Plant
Science

*A Series of Books
in Agricultural Science*

Plant

Second Edition

Science

AN INTRODUCTION TO WORLD CROPS

Jules Janick
PURDUE UNIVERSITY

Robert W. Schery
THE LAWN INSTITUTE

Frank W. Woods
THE UNIVERSITY OF TENNESSEE

Vernon W. Ruttan
AGRICULTURAL DEVELOPMENT COUNCIL

W. H. FREEMAN AND COMPANY
SAN FRANCISCO

The illustrations on the cover are reproduced through the courtesy of The Metropolitan Museum of Art.

Library of Congress Cataloging in Publication Data
Main entry under title:

Plant science.

Includes bibliographies.
1. Agriculture.   I. Janick, Jules, 1931–
SB91.P55 1974   630   73-13921
ISBN 0-7167–0713–6

1   2   3   4   5   6   7   8   9

Preface
TO THE *Second Edition*

This text was written to give attention to the scientific, technological, and economic foundations of world crop production. The sciences dealing with plants and the technology and economics of crop production and marketing have become irrevocably entwined. Each contributes relevance and meaning to the other. Although the origins of the plant sciences can be traced to the seventeenth and eighteenth centuries, they are in many respects truly twentieth-century disciplines. Some crop production practices, however, are older than written history, and form an integral part of man's legends and rituals. It is no accident that all people celebrate festivals that are based on the agricultural calendar. It is our purpose here to blend the twin themes of plant science and crop agriculture into a single narrative. The formal disciplines that have been established in agriculture (agronomy, horticulture, forestry), botany (systematics, physiology, genetics, ecology, etc.), and the social sciences (economics, sociology, political science, history) must be interrelated to provide a full understanding of the relationship between men and plants.

This text was planned for use primarily as an introductory, university-level course in plant or crop science. A number of universities have developed such a course by consolidating material from some combination of the beginning courses in agronomy, horticulture, and forestry.

In view of the fact that this book was written to survey an area in which there is no long-established course, it encompasses more than is usually found in an introductory text. Nevertheless, the entire book could be covered in a two-semester sequence, Parts I to IV in one semester, Parts V and VI in the second. Under the quarter system various combinations of chapters and parts could be used, the choice depending on the particular objectives of the instructor. For example, Part V could form the basis for a quarter course in Economic Crops.

Parts I, *Plants and Men,* introduces the relationship between crops and civilization in both historical and biological terms. Part II, *Nature of Crop Plants,* is a formal presentation of plant relationships, structure, and development. Part III, *Plant Environment,* is an introduction to the ecology of crop plants. Part IV, *Strategy of Crop Production,* features technological aspects of agricultural practices. Part V, *Industry of Plant Agriculture,* surveys individual crop species. Finally Part VI, *The Marketplace,* explores the relation between crops and the economic community.

In the second edition two new chapters have

been added: Chapter 5, "Agriculture, Pollution, and the Environment," and Chapter 27, "The Organization of Agricultural Research Systems." Also, those changes that have occurred since 1969, when the first edition was published, have been incorporated.

Each of the four authors who collaborated on this book has had specific responsibility for particular chapters: Jules Janick, 1, 3, 7–9, 18, 19; Robert W. Schery, 6, 15, 20–23; Frank W. Woods, 2, 5, 10–14, 16, 17; Vernon W. Ruttan, 4, 24–29.

We all, however, bear collective responsibility for the book as a whole. We wish to give thanks to our families, colleagues, and friends who have suffered through this endeavor with us. We gratefully acknowledge the help of Dr. Arthur H. Westing, who read through the entire manuscript.

January 1974

Jules Janick
Robert W. Schery
Frank W. Woods
Vernon W. Ruttan

Contents

PART I

PLANTS
AND MEN

CHAPTER 1

Crop Plants and World Affairs

Earth is a plant-oriented planet. The green plant is fundamental to all other life. Were man to perish tomorrow, vines would destroy his mighty temples and grass would soon grow in the main streets of the world. In contrast, the disappearance of plants would be accompanied by the disappearance of man along with every other animal.

The importance we attribute to any product is usually related to the probability or actuality of a shortage. Thus, those that are plentiful and readily available are often held in low esteem, even though our very existence may depend upon them. The oxygen we breathe, the nutrients we consume, the fuels we burn, many of the most important materials we use, are all related to plant life.

Plants have a wide spectrum of uses. The most obvious is for human sustenance. Plants supply all of man's food, either directly or indirectly as feed for animal intermediaries. They are also utilized as a source of structural support, as a construction material, and as the raw material in the manufacture of fabrics and paper and such "synthetics" as plastics and rayon. We have come to depend upon many of the complex substances that plants produce —dyes, tannins, waxes, resins, flavorings, medicines, and drugs. Living plants, besides having a direct effect on the ecological position of man, are used to control erosion by water and wind, to provide a setting for recreation and sports, as landscape materials, and to satisfy man's desire for beautiful objects.

The story of man is largely a chronicle of his struggle for dominion over his environment. The efficiency of this control is thought of in terms of civilization, or culture. To a great extent, controlling the environment means controlling plant life. The failure to provide a high standard of living for the population as a whole in so-called under-developed countries is generally associated with an inefficient system of crop production and distribution. Similarly, the decline of developed societies can in many instances be associated with a disruption in the basic system of resource utilization. Crop production, the management of useful plants, is the very basis of our civilization.

CROP PLANTS

In its broadest sense, the term "crop plant" means any plant utilized by man for any purpose. In the restricted but common sense of the term, crop plants are those useful plants that fit economically into the scheme of man's work and existence. They are plants that are managed to some degree. This may involve merely a systematic type of harvest, as in some types of forestry, or completely artificial culture throughout the entire life of the plant, as in greenhouse fruit production.

The number of species considered as crop plants depends on the definition employed. For example, in Bailey's *Manual of Cultivated Plants*,° a well-known compilation for the United States and Canada, 5,347 species are listed. A world-wide compilation would number between 10,000 and 20,000, and would include many plants that can be classified as crops only under the broadest definition. The number of plants that fit economically into man's activity is probably between 1,000 and 2,000 species. Those of major importance in world trade are relatively few, perhaps between 100 and 200. Fifteen species provide the bulk of the world's food crops: rice, wheat, corn, sorghum, barley; sugar cane and sugar beet; potato, sweet-potato and cassava; bean, soybean, and peanut; coconut and banana. The course of history has been changed—and continues to be changed by only a few plants, out of an estimated world total of 350,000 species. Man has struggled to obtain them and to cultivate them. These are the plants of legend and literature, of song and sacrifice; they are the plants with which this book is largely concerned.

DEVELOPMENT OF AGRICULTURE

The discovery of fire and the development of agriculture are the two innovations that form the basis of civilization. Fire is so basic to our existence that it is hard to imagine man without it. Man's use of fire not only marks the beginning of social living but eventually gave birth to a whole series

° Rev. ed. Macmillan, 1949.

of related technologies that would otherwise be inconceivable. The most important immediate result of fire was the expansion of food supply, for a great number of foods are inedible, unpalatable, or unsanitary unless they are cooked. The continued development of any society hinges on the availability of an adequate source of food. In primitive societies based on food gathering or hunting, each individual must be totally involved with the urgencies of securing sustenance. Abundance is temporary and exceptional. The solution to this problem came with the "invention" of a series of complex and related technologies involving an intimate relationship between crop plants and domestic animals—the development of **Agriculture.**

Agriculture is a relatively recent innovation when considered in relation to the history of mankind, for man has been a mere collector of food for the greatest part of his existence. The first production of food by crop cultivation and actual domestication dates back 7,000 to 10,000 years, to the Neolithic age. Agriculture seems to have developed independently at widely separated times in a number of parts of the world. The gradual development of agriculture brought forth a dependable surplus. Such a surplus releases from food production those individuals with skills in other specialties. The development of new specialists is possible only when increased agricultural efficiency allows the exploitation of newly acquired leisure. This is still true. The resultant increases in the standard of living are measured by the manner in which former luxuries become everyday necessities.

The origins of civilization can be traced to the discovery that a plentiful food supply could be assured by the planting of plant parts or seed. Plants that grow rapidly and produce a crop within a season were probably the first cultivated. The technology involved in the cultivation of long-lived plants such as fruit trees is time-consuming and requires a higher order of technology; as a result, these crops were probably gathered from the wild. Some crops are still harvested from the wild. The chicle industry of Central America and the blueberry industry of Maine depend on a source of wild plants.

Practically every one of our present-day crop plants (as well as domestic animals) was developed in prehistoric times. This was achieved by means of two distinct processes: **domestication,** the bringing into cultivation or management of some wild species; and **selection,** the differential reproduction of these species. Primitive man displayed uncanny ingenuity in the process of domesticating wild plants and preparing them for food. For example, the cassava, a major source of starch in tropical South America, contains a deadly poison (hydrocyanic acid), but somehow it was long ago learned that the poison could be removed through the cooking process. Such techniques are not easily come by! Selection sometimes led to the creation of new types, and was extremely effective for many plants. Most of our cultivated crops differ markedly from their wild ancestors, and many are so changed that precise lines of descent have become obscure. Early man was a plant breeder, and a very effective one, without any knowledge of genetics.

Crop plants have accompanied man in his wanderings and migrations. The introduction of species into new habitats has been one of the most important features in world agricultural development. Today the centers of production of nearly all agricultural crops are far removed from their centers of origin, as is shown in Table 1-1.

Civilization rests on discoveries made by people little remembered by history. Precisely where the first plants were cultivated is unknown. According to archeological evidence, however, agriculture had its origin 7,000 to 8,000 years ago, somewhere in the then well-watered highlands of the Indus, Tigris, Euphrates, and Nile Rivers. The antecedents of this agriculture undoubtedly date back many more thousands of years. Southeast Asia, with its diverse geography and consequent diversity of vegetation, its mild climate, and its capacity to support a stable population from a fishing-hunting economy, has been proposed as a likely location for the birth of primitive agriculture. This area is especially rich in asexually propagated crop plants, and perhaps the planting of vegetative parts preceded the planting of seed. Primitive agriculture may have originated independently in a number of parts of the world, and developed through the dispersal and divergence of new plant forms

Table 1-1   THE CENTERS OF WORLD PRODUCTION AND CENTERS OF ORIGIN FOR SELECTED WORLD CROPS.

CROP	CENTERS OF PRODUCTION	CENTERS OF ORIGIN
Cacao	Africa	Brazil
Coffee	Brazil	Abyssinia
Maize	Midwest U.S.	Tropical America
Monterey Pine	Australia	California
Pineapple	Hawaii	Brazil
Potato	Eastern Europe	Peru
Wheat	North-central North America	Central Asia

in new environments. As agriculture moved into more severe climates, seeding replaced planting of vegetative parts as the dominant technique.

When agriculture came to the Old World there was a movement to river valleys, in which the twin obstacles of drought and flood had to be overcome. Gigantic changes were catalyzed by the innovations required by cereal culture and irrigation. The new technology increased the need for a higher level of social order; great works are required to make rivers benefit rather than harass man. The success of this technology can be measured in terms of the ever-increasing populations that were supported. More than 5,000 years ago a stone age culture was transformed to the full-fledged urban civilizations of Egypt and Sumeria.

NEAR EASTERN ORIGINS

The birthplace of many early civilizations lay in the territory now referred to as the Near East. Particularly important was Mesopotamia—"the country between the rivers." The superlative juxtaposition of habitats that exists there supports a diversity of wild species. Such diversity is likely to foster change and stimulate speedy adaptation under the pressure of early selection. Important plant discoveries and early domestications were presumably made in the lower elevations of the Zagros mountains and brought down to the fertile Tigris-Euphrates flood plain for full domestication under management. The earliest traces of the transition from food-gathering to food-producing economics are found here. Modest genetic change (an all but inevitable consequence of selection for harvestable types) would encourage rapid adaptation to intensive cultivation in river valleys. It is likely that these first crops, mainly grains, were selected for high yield and valued principally as high-energy foods. Vegetable oils were no doubt secured by gathering fruits from the wild; other requisites for a balanced diet were obtained by catching game. Acorns, pistachio nuts, olives, and hazel nuts abounded in the region.

Figure 1-1 is a map of the Fertile Crescent, where so many important civilizations have arisen.

The high plateau east of the Zagros Mountains is inhospitable, and probably only the deltas of the mountain valleys were habitable in primitive times. But west of the Zagros are fairly moist uplands, below them somewhat drier steppes, and even drier alluvial bottoms below the steppes. Irrigation of the deep, rich soils of the alluvial bottoms would make for quite remunerative agriculture. Ancient settlements were amazingly abundant on the western foothills of the Zagros, indicating that these foothills once had a much greater carrying capacity than seems possible today. Very likely, settlements ranged from essentially permanent ones in the lowlands, to temporary camps of migratory tribes that followed game (and eventually

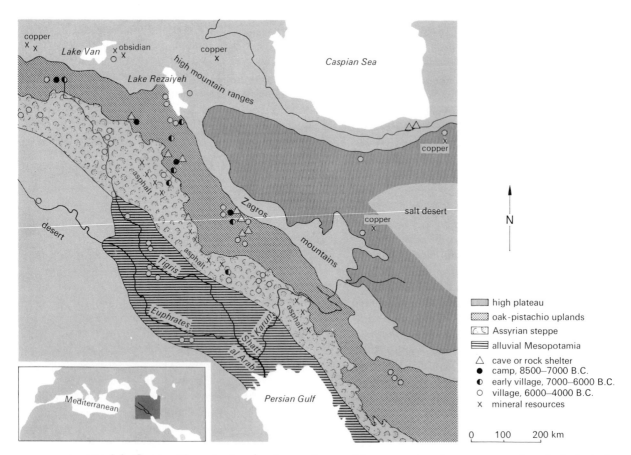

FIGURE 1-1. *Greater Mesopotamia, showing environmental zones, mineral resources, and archeological sites.* [*After K. V. Flannery, "The Ecology of Early Food Production in Mesopotamia," Science, Fig. 2, Vol. 147, pp. 1247–1255, 12 March 1965. Copyright © by the American Association for the Advancement of Science.*]

herded livestock after its domestication), to seasonally suitable grazing and browsing land in the mountains. Itinerants may have carried on trade with peoples of the Iranian plateau, exchanging copper at the more permanent settlements for food and seasonal accommodations.

Preagricultural subsistence in Mesopotamia relied greatly upon the chase. The wild sheep, goat, pig, gazelle, deer, ass, fish, birds, and small game were all hunted. Wild fruits and seeds of the vicinity were gathered, too. Wild alfalfa and other legumes, and of course the primitive grains, were repeatedly utilized. It is not unlikely that a migrating tribe might harvest one crop on the steppes in March, another higher in the foothills in April or May, and still another at higher elevations in summer. A region with so richly varied a natural fare would offer many candidate species for selection and trial for adaptation and storage qualities in new environments.

The first crop plants would not be acceptable today. Wild grains shatter easily, and their tough husks make threshing difficult. Perhaps unconsciously, early farmers overcame these difficulties by selecting grains with a tough rachis (almost automatic under a harvesting system), and by roasting and grinding husked seeds. The requirement of intensive agriculture brought about directed selection to improve the crop and the preservation of other genetic change. The improved crop, entirely an artifact of human manipulation, and perhaps unable to persist in the wild, is the culminatation of a long process involving changes in the ecological relationship between man and locally available plants!

As in modern agriculture, early farmers would endeavor to secure greatest yield with least labor. Concentrating the crop by sowing seed in areas where weeding and protection against pests were possible would have afforded a definite advantage over collection from chance wild stands. This practice was established very early for both wheat and barley. Barley may have become domesticated quite fortuitously, as an impossible-to-eliminate weed in wheat (rye and oats were apparently introduced into European agriculture in this way from the Near East.) Certain millets, field peas, lentils, vetches, chickpea, horsebean, and flax (lin-

seeds) were also introduced from the Near East. The wine grape was cultivated at least four millennia before Christ, and gradually the olive, date, apple, pear, cherry, fig, and other fruits were managed in orchards rather than sought in the wild.

AGRICULTURE AND ANTIQUITY

Mesopotamia, aptly called the cradle of civilization, was to influence all of antiquity. The first urban economy to develop here was based on an agricultural technology that led to temples and priests, granaries and clerks. The creation of a "social surplus" made warfare and slavery economic institutions. The administration of the stored surplus pressed the need for a system of accounts. The solution to this problem came 6,000 years ago with the invention of writing, which marks the initiation of civilization in its present-day meaning. The Mesopotamian cultures existed for thousands of years under many different regimes. Their influence, although difficult to define precisely, radiated to Syria and Egypt and perhaps to India and China.

The backbone of Mesopotamian agriculture consisted of crop plants that are still important to the world's food supply: wheat and barley, date and fig, olive and grape. The ancient cultures of Mesopotamia—Sumerian, Babylonian, Assyrian, Chaldean—developed an increasingly complex and integrated agriculture. Their ruins reveal the remnants of irrigated terraces, parks, and gardens. Four thousand years ago burnt-brick-lined irrigation canals with asphalt-sealed joints helped keep 10,000 square miles under cultivation to feed 15 million people. By 700 B.C. an Assyrian Herbal was compiled that contained the names of 900 plants. The books of the old Testament are rich in allusion to the agriculture of this era.

Our knowledge of ancient agricultural development is nowhere greater than in Egypt, where the drifting desert sands have preserved the records of an astonishing era. Although the Nile Valley had supported man for at least 20,000 years, its agricultural development is believed to have stimulated the transformations that occurred in the Mediterranean area. The great Egyptian civilization, to which our own culture owes so much, flourished on the agricultural abundance made possible by the constant renewal of fertility from the overflow of the Nile. By 3500 B.C. Egypt had established a centralized government with Memphis as its capital, and by 2800 B.C. had developed to a level of sophisitication capable of such gigantic works as the construction of the pyramids.

The Egyptians were masters at developing techniques of drainage and irrigation. Drainage, the removal of excess water, is a necessity in areas like the Nile Valley, in which flooding was an annual event and of concern to municipal as well as agricultural centers. Drainage requires both the development of the slope of the ground and the establishment of efficient conduction systems. Irrigation, the artificial supplying of water to crops, involves the procurement, conduction, and application of water. The problems of drainage and irrigation are intertwined; the Egyptian solution was to construct a series of dikes for storing water and canals that served both purposes. The Egyptians developed techniques for raising water that are still being used! The most basic invention was the **shaduf,** a suspended container balanced with a counterweight (Fig. 1-2). With this innovation, it was possible to raise 600 gallons of water as much as six feet per man-day.

Land preparation technologies can be traced through refinements in the simple hoe. The hoe was originally a pointed, crotched stick and was

FIGURE 1-2. *Irrigating with a shaduf. The farmer has raised water from a pool with the aid of a clay counterpoise to irrigate a palm garden. From a tomb at Thebes (ca. 1500 B.C.). [From Singer, Holmyard, and Hall (editors),* History of Technology, *Vol. 1. London: Oxford Univ. Press, 1954. Courtesy The Metropolitan Museum of Art.]*

FIGURE 1-3. *Agricultural implements and land reclamation in ancient Egypt. From a tomb at Thebes (ca. 1420 B.C.). [From Singer, Holmyard, and Hall (editors),* History of Technology, *Vol. 1. London: Oxford Univ. Press, 1954. Courtesy The Metropolitan Museum of Art.]*

used (as it is today) with a chopping motion. The early plow was merely a hoe drawn by man (later by animal) to scratch the surface of the soil (Fig. 1-3), and is still in use in many parts of the world. Later the plow was improved through the reinforcement by iron of parts in contact with the soil and more substantial and efficient construction. The modern, or "soil inverting," plow is a more recent European invention that dates back to the eleventh century. The Egyptians employed various cutting tools in harvesting, of which the curved metal sickle was the most refined.

The Egyptians developed the various technologies associated with the culinary arts—the ceramics industry, baking, wine making, and food storage. Their methods of preserving food included fermentation and pickling, drying, smoking, and salting. Many plants were cultivated for fiber, oil, and other "industrial" uses. These included papyrus for paper, date palm for fiber, castor bean for oil, pines for resins. The Egyptians created the first pharmacopoeia, a collection of drug and medicinal plants, and the first spice, perfume, and cosmetic industries.

Along the Nile great formal gardens were created, replete with exotic ornamentals and pools containing fish and water lilies. In orchards the date, grape, fig, lemon and pomegranate were cultivated. Vegetable gardens included cucumber, artichoke, lentil, garlic, leek, onion, lettuce, mint, endive, chicory, radish, and various melons. (See Fig. 1-4.)

Egyptian civilization persisted for an incredible 35 centuries. It produced not only a remarkable technology but a magnificent art, and a complex,

FIGURE 1-4. *Fruit- and vegetable-growing in Egypt.* (*Top*) *Picking figs. From a tomb at Beni Hasan (ca. 1900* B.C.). (*Middle*) *Collecting grapes from a round vine arbor and treading to express the juice. From a tomb at Thebes (ca. 1500* B.C.). (*Bottom*) *Irrigating a vegetable garden.* [*Top and middle figures from Singer, Holmyard, and Hall (editors),* History of Technology, *Vol. 1. London: Oxford Univ. Press, 1954. Courtesy The Metropolitan Museum of Art. Bottom figure from Gothein,* History of Garden Art, *Vol. 1. New York: Dutton, 1928.*]

cause of a combination of uncertain rainfall and shallow soil. Consequently, the strength of Greek civilization was destined to be based on sea power and to be nourished by trading and conquest.

GREECE AND ROME

The "Greek miracle," the extraordinary achievement wrought some 2,500 years ago by an ancient civilization, was the result of a combination of factors that is still difficult to explain. Because our cultural heritage in art, literature, and ethics is in large part traceable to Greek influences, it is easy to overemphasize the technological level of Greek culture (Fig. 1-5). The achievements of the Hellenic world owed much to the established technologies of Mesopotamia and Egypt; they were developed when the pyramids were more than a thousand years old. The rise of Greece, and later of Rome, is in some ways comparable to the still later rise of the Huns and the Goths—a victory of barbarism over advanced but tired cultures. The technology of the invader was often inferior to that of the invaded and subdued. Although the Greeks added only incidentally to the practical, their curiosity and their analytical attitude concerning the nature of things were to have profound effects on the course of future technological advance. For

FIGURE 1-5. *Greek plough. From a black-figured Nikosthenes cup (sixth century* B.C.). *Note the similarity to Egyptian implements which, at that time, had been in use for 1,000 years.* [*From Singer, Holmyard, and Hall (editors),* History of Technology, *Vol. 2. London: Oxford Univ. Press, 1956.*]

if bewildering, theology. By the time Egypt had become a Roman province (3 B.C.) its influence had already permeated the ancient world—an influence that is still felt today.

In approximately 1000 B.C. the introduction of iron and the rise of the Indo-European invaders were to create a period of turbulence in the history of civilization and cause a shift to the Aegean—the bridge between Europe, Asia, and Africa. Phoenician sailors passed on the technological heritage of Mesopotamia and Egypt to the emerging Greek Isles. The agricultural products of the Eastern Mediterranean region—wheat, barley, grapes, figs, and olives—were never in abundant supply be-

example, the Greeks, while contributing little to practical agriculture, made great contributions to the formal study of plants. The science of botany originated in the Greek mind.

Theophrastus of Eresos (372–288 B.C.), brilliant student and disciple of Aristotle (whose botanical works are lost), appears to have been among the first to become interested in plants for their own sake. His two treatises, *History of Plants* and *Causes of Plants*, have earned him the accolade "Father of Botany," and influenced botanical thinking until the seventeenth century. These writings deal with such diverse topics as morphology, classification, seed and vegetative propagation, geographic botany, forestry and horticulture, pharmacology, insect pests, and plant flavors and odors. Theophrastus treats over 500 kinds of plants, both cultivated and wild. He distinguishes between angiosperms and gymnosperms, between monocots and dicots, and discusses the formation of annual rings and methods of collecting resins and pitch. He even discussed the "dusting" of the fruit-forming "female" date palm with flowers of the nonfruiting "male" to effect fruiting. This straightforward account of sexuality in plants became virtually lost for 2,000 years!

The Greek genius proved to be incapable of political survival. The intercity struggles and rivalries paved the way for complete destruction by the armies of Macedonia. There are some who trace the downfall of Greece to the effect of increasing populations on declining resources as much as to wars and internal decay. It appears that the agricultural base of Greece was not sufficient to support an ever-growing civilization.

Greek civilization was absorbed by a new breed to the west. The Roman empire, unlike the Greek, was built on a firm and powerful resource base. In contrast to the Greeks, the Romans were extremely interested in the practical aspects of agriculture. Agriculture was a vital part of the economy and of real concern. The Romans' main source of revenue was land taxes; their most important legislation dealt with agrarian planning. Great fortunes were invested in farm land. Rome grew to grandeur on the foundation of a sound and functional agricultural technology. As they conquered, they built a culture that was Greek in origin but Roman in application. Although the Romans can be credited with relatively few original ideas, it must be recognized that they did improve upon what they found. Their enduring trademarks were roads and aqueducts. The Romans were of a modern mind—civilized and urban, yet bound to the land through business and inclination. The combination of their empiricism and their genius for political management produced the largest contiguous empire the world has seen—an empire destined to last a millennium.

Roman agricultural practices are well recorded. The earliest of the agricultural writings in Latin is the terse *De agri cultura* of Marcus Porcius Cato (234–149 B.C.), who writes of the practical aspects of crop and livestock management, and especially of profits—slaves should be fed only the barest minimum. The origins of a rural philosophy are found in his conclusion that farmers are not only the best citizens but make the best soldiers. About a hundred years later a longer, better organized, and more complete work was produced by Marcus Terentius Varro (116–28 B.C.). Varro's *De re rustica libri III* (three books on farming) emphasized the dependence of the commonwealth on a sound agriculture. The *Georgics* of Virgil (70–19 B.C.), one of the era's literary masterpieces, speaks not only of the details of husbandry but extols the virtues of rural life. In the second half of the first century A.D. a number of works dealing with agriculture were compiled. These include a twelve-book treatise by Columella and the encyclopedia *Historia naturalis* of Pliny the Elder (A.D. 23–79), a monumental conglomeration of science and ignorance. It is from these and other works that the agrarian history of Rome has been deduced.

During the early period of Roman history (up to 200 B.C.) the basic agricultural institution was the village community. Early Rome has been described as a market place that served a hamlet of truck gardeners. Individual holdings were small, ranging in size from one to four acres, and intensively managed. (To wish for more was considered dangerous.) But as the Roman state expanded in territory and acquired a slave-labor force through conquest, a larger production unit, the *latifundium*, came into being. These large-scale holdings were carved out of public lands and distributed by the state. The resultant plantation system was to foster the growth of enormous individual fortunes and to encourage graft and corruption, which spread virulently. The abundance of low-cost labor that resulted from the increasing use of captive "barbarians" and the expanding size of individual holdings resulted in a disequilibrium in the social system. The soldier-farmer-citizen lost his place as a stabilizing force in Roman life. Absentee ownership, slavery, and soil exhaustion resulted in ever-declining agricultural productivity. In addition, continual warfare had the end result of reducing the small landholder class by attrition, only to increase the holdings of the remaining few. Tribute from conquered states in the form of grain and other food products relieved temporary shortages, but in the long run probably contributed to a decline in agricultural productivity on the Italian peninsula.

An alternative system of agricultural production became prominent about A.D. 200 with the substitution of tenancy for slavery. Rents rather than produce became the return from investment in agriculture. The increase of an indebted tenant class in the rural areas and a landless, discontented "proletariat" in the cities were to have unhappy consequences. With a decline in land productivity the bond between landowner and tenant deteriorated from a symbiotic to a parasitic relationship and was accompanied by the gradual erosion of freedom of the erstwhile renter. For example, it became a crime for the son of a farmer to leave the farm on which he was reared, and soon "free" men as well as slaves were bound to the land they cultivated. Such legislation was no doubt originally enacted to increase lagging agricultural productivity, but it failed badly in this respect. A declining population in Europe followed the declining productive capacity of the land. The surplus necessary to support cities evaporated, and the decimation of the metropolitan centers was followed by a lowering in the quality of civilization.

The tenant was systematically reduced from a citizen to a serf, and the slave retained his yoke under a new name. The renter who traded his labor for land (and production) was in fact bound to it—originally through the machinery of law but soon even more effectively through force of custom and habit. The resultant manorial system (named after the great houses or manors of the owner) became a stable system, and signalled a return to a village economy. It lasted in Europe as a dominant condition for well over a thousand years and it still exists in many areas of South America. It was the general system of life in the Middle Ages.

The agricultural technology of Rome at its height, although probably inferior to the Egyptian, was not equalled in Europe until the Renaissance. Agricultural writings note grafting and budding, the use of many kinds and varieties of fruits and vegetables, legume rotation, the use of manure and marl, soil fertility appraisals, and even cold storage of fruit. Mention can be found of a prototype greenhouse (*specularium*) that was constructed of mica for vegetable forcing.

It was in Rome that ornamental gardening was developed to a high level. The good life was that of a gentleman farmer, his expression of conspicuous consumption being the country estate, or *suburbanum*. It contained fruit orchards and flower gardens. Splendid mansions were enclosed by frescoed walls, statuary, fountains, and pools. The look of luxury was artfully conceived.

The rise and fall of the political fortunes of imperial Rome were paralleled by similar trends in agriculture. Which was the cause and which the effect is not clear; probably they were interrelated. The strain of supporting and defending an overexpanded empire undermined the agricultural foundations; an unstable and exhausted agriculture reduced the capacity for further growth and development. By the third century A.D. Rome had become an economic, political, and military failure. Nevertheless, the remarkable fact remains that Rome, an amalgamation of diverse peoples and customs, survived for a millennium. The end was rather inglorious: the empire retreated to the villages from which it had been created. Western civilization and agriculture were to take a step backward into ignorance and misfortune.

MEDIEVAL AGRICULTURE

With the decline of Rome and the West, the mantle of technological superiority passed to the

Near and Far East. During the thousand years between 500 and 1500, it was the East that provided the most skilfully crafted products. The rise of Islam after A.D. 700 provided a repository for Greek culture and preserved the heritage of antiquity. It is intriguing to note that the Eastern civilizations were at the peak of their achievements when Western Europe was at its low ebb.

During the early decline in technological progress much of the agricultural art survived in the monastic gardens. Gardening became an integral part of monastic life, providing food, wine, decoration, and medicinals. The job of gardener (*Hortularium,* or *Gardinarium*) became a regular monastic office. Fruit varieties and vegetable strains were preserved and improved. The few botanical and agricultural writings of this period are derived from the church, the guardian of literacy, but are for the most part compilations traceable to Roman sources, especially Pliny's *Natural History.* It was to be many centuries before the technology of antiquity was equalled.

The Byzantine empire, the eastern heir of Rome, was to be the filter through which passed techniques and ideas, as well as silks and ivories from Persia, India, and China. Slowly developing contacts with the East brought many things that were to have a profound influence on the future course of history. Gunpowder, paper making, the compass, and what we know as arabic numerals all come to us from the Far East. In the field of agriculture and crop production many important products of Eastern origin were continually introduced. Eastern crop plants such as the peach were known to the Chinese before 2000 B.C., and had reached the Mediterranean at about the height of Greek civilization, and was described in certain writings as a Persian fruit, a misconception that survives in the present botanical binomial, *Prunus persica.* Sugar cane, native to eastern Asia, although known to classical antiquity, was not commonly used in Europe until it was introduced by the Arabs into Palestine, Sicily, Spain, and the Greek Islands. The soybean is the most recent of a long list of important crop introductions from the Far East. One of the most important inventions brought from the East was an efficient collar that did not choke the horse under strenuous pulling.

The development of such an efficient harness made possible the utilization of the full power of the horse in agriculture. The padded collar, known to have been used in China from the third century, was unknown in Europe until the ninth century.

The lure of the Orient and its wealth had long been a determining force in Western civilization. It was the search for new routes to the spice-rich East that gave impetus to the exploits of the daring navigators of Italy, Spain, and Portugal. The discovery by Columbus of the New World in 1492 marks a convenient if inaccurate date to assign to the beginning of the Modern Age.

In the Middle Ages the manor, or estate, was the basic unit of agricultural production, and the smallest component of the feudal state. The estate, consisting of small tenant holdings and the lord's land, was characterized by a high degree of self-sufficiency. This very high degree of independence retarded communications and kept progress down to a low level.

Crop growing was based on extensive production in large fields and intensive production in small gardens. The tenant had the use of strips in a number of fields, the right to a share of hay and firewood, and the right to graze his cow on a common pasture (the commons). The usual system of crop production was a **fallow system,** in which part of the land was kept unplanted to restore soil fertility. Fields were rotated among the tenants, a practice that ensured an equitable distribution of the good land but probably removed any incentive to improve the soil.

The present classification of crop agriculture into **agronomy, horticulture,** and **forestry** is medieval in origin, and traces back to the system of land distribution of the manor. The practice of planting extensive areas in grains and forages gave rise to the specialized discipline of crop agriculture now called agronomy. Horticulture developed from the practice of cultivating intensively managed kitchen gardens that provided fruits and "herbs" as well as the increasing degree of ornamental planting on the lord's manor. Forestry developed from the protection of the wild lands that produce timber and support game. The relationship between forestry and game management is an old one.

A decline in feudalism came about with the growing rise of towns and the emergence of strong national states. The re-emergence of towns offered an alternative choice for the peasant. The rise of the modern state was based on the destruction of power of the manorial lords. Feudalism's breakdown was forewarned by a series of unsuccessful peasant revolts as early as the fifteenth century, yet the complete disappearance of feudalism was slow to come. Although it was gradually eliminated in England in the sixteenth century, serfdom was not legally abolished in many countries until very recently. It was legally abolished in Prussia in 1807, in Russia in 1861, and in Japan in 1874.

The recovery of Europe from the backwardness of the so-called Dark Ages was a slow but ever-continuing process. It came with the increase in trade and economic activity. With the decline in feudal ways and the accumulation of technology (incorporated from Islamic cultures in Spain, old Roman techniques preserved by the monasteries, and the ever-increasing appearance of new processes and inventions), agricultural methods, tools and products improved. The rising population of towns and the expansion of trade and finance also enriched the rural economy. New industries created a market for industrial crops—sugar cane, hemp, flax, vegetable oils, and dyes. The production of wine became a key industry and vinyards began to flourish in favorable regions. Strip farming came to be abandoned in favor of an enclosed system in which individual farmers took continuing responsibility for their own land. Crop rotation and improved plowing techniques increased in usage. Water mills, although known to the East, began to appear in Europe toward the end of the twelfth century. From the eleventh to the fourteenth centuries the great dikes (the "Golden Wall") were constructed to hold back the sea from inundating Belgium and Holland. At the end of the Middle Ages the agricultural land of Europe was cleared, drained, and cultivated to provide the foundations for the eventual triumph of European civilization.

As feudalism gave way to trade, a real rise in the standard of living was produced. The diet of the peasant class improved from one consisting mainly of cereals to one that included garden products and meat. By the thirteenth and fourteenth centuries orchards and gardens were common outside of the monastery. The revival of learning known as the Renaissance had a great impact on agriculture, and a horticultural revival spread from Italy to France and then to England. The peak of Renaissance horticulture is to be found in the magnificent gardens of LeNôtre (1613–1700), the most famous being those at the Palace of Versailles, built for the French monarch Louis XIV.

THE NEW WORLD

The discovery of the New World raised great expectations in Europe for such riches as gold, spices, and new foodstuffs—expectations that were to be fulfilled on all counts. Gold and silver in quantities the world had never seen were to fill the coffers of indebted sovereigns. Pizarro returned to Seville from Peru with a gold cargo of over 17 tons, most of which consisted of artistic works melted down into ingots. This was the cargo of but one ship! The foodstuffs were found in the form of strange and wonderful plants. Man transposed to the New World had brought under cultivation practically all the indigenous plants we now use: crops such as maize,° potato, tomato, sweet-potato, squash, pumpkin, peanuts, kidney and Lima beans; cranberries, avocados, Brazil nut, cashew, black walnuts, pecan, pineapple; cacao, vanilla, chili, quinine, coca, rubber, and tobacco.

The metal treasure is all gone—hardly a single piece of the golden objects shipped to Spain remains. The gold was not only responsible for the annihilation of the Aztec, Mayan, and Inca Empires but led to the ruin of the nations that expropriated it. As Von Hagen has said of the mighty cultures in America, "what was not destroyed by man was overwhelmed by the insults of time." The inheritance of the greed that conquered a continent is the remnant of feudalism that still lingers in many parts of Latin America. But the plants of America have produced, and are still producing,

---

° Maize (*Zea mays*), also "corn," originally "Indian corn" in the United States. The term "corn" originally referred to any seedlike fruit, especially the seed of cereals—wheat, rye, oat, barley.

14

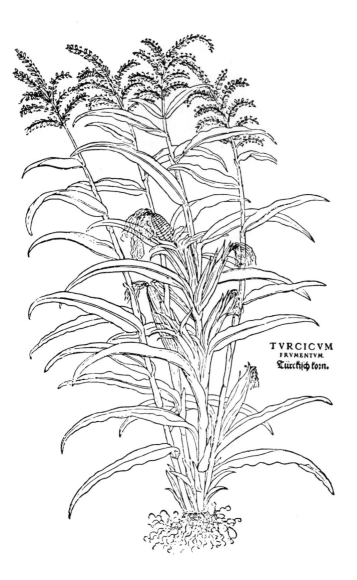

TVRCICVM
FRVMENTVM.
Türckisch korn.

FIGURE 1-6. *The first woodcut of maize, from Leonhard Fuchs' herbal* De historia stirpium, *published in 1542. Fuchs erroneously stated that maize was introduced into Germany by the Turks, hence his name* Turcicum Frumentum, *or "Türckisch korn." Maize was unknown to the Old World until after the time of Columbus.*

wealth that pales into insignificance all the golden booty (Fig. 1-6).

The exact time that man came to the Americas is not known. It is generally believed that the first arrivals reached North America at least 40,000 years ago from Asia over a once-existing land bridge across the Bering Straits. It seems that few cultural traditions were carried along; the development of civilization in the Americas was mostly home-grown. Similar needs inspire similar inventions. New World agriculture paralleled many developments in the Old World, although there were important differences. The wheel, for one, was practically unknown, except as a child's toy, and there was no "iron age." The conquistadores came upon civilizations in the age of stone. The basic agricultural tool was the digging stick.

Some of the American cultures never developed beyond the Paleolithic mold. For example, the Plains Indians of North America never completed the transition to an agricultural economy. The men hunted; only the women planted. Until the horse was introduced into North America by the Spanish the nomadic Plains Indians were able to hunt buffalo only with great difficulty; the horse gave them ready access to the tremendous herds and completely altered the course of their cultural development.

To the south, however, a different situation emerged. By 1000 B.C. the great migrations that had filled the continent were over, and the agricultural patterns were formed. Temple-cities began to rise as an agriculture based on the culture of maize bestowed a sufficient surplus of food to free some individuals from laboring in the fields. A series of cultures developed, culminating in the Aztec in Mexico, the Mayan in Central America, and the Inca, which spread from Peru throughout most of the Andes.

In the New World the sequence of developmental steps to a sedentary agriculture has been particularly well worked out for the Tehuacan Valley, some 150 miles south of the site of Mexico City. Maize, one of the world's most important crops, appears to have been brought into cultivation there, along with beans, squash, and other prime food plants. The detailed record unearthed in the Tehuacan Valley is a tribute to the efforts of skilled archaeologists.

As in the Fertile Crescent in the Near East, the Tehuacan Valley is ringed by foothills and mountains that provided a wealth of environmental diversity. The valley climate is dry, preserving well the archaeological evidence in the middens. Even desiccated feces have been found, providing some direct evidence of diet. The trash accumulations of ancient habitations are sufficiently well preserved to shed light on such details as customs of the occupants, months of occupancy, and climatic change. Sufficient material for accurate carbon-14 dating is available.

By 7200 B.C. some form of primitive agriculture had been in existence for three or four millennia. The valley was sparsely occupied by small, nomadic families or tribes, which changed camps several times yearly and subsisted chiefly on wild plants and hunting. Pollen studies indicate that the valley was sufficiently dry to be grassland. Its occupants used flint tools, but apparently had no devices for cultivating the soil.

Within the next 2,000 years the small nomadic groups gathered at times in base camps, but dispersed to follow the seasons from valley floor to uplands, as did the primitive farmers in the Near East. The first crude planting of seeds seems to have begun by about 5000 B.C., with at least squash, chili, and avocados having been domesticated; cotton and maize were probably still gathered from the wild. There were tools for grinding grain, and no doubt primitive digging sticks for planting seeds.

During roughly the next 2,000 years, up to 3400 B.C., valley populations increased tenfold and became still more sedentary (although small bands dispersed seasonally for plant collection and hunting). Within another 2,000 years maize, several beans, several species of *Amaranthus*, yellow sapote (*Lucuma salicifolia*), certain squashes, and perhaps pumpkin were added to the rolls of domesticated plants. Their agriculture was clearly efficient, and their settlements gradually increased in size. But fecal deposits indicate that wild animals and undomesticated plants still accounted for over half of their diet.

By the last millennium B.C., the people of the valley were essentially full-time agriculturists, supporting developing civilizations; one digging shows a sequence of 17 consecutive floors. Improved maize was by then a mainstay, and doubtless there were choice slections of the many plants already mentioned. By the birth of Christ wattle-and-daub houses were grouped in villages that sported various ceremonial structures, the occupants being full-time farmers who employed irrigation. New plant introductions included tomato, peanut, lima bean, and guava. Agriculture continued to expand until the Spanish conquest in 1540.

Archaeological findings indicate pretty well what the earliest forms of maize were like. With such a model in mind, Mangelsdorf has reconstructed an "ancestral" maize by crossing and backcrossing podcorns and popcorns. The resulting plant is quite unlike modern maize; its seeds are not enclosed by husks and are borne on fragile branches of the tassel as well as on ears at high positions on the stalk. Wild maize similar to this was an important food item until the third millennium B.C. in the Tehuacan Valley. Evidently it was brought down to the valley floor for cultivation, to an environment too dry for agriculture until irrigated. Planting techniques probably involved nothing more than dropping a seed into a hole poked in the soil with a pointed stick.

From about the fifth millennium B.C. until nearly the time of Christ, a primitive cultivated maize was much utilized. About 1500 B.C. improved types began to appear, leading to modern maize. The biggest genetic advance may have come as early as several millennia B.C. through natural hybridization with the related plant teosinte (*Euchlaena mexicana*), itself perhaps the result of a cross between maize and gama grass (*Tripsacum dactyloides*). Wild maize apparently became extinct before the time of Christ, perhaps genetically "swamped" by cultivated types. Undoubtedly the natural habitat of wild maize had become thoroughly replaced by cultivated fields.

Here, as in the Near East, human cultures gradually adopted techniques found useful by trial and error to reduce need for migration in search of wild plants and animals. New crop originations were profuse in the Tehuacan Valley, but certainly a number of other centers must have domesticated particular species independently. A crop so important to New World civilization as maize undoubtedly gained great renown and became widely spread from its center of domestication. It was grown throughout Mexico and in parts of North America 4,000 years ago. Fossil maize pollen over 2,000 years old has been identified in swamp deposits as far north as Virginia.

Agriculture in the New World was developed by the genius of the Indian to conform to the diversity of conditions. For example, the Aztecs developed unique floating gardens in watery environments. Reed baskets eight feet in diameter were filled with earth and set in lake waters to become stationary with the anchoring of roots. The Mayans, using the corn stalk as a support for the bean, developed (perhaps unwittingly) a basis for a balanced cropping program and a balanced diet that still sustains the Indians of this region. The form of agriculture perfected by the Incas was uniquely adapted to their mountain terrain. It was based on the potato, which produced abundantly in high altitudes and was preserved by dehydration, which required repeated freezing and crushing. Deep narrow valleys were irrigated through the long-distance channeling of streams and the creation of stone-laid reservoirs. Stone terraces, many still used, prevented erosion. More people were maintained on a higher standard of living 500 years ago than at present in the forbidding Andes region.

THE RISE OF SCIENCE

Philosophy, literally "love of knowledge," denotes the orderly arrangement of information—information dealing with abstract ideas or with the substance of the real world (natural philosophy). Such an orderly arrangement of information can be built up in various ways. Beginning with a reasonable premise, and drawing conclusions (generalizations) in a logical manner is one very attractive method. This method, called **deduction,** was formalized by the Greek philosophers, who formulated rules of logical reasoning designed to extract true inference from generalizations. The flaw in this method of reasoning is that it depends on the correctness of the original premise, which in this method of reasoning is not subjected to test. An alternative method of reasoning is **induction.** In this method generalizations are derived from observations. The generalizations must then be continually tested against new observations and, if possible, against specially devised observations to test the validity of the assumption. Most important, the generalization is considered an *approximation* to the "truth," subject to change, if necessary, to fit any new observations that challenge the validity of the assumption. A generalization, to be useful and valid, must not only fit each and every observation, but must prove useful in prediction. The concept of the testing of generalizations developed a new natural philosophy known as "**science**" (from the Latin "to know"). The word science is now used in place of the older term "philosophy" in reference to the body of knowledge of a subject (for example, plant science). Science is in fact a method of inquiry for the organization of information. **Science** is the **scientific method.** The scientist is one who contributes to this body of information by making generalizations, performing observations, and testing one observation against the other. Experimentation is his most powerful tool. The creative part of science lies in the devising of experiments to test conclusions and the synthesis of generalizations. The discoveries of science are often intuitive or accidental, sometimes removed from actual experimentation except as they are used in subsequent testing. They are a combination of intelligence and insight—from the ability of some to look at old problems in fresh ways. In the creative process, experimentation aids but is not necessarily a guarantee of discovery. "Experiment is experience sharpened to a point—useful as a digging stick but not as a divining rod."[*]

In the ancient world technology gave rise to science; only recently has science been the source of technology. The early contact between science and technology had been only marginal. Philoso-

[*] H. S. Harrison, "Discovery, Invention, and Diffusion," *in* C. Singer, E. J. Holmyard, and A. R. Hall (editors), *History of Technology* (London: Oxford Univ. Press, 1954), vol. 1, p. 61.

phers felt little responsibility for making practical use of their findings, and the artisan understood little beyond his "craft mysteries"—the inherited, traditional practices. The concept that science and technology were interrelated, both being concerned with natural phenomena, developed very slowly. The fusion of craft mysteries with the new technology of science led to an industrial society that was to engulf the Western World. The technological superiority of the West over the East became evident only at the close of the Middle Ages and then only in military technology. Conquest and economic intrusion were the signs of the vigor of Western civilization.

The accumulation of information through observation (or experience) and experimentation began to reach significant proportions in the sixteenth century and reached maturity in the awesome discoveries in astronomy and physics of DaVinci (1452–1518), Galileo (1564–1642), and Newton (1642–1727). Renewed interest in botanical studies accompanied the revival of learning, as evidenced by the increasing appearance of herbals—descriptive treatises of plant material. The development of an appetite for scholarly works coincided with the European invention of printing from movable type about 1440 (printing from wood blocks was known earlier in China), which had tremendous impact on the spread of ideas.

Fundamental studies of botanical matters began to appear in print in the seventeenth century, stimulated by the invention of the microscope (1590), just as the invention of the telescope (1608) was to be the generating factor in the advance of astronomy. Robert Hooke (1635–1703), with the aid of the microscope, described "cells" (the name still in use) in cork and extended these findings to other plants. The studies of Antoni Van Leeuwenhoek (1642–1723) in observing and describing protozoa and bacteria are well known. The microscope also influenced the fundamental studies on plant anatomy of Marcello Malpighi (1628–1694) and Nehemiah Grew (1628–1711), research unsurpassed for a century.

The importance of experimentation as a tool in biology increased during the seventeenth and eighteenth centuries. Experimental studies made by the Dutch botanist Rudolph Jacob Camerarius (1665–1721) clearly established the nature of sexual reproduction in plants. The clarification of this basic concept, together with the later hybridization experiments of J. C. Koelreuter (1733–1806), were to lay the foundation for the brilliant experiments on inheritance in the garden pea by the Austrian Abbot Gregor Mendel (1822–1884).

Plant taxonomy was to mature in the eighteenth century. During the more than twenty centuries following Theophrastus, countless lists of plants were developed. The usefulness of these compilations was limited by the lack of a functional and consistent method of classification. This problem became increasingly evident as the number of known plants and plant products increased. For example, the famous medical herbal compiled by the Greek Pedanius Dioscorides in the first century A.D. lists 600 plants; a later Arabic work, The *Kitah al-Jami* of Abu Muhammed ibn Baitat (1190–1248), lists 1,400 plants, foods, and drugs; the *Pinex theatri botanici* (1623), written by Kaspar Bauhin (1590–1624), describes 6,000 species! The Swedish physician Carl von Linné (1708–1778), better known as Linnaeus (the Latinized form of his name) developed a workable system of classifying plants and animals. The system he devised is based on structural differences and on similarities between the morphology of reproductive parts, the organs least likely to be influenced by environmental conditions. Linnaeus' fame further rests on the development, with his students, of the binomial system of naming organisms according to genus and species—a system still in use.

The eighteenth century also saw the start of fundamental studies in plant physiology. The Englishman Stephen Hales (1677–1761), greatly influenced by Harvey's discovery of the circulation of blood (1628), was the first to use careful experimental techniques and quantitative methods. He pioneered studies of transpiration and the movement of sap. Another Englishman, Joseph Priestley (1733–1804), the discoverer of oxygen, was the first to recognize the analogy between combustion and respiration. He demonstrated that growing plants "restored" air that had been respired or had candles burnt in it. It was not for many years after this, however, that the significance of his findings came to be understood.

AGRICULTURAL REVOLUTION

The term "agricultural revolution" refers to the transition from medieval to modern practices. The emergence of new agricultural techniques in the seventeenth and eighteenth centuries can be traced to new crops and economic changes. As serfdom disappeared, so did the rights of the peasant to graze livestock on the lord's common field, to hunt and fish on his lands, or to cut wood in his forest. The system in which the cultivated land of the village was divided up into different regions, with each peasant managing one or more strips, gave way to the enclosure of fields to be cultivated in one large-scale operation to increase efficiency. The eventual disruption of rural life was cruel, but it was on these expanded farms that major advances in agricultural technology originated. During the eighteenth century the value of good farmland rose tenfold. The increased productivity was responsible for the growth of the urban population, and in effect allowed the industrial revolution to happen.

The practices of crop agriculture were greatly affected by the acquisition of new foods during the Age of Exploration. Three of the world's great food plants—potato, maize, and rice—entered European agriculture. The potato, introduced into Spain before 1570, was to become one of the most important food and industrial (starch and alcohol) crops in Europe. Introduced into Ireland in the late sixteenth century, it became in time virtually the only source of food for most of the population. When the potato crop failed several years in succession in the mid-nineteenth century as a result of the late blight disease, the Emerald Isle was depopulated by death and migration. Maize, another New World introduction, became an important crop in southern Europe. Rice, although introduced into Spain in the eighth century, was not cultivated in Italy until the sixteenth century.

The demand for three new beverages: tea (from China), coffee (from Abyssinia via the Moslem world), and cocoa (New World) also gave great importance to sugar as a sweetening agent. Sugar cane was imported to the New World, and the West Indies soon provided the main source of Europe's supply in the seventeenth and eighteenth centuries. Beet sugar did not become important until about 1800. The exploitation and taxation of these crops were to have a profound effect on the body politic.

The problem of insufficient feed during the winter in the temperate countries of Europe was alleviated by such new winter crops as clovers, which did not deplete soil fertility, and turnips. The new farming systems, which made use of enclosed fields, made it possible to regulate animal populations, and the more intensive rotations led to increased development of animal agriculture. Pasture land that had been inefficient under the old practices tended to decrease. The relation between pasture for sheep and arable land for cereals was kept in balance in England by the changing prices for wool and grain. In the newly developing systems the cattle were fed in stalls from fodder grown in the fields. The system required considerable labor, but was very efficient in land utilization. Moreover, it provided large quantities of manure, which previously had been largely wasted but could now be used to fertilize the soil. Fertilization practices also improved with the use of marl—a clay-like material rich in calcium carbonate.

The old fallow system, in which one-third of the land was unplanted to preserve and restore fertility, gave way to improved, "scientific" methods of crop rotations that permitted great increases in production through more efficient utilization of the land. Although these methods were not new in themselves, the idea that crop rotation could be used to bring about several desirable results was new. The advantages of crop rotation (although not entirely understood) were (1) the control of diseases and weeds that tend to become severe problems under continual one-crop plantings, (2) convenience and efficiency, (3) fertility improvements by the use of nitrogen-fixing legumes, (4) erosion control by the use of winter cover crops, and (5) spreading the risk by increasing the number of crops.

The beginnings of mechanization were the preview of a full-fledged revolution that was to change the course of agriculture completely. The development of improved flour-milling techniques heralded the use of mass production for the process-

FIGURE 1-7. *Farming in France, 1735. Ploughing, broadcast-sowing by hand, and harrowing-in the seed.* [*From Singer, Holmyard, and Hall (editors)*, History of Technology, *Vol. 4. London: Oxford Univ. Press, 1958.*]

ing of foodstuffs. The increases in long-distance trade of agricultural crops and products necessitated testing and evaluation procedures, and the resulting establishment of standards led to a rise in the quality of agricultural equipment. Farm machinery was at first relatively simple, and was derived more from the ingenuity of local craftsmen than from specialized knowledge (Fig. 1-7). New plows, seed drills, threshing and harvesting machines, however, soon became increasingly complex. Rising industrialism was to find a real market in agricultural tools, and by 1850 much of the agricultural machinery in Britain was furnished by factories.

NORTH AMERICAN AGRICULTURE

The North American colonists, although lacking wealthy empires to plunder, found rich agricultural resources to develop. Nearest at hand was the forest, which immediately provided two exportable surpluses, timber and fur. Practically the

entire Eastern seaboard was continuous forest. While the development of farming necessitated land clearing, the forest was to provide riches in the form of fuel; building materials, posts, rails, furniture, implements, staves, ship frames and masts; tar, pitch, and turpentine; potash for fertilizer and soap, and hemlock bark for tanning. The early development of colonial manufacturing such as ship building was based on timber resources; the forest provided the chief source of foreign exchange. The "Broad Arrow" policy of England (the three blazes marking trees represented the broad arrow, symbol of the British Navy), which reserved to the Crown all trees over two feet in diameter to ensure a supply of mast timber, was to contribute to the growing disenchantment between Britain and her American colonies.

Farming practices in North America were developed out of the experiences of the Old World. Agriculture in the first colonies resembled the early agriculture in Europe. The potential of maize was realized immediately, however, and it soon became the dominant grain crop. The rise of tobacco and, soon after, rice, as export "cash" crops, stimulated the development of slavery, for both crops require high labor inputs. (The decline in the price of tobacco decreased the virulence of slavery; in 1778 Virginia prohibited the importation of slaves.) With Eli Whitney's invention in 1793 of a machine that could separate the cotton fibers from the seed, cotton became an extremely profitable crop; Whitney's gin did the work of 100 men. Exports increased from 275 bales in 1797 to 36,000 in 1800. King cotton was to establish the plantation system as the basic agricultural pattern in the southern United States. The slavery issue and the future of the plantation system, which depended upon slaves, were eventually to be decided by a civil war.

The young United States was primarily an agricultural nation. In 1800, 90 percent of the working population was engaged in farming. The early agricultural pattern in the United States was strongly affected by the seeming limitlessness of new lands. It was new soils that were most responsible for the unprecedented levels of productivity.

The improvements in European agricultural practice were followed closely in the United States, particularly in the field of mechanization. The mechanical reaper of Cyrus McCormick, invented in Virginia in 1831, was the first to be generally adopted. Labor shortages and high grain prices during the Civil War accelerated the adoption of machines, and the leadership in mechanization was to pass to the United States, a fact that does much to explain the technological progress in American crop production in the twentieth century (Fig. 1-8).

M'CORMICK'S PATENT VIRGINIA REAPER.

FIGURE 1-8. *Development of the reaper. (Top) Cyrus McCormick's first reaper, 1831. (Middle) The 1851 McCormick reaper. (Bottom) The twine binder (1881) reaped and tied sheaves of grain in one operation.*

BEGINNINGS OF THE AGRICULTURAL SCIENCES

The changes in agriculture brought into sharp focus many problems that traditional practices could not cope with. In the forefront were the difficulties associated with plant and animal nutrition and disease control. Until the sixteenth century little had been added to man's knowledge of plant nutrition since Aristotle had formulated his hypothesis that plants absorbed foods from the soil. Malpighi had mentioned that leaves might have a role in the elaboration of food, and Nehemiah Grew had suggested that light absorption might play a role in this process. Various eighteenth-century researchers had worked out some notion of the relation between oxygen, carbon dioxide, and soil nitrogen, but in spite of this the main theories of plant nutrition accepted at the beginning of the nineteenth century were grossly incorrect. It was felt that the carbon source of plants was decomposed soil organic matter ("humus") and that the inorganic salts in the soil were little more than growth stimulants. The beginnings of modern plant nutrition and agricultural chemistry are found in the work of the German chemist, Justus von Liebig (1803–1873), who in 1840 presented evidence that contradicted the "humus" theory of plant nutrition. Although he erroneously concluded that plant nitrogen is derived only from the ammonia of the air, he initiated the correct view that plants obtain carbon from carbon dioxide and mineral constituents (potassium, calcium, sulfur, and phosphorus) from the soil. Pasteur's suggestion in 1862 that microorganisms might be involved in soil nitrification was confirmed 15 years later. Studies in plant nutrition advanced at a rapid rate at the end of the nineteenth century, and were paralleled by the growth of the artificial fertilizer industry.

The understanding of the biologic nature of disease was to have a profound effect on agricultural practices. The studies of early mycologists provided the foundations but did little to eliminate the ancient belief that disease was a scourge descended from the demons. The results of experiments of Matthier du Tillet in 1755 (stimulated by a prize offered by the Academy of Bordeaux) clearly showed the disease-producing potential of the "black dust" from bunted wheat and described treatments of the grain with lye and lime to control the disease. It remained for Benedict Prevost (1755–1819) to demonstrate that the black dust of bunt was in fact spores of the fungus *Tilletia tritici*. Prevost described the fungicidal properties of low concentrations of copper sulfate and its use in seed treatment. These early studies were generally ignored until Louis Pasteur (1822–1895) established that microorganisms are responsible for rather than engendered by disease. A rational explanation of the diseases of plants was offered by the German botanist Heinrich Anton de Bary (1831–1888) through his classical work on the late blight disease of potato. He established the concept that specific plant diseases are caused by specific microorganisms, and defined the concepts of parasitism, infection, and resistance. The control of plant disease as an agricultural practice traces to the discovery by Millardet in 1882 that combinations of copper sulfate and lime (Bordeaux mixture) when applied as a spray to grapes would control mildew, a ruinous disease of the French grape industry.

DEVELOPMENT OF AGRICULTURAL EDUCATION AND RESEARCH

Scholarly writings in agriculture, although at first infrequent, appeared continuously in Western writings. The important work *De Re Rustica* (1594), by the Frenchman Charles Estienne, was to appear in many editions, including English translations. One expression of the agricultural revolution in the eighteenth and nineteenth centuries was the increasing appearance of works on agriculture. The emergence of agricultural professionalism was characterized by the appearance of agricultural societies and agricultural publications. The Danish Agricultural Society was founded in 1769; the establishment in 1785 of the Philadelphia Society for the Promotion of Agriculture was soon followed by the formation of many similar societies in the United States. The Royal Agricultural Society in England was incorporated by Royal Charter in 1840. A Board of Agriculture established in England in 1793 was the first of a long and continuous series of governmental agencies to deal with the problems of agriculture.

Formal educational training was another consequence of the increasing stature of agriculture. A school of veterinary medicine was founded in Copenhagen in 1773. Chairs of agriculture were established at the leading universities—Edinburgh, 1790; Oxford (rural economy), 1796; and Yale (agricultural chemistry), 1847. In the early 1800's a number of agricultural institutions were founded in England and on the continent. The beginnings of the agricultural experiment station trace to the creation in 1834 by John Bennet Lawes (1814–1901), with the later collaboration of Joseph Henry Gilbert (1817–1901), of a private laboratory dedicated to research in agriculture at Rothamsted. This famous institution, which is still in existence, was to be extremely productive. In addition to the early work on fertilizers and fertilization practices, work on experimental technique was to have important consequences for statistics. In 1842 Lawes patented superphosphate, formed by treating phosphate rock with sulfuric acid, and in effect initiated the chemical fertilizer industry.

The increasing interest in agriculture in the United States resulted in the establishment in 1862 of the Department of Agriculture, formerly a division of the Patent Office. In the same year two other famous pieces of agricultural legislation were signed by President Lincoln: the Homestead Act, which provided for staking farms on governmental lands, and the Land-Grant College (Morrill) Act, which established educational institutions in "agriculture and the mechanical arts." Iowa became the first state to accept the provisions of the Land-grant College Act. This legislation was to have a tremendous impact on agriculture and technology as well as on the American educational system. Whereas higher education had previously been the privilege of a few, the new institutions established the principles of people's colleges. The state agricultural experiment stations were to develop a vast pool of agricultural information—information on the classification of soils, diseases and insects, and the development of improved variety collections. It was this body of knowledge that provided the base for the great strides made in American agriculture in the twentieth century.

TWENTIETH CENTURY TECHNOLOGY

Although the population of the United States is continually increasing, the country is under no immediate internal pressure to increase food supplies, even though its farm force is decreasing and its farm acreage declining (Figs. 15-7, 28-7). In 1910 each farmer produced for himself and seven others; in 1967 he "supported" more than 50 people. This increase in labor efficiency is the result of improved technology. The magnitude of this increase has been such that the recent agricultural problems have tended to be due to overproduction and low prices. Although the new technology in agriculture had no definite beginnings, its effects became noticeable in the late 1930's, when there was a sharp increase in the rate of technological change (see Chapter 28).

The increase in the United States' agricultural production from 1880 to 1920 was largely a reflection of the increase in expenditures—more land and more labor. Returns per acre remained relatively constant. At the outbreak of World War I the number of people engaged in farming was the highest in history (even though the percentage was declining), as were the number of acres under cultivation and the number of horses and mules (25 million). The replacement of animal power by the gasoline engine in the 1920's was the first great step in the twentieth-century technological revolution. The release of the land formerly used for feed stabilized the theretofore expanding farm force and acreage during the decade following World War I. In the following decades the early pioneering work of the agricultural experiment station began to pay off in the form of improved agricultural practices. The resurgence of genetic investigations in the quarter century since the rediscovery in 1900 of Mendel's paper on inheritance was translated into new and better crop varieties. The development of hybrid corn and its rapid adoption (Fig. 28-4) in the corn belt was the most spectacular of these achievements. These improved genetic stocks, in combination with the increased use of inorganic fertilizers, accounted for a large part of the tremendous increase in production necessitated by World War II.

A whole new set of technological advances was initiated in the late 1940's, made possible by the research in the preceding decades. Agricultural chemicals in the form of weed killers, organic fungicides, and insecticides quickly followed the spectacular commercial success of the insecticide DDT and later the broadleaf weed killer 2,4-D. These improvements increased yields per unit area as well as conserved labor. Mechanization increased in the 1950's to include even "chore" tasks. The increasing use of supplemental irrigation in the United States made economical the use of additional fertilizer. The net result has been a steady increase in agricultural efficiency.

In the late 1950's a new trend in research developed when the Federal government decided to support basic research on a very large scale. Although the amount of research funds provided to support applied agricultural research has not diminished, the percentage of funds has declined sharply over the years. In essence a cycle is being repeated: the freedom from the exigencies of agricultural problems has enabled a concentrated effort in the "luxury" areas of research. Present studies of the basic mechanism of plant growth and development cannot but have a profound effect on the future technologies of crop production.

In the 1960's two approaching crises impinged on world consciousness. One was an emerging world food shortage caused by an unchecked population increase in the tropical world. The second, first brought into prominence by Rachel Carson's *Silent Spring* (1962), was the polluting consequences of technology itself. The two problems were soon shown to be part of a single issue—the distribution of the earth's renewable and nonrenewable resources. Ironically, as man raced to conquer the moon he discovered the earth. Whereas our plant planet had traditionally been thought of as "Mother Earth," the source of bounty and plenty, the new concept, of a "Spaceship Earth" stressed the finiteness of her resources. The new earth was seen as if from the vantage of the man on the moon: small and isolated, yet beautiful and precious as never before. It soon became apparent that there was indeed a limit to earth's resources and that increasing population acted as a growing divisor of the world's wealth.

Worst of all was the realization that the narrow application of technology, geared only to short-term increases in production, was leading at best to a fouling of the nest, and at worst to mass poisoning.

The social consequence was the rise of the ecology movement. Conceived in fear and born in crisis, the movement inspired religious fervor and struck a responsive chord in the United States during the 1960's. Although its message has been strident and its thrust has sometimes veered uncomfortably toward anti-intellectualism and anti-science, the ecology movement has proved itself to be a force to be reckoned with. The banning of DDT, its *cause célèbre,* was the most notable of its victories. Recently the ecology movement has spread to the field of social problems as concern for the displaced agricultural worker, the human jetsam of agricultural progress, has risen.

The major environmental problems of that decade have not been solved as of this writing (1973), but events of the past few years have led to increased optimism. In the late 1960's a dramatic increase in the production of major food crops (rice, wheat, maize, as well as grain sorghum and potato) occurred in the developing world. A central breakthrough occurred in plant breeding—the creation of high-yielding, fertilizer-responsive cultivars of wheat and rice suitable for the tropics. At the same time more emphasis was placed on the use of agricultural technology in developing countries, such as India, for example. Increase use of technology in rural areas plus the "miracle" grains, then, introduced what is known as the **"green revolution,"** and advanced a number of tropical countries a small step toward self-sufficiency in foodstuffs. Progress in population control has also been encouraging in at least some countries. The clamor over pollution has made agricultural technologists in the United States more aware of the long-range implications of their action. The feeling now exists that technological progress combined with statesmanship can reverse the trend toward environmental deterioration and lead to not only sustained but increased agricultural production.

One result of the new technology in the United States is the breakdown of the differences that once separated the disciplines of agronomy, horticulture, and forestry. These divisions of plant agriculture had been based on a combination of intensity of agricultural production and tradition. Thus forestry was concerned with extensively managed timber lands, whereas horticulture dealt with intensively managed "gardens." The acceptance of the horticultural concept of intensive care throughout the life of the crop plant and the growing use of mechanization and agricultural chemicals have obscured these differences. Thus, for all crops, there is a growing concern with each phase of development, a decrease in labor input, and at the same time an expanding scale of production—trends that emphasize the increasing role of management and science. There are few differences in the modern Midwest production of cannery tomatoes and field corn. Pruning, once confined to horticulture, is now an established practice in the production of timber; the nursery end of forestry is a true horticultural operation. Similarly, the large-scale practices of field crop production are being adapted to the production of fruit, vegetables, and ornamentals. The traditional differences between the disciplines—differences based on species—were never completely satisfactory, and each new crop raised new difficulties and jurisdictional disputes. Tobacco and potatoes, for example, are considered horticultural crops in some locations and agronomic crops in others. None of the tropical crops fit neatly into either category.

A further indication of the growing unity is the speed with which practices initiated in one discipline are quickly adopted by the other. Growth regulators most widely used in horticulture are rapidly being explored for agronomic crops. The use of computer programming, first accepted in forestry, is being adapted to all cropping practices. Tradition remains the main factor separating the three crop disciplines.

A striking feature of the new agriculture is not only the magnitude of results but the increasing rapidity with which these discoveries are made. No invention comparable to the gin affected cotton production for a century and a half. Recently the mechanical cotton picker, an innovation of similar magnitude to Whitney's gin—one machine re-

24

FIGURE 1-9. *The mechanical cotton picker is the most sophisticated present-day farm machine.* [*Courtesy John Deere Co.*]

places 100 pickers (Fig. 1-9), chemical weed control, defoliants, and hybrid cotton are completely transforming cotton production.

The future course of agricultural technology cannot be charted with certainty. Many of the basic innovations have not been conceived, but the trend in the United States is unmistakable. The increased development and use of chemical herbicides points to the elimination of most cultivation. Plant growth in a weed-free environment will change many of our present agricultural practices, such as field preparation and planting. Pesticide development and the incorporation of genetic resistance can be expected to further reduce losses due to pests and diseases. Many crops will be improved genetically through the utilization of hybrid vigor. Mechanical harvesting will extend to hard-to-pick fruit crops. The decline in the farm labor force as well as increased yields per acre will continue.

In addition to the refinement of present agricultural practices, new methods of producing food have been seriously considered since the close of World War II. These vary from the extraction of plant proteins to the culture of algae or microorganisms. Recent stimulation has come from aerospace research on algae as a means of oxygen renewal in space flight. The mass growth of microorganisms is of course a highly developed present-day technology. The end product has a high value per unit weight as in antibiotic production or the fermentation industries. However, the claims that some new method of food production would quickly arise to solve the world's fast approaching food crisis appear to have been exaggerated or at least premature.

The future course of the plant sciences promises to be an exciting journey. Merely to provide a sustaining diet for the present world population, much less an expanding one, is a herculean task. The goal of proper nutrition for all people and the judicious exploitation of plant life will require the joint effort of creative scientists and wise statesmen. It is an endeavor worthy of the best in men.

Selected References

Gras, N. S. B. *A History of Agriculture* (2nd ed.). F. S. Crofts, New York, 1940. (Development and history of rural life in Europe and the United States.

Reed, H. S. *A Short History of the Plant Sciences.* Ronald Press, New York, 1942. (An excellent review.)

Sarton, George. *A History of Science* (2 vols.). Harvard University Press, Cambridge, 1952, 1959. (The development of science up to the "golden age" of Greece. Only the first two volumes of a projected eight-volume treatment of the history of science were written before the author's death. Consult his monumental *Introduction to the History of Science* for a complete and annotated bibliography.)

Sauer, Carl O. *Agricultural Origins and Dispersals.* American Geographic Society, New York, 1952. (An account of the prehistory of agriculture in the Old and New Worlds.)

Singer, C., E. J. Holmyard, and A. R. Hall, (editors). *History of Technology* (5 vols.). Oxford University Press, London, 1954–1958. (A monumental work on science and technology up to 1900. Agriculture and related fields are well documented.)

Von Hagen, Victor W. *The Ancient Sun Kingdoms of the Americas.* World Publishing, Cleveland, 1961. (A popular history with much technological information on the Aztec, Maya, and Inca civilizations.)

Wright, R. *The Story of Gardening.* Garden City Publishing Co., Garden City, 1938. (A popular history of horticulture and gardening.)

CHAPTER **2**

Energy and Crop Production

Our industrialized society feeds greedily and insatiably on energy made available by burning fossil fuels—coal, petroleum, gas. This stored energy was captured from the sun eons ago by plants whose remains became buried deep in the earth's crust. These fossil fuels are of course limited, and the efficiency with which we use them, which depends on our level of technology, will determine our standard of living for some time to come. Although we obtain some energy from water power, and small but increasing amounts from nuclear reactors, the amounts obtained from these sources are still insignificant on a world basis. The world energy use is equivalent to the consumption of 4.2 billion tons of coal annually. Dividing this figure by the earth's estimated population of 3.5 billion gives an energy equivalent of 1.2 tons of coal per capita per year.

Impressive as these figures may be, they are dwarfed by the present-day biological energy transformations that sustain us. Each day every human consumes, on the average, the equivalent of 1.8 pounds of plant material containing 0.7 pounds of carbon. This is about 260 pounds of carbon per person annually, which amounts to a

world total approximately 0.46 billion tons. Practically all of this food is produced from cultivated land, but because of the inefficiencies involved 5 billion tons of carbon must be produced each year (Table 2-1) to supply the amount actually consumed. The difference between the amount consumed and the amount produced is due to a number of factors. On the average only about 20 percent of the cultivated plant is eaten—for example, the grain constitutes only 28 percent of the dry weight of the wheat plant. Further, one-half of the plant material produced on cropland is consumed by animals, and only 3 percent of this stored energy ends up as human food. Losses to pests and diseases account for one-third of the total production. As we shall see, only 1–2 percent, or less, of the total energy from the sun is utilized in fixing the carbon in plants. In the end, to produce the 0.36 billion tons of carbon per year needed at present to sustain man, 50 to 100 times as much energy is required as is used for all other purposes by today's world population.

Basically, agriculture is concerned with the conversion of solar energy to energy usable by

Table 2-1 ANNUAL WORLD FIXATION OF CARBON BY PLANTS, AND THE FATE OF CARBON FROM CULTIVATED LAND.

SOURCE OF CARBON	AVAILABLE CARBON (BILLIONS OF TONS PER YEAR)	POPULATION THAT COULD BE ADEQUATELY SUPPORTED (BILLIONS)
Fixed by photosynthesis		
World total	150	1160
Ocean	134	1040
Land	16	120
Cultivated land	5	38
Fate of carbon from cultivated land		
Plants consumed by humans ($\frac{1}{2}$), 20 percent of plant used	0.5	
Plants consumed by animals ($\frac{1}{2}$), 3 percent converted to human food	0.08	} 4.4
Plant material available after $\frac{1}{3}$ losses due to pests, including disease	0.38	2.8

Source: Adapted with permission from Brown, Bonner, and Weir, *The Next Hundred Years*, New York: Copyright © 1957 by The Viking Press, Inc.

man—for food and fiber. Sadly, the average technology employed in the world's agriculture is insufficient to meet the immediate needs of our expanding populations. Fortunately, some cultures have developed intensive techniques so that a few men are able to produce the food required by hundreds of people. To increase agricultural production and efficiency, we need to be concerned with all phases of energy transformation—from the sun to the supper table.

CROP ENERGETICS

The problem of understanding and quantifying the total energy balance at the surface of the earth has been of concern to meteorologists for many years. They have become interested in how much of the sun's radiant energy is used for the evaporation of water, how much is reflected back to the sky, and how much is diverted into various other channels (Fig. 2-1). This accounting process is a matter of "balancing the energy budget," and is a bookkeeping procedure in every sense of the word, although much more complex than the business accountant's double-entry system. Precise and ingenious instrumentation has been devised to measure the energy flux at the surface of the earth and to determine how the total energy budget balances. This objective of meteorology is of more than academic interest, for it relates directly to water availability for human needs. Unfortunately, the amount of energy captured by plants is so small that meterologists seldom take it into account. The total amount used in photosynthesis usually amounts to only 1 or 2 percent of the total solar energy input, which is within the limits of computational error for the total input. Energy requirements have been calculated for various physical processes that take place in croplands. For example, it has been estimated that about two-thirds of the net radiation falling on vegetation is used in the evaporation and transpiration of water.

28

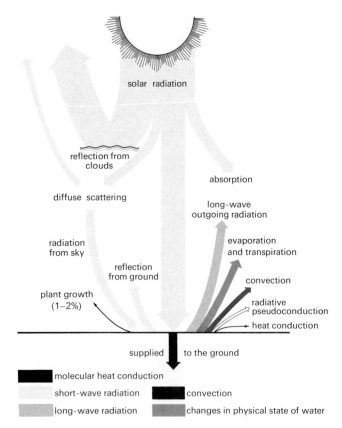

solar radiation

reflection from
clouds

diffuse scattering

absorption

long-wave
outgoing radiation

radiation
from sky

evaporation
and transpiration

reflection
from ground

convection

plant growth
(1–2%)

radiative
pseudoconduction

heat conduction

supplied   to the ground

molecular heat conduction

short-wave radiation

convection

long-wave radiation

changes in physical state of water

FIGURE 2-1. *Energy exchange at noon on a summer day. The arrow width indicates relative amounts of energy transferred. Note that plant growth accounts for a very small part of the total energy budget.* [*After Geiger,* The Climate Near the Ground. *Cambridge: Harvard Univ. Press, 1950.*]

In recent years considerable work has been done by biologists who are trying to quantify the productivity of plants. This field of research, called **production ecology,** is concerned with the capture of radiant energy in photosynthesis, its conversion into chemical energy, and its flow through plant and animal communities.

The energy captured in photosynthesis is conveniently represented in part by the total biomass on a unit area of the earth's surface at any given time. **Biomass** merely refers to total living organic matter, usually on an oven-dry-weight basis, but for some purposes it may also include nonliving materials, such as bark, cuticle, and resinous deposits.

Even though a crude measure, biomass is useful for making comparisons of different crops and different land areas. The amount of organic matter present is only a partial measure of production, because respiration requires a high proportion of carbohydrates, which are part of the total or gross production.

Photosynthetic efficiency

Every year, an average of 263,000 langleys° of solar energy is received at the outer edge of the earth's atmosphere. Of this, approximately 123,000 langleys is either absorbed or reflected back into the atmosphere by molecules, water vapor, and dust, while the balance of 140,000 langleys actually reaches the surface of the earth. In terms of energy values one square meter of the earth's surface will intercept in one day the amount of energy required daily by an active person. This tremendous input of energy is potentially available for use by plants and animals. But solar radiation is not uniformly distributed over the surface of the earth. It varies with cloudiness, the amount of dust in the atmosphere, latitude, altitude, and local topography, season, and time of day (Fig. 2-2).

It is interesting to compare the input of solar energy per unit area of land surface at various places on earth. In the United States the average amount of solar radiation received per day is 300 langleys and varies from about 100 to 800 langleys depending on season and region. At the University Experiment Station in Alaska, the total annual input of radiation is about 17,920 langleys during the course of a 64-day frost-free season. In Miami, Florida, where the growing season lasts about 10 months, more than 132,300 langleys is available. The great difference in solar input between tropical and temperate regions is not indicative of the actual relative productivity, however, for many other factors must be taken into account. The effects of air pollution must not be overlooked in heavily populated or industrial areas. Total sunlight reaching the ground in and around London,

°1 langley = 1 gram calorie per square centimeter. One gram calorie is the amount of heat energy necessary to raise the temperature of 1 g of water 1°C (specifically from 14.5 to 15.5°C); 1 kilocalorie (or 1 Calorie) = 1000 gram calories.

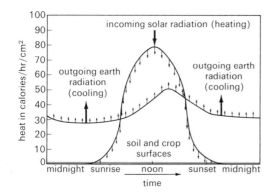

FIGURE 2-2. *Incoming energy is balanced by losses. Gains exceed losses during the day; the reverse is true at night.* [*After Newman and Blair,* Crops and Soils Magazine, *June–July 1964.*]

for example, is reduced by more than half by pollution and fog.

Of special interest to agriculture is the radiant energy in the visible part of the spectrum. Our biological bookkeeping is concerned mainly with this part of solar radiation, for green plants are able to convert light energy to chemical energy through the process of photosynthesis. This important reaction has been the subject of intensive study for many years, and in 1961 Dr. Melvin Calvin received a Nobel Prize for his contributions to the elucidation of the physiological processes involved.

The raw materials and the end products of photosynthesis can be summarized in the following chemical equation:

$$6CO_2 + 12H_2O \xrightarrow[\substack{\text{in the presence of} \\ \text{chlorophyll, certain} \\ \text{enzymes, and cofactors}}]{\substack{673,000 \text{ calories of} \\ + \quad \text{light energy}}}$$

$$C_6H_{12}O_6 + 6O_2 + 6H_2O,$$

which merely says that 6 moles of carbon dioxide, 12 moles of water, and 673,000 calories of energy will yield 1 mole of glucose, 6 moles of molecular oxygen, and 6 moles of water.° This statement does not show the true complexity of the process, for dozens of contributory biochemical processes and energy exchanges take place during the photosyn-

thetic process. The 673,000 calories in the equation is released as heat when the mole of glucose is burned. Because of the inefficiencies of the numerous reactions involved in photosynthesis more than 2 or 3 million calories is required for each mole of glucose produced. Under special laboratory conditions experiments on the efficiency of the cholorphyll molecule in converting light into chemical energy have shown efficiencies as high as 75 percent, whereas estimates based on short-term experiments with growing plants show efficiencies that range from 15 to 22 percent. (In diffuse light cultures of algae have reached efficiencies of 20–50 percent but this range drops to 2–6 percent when large tanks are used). Under field conditions the overall efficiency of crops over extended time periods is only a few percent. Under average conditions less than 1 percent of the total radiation received from the sun is fixed by plants.

The solar energy used by plants is derived only from wave lengths between 0.4 and 0.7 microns, the portion of the spectrum that we perceive as light. When captured by plants the energy is incorporated in the molecular bonds of many different kinds of compounds. Because all chemical bonds are not of equal strength, different compounds may possess different energy levels. Some, such as cellulose, are terminal structural products. Others, including starches and oils, function as intermediate storage compounds and may be converted to other forms, subject to demand by metabolic "messengers." Sugars are not only mobile transfer products, but are used almost immediately in respiration, the biological combustion of energy-rich carbon compounds. In a gross manner, respiration can be characterized as the reverse of the simplified photosynthetic equation. Of the energy released and made available by respiration, part

°Although photosynthesis usually involves the reduction of carbon dioxide and the splitting of water to form sugars, some primitive plants carry on variations of this process. Purple sulfur bacteria, for example, use H_2S rather than H_2O and produce elemental sulfur rather than oxygen. The equation for photosynthesis, therefore, can be generalized as follows:

$$6CO_2 + 12H_2X + \text{light energy} \rightarrow C_6H_{12}O_6 + 12X + 6H_2O.$$

Here H_2X is a compound that donates electrons; X may be either oxygen or sulfur. Light supplies the energy required to separate hydrogen from its donor.

is lost as heat and part is utilized in biosynthesis and chemical work. To a large extent the net amount of energy incorporated into organic matter represents the difference between photosynthesis and respiration. Transfers of energy from one compound to another within the plant are handled by special energy carriers such as adenosine triphosphate (ATP). This flow of energy is a specialized part of the energy concept, and is receiving much attention from biologists.

One important factor which determines the actual photosynthetic efficiency of a leaf is the manner in which the rate of photosynthesis changes with light intensity. The rate of photosynthesis increases with light intensity only up to a certain intensity. At this intensity it is convenient to say that the leaf is **light-saturated.**

Although photosynthesis increases with increasing light, it does so at a decreasing rate, and the efficiency of light capture consequently decreases. In fact, maximum efficiency can be obtained only under relatively low light intensities. When intensities are high, relatively more light passes through leaves and is reflected from them. At low intensities a high proportion of light may be absorbed and used. For short periods of time efficiencies of 7–10 percent are possible for some crop plants, but efficiencies of 2 to 3 percent are probably the most that can be expected over extended periods of time, even if temperature, carbon dioxide, water, and mineral elements are optimal.

Even though less efficient, total production is still greater at high intensities. Although photosynthesis in a single leaf may level off at about 3,000 footcandles, the rate of photosynthesis in the whole plant may continue to increase up to 10,000 foot-candles because more light reaches the lower leaves, which are shaded.

Light may pass through only a few layers of leaves in low crops but may pass through as many as 15 or 20 layers in tropical forests, where 95 percent of the light may be absorbed before reaching the ground. Sugar beet leaves have a vertical distribution of about a foot, whereas in forests the distribution may extend more than 300 feet. In natural forest stands the amount of light reaching the lower leaves is below the **compensation point** —that light intensity required to maintain a rate

of photosynthesis equal to the rate of respiration. Lower branches may actually persist at the partial expense of the upper ones, which supply them with some carbohydrates and thus have been called "negative branches."

Low crops, with leaves in a relatively narrow zone close to the ground, seem to be more efficient producers than forests, in which leaves are spread over a wide range of heights. This is probably due to the fact that in low crops a smaller percentage of leaves is exposed to light below the compensation point. Moreover, trees have a more extensive transportation system through which the compounds move, and the construction and maintenance of the "plumbing system" and the movement of compounds requires much energy.

The growth of plants is a function of the efficiency with which they produce dry matter. This involves the efficiency with which they capture light energy and the efficiency with which they transform it into organic matter. The combined efficiency of these two processes in our best cultivated crops—sugar cane, sugar beet, wheat, and rice—is between 1 and 2 percent. But since only a portion of the crop plant is edible (28 percent of dry weight of cereals, 55 percent of the root of sugar beet), the net, or overall, efficiency is much lower. Furthermore, there are losses, as far as man is concerned, due to diseases, pests, and fires.

Utilization efficiency

One of the most troublesome problems, to those concerned with supplying the necessities of life to our expanding population, is the inefficiency with which consumers use food and other plant products. Plants usually function with an average efficiency of no more than 1 or 2 percent in storing the energy that arrives from the sun. Animals and other organisms that feed on the green plants are equally inefficient. (Some of us, judging by the shape of things, are obviously much more efficient than others.) And so it goes, with each successive consumer utilizing only a part of the food it consumes. Consumption is, of course, a means of transfering energy and building materials from one organism to another. Such a transfer of materials takes place through a **food chain** or **food web**, depending upon the linearity of transfer, to different kinds of organisms (Fig. 2-3). Eventually, all of the original energy produced by plants, and not dissipated by fire, is depleted by respiration (Fig. 2-4). We humans are primarily concerned with the food chain from plants to ourselves. Food webs, however, may be extremely complex, and may involve various degrees of predation among lower forms.

It is easy to see that the **ecological efficiency** of the various organisms in the food chain is of the utmost importance. Ecological efficiency may be either a measure of the effectiveness with which (1) solar radiation is captured or (2) the biomass of one organism, either plant or animal, is converted to the biomass of another. Respiration and the production of indigestible materials such as hair, hide, and bone all detract from the efficiency of conversion.

The flow of energy through food chains and food webs is essentially a one-way process. It has an analogy in the second law of thermodynamics, which says that the spontaneous transformation of energy can take place only from concentrated form to a more diffuse form. Each time a transfer of materials occurs some energy is lost and never recovered. It has been estimated that 100,000 pounds of algae are required to produce one pound of codfish! All the remaining energy incorporated into the algae is lost in respiration, somewhere in the food chain between the algae and the cod, and even the codfish doesn't last but a few years before it too is reduced into carbon dioxide, oxygen, a few minerals, and heat energy. The heat energy is lost, however, for it cannot be recovered and incorporated directly into the biomass, even though heat from biological combustion does affect the rates of reactions that take place in living organisms.

Suppose, for a moment, that the world population has increased and the demand for food is so great that every possible means at hand must be applied to avoid starvation. When we eat animals we lengthen the food chain from plant to human; to shorten this chain means less steak and more starch. Shortening the food chain precludes the great loss of energy resulting from each conversion

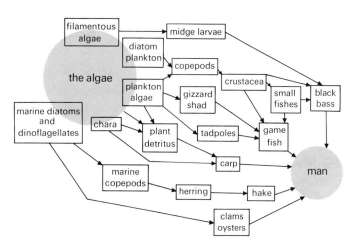

FIGURE 2-3. *Aquatic food web from algae to man. Each time a transfer of energy is made, 90 to 95 percent loss occurs.* [After Transeau, Samspon, and Tiffany, Textbook of Botany. *New York: Harper & Row, 1940.*]

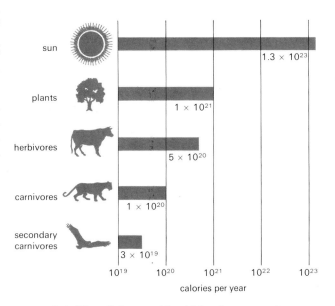

FIGURE 2-4. *The efficiency with which solar energy is used decreases with each step along the food chain. Plants use only a small fraction of the energy that reaches them; herbivores indirectly use only a part of this energy, and carnivores still less.* [From Cole, "The Ecosphere." Copyright © 1958 by Scientific American, Inc. All rights reserved.]

to the next link. In time of shortage the strategy must be to get as close to the primary producers as possible, eliminating the intermediaries. We have eaten primary plant producers in part, or in their entirety, for many thousands of years, and will continue to do so as a matter of choice. The fact is that many plant food sources are extremely palatable. In the Orient the diet of most people consists almost exclusively of plant materials, whereas in the United States we have one of the highest per capita consumptions of meat of any nation.

Neglecting dietary custom and limitless resources, it is pertinent to know what is the most efficient method of providing a hungry world with food. It has been demonstrated, in a laboratory, that algae can produce biomass on a dry-weight basis of 40 tons per acre per year. Of this production, 20 tons would be protein, 3 tons would be fat, and 17 tons would be carbohydrates. No present crop produces organic matter so efficiently.

The culture of algae as a crop is possible with a one-celled species called *Chlorella*. Much is known about its physiology and reproductive habits, and it would be a simple matter to get it into production should economic factors become favorable. Perhaps few readers would find algae soup as desirable as sirloin, but it is at least as rich in some vitamins and amino acids. One group of people ate algae soup for an extended time and found it to be palatable and nutritious. Algae powder could probably be incorporated into many of our foods with little apparent change in quality or flavor. If it is not acceptable as human food, the *Chlorella* might be rendered into feed for animals, thereby allowing us to maintain a more conventional diet. Many other kinds of algae are eaten by people all over the world (Fig. 2-5).

In a food crisis even wood can be made edible! By treating woods with a strong acid, molasses can be made that is easily purified for human consumption. How many years would a giant redwood sustain a man if converted into human food? Cultivating yeast on wood molasses could produce a complete, well-balanced food. In Germany, during World War II, fats were manufactured from coal. This, however, is a complicated process that can be used only when expense is no object.

Woods that cannot be used as structural or food

FIGURE 2-5. *Intensive agriculture can be conducted even in the oceans. Posts and nets support a crop of edible seaweed,* Porphyra, *in the Inland Sea of Japan.* [*From Holt, "Food Resources of the Ocean." Copyright © 1971 by Scientific American, Inc. All rights reserved.*]

materials could be used for the production of gaseous and liquid hydrocarbons. Even though the costs of conversion are high, some countries do find these techniques economically feasible. Alcohol power plants are already in use in some countries where fossil fuels are too expensive. But depending upon the efficiency of conversion, the conversion of wood to gasoline is more likely to become generally economic than the conversion of wood to alcohol.

Even petroleum may become a source of highly proteinaceous food. It has been found that certain bacteria can use petroleum hydrocarbons as a source of energy and building materials.

THE ENERGY FLOW CONCEPT IN AGRICULTURE

The concept of agriculture as a technology that directs the flow and concentration of energy has a number of potential contributions. This point of view gives a common denominator, energy, by which crop production can be measured, and provides an absolute standard for equating the effects of different cultural treatments and practices. Considering crops as traps that can be manipulated to capture, transfer, and store energy provides insight into the means by which we can improve and increase agricultural efficiency and productivity.

Measuring productivity

The fundamental objective of agronomists, horticulturists, and foresters is to increase the efficiency with which solar energy is converted to useful products. Although the products are sold in arbitrary units (pounds of potatoes, bushels of corn, gallons of cider, or cords of wood), much better and more sophisticated units for measuring energy conversion are available. How much air space is there in a bushel of apples? How large is a head of lettuce? It is easy to see why bushels, pecks, and cords are not adequate for scientific use, and not even so for highly intensive agriculture. Furthermore, it is often desirable to measure rates of production of standing crops in the field. What are some of the possibilities for more precise measurement?

Plant productivity can be precisely estimated by measuring either the oxygen released or the carbon dioxide used in photosynthesis. Since the amount of carbon in CO_2 is directly proportional to the amount of carbon fixed in sugars during photosynthesis (Fig. 2-6), productivity can be estimated by the rate of disappearance of CO_2 from its environment. This is a straightforward task in a small growth chamber, but is difficult in the field, where the apparatus for collecting and measuring gases must be disturbingly complex.

It is also possible to estimate production by determining the amount of chlorophyll present in a given amount of vegetation on a given area of land. To do this a sample of leaves of known weight is collected, and the chlorophyll extracted in boiling alcohol. By knowing (1) the total weight of leaves from which the chlorophyll was extracted and (2) the efficiency of chlorophyll in photosynthesis, the total photosynthetic efficiency of vegetation can be estimated. This technique is sometimes used for forest vegetation.

The use of energy to measure productivity gives a degree of precision not heretofore possible. It offers a single unit, the calorie, which is equally useful from the time that light energy is captured by plants until it is incorporated into consumer products. For example, a fairly active man requires approximately 3,000 kilocalories of energy each day, a yearly requirement of 1,100,000 kilocalories. This energy theoretically could be supplied by about 2 tons of potatoes (fresh weight) or 750 lb of wheat—a filling if incomplete and uninspiring menu (Table 2-2). The energy unit of measurement is also useful for expressing the production of fiber crops such as wood. Paper mills are already buying wood on a weight basis. Even though few foresters regard boards or sticks of pulpwood as bundles of energy, this is precisely what they are. The specific energy content of a woodpile or a bale of cotton depends upon the total amount of cellulose present. Energy levels of crops can be used precisely as expressions of quantity regardless of volume or specific gravity.

The process of estimating energy values in terms of caloric equivalents is a fairly complicated process, and must be done by statistically controlled sampling of the biomass. The energy values of organic materials are determined by burning known quantities of materials under carefully controlled conditions and determining how much heat is given off.

It may well be that many agricultural products which we enjoy today will not be produced on the farm, but will be manufactured chemically from component parts. If people could become conditioned to eat and enjoy reconstituted foods and make other such substitutions, agricultural production could be channeled into crops that offer maximum productivity of one or more of the basic requirements of the human diet. For example, certain crops would be grown primarily for their high caloric content, whereas others would be grown for their content of particular organic compounds, such as amino acids. The basic materials, after being synthesized by plants, could be subsequently elaborated into foods with a wide variety of flavors, tastes, and textures, thus satisfying the psychological need for a varied diet.

Increasing crop production

One of the present-day anomalies is that some areas of the world are blessed (?) by agricultural surpluses while others are plagued by persistent and agonizing shortages. There is, of course, a

34

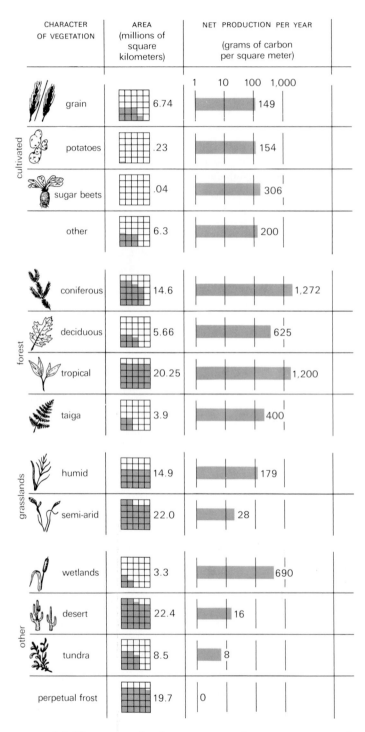

CHARACTER OF VEGETATION		AREA (millions of square kilometers)	NET PRODUCTION PER YEAR (grams of carbon per square meter)
cultivated	grain	6.74	149
	potatoes	.23	154
	sugar beets	.04	306
	other	6.3	200
forest	coniferous	14.6	1,272
	deciduous	5.66	625
	tropical	20.25	1,200
	taiga	3.9	400
grasslands	humid	14.9	179
	semi-arid	22.0	28
other	wetlands	3.3	690
	desert	22.4	16
	tundra	8.5	8
	perpetual frost	19.7	0

FIGURE 2-6. *The amount of carbon incorporated in organic compounds is a measure of the amount of organic matter that is produced. Within any given area, cultivated vegetation is less efficient than forest vegetation, but greater than for grassland. Increasing the productivity of the vast amount of desert area on the surface of the globe may be man's greatest scientific challenge!* [From Deevey, "The Human Population," Copyright © 1960 by Scientific American, Inc. All rights reserved.]

marked difference in the intensity and efficiency of crop production practices used throughout the world. Consequently, appropriate methods for increasing production will depend on the existing level of technology.

In underdeveloped areas the productivity of land can be increased by many agricultural techniques that are now routine in developed countries. In an area saturated with plants, some environmental factor soon becomes scarce. Light, soil moisture, and nutrients are frequently the factors in short supply. These are some of the well-known growth-limiting factors with which agriculturists concern themselves. The technology necessary to identify and overcome these limiting factors and to control crop hazards, such as pests, disease, and fire, is the subject of a large part of this book. This is the direction that agricultural improvement in underdeveloped areas must take.

As limiting factors are recognized and corrected it becomes increasingly difficult to obtain further gain. Nevertheless, there is much room for improvement in the underdeveloped areas of the world. Fertilization is one of the most powerful tools for increasing crop production, yet in underdeveloped agricultural areas little or no fertilizer is applied. Large increases in efficiency can be obtained with crop improvement through genetics and breeding. The ruinous losses by pests and diseases in underdeveloped areas can be prevented with better control methods.

In the developed agricultural areas the immediate, practical problems of production—the amount of fertilizer and the method of applying it, the control of crop pests, the optimum depth to which plowshares should be set, the length of the rotation period, the choice of weed killers— have received much attention. These types of problems have often been solved on an experience and expediency basis, but they have also been tackled by organized research groups using experimental approaches. Many problems can be now handled routinely by "cook-book solutions." The surplus of farm and forest products produced each year in the United States bears witness to the efficiency of this system.

In the United States most of the small, inefficient farmers and timber growers are being forced, by economic pressures, either to leave their farms or to seek supplemental employment in nearby cities. Consequently, the production of farm and forest products has increasingly become the privilege of a relatively few large producers. Even though small applied research projects have been useful in years past in providing immediate answers to many problems, this approach no longer works in advanced agricultural economies. Most of the remaining problems are of a basic nature. To solve them we need a hard core of biological and economic facts. Crop production is taking on a new personality. It is big business, and has become more nearly a science than an art. It is only prudent to assume, in the face of increasing national and world populations, that local overproduction will cease to be a problem. Should international crop sharing become a complete reality, the day of

Table 2-2 YIELD PER ACRE AND EQUIVALENT CALORIC VALUES FROM SELECTED CROPS PRODUCED IN ONE GROWING SEASON.

CROP	U.S. YIELD (1959–1963 AVERAGE) PER ACRE	YIELD IN LB/ACRE	KILOCALORIES/LB	MILLIONS OF KILO-CALORIES/ ACRE	DAY EQUIVALENTS OF ENERGY/ PERSON/ACRE[1]
Potato	19,200 lb	4,200[2]	1,270	5.4	1,800
Rice	3,582 lb	3,600	1,460	5.3	1,800
Corn	60.3 bu	3,400	1,450	4.9	1,600
Soybean	24.2 bu	1,500	1,600	2.4	800
Wheat	24.5 bu	1,500	1,470	2.2	700

[1] At rate of 3,000 kilocalories per day per person.
[2] Dry weight (22 percent).

underproduction in the United States is likely to face us sooner than we realize. To prepare for this almost certain eventuality, it is essential that we learn to improve the efficiency of each phase of the energy flow system of agriculture.

Residue utilization

The first approach to increasing utilization efficiency should be to use a greater part of current production. This is partly a simple matter of thrift. In some countries straw, cobs, husks, shells, and manure are carefully collected and used either for feed or for fuel. But in other, more prosperous countries they are usually discarded or, at best, composted. In the forests of Europe, even the smallest branches and twigs are carefully collected, while in the United States, tops, bark, slabs, and sawdust are usually burned or left to rot. The cellulose in these discarded parts of trees is potentially as good as that in the wood of the finest firs and pines, and the bark has many unusual and useful properties that have infrequently been exploited. Furthermore, many forest species are totally rejected for use because of their small size or poor form.

The extraction of usable fibers from plant residues is a field that holds much promise. Of the many kinds of residues, grain straw, corn stalks, sugar cane bagasse, cottonseed hulls, cotton stems and pods have been utilized to the greatest extent. Strawboard is a commercial product; ground walnut and pecan shells are used as fillers in plastic molding powders; sugar cane bagasse pith is useful as a filler in low-density dynamite. Fine quality papers of the kinds required for cigarettes, fine books, and stationery are made from the fibers of hemp and flax. Sawdust is collected at sawmill sites for manufacture into pulp, an operation which has greatly increased the utilization of wood. Sawdust does not make paper equal to that made from wood chips, but nonetheless it has a number of excellent applications. Someday, a widespread commercial use may be found for the enormous quantities of lignin that are usually dissolved and discarded in the manufacture of paper.

Many industries based on the utilization of discarded plant products could be established. Tops of plants produced mainly for their starchy roots and tubers could be used as sources of protein. Proteins from leaves are now used for animal feed and may someday be extracted for human consumption.

Energy capture

There have been many proposals to broaden the base of energy capture. Most of the photosynthesis on the surface of the earth takes place in water, mainly in oceans and seas (see Table 2-1). Only an insignificantly small fraction of this production is used as food for humans. By systematic fertilization and management, it might be possible to "farm" certain areas of the seas and oceans to raise their level of productivity even higher. Can we produce fiber as well as food in this manner? Before planning intensive ocean management, we must show better extensive use of the resources of this vast reservoir.

In some places it may be feasible to extend agricultural production to deserts and other unfavorable climates. Many desert soils are extremely fertile and only need carefully controlled irrigation to become highly productive.

One obvious method of increasing energy capture is to increase the length of the growing season. Artificial extension of the growing season by using greenhouses makes it possible to produce crops where the growing season is very short. But this technique has limited application in today's economy because of the high cost of building and maintaining greenhouses, and generally has been restricted to the production of flowers and high-value food crops such as tomatoes and grapes. In the future, though, solar heating may be used more extensively, even to the point of making possible the production of relatively low-value carbohydrate sources on a 12-month basis.

These proposals require extremely high inputs of technology and capital. The process of energy capture in present agricultural areas offer opportunities for improvement. For example, in many situations of intensive production the factor that limits energy capture is the availability of carbon dioxide. Producers of greenhouse crops can increase yields by releasing CO_2 gas from storage cylinders to "fertilize" the air.

Air movements affect photosynthesis by influencing CO_2 exchange between leaves and the surrounding atmosphere (Fig. 2-7). Thus the orientation of rows with respect to prevailing winds may influence productivity. Where the prevailing winds are negligible to moderate, corn yields can be increased if the rows are oriented at right angles to the wind. The tops are blown back and forth and there is greater turbulence and mixing of air.

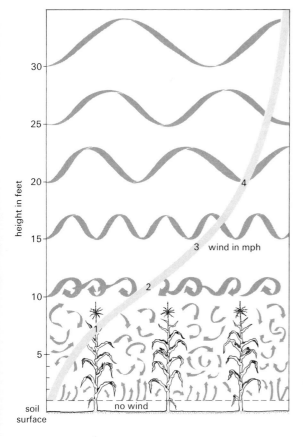

FIGURE 2-7. *The turbulence of wind increases with distance above the soil surface. Crop spacing can increase turbulence at lower elevations, thereby increasing carbon dioxide exchange.* [*After Neuman and Blair,* Crops and Soils Magazine, *June–July 1964.*]

Plant morphology also has a relation to energy capture. Such characters as plant type, leaf shape, and branch angle influence light absorption. Redesigning crop plants to increase the capture and use of energy will be the future course of the technology of plant improvement.

The energy-flow concept will be particularly applicable in forestry. The present production of lumber and pulpwood may well give way to a forest economy based on two commodities, alpha cellulose for structural purposes and degradation products of lignin for stock feed. The primary need, however, will be for molecules of cellulose. Such a change in the objectives of production seems logical. In the form of dissolving pulp, a product of alpha cellulose, wood can be exploded, foamed, extruded, moulded, shaped, and combined with other materials to form products that are dimensionally stable; resistant to fire, insects, decay and rot; and highly adaptable to many uses. Unwoven fabrics are already finding a good market, and it may not be too long before fiberless paper is in use.

Since straight tree trunks and a high degree of structural stability are not important to cellulose production *per se,* tree form would be of no consequence. The form and structure of the forest stand would receive primary consideration. Forests would be designed to make energy capture and storage most efficient. It is probably safe to assume that harvesting and processing techniques would keep pace with forest management and wood utilization, so that even the shrubs, saplings, and suppressed trees of multilayered forests would be used profitably. Under such conditions as these, there might well be a change to a more naturalistic silviculture (tree culture) based on the hypothesis that vegetation types with many layers are photosynthetically more efficient than those with only a single layer.

Programming crop production

We have stressed the capture of light energy by plants, and its flow through plant and animal communities. The energy concept does not alter the objectives of crop producers. Nor would many practices necessarily be changed, for we have learned through experience to manipulate our crops to make good use of available energy. We will, however, be able to control crop production with much greater precision than is now possible.

Business uses computers to "plan" the most economical and profitable succession of events for particular purposes. One of the important mathematical decision-making devices employed is called linear programming. It has been accepted by American industry and become a routine management tool. But it is only usable because all of the different kinds of materials and efforts involved have a common denominator. Time, materials, and manpower can be equated in dollars and cents.

By convention, biologists describe the ecological environments of organisms in terms of rainfall, temperature, soil moisture, wind, sunlight, available nutrients, soil conditions, composition of the atmosphere, pollutants, and living organisms. Although these are adequate parameters, they must be used correctly and on a comparable basis if they are truly to be understood. The most suitable common denominator for crop production may be energy. Measurements of input and of output can both be compared by making use of caloric equivalents. Thus the energy concept will allow more precise analysis, computerized if necessary, in crop production.

In order to maximize the capture and storage of energy, the specialty of crop energetics will involve not only the standard cultural practices, but many factors that have received little attention in the past, such as plant density and distribution, leaf coloration, and height of individual plants.

Scientists are quantifying the chemistry and physics of plants in terms of their metabolic requirements. The water relations of plants and soils are especially vulnerable to this approach, and much is being done to explain and describe the movement of water in the environment on the basis of energy gradients, using absolute mathematical expressions. Sooner than we realize, most of the biological aspects of crop production will yield to mathematical quantification.

Selected References

Asimov, I. *Life and Energy*. Doubleday, New York, 1960. (A broad, popularized treatment of fundamental concepts of energy relationships.)

Brown, H., J. Bonner, and J. Weir. *The Next Hundred Years*. Viking Press, New York, 1957. (An exciting and stimulating discussion of the earth's natural resources in relation to man and technology.)

Dinauer, Richard C. (editor). *Physiological Aspects of Crop Yields*. American Society of Agronomy, Crop Science Society of America, Madison, Wisconsin, 1969 (Proceedings of an international symposium on yield-related aspects of crop physiology and energetics.)

Ehrlich, P. R., J. P. Holdren, and R. W. Holm (compilers). *Man and the Ecosphere: Readings from Scientific American*. W. H. Freeman and Company, San Francisco, 1971. (A collection of papers, some of which deal with certain aspects of energy-flow systems in nature.)

Gates, D. M. *Energy Exchange in the Biosphere*. Harper & Row, New York, 1962. (The title describes the book completely.)

Gates, D. M., and L. E. Papian, *Atlas of Energy Budgets of Plant Leaves*, Academic Press, New York, 1971. (A review of the factors—that is, energy balance—needed to keep a leaf viable, and therefore a crop growing and productive. The book discusses leaf temperatures, transpiration rates, etc. as they relate to wind speed, relative humidity, radiation, leaf shape, and so on.

Margaleff, R. *Perspectives in Ecological Theory*. University of Chicago Press, Chicago, 1968, (An excellent small book dealing with the energetics of ecosystem structure and management.)

Odum, E. P. *Fundamentals of Ecology*. W. B. Saunders, Philadelphia, 1971. (A textbook on ecology with an excellent section on the energetics of ecosystems.)

CHAPTER 3

Food and Human Needs

Food is any material that provides an organism with energy and nutrients. Green plants utilize the sun's energy to synthesize their food supply directly from atmospheric carbon dioxide and inorganic substances of the soil. (As soil nutrients do not supply energy they are only incorrectly referred to as "plant-food.") Some microorganisms have very simple requirements and need only an energy source (such as sugar) and inorganic nutrients. Animals, however, require their food "ready-made" in the form of highly complex molecules. Their diet may be quite specialized, but they ultimately depend on the energy captured in photosynthesis. Man is omnivorous, and obtains food directly from both plants and from animals.

Insufficient food leads to intolerable consequences. Most of us experience great discomfort if we do not eat every five or six of our waking hours. A person cut off from all food for 7 to 8 weeks faces certain death. There are also qualitative aspects to food. Individuals who go for extended periods without the proper kind of food suffer irreparable damage; the consequences are especially severe in the young. Further, a relationship exists between proper diet and mental competence and achievement. In addition to the biologic requirements for energy and nutrients, there is a cultural dimension to food. All peoples hold strong emotional preferences in their choice of diet. Dietary customs become deeply ingrained and difficult to change. Thus the food we depend upon for nourishment and sustenance greatly influences our style of living.

Improper diet leads to starvation, malnutrition, obesity, poisoning, disease, or loss of life itself. The problems of food, from production to ultimate use, are the proper concern of all.

NUTRITION

Nutrition refers to the processes by which living things ingest and assimilate food. Proper growth, maintenance, and functioning of the body are dependent on the choice, amounts, and combination of food materials. The nutritional requirements of man are not perfectly known, partly because of the inevitable hazards involved in attempting to use man as a test organism. Many nutrients are required only in minute amounts, and some of these are present in almost all natural foods. Furthermore, precise quantitative determinations are difficult to make because bacteria residing in the intestinal tract may supply some nutrients (e.g., biotin, folic acid, vitamin K). Consequently, we obtain much of our nutritional information from work on other animals (particularly rats), and much more is known about the nutritional requirements of chickens or pigs than of humans.

The nutritive value of food must be discussed in terms of chemical and energy units. (Food as an energy source will be discussed in a later section.) The chemical substances required by man include inorganic materials (water and certain elemental minerals) and organic materials (amino acids, fatty acids, and accessory factors popularly grouped as vitamins). Those nutrients that cannot be synthesized from other constituents of the diet are considered essential (Table 3-1). A deficiency of any essential nutrient generally results in a specific physiological response (symptom) for each species; the most common first symptom in children is decreased growth. When all nutrients but one are present in adequate quantities, growth is usually proportional to the supply of the limiting nutrient. Essential nutrients are usually not consumed in pure forms but as components of foods. Although our nutritional requirements can be listed rather specifically in terms of essential nutrients, nutritionists prefer to list them in terms of the broad classification of nutrients into carbohydrates, fats, proteins, vitamins, and minerals.

Nutrients

CARBOHYDRATES   The carbohydrates consist of a large group of compounds composed of the elements carbon, hydrogen, and oxygen. Because there are usually two hydrogen atoms for every oxygen atom, they incorrectly appeared to early chemists as "hydrates of carbon" ($C \cdot H_2O$), and thus the epithet. Carbohydrates make up the bulk of the dry weight of plants. Although there are many kinds, the main plant carbohydrates are sugars, starches, and the various celluloses. Animal products contain only small amounts, with the

Table 3-1   NUTRIENTS REQUIRED BY MAN.

ESTABLISHED AS ESSENTIAL	PROBABLY ESSENTIAL
INORGANIC	
Water	
Minerals	
Macroelements:	
Calcium	
Chlorine	
Magnesium	
Phosphorus	
Potassium	
Sodium	
Sulfur	
Microelements:	
Copper	Fluorine
Iodine	Molybdenum
Iron	Selenium
Manganese	Zinc
ORGANIC	
Amino acids	
Isoleucine	Histidine[1]
Leucine	Arginine[1]
Lysine	
Methionine	
Phenylalanine	
Threonine	
Tryptophan	
Valine	
Fatty acids	
	Linoleic acid
	Arachidonic acid
Vitamins	
Ascorbic acid	Biotin
Choline[2]	Pantothenic acid
Folic acid	
Niacin[3]	
Pyridoxine	
Riboflavin	
Thiamine	
Vitamin B_{12}	
Vitamins A, D[1], E and K	

Source: After White, Handler, and Smith, *Principles of Biochemistry*, 3rd ed. New York: McGraw-Hill, 1964.

[1] Indicated to be unnecessary in adults in short-term studies but probably necessary for normal growth of children. (Vitamin D required by pregnant women; requirement can be met in children by exposure to sunlight.)

[2] Requirement met under circumstances of adequate dietary methionine.

[3] Requirement may be provided by synthesis from dietary tryptophan.

exception of milk, which contains lactose, a sugar.

These carbon-containing compounds exist in many different forms, each with a molecular architecture of its own. Sugars are simple carbohydrates that form more complex materials (polymers) when linked in chain-like fashion. Sucrose, the common table sugar, is composed of two linked simple molecules: glucose and fructose. Linked chains of hundreds of glucose molecules make up starch, a storage form of carbohydrate found in fleshy tubers and roots, and the principal reserve carbohydrate of many plants. Inulin, a polymer of fructose, is the principal reserve carbohydrate of the Jerusalem artichoke (*Helianthus tuberosus*). Inulin has interesting medical uses because, unlike the glucose-containing carbohydrates, it can be metabolized by diabetics. Cellulose is formed from interwoven and interconnected strands of a polymer of glucose, but differs from starch in its branch-like linkage relationships. Cellulose is very insoluble and is not digested by man, but can be broken down by microorganisms. Some plant-eating animals can utilize cellulose because of the fermentation action of microorganisms in their gastrointestinal tract. This is accomplished very efficiently within ruminant animals (cattle, sheep, goats) and those species with a large caecum, such as the horse and the rabbit.

Carbohydrate foods are commonly classified on the basis of digestibility into two components: **crude fiber,** the coarse cellulose-like portion, and **nitrogen-free extract,** principally starches and sugars. The crude fiber is not easily digested and yields a lesser proportion of its energy than does the nitrogen-free extract. Although crude fiber is of little direct use to man, it is the basis of many animal rations; the average animal ration is 75 percent carbohydrates.

With the exception of vitamin C (ascorbic acid), carbohydrates are not in themselves essential as a human food. Man, like other animals, can be maintained on a carbohydrate-free diet by substituting lipids as an energy source, which the Eskimo does. But the universal use of carbohydrates in the form of sugars and starches as an inexpensive form of energy renders them practically indispensable as a human food. For example, about half of the calories in the diet of the average American are provided by carbohydrates. In other countries carbohydrates supply an even greater fraction of calories in the diet.

FATS Fats and fat-like substances are grouped under the general term **lipids.** They are soluble in fat solvents such as ether or chloroform. Fats are esters of long-chain organic acids (known as fatty acids) and alcohols such as glycerol. Fats contain carbon, hydrogen, and oxygen, but unlike carbohydrates, the proportion of oxygen is very low.

Fats occur in small amounts in all living protoplasms. In plants they generally accumulate in seeds.

Crop	Percent crude fat in seed
coconut	65
peanut	40–50
flax (linseed)	30–35
soybean	15–20
cotton seed	15–20
maize	5

There is an inverse relationship between the amount of carbohydrate and the amount of fat in seeds. The amount of fat in the seed may be altered by selection. Long-term selection in maize has produced varieties with high (>14 percent) and low (< 2 percent) oil contents (see Fig. 19-10). Some fruits (olives, avocado) also have high fat contents.

The differences in fats are related to the kind and arrangement of their fatty acids. Fatty acid molecules differ in their length (usually from 12 to 26 carbons) and in the degree of hydrogen saturation—that is, number of double bonds. The degree of saturation affects their stability and firmness. The distinction between oil and fat is made on the basis of whether the substance is solid or liquid at room temperature. Oils have large numbers of double bonds; that is, they do not carry all of the hydrogen possible. Plant oils may be changed to fats by the addition of hydrogen (hydrogenation).

Fats are insoluble in water and must be rendered soluble (digested) before they can be utilized by the body: digestion splits the fatty acids from glycerol. The position at which a fatty acid is attached to the glycerol affects digestibility, because a fatty acid attached at mid-position is removed last if at all. Digestibility depends also on the length of the fatty acid molecule; the short-chain fatty acids are most easily digested.

Fats, the most concentrated energy source, have always been highly valued as a source of food. Some of the unsaturated fatty acids are essential. The most important is linoleic acid, an 18-carbon unsaturated fatty acid with reactive double bonds at the 9 and 12 positions. Fats aid the absorption of certain fat-soluble vitamins and are responsible for the feeling of satiety experienced after eating.

In the U.S., fat contributes 40–50 percent of the calories in the average diet; in some less prosperous parts of the world, fat contributes only about 20 percent. It is not clear with whom the advantage lies, for in some individuals the level of cholesterol—a natural fatty material in the blood—increases to dangerously high levels with high intakes of saturated fats. High cholesterol levels have been associated with coronary disorders, but the precise relationship between fat intake and cholesterol levels is not at all clear. Much remains to be learned about fat metabolism and its relation to human nutrition.

PROTEINS   Protein (the term applied to proteins collectively) is a major constituent of living material. Next to water protein is the most abundant substance in our bodies, accounting for about half the dry weight. Protein is the characteristic material of animals. On a dry-weight basis, muscle (beef-steak) is 47 percent protein, milk 27 percent, and eggs 50 percent. If cellulose is excluded (and lignin in woody plants), protein also makes up the greatest proportion of the dry weight of most plant cells.

Proteins are complex molecules made up of different combinations of amino acids. Amino acids, the basic unit of proteins, are synthesized in plants (and microorganisms) from carbohydrate fragments and from the nitrogen in ammonium ions (NH_4^+). Although proteins contain carbon, hydrogen, and oxygen, nitrogen is the characteristic element. Since most proteins average 16 per-

cent nitrogen, the protein content of foods may be estimated by multiplying the nitrogen content by a factor of 6.25. Moreover, protein utilization is studied by measuring the nitrogen entering and leaving the body. Of the essential elements, sulfur is found in two of the amino acids; other essential elements may be found in protein in small quantities.

Each of the amino acids contains a carboxyl group (—COOH) and an amino group (—NH_2). Although there are many amino acids, only about twenty-two are found in protein. Amino acids differ in the number of carbon atoms and attached groups. The amino acids combine at their amino and carboxyl groups to form peptide linkages. A protein can be defined as a group of amino acids held together by peptide linkages. Proteins may be very large. The molecular weight of gliadin, one of the wheat proteins, is 27,500 and that of glutenin, another wheat protein, is in the millions.

There are many kinds of proteins. In plants, proteins are commonly classified on the basis of their origin. Seed proteins are the primary reserve forms. The proteins found in the endosperm of wheat (commonly referred to as gluten) are responsible for the characteristic elastic property of dough. Proteins in plant tissue occur primarily in the cytoplasm and chloroplasts. Nucleoproteins, proteins that contain nucleic acid, make up the skeleton of the chromosomes. Enzymes, the major regulatory material of the living organisms, are protein.

Animals require amino acids rather than protein *per se*, and since they cannot synthesize all amino acids, they ultimately depend on plants or microorganisms for those they cannot synthesize. Cattle, sheep, and goats can obtain their amino acids from bacteria that reside in their rumen, and in this way can exist on a diet of purified carbohydrates and simple nitrogen compounds. It is thus possible to produce milk from the cellulose in corn cobs plus urea and ammonium salts as a source of nitrogen. As proteins differ in the kind and proportion of amino acids, it follows that proteins differ in nutritive value. High-quality proteins are those that supply a proper balance of essential amino acids, and this proper balance is required in concert. Rats fed the necessary amino acids in sequence, three hours apart each day, fail to grow.

When proteins are digested, specific enzymes in the gastrointestinal tract break the peptide linkages. The amino acids are reassembled in the tissues to form new proteins required by the particular species. Amino acids left over are not stored but are stripped of the amino group by the liver. The remainder of the molecule is used to provide energy or is converted to fat and stored.

Many animal proteins have a greater biological value as human food than proteins from single plant sources (Fig. 3-1). This is because the assortment and proportion of amino acids in animal protein are similar to those required in the human diet. The amino acids that occur in proteins are shown in Figure 3-2. Cereal grains are likely to be relatively low in lysine, as well as methionine and tryptophan, whereas proteins from oil-bearing seeds are likely to be low in the sulfur-containing amino acids cystine and methionine. Plant proteins also tend to be less digestible. Although proteins are generally found in low concentration in plants as compared to animals, adequate protein levels can be supplied by the skillful complementation of different plant protein sources. Further, it has been found that the amino acid composition of plant proteins may be improved. The *opaque-2* gene greatly increases the biological value of corn protein by increasing lysine and tryptophan content (Fig. 3-3).

Protein is in short supply in the world. It is the most prized and the most expensive of the major nutrients. The world food problem is largely one of insufficient protein. Insufficient protein in the diet of children leads to a deficiency syndrome characterized by growth retardation. The syndrome was originally called **kwashiorkor** ("displaced child"), from its occurrence in Central African infants deprived of their mothers' milk (Fig. 3-4).

VITAMINS   The relationship between diet and disease is very old. The curative power of liver on night blindness was recorded by Hippocrates. The use of lime juice by British sailors to prevent scurvy (hence the term "limies") has been traced

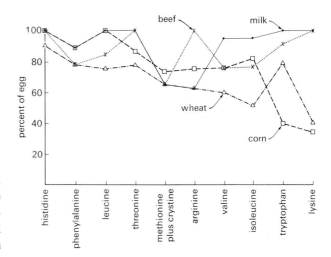

FIGURE 3-1. *Evaluation of protein as compared to egg. The egg ratio is the percentage by which the amino acid content of a protein departs from that of whole egg protein, with excesses taken as 100 percent.* [*After Crampton,* Applied Animal Nutrition. *San Francisco: W. H. Freeman and Company, © 1968.*]

to a suggestion by James Lind in 1757 on the beneficial value of fresh fruits and vegetables. Early nutritionists found that animals fed on rations of purified forms of known foods sickened and died, yet thrived when small amounts of certain substances, such as water-soluble extracts from yeast or wheat embryos (germ), were added. In 1912, Frederick G. Hopkins and Casimir Funk suggested that specific human diseases, such as beriberi, rickets, and scurvy, were caused by the absence of certain nutritional substances in the diet. These were termed vitamines ("vital amines"), as the first active preparation was an amine. When other substances were shown to be involved, the term **vitamin** (minus the "e") was retained to refer to essential growth factors required in very small amounts. Many essential vitamins have been discovered since that time. Although the functions of some vitamins are unknown, many have been shown to be coenzymes, which are required for the function of particular enzymes. The water-soluble B vitamins tend to play a part in energy transfer.

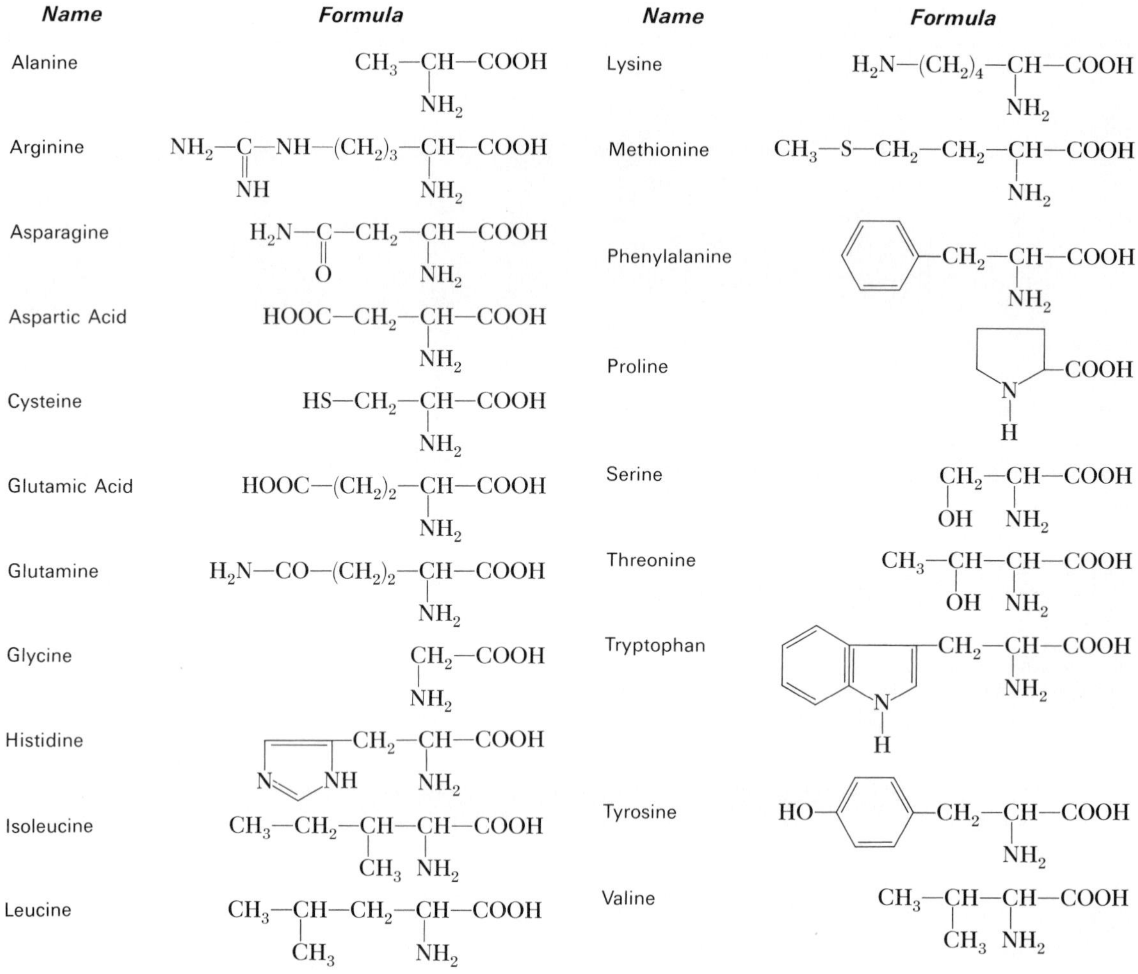

Name	Formula	Name	Formula
Alanine	$CH_3-\underset{\underset{NH_2}{\mid}}{CH}-COOH$	Lysine	$H_2N-(CH_2)_4-\underset{\underset{NH_2}{\mid}}{CH}-COOH$
Arginine	$NH_2-\underset{\underset{NH}{\parallel}}{C}-NH-(CH_2)_3-\underset{\underset{NH_2}{\mid}}{CH}-COOH$	Methionine	$CH_3-S-CH_2-CH_2-\underset{\underset{NH_2}{\mid}}{CH}-COOH$
Asparagine	$H_2N-\underset{\underset{O}{\parallel}}{C}-CH_2-\underset{\underset{NH_2}{\mid}}{CH}-COOH$	Phenylalanine	(ring)$-CH_2-\underset{\underset{NH_2}{\mid}}{CH}-COOH$
Aspartic Acid	$HOOC-CH_2-\underset{\underset{NH_2}{\mid}}{CH}-COOH$	Proline	(ring with N–H)$-COOH$
Cysteine	$HS-CH_2-\underset{\underset{NH_2}{\mid}}{CH}-COOH$	Serine	$\underset{\underset{OH}{\mid}}{CH_2}-\underset{\underset{NH_2}{\mid}}{CH}-COOH$
Glutamic Acid	$HOOC-(CH_2)_2-\underset{\underset{NH_2}{\mid}}{CH}-COOH$	Threonine	$CH_3-\underset{\underset{OH}{\mid}}{CH}-\underset{\underset{NH_2}{\mid}}{CH}-COOH$
Glutamine	$H_2N-CO-(CH_2)_2-\underset{\underset{NH_2}{\mid}}{CH}-COOH$	Tryptophan	(indole ring)$-CH_2-\underset{\underset{NH_2}{\mid}}{CH}-COOH$
Glycine	$\underset{\underset{NH_2}{\mid}}{CH_2}-COOH$	Tyrosine	$HO-$(ring)$-CH_2-\underset{\underset{NH_2}{\mid}}{CH}-COOH$
Histidine	(imidazole ring)$-CH_2-\underset{\underset{NH_2}{\mid}}{CH}-COOH$	Valine	$CH_3-\underset{\underset{CH_3}{\mid}}{CH}-\underset{\underset{NH_2}{\mid}}{CH}-COOH$
Isoleucine	$CH_3-CH_2-\underset{\underset{CH_3}{\mid}}{CH}-\underset{\underset{NH_2}{\mid}}{CH}-COOH$		
Leucine	$CH_3-\underset{\underset{CH_3}{\mid}}{CH}-CH_2-\underset{\underset{NH_2}{\mid}}{CH}-COOH$		

FIGURE 3-2. *Structure of the twenty amino acids that occur in proteins. Isoleucine, leucine, lysine, methionine, phenylalanine, threonine, tryptophan, and valine are essential to humans; arginine and histidine are required by children but probably not by adults.*

Letters of the alphabet were first used to describe the mysterious nutritional factors. These letters (A, B, C, D, E, K, and others) have persisted. It was found, however, that some factors involved more than one substance. The original B factor has been shown to consist of more than a dozen entities. These factors are now designated as specific substances: for example, thiamine (B_1); riboflavin (B_2); pantothenic acid (B_3); three related substances, pyridoxine, pyridoxal, and pyridoxamine (B_6); and niacin (nicotinic acid). Because these commonly occur together they are referred to as the B complex.

Vitamins are classified as fat soluble (A, D, E, K) or water soluble (B complex, C), and are often described in terms of their deficiency effects. For example, a deficiency of Vitamin C results in loose teeth and in cracked and inflamed mouth tissues; the syndrome is called scurvy.

Nutritional disorders due to vitamin deficiencies are often a result of ignorance as well as economic conditions. For example, vitamin A deficiency is

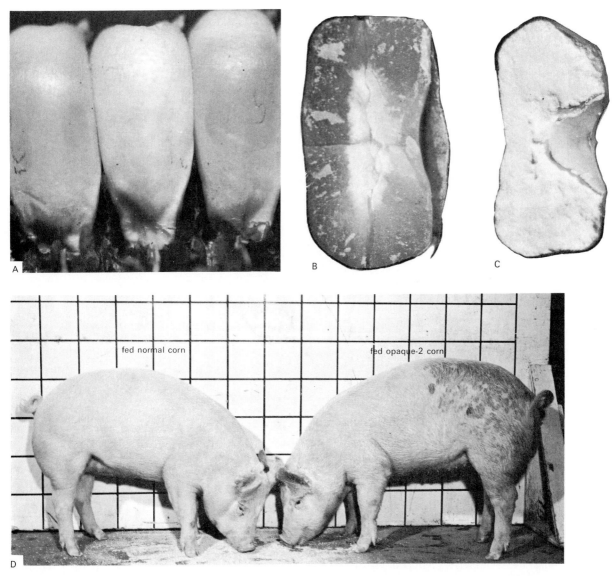

FIGURE 3-3. (A) *Kernels on an ear of maize, showing* opaque-2 *mutant (center) and normal kernels. The* opaque-2 *endosperm contains 69 percent more lysine than the normal endosperm. Cross sections through a normal kernel (B) and an* opaque-2 *kernel (C) show a reduced horny endosperm associated with the* opaque-2 *gene. (D) The superior feed value of* opaque-2 *corn is due to its higher protein quality. Because lysine is the limiting amino acid in corn the* opaque-2 *mutant almost doubles the feed value of corn for monogastric animals, such as the rat, the pig, and man.* [*Courtesy Purdue Univ.*]

frequent in tropical regions even though it is readily available in native plants that could be, but are not, utilized as food because of dietary preferences. The deficiency may be most marked when food is plentiful, and may actually decrease during severe shortages, when people are forced to consume "unacceptable" but vitamin-A-rich food.

MINERALS   Ninety-six percent of the weight of the human body is made up of four elements: oxygen, carbon, hydrogen, and nitrogen. The remainder consists of essential elements. On the basis of the amounts required, the essential elements (commonly referred to as **minerals**) can be divided into macroelements (calcium, phosphorus, potassium, sulfur, sodium, chlorine, magnesium) and microele-

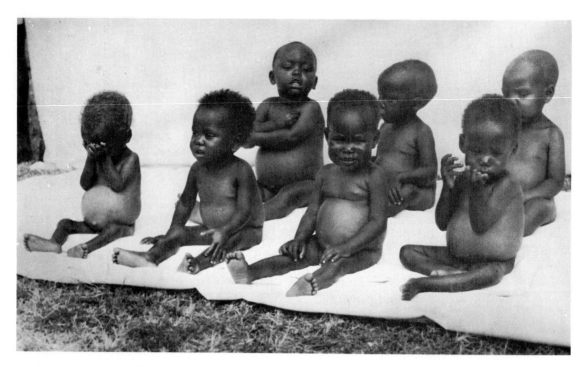

FIGURE 3-4. *Symptoms of kwashiorkor (protein deficiency) in African children.* [*Courtesy E. Wilson.*]

ments, or trace elements (iron, manganese, copper, iodine, and perhaps others). Of the macroelements, calcium and phosphorus are required in greatest amounts. They make up 70 percent of the total ash and are found mostly in the bones and teeth. Foods vary widely in their content of these elements. Milk, an especially good source, is therefore a very valuable food for children, whose need for calcium and phosphorus is especially great. Inorganic forms of calcium and phosphorus may be added to the rations of animals. Deficiencies of the remaining macroelements are rare in the mixed diets of humans.

Of the microelements, iodine is the most commonly deficient. Deficiency symptoms of enlarged thyroid (goiter) are common in geographic areas in which the diet is low in iodine. Supplementary iodine is usually provided in inorganic form, as potassium or sodium iodide, and added to common table salt. Trace element deficiencies are more common in livestock and poultry than they are in human beings.

Food as an energy source

Food must supply the fuel for the living machine. The energy requirement depends on the size of the machine and the amount of work performed. The potential chemical energy in food, in calories, is obtained by burning it and measuring the amount of heat released. This is done in a device called a "bomb calorimeter." The kilocalorie, or Calorie (see p. 28), is the unit used for measuring the energy produced by food oxidized in the body.

For a given quantity of food the amount of calories actually utilized by the body is somewhat less than that measured by the bomb-calorimeter method, because some food may not be digested, some may be stored in the body in the form of tissue, and some components (such as protein) are not completely oxidized in the body.

Fats and carbohydrates are metabolized in the body to carbon dioxide, water, and heat. Protein, however, leaves residues (such as urea), which are eliminated in the urine.

Nutrient	Kilocalories/g	Kilocalories utilized by the body
carbohydrates	4.1	4
fats	9.5	9
proteins	9.7	4

Basal metabolism refers to the amount of energy expended (while awake and at rest) of all involuntary vital processes (respiratory mechanisms, the production of body heat, etc.) under uniform environmental conditions. Although basal metabolism varies with age and sex, it tends to be the same for individuals of similar size, differing only from 5 to 10 percent. Extreme changes in basal metabolism are then an indication of some pathological metabolic disturbance such as an overactive or underactive thyroid.

The total energy requirement above that required for basal metabolism depends on muscular activity and adjustments to the environmental temperature. Muscular activity requires energy. In addition, the body must eliminate the heat pro-

duced from the combustion of food. Under the comfortable temperature of 21°C (70°F) about three-quarters of the heat produced is eliminated by radiation and conduction and one-quarter by the evaporation of perspiration. More energy is required to perform a given task in hot, humid weather, because all heat must be eliminated by perspiration—an energy-requiring process. During cold weather we can warm ourselves by exercising. Heat can also be produced involuntarily, as it is during the involuntary reflex that we call shivering.

Caloric requirements depend on activity. A 145-lb person leading a sedentary existence requires approximately 2,500 kilocalories per day, but as much as twice that amount if extremely active. The human requirements for calories and other nutrients are summarized in Figure 3-5.

Energy requirements take first call on either food or reserves. Excess energy expenditure over receipts takes place at the expense of stored fat or even body tissues. The short-term response is loss of weight. Longer exposures to a deficit of

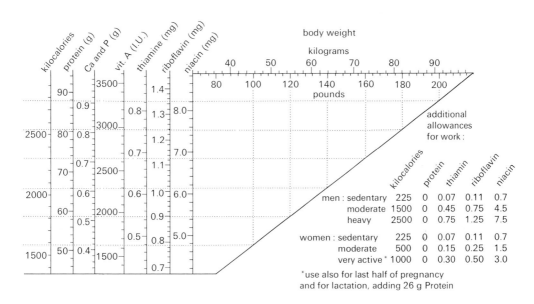

FIGURE 3-5. *Standard allowances for maintenance of adults (based on dietary standards for Canada, 1948). To find recommended amounts, read down from point showing body weight to diagonal. Follow across to scales on left, and add an allowance for work.* [*From Crampton and Lloyd,* Fundamentals of Nutrition, *San Francisco: W. H. Freeman and Company,* © *1959.*]

energy foods leads to reduced vitality, physical exhaustion, and ultimately death. Our need for energy is the basis of the hunger urge. Appetite in most people serves to regulate the balance of energy intake with energy output. When intake exceeds output the body stores energy in the form of fat.

Diet

Nutrients and energy-rich substances may be consumed in many forms. Most of us, however, have a pattern of eating that involves only certain selected foods. Food consumption patterns are a result of many factors—economic, technological, geographical, cultural. Our diet is inextricably bound up in our pattern of living.

Although sustenance may be derived from a seemingly limitless supply of other life forms, relatively few are actually used to satisfy man's nutritional requirements. Nevertheless, there are distinct differences in diets. These differences portend important consequences in the quality of life for mankind and for individual man. Food, unlike the air we breathe, is obtained only with a great deal of effort. In many areas of the world food is expensive, and a high percentage of mankind trades the greater proportion of its productive labor to purchase it. Food, an absolute necessity, comes first.

From a nutritional standpoint, the important components of diet reduce to an accounting of calories, proteins, and vitamins. Dietary differences reflect the way these needs are satisfied. Ultimately, the key factors in food supply are availability of suitable land per person and level of technology.

The energy food that mankind requires has been made plentiful through the discovery of agriculture. Certain plants are capable of high calorie output per acre in convenient edible form, including the grains (especially wheat, rice, and corn), and potato, cassava, and sugar cane. All great civilizations have been able to cultivate at least one of these important calorie producers in abundance. These foods are versatile and are readily transformed to such products as bread, spaghetti, tortillas, rice cakes, and poi.

Our demand for protein can also be satisfied in a great many ways. Although the choice is seemingly dependent on many factors, there is one underlying economic principle: the further that man gets from plants in the food chain, the higher is the cost of his food. Only where there is a land surplus, and sufficient technology to exploit it, can man afford the luxury of converting grain into meat.

The use of animal sources of protein has always been very attractive. There is a firm biological basis for this because animal proteins are complete; that is, they contain all eight or nine essential amino acids in amounts fairly close to our needs. The beneficial qualities of meat also include minerals and vitamins. This is reflected in man's preferences toward meat, especially in combination with fat. In this combination, the protein and energy are supplied together. A few cultures exist entirely on animal protein and fat; for example, the Eskimo's diet consists chiefly of fish and blubber, and the staples in the diet of the Masai in Africa are the blood and milk of cattle.

Animals, however, are inefficient converters of feed to meat. Only 3 percent of the total energy in plants is converted to potential human food. Not all of the plant is consumed by animals; of the part that is consumed, some is "walked off," or wasted. Moreover, much of the animal is not consumed by man. As a result, when meat is expensive, man has shifted to eating that plant food formerly used as feed. In highly developed societies the convertibility of feed into meat proteins provides an ever-present variety of foods, one's choice being governed by price. If meat is scarce, high prices merely mean a shift to plant protein. Where such an abundance of food is available to feed animals, there is seldom any real threat of famine.

In parts of the world where the pressure of human population on land resources is great, animal protein satisfies only a small fraction of the human requirements: the plant food produced is virtually all consumed by man. When shortages occur there is no other immediately available supply of food. Under such circumstances the specter of famine and starvation can be expected to recur.

The special value ascribed to animal protein is

reflected in the many cultural attitudes toward the consumption of animal flesh, in contrast with the few toward plants. This appears to be manifested in religious prohibitions of various types. For example, Catholics still observe a number of meatless days as penance. Most of the complex dietary laws in Judaism govern the eating of animal products. This includes the species and parts of the animal that may be eaten and the kinds of food that may be combined (mixing of milk and meat is prohibited). Devout Moslems and Jews do not eat pork. Hindus do not eat cattle, for they regard them as sacred, and some sects abstain from meat in any form. All of us scoff at the "taboos" of others, but strongly defend our own "preferences." In the United States many will not eat rabbit, and most draw the line at horsemeat or squirrel.

FOOD STORAGE AND PROCESSING

Although the demand for food is continuous, its production is not. The growth of many crops conforms to a seasonal cycle and the harvest becomes available from particular regions only once each year. In practically all areas there is a period during which crops will not grow—winter in temperate regions and periods of drought or excessive moisture in the tropics. To insure a steady supply without recourse to nomadism it is essential that food be preserved.

In order to extend the supply of harvested foodstuff a number of destructive forces must be circumvented. These include natural physiological deterioration, decay by microorganisms, and losses due to pests, such as rodents and insects. The methods of food preservation are many in number, and the choice of method depends on the product, its use, the destructive agencies, dietary preferences, and the length of time the product must be preserved.

Acceptable quality of some products is a transient and ephemeral thing. A fresh apple is alive when eaten. Thus the preservation of some foods involves extending the living state. For the long-term storage of most food crops, however, life must be terminated and microbial action prevented. Edibility is maintained even though the processes involved materially alter the food from its original condition. Often this results in a new product with utility beyond its preservation function—for example, milk to cheese, tomatoes to ketchup.

Storage of perishables

Plant products, such as fresh fruits and vegetables, and the great bulk of animal products—meat, milk, fish—have a usable period measured in days. Living plant products deteriorate largely through respiration and excessive loss of water. Animal food products are subject to rapid microbial decay. Short-term storage without extensive changes in form by processing may be accomplished by controlling the environment—specifically, by controlling temperature, humidity, and oxygen level.

Temperature is the critical factor in storage. It affects the respiration rate of the living commodity and the growth rate of the microbial population. The lower limit must be determined for each commodity. Storage temperature is usually kept above the point at which the commodity freezes. For example, apples store best at about $-1°$ to $0°C$ ($30°$ to $32°F$). Some tropical fruits, however, are injured at temperatures below $10°C$ ($50°F$). The evolution of heat by respiration in living materials must be considered in determining the proper storage temperature. Animal carcasses and products such as milk are originally at high temperatures—$37°$ to $40°C$ ($98°$ to $104°F$)—and must be rapidly cooled. Year-round consumption of fresh fruits and vegetables and animal products requires a high level of technology, which is manifested in today's refrigerated transport and storage equipment.

Humidity affects the perishability of many plant products. In general, low humidity results in dessication and wilting. High humidity, however, favors the development of decay, especially if temperatures are too high.

In addition to being temperature-dependent, respiration is also directly affected by the amounts of oxygen and carbon dioxide in the storage atmosphere. Reducing the amount of oxygen slows the respiration rate. Controlling the amount of oxygen is known as modified atmosphere storage. In gastight storage space filled with freshly harvested living produce, oxygen is consumed and equal concentrations of carbon dioxide released. By controlling the introduction of air, oxygen levels may be kept low enough to reduce respiration and yet not low enough to permit anaerobic respiration, which results in the formation of alcohol. This has been accomplished quite simply with the use of polyethylene film liners, which are differentially permeable to carbon dioxide and oxygen. Oxygen may be removed rapidly by flushing with the inert gas nitrogen. The combination of low temperature and modified atmosphere makes it possible to keep fresh fruit for periods of months.

Food processing

Relatively long-term preservation of food may be achieved by physical and chemical processes that sterilize the food or render it incapable of supporting the growth of microorganisms. These processes include drying, canning, freezing, the use of preservatives, and irradiation. Food processing has become one of the major agricultural industries.

DRYING   Drying is the most ancient method of food preservation. Removing water from the tissues results in a highly concentrated material of enduring quality. The natural deterioration of the product by respiration is stopped because of enzyme inactivation, and the lack of free water protects the dried product from decay by microorganisms. Drying may be achieved naturally (**sun drying**) or artificially (**dehydration**) by passing warm air over the product. Freeze-dehydration, the drying of quick-frozen food in a high vacuum, holds promise of increased usefulness because the quality of the reconstituted product is much higher than that achieved by ordinary dehydration.

Drying is the natural means of preservation for grains. They are harvested when nearly dry, and consequently can be stored for extended periods of time. Drying is also useful for many other products; much fruit is preserved by drying, includ-

ing grapes, plums, apricot, date, fig, banana, peach, apple, and pear. Meat and fish can also be preserved in this manner. In tropical areas sun-dried meat (**charqui** or "jerked" meat) is widely used. For these products salt may be used in combination with the drying process.

Dried, concentrated, processed foods such as flour, protein supplements, powdered milk, and powdered eggs can be stored for long periods if protected from rodents and insects. The storage life of dried materials is extended at cool temperatures. High humidities encourage the growth of molds.

Drying, the most ancient of processing methods, appears to be the technique of the future. It has many advantages: by removing water it reduces weight and, in some foods, bulk. Food preservatives are not required for the process, and packaging problems are minimal. Storage life is long, and nutrients are virtually unaffected. Loss of quality, the limiting factor in dried foods, is being overcome by advances in technology.

CANNING   **Canning** consists of heat-sterilizing food and sealing it in an airtight container. The process destroys both the pathogenic and the food-spoilage microorganisms and inactivates the enzymes that would otherwise cause decomposition during storage. The sealed container prevents reinfection after the food has been sterilized and prevents gaseous exchange.

The application of sufficient heat to sterilize food and inactivate enzymes results in alteration of color, flavor, texture, and nutritive value. Although the transformations are similar to those that occur in the normal cooking process, the treatment must be more extreme for many foods. Consequently, loss of quality is one of the limiting factors of the canning process. Low-acid foods require relatively severe heat treatment, since they can support growth of *Clostridium botulinum*, a bacterium that causes food poisoning. A millionth of a gram of the toxin produced by this organism will cause death. Thus, all foods capable of sustaining growth of this organism are processed on the assumption that the organism is present.

Although highly efficient systems have made the

cost of canning reasonable in highly developed economies, canned foods are still out of the reach of a large part of the world's population. In addition, the finished product is bulky and heavy, and unless proper precautions are taken is not suitable for extended storage under all conditions. At storage temperatures above 49°C (120°F) certain heat-loving bacteria not ordinarily killed by the sterilization process will continue to grow. In humid regions, and especially in coastal areas where salt concentrations are high, storage life is limited by corrosion of the metal containers. Nevertheless, in developed countries, canning is still the most important processing method, largely because of convenience.

FREEZING Freezing protects food from spoiling because most microorganisms cannot grow at temperatures below 0°C (32°F). The freezing process stops most enzymatic activity and is not in itself destructive to nutrients. Some products, however, require heat treatment before freezing, in order to inactivate enzymes that affect flavor and color during storage.

The basic principle in freezing food is the speedy removal of heat. This is achieved by cold air blasts, direct immersion in a cooling medium, contact with refrigerated plates, or liquid air, nitrogen, or carbon dioxide. Freezing in still air is the slowest method. The rate of freezing has a large effect on quality. In slow freezing relatively large ice crystals develop and fracture the tissue cells, resulting in the loss of cellular fluid and an undesirable soft texture when the food is thawed. If freezing occurs rapidly, the crystals formed are about one-hundredth the size of those formed at slow rates. Because these crystals are tightly packed, fewer cells are ruptured. In quick freezing the refrigerant is kept at temperatures of −29° to −40°C (−20° to −40°F), and the water in food is completely frozen in thirty minutes or less.

The success of freezing as a method of preservation depends on the continuous application of the process. Thus the method is only suited where the population has refrigeration equipment. Because of the high quality of frozen foods as compared to canned foods, freezing has taken over the mar-

ket for many products. For example, frozen concentrate juice now accounts for the greatest use of the U.S. orange crop.

PRESERVATIVES There are many materials that extend the usable life of foods, including salt, sugar, vinegar, spices, and a number of synthetic chemicals.

Common salt, because of its action in controlling bacterial growth, is widely used for preservation, especially for fish, meat, and meat products. "Curing" meats with salt is an ancient art—the term "corned" is derived from an old word for granulated salt. In addition, nitrate and nitrite salts, and sugar and other materials, may be added for color and flavor. Smoking, often used in combination with curing for adding flavor, also acts as a meat preservative. Sugar, like salt, preserves foods by reducing the availability of free water. Acid fruits concentrated to at least 69 percent solids are commonly preserved in the form of jelly (concentrated juice), jam (concentrated fruit pieces) or preserves (concentrated whole fruit), butter (a semisolid product), and glacéed, or candied, fruits.

Many preservative materials are obtained naturally through **fermentation,** the anaerobic microbial decomposition of carbohydrates. Fermentation may be accomplished by a number of different organisms. Among the variety of end products are: carbon dioxide and water (complete oxidation), acids (partial oxidation), alcohols (alcoholic fermentation), and others. The buildup of some of these products to sufficient concentrations creates unfavorable conditions for microorganisms, including the original one. In addition, fermentation may impart certain flavors and secondary effects that are regarded as desirable. Fermentation may be controlled by conditions that favor the growth of one type of organism through the regulation of pH, oxygen, temperature, and the use of salt (pickling refers to fermentation in combination with salting) and other curing materials.

Fermentation plays an important role in processing many kinds of food. Cheese is a fermentation product; fermented vegetables make highly prized relishes. The fermentation of grain or fruits yields the alcoholic beverages: beer, wine, and spirits. Ensilage, fermented fodder, is an important animal feed.

A number of chemicals are used to retard deterioration, usually in conjunction with other methods of preservation. These include inorganic agents, such as sulfur dioxide and chlorine, and organic agents, such as benzoic acid, certain fatty acids, sorbic acid, ethylene and propylene oxides, and various antibiotics. Chemical preservatives and other food additives have a legitimate place in food processing. They are unacceptable, however, if used to deceive or if there is any associated health hazard.

IONIZING RADIATION   Radiation has, at present, a limited use in preserving some foodstuffs. Gamma radiation extends the storage life of perishables by inhibiting growth processes, such as sprouting in potatoes. It is also used in the decontamination of food by killing insects or parasites, and in certain unit operations as meat tenderizing. There is evidence that radiation changes the flavor, texture, appearance, and nutrient value of food, as do other forms of processing, such as heat treatment. Although there is no direct evidence of toxic effects, the radiation of foodstuffs for preservation purposes must be considered as largely experimental. Food preservation by irradiation has not yet lived up to early expectations.

PROVIDING WORLD FOOD NEEDS

A large proportion of the world—estimates run as high as half—is underfed or malnourished (Fig. 3-6). More distressing, the situation is deteriorating; as it is inextricably bound up with the "population explosion" (see Chapter 4). Although feeding the present world's population seems difficult, but conceivable, even to the unduly pessimistic, the problem of nourishing a population twice the present size (to be reached by the year 2000 according to most projections) is staggering. Because of the present age distribution and other factors, the world's population will surely continue to rise in the next several decades. At best, only the rate of increase can be slowed. The population must at some point decline, however, either because of its impossible demands on resources or because man will have found some way to control

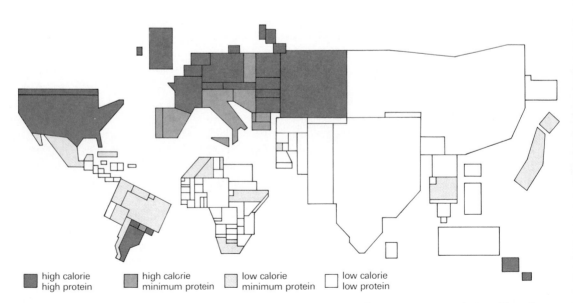

high calorie high protein   high calorie minimum protein   low calorie minimum protein   low calorie low protein

FIGURE 3-6. *Nutritional status of the world. The land areas are distorted to represent population. Note that the malnourished constitute more than half the world's population.* [*From Mattson,* International Science and Technology, *December 1965.*]

his multiplication—by regulating births or regulating deaths! The solution to better nutrition for the world's inhabitants must therefore involve both population control and increased food supply. The upshot of failure will be incomprehensible misery.

The suggestions for increasing world food production may be simply summarized: more land, new foods, greater efficiency. The simplicity of the suggestions, however, obscure the serious problems that each entails. The potentials and limitations of each must be carefully considered.

Developing productive lands

A map of the world's population is in large part a map of the world's productive areas. Man has already filled to overflowing most of those areas capable of supporting a large population easily. This is not to imply that all the usable land is gone, but the large areas of land presently unfarmed are generally unsuited for agriculture unless modern technology can be employed on a massive scale to remedy their deficiencies. Among these deficiencies are low soil fertility, excess or lack of rainfall, and unsuitable terrain and climate. Making such lands agriculturally productive is very expensive and in many places prohibitive. For example, the recent cost of opening new lands in the Western United States was about one or two thousand dollars per acre, much more than the value of the crops for many years. If production is the main concern, these capital resources could, in general, have been used more efficiently to increase the yield from land already cultivated. Furthermore, because the unused lands of the world are outside of populated areas, the problems of distribution would be very great. The exploitation of new lands (and new seas?) does not seem to provide the key to the world's present food problem.

New foods

The suggestion that some new food will appear to alleviate man's problems is a recurring and wistful theme—from heavenly manna to Al Capp's fanciful "schmoo." In contrast to the change taking place everywhere around us, little or no change

seems to be taking place in our diet: our present foods have been with us interminably long. Many scientists are quick to point out that more efficient food sources are available and need only be exploited (not for those of us who are quite content with our present fare, but for others). And they are right. The culture of microorganisms or such aquatic plants as *Chlorella*, or the extraction of protein from forages, or the fermentation of petroleum fractions would certainly be more efficient—at least on theoretical grounds—than the present food sources. The culture of microorganisms requires little space, is independent of climate, and takes place extremely rapidly. The use of petroleum for culturing microorganisms and its direct conversion to food are distinct possibilities. The substance 1,3-butanediol, a cheap industrial chemical derived from natural gas, can be produced in very large quantities. It is completely digestible and is a good energy source. It does, however, have an objectionable odor. In the future, processes of this kind may very likely be used economically for producing food or feed. Unfortunately, because of shortages of time and capital, such processes do not seem to be ready solutions to the world's immediate requirements. Further, there are many special problems inherent in the adaptation of this advanced technology, in addition to those of acceptance.

Even if the processes were practicable, the acceptance of the new foods would remain a serious problem. New foods more readily find acceptance in highly developed countries, where they are least needed. In the United States, for example, soybean products are being substituted for a whole spectrum of dairy products (butter, whipped cream, ice cream), and soy protein is beginning to find a place in the market as a binder for meats. Synthetic chicken and ham are being marketed. Vitamin enrichment of such foods as bread, milk, breakfast foods, and fruit drinks is commonplace and often legally required. New foods typically become acceptable when they mimic highly acceptable familiar products. The cost of educating populations even about *acceptable* new products is difficult at best, as any of our large food processing firms will verify. It is a certainty that any preparation of algae (or bacteria or yeast) will not be used as some green, bitter slurry, but rather to fortify familiar products, very likely different ones in each area, or as livestock feed. Another approach is the use of cheaper plant or fish proteins to enrich less nourishing local products. But if the new foods come from outside of the food deficit area, the related costs of packaging, distribution, and development would probably make them too expensive in the area where needed most.

Improving agricultural practices

Remarkable increases in productivity have been obtained merely by applying new techniques to present-day agricultural systems and using land presently under cultivation. Employment of these techniques in underdeveloped areas offers the most acceptable means of increasing the world's food production in the short run. Food production should be increased in precisely the areas where the food will be consumed.

There are, of course, economic difficulties in this approach that cannot be ignored. It has been estimated, for example, that the capacity of the chemical fertilizer industry must double if productivity is to double. Similarly, a great expansion of the food preservation and storage industries is essential. The accomplishment of these ends requires heavy investments of capital. Further, and contrary to what is generally believed, the developments that have increased production in temperate regions are not directly transferable to tropical regions, where most of the world's food problems exist. For example, such improvements as genetic gain can be spectacular, but varieties must be developed in large part for specific areas of adaptation—a time-consuming and expensive process. The most important requirement is "pay-off" technology—that is, technology that produces immediate gains. This is absolutely necessary because the sustenance-level farmer has essentially no surplus capital to risk. Results must be guaranteed. This seemingly obvious fact is too often overlooked and little appreciated. This kind of technology, if not developed in the area of application, must at least be refined and adapted for use in the area of application. Thus, technology and technologists must be improved together. Still further, agriculture is an evolving system in which one improvement depends on another. New pesticides, to be efficient, require efficient application equipment. Increased production requires added storage capacity. Agricultural development cannot be conducted on a piecemeal basis.

It has been said that an underdeveloped country can more easily obtain a modern steel industry than a modern system of agriculture. This is because of the diffuse nature of the agricultural industry—spread over large areas and involving whole populations. Improving the agriculture of an underdeveloped country involves improving the whole structure of society. Thus, the development of a modern agricultural system depends as much on improving education and communication channels as it does technology. Whole philosophies must be created to change the concept of agriculture from an extractive, feudal system to an investment business enterprise. Man will face extinction if he shirks his obligation to project, to foresee, to take measure of the calamities that are fast approaching.

Selected References

Bogert, Jean, George M. Briggs, and Doris Howes Calloway. *Nutrition and Physical Fitness* (9th ed.) W. B. Saunders Company, Philadelphia, 1973. (The principles of human nutrition.)

Borgstrom, Georg. *Principles of Food Science.* (2 vols.) Macmillan, New York, 1968. (An introduction to food science: volume 1 covers food preservation; volume 2 covers food chemistry, microbiology, marketing and utilization.)

Crampton, E. W., and L. E. Lloyd. *Fundamentals of Nutrition.* W. H. Freeman and Company, San Francisco, 1959. (An integrated treatment of nutrition and metabolism.)

Desrosier, N.W. *The Technology of Food Preservation* (3rd ed.). AVI Publishing Co., Westport, Conn., 1970. (A general textbook.)

Hoff, Johan E., and Jules Janick (compilers). *Food: Readings from* Scientific American. W. H. Freeman and Company, San Francisco, 1973. (A collection of brief but stimulating articles on nutrition and food production.)

Mayer, Jean (editor). *U.S. Nutrition Policies in the Seventies.* W. H. Freeman and Company, San Francisco, 1973. (The causes, effects, and possible remedies for hunger and malnutrition in the United States based on the 1969 White House Conference on Food, Nutrition and Health.)

U.S. Department of Agriculture. *Food.* Yearbook of Agriculture, 1959. (The story of food, written for the layman.)

Wohl, Michael G., and Robert S. Goodhart (editors). *Modern Nutrition in Health and Disease* (4th ed.). Lea & Febiger, Philadelphia, 1968. (An advanced treatise.)

CHAPTER **4**

World Population

More than one million identified species of animals and plants inhabit the earth. Of these only man can willfully control and modify his environment. Because of this ability he now dominates the earth to an extent that has not been achieved by any other species. This domination is relatively recent in human history. It has given rise to a new dimension in population growth with which man is now only beginning to cope.

The evolutionary process has endowed all species of plants and animals with a reproductive potential that, if unchecked, would overpopulate the earth within a few generations. This reproductive potential has been and still is controlled by diseases, limitation of food supply, and competition among species in the struggle for existence. No species has yet been able to free itself from the biological regulation of growth in large numbers over any significant time period. Man cannot yet lay claim as an exception to this rule. Man's failure to achieve an optimal balance between food supply and rate of growth of population is increasingly due to limitations of social organization rather than technical feasibility.

THE EMERGING POPULATION CRISIS

The length of time that man has inhabited the earth is now estimated at between one and two million years, although some estimates indicate an even earlier origin. The total population of mankind probably had not reached a quarter of a billion until approximately 2,000 years ago (Fig. 4-1). World population did not reach half a billion until the middle of the seventeenth century. Only two more centuries were required to again double the world population to slightly more than one billion. The time required to add an additional half billion has successively shortened. At the rate of growth now being achieved, during the third quarter of the twentieth century only six or seven years is required to add a half billion to the world's population.

The "population explosion" of the last several decades has given rise to two sharply opposed viewpoints regarding its implication for human welfare. On the one hand there is a small group of alarmists, often termed neo-Malthusians after the famous clergyman-scholar who around 1800

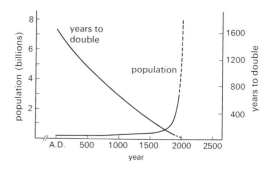

year (A.D.)	population (billions)	number of years to double
1	0.25(?)	1,650(?)
1650	0.50	200
1850	1.1	80
1930	2.0	45
1975	4.0	35
2010	8.0	?

FIGURE 4-1. *Estimated population of the world and the number of years required for it to double. [After Dorn, "World Population Growth," in Hauser (editor), The Population Dilemma. © 1963 by the American Assembly, Columbia University, New York. Reprinted with permission of Prentice-Hall, Inc., Englewood Cliffs, N.J.]*

first drew attention to the economic implications of rapid population growth. The neo-Malthusians have coined such phrases as "population bomb" to express their concern about the implication of a continued growth of population at the present rate. They would pose population policy as the major issue facing mankind in the next several decades.

At the other extreme are the "technocrats," who have faith that the possible application of science and technology to the production of crops and other raw materials will make possible a rising level of living for most of mankind for an indefinite future. Under these conditions they would expect a slowly declining birth rate to solve the population problem.

A careful sifting of the evidence implies that advancing science and technology could permit an adequate life for the world's present population, and for anticipated population growth over the next several decades. This does not imply, however, that

concern with population policy is not of immediate significance. Indeed, to cast the problem in terms of whether science and technology can meet the requirements for food, fiber, and other forms of energy implied by current or anticipated population growth is highly misleading.

Current population growth rates in countries with high "medieval" birth rates and low "modern" death rates are in the range 3.0 to 4.0 percent per year. Under these circumstances a high rate of investment in agricultural development and in educational and health facilities and services is required simply to prevent declines in individual levels of consumption and welfare. The resource demands of rapid population growth can probably be met, but only at the expense of qualitative improvements in the health, nutrition, education, and other dimensions of human welfare.

It would appear that the warnings and the concerns of both the neo-Malthusians and the "technocrats" are relevant to current population policy. Science and technology must be applied to the fullest extent possible if improved standards of living are to be achieved for the people who are now alive. At the same time, the rising population growth rates of the last several decades must be reversed if economic development is to become a continuing process for most regions of the earth.

The analysis needed to support this conclusion will be discussed in greater detail in the rest of this chapter. Attention will first be given to the principles of demographic changes and then to the problem of expanding crop production.

POPULATION DYNAMICS

Inferences drawn on the basis of world population growth rates are likely to be quite unrealistic for any particular country or region in the immediate future. The world is not a single economic unit. The earth's resources and man's technological knowledge are not reservoirs available to all. Population growth rates vary widely among countries and regions. In general, technological knowledge is expanding the availability of resources most rapidly in exactly those regions or countries where population growth rates are lowest.

In Europe, and in the USSR, North America, and Oceania, population growth rates are typically low—2.0 percent per year or less, and declining. In most countries of Africa, Asia, Central America, and South America—South Africa and Japan being exceptions—population growth rates range from 2.0 percent to close to 4.0 percent per year. Most of the countries with relatively low population growth rates have experienced a transition from high to low population growth rates within relatively recent history. This transition to relatively low population growth rates, based on both low birth rates and low death rates rather than high birth rates and high death rates, is referred to as the **demographic transformation** (Box 4-1).

Will the rapidly growing population of Asia, Africa, and Latin America go through the same demographic transition that has been occurring in Western Europe, North America, and Oceania? A purely biological theory of population would imply a negative answer to this question.

Malthusian population growth

In a population system in which growth rates are determined only by biological or "natural" factors, population growth would be determined by (1) the biological capacity of women to produce children (fecundity); (2) the natural length of life under optimum conditions (longevity); and (3) the natural ecological conditions that affect food availability and health and reduce the actual number of children born (fertility) to below the biological maximum and reduce the average length of life (mortality) to below the natural optimum.

The best known biological or natural theory of population growth was that attributed to Thomas Malthus (1766–1834), an English clergyman-scholar. Malthus concluded from the historical and scientific evidence of his time that food production had a tendency to increase in an arithmetical ratio (1, 2, 3, 4, . . .) while human population increased in a geometrical ratio (1, 2, 4, 8, . . .) over time. He concluded that since human population growth is ultimately limited by the amount of food the world can produce, the rate of human population growth would of necessity be brought into balance with the rate of growth of food production only

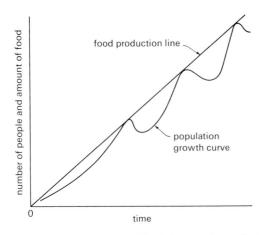

FIGURE 4-2. *Diagram of Malthus' theory of population growth.* [*After Snodgrass and Wallace,* Agriculture, Economics, and Growth, *Fig. 5–3. New York: Meredith Publishing Company. Copyright © 1964. With permission of Appleton-Century-Crofts.*]

by the natural checks of famine and pestilence.

The Malthusian theory of population growth was not, however, a completely biological theory. Malthus noted that population growth could also be limited by such factors as delayed marriage (restraint) or prostitution (vice). He did not regard these factors as having much practical significance, nor did he visualize the potential impact of technological change, resulting from advances in the natural sciences, on man's capacity to increase the production of food and fiber. Malthus visualized a pattern of population growth and decline like that shown in Figure 4-2. A similar pattern is suggested in a recent study by Meadows and associates in *The Limits to Growth.*

There have also been other biological or natural theories of population growth. Some have emphasized the negative effects of population density or complexity of life on the capacity to reproduce. The most recent to gain widespread attention was put forth by the Brazilian biologist Josué de Castro. He suggested in *The Geography of Hunger*[*] that improved diets, particularly those that are high in protein, may impair the ability to repro-

---

[*] Josué de Castro, *The Geography of Hunger,* Little, Brown, and Co., Boston, 1952.

duce. His suggestion that hunger is the cause rather than the consequence of population growth is refuted, however, by evidence that well-fed American, British, and Scandinavian wives would conceive at rates that are comparable to the rates in low-income countries if they did not practice contraception.

Sociological explanation of population growth

The demographic transformation in Western Europe, which began there in about 1875, when birth rates were in the middle and high 30's (per 1,000 population), continued until the late 1930's, when birth rates had declined to between 15 and 20; this decline in birth rates is generally taken as a refutation of at least the crude hypothesis proposed by Malthus. Today it is generally accepted that the rate of population growth, even in the high-growth-rate countries, is more of a reflection of social and cultural customs and institutions than of natural or biological factors. Population growth is thought of as a social rather than a natural phenomenon. Explanations for differences in rates of human population growth are sought in terms of sociological or institutional influences, such as religion and law; technological influences, such as medicare and public health; economic factors, such as income and the rising economic value of women's time; and cultural factors, such as attitudes toward the role of women in society.

These factors affect the rate of population growth through their separate effects on the birth rate and on the death rate (see Box 4-1). During the Western demographic revolution the factors that tended to decrease death rates were more influential than the factors that tended to decrease birth rates. Better control of disease by means of improved housing, better food and water supplies, adoption of sanitary measures, advances in preventive medicine, and, more recently, development of insecticides and antibiotics have all contributed to reducing the death rate. Furthermore, rapid adoption of these advances for the purpose of reducing infant mortality or prolonging average life span has not provoked any serious moral or social conflicts.

Box 4-1 BIRTH AND DEATH RATES, 1770–1970*

In the industrialized nations in the last century, the declining death rate was accompanied by a decline in the birth rate. From a rate of about 40 per 1,000 in 1875, the birth rate in developed countries went down to about 20 per 1,000 by 1970. Demographers refer to this gradual shift in both birth and death rates as the "demographic transition."

The difference between the birth rate and the death rate is a measure of how fast the population is increasing. For developed countries, this difference is about 10 per 1,000 population, or about 1 percent. This means that despite going through the demographic transition, developed countries continue to grow in population.

In underdeveloped countries, the sharply dropping death rates has not been accompanied by a significant drop in the birth rate. In these countries, the birth rate has continued to be close to 40 per 1,000 while the death rate has fallen to about 15 per 1,000 in 1970. The result is an increase in population of about 25 per 1,000 population each year, or 2.5 percent.

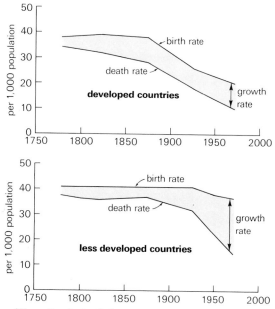

*From Population Reference Bureau, Inc., *The World Population Dilemma*, Columbia Books, Inc., Washington, 1972.

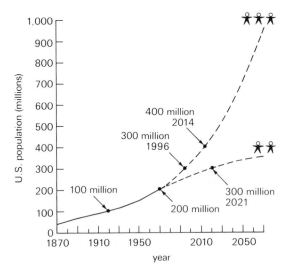

FIGURE 4-3. *Population growth and family size.* [*After Commission on Population Growth and The American Future,* Population and the American Future. *Bergenfield, New Jersey: The New American Library (Signet), 1972, p. 19.*]

The adoption of practices intended to reduce birth rates required much more complex changes in personal and social motivation. Religious beliefs and social values—for example, perpetuation of large families—were often in conflict with new values concerned with the improvement in welfare and opportunities that could be gained by maintaining small families. Only after the new values were widely accepted did the control of family size through some combination of delayed marriage and contraceptive practice become extensive enough to reduce birth rates.

The lag in the decline in birth rates behind the decline in death rates has resulted in exceptionally rapid population growth during the middle years of the demographic transformation in most countries. Even the seemingly modest change from a three-child family to a two-child family can have a tremendous impact when the change occurs over several generations. In the United States an average of three children per family would, for example, result in a population of about 320 million in the year 2000 and almost one billion by 2070 (Fig. 4-3). An average of two children per family would, in contrast, result in a national population of about 270 million in 2000 and a leveling off of the population at approximately 350 million by 2070. The pattern of fluctuations in fertility rates in the United States throughout the past several decades indicates that considerable caution should be exercised in attempts to forecast the fertility behavior of American families. Recent data suggest, however, the U.S. fertility rate is becoming one that is consistent with a stable, or nearly stable, population (Fig. 4-4).

A Neo-Malthusian perspective

The continuum described above was, until recently, regarded as a transition through which all nations would pass. All nations and people seemed to be moving through the demographic transformation, although at different rates. Current developments do not, however, support this optimistic view. There appears to be an increasing polarization into high-income, developed nations in which the population increases slowly and low-income, underdeveloped nations in which the population increases rapidly.

This polarization appears to be related to the extremely rapid decline in infant mortality and to an increase in average longevity. For example, the death rate among the Moslem population in Algeria declined more in the eight years between 1946 and 1956 than the death rate in Sweden did during the period 1775 to 1875. The result is that

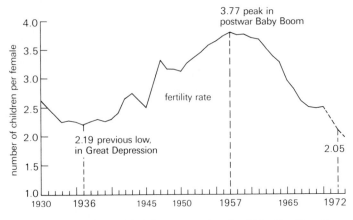

FIGURE 4-4. *Fluctuation in the U.S. Fertility Rate, 1930–1972.*

population growth rates in the most underdeveloped countries now exceed those in the developed countries even during their periods of most rapid population growth.

In the Philippines, for example, the population growth rate exceeded 3.25 percent per year in the late 1960's. It could rise to even beyond this level in the 1970's. Rates of more than 3.0 percent per year are entirely outside the range of historical experience in western countries. In the West population growth rates have rarely exceeded 2.0 percent.

POPULATION AND FOOD

In a Malthusian economy the relationship between population growth and food production is relatively simple: the rate of increase of food production limits the rate of population growth. Wars and epidemics or the discovery of new land resources might ease the food situation for a time, but population growth would ultimately be limited by the difficulty of increasing food production as cultivation extended to the less productive soils and as intensity of labor use increased. In the Malthusian system food production is the independent and population growth the dependent variable.

In a world in which countries are passing through the several stages of the demographic revolution, the relationship between population and food production is much more complex than in a simple Malthusian system. As a nation undergoes the transition from high birth and death rates to low birth and death rates there is a decrease in the proportion of national resources, or of personal income, spent on food. During the early stages of development 75 percent or more of the income of a population may be spent on food. By comparison, less than 25 percent of the income of consumers is typically spent on food in developed countries of today.

The demographic transformation, if successful, is accompanied by a transition from a situation in which the rate of population growth depends on the rate of growth in food production to one in which the rate of growth of food production depends on population growth. This transition is from a situation characterized by the "pressure of population on food supplies" to one characterized by the "pressure of food supplies on population." In economies like India's, population growth is highly dependent on growth in food production (plus food imports). In the United States the growth of food production is restricted by the growth of population (plus food exports).

During the intermediate stages of the demographic transformation the population-food relationship is even more complex. During the initial stages of the transformation death rates decline more rapidly than birth rates. This alone would result in a rise in the food needs. If per capita income is also rising, this adds further to the demand for food. As incomes rise from levels of $50 to $100 per capita per year much of the increase in income is spent in increasing the quantity and the quality of food consumed. The result is a rise in the demand for food that is substantially greater than the increase in population. Table 4-1 shows the figures for these increases in 40 countries (see also Chapter 25).

Before death rates begin to decline substantially, population growth rates and the demand for food may both be increasing at little more than 1.0 to 1.5 percent per year. As death rates continue to decline and birth rates remain high, the rate of population growth may rise to well above 3.0 percent per year, as in the Philippines and Malaya and a number of countries in Latin America. If rapid population growth is accompanied by an increase in per capital income of 2.0 to 3.0 percent per year, the demand for food may increase by 5.0 or 6.0 percent per year.

An increase in the demand for food in this range is completely outside the experience of the developed countries. In these countries death rates declined slowly and birth rates began to decline while death rates were still relatively high. The question that remains to be answered, for countries experiencing such high rates of population growth and food demand, is whether this phase will be followed by a decline in the rate of population growth as a result of lower birth rates, along the lines of a "classical" demographic transformation, or by a decline in the rate of population growth as a result of a new "Malthusian" equilibrium.

Table 4-1 ANNUAL RATE OF GROWTH (IN PERCENT) OF POPULATION, DOMESTIC FOOD DEMAND, CROP OUTPUT AND AGRICULTURAL PRODUCTION IN 40 DEVELOPING COUNTRIES, 1950–1968.

| COUNTRY | POPULATION GROWTH RATE | RATE OF GROWTH OF | | RATE OF GROWTH IN DOMESTIC FOOD DEMAND | SURPLUS OR DEFICIT IN AGRICULTURAL PRODUCTION |
		CROP OUTPUT	AGRICULTURAL PRODUCTION		
Group I (rapid economic growth; gross national product, 6% or above)					
Greece	0.8	4.1	4.6	3.0	1.6
Bulgaria	0.9	3.0	4.0	4.4	−0.4
Yugoslavia	1.1	5.4	4.6	4.6	0.0
Poland	1.5	3.1	3.2	3.7	−0.5
Jamaica	1.8	3.4	2.7	3.8	−1.1
UAR	2.5	2.9	2.4	5.4	−3.0
South Korea	2.8	4.3	3.7	5.1	−1.4
Iran	2.9	NA	2.7	4.7	−2.0
Panama	3.1	3.1	3.6	4.5	−0.9
Taiwan	3.2	3.8	4.4	6.6	−2.2
Mexico	3.3	5.7	5.1	4.4	0.7
Thailand	3.3	4.5	4.5	5.3	−0.8
Venezuela	3.7	5.4	5.1	4.5	0.6
Group II (medium economic growth; gross national product, 4–6%)					
Spain	0.8	1.6	2.9	2.6	0.3
Burma	1.9	1.6	1.3	3.8	−2.5
Bolivia	2.2	3.5	2.8	3.7	−0.9
India	2.2	2.8	2.6	3.5	−0.9
Peru	2.7	3.3	3.1	4.1	−1.0
Turkey	2.7	2.2	3.4	3.7	−0.3
Iraq	2.8	2.3	2.0	4.7	−2.7
Sudan	2.9	NA	4.1	4.3	−0.2
Brazil	3.0	4.3	3.8	4.2	−0.4
El Salvador	3.0	5.4	4.0	4.0	0.0
Malaysia	3.1	3.7	4.1	4.3	−0.2
Nicaragua	3.1	7.9	5.9	4.2	1.7
Colombia	3.2	3.2	3.3	3.9	−0.7
Ecuador	3.2	4.5	6.0	3.8	2.2
Honduras	3.2	NA	3.2	4.2	−1.0
Philippines	3.3	4.1	3.7	4.8	−1.0
Guatemala	3.4	6.7	5.0	4.1	0.9
Dominican Republic	3.6	NA	1.6	4.0	−2.4
Costa Rica	3.8	3.7	4.2	4.5	−0.3
Group III (slow economic growth; gross national product, 4% or below)					
Uruguay	1.4	−0.5	−0.1	1.1	−1.2
Argentina	1.7	2.5	2.0	2.1	−0.1
Tunisia	2.1	−1.9	0.9	2.4	−1.5
Indonesia	2.2	1.9	2.3	2.1	0.2
Ceylon	2.4	3.2	2.9	3.0	−0.1
Chile	2.4	2.2	2.2	3.0	−0.8
Pakistan	2.4	2.7	2.9	3.4	−0.5
Paraguay	2.6	NA	2.4	3.0	−0.6
Morocco	2.8	−0.2	1.1	2.1	−1.0
United States	1.5	1.7	1.9	1.6	0.3
Japan	1.0	2.4	3.3	3.6	−0.3

Source: After USDA, *Economic Progress of Agriculture in Developing Nations, 1950–68,* For. Agric. Econ. Report No. 59, Washington, 1970.
NA = Not available.

For less developed nations, the rate of growth of domestic crop production has clearly been behind the rate of growth in demand during most of the period since World War II (Table 4-1). If these countries are to escape the Malthusian "trap" and proceed along the path of demographic transformation, either the rate of growth of food production must expand more rapidly or the rate of population growth must decline. Recently, a few countries have begun to experience changes. An agricultural "green revolution" has resulted in more rapid growth of grain production in Mexico, Pakistan, the Philippines, and a few other countries since the mid-1960's. The rate of population growth has begun to recede in a few countries such as South Korea and Taiwan. By and large, however, the institutional capacity of agricultural and medical science for the dissemination of knowledge and materials, needed to speed the rate of growth of food production or reduce the rate of population growth, either is lacking or is just now being established.

POPULATION POLICY

None of the problems of population growth can be solved without sustained increases in both agricultural and industrial production. The fundamental problems that must be solved to achieve sustained increases in agricultural production are dealt with throughout this book.

It is also clear that population growth itself represents an obstacle to the achievement of sustained growth in agricultural and industrial production. High rates of population growth impose heavy demands on technical progress and capital formation simply to maintain the existing per capita rates at which food, fiber, and other forms of energy are consumed. Rapid population growth is, therefore, competitive with improvements in the quality of human existence.

There must, therefore, be a dual approach to the problems of population. One involves technological, social, and economic development to improve life for the people who are now alive. The other approach is demographic. There must be declines in birth rates if economic development is to become a continuing process. The problem,

then, is not whether to speed up the rate of technological development or to reduce birth rates. Rather, for each country, the problem is to determine what combination of policies should be followed to achieve agricultural development and reduce population growth.

Adoption of an effective population policy is complicated by a number of social and biological factors. The social problem centers primarily on the ethical issues related to the use of contraceptive devices and the development of motivation for the control of family size. The biological problem centers on the development of low-cost, effective, socially acceptable contraceptive devices or techniques.

Ethical problem of population control

Most of the world's major religions have no generally accepted or firmly enforced objections to birth control. Individual Moslems, Hindus, Buddhists, and Confusians may be found to favor or oppose birth control, but there is no "official" theological or ethical rationalization of either position.

The Roman Catholic position does represent a more difficult theological obstacle to population control. The Catholic position does not oppose population control as such. Rather, it relates to the means by which fertility is regulated. Catholic doctrines on marriage and procreation have been significantly modified during the twentieth century. Periodic continence as a method of regulating births has been endorsed. There has also been increased emphasis on personal and social values in marriage. Most observers anticipate further change in Catholic doctrine that will facilitate the development of policies designed to lower the birth rate in Catholic families and in predominantly Roman Catholic nations.

Social factors

Social inertia has, in the past, been at least as important an obstacle to the decline of birth rates as have religious and ethical values. The changes in the status of women, aspiration for social mobility, and the growth of the social and economic importance of formal education have contributed to declines in birth rates in both Catholic and non-Catholic countries of Europe. Where motivation to control family size was strong, birth rates began to decline even before the development of modern contraceptive devices.

Among the important social, or economic, factors leading to the decrease in family size in the higher-income countries is the rising value of female labor. As employment opportunities for women have risen, the time devoted to child bearing and rearing have become increasingly competitive with other goals. Concurrently, the material worth of children, at least in large numbers, has lessened. Urbanization has reduced their economic value in production, and the development of social security and other institutional support for the elderly has reduced their importance as a source of security in old age.

For thousands of years man has sought reliable means of avoiding unwanted children by contraception—preventing the male sperm from fertilizing the female egg during sexual intercourse, or in other ways preventing the fetus from forming. Progress has not been as rapid in this field of medical science as in many others, partly because less effort has been devoted to methods of birth control than to improving the methods of extending life by deferring death. Furthermore, until very recently the available methods of birth control were not well adapted to meet the needs of poor, illiterate couples lacking either privacy or adequate sanitary facilities.

The principle methods of contraception include jellies and foams to kill the sperm; condoms and diaphragms to block the sperm; oral contraceptives to make the women temporarily infertile by chemical means; and the IUD (Intra-Uterine Device) to make the woman temporarily infertile by the insertion of a plastic or metal device into the uterus. In addition to contraception, surgical sterilization has achieved growing popularity, particularly among mature adults who wish to curtail their family size. The usual operation is either a vasectomy for the man, or a tubal ligation for the woman. Both are virtually 100 percent effective. The abortion is another surgical method of reducing unwanted births. Its use tends to be used most extensively where the more effective contraceptive technologies are not available.

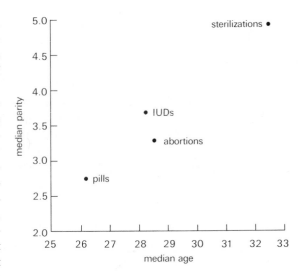

FIGURE 4-5. *The choice of birth control method is greatly influenced by age and parity status (number of children). These data have been obtained from postpartum family planning programs and hospitals throughout the world.* [*After* Population Program Assistance, *Agency for International Development, Washington, D.C., December, 1971.*]

All of the methods discussed above have a place, depending on a particular country's cultural traditions and stage of medical development. Also, among individuals, variation in birth control methods is substantial, the one empolyed often depending on the age of the woman and the number of children (parity status) in the family (Fig. 4-5).

Population programs

There has been widespread interest in the development of population policies. Both Mainland and Nationalist China have active programs to reduce birth rates. South Korea has a strong government-backed program to spread family planning. Japan has achieved one of the world's lowest birth rates. India and Pakistan have initiated strong birth control programs. Tunisia and Egypt are both developing programs to spread family planning. Even in Latin America there is widespread concern about the "population explosion," and a few Latin American countries make contraceptive services available in government hospitals.

The fertility control efforts in Taiwan have been particularly interesting because they have been related to a well-planned research program designed to provide information on the efficiency and acceptability of different methods and devices.° The Taiwan preprogram survey indicated that most women were having more children than they wanted and approved of the idea of limiting family size. Not only were they aware that limiting family size would improve the welfare of their families, but they believed that the number of children should not be left to "fate" or "providence." Their attitudes were, in general, more advanced than many officials believed them to be.

Similarly, a study in the Philippines showed that rural wives would like to limit the size of their families. The study also showed, however, that the wives did not know that effective methods were available. In spite of their lack of knowledge a very high percentage of the women studied were trying desperately, to the best of their ability, to prevent themselves from becoming pregnant by the use of the relatively ineffective practices available to them.

Prospects for the future

There is every reason to anticipate that birth rates in many of the newly developing countries can be reduced more rapidly than might have been expected from the earlier experiences of Western Europe. Possibilities appear even more favorable than they did in the early 1960's. "Religious, ideological, and nationalistic disputes about birth control have subsided; governments are progressively aware of the need to reduce the birth rate, much of the public is already anxious to control its childbearing; and there is a new method of contraception that requires no attention after it is first obtained and which is cheap, safe, and

effective for the large majority of the childbearing population."°

It is now possible to think that the world will—by 1980—be well on the way toward finding positive solutions to its problem of excess reproduction and population growth. This optimism is based on a number of favorable developments.†

1. The concept of fertility control, and its individual and social benefits, has been broadly disseminated in an increasingly receptive world.
2. Many countries have removed or modified laws that restricted the availability and use of the most effective means of birth control.
3. Resources available for the development of population and family planning programs have increased rapidly.
4. International organizations have greatly increased their activities in the population and family planning field.
5. Key breakthroughs in fertility control technology have occurred or now seem feasible.
6. Many countries are developing increasingly effective family planning programs.
7. Recent census and other demographic data indicate a more favorable world population figure than has been projected from earlier data.

What has been accomplished is, however, only a beginning. If population growth rates are to decline fast enough to allow savings and capital to be diverted from simply meeting food needs to improving the quality of life, positive government programs will be required. At least part of the resources that would otherwise be devoted to expanding national agricultural and industrial potential will have to be devoted to the extension of public education and medical services directed to reducing the birth rate.

° Bernard Berelson and Ronald Freedman, "A Study in Fertility Control," *Scientific American*, Vol. 210, No. 5, May 1964, pp. 29–37.

° F. W. Notestein, "World Population Determinants in the Future," *in World Population and Food Supplies, 1980*, American Society of Agronomy, Madison, 1965.

† Agency for International Development (AID), *Population Program Assistance*, U.S. Government Printing Office, Washington, D.C., December, 1971.

Selected References

Commission on Population Growth and The American Future. *Population and the American Future.* The New American Library (Signet), Bergenfield, N.J., 1972. (An authoritative analysis, by a commission proposed by President Nixon and authorized by the Congress, of population trends and of the implications of future population growth for the United States.)

Ehrlich, Paul R., and Anne H. Ehrlich. *Population, Resources, Environment: Issues in Human Ecology* (2nd ed.). W. H. Freeman and Company, San Francisco, 1972. (An ecological perspective on population growth in relation to resource constraints.)

Farmer, Richard N., John D. Long, and George J. Stolnitz (editors). *World Population—The View Ahead.* Bureau of Business Research, Indiana University, Bloomington, 1968. (The report of a 1967 Conference on World Population Problems held at Indiana University. Careful analysis of the outlook for world population growth; the technological, economic, and sociological basis and implications of population growth; the biological and technical aspects of birth and death control. A sober rather than an alarmist perspective.)

Meadows, Donella H., Dennis L. Meadows, Jorgen Randers, and William W. Behrens, III. *The Limits to Growth.* Universe Books (for Potomac Associates), New York, 1972. (A neo-Malthusian exposition of the relationships between resources, technology, and population growth. Utilizes systems simulation methodology and rhetoric to project a world population crisis during the second half of the twenty-first century.)

Population Reference Bureau. *The World Population Dilemma.* Columbia Books, Washington, D.C., 1972 (A very useful introductory treatment of population problems. Sections on the origins of the population explosion, on population in relation to material and environmental resources, and on population control technologies and policies. The Population Reference Bureau also publishes the bimonthly *Population Bulletin*, each issue of which is devoted to a specific population problem or issue.)

United Nations. *A Concise Summary of the World Population Situation in 1970.* Population Studies No. 48, New York, 1971. (An authoritative review of historical patterns of population growth; population projections for 1970–2000.)

CHAPTER 5

Agriculture, Pollution, and the Environment

Pollution is a general term applied to undesirable changes in the environment. Such changes may be due to substances introduced in soil, air, or water, or to natural substances modified so that they become undesirable. Pollutants may be biological in nature, such as pests or pathogens, or forms of unwanted energy, such as waste heat, noise, and radiation. The term has also been expanded to include visual or esthetic insults to the environment. Although everyone agrees that pollution is undesirable, there is considerable argument over who is responsible and what can be done about it. In this chapter pollution as it affects agriculture and the environment, and the ways in which it relates to present-day circumstances will be considered. It is a particularly important subject now, since the issue of minimizing pollution and maximizing production has become legal as well as moral.

MOTHER EARTH VERSUS SPACE SHIP EARTH

Early humans usually spent their lifetime, averaging perhaps 20 or 25 years, within a day's walking distance of their birthplace. They did not need to go farther. Using the rough comforts of generations past, a cave or some other well-concealed spot, they secured their food by hunting and gathering. But as the populations of the world increased, it became necessary to live in small, more immobile groups for mutual aid and protection. The hunting and foraging parties of these groups would deplete their immediate vicinities and were obliged to range increasingly greater distances to secure food. With the development of agriculture the proportion of food obtained from cultivation increased. Through time, and countless trials and errors, mankind learned to husband the soil, and to respect it (Fig. 5-1).

Thus, **Mother Earth** became the great provider for the cultivator, too, although her bounty was often grudgingly given. Throughout the centuries, the reverence for "nature" as a sustainer of life has endured. Even today we assign the properties of life to soil itself and describe it in terms we would use for living things—for example, "beautiful," "productive," "dead," "mature," "old," and "young." The forces of nature were thus personified and deified, and we worship them still.

With the passage of time, our sense of direct

FIGURE 5-1. *Man and nature can coexist in harmony and peace, creating an environment in which there is abundance for all.* [*Courtesy TVA.*]

dependence on the land became obscured and reverence for the soil and the environment faded. Today, one person can produce food for 50 people, and a city dweller may never see a field of corn planted or a wheat harvest. Most planners and engineers disregard the fact that the supply of rich tillable land is limited, and they continue to cover our most productive soils with cement and asphalt. The millions of people who must work in cities, and escape each day to the more pleasant surroundings of suburbia, have tempted developers to transform the countryside into a plethora of squalid look-alike housing developments and trailer parks. Even the law in many parts of the United States discourages the use of farmland for crop production, and increasingly more land is being taxed according to its "highest use," which means its greatest potential for residential, industrial, or other commercial exploitation. Present tax structures, then, make it economically infeasible to raise agricultural or forest crops on much of the land—assessed at $2,000 to $10,000 an acre. These materialistic and impersonal values we have placed on the soil that sustains us must ultimately change if this nation's traditional self-sufficiency in agricultural production is to endure.

Until recently, little public concern was expressed over the prevailing methods of waste disposal; refuse in general—including the wastes from manufacturing industries such as paper mills and cannery plants, power industries, and city sewers, as well as those directly from agriculture itself—was assumed to be harmless, until contrary evidence was produced by some regulatory body. The rationale that allowed this permissive attitude to develop in the United States had its roots in the early history of this nation at a time when resources seemed almost unlimited and the pressures to industrialize and develop a greater national economic base dominated nearly all other aspects of life. Most people welcomed new industry, regardless of its ravages to the countryside, and disregarded the evidence that streams were becoming polluted and vast areas of land disfigured and ruined (Fig. 5-2). Unfortunately, the same robber-baron philosophy still prevails today, and many developing nations have decided to ignore pollution in their efforts to industrialize as rapidly as possible. We can hardly censure them, for they are only following our example.

To be sure, many may still pay verbal homage to Mother Earth, but they continue to assume that

FIGURE 5-2. *When the conservation measures are neglected, destruction often ensues.* [*Courtesy TVA.*]

FIGURE 5-3. *The finite nature of space-ship earth is most apparent when viewed from the moon.* [*Courtesy NASA.*]

awakened to a growing apprehension that much of what we cherish might be lost forever. This revolution in thinking, known as the environmental or ecology movement, found an especially receptive audience in the youth, a generation at least partially jaded by abundance, as well as confused and exasperated by an unpopular war. A series of dramatic events and findings quickly unfolded: the publication of Rachel Carson's *Silent Spring;* the thalidomide disaster; military use of defoliants; an association of DDT with reduced eggshell thickness and the potential extinction of a species; the linkage of smoking and cancer. The astronauts hurtled through space to the moon, and rediscovered the earth; and the rest of us, too, saw it anew, as a lonely, tiny craft, with limited resources: **Space-ship Earth** (Fig. 5-3). Environmental pollution became a cause with which all could identify.

This rising concern for environmental quality has come from two major sources. The first is the increased demand for the environmental assimilation of residuals derived from growth in commodity production and consumption, and from the energy production and transportation services associated with this growth. The second, and related, source is the rising demand for outdoor "retreats," where such amenities as the absence of pollution and congestion are present. The increased competition between the demand for environmental services for disposal of residuals and that for untouched natural, or wilderness, areas resulted in the 1960's in a heightened awareness of environmental problems in industrial countries.

At the same time, it is useful to emphasize certain considerations that have frequently been ignored in the heat of the challenge by the new environmental concerns. First, it is well to remember that throughout history man has been continuously challenged by the dual problem of (a) how to provide himself with adequate sustenance and (b) how to manage the production and disposal of what is now increasingly referred to as residuals or, in less elegant language, as garbage. Failure to make balanced progress along both fronts has at times imposed serious constraints on society's growth and development. Even in poor communities, whose people are living close to

her resources are infinite: she is expected to be forever tolerant of the abuse of her erring and spoiled children. Although concern for the environment had long been championed by eloquent and forceful exponents (such as Henry Thoreau, Gifford Pinchot, and John Muir) their combined voices never really touched the conscience of America whose heroes were frontiersmen rather than Indians, cowboys rather than farmers, and industrialists rather than craftsmen.

However, in the early 1960's a new awareness of environmental issues emerged in both the scientific community and in society generally; we

nature, use of energy to dispose of residuals is directly competitive with the use of energy to provide for sustenance.

It is also worth remembering that the capacity of a society to solve either the problem of sustenance or the problems posed by the production of residuals is inversely related to the rate of population growth and positively related to the society's capacity for innovation in science and technology and in social institutions. Continued technological advance is essential for further advances in both the materials and esthetic dimensions of consumption. The fundamental significance of advances in science and technology is that they permit the substitution of knowledge for resources. It is clear, then, that scientific and technical effort must now be redirected toward reduction of environmental stress.

Technology, however, its potential benefits notwithstanding, was one of the first to be identified as a major pollutant, and those who did so assumed the roles of heroes.

Agriculture, the self-proclaimed friend of the soil, was not to be spared as a "special case," and its technological aids were considered as harmful as those of industry. Although we may indeed be able to affix the cause of pollution to a few major sources, and although all of us may identify with the cause of a "clean environment," it is also clear that each of us contributes to its degradation. We pollute every time we drive our automobile, wash our clothes, or turn on the light. If cars, detergents, the power industries, or agricultural production practices are recognized polluting agents, then so are we who make use of them.

AGRICULTURAL PRACTICES AND THE ENVIRONMENT

Modern agriculture is presently identified as both object of pollution and pollutor itself. However this dual role is particularly ambiguous because of the continuing fluctuations in the interrelationship between agriculture, the chemical revolution, and the population explosion. The growth of cities and the increase of population has resulted in the continued encroachment of asphalt and steel on

farmland, even as the numbers of mouths to feed have increased. Cities and their industrial complexes have preempted still more valuable space for land fills for their refuse, and noxious gasses are constantly spewed from smokestacks, burning dumps, and automobiles. However, just as the urbanite relies on new artificial fibers and medicines, so has agriculture turned to the chemical industries for mineral fertilizers and pesticides to control weeds, fungi, and insects. As a consequence of its search for increased efficiency, agriculture, too, has helped to pollute the environment with pesticide residues and leached nutrients. Nevertheless, as we seek solutions, the basic purpose of agriculture—to produce food and fiber— must be clearly kept in mind: to eliminate pollution and at the same time create a food shortage will not better mankind.

That agriculture contributes to pollution in various ways (Fig. 5-4) can be most clearly illustrated by descriptions of two cropping situations—corn production practices in the Midwest and forest clear-cutting.

Corn production

In the Midwest, approximately 120 days are needed to grow a crop of corn from the time seed is planted. During the other 245 days of the year, the soil is bare and subject to erosion, since it is not protected by a mantle of vegetation. In some parts of the world a second crop is grown which will protect the soil, but in the climate of the United States Corn Belt there is no respite from winter.

Corn is generally planted on a bare site. As the seedbed is prepared for planting, dust is sent flying into the air and blown away. Soil is lost in a similar manner while the crop is growing and during the fallow period between crops.

To grow corn most efficiently one must apply pesticides, starting before or at the same time the seed is being planted. These chemical materials can infiltrate into the environment by scattering and blowing, vaporization, leaching, and adsorption onto soil particles, and incorporation into the crop and crop residues. The fate of these pesticides is now considered to be one of agriculture's

72

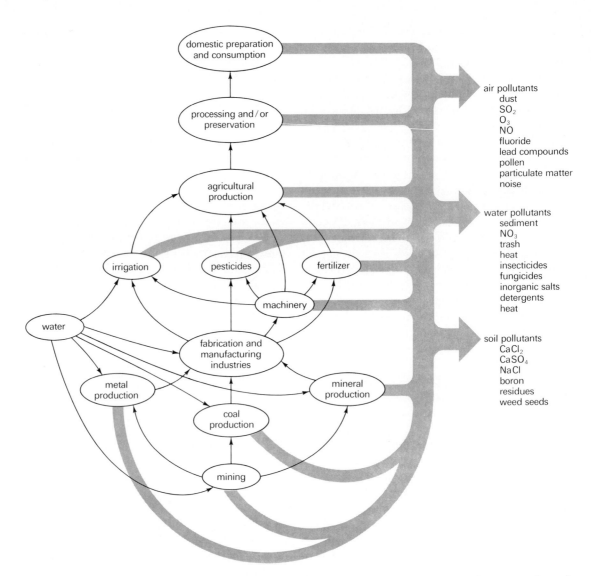

FIGURE 5-4. *At every stage in the production of food for modern man, pollution results. From the mining of metals to make machinery with which to work the fields, to the ultimate disposal of human wastes, there is a pollution syndrome which overlaps, feeds back, and loops, so that the exact sources of pollution are not always clear. This diagram is much too simple to indicate the complexity and immensity of the situation. The transportation system, carrying materials from one place to another in the production network, creates a staggering amount of pollution that must also be accounted for.*

major problems. They will eventually break down chemically or biologically, although some chemicals may take years to do so. However, if the amounts applied are greater than the amounts lost, an accumulation will ensue.

Unless the fields are contour plowed, rainfall drains almost immediately into waterways and streams. Hence, runoff frequently contributes to floods and soil erosion. A conservative estimate of annual soil loss by erosion from a well cared for cornfield with little slope, would be approximately 18 metric tons per hectare (8 tons per acre). Since the total weight of soil in the upper 15 cm (6 inches) of soil is approximately 2,200 metric tons per hectare (1,000 tons per acre), such a loss is hardly significant in any one year. However, the annual loss can be four times as great, and is sometimes greater by a factor of 10. Unfortunately, most of the soil that is lost goes directly into streams and ponds, creating sedimentation problems by silting (Fig. 5-5).

If production in the U.S. Corn Belt is to be profitable, grain yields of about 100 bushels per acre (5 tons per hectare) are necessary. If these yields are to be obtained, the nutrients that have been removed by the growing crop must be replaced, the usual method being the application of mineral fertilizers. However, those nutrients which are soluble in water, or which do not become tightly affixed to soil particles, are lost in the soil water runoff, thereby creating the possibility of excess aquatic vegetation from an over-enrichment of nutrients in streams and lakes, and eventually fish kills from the depletion of dissolved oxygen. This process is called **eutrophication**. Although phosphorus is a source of water deterioration, the amount lost from the soil because of agricultural practices is probably insignificant, because phosphate ions are tightly bound to soil particles; rather, most phosphorus comes from household detergents. However, nitrogen loss from the soil is much greater, because nitrates are highly soluble and are readily leached from soil. Consequently these compounds may be the main elements contributing to eutrophication of many lakes and farm ponds. A critical problem can also occur when nitrogen compounds find their way into underground aquifers, because excess nitrate (NO_3) or nitrite (NO_2), when ingested, may be hazardous, especially to the young. A recent study, however, has indicated that present patterns of fertilizer use pose no present global danger from these compounds.

Corn is a row crop and, as usually managed, requires intensive cultivation; however a new system of called "minimum tillage" has been developed which reduces erosion but relies on greater use of herbicides (Fig. 5-6). Tractors, used to cultivate corn, produce fumes containing hydrocarbons. These fumes are not a major source of pollution in rural surroundings but they must

nevertheless be counted as a contributing factor.

Smut, a common fungus that attacks corn, releases spores that can cause severe irritation to some persons. Similarly pollen from ragweed growing at the uncultivated edges of fields is a cause of asthmatic attacks. Corn pollen is irritating to some, but is usually no more than a local nuisance to those working directly with the crop.

Under modern systems of production most crop residues are returned to the soil, but some are burned in the field. Until recently, before the large-scale adoption of picker-shellers, cobs were harvested along with the grain and disposal was a problem. Thus burning contributed, and still does to some degree, to air pollution.

Clear-cutting in forestry

Clear-cutting is a practice in which all the merchantable timber is removed from an area at one time, usually leaving the soil essentially treeless. Once the forest canopy is removed by clear-cutting, precipitation falls directly on the soil and a greater proportion of the water runs off immediately. Many of the environmental problems that result from corn production are also created by clear-cutting, though they are not as severe. Natural fertility is lost. Pesticides are introduced into the ecosystem. Machinery that is used to harvest the crop releases hydrocarbons into the atmosphere. The humus in the topsoil is exposed to the sun so that it decomposes faster, thus reducing the overall quality of the site. Smoke from fires used to burn brush piles pollutes the atmosphere. Esthetically, a clear-cut is nearly always displeasing. One can rationalize all of these objections by saying that clear-cutting is necessary to grow certain kinds of trees which are especially light demanding, especially in their juvenile stages, and that it has a minimal negative effect in certain environments. However, this is a weak defense against the onslaught of preservationists, many of whom would prefer that no trees at all should be cut.

Agriculture as a pollutor

As indicated in the discussions of corn production and clear-cutting practices, agriculture does indeed contribute to the deterioration of the environment, as does industry. In general, those spe-

74

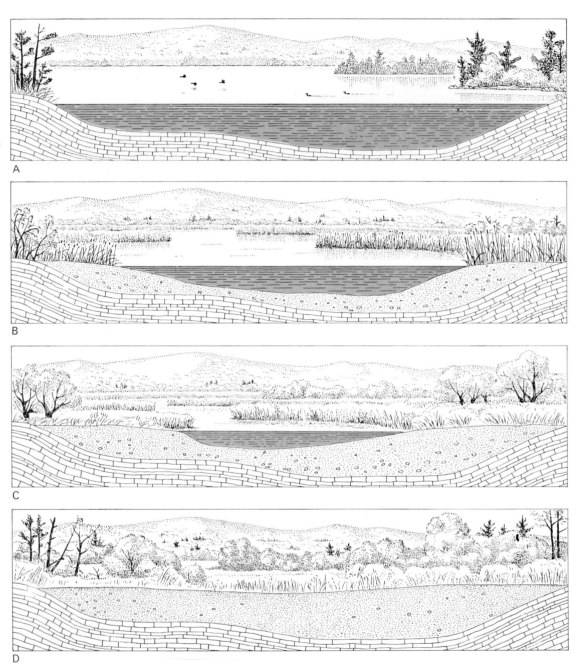

A

B

C

D

FIGURE 5-5. (A) *Obliteration of lakes and other kinds of impoundments is a process that starts at the edge of the water: a few bog-adapted conifers rise in a forest of hardwoods. (B) Next the debris of shallow-water plants turns the lake margin into marsh that is gradually invaded by mosses and bog plants, bog-adapted bushes and trees such as blueberry and willow, and additional conifers. (C) Eventually the lake, however deep, is entirely filled with silt from its tributaries and with plant debris. (D) In the final stage the last central bog soon grows up into forest.* [From Powers and Robertson, "The Aging Great Lakes." Copyright © 1966 by Scientific American, Inc. All rights reserved.]

FIGURE 5-6. *By growing crops such as corn in soil without tillage (so that the soil has a perennial cover of sod), soil losses are negligible. This system of production will be used much more in the future.* [*Courtesy Ohio Agriculture Research and Development Center.*]

cific practices most likely to cause pollution are soil erosion, the improper use of plant nutrients, and pesticide usage.

Erosion has long been an agricultural problem, and soil conservation has always been the byword of progressive agriculture. Long before it became fashionable to wave the banner of environmentalism, agriculture had been concerned over this problem—because long-range maintenance of productivity was in agriculture's obvious self-interest. Consequently, a discussion of soil and soil conservation is necessarily included in any basic coverage of agricultural practices (see Chapters 12 and 16).

In the past, the excessive use of plant nutrients has usually been considered an agricultural problem only in relatively isolated situations. These include arid areas where the penalty of overfertilization is severe, leading to salinization of water supplies and high salt content of soils. Overfertilization of greenhouse soils, has similarly generated an appreciation of the high-salt problem that can result from this practice. However it is fair to say that in the past underfertilization has received the greatest attention, and the problems of excess nutrients in humid regions have not been considered except for economic purposes: that is, overfertilization has always been unprofitable, and it has been cautioned against only for that reason.

The pesticide problem appears to have been underestimated, even though problems of overuse were recognized long ago. The discovery of the new "miracle" pesticides, such as DDT, seemed to provide a welcome solution because their concentration appeared infinitesimally small in comparison with lead arsenate, for example, and their potential polluting consequences were largely overlooked in agriculture. Today, even after the "banning" of DDT, the solution is still not clearly apparent. We now find that some of the replacements, the so-called "hard" or persistent pesticides, are quite dangerous, particularly for those applying them. It also appears that the toxic effects of such substances as DDT may have been exaggerated. Many have felt that this banning was more emotional than rational, and the result of the need to identify a scapegoat that would be symbolic of our general pollution problems.

Agriculture has never relied exclusively on chemical control of pests. Genetic resistance has been a key objective in crop-breeding programs, and biological control is not at all a new agricultural practice.

Finally, it would be folly to attempt to grow food and fiber crops at a "zero" level of pollution. The cost of "organically grown" foods is approximately twice that of food produced by conventional practices. We could eliminate farm machinery and return to primitive ways, but it would require a mass movement back to the land, a chilling prospect to all but the romantic. In other words, over fastidiousness is not the solution. This is not to imply that we should not use conservation practices with sound ecological bases. We must! But we must determine how we can feasibly produce sufficient amounts of food, given the economic realities, and the necessity to safeguard public health and welfare. Our goal must be not only sustained but increased yield, achieved with minimum damage to the environment.

INDUSTRIALIZATION AND AGRICULTURE

As a result of the trend toward urbanization and industrialization, the threat to agriculture, primarily from air pollution, has increased. Figure 5-7 is a diagram of the global circulation patterns of certain atmospheric pollutants.

Air pollution is not only esthetically offensive but also a genuine health hazard to animal and man. It can be extremely injurious to vegetation, so that agricultural industries located around population centers, such as market gardens, are especially endangered. Air pollution is often harmful to urban greenery in parks, medians of avenues, and residential plantings. Plants are injured by particulate matter as well as by gasses such as sulfur dioxide and ozone.

The principal sources of air pollution are automobiles (60%), industries (19%), electric power generating plants (12.5%), space heating (6%), and refuse disposal (3%). In the United States, these sources generate about 125 million tons of air pollutants annually. The responsible agents are mainly carbon monoxide, oxides of sulfur, ozone,

hydrocarbons, and particulate matter. As a result of the urban concentration of these polluting sources, air in urban areas is on the average three times as polluted as rural areas, but the gradient may be 50-fold or greater.

The principal phytotoxic substances are discussed below.

OZONE   This compound was recognized as a major phytotoxicant in 1958. Since then, ozone damage has been observed on ornamental, agronomic, and forest crops of many kinds.

Ozone is most abundant in urban environments where there are large quantities of the chemicals necessary for its formation. Automobile engines are the chief source. Nitric oxide (NO) is formed in the engine where it passes into the atmosphere to oxidize quickly, forming nitrogen dioxide (NO_2). The energy in sunlight then splits the NO_2 into NO and molecular oxygen (O). The O then combines with atmospheric oxygen (O_2) to form ozone (O_3). The reverse reaction would take place except for the fact that hydrocarbons then react with the NO, with the resultant accumulation of ozone. While 1 and 2 pphm° is a normal level of ozone, the daily maximum in the Los Angeles basin has reached 100 pphm. Daily peaks frequently reach 30 pphm in the summer and fall. But ozone reacts quickly, once in the atmosphere, so that photochemically produced ozone disappears soon after it ceases to be produced.

Ozone has various effects on plants (Fig. 5-8). Photosynthesis is decreased as long as plants are exposed, and growth is thus impaired. Respiration may be increased, resulting in a reduction of carbohydrates. The stomata, the openings that permit gaseous exchange, close when ozone is in the atmosphere, limiting gas exchange and all other processes which depend on CO_2 and O_2 exchange.

Ozone injury first presents a shiny, oily appearance on upper leaf surfaces. If the exposure is sublethal and is soon halted, the leaf will return to normal. In tissues that have been killed, the shiny areas take on a dull, gray water-soaked appearance. The groups of cells subsequently turn white, giving the leaves a stippled appearance. In

° pphm = parts per hundred million.

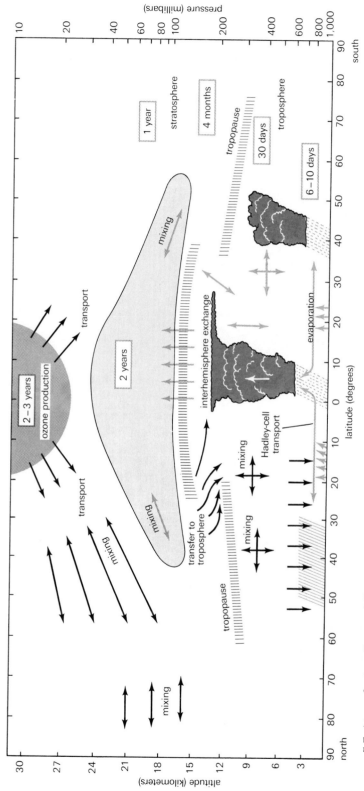

FIGURE 5-7. *Atmospheric pollutants follow global circulation patterns. The numbers in the boxes indicate the period of residence for aerosols, one of the forms in which pesticides are applied to crop plants.* [*From Newell, "The Global Circulation of Atmospheric Pollutants."*]]

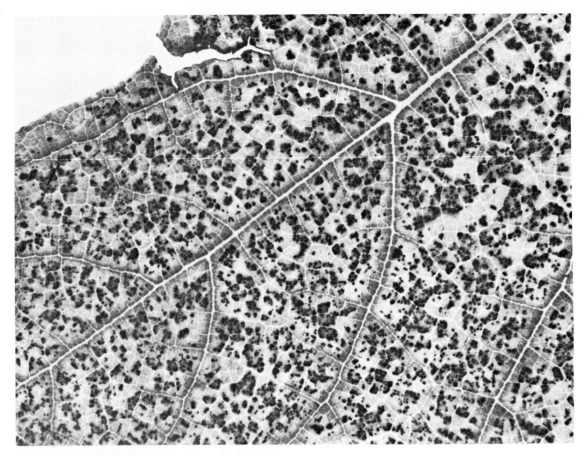

FIGURE 5-8. *Ozone has varied effects on plants.* (Top)
*Grape leaves develop lesions, which are distributed
interveinally over the entire leaf surface.* (Bottom) *Injury is concentrated in the terminal leaves of tomatoes*
(var. Manapal), *which become yellow near the tip and
die gradually as the damage extends from the tip toward
the base.* [*Courtesy USDA, EPA, and Walter Kender.*]

monocotyledons, such as corn, wheat, rye and oats, there is a chlorotic stipple between the veins. Leaf bronzing and early senesence occur in some woody ornamentals.

Some species are so sensitive that they need no urban source to produce symptoms of ozone damage. White pine is one such plant, being subject to a disorder called white pine needle blight or emergence tip burn. Stipple is an ozone-induced disorder of grapes, in which small brown or black lesions appear on young leaves, and ultimately leads to bronzing and defoliation.

SULFUR DIOXIDE SO_2 is released from burning coal. Power plants, refineries, and home furnaces especially contribute to the SO_2 content in the air. Additional SO_2 is cast into the air when some ores such as nickel, copper, and iron are smelted.

This compound can kill vegetation and threaten the health of man and animals. It enters leaves through stomata, passing into the intercellular spaces of the mesophyll where it is absorbed on the wet cell walls. The mechanism by which it damages plants is still a matter of speculation, but one theory suggests that it interferes with protein synthesis. Sulfur damage can also result if accumulations are great enough to be directly toxic to cells. Sulfur causes cells to shrink and collapse, and chloroplasts to disintegrate.

The external symptoms of SO_2 damage are varied (Fig. 5-9). Alfalfa and red clover develop white spots on leaves, whereas white pine tips lose color. Both the spots and the pine tips will become green again in a few days if the source of SO_2 is removed. If not, they become necrotic.

Studies with alfalfa, barley, and wheat have shown that crop yield is directly proportional to the degree of leaf necrosis. The height growth of white pine, an especially sensitive species, may be depressed by SO_2. Other trees that can be seriously retarded include Douglas-fir and lodgepole pine.

One of the major examples of SO_2 damage to an ecological system occurred in the copper basin of Tennessee. In the early days of smelting, the ore was placed on large pyres of logs which were ignited. The copper melted and flowed to the bottom of the fire, where it solidified and was picked out after the ashes had cooled—a crude

but effective smelting operation. Not only were trees cut to build the pyres, but others were killed by the large amounts of SO_2 released in the smelting operation. Over 2,500 hectares (7,000 acres) were completely denuded. Similar disasters have occurred in British Columbia and Montana. Even today the SO_2 emission from modern copper and sulfuric acid plants restricts vegetation growth. Because all species of plants are not equally affected, SO_2 changes the composition of native vegetation where it is constantly present, even at low levels.

FLUORIDE   When ores containing minerals such as mica, hornblende, and cryolite are smelted, fluoride escapes, as it does when fluoride compounds are manufactured or used for catalysts and fluxes. Both gases and particulate forms are released.

Fluoride enters plants by diffusing through stomata to intercellular spaces, where it is absorbed into the cells and transported through vascular tissues to the tips and margins of leaves, and it then accumulates in chloroplasts, cell walls, nuclei, and cytoplasm. Although plants may normally have a fluoride content of from one to 20 ppm, those in industrial areas may contain 100 to 1000 ppm. A few plants can accumulate large amounts of fluoride even in areas that have no industrial pollution. For example, tea may contain as much as 1300 ppm without apparent damage.

The exact processes through which fluorides injure plants are not known, but they seem to interfere with enzymes that affect many metabolic processes. It is thought, too, that fluoride may precipitate calcium within the plant, thus causing a calcium deficiency.

Fluoride causes mottling and necrosis at the tips and margins of leaves of sensitive broad-leaved species. On grasses, tips are killed, and stippling or mottling is found along the margins. On conifers, necrosis starts at the tips of the current year's needles.

The fruits of some species are extremely sensitive, even more so than the leaves. In peaches, fluoride causes the soft-suture disease, a premature ripening along one or both sides of the suture line. This area overripens or rots before the rest of the fruit matures.

FIGURE 5-9. (*Left*) *Sulfur dioxide fumes have damaged this white birch leaf.* (*Right*) *This leaf is from a tree grown in filtered air.* [*Courtesy USDA.*]

Fluoride emanations do not always damage plants. Conifers, aspen, grapes, and orange trees may actually increase in growth upon exposure to small dosages of atmospheric fluoride. But if the amounts are large, crop production is decreased and natural vegetation may be changed in composition.

SMOG This term was coined in London in 1905, meaning a combination of smoke and fog. Los Angeles, the smog "capital" of the world, established an Air Pollution Control Board in the late 1940's to study the problem and prescribe a course of action leading to its control. Research has shown that smog contains dusts, oxides of nitrogen, hydrocarbons, ozones, sulfur dioxide, aldehydes, and numerous other compounds. The components of smog most damaging to plants seem to be oxidants, which include ozone, nitrogen oxides, and peroxides (Fig. 5-10).

Dr. Aarie Haagen-Smit, a pioneer in the study of the atmosphere in and around Los Angeles, found that the hydrocarbons and nitrogen oxides from automobile exhausts react in a normal atmosphere in the presence of sunlight to yield toxic gases, which cause plant damage and eye irrita-

FIGURE 5-10. *Petunia plants react very strongly to smog, which has become a natural component of the atmospheric environment of cities. [Courtesy California Statewide Air Pollution Research Center.]*

tion. The end products from this series of reactions are a group of compounds called peroxyacetal nitrate (PAN). PAN and ozone (photochemical oxidants) cause great damage to plants (Fig. 5-11). Even if exposure is as slight as 5 ppm for 10 minutes, some sensitive species are damaged (Fig. 5-12).

PAN causes several types of injury to plants. Glazing and bronzing occur on spinach, lettuce, endive, and sugar beets.

Additional contaminants include ammonia, chlorine, and other gases that have been accidentally spilled into the atmosphere and have caused local problems. Even ethylene, a natural plant product, will injure plants if it is present in too great a quantity (Fig. 5-13).

SUSPENDED PARTICLES Another source of pollution injurious to plants, these particles can come

FIGURE 5-11. *These pinto bean plants were exposed to 5 ppm PAN for ten minutes. The plant on the left was in the light during fumigation while the plant on the right was in the dark. The variation between the two illustrates the photochemical reaction of PAN in producing phytotoxicity. [Courtesy EPA.]*

FIGURE 5-12. *Radish var. 'Cherry Belle' grown in un-filtered air (left) and filtered air (right). Damage was caused by various oxidants.* [*Courtesy USDA.*]

FIGURE 5-13. *Ethylene can injure greenhouse crops such as orchids.* [*Courtesy California Statewide Air Pollution Research Center.*]

from smelters, smokestacks, mining operations, cement plants, quarries, strip mines, mineral processing operations, and other industries, as well as from various agricultural practices.

Dust, when it is most damaging, can sandblast seedlings. The epidermis of cotton, tobacco, and even young pine seedlings can be eroded away at the soil surface. Such damage occurs frequently in the coastal plain of the southeastern United States.

When particulate matter coats leaves, photosynthesis can be impaired. Finer particles may obstruct stomatal openings so that gas exchange is inhibited. When airborne particles settle on some crops, such as lettuce and celery, the quality of these vegetables may be impaired, as may that of fruits and ornamentals.

Perhaps the most subtle of all damage by dust may be indirect: dust particles carry pesticides. Even dusts that have been airborne for as much as 1,000 miles have been found to have a coating of pesticides. This means, then, that the spraying practices of a corn farmer in Kansas could have a direct effect on, for example, grape production in New York.

Another kind of potentially damaging particle comes from the smoke resulting from controlled burning. Foresters have for a long time burned ground cover and low-shrub vegetation of certain forest types. These creeping ground fires are used for several purposes: (1) fuel reduction to prevent an accumulation of dangerous amounts; (2) destruction of pathogen spores (such as the fungus carrier of the brown-spot needle blight); (3) improvement of visibility. Unfortunately, such controlled burns produce much objectionable smoke (Fig. 5-14). However, an alternative solution for the residue problem is not yet apparent.

Dusts from various sources, aerosols, and other kinds of particulate matter, including smoke, all cause injury to shade trees.

SOLUTIONS TO AGRICULTURE'S POLLUTION PROBLEMS

Many approaches are necessary to solve the problems of pollution—to reduce agriculture's own

FIGURE 5-14. *Controlled burns have been used for many years in routine forest management. Designed to reduce ground cover and prevent litter accumulation, they are "cool" fires which are managed so that they spread into the wind. Consequently, much objectionable smoke is generated.* [*Courtesy St. Regis Paper Company.*]

contribution as well as to alleviate the ill effects of pollution on agriculture. Many of the cures are already well known, such as soil conservation. Others are new because many of the pollution sources, such as mercury, did not previously exist or remained unrecognized or underestimated (Fig. 5-15). It should also be remembered that there is almost always a trade-off between alternative approaches to environmental stress. For example the use of fertilizer and pesticides has prevented much of soil erosion. Corn acreage has remained at about 100 million acres since the mid-1920's in the United States, while output has more than doubled. If yield increases had not occurred, it would have been necessary to push corn production onto the more fragile margins—the steeper slopes and the poorer soil areas. Thus, the control of one pollution problem, soil erosion, is achieved in part at the expense of chemical pollution.

Improving agricultural practices

Because many of the solutions for pollution have a negative short-term economic value, there is no immediate profit incentive, the usual motivating force behind capital investment and action. The

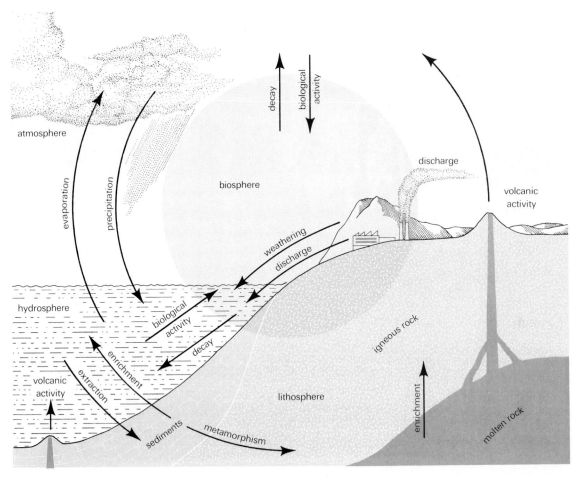

FIGURE 5-15. *Mercury is present in the lithosphere, the hydrosphere, and the atmosphere in trace amounts; however, it becomes concentrated in organisms by means of various biological processes.* [*From Goldwater, "Mercury in the Environment." Copyright © 1971 by Scientific American, Inc. All rights reserved.*]

following consideration of the possibilities for cleaning up agricultural pollution will indicate the complexity of the task.

INCREASED USE OF AGRICULTURE RESIDUES Rather than discard waste products from agriculture, it would be infinitely wiser to find uses for them. Attempts have been made, but the primary motivation has usually not been the reduction of pollution but rather the possibility of increasing profits. The incentive to "sell" a material that has posed an expensive disposal problem is very strong.

Organic waste from processing plants may be used in the future as a soil additive. Experimental trials have already begun and feasible technology is known. However, if these wastes are to be successful, soils must be porous enough to accept the large volumes of water, and transportation costs from factory to field must not be prohibitive.

Disposal of waste with a high content of cellulose is especially difficult. Cellulose wastes of many kinds could be converted to protein if they were broken into sugars and fermented in the presence of nitrogen and phosphorus. Although not a profitable venture at this time, it could become so in the future. Termites can use wood as an energy source because microorganisms in their intestine break down the cellulose into sugars. In a controlled environment, which would favor growth of such microorganisms, raw sawdust could be biologically converted to produce sugars. Acid hydrolysis is another available technique for transforming cellulose to sugars.

The amount of utilizable wastes from food and food processing in the United States is nearly 50,000 metric tons. Under controlled conditions the potential yield of methane gas from the breakdown of these wastes is almost 30,000 cubic meters (1 trillion cubic feet) each day—hardly an amount of energy to be ignored. There are several methods by which this can be done, but one of the most promising is a microbiological two-step fermentation. Acids, alcohols, and aldehydes are produced in the first step; methane and carbon dioxide are among the products in the second step. Although pilot plants for this process exist, problems of garbage sorting and transportation remain to be solved.

Many other uses of agricultural wastes are technically feasible. Citrus wastes are now commercially processed into a nutritious feed. Paper can be made from rice and cereal straw. Uses of residues from forestry operations have been discussed in Chapter 2.

WATER AND SOIL CONSERVATION   Water conservation is essential at all stages of crop production. This problem is extensively discussed in Chapters 13 and 17. Industry uses much less water than agriculture, but it still requires vast amounts of fresh, clean water for cooling and/or processing, and returns most of it, frequently in a highly polluted condition, to its original source. In contrast, most of the water that is used for agriculture is returned into the atmosphere in a "pure form" by transpiration and evaporation, though some of the remainder, which is returned to the original source, is polluted with dissolved salts and particulate matter. The key to better water conservation lies in more diligent application of known water and soil conservation practices.

Soil conservation, discussed at length in Chapters 12 and 16, cannot be separated from water

management (Fig. 5-16). The sediment from surface erosion constitutes, by far, the greatest volume of waste from forestry and farming operations.

In Chapter 16, the use of mineral fertilizers is discussed in detail. The pollution problems that have resulted from the practice of fertilization have, in recent years, been largely due to overuse and misuse. Three important principles of fertilizer use have emerged: (1) apply fertilizers only if and when they are needed; (2) avoid overuse; (3) place them precisely for immediate uptake. Adherence to these principles, rather than the use of "shotgun" methods, reduces the cost of fertilization as well as the attendant pollution problems. The proper use of fertilizer materials requires an exact understanding of what nutrients are required and when they are needed.

REDUCING PESTICIDE HAZARDS   Minimizing the hazards associated with pesticide usage is a complicated issue without easy solutions. In assessing the effects of pesticides, consideration must be given to the protection of the applicator and the direct consumer, and the possible consequences of those residues that may pass into the ecosystem to become worldwide contaminants. Still, if pesticides were completely eliminated, the alternatives would be severe crop losses or failures, increased filth from pest infestation, and problems resulting from disease-associated toxins (mycotoxins)—all creating, of course, totally unacceptable situations for the human population.

Feasible solutions include a variety of measures: more precise pesticide monitoring, greater selectivity of nonpersistent and biodegradable materials, and increased efficiency. Finally the reliance on chemicals alone must be decreased. This goal may be accomplished by integrating other systems of pest control.

Pesticides have undergone several stages of development. The early insecticides consisted of quite simple and crude chemical poisons, such as nicotine sulfate for use against sucking insects, lead arsenate to kill chewing insects, and kerosene, which was spread on surfaces of ponds to destroy larvae. For these materials to be effective, large amounts were required, and consequently serious

FIGURE 5-16. (Top) *Even the most severe soil erosion can be corrected. The owner of this gullied land in Buncombe County, North Carolina, planted the area with TVA-produced pine and locust seedlings in the late 1930's. (Bottom) This is the same tract nearly 20 years later. The Tennessee Valley Authority assists local forestry and agricultural agencies in reforestation for timber production and watershed protection. [Courtesy TVA.]*

problems occurred, not the least of which was the inefficacy of control in many instances. The so-called second-generation insecticides, first used in World War II, were fantastically potent in very minute amounts. DDT, the first of the chlorinated hydrocarbons, can be credited with staving off the plague during the period of reoccupation in southern Europe and has been used throughout the world for malarial control. Many other "miracle" insecticides followed: for example, parathion and malathion. Three major problems soon became obvious: (1) many of the materials were broad spectrum, killing both harmful and beneficial organisms; (2) the development of resistance soon reduced efficacy; (3) problems of residues and their subsequent accumulation through magnification in the food chain (Fig. 5-17) occurred.

We now understand that we must require pesticides to be selective as well as safer. Either they must be **biodegradable** so that they can be readily decomposed by the action of resident microorganisms of the environment into which they are introduced, or they must be chemically degradable. A "third generation" of insecticides, which would overcome present deficiencies is presently being developed. Some achieve selectivity through their effects on specific phases of insect development. For example, materials that mimic the juvenile hormone of specific insects prevent larvae from maturing, or render adult forms permanently sterile. One juvenile hormone, juvabione, has been found to kill bugs in the family Pyrrhocoridae, which includes several extremely destructive cotton pests in Asia.

But the problem does not end with the selection of a pesticide that seeks out only a specific target. It has been calculated that only one pesticide molecule out of every million released into the environment "hits" its intended target. More effective placement would greatly reduce the minimum amount required. For example, some species of insects produce chemical attractants (pheromes) that attract males for great distances. The isolation and synthesis of these sex attractants suggests their use as lures to achieve more effective destruction with very reduced quantities of the active insecticide.

Other notable alternatives to chemical control

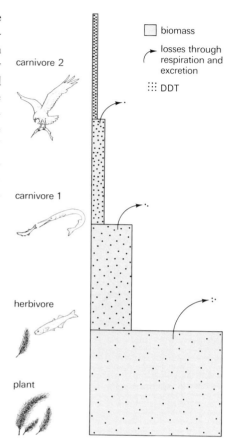

FIGURE 5-17. *As the "biomass," or living material, is transferred from one link to another along such a chain, usually more than half of it is consumed in respiration or is excreted (see arrows); the remainder forms new biomass. The losses of DDT residues along the chain, however, are small in proportion to the amount that is transferred from one link to the next. For this reason high concentrations of DDT residues occur in carnivores.* [From Woodwell, "Toxic Substances and Ecological Cycles." Copyright © 1967 by Scientific American, Inc. All rights reserved.]

are the various "biological methods." These include the use of genetic resistance and the introduction and control of natural predators, diseases, or genetic defects of the pests themselves. The principles of biological control are not new (see Chapter 18 and 19). Although biological control is not adaptable to all situations, the use of these methods must increase, and a more stringent regulation of agricultural chemicals must be put into effect.

Perhaps the best hope for effective and safe pest control lies in a combination of methods that depend on minimal use of chemical pesticides coupled with other physical and/or biological methods. This **"integrated control"** appears to be more realistic from an ecological point of view for it relies on a number of complimentary methods. It assumes that control, though perhaps not perfect, will be at least sufficient, and it assures that the cure will not be more hazardous than the malady.

Protecting plants from pollution injury

The obvious solution to the ill effects of industrial pollution on agriculture is to reduce pollution at the source. Unfortunately, although many give verbal support to such a measure, few would willingly pay for the consequences. There is no evading the fact that the reduction of pollution must increase the cost of products we all use, whether it be the cost of buying or running our automobiles or that of turning on the light. But zero level of pollution is neither a realistic nor a necessary goal. The aim must be to reduce pollution to a level that will not deteriorate our environment in the course of an indefinite period of time. This goal is essential, not only for our direct health and well-being but also for plant life. Materials that pollute the air, water, and soil tend in various ways to affect plant growth, productivity, and quality adversely.

Although plants may themselves detoxify the environment, this effect appears to be minor and therefore provides no excuse for us to relent in efforts to reduce pollution. In fact the evidence is clear that in many areas, present levels of pollution are highly toxic to plant life. Growing plants in the filtered environments of greenhouses is not a practical possibility. While a number of studies have indicated that environmental factors such as light and nutrition affect plant reaction to pollutant injury, this does not seem to offer a viable method of control. There do, however, appear to be genetic differences in the ability of plants to

react to many of the common forms of pollutants.

At present, the two basic mechanisms for pollution "resistance" are these: (1) physiological systems that allow the plant to tolerate excesses of pollutants; (2) structural and physiological mechanisms that tend to exclude pollutants from entering the plant. The second system is the better understood. Air-borne pollutants enter the leaves largely through stomatal apertures. Resistance to ozone injury has been related to the sensitivity of guard cells, which close in the presence of ozone, just as drought resistance in many species is a function of stomatal response to water stress. Closure by these cells prevents pollutants from entering the plants just as it prevents excessive water loss by preventing transpiration. Significant differences in stomatal sensitivity occur between species and within species.

In areas of high pollution one can select species that resist pollution injury through their inherent protective mechanisms. For example, deciduous trees are much more resistant to air pollution than evergreen trees. Pollution injury in Germany is almost entirely restricted to conifers, and in the highly industrialized regions of that country foresters are in the process of replacing spruces and firs with hardwoods and resistant conifers such as Austrian pine and Corsican pine. Even within the same stands, it is possible to find variation from tree to tree in resistance and susceptibility. Eastern white pine is highly variable in resistance to oxidants and to coal smoke and gas from power plants. Propagation with the use of cuttings from resistant trees has made it possible to produce plant stock for areas subject to these pollutants.

Similar differences in susceptibility have been found in lines and cultivars of such plants as grape, onion, tomato, and corn. The precise basis for these variations in resistance is unknown but genetic resistance appears to be a viable possibility in the continuing efforts to combat pollution injury.

Recently, evidence that pollution resistance can be obtained by chemical treatment has been discovered. Succinic acid-2,2-dimethylhydrazide provides petunia with resistance to ozone injury. Various materials that affect stomatal apertures are also known. This new area of research offers another means for the amelioration of pollution injury.

POLLUTION AND THE LAW

The National Environmental Policy Act (NEPA) of 1969 places the responsibility for assurance that any Federal activity will not be significantly detrimental to the quality of the human environment directly on the Environmental Protection Agency. Furthermore, in some states, the courts have decided that private individuals and industries also will be subject to the provisions of similar laws established by the governments of those states. Actions that take place in one state, and that affect the environment of people in other states will be subject to federal scrutiny too. For example, if pesticide sprayed in one state drifts into another, federal action could be invoked.

Agribusiness will be subject to laws such as NEPA. The provisions of this act call for the presentation of an Environmental Statement, popularly called **Impact Statement**, prior to initiation of new federal projects. The statements must include consideration of the following items:

(a) A description of the proposed action including information and technical data adequate to permit a careful assessment of environmental impact by agencies, institutions, and individuals.

(b) The probable impact of the proposed action on the environment, including impact on ecological systems such as wildlife, fish, and marine life. Both primary and secondary significant consequences for the environment should be included.

(c) Any probable adverse environmental effects which cannot be avoided.

(d) Alternatives to the proposed action. Sufficient analysis of such alternatives and their costs and impact on the environment should accompany the proposed action through the review process in order not to foreclose prematurely the options which might have less detrimental effects.

(e) The relationship between local short-term uses of man's environment and the maintenance and enhancement of long-term productivity.

(f) Any irreversible and irretrievable commitments of resources which would be involved in the proposed action, should it be implemented.

(g) Discussion of problems and objectives raised by Federal, state, and local agencies and by private organizations and individuals in the review process and disposition of the issues involved.

The statement is routinely sent to various agencies, organizations, and individuals for review, and to others on request. If one is not satisfied that the environment will remain unharmed, court action in the form of an injunction may be taken that can delay construction or other action until the statement is satisfactory to all concerned.

NEPA applies particularly to the large impacts on the environment. Dams, power plants, and canals are clearly within the scope of the law. It is reasonable to expect that, in the future, those ongoing projects having even less impact will be subject to scrutiny. Even individual crop production units, such as farms, may be reviewed, and owners forced to implement the best possible conservation practices, a measure that would increase the cost of crop production.

In 1972, the Federal Environmental Pesticide Control Act was passed. This Act will have a strong effect on American agriculture because it provides that private citizens who wish to apply restricted-use pesticides must be certified and have a permit even to purchase them. It also delegates the responsibility for a plan to carry out the provisions of the Act to the individual states. The 1972 Act also specifies a list of pesticides for which no permit is requested.

The laws on the use of pesticides are complex, and they are constantly changing. Some, such as DDT, sodium fluoroacetate, and organic mercury fungicides have been permanently banned. Others, including parathion, paraquat, and 2,4-D, are presently restricted to use by persons with an appropriate permit. A few, such as endrin, are banned from use only on food crops. Those that present very little or no danger to man, such as pyrethrum, malathion, and rotenone, are unrestricted. It seems likely that additional changes in legislation and in administrative rulings will be necessary in order to achieve effective safety standards in the marketing of pesticides and at the same time continue to provide incentives for the development and marketing of more effective pesticides.

AGRICULTURAL PERFORMANCE

Ecology is the study of the interactions of organisms with one another and with their environments, in time and space (Fig. 5-18). The task of the ecologist is to investigate, understand, and explain such phenomena as forest succession, the growth of cities, and the conditions that must be maintained for sustained crop yield. The ecologist

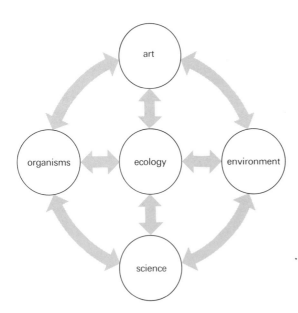

FIGURE 5-18. *Ecology is neither a science nor an art; rather, it is a way of considering all of the relationships between organisms and their environments in both space and time. The value judgements that must be made regarding the interplay of organisms and environments require perspectives ranging from the objectivity of science to the subjectivity of art.*

does not necessarily make economic decisions on the maximization of profits, nor is he expected to provide a social commentary on the benefits of his thoughts to society. Rather his role is to understand the functioning of natural and manmade ecosystems and to predict what will happen when they are in too great a state of imbalance.

Man-induced upheavals of the balance of nature are not limited to the twentieth century. In the course of the past millennia, we have either caused or permitted countless ecological disasters through our various efforts to circumvent or thwart nature. A few examples illustrate this point.

In the United States, the Great Plains region has always been subject to periodic drought. As long as they were covered with short sod-forming native grasses, the dry spells did little damage. But in 1889, the Oklahoma territory was opened for homesteading, and the number of settlers increased from a few thousand to 60,000, as more and more acres of protective grass were plowed under for farmland. This took place during a period of abundant rainfall, and by 1900, the rich fertile soils of Oklahoma supported 390,000 persons. In 1910 a dry spell reoccurred, but by 1914 the soils were, once again, back in full production. In 1924 there came a prolonged drought, and the topsoil of Oklahoma was blown even as far as New York City and into the Atlantic Ocean. The resultant emigration toward the West Coast is documented in John Steinbeck's moving epic *The Grapes of Wrath*.

In the Middle East, engineers had long contemplated building a great dam across the Nile to impound water from the spring rains for irrigation during the hot summers when the water level subsided. The Aswan dam, completed in 1971 was built to fulfill this objective as well as to provide hydroelectric power. Water storage was begun in 1964, in the partially complete dam, and a river that had flown freely for thousands of years was suddenly trapped and harnessed. Some of the anticipated benefits of the dam have materialized. Power is being produced, and higher yields of cotton, grain, fruits, and vegetables have been obtained in some locations. But at the same time, many problems have resulted. The sediment that was formerly washed out to sea carried mineral nutrients with it and was important to the food web of offshore life. Consequently, the annual catch of sardines decreased by 18,000 tons. Each year the flooding waters carried away salts that would otherwise have accumulated in the soil. Once the flooding stopped, salts accumulated and salinity is now a threat to productivity. Flooding also carried the silt and its nutrients into the fields downstream from the dam. Now that the nutrients are no longer being added to the soil naturally, chemical fertilizers must be added. Other types of problems have arisen too. During the dry periods, a snail fluke, called bilharzia, was kept under natural control. After the irrigation canals were built, that natural annual control ceased to exist, and this blood fluke, which spreads easily to man, now infects 80 percent of the people who must work in the canals.

In parts of India today, overgrazing is destroying ground cover while goat herders, using sickles with 20-foot-handles, cut the limbs of trees for their goats. The trees then die and the unprotected soil blows away, so that the sites are no longer productive.

Each of these case histories is an example of **ecological backlash**—an unforseen detrimental consequence of a modification of the environment that cancels out or overcompensates for long-term net gains that were expected. Poor conservation was practiced.

In plant agriculture, conservation means *wise use,* the aims being to insure not only a sustained but also an increasing yield of useful plant products, while maintaining an environmental quality that is compatible with esthetic needs, recreational opportunities, wildlife, and all other uses to which the land might be put (Fig. 5-19). It implies planning and foresight. It demands the application of ecological principles. The above examples of land misuse are not necessarily caused by faulty planning, but primarily by a misunderstanding of ecological principles!

ECOLOGY AND CHANGE

Perhaps the dominant ecological principle is that of change: change is inevitable. Whether or not we inflict ourselves and our artifacts on an ecosystem, all natural systems are constantly changing. A mature forest ecosystem, such as a deciduous forest of beech and maple, perhaps the most stable of all kinds of vegetative communities, undergoes fluctuations, compensations, and counterbalances that take place in a dynamic equilibrium. The components are constantly changing in intensity, numbers, kinds, and complexity.

A second important principle is that change is directional—that is, always toward greater complexity, stability, and maturity. In plant succession, various kinds of plants give way to succeeding ones until, at last, a condition in which biota and environment are in harmony is reached, and the system is in a state of **homeostasis** or self-regulation. Such an ecosystem will maintain itself for long periods of time, or until some major change occurs to upset the balance. So it was with the construction of the Aswan Dam. Man, unable and perhaps unwilling to make great changes in the environment, had learned to live in harmony with the system of the great Nile River, and this system existed for thousands of years—until the dam was built.

When we inflict ourselves on a successionally mature and stable forest by cutting a tree or by clearing it for pasture or for farming, an earlier successional stage of plant life occurs. In a cornfield, our weed problems are not caused by the trees and shrubs of the mature forest that would eventually occupy the cornfield, but by invading herbaceous weeds of an earlier successional state that are competing for soil moisture and nutrients.

Nature does not necessarily give us trouble when we manipulate it wisely, but our actions must be planned, so that we take into account the checks and balances that have evolved in the course of time. Common sense dictates that we should try to predict the consequences of our actions. Realistic assessments of what the consequences might be, if a dam is built, the soil plowed, pesticides applied, or a forest cut, are completely within the capabilities of the human mind. If we fail to use ecological principles in producing crops, then we can expect more and even greater disasters than in the past.

FIGURE 5-19. *In the early days of American Forestry, the primary emphasis was on fire protection. Today, foresters are expected to provide not only wood, but recreation opportunities, a continuous supply of good water, wildlife, and pleasant surroundings.* [*Courtesy U.S. Forest Service.*]

RATIONAL CHOICES

Through history, man has always feared, condemmed, and destroyed or banished not only other human beings but also various animals, plants, and objects—even ideas and concepts—that it was thought might possibly cause injury or discomfort. Today we still launch inquisitions and attacks on people, and even on various inanimate objects. For example the banning of DDT would appear to have been the result of this kind of decision making. The evidence indicates that DDT, used properly, can make the quality of life better for people throughout the world. Used improperly, it may be a true hazard in the environment (Fig. 5-20). A rational approach to this problem would be to control and restrict its use to situations in which it will not create an environmental hazard. However, the emotional approach—to ban it completely because it poses a potential threat—has prevailed. If the same approach were applied to other compounds and products, we might very well forbid the use of such everyday items as sugar. Eaten in too great a quantity, sugar will cause obesity, cardiovascular problems and shortened life spans. Yet, most of us would never think of eliminating sugar from our diet, although wise use restricts its intake. Every age has had its witch to attack, and even an enlightened age such as ours has not denied itself this pleasure. DDT has been one of the objects offered for our ritual sacrifices.

In societies that are underdeveloped or that

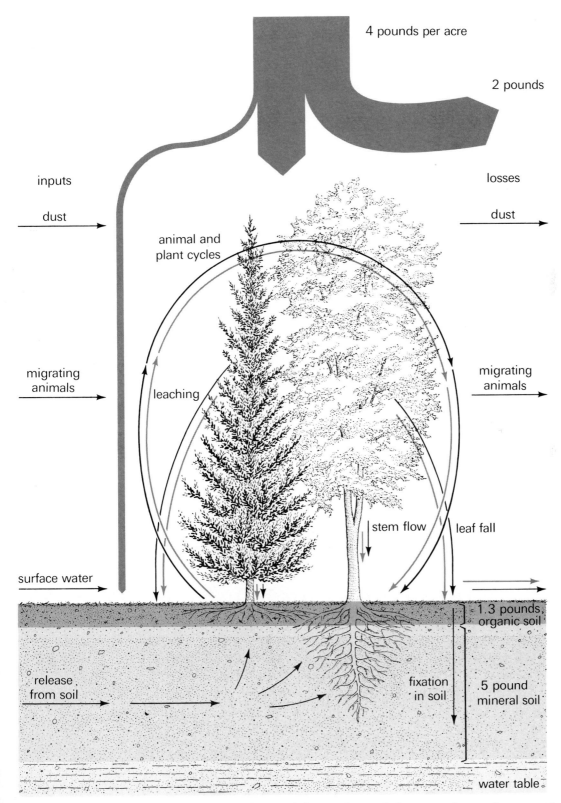

4 pounds per acre

2 pounds

inputs

losses

dust

dust

animal and
plant cycles

migrating
animals

migrating
animals

leaching

stem flow

leaf fall

surface water

1.3 pounds
organic soil

release
from soil

fixation
in soil

.5 pound
mineral soil

water table

FIGURE 5-20. *A forest community is an integrated array of plants and animals that accumulates and reuses nutrients in cycles. DDT also participates in cycles. Three years after a New Brunswick forest had been sprayed with DDT (2 pounds per acre in the course of seven years), the DDT residues were as shown (in pounds per acre).* [*From Woodwell, "Toxic Substances and Ecological Cycles." Copyright © 1967 by Scientific American, Inc. All rights reserved.*]

have few resources, pollution is not a problem. When Eskimos kill a seal, practically every part of the animal is used for food, clothing, or tools; no part that has a potential use is discarded. Even if it were, the population density is so low that no problem would arise. When an Indian in the Amazonian jungle grows a crop of manioc and bananas, the detritus from the unused parts of the plants does not accumulate to create a pollution problem because it decays rapidly in the heat and humidity of the jungle. In such societies, a bottle or a can, once emptied of its original contents, is likely to be coveted and used again and again.

Unfortunately the same cannot be said of our own enlightened and highly developed society. Our monumental garbage dumps are dwarfed only by the problems we encounter in trying to find locations for new and even grander ones. Our roadsides have become convenient receptacles for unwanted cans, bottles, paper, garbage, and discarded car bodies. An advanced society or community "learns" to perpetuate itself by recycling the materials essential for continued growth—wasting little, reusing most. While we think of ourselves as a "developed country," in many ways we behave as a people in a developmental stage, wantonly discarding our resources.

Frequently, pollution is a problem of concentration. The wastes that a herd of beef cattle produce on an open range or pasture usually present no problems, but actually contribute to the system. The same herd, in a feed lot, next to Kansas City, will cause very serious problems! Similarly, a small portable sawmill operating in a farmers woodlot for four to six weeks before it is moved on does not produce a pile of sawdust so large that it cannot be leveled and scattered about with a farm tractor. It may even be used sometimes as a soil amendment, with appropriate additions of mineral fertilizers. But a large saw-

mill, operating for many years in a single location will create problems of sawdust and bark disposal. In years past the sawdust was burned, either to help supply the power requirements of the mill, or in a special disposal furnace. Today, sawdust is increasingly used in paper, as a filler in many products; and even bark, a by-product once thought to be absolutely without a positive economic value, is being used as a mulch for ornamental plants.

Still, we cannot escape the fact that we must convert enough land, produce enough fertilizers, cut enough timber, mine enough minerals, plow enough land, process enough food, and dispose of enough waste to feed, house, and clothe a population that could double within the next 30 years. Our problems have only begun! If we wish only to maintain our present standards of living, production still must be substantially increased. If we wish to continue to increase our living standards, then production must expand even more. Consumers must ultimately make the decision on what is to happen, for it is they who will pay the costs. Good conservation practices, in which high quality is maintained while pollution is abated, will increase prices for everyone and we must be prepared to pay.

It is folly to believe that completely unguided individual action will result in greater food production and cleaner environment simultaneously. Past events suggest that the nature of the human species is aggressive, acquisitive, and selfish, and controllable only by mutual coercion. Given the freedom of choice, we dump wastes in rivers because it is cheap and easy. We stop only if our laws say that we must do so. Thus, our laws must be so structured that individuals and groups are motivated to protect the quality of the environment. The regulations must be feasible, enforcable, and enforced!

Selected References

Benarde, M. A. *Our precarious habitat.* W. W. Norton, New York, 1970 (This book contains material on the problems of pollution and food contamination as they affect contemporary food production.)

Brady, N. C. (editor). *Agriculture and the quality of our environment.* A.A.A.S. Publication 85, Washington, D.C., 1966. (An important collection of technical papers dealing with many aspects of plant agriculture, as it affects and is affected by pollution.)

Ehrlich, P. R., and A. H. Ehrlich. *Population, Resources, Environment: Issues in Human Ecology* (2nd ed.). W. H. Freeman and Co., San Francisco, 1972. (A popular text on the environmental problem.)

Hayes, J. (editor). *Landscape for living.* USDA Yearbook of Agriculture, Washington, 1972. (The use of plant materials to improve the quality of the human environment.)

Helfrich, H. W. *The Environmental Crisis.* Yale University Press, New Haven, Connecticut, 1970. (A collection of essays on environmental quality, including several on food production.)

Hellman, H. *Feeding the world of the future.* M. Evans and Company, New York, 1972. (A short, interesting book about human food production and the problems resulting therefrom.)

Jacobson, J. S., and A. C. Hill (editors). *Recognition of Air Pollution Injury to Vegetation: A Pictoral Atlas.* Informative report No. 1, Air Pollution Control Association, Pittsburgh, 1970. (An excellent collection of photographs showing most types of air pollution injury to plants.)

Ludington, D. C. (editor). *Agricultural Wastes: Principles and Guidelines for Practical Solutions.* Cornell University Conference on Agricultural Waste Management, Syracuse, N.Y., 1971. (A collection of papers dealing with disposal of agricultural wastes as well as with alternative uses.)

Matthews, W. H., F. E. Smith, and E. D. Goldberg (editors). *Man's Impact on Terrestrial and Oceanic Ecosystems.* The MIT Press, Cambridge, Mass., 1972. (Discusses the influence that pollution may have on cropping productivity, as a result of the disruption of ecosystems, by influencing living organisms either directly, or indirectly by alterations of climate).

National Academy of Sciences, Committee on Nitrate Accumulation. *Accumulation of Nitrate.* National Academy of Sciences, Washington, D.C., 1972. (Survey of nitrogen sources in the biosphere, and the hazards and problems resulting from its use.)

Nicholson, M. *The Environmental Revolution.* McGraw-Hill, New York. 1970. (A comprehensive book on various aspects of environmental conservation.)

Owen, O. S. *Natural Resource Conservation.* Macmillan, New York, 1971. (A textbook in conservation, with chapters on many of the problems of agricultural production.)

Smith, R. L. (editor). *The Ecology of Man: An Ecosystem Approach.* Harper and Row, New York, 1972. (An excellent collection of essays, many of which are relevant to the problems of modern agriculture.)

Spilhaus, A. (editor). *Waste Management and Control.* Publication 1400 National Academy of Sciences. Washington, D.C., 1966. (A general statement of the pollution problem in its broadest context.)

Treshow, M. *Environment and Plant Response.* McGraw-Hill, New York, 1970. (The effects of environmental factors on plant growth and productivity.)

Turk, A., J. Turk, and J. T. Wittes. *Ecology, Pollution, Environment.* W. B. Saunders, Philadelphia, 1972. (A small book that discusses agriculture and its problems from an ecosystem approach.

Wadleigh, C. H. *Wastes in Relation to Agriculture and Forestry.* USDA Misc. Pub. No. 1065, Washington, D.C., 1968. (A valuable book on the problems of pollution—those affecting agriculture and those created by it.)

White-Stevens, Robert (editor). *Pesticides in the Environment* (2 vols). Marcel Dekker, Inc., New York, 1971. (A comprehensive collection of papers on the chemistry, analysis, toxicology, and pest resistance of most pesticides in current use.)

PART II

NATURE OF CROP PLANTS

CHAPTER 6

Origin and Classification

Man and the plants upon which he is dependent are both products of evolutionary development. During the first billion or so years, life forms sustained themselves mainly on what chanced to be at hand. Not until the last few tens-of-thousands of years has there existed any organism both dexterous enough and intelligent enough to oversee other forms of life for its benefit—not until an especially advanced primate, early man, slowly learned to manage his environment sufficiently to maintain for his own use certain of nature's wealth of species. In the past few thousand years the tempo of change and exploitation has increased exponentially, as man, with his bent for technology, set about literally to alter the face of the earth and to populate it everywhere. How has this been possible? What is the story behind the hundreds of kinds of plants that help form the basis upon which modern societies rest? And how does modern man cope with the need for ever more useful and efficient plant types to sustain the tremendous superstructure of civilization? Evolution and genetics help clarify the origin. Classification supplies the inventory.

THE EVOLUTIONARY SEQUENCE

Although intelligent speculations can be offered, no one can be sure just when, why, or how life first started. Indeed, it is impossible to characterize life exactly, impossible to distinguish it clearly in some of its simpler and more ancient forms from the nonliving. Two chief criteria characterize earthly life: presence of carbon, typically in large organic molecules, and reproduction or self-duplication—the ability of life to make new life from the non-living substances in its milieu.

Simple forms of life, probably quite similar to bacteria-like plants, existed two to three billion years ago (Fig. 6-1). Their remains have been found in rocks that old, and there is supplementary evidence of their influence on their ancient environment—for example, abundantly oxidized beds of iron in what seems otherwise to have been a reducing environment. This fairly elaborate form of life must itself have evolved over a long period of time in which simple molecules linked to form progressively larger molecules, some of which

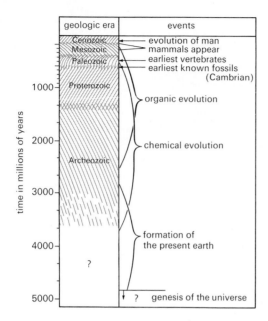

FIGURE 6-1. *Evolutionary time scale.*

were able, eventually, to duplicate themselves by assembling appropriate ingredients found in the primordial sea and atmosphere. Even in space, amino (nitrogen–hydrogen) compounds apparently occur. The newly formed earth must have contained such simple ingredients as methane, ammonia, and carbon dioxide. Laboratory tests show that under certain not-too-extreme conditions of heat and radiation these can combine into large molecules. The Russian biochemist A. I. Oparin hypothesized a sort of "colloidal broth" for the beginnings of life, and Oparin's followers have elaborated quite plausible explanations of how early life may have started.

Although circumstantial evidence indicates that life was created but once, and that from this creation has arisen the whole complexity of the modern living world, life could have been created many times; we cannot be sure. But it is likely that once created, life so altered the environment that suitable conditions for our hypothesized beginnings no longer existed. For example, it is very likely that the early atmosphere contained little if any free oxygen; it is probable that green plants,

through photosynthesis, have altered an original atmosphere rich in carbon dioxide to one quite impoverished of it and instead rich in oxygen. Oxygen, a relatively reactive gas, is inimical to the reducing atmosphere seemingly necessary for the slow buildup of compounds leading toward the first life. And with life came such organisms as today's bacteria, pirating for their sustenance large molecules as a source of energy ("food"). No longer could free organic compounds exist for any length of time, awaiting chance concatenation in an organic soup; instead they would be immediately utilized by life itself. Though it is possible in the laboratory to assemble proteinaceous molecules into something at least very akin to life, it is unlikely under present-day natural conditions that there would be either time enough or the proper "mix."

All of this, of course, is quite speculative, and in a sense incidental to our story. But from some such beginning have come the myriad of life forms with which classification concerns itself. How new organisms arise and evolve are equally of intellectual concern and essential to providing ever more useful forms for man's utilitarian purposes.

Origin through natural selection

The evidence that evolution has occurred is overwhelming, yet the mechanisms by which it proceeds are still only partly clarified. Not until the publication of Charles Darwin's *Origin of Species* (1859) was any scientifically credible theory formulated. Alfred Wallace, at the opposite end of the earth (Darwin in England, Wallace in Malaysia), arrived independently at roughly the same conclusions at about the same time. But Darwin's work remains the landmark; it is Darwin who is credited with ushering in one of biology's momentous concepts—evolution by natural selection.

In essence, natural selection—popularly, "survival of the fittest"—views all life as continuous back in time; all present forms are closely or remotely related, depending upon how long ago and with what rapidity they diverged from a common ancestor. Life is a huge "tree," in which the present forms are represented only by the tips of the many branches; back in time these interconnect. What are now discrete forms arose from similar forms. Only those best adapted to the environment of their times survived.

Organisms are capable of reproducing far beyond the capacity of the environment to sustain all of their offspring. Only a few of the best adapted progeny will survive the competition and yield future generations. Progeny vary, although often only subtly. Those variations that best equip an organism to survive and reproduce its kind are the ones that will be perpetuated by natural selection. Thus did nature, over billions of years, gradually evolve the wide array of organisms inhabiting the modern environment.

Through the ages continents have risen and subsided, seas have come and gone, glaciers have formed and melted, desert has superseded rain forest and vice versa. With unceasing change there is certainly opportunity for new forms better adapted to changed conditions to survive. Thus, groups that were once great are now extinct, like the dinosaurs that dominated the earth for many millions of years. A relative upstart, man, now has his turn. He like all other life before him must abide by nature's one pervasive rule—living organisms must come to terms with their environment. To fail means extinction, and millions of extinctions mark the trail from early life to the present. Extinction can come from overaggressiveness that exhausts the resources upon which an organism depends, as well as from inability to meet the challenge of change. Master of his surroundings to a degree that no organism before him has been, man still must prove that he can harmonize his procreation and exploitation with his finite environment!

Origin through crop breeding

Changes in the genetic material (mutations) form the basis for variation in living things. Mutations result from the alteration or displacement of the genes. Along with the considerable variety of gene combinations, mutations are the grist for the mill of natural selection. Evolution can be thought of as occurring by changes in "populations" of genes, from which new harmonious combinations are isolated.

It is evident that mutations that confer some increase in fitness might have a good chance of becoming fixed, at least in a small population. Most, however, tend to be disadvantageous. Low adaptiveness of a mutant tends to lead to its extinction, whereas high adaptiveness of a mutant tends to lead to disappearance of the ancestral type. High adaptiveness in some environments and low adaptiveness in others—a more usual condition—leads to selection of independent populations. The more progeny that survive, the greater the frequency of particular genes. This in essence is the mechanism operative in natural selection. Similarly, selection of certain progeny for perpetuation has always been the basic method of the plant or animal breeder.

Most animal life is dependent upon the seed plants, which evolved some hundreds of millions of years ago apparently from fern-like ancestors. Man himself is relatively recent, perhaps a million years on the scene. Through the ages he learned to save some of the wild seed he found, and to sow this to ensure future supplies. Such was the beginning of agriculture, and the emergence of man from a life of constant roaming in search of sustenance. As was seen in Chapter 1, only when man became sedentary and began living in relatively permanent communities could civilization begin. Even in the most complex of modern societies, the domesticated seed plants and the animals that feed upon them are still the foundation of man's existence.

Man-made alterations in wild plant populations began with domestication—the first step toward crop breeding. It is a marvel that the art could be so cleverly practiced before the dawn of history, and without genetic understanding. The present-day improvement of crops is effected by genetic manipulation in addition to selection. Variable populations of known potentiality are created from whatever useful forms may be selected (see Chapter 19). In the future it is theoretically possible that the breeder will be able to control which gametes survive and which do not.

The plants (and animals) that primitive man found useful were available just as natural selection delivered them. To this day we still use many wild plants, unchanged from the way they occur in nature. Familiar examples are persimmon, hickory, black walnut, Brazil nut, rubber, the various medicinal plants, and the many trees used for lumber, pulp, and firewood.

For most of today's crop plants, however, wild plants were but the starting point for further selection and alteration by man. In short, man has superimposed upon natural selection his will and his ability to exercise control over his environment in order to hasten the evolution of forms he wants. He, rather than nature, now decides which forms will survive and be perpetuated. By undertaking selection, unconsciously or knowingly and with purpose, man has caused the evolution of plants to take place far more rapidly than is normal in nature. As a matter of fact, so complete have been the evolutionary changes that ancestral forms of many domesticated plants are not even known. Such plants that occur only under cultivation—cabbage and corn are examples—are known as **cultigens.** Improvements in yield and quality, achieved through plant breeding, make possible a modern and efficient agriculture.

PLANT SYSTEMATICS

The results of evolution—the plants and animals sharing the earth with man today—have been classified into discrete and identifiable units termed **taxa** (singular **taxon**). Some taxa include members that are only physiologically distinguishable; others manifest unique clearly visible structural form. An example of the former might be two soybean varieties that look quite alike but are differentially hardy; an example of the latter might be the giant sequoia, which has few close relatives.

Science must concern itself with some sort of method of classifying life forms. One logical way of arranging the plant kingdom is according to evolutionary relationships. Very closely related taxa are grouped as subunits of the same species; more distantly related and distinctive taxa fall into groups of higher rank. The name identifying an organism not only pigeonholes it, but relates it to all others.

The basic taxon used in identifying living organisms is the **species** (singular and plural). This was presumed by early botanists to be the unit of creation, the distinctive entity. The results of Darwin's concepts and, later on, the results of genetic studies have cast some doubt about that, but the species is still the universally accepted typification that is defined and visualized for practical classification. Whether a species is an arbitrary unit on a long intergrading continuum, or whether evolution takes place in quantum-like jumps to make discrete units of appreciable magnitude can be argued pro and con. For our purposes a species is best visualized as a population of related and interbreeding forms, which under normal circumstances remain reasonably distinct from similar populations that constitute other species.

Naming plants

To identify and discuss important plants and animals, some name is needed. Orderly trade and commerce depend upon widely recognized names for such familiar plants as bean, maize, white oak, and so on. Our primitive ancestors undoubtedly gave names to the wild plants and animals that they depended upon for food. In fact it appears that primitive peoples, such as the American Indians and the early Micronesians, were much better acquainted with plants and more adept at knowing them by name than is the average highly educated and "civilized" man of today.

Plant identification has been an evolutionary process in itself, starting perhaps with simple names designating use, and ending today in highly codified systems so abstruse as to be intelligible only to taxonomic specialists. (**Taxonomists** are scholars concerned with properly identifying and naming life forms according to internationally accepted rules of procedure, and in the process surmising evolutionary or natural relationships.) As early as the Middle Ages it was recognized that a universal scholarly language must be used that could be understood by learned men everywhere, no matter what their native tongue. The important treatises were written in Latin, and to this day Latin remains the accepted language for the definitive or scientific naming of organisms. Latin is a dead language, unchanging, understandable to scholars everywhere, hence eminently suitable for such universal communication.

The internationally accepted rules for naming plants prescribe that each plant be distinguished by two primary identifying names—the **generic** name (genus) and the **specific** name (species), in Latin form. (This is often called the **binomial system.**) The original description of the plant must be given in Latin, too. The rules further prescribe that (with a few expressly indicated exceptions) the earliest properly published name after 1753 (the date of Linnaeus' *Species Plantarum*) be the one accepted to identify the plant. If two authors unbeknownst to each other describe the same species and assign it different names, the earliest published name has priority and is the correct designation. Even if a species is placed in an incorrect genus and is later transferred to another, the original specific name must be used if it was the earliest to properly identify the species. The author of a plant name is often listed after the specific name; the names of well-known taxonomists are indicated merely by an initial or an abbreviation. Thus L. stands for Linnaeus, as in the Latin name for wheat, *Triticum aestivum* L.

Keep in mind, however, that even Latin naming according to accepted rules has shortcomings. Sometimes botanists don't agree on what constitutes a species; or, as mentioned, were unaware of prior publication. Critical study often reveals that two or more names (**synonyms**) apply to the same species. Thus the Latin names in this book won't necessarily agree with every author's choice. Technically, only one name can be valid, but until the plant group is thoroughly studied by a competent taxonomist, one can't be sure which name applies to exactly what entity. Nevertheless, the combination of common name and Latin name should certainly make clear enough the identity of the plants discussed in this book.

Species may be subdivided into **subspecies** or **varieties,** for additional identification. The names of botanical subspecies are also written in Latin. Popular names given to **cultivars** or **clones** ("agricultural varieties"), are sometimes added, but are not Latinized—for example, *Rosa foetida* 'Austrian Copper.' *Rosa* is the generic name, *foetida* the specific name, Austrian Copper the name of the cultivar ("variety").

Until publication of Linnaeus' *Species Plantarum*, the name of a plant was not limited to two primary words, the genus and species. The medieval herbals had employed whole phrases or sentences, **polynomials,** often several lines long. As more and more kinds of plants were discovered, the discoverers simply added additional Latin adjectives to distinguish new plants from their relatives. The times were certainly ripe for the simpler binomial system finally established by the students of Linnaeus.

Linnaeus is regarded as the father of modern taxonomy. As ingenious as he was he nevertheless had two centuries of the accumulated experience of many herbalists to build on. Of Linnaeus' predecessors, Vallerieus Cordon (1515–1544) in particular stands out as one of those whose work foreshadowed that of Linnaeus. The descriptions and organization in Cordus' *Historia Stirpium* (published posthumously) have a modern ring to them. With the rise in world exploration, during Linnaeus' lifetime, he had unparalled opportunity to systematize the then expanding biological world. During Linnaeus' time botanists and zoologists were chiefly concerned with gathering and naming new forms, an occupation at which Linnaeus was without peer. So great was his renown, and so widely accepted his new system (based chiefly upon reproductive structures of plants) that explorers the world over sent collections to him and rightly regarded him as the top authority of his time. Few men can match Linnaeus in the quantity of species named, or in the intuition with which he systematized their cataloguing.

Classifying plants

Naming plants as they are discovered is essential for identification. But equally important is the systematic pigeonholing of them for orderly reference—classification. There are many ways in which plants can be classified. An obvious way is to designate them according to their use by man: food, fuel, lumber, feed for grazing stock, and so on. They might be categorized according to where they grow: in the sea, in fresh water, on swampy ground, on high ground, on the prairie, in the forest, in the jungle, on the tundra, and so on. Or they could be grouped according to general appearance and stature: leafy herbs, shrubs or bushes, vines, trees, and so forth. We actually group some plants by a combination of appearance and use; examples are vegetables, fruits, roots, and nuts. These groupings are not very precise, but are adequate for certain everyday purposes. They are not adequate, however, for communication between people of different cultures: one man's fruit is another man's vegetable. Rather, the traditional basis for botanical classification has been based primarily upon appearance (morphology), especially of the reproductive (flower) structures (which are not so prone to change in response to the environment, and thus ordinarily reflect natural relationships more accurately than do vegetative characteristics, many of which alter rapidly to adapt to the environment—for example, all desert plants develop water-conserving forms whether or not they are related).

DESCRIPTIVE CLASSIFICATIONS

Growth habit

The practice of distinguishing plants as annuals, biennials, or perennials goes back at least to the early Greeks. **Annuals** complete their growth cycle in a single growing season, and are perpetuated by seed. The major crop plants of the world are annuals, including all of the grains and many of the legumes, such as soybean, pea, and bean. **Biennials** are plants that take two seasons or years to complete their growth cycle. Usually during the first year the plant accumulates food reserves in storage organs, and during the second produces reproductive flowers and seed. Some of the root crops, such as beets, carrots, and parsnips, are biennials, as are such vegetables as onion and cabbage and a few of the ornamentals, such as hollyhock. By and large, biennials are of secondary importance compared to annuals. **Perennials** are plants that continue growing indefinitely. Some die back to the ground each winter, but revive from the roots the following spring; this habit is typical

of most forage grasses and alfalfa, plus some garden plants, such as rhubarb. Most perennials simply add new growth each year, with little loss or destruction of aboveground parts. Some species—especially tropical forms, such as tomato—are perennial in one climate, but are cultivated as annuals in another. The same is true of cotton: tropical forms are often grown as perennials, whereas temperate forms are usually annuals.

Structure and form

Herbaceous plants are soft and succulent, with little or no secondary tissue, Conversely, **woody** plants are those that develop secondary stem tissues and abundant xylem. Herbaceous plants and plant parts are much more likely to be edible than woody ones; woody plants are more suitable for building materials or as a source of such substances as cellulose.

Herbs are of course herbaceous plants, but in a limited popular sense an herb is a small aromatic species primarily used for scent and flavoring. **Vines** are trailing or climbing plants that are not sufficiently woody to support themselves (and many are entirely herbaceous). Large vines with somewhat woody stems, particularly of the tropics, are **lianas**. The wild grapevine of temperate climates would be a liana, if we were accustomed to using that term in temperate regions. **Shrubs** are low, woody plants with stems that are rigid and strong enough to sustain considerable erect growth; shrubs typically have several stems of essentially equal size and rank. **Trees** are plants that are abundantly woody, usually have a single stem or trunk, and typically grow quite tall.

Leaf retention

Although almost all plants drop their leaves sooner or later, those that do so annually, and are leafless for extended periods, usually winter, are said to be **deciduous**. Familiar hardwood trees, such as oaks and maples, are of this type. Plants that maintain green leaves throughout the year are said to be **evergreen**. Needle trees, such as the pines and spruces, are of this type, as are certain broadleaf plants (Dicotyledonae), such as holly and rhododendron.

Climatic adaptation

This sort of classification is essentially self-explanatory. **Tropical** plants grow in warm climates where freezing rarely if ever occurs. Most shed their leaves once a year in response to changing season, but the period during which old leaves are replaced by new ones is relatively brief, and many are essentially "evergreen." **Temperate** plants grow where there is a marked winter season, typically with considerable freezing. Their climatic range is great—from the arctic, where the growing season (summer) is brief and the winter severe, to approximately the latitude of northern Florida, where the growing season is prolonged and winter is mild. The climate in such places as Florida is often termed **subtropical.** Climates in the colder part of the temperate zone are often referred to as **boreal.** Most temperate species not only withstand cold well, but require winter weather for full vigor and flowering. Kentucky bluegrass, for example, will not set seed unless freezing temperature combines with short day length to trigger this response.

Related is plant hardiness and tenderness. Plants are termed **hardy** if they will survive the winter climate of a particular region. Plants termed **tender** will not survive winter without special protection. Whether a plant is hardy or tender depends on where one wishes to grow the plant. This is true of many fruit trees and ornamental shrubs. The terms "hardy" and "tender" are also used with reference to a plant's ability to tolerate lack of water. In general, the diversity of plants is greatest in humid, tropical climates, and decreases progressively toward the arctic and antarctic regions or toward drier environments. Relatively few kinds of plants are hardy enough to withstand the effects of freezing and desiccation that plants are subjected to in arctic climates.

Usefulness

The distinctions between useful, useless, and harmful plants were among the first that primitive man had to make. We can suppose that species faced extinction when any considerable portion of their environment was toxic. By the same token, evolutionarily successful man is harmfully affected by very few plants or plant parts. Of course, almost anything used immoderately can be discomforting, even lethal. But in general only a few kinds of bacteria, some mushrooms, and very occasional seeds or parts from higher plants are deadly **poisonous.** Discomfort can come from a number of stinging or irritating plants, such as nettles and poison ivy. And of course many plants and parts of plants are **inedible,** mostly because they have unappealing taste, little nutritive value, or are too tough and woody even to chew.

Plants can be useful in many ways. Those that provide utilizable substance are **crops.** Those providing esthetic satisfaction are **ornamentals,** or ornamental crops. **Forages** are crops that animals graze upon.

Classification of plants according to the edible parts are often based more upon custom than botanical accuracy, and the delimitations are vague. Horticulturists divide edible garden plants into **fruits** and **vegetables.** Botanically, fruits are seed-bearing structures derived from flowers. As the word is popularly used, "fruits" are pulpy, juicy, and tasteful, and generally but not always come from woody perennials, often trees or shrubs—for example, apple, peach, orange, and quince. The strawberry and banana plant, although herbaceous, are grouped by horticulturists as fruit plants. Herbaceous plants that are usually but not always annual or biennial and have edible parts (including, confusingly, the botanical fruit) are termed vegetables. Common examples of vegetables include lettuce (edible leaf), asparagus (edible stem), beet (edible root), cauliflower (edible flower), tomato (edible fruit), and pea (edible seed). No precise distinctions can be made between "fruit" and "vegetables." With grasses, the edible portions are the dry, one-seeded fruits ("seeds") called **grains**—such as rice, wheat, oats, barley, millet, sorghum, and many others. **Root crops** are those with swollen underground stems or roots, such as the beet, sweet-potato, and manioc (roots); or the potato (stem and tuber). There are many other general classifications based on plant parts such as **nuts, berries, bulbs,** and **corms.**

THE MODERN CLASSIFICATION

A number of attempts were made before Linnaeus' time to catalog new plants methodically as they were discovered. With Linnaeus came the great practical step forward. Later botanists all strived for logical order in arranging species. They recognized that general appearance was not fundamental, and that completely unrelated species might have similar characteristics as a result of response to a given environment. For example, many desert plants are waxy or fleshy, many rain forest plants are tall trees, no matter what their botanical affinities. General morphology (shape, form) is essential to the description of a species, but the reproductive structures (flowers) give a truer, more constant reflection of broad relationships. The basic flower structure of a species is relatively constant, and less subject to environmental variation than vegetative structures.

As noted earlier, Darwin's *Origin of Species* provided breath-taking illumination of the evolutionary processes. Reasons for degrees of relationship suddenly became clear. Ever since, taxonomists have considered one of their main obligations to be the decipherment of "natural relationships"—that is, the presumed evolutionary background of the groups they are studying. Because of the immensity of the plant kingdom, intense study of this kind is possible only for relatively few plant groups. Many taxonomists spend a lifetime specializing within a certain group, such as a family, but the great majority of plants have not received such attention.

It is remarkable that the early botanists, who had little more than morphology and a bit of geography to go on, were for the most part intuitively correct in their grouping of plants according to natural relationships. It is apparent now that most ancestral life forms have become extinct, and that the "missing links" between related groups are not likely to be found except as occasional fossils. The gradual accumulation of many lines of evidence make it possible to piece together evolu-

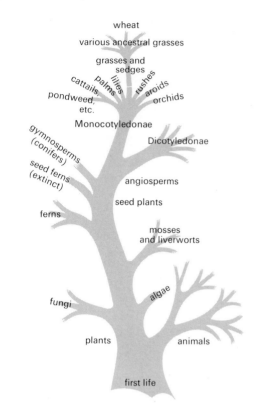

FIGURE 6-2. *Hypothetical evolutionary sequence leading to wheat.*

tionary sequences such as that roughly outlined in Figure 6-2.

We will regard systems adopted since Linnaeus as "modern." Linnaeus' species concept has remained intact, although many groupings of plants have been changed according to various authors' beliefs as to what constitutes a natural relationship. Surprisingly, there is good general agreement between systems conceived by eighteenth and nineteenth century botanists and those of more recent date. As taxonomists acquire better tools for deciphering relationships (for example, chromosome counts, serological and chromatographic tests, improved chemical analysis and computerized investigations), one would expect that taxonomic changes would be based upon a more universally accepted concept of just what constitutes a natural relationship. However, evidence from these approaches generally confirm the process of evolution, and are valuable support for the traditional morphological and paleontological approaches. By and large their usefulness is greatest within a relatively small group, such as a family or a genus. With good microscopes and modern techniques for easily staining cell nuclei, chromosomes have been counted and catalogued for a great many species. The information is especially helpful in suggesting hybrid relationships and polyploid series (see Chapter 19).

Serums made from plant extracts have been used to test affinities between plants. A protein extract from one plant is injected into an animal to "immunize" it to the extract. The degree of precipitation in the reaction between the antiserum obtained from the animal and the extract of another plant is presumably an index of their similarity, and hence of their relationship. A similar technique consists in comparing the genetic material, DNA, from different taxa to obtain a measurement of relatedness on the gene level. Extracted DNA from one taxon is inoculated with radioactive-tagged DNA from another. Recombination occurs in proportion to the similarity of the DNA. Comparisons have been made between the DNA from several higher organisms—man, monkey, mouse, hamster, guinea pig, rabbit, cattle, and salmon. The results show that the DNA of man has much in common with that of the monkey and little in common with that of the salmon, and that the hamster is more closely related to the mouse than to the guinea pig. This same sort of comparison is possible with such crops as wheat, rye, oats, and their various wild relatives.

More recently, chemical investigation has been used to analyze complex biochemical compounds. Chromatography has proved to be a powerful tool of biochemical taxonomy. It is presumed that similarity of compounds indicates similar biochemical pathways under control of genes. Chemical analysis has been effective in elucidating relationships in *Eucalyptus* (by essential oils), *Pinus* (by terpenes), *Senecio* and *Veratrum* (alkaloids), various crucifers (isothiocyanates), and many legumes (various organic compounds). Often the results of these techniques support or cast doubt on long-accepted

relationships established on the basis of morphology alone. For example, betacyanins occur in all families of the order Centrospermae except the Caryophyllaceae. Is the Caryophyllaceae improperly included in the Centrospermae?

Judging from the results of such studies, it is safe to predict that much of the structure of the system of classifying the plant kingdom will remain essentially as it is now, differing little from its organization according to natural relationships as primarily perfected by German taxonomists of the late Nineteenth Century. Surely the Linnaean concept of species in a genus will be retained, and so will the hierarchy now so well recognized: genera (plural of genus) in a family, families in an order, orders in a class, classes in a division (the "phylum" of the zoologist). Various "sub" designations (subclass, suborder, etc.), or other in-between groupings ("section," "tribe"), can always be employed as convenience dictates. Just how one may care to delimit and by what name identify these categories is often a matter of preference.

Because the heyday of plant collection and initial classification coincided with the tremendous surge of world exploration in the Nineteenth Century, many of the important type specimens (the preserved plant from which the original description was derived) are ordered in the great herbaria of the world according to a system that originated in Germany before the turn of the century and was brought to perfection in the monumental work of Engler and Prantl, *Die Natürlichen Pflanzenfamilien*. Although its details are not necessarily up to date, there is obvious convenience to maintaining the Engler-Prantl classification as a standard. As an example, the pigeonholing of wheat, *Triticum aestivum*, would go about as follows:

Kingdom—Plant (as distinct from animal)
Division (or Phylum)—Spermatophyta (seed plants)
Class—Angiospermae (seeds in fruit)
Subclass—Monocotyledonae (seed with single seed leaf)
Order—Graminales (grass-like families)
Family—Graminae or Graminaceae (the grasses)

Genus—*Triticum* (the wheats)
Species—*aestivum* (bread wheat)

A reading upward from the smallest taxon efficiently categorizes the plant in the overall scheme of things. It is wheat, in the grass family, related to sedges (also in the Graminales) and more distantly to the palms (Monocotyledonae), and of course even more distantly to other seed plants. If this were drawn as a "family tree," the ancestral genealogy would look something like Figure 6-2.

Major and minor taxa

The categories of division, class, order, and family are referred to as **major taxa** and from genus below as **minor taxa.** The family, a group consisting of related genera is a distinction that most botanists can agree upon. Certain families stand out as mainstays of modern agriculture. Perhaps the most important of all is the grass family, **Graminae,** source of most of the important grains, staff of life for the world. Without the easily stored and handled grains, human feeding would be exceedingly cumbersome. Almost equally as important as grain in many parts of the world is grass for the grazing of livestock. Sugar cane, too, is a grass, as are corn and sorghum. And in many parts of the world various bamboos, which are grasses, are used for structural purposes. Everywhere grasses protect and build soil.

Perhaps next in importance is the legume family, the **Leguminosae,** which includes beans, peanuts and such nutritious forages as clover, vetch, and alfalfa. In the tropics, trees of the Leguminosae are utilized for their fine timber as well as their fruit (pods and seeds). The Leguminosae live in symbiotic relationship with certain bacteria; nodules on the roots house these bacteria, and they trap gaseous nitrogen and convert it to nitrate. Nitrate is an essential nutrient that the legumes of the world had long been producing before there was ever a fertilizer industry. With usable nitrogen so readily available, it is not surprising that legumes are generally rich in protein. Many tropical populations get along largely without animal protein by substituting proteinaceous beans instead.

The palm family (**Palmae**), important in the tropics, yields food from its fruits (coconut, date), structural materials from its stems and leaves, and various waxes and oils from its leaves, fruits, and seeds. The rose family (**Rosaceae**) not only contributes many ornamentals, but is important for many familiar fruits (apple, pear, peach, apricot, plum, cherry, strawberry, raspberry, and others). The melon family (**Cucurbitaceae**) was perhaps more important to primitive civilizations than it is to urbanized man, but is prized as the source of squash, pumpkins, gourds, watermelons, muskmelons, and cucumbers. The mustard family (**Cruciferae**) yields the cole plants (the cabbage group, including broccoli, cauliflower, Brussels sprouts, and so on), and along with the celery, or parsnip, family (**Umbelliferae**) is the source of many oil-bearing seeds. The pine family (**Pinaceae**) is, of course, an outstanding source of forest products.

The minor taxa are those below the genus. The genus is usually a group small enough for intensive study. The generic concept is an ancient one; most genera were established before the concepts of evolution were known and are based primarily on morphological distinctions. As a result, many well-known genera differ considerably in complexity and require extensive revision. This is especially true for those of economic importance, such as *Sorghum*, a genus whose genetics have been muddled by man through selection.

The species, the basic unit of all taxonomy, is also subject to subdivision by various morphological criteria. The traditional subdivisions of the species are, in decreasing order of complexity, **subspecies, variety,** and **race** (form). But the definitions for these terms are vague and largely a matter of individual preference. Consequently modern taxonomy, making greater use of genetics and ecology, has developed new terms with more precise definitions that are subject to experimental verification. These include the following:

Comparium: Integrity maintained entirely by genetic barriers; often equivalent to the traditional genus.

Cenospecies: Integrity maintained by genetic barriers reinforced by ecological barriers; often equivalent to the traditional genus or species.

Ecospecies: Integrity maintained by ecological barriers reinforced by genetic barriers; often equivalent to the traditional species.

Ecotype: Integrity maintained entirely by ecological barriers; often equivalent to the traditional subspecies, variety, or race.

Identifying higher plants

Until one has mastered taxonomy sufficiently to spot most taxa (at least to the family level), usually on the basis of floral similarity, the theory behind natural classification is fine, but of limited practical value. In order to facilitate identification **keys** (or "analytical keys," also called "dichotomous keys") are utilized, and these are of value to both the uninitiated and the expert in finding out the name of a particular plant. The principle of these keys is the selection between alternatives (usually only two choices are given for each characteristic), and one continues progressively into finer subdivisions of family, genus, species, or subspecies.

A key to all plants, everywhere, would of course be so complex, cumbersome, and time-consuming that the identification effort would be completely frustrated. Rather, most keys are confined to limited geographical areas, or to certain categories of the flora, as are, for example, *Spring Flora of Missouri, Illustrated Guide to Trees and Shrubs* [of the Northeastern United States], or *West Virginia Grasses,* in which the subject coverage is restricted to manageable proportions.

To use a key effectively, one ordinarily needs a specimen of the plant being identified that exhibits both vegetative and floral characteristics (e.g., a twig with leaves and flowers); especially if the plant is a large one, the specimen should be accompanied by information indicating size, habitat, growth habit, flower color, fruit characteristics (if available), and so on. Ordinarily it is impossible to "key out," or identify, a native plant unless a flowering (or fruiting) specimen can be obtained, although some attempts have been made to devise keys for such flora as lawn weeds or small groupings of cultivated plants, based entirely on stem and foliage characteristics. A magnifying lense or binocular microscope will probably be needed to determine such characters as type of pubescence (hairiness), position and shape of stamens and pistil (and similar features determined through dissection of a flower)—characteristics often utilized for making final distinctions.

Dichotomous keys typically identify the two (occasionally more) alternatives to a characteristic by the same code letter or number, with subsequent alternatives (after a choice has been made) following sequentially, and usually indented. A very simple example follows:

A—Colored flowers, not white
    B—tree ....taxon 1
    BB—herb...taxon 2
AA—white flowers
    B—leaves pubescent (hairy)
        C—stamens 4 ....taxon 3
        CC—stamens 8 ....taxon 4
    BB—leaves glabrous (smooth) ....taxon 5

If the specimen has white flowers, alternative AA is chosen and all subheadings under A are passed over; if its leaves are pubescent, alternative B (under AA) is selected; then if eight rather than four stamens occur in the flowers, alternative CC (under AA, B) is chosen. If no further alternatives are listed the name of the taxon is given, in this example "taxon 4". Sometimes rather than being sequentially indented alternatives are progressively listed by number, as follows:

1. Trees or shrubs...............................2
1. Herbaceous.....................................4
    2. Leaves scale-like .... Taxon A
    2. Leaves flat and expanded........3
3. Leaves opposite .... Taxon B
3. Leaves alternate............................. Taxon C
    4. Leaf margins serrate
        (toothed)...............................5
    4. Leaf margins entire
        (not toothed) .... Taxon D

· ·

If the specimen is an herb with serrate leaves the progression is 1, 4, 5, and so on.

Most identification keys are designed for professional use and their terminology is unintelligible outside of botanical circles; to employ them effectively requires mastery of terms used to describe plant form precisely. The brief key that follows is taken from the *Flora of Panama*, for the identification of the species of *Albizzia* (trees of the Leguminosae family) growing there.

a. Leaflets small and narrow (5–11 mm. long and about 2 mm. wide); numerous (9–50 pairs per pinna).
  b. Leaflets less than 30 pair per pinna, pubescent; mature peduncles 2–3 cm. long; flowers relatively large (calyx about 2.5 mm. long) ....1. A. CARBONARIA
  bb. Leaflets usually 30 or more pairs per pinna, glabrous; peduncles less than 1.5 cm. long; flowers small (calyx about 1 mm. long)................................2. A. CARIBAEA
aa. Leaflets larger and broader (mostly 15–40 mm. long and 10–20 mm. wide), few or several (2–9 pairs per pinna).
  b. Leaflets tapered and bluntly acute apically; flowers sessile ..3. A. ADINOCEPHALA
  bb. Leaflets obtuse or rounded to emarginate apically; flowers pedicellate.
    c. Flowers all alike, short-pedicellate (pedicels mostly about 2 mm. long); mature leaflets, young twigs, and legume sparingly pubescent or glabrous..................4. A. LEBBECK
    cc. Flowers dimorphic, all except center one long-pedicellate (pedicels 8–15 mm. long); mature leaflets, young twigs, and legume tomentose to moderately pubescent ....5. A. LONGEPEDATA

One must know what "pinna," "peduncle," "sessile," "pedicellate," "acute," "obtuse," "glabrous," and "tomentose" signify to use even so unelaborate a key as this. The lengthier key for determining the genera of the Leguminosae in the same publication (by which one determines, to begin with, whether the specimen is *Albizzia* or some other genus) requires additional expertise with taxonomic language, and if one did not have enough experience to seek the plant in the Leguminosae, the use of a comprehensive manual to determine family would be even more complicated and probably a task for someone with taxonomic training.

The cultivated variety, or cultivar

In agricultural terminology the term "variety" refers to a named group of plants within a particular cultivated species that is distinguished by a character or a group of characters. Examples are the 'Bartlett' pear, 'Kenya Farmer' wheat, 'Ranger' alfalfa, and 'U.S. 13' corn. To avoid confusion with the botanical variety, the term **cultivar** (a contraction of *culti*vated *vari*ety) has been suggested as the preferred term. The cultivar is the taxon that is most often of relevance to the individual crop producer.

The cultivar, or agricultural variety, biologically speaking, can refer to one of several entities, often depending on the method of reproduction involved in the particular crop. Thus with reference to crops that are generally asexually propagated, such as potato and various fruits, the term means a particular **clone,** the product of a single individual. For example, the names 'Bartlett' pear, 'Russet Burbank' potato, and 'Senga Sengana' strawberry refer to unique genotypes perpetuated by vegetative propagation. They are often referred to as clonal cultivars. Many improved clonal cultivars are the result of mutations (known as sports) that produce some advantageous change, such as the red pigment in the skin of sports of the 'Delicious' apple. In a few plants clonal propagation can be made by seed that does not develop sexually and is identical with the mother plant—a condition known as apomixis. Kentucky bluegrass (*Poa pratensis*), favorite for lawns as well as a prime pasture species, is largely apomictic.

With reference to self-pollinated plants, such as wheat or tomato, the term "cultivar" means a particular inbred or pure line that breeds true naturally. These cultivars are maintained by avoiding contamination and mixing. In cross-pollinated crops, such as alfalfa, the term refers to a population of plants distinguished on some morphological or physiological basis and maintained by selection and isolation. These are referred to as open-pollinated cultivars. Finally, the term "hybrid cultivar" may refer to a particular combination of inbred lines. Thus it can be seen that the meaning of cultivar in the agricultural sense depends on the particular crop in question. The improvement of crop cultivars is the goal of plant breeding.

AN OVERALL VIEW

Although the chief domestications were complete before recorded history, man is still selecting new plants for his use, and improving old. The breeder of a particular crop usually works with subdivisions of lesser rank than the genus. By and large, species of different genera will ordinarily not hybridize, but many species naturally cross with others of the same genus. Indeed, this seems to be one of nature's means for creating new combinations for natural selection. The plant breeder selects from the progeny of his crosses those individuals having advantage. Useful selections become the cultivars of modern agriculture.

As improved cultivars are created, older ones are abandoned. Scientists have begun to wonder—as uniform high-yielding cultivars are now being pressed upon "underdeveloped" parts of the world, in place of traditional locally-adapted types—whether we are not losing germplasm that might prove immensely valuable in some future age under different circumstances. By allowing wild plants to become severely decimated, even extinct, and by not perpetuating "obsolete" cultivars, this could happen. Efforts are now being undertaken, perhaps just in the nick of time, to preserve a few fits and snatches of natural habitat ("living museums"), and a Federal Seed Storage

Laboratory (germplasm bank) has been created at Fort Collins, Colorado, to house stocks of older cultivars that might someday prove of use to future plant breeders.

As pedantic and uninspiring as classification may seem to those not engaged in its research, it is apparent that some understanding of it and the terminology that goes with it is fundamental to discussions involving numerous plants. Classification is essential in order to identify—that is, to "key out"—plants; to determine whether they are known or unknown. The system is by no means perfect, even for the more easily categorized wild plants. Applied to modern cultivars and hybrids, it does not always work well, and there is not always agreement about how to categorize such cultivars as those that originate through crosses between species (often such cultivars are given a formal name prefixed by ×, as × *Magnolia soulangeana* for the Saucer Magnolia). For the lower organisms, such as bacteria, morphology offers little to go on (even with the aid of the best microscopes), and consequently we must rely upon physiological reactions, chemical by-products, and colony formation. Nevertheless, a plant name should bring to mind diverse information about it. In this sense the name of a plant is a summary of it; it is an epitomization of its genetics, physiology, uses, importance to man, and so on. Regarded in this way, classification becomes one of the broadest and most dynamic fields of study.

Selected References

Bailey, L. H. *Manual of Cultivated Plants* (rev. ed.). Macmillan, New York, 1949. (Classic compendium for identifying cultivated plants, especially those having ornamental value, by the outstanding horticultural taxonomist of the century. For reference rather than reading, but worth examining to become aware of how plants are practically catalogued and how keys are used for their identification. The book contains comprehensive glossaries of Latin names, technical terms, and taxonomic authorities. The manual follows the classical concept of Engler and Prantl.)

Blake, S. F., and Atwood, A. C. *Geographical Guide to Floras of the World; Annotated List with Special Reference to Useful Plants and Common Plant Names. I. Africa, Australia, North America, South America; Islands of the Atlantic, Pacific, and Indian Oceans. II. Western Europe; Finland, Sweden, Norway, Denmark, Iceland, Great Britain with Ireland, Netherlands, Belgium, Luxembourg, France, Spain, Portugal, Andorra, Monaco, Italy, San Marino, Switzerland.* USDA, Miscellaneous Publications 401 and 797, Washington, D.C., 1942, 1961. (This exhaustive bibliography, containing 9,866 titles, is the only one of its kind. It covers more than half the world, but does not include

Germany, central and eastern Europe, or Asia and associated islands.)

Blunt, Wilfrid, with the assistance of W. T. Stearn. *A Life of Linnaeus.* Viking, New York, 1971. (A readable biography of Linnaeus from his boyhood to his attainment of recognition as the leading botanical taxonomist; the book reveals the many facets of knowledge that prevailed at that time and with which Linnaeus coped in his rise to fame.)

Core, Earl L. *Plant Taxonomy.* Prentice-Hall, Englewood Cliffs, N.J., 1955. (Suggested as an easily readable reference that relates classical taxonomy, based upon morphology, with some of the more modern techniques. The author holds to the traditional approach of first summarizing the plant kingdom and later describing the plant families and the important genera in them. This more or less traditional approach has been abondoned in some later taxonomy textbooks.)

Davis, P. H., and V. H. Heywood. *Principles of Angiosperm Taxonomy.* Van Nostrand, Princeton, 1963. (An excellent, detailed summary of most of the subject matter in this chapter, by English authorities.)

Grant, Verne. *Plant Speciation.* Columbia University Press, New York, 1971. (This volume contains many examples of evolutionary development of species, and is concerned with the difficulties of adequately defining what a species is both taxonomically and biologically.)

Harborne, J. B. (editor). *Phytochemical Phylogeny.* Academic Press, New York, 1970. (Surveys the impact of phytochemical research on our concepts of the origin of plant life and the evolution of modern plants. At this 1969 meeting, various authors presented papers on subjects ranging from palaeobotany through chemical composition of plant parts and secretions to discussions of the environment and enzyme evolution in plants.)

Hawkes, J. G., (editor). *Chemotaxonomy and Serotaxonomy,* Academic Press, New York and London, 1968. (Discussion of serological and biochemical techniques for establishing "chemical" systematics. Proceedings of a symposium on the subject held at the University of Birmingham.)

Heywood, V. H. (editor). *Modern Methods in Plant Taxonomy.* Academic Press, New York and London, 1968. (A series of review papers given at a conference of the Botanical Society of the British Isles and the Linnean Society of London in 1967, resulting in a comprehensive review of the "state of the art" in plant taxonomy as of that date.

Chemosystematics and numerical taxonomy are aspects discussed.)

Jardine, Nicholas, and R. Sibson. *Mathematical Taxonomy.* Wiley, New York, 1971. (A review of taxonomic methodology, drawing upon widespread European references, and including a section that discusses the principles of biological classification in a numerical context.)

Lawrence, G. H. M. *Taxonomy of Vascular Plants.* Macmillan, New York, 1951. (A standard text in systematic botany. Part I covers both the theoretical and practical aspects of classification, nomenclature, and identification. Part II describes all tracheophyte families known to be native to or introduced into the United States. The appendix contains a detailed glossary of botanical terms.)

Oparin, Aleksandr Ivanovich. *The Chemical Origin of Life* (English translation from the Russian of 1936 by Ann Synge.) Charles C Thomas, Springfield, Ill., 1964. (American Lecture Series Publication 588.) (Tedious for thorough reading, but worth scanning as a landmark that ushered in modern concepts about the creation of life.)

Ponnamperuma, C. *The Origins of Life.* Dutton, New York, 1972. (Discussions suited to beginning students on the relationship between living things and their cosmic environment. Stresses "chemical evolution" and the antiquity of life.)

Porter, C. L. *Taxonomy of Flowering Plants* (2nd ed.) W. H. Freeman and Company, San Francisco, 1967. (An excellent elementary text).

Scagel, R. F., G. E. Rouse, J. R. Stein, R. J. Bandoni, W. B. Schofield, and T. M. C. Taylor. *An Evolutionary Survey of the Plant Kingdom.* Wadsworth, Belmont, Calif., 1965. (A thoroughgoing survey of the plant kingdom, integrating information from the fields of physiology, cytogenetics, and so on; a compendium of information suitable for students planning advanced study in the subject area.)

Stebbins, G. L. *Chromosomal Evolution in Higher Plants.* Addison-Wesley, Reading, Mass., 1971. (A review integrating molecular biology, cytology, and population studies of plants. Of interest for concepts of evolutionary mechanisms, although slanted more to the expert than to the beginner).

Swain, T. (editor). *Chemical Plant Taxonomy.* Academic Press, London, 1963. (A thorough consideration of chemical and allied techniques for determining relationships among plants.)

Valentine, D. H. (editor). *Taxonomy Phytogeography and Evolution.* Academic Press, New York, 1972. (Comprehensive background discussions relating to origin and classification.)

CHAPTER 7

Structure and Function

Flowering plants exhibit tremendous diversity in size and structure. Plant organs—stem, leaf, root, flower, fruit, and seed—show extreme modification; crop plants especially have unusual features. Although many structures are superficially very different (for example, the air-borne root of the orchid and the swollen root of the sweet-potato), they can be shown to be functionally and morphologically related. The present discussion must necessarily be a balance between generalizations stressing the similarities of structure and function among different plants, with specific emphasis on the differences that distinguish individual crop plants.

THE CELL AND ITS COMPONENTS

The structural unit of plants, as well as of animals, is the **cell.** Cytology, the study of cells, is concerned with the organization, structure, and function of cells. The concept that the cell is the basic unit of all living things is still accepted by biologists as dogma. The complex multicellular organism is an integrated collection of living and nonliving cells. The high degree of synchronization and coordination, however, in the total organism creates an entity that is in effect greater than the sum of its parts.

Plant cells vary in shape from spherical, polyhedral, and ameboidal to tubular. Most cells are between 0.025 mm and 0.25 mm in size (0.001–0.1 inch), but some fibers (long, tubular cells) are as much as 2 feet long. The concept of the typical, or generalized, plant cell is a useful one (Fig. 7-1). The most prominent components visible under the light microscope are the densely staining nucleus, and the more or less rigid cell wall enclosing the cytoplasm. The cytoplasm contains a number of structural bodies, or organelles, such as plastids, mitochondria, and vacuoles and various other entities such as crystals, starch grains, and oil droplets. A distinction is often made between living and nonliving substances in the cell, but such distinctions have little meaning.

The **cytoplasm** (less precisely termed protoplasm) is an extremely complex substance, both physically and chemically. It is composed of 85

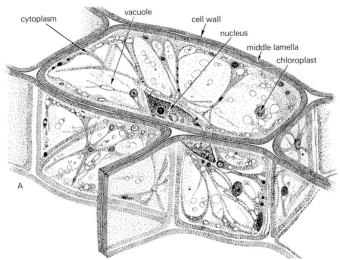

light microscope image   electron microscope image

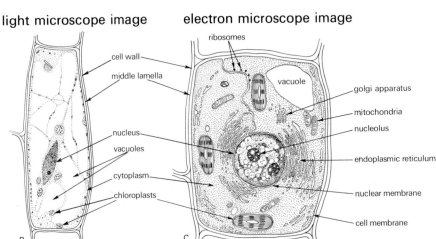

FIGURE 7-1. *Three views of a plant cell and its component parts. Recent electron microscope studies have produced a new concept of the ultrastructure of the cell.* [*Parts A and B after Esau,* Plant Anatomy. *New York: Wiley, 1953. Part C after Brachet, "The Living Cell." Copyright © 1961 by Scientific American, Inc. All rights reserved.*]

to 90 percent water (by fresh weight); the remaining 10 to 15 percent consists of organic and inorganic substances that are either dissolved (salts and carbohydrates) or in the colloidal state (proteins and fats). The physiological activity of the cell occurs in the cytoplasm. Although the living state appears to be dependent on the cytoplasm remaining within the structural confines of a "cell," various so-called vital processes (such as photosyn-

thesis) have recently been found to be possible outside of the cell. The concept of what constitutes a vital process changes as biological techniques increase in sophistication.

Surrounding the cytoplasm is the **plasma membrane,** which lies against the cell wall. The cell wall is more or less completely permeable to all solutes and solvents, whereas the plasma membrane is semipermeable. The plasma membrane is composed of lipoproteins, which account for its elasticity and high permeability to fatty substances. Although it is ultramicroscopic in thickness, the plasma membrane can be defined in plasmolyzed cells. The electron microscope has revealed that the plasma membrane is not a mere boundary to the cell but has structural definition. The cell appears permeated by internal membranes that, along with the external membrane, are thought to constitute a system called the **endoplasmic reticulum.**

The **nucleus,** a dense, usually spheroidal body, is located within the cytoplasm. The close affinity of the nuclear material for many dyes makes it the most conspicuous feature of a stained cell. The nucleus is, in effect, the control center of the cell, since it contains the **chromosomes** (*chromo,* colored; *soma,* body). The chromosomes contain deoxyribonucleic acid (DNA) associated with protein. The arrangement of the DNA contains genetic information, somewhat analogous to the punch cards in an electronic computer. The information here is then relayed to the cytoplasm in the form of RNA (ribonucleic acid), a substance similar to DNA. The information contained in DNA affects the machinery of the cell through the control of protein synthesis. The actual site of synthesis is carried out not in the nucleus but in small particles in the cytoplasm called **ribosomes.** The information contained in DNA provides the basis for the physiological function of the cell and determines the metabolic and morphological features of the organism. It is also a major part of the hereditary bridge between generations.

The cell wall, one of the structures that distinguishes plant cells from animal cells, is usually thought of as a deposit, or secretion, of the cytoplasm. The plant cell wall is composed of three basic groups of constituents: cellulose, lignin, and pectin compounds. **Cellulose** and a number of closely related compounds, such as hemicellulose, make up the firm layer of the cell wall. These compounds are, basically, branching long-chain units (polysaccharides) formed from the simple sugar glucose (see p. 41). Hemicellulose is related to cellulose but contains sugar and nonsugar components. Cellulose and its derivatives are highly combustible materials. **Lignin** consists of a complex mixture of chemically related compounds— specifically, polymers of phenolic acid. The deposition of lignin (lignification) hardens the cellulose walls into an inelastic and enduring material resistant to microbial decomposition. Lignins cause yellowing in paper and must be dissolved from the wood pulp used in making high-quality papers. **Pectin compounds** are water-soluble polymers of galacturonic acid that form sols and gels with water. The familiar substance pectin is used as a solidifying agent for jams and jellies.

The **cell wall** is laid down in distinct layers, and its thickness varies greatly with the age and type of cell. Three regions of the cell wall are generally distinguished:

1. The **middle lamella,** a pectinacious material associated with the intercellular substance. The slimy nature of rotted fruit is due to the dissolving of the middle lamella by fungal organisms.
2. The **primary cell wall,** the first wall formed in the developing cell. It is composed largely of cellulose and pectic compounds, but closely related substances and noncellulose compounds may also be present, and it may become lignified. The primary cell wall is the wall of dividing and growing cells; in many cells it is the only wall.
3. The **secondary cell wall,** which is laid down inside the primary cell wall after the cell has ceased to enlarge. It appears to have a mechanical function, and is similar in structure to the primary wall, although higher in cellulose, and often contains some lignin.

The cell wall in plants is not continuous. It appears to be pierced by cytoplasmic strands (**plasmodesmata**) that provide a living connection between cells. Furthermore, thin areas called **pits** occur in the walls of some cells.

Plastids are specialized, disc-shaped bodies contained in the cytoplasm, and are peculiar to plant cells. They are classified on the basis of the presence or absence of pigment into leucoplasts (colorless) or chromoplasts (colored). Leucoplasts occur in mature cells that are not exposed to light, and some types are involved in the storage of starch. Of the colored plastids, those containing chlorophyll (**chloroplasts**) are the most significant, for they are the complete structural and functional

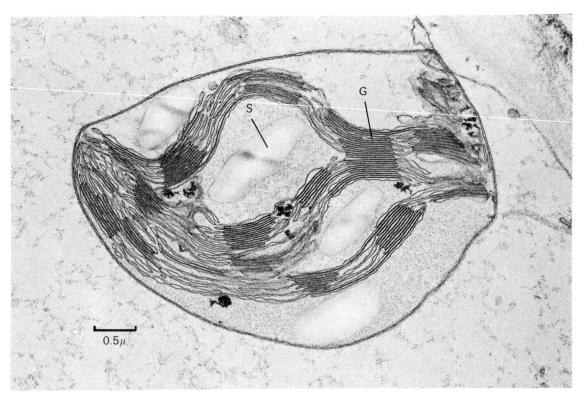

FIGURE 7-2. *An electron micrograph of the chloroplast from tobacco leaf. Note granum (G) and starch grain (S). [Courtesy W. M. Laetsch.]*

unit of photosynthesis, the process by which carbon dioxide and water are transformed to carbon-containing compounds. The reactions result in the formation of starch grains within the chloroplast.

There are about 20 to 100 chloroplasts in each chlorophyllous cell of a green leaf. Electron microscope photographs have revealed the internal structure of chloroplasts (Fig. 7-2). Structural units that resemble stacked coins called grana contain the chlorophyll and are the receptors of light. The actual transformation of carbon dioxide to carbon-containing compounds occurs in the surrounding material, called stroma.

Chloroplasts have a kind of independent existence in the cell. Although they are under the influence of nuclear genes, the chloroplasts are self-replicating and apparently arise only from pre-existing plastids.

Mitochondria are the cell's power plant. They appear as small, dense granules under the light microscope, but electron microscopy has revealed that they have a complex involuted internal structure. The mitochondria are made up of proteins and phospholipids. All of the known functions of mitochondria are related to enzymatic activity connected with oxidative metabolism. This activity occurs through the formation of the energy-carrying substance called adenosine triphosphate (ATP).

Vacuoles are membrane-lined cavities located within the cytoplasm, and are filled with a watery substance known as the cell sap. The cell sap contains a number of dissolved materials—salts, pigments, and various organic metabolic constituents. In actively dividing cells the numerous vacuoles are small, whereas in mature cells the vacuoles coalesce into one large vacuole that occupies the center of the cell, pushing the cytoplasm and the nucleus next to the cell wall.

A number of complex materials may be found in the cytoplasm; among these are crystals, starch grains, oil droplets, silica, resins, gums, alkaloids, and many organic substances. Many of these compounds are reserve or waste products of the cell.

TISSUES AND TISSUE SYSTEMS

Although the plant ultimately originates from a single cell (the fertilized egg), the marvels of cell division and differentiation produce an organism composed of many different kinds of cells—different structurally and physiologically. It is this difference in cell morphology and cell arrangement that results in the complex variation among different plants and within an individual plant.

Plants are made up of groups of similar types of cells in a definite, organized pattern. Continuous, organized masses of similar cells are known as **tissues**. Tissues can be classified in several ways, but no single system of classification is universally used. The following system retains many of the customary botanical terms.

Meristematic tissue: actively dividing, undifferentiated cells
Permanent tissues: nondividing, differentiated cells
    Simple tissue: composed of one type of cell
      parenchyma: simple, thin-walled cells
      collenchyma: thick-walled, supporting tissue
      sclerenchyma: thick-walled, highly specialized elements
    Complex tissue: composed of more than one type of cell
      xylem: water-conducting tissue
      phloem: food-conducting tissue

Meristematic tissue

Meristematic tissue is composed of cells actively or potentially involved in cell division and growth. The meristem not only perpetuates the formation of new tissue but perpetuates itself. Since many so-called permanent tissues may, under proper stimulation, assume meristematic activity, no strict line of demarcation exists between meristematic and permanent tissue.

Meristems are located in various portions of the plant (Fig. 7-3). Those at the tips of shoots and roots are known as apical meristems. The apical meristem of the shoot is also known as the growing point. The increase in girth of woody stems results from the growth of lateral meristems, collectively referred to as the **cambium.** The meristematic regions of grasses become "isolated" near the nodes, and are called intercalary meristems. Thus the mowing of lawns does not interfere with the growing portion of the grass plant. Tissues differentiated from apical meristems are referred to as **primary tissues.** Others, especially tissues formed from the cambium, are **secondary tissues.**

Although there are exceptions, meristematic cells are usually small, roughly spherical to brick-

FIGURE 7-3. *Diagrammatic longitudinal section of a grass plant, indicating the location of the meristems. These shaded areas are the youngest parts of the plant.* [*After Eames and MacDaniels,* An Introduction to Plant Anatomy. *New York: McGraw-Hill, 1947.*]

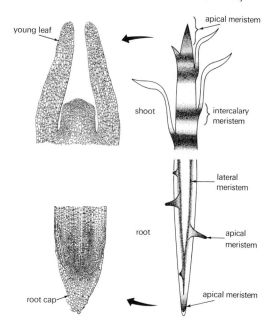

shaped, with thin walls, and inconspicuous vacuoles. In preparation they appear darkly stained, owing to the relatively small amount of cytoplasm in relation to the size of the nucleus.

Simple tissue

Permanent tissues are derived directly from meristems. They are referred to as simple tissues when composed of one type of cell (Fig. 7-4).

Parenchyma is unspecialized tissue. It makes up a large proportion of many plant organs, such as the fleshy parts of fruits, roots, and tubers, and is the seat of activities that depend on living protoplasm, such as photosynthesis, assimilation, storage, and secretion. It is relatively undifferentiated, thin-walled, living tissue, capable of growth and differentiation.

Collenchyma tissue is characterized by elongated cells with unevenly thickened primary walls. Collenchyma cells commonly occur in strands or continuous cylinders and function largely as mechanical support in early growth. The strands of the outer edge of celery petioles are collenchyma.

Sclerenchyma cells are thick-walled, have small cavities, and occur individually, in small groups, or in continuous masses. They are often heavily lignified. Sclerenchyma cells, in contrast to parenchyma and collenchyma, are nonliving when mature. When these cells are long and tapered, they are usually referred to as fibers. Fibers tend to interlace and interlock to form sheets of tissues. Short or irregularly-shaped sclerenchyma cells are called sclereids. Clusters of sclereids, or "stone cells," are responsible for the gritty texture of the pear fruit. In masses, sclereids produce the hardness of walnut shells and of peach and cherry pits.

In common usage the term "fiber" refers to thread-like strands of any substance. The fibers of commerce may be derived from plants (flax, hemp), animals (silk, wool), or minerals (asbestos). Many plant fibers are derived from tough, flexible strands that occur in stems, leaves, or roots. Such fibers are generally composed largely of sclerenchyma, although other tissues such as vascular bundles may be involved. Some well-known fibers are morphologically unrelated to sclerenchyma; cotton

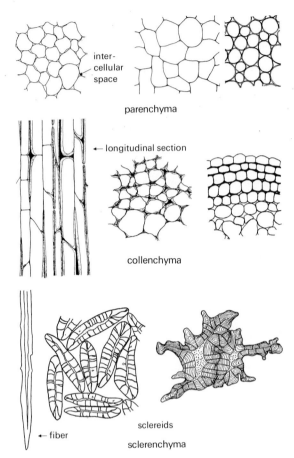

parenchyma

← longitudinal section

collenchyma

← fiber

sclereids

sclerenchyma

FIGURE 7-4. *Simple tissue.* [*After Eames and Mac-Daniels*, An Introduction to Plant Anatomy. *New York: McGraw-Hill, 1947.*]

"fibers" are in fact epidermal hairs of the seed coat and are almost pure cellulose. Plant fibers may be classified, on the basis of their origin, as bast fibers (phloem, pericycle, or cortex), wood fibers, vascular bundle fibers, or surface fibers (seed hairs).

Complex tissue

Complex tissue is composed of combinations of simple and specialized cells and tissues. The two major types of complex tissues, xylem and phloem, are continuous throughout the plant; in addition to supporting the plant, they function in the conduction and storage of food and water.

XYLEM The principal water-conducting tissue of woody and herbaceous plants is **xylem,** an enduring tissue that consists of living and nonliving cells (Fig. 7-5). The basic element of xylem is a specialized nonliving cell called the **tracheid.** These are typically angular, elongated, tapered cells having walls that are hard and usually lignified although not especially thick. The cell is devoid of contents and is adapted for its principal function of water conduction, although the rigid walls also function in support. Water and its dissolved contents moves readily through the empty tracheid, flowing from cell to cell through the numerous pits. The tracheid cell walls are often unevenly thickened or sculptured in a spiral fashion, which makes them easy to distinguish in longitudinal section.

In the primitive condition the tracheids constituted the principal cell type of xylem, but the xylem has evolved into a more complex and specialized tissue in which specialized fibers have taken over the role of structural support and in which efficient specialized series of cells (tracheae), called vessels, function in water conduction. **Vessels** are formed from meristematic cells that are lined up end to end and whose end walls have been dissolved. The cells may be very large in diameter, and the vessels may be many feet long. In highly specialized xylem thin ribbon-like structures of parenchyma tissue, called medullary rays, run inward toward the center of the stem at right angles to the vessels. These rays serve for lateral conduction of water and for food storage.

Primary xylem is formed by differentiation of the apical meristems of root and shoot. In perennial woody plants, secondary xylem is also formed from the cambium. The differential seasonal growth of xylem produces the familiar annual rings. Spring wood consists of larger cells with thinner walls and appears lighter, or less dense, than summer wood.

Wood and lumber. All of the xylem of vascular plants—including the veins of leaves or the string-like vascular bundles of palms and bamboo—is properly called wood. The name xylem is in fact derived from the Greek word for wood. The lumberman's terms "softwood" and "hardwood" have little relation to actual hardness but refer to the two great classes of trees, gymnosperms and angiosperms (specifically, the dicotyledons), which provide lumber. The wood of gymnosperms (such as pine, spruce, and larch) is called softwood, or nonporous wood; the wood of angiosperms (such as oak, mahogany, and teak) is called hardwood or porous wood (Fig. 7-6). In hardwoods water conduction is carried on in vessels, some of which are large enough to be seen with the naked eye. In softwoods water conduction is carried on in tracheids, which also provide support. Many softwoods contain resin, either in very thin tubes (resin ducts) parallel to the tracheids or in short cells like those of the California redwood.

In general the central portion of the trunk of a tree is darker than the outer portion. The darker center is called heartwood; the lighter outer portion, sapwood. The change from sapwood to heartwood occurs when tannins, gums, resins, and other substances accumulate in aging xylem cells, and is associated with the cessation of their functioning as liquid-conducting elements.

The various odors of different woods are a result of aromatic gums, resins, and essential oils that accumulate in the parenchyma cells. The grain of wood is due to the orientation of the structural

FIGURE 7-5. *Cells of the xylem.*

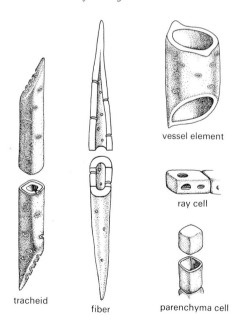

vessel element

ray cell

parenchyma cell

tracheid

fiber

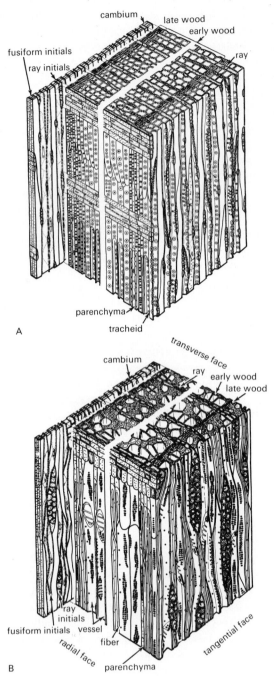

FIGURE 7-6. *Block diagrams of the cambium and secondary xylem of (A) white cedar* (Thuja occidentalis), *a gymnosperm (softwood), and (B) tulip tree* (Liriodendron tulipifera), *an angiosperm (hardwood).* [*From Esau,* Plant Anatomy. *New York: Wiley, 1954.*]

elements. Interesting patterns arise because of irregularities of the grain or the orientation of the grain in relation to the sawed or sliced surface. Knots are the basal ends of branches that become embedded in the thickening trunk. Wood's anatomical components and arrangement are characteristic for each species and can be used as a means of identification.

The remarkable properties of wood—its strength and lightness, its elastic properties, and its relative softness—are a consequence of the physical nature of the wood material and its anatomical structure. Factors such as density, moisture content, and defects influence the mechanical properties of lumber.

PHLOEM   The basic components of the **phloem** are series of specialized cells called **sieve elements.** It is through these elongated living cells with thin cellulose walls that the food-conducting function of the phloem is carried out. Upon maturity the nucleus of the sieve tube element disappears. In the angiosperms, however, specialized parenchyma called **companion cells** are in intimate association with the sieve cells (Fig. 7-7). In addition to these,

FIGURE 7-7. *Cells of the phloem.*

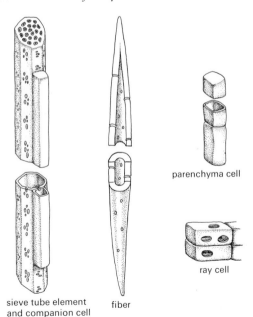

fibers and sclereids may be present. The fibers of hemp and flax are derived from the phloem tissue.

The phloem, like the xylem, is formed both by the apical meristem and by the cambium. The phloem, however, is not enduring, and the old phloem gets crushed and largely obliterated in woody stems. It is protected by special meristematic tissues (cork cambium) that produce parenchymatous tissue. The phloem, the corky tissue, and the other incidental tissues constitute **bark.**

ANATOMICAL REGIONS

Plant tissues can be classified in terms of structure and function into anatomical regions. These include the vascular system, cortex, and epidermis (Fig. 7-8). These regions are interrelated, and it is sometimes hard to decide where one ends and the other begins. The pith, pericycle, endodermis, and secretory glands are components of one or more of these regions.

FIGURE 7-8. *Anatomical regions of the mint stem.*

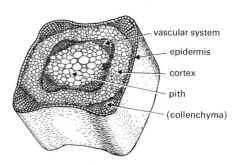

vascular system

epidermis

cortex

pith

(collenchyma)

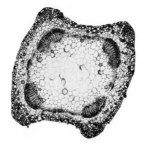

The vascular system

The vascular system consists principally of the xylem and phloem tissues. It serves primarily as the conduction system of the plant, but because it may also serve in support it may be compared to both the circulatory and skeletal systems of animals. Although xylem and phloem are usually found in association, there are differences in their structural relationship to each other. Typically, the vascular system forms a continuous cylinder in the stem, in which the outer portion is phloem and the inner portion is xylem, which surrounds an area of parenchymatous tissue known as the pith. In some plants—for example, the monocotyledons—the vascular system is discontinuous, and appears as a series of strands in longitudinal section, and as bundles in cross section (Fig. 7-9). In other plants the primary vascular system is discontinuous and forms a series of strands, and the secondary vascular system forms a continuous cylinder. In some plants the vascular system makes up the greater part of the root, the amount of pith being minimal.

In roots the vascular system is separated from the cortex by specialized tissues called the pericycle and the endodermis. The **pericycle** encircles the vascular system. It is composed of parenchymatous tissue and is the source of the branch roots and stems that arise from the root. The **endodermis** is commonly a single sheet of cells separating the vascular system from the cortex; it is not absolutely clear whether it is part of the vascular system or the cortex. It appears to have a protective function. The pericycle and endodermis are usually absent in the stem.

The development of the shoot is accompanied by the expansion and elongation of the vascular system. The part of the vascular system that is derived from the apical meristem is called the primary vascular system. When further increase is derived from the cambium, the resultant growth is referred to as the secondary vascular system.

In gymnosperms and dicotyledons the secondary vascular system is derived from the cambium, a cylinder between the xylem and the phloem. The cambium produces secondary xylem toward the inside and secondary phloem toward the outside.

122

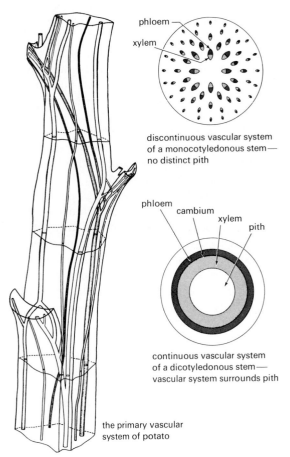

phloem

xylem

discontinuous vascular system
of a monocotyledonous stem—
no distinct pith

phloem cambium

xylem

pith

continuous vascular system
of a dicotyledonous stem—
vascular system surrounds pith

the primary vascular
system of potato

FIGURE 7-9. *Discontinuous and continuous arrangements of the vascular system. The primary vascular system of the potato stem initially appears as separate bundles but becomes embedded in secondary tissue as the stem matures. The vascular system of the mature stem is continuous.* [*After Eames and MacDaniels,* An Introduction to Plant Anatomy. *New York: McGraw-Hill, 1947.*]

Not all plants have a cambium; in those herbaceous plants that do not, the increase in diameter is the result of enlargement of cells of the cortex and vascular system.

Monocotyledonous plants generally lack secondary growth, but they may produce very large stems, as do the palms, by a special thickening meristem. The thickening meristem usually ceases activity close to the apex of the plant, but in some

monocots it remains active and produces secondary growth, and in fact is referred to as cambium.

Cortex

The cortex is the region between the vascular system and the epidermis. It is made up of primary tissues, predominantly parenchyma. In old woody stems, the formation of cork in the cortex, with the subsequent disintegration of the outer areas, tends to obliterate the cortex as a distinct area. Cork is formed when mature tissue becomes meristematic and forms cells with walls that contain a waxy substance called suberin. This corky protective sheath that replaces the epidermis when it is killed or sloughed away is known as the **periderm.** Callus tissue, which is formed as a result of wounding, may also become corky. The cork industry depends on the large amounts of this tissue produced by the cork oak (*Quercus suber*).

Differentiated portions of the periderm that form ruptured, rough areas are referred to as **lenticels.** These are particularly conspicuous in the bark of young woody stems of some species. The "dots" on apple skin are lenticels. This loose arrangement of cells tends to tear the epidermis, and fruit lenticels become subject to the entrance of decay organisms and act as points of water loss.

Epidermis

The epidermis is a continuous cell layer that envelops the plant. Its main functions include mechanical protection, water conservation, gaseous exchange (in shoots), water absorption (in roots), and other functions that predominate in other regions, such as secretion. Except in older stems and roots, where the epidermis may be destroyed, it sheaths the entire plant. The structure of the epidermis varies, and may be composed of different kinds of primary tissue. The epidermal cells located a short distance from the root tips form tubelike extensions called root hairs, which function in the absorption of water and inorganic nutrients. Hairs are also found in epidermal cells of the shoot, and may be of complex multicellular structure. Cotton fibers are extremely long epidermal hairs on the seed. The velvety feel of rose petals is due to the uneven surface of their upper epidermal cells. Guard cells, which form **stomata,** openings in the leaf that permit gaseous exchange, are modified epidermal cells.

A significant feature of epidermal cells is the **cuticle,** a waxy layer that appears on the exposed surface of the cell. The waxy material **cutin** acts as a protective covering that prevents the desiccation of inner tissues. It is particularly noticeable in many fruits—for example, apple, nectarine, and cherry, where its accumulation results in a blush that polishes to a high gloss. Carnauba wax is derived from the leaves of a native Brazilian palm (*Copernicia cerifera*).

Secretory glands

A number of morphologically unrelated structures—some very simple, others quite complex—secrete or excrete complex metabolic products. These substances are often exuded in viscous or liquid form. Many of these substances are of economic importance—for example, resin, rubber, mucilages, and gums.

Secretory functions are carried out by many multicelled, hair-like epidermal appendages called trichomes. Moreover, complex secretory structures called glands develop from subepidermal as well as epidermal tissue in various parts of the plant body. The fragrance of flowers is produced in glands called nectaries. Some specialized glands secrete essential oils (see p. 140). In citrus fruits such glands occur in the peel, as can be verified by squeezing a portion of the rind next to a flame. Rupturing the glands releases the volatile and inflammable essential oils. Gums, mucilages, and resins are formed in ducts and canals, intercellular spaces between groups of specialized cells. Often these substances are secreted in response to wounding, as in the stem of peach and cherry.

Latex, a milky, viscous substance containing various materials, particularly gums, is found in some angiosperm families. It may be formed in ordinary parenchyma or may be formed in a complex series of branching tubes called laticifers. They are not considered distinct tissue but are associated with other tissues, commonly the phloem. Latex is released under pressure when

these cells are punctured. Rubber is a constituent of some latex-forming plants. The content varies greatly, but in the rubber tree (*Hevea brasiliensis*) 40 to 50 percent of the latex is rubber. The function of the latex duct network is not clear, but it is felt that it probably serves as an excretory system for the plant and plays some role in wound healing.

MORPHOLOGICAL STRUCTURES

The flowering plant consists of two basic parts: the root, the portion that is normally underground, and the shoot, the portion that is normally above ground. The shoot is made up of stems and leaves, although these structures are not morphologically distinct in all plants. Leaves arise from enlarged portions of the stem called nodes. The flower may be thought of as a specialized stem with leaves adapted for reproductive functions. Buds are miniature leafy or flowering stems. The fundamental parts of monocots and dicots are presented in Figure 7-10.

The root

The root, although visually inconspicuous, is a major component of the plant, both in terms of function and absolute bulk. It usually consists of perhaps one-third of the dry weight of the entire plant body. The root is structurally adapted for its major function of absorption, anchorage, and storage. Owing to its complex branching, which occurs irregularly rather than at nodes as in stems, and to its tip area of root hairs, the root presents a very large surface in intimate contact with the soil. The process of absorption of water and nutrients in most plants is carried out mainly in the root hairs, which are constantly renewed by new growth. In most woody plants and in some herbaceous plants, absorption is further increased by mycorrhizae—fungi that live in symbiotic association with the roots. The growth of the plant is in part limited by the extent of its underground expansion. This vast network of roots anchors the plant and supports the superstructure of food-producing leaves. The older roots also serve as a storage organ

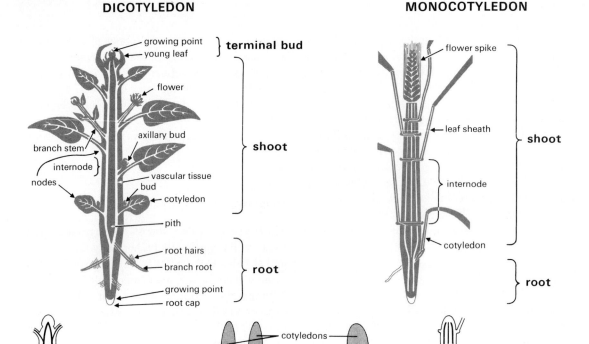

FIGURE 7-10. *The fundamental plant parts.*

for elaborated foodstuffs, as in the sweet-potato and the sugar beet.

The original seedling root, or primary root, generates the root system of the plant by forming various branching patterns (Fig. 7-11). When the primary root becomes the main root of the plant, the network is referred to as a taproot system, as in walnut, carrot, and alfalfa. In many plants, however, the primary root ceases growth when the plant is still young, and the root system is taken over by new roots that grow adventitiously out of the primary root, or, as in many of the grasses, out of the stem, forming a fibrous root system. In addition, many taprooted plants form an upper network of fibrous-like feeder roots, as in apple. This permits deep anchorage and a more reliable water supply while providing absorptive capacity at the more fertile upper layers of soil. A fibrous root system may be formed artificially by destroy-

ing the taproot. This is accomplished by transplanting or undercutting and is a standard practice with shrubs and trees. Nurserymen build up a fibrous root system concentrated in a "ball" below the plant. This permits even relatively large plants to be successfully transplanted.

In general, plants having a fibrous root system are shallow-rooted in comparison with taprooted plants. Shallow-rooted plants will, of course, be more subject to drought and will show quicker response to variations in fertilizer. Shallow-rooted species depend mostly on the rain that falls during the growing season; deep-rooted species depend mostly on cumulative annual rainfall.

The anatomical structure of the root is shown in Figure 7-12. The arrangement of its vascular system differs from that of the stem. Except in monocots, the root lacks pith. The pericycle and the endodermis surround the vascular system. A

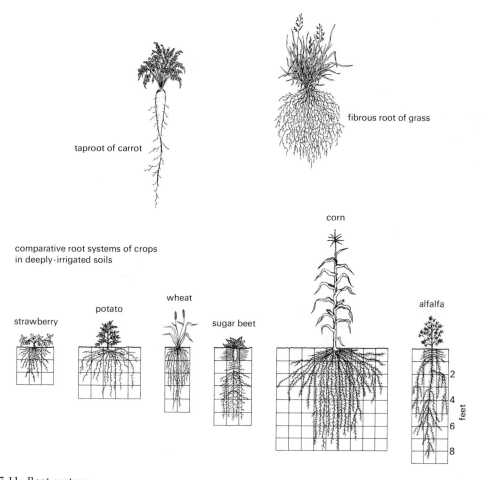

taproot of carrot

fibrous root of grass

comparative root systems of crops in deeply-irrigated soils

corn

strawberry

potato

wheat

sugar beet

alfalfa

2

4

feet

6

8

FIGURE 7-11. *Root systems.*

cambium serves to increase the girth of roots, just as it does in the stem in some plants.

The root displays considerable structural variation. The names for various roots are descriptive of their functional adaptations—for example, storage, aquatic, brace, aerial, holdfast. Many plants are grown as crops because of their root modifications. Roots of certain species become swollen and fleshy as a result of the storage of food in the form of starches and sugars. Some of these storage roots, such as the carrot, cassava, and sweet-potato, are edible. The ability of some roots—for example, the sweet-potato and the Dahlia—to form adventitious shoot buds renders them important in propagation.

The shoot

The shoot has been described as a "central axis with appendages." The "central axis"—that is, the stem—supports the food-producing leaves and connects them with the water- and nutrient-gathering roots. The stem also is a storage organ, and in many plants its structure is greatly modified for this function. Young green stems also have a small role in food production. Plants assume extremely varied forms, ranging from a single upright shoot, as in the date palm, to the prostrate branched "creepers." It is the structure and growth pattern of the stem that determines the form of the plant. Basic structural and anatomical features

126

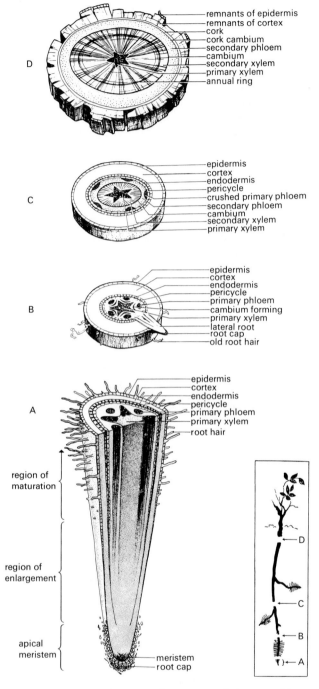

Labels for figure D (top):
remnants of epidermis
remnants of cortex
cork
cork cambium
secondary phloem
cambium
secondary xylem
primary xylem
annual ring

Labels for figure C:
epidermis
cortex
endodermis
pericycle
crushed primary phloem
secondary phloem
cambium
secondary xylem
primary xylem

Labels for figure B:
epidermis
cortex
endodermis
pericycle
primary phloem
cambium forming
primary xylem
lateral root
root cap
old root hair

Labels for figure A:
epidermis
cortex
endodermis
pericycle
primary phloem
primary xylem
root hair

region of maturation

region of enlargement

apical meristem
meristem
root cap

FIGURE 7-12. *Anatomical structure of the root.* [*From Muller*, Botany: A Functional Approach. *Reprinted with permission of The Macmillan Company. Copyright © 1963 by Walter H. Muller.*]

of herbaceous and woody stems are shown in Figure 7-13.

The upright growth of plants that have one active growing point and a rigid stem is considered the normal, and our descriptive terms are used to differentiate other growth patterns. Typical shrubby or bushy growth is due to the absence of a main trunk or central leader. Plants with this pattern of growth are characterized by a number of erect or semierect stems, none of which dominates. The distinguishing feature is form rather than size. Similarly, slender and flexible stems that cannot support themselves in an erect position are known as vines or creepers. Vines will trail unless an independent support is made available to permit them to attach and grow upright. Some vines are herbaceous (morning glory, pea), and others are woody (grape), in which case they are called lianas.

BUDS   The stem is divided into mature and actively growing regions in which growth and differentiation take place. The "embryonic stem" is referred to as a bud. All buds do not grow actively; many assume an arrested development, or become dormant, but are nevertheless potential sources of further growth. Overwintering is one of the main functions of buds. Although buds may be embedded in the stem tissues, and relatively inconspicuous, they may become quite structured. The form, structure, and arrangement of buds prove to be useful guides in describing woody plants, even when the leaves are absent, as in winter. Growth may originate from a single terminal bud or from lateral buds, which occur in the leaf axis. In addition, buds may be formed in internodal regions of the stem, leaves, or even roots, often as a result of injury. These are called **adventitious** buds.

Buds may produce leaves, flowers, or both, and are referred to as, leaf, flower and mixed buds, respectively. When more than one bud is present at a leaf axis, all but the central or basal bud are called accessory buds. The arrangement or topology of buds on a stem (their phyllotaxy) corresponds to the leaf arrangement. If leaves are opposed to each other at the same level, the leaf (and bud) arrangement is said to be opposite or whorled. When single leaves grow at different levels, they are arranged in a spiral and are said to be alternate. The spiral pattern of leaf arrangement is varied and is expressed as a fraction ($\frac{1}{2}$, $\frac{1}{3}$, $\frac{2}{5}$, $\frac{3}{8}$), where the numerator is the number of turns to get to a leaf directly above another and the denominator is the number of buds passed. Phyllotaxy has some taxonomic significance, since it is often the same throughout a genus and often even applies to a whole family.

STEM MODIFICATIONS The stem may be greatly modified from a basically upright cylindrical structure (Fig. 7-14). Some of these alterations may appear quite bizarre; yet upon close analysis (particularly of their ontogony) these modifications can be shown to be basically stem-like in structure; that is, they have nodes, leaves, or similar scale-like structures, and they function in transport or storage. Stem modification can be divided into aboveground forms (crowns, stolons, spurs) and belowground forms (bulbs, corms, rhizomes, tubers). Since many stem modifications contain large amounts of stored food, they are especially significant in propagation, and, in potatoes, are important as a source of food.

Aboveground modifications. The crown of a plant is, in general, that portion just above and just below ground level. (The forester's term "crown" refers to the branched top of a tree.) This portion of the plant may become greatly enlarged, as in baldcypress (*Taxodium distichum*). Crowns may be thought of as "compressed" stems. The structure of the strawberry crown can be clearly seen by artificially elongating it through treatment with gibberellins. Leaves and flowers arise from the crown by buds, as they do in ordinary stems. Fleshy buds from crowns may produce new plants, referred to as crown divisions. The crown may be modified into a food storage organ, as it is in asparagus.

Short, many-sided, horizontal branches that grow out of the crown and bear fleshy buds or leafy rosettes are referred to as offsets, slips, suckers, pips, and so on. These stem modifications, which can be collectively termed **offshoots,** are important for propagation.

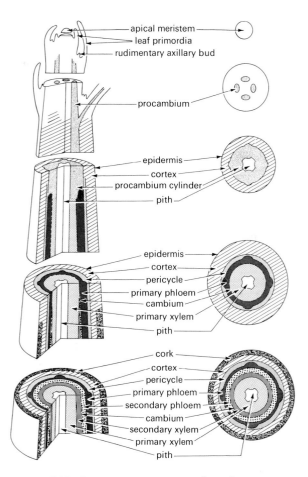

FIGURE 7-13. *Diagrammatic sections through a stem.* [*After Holman and Robbins,* A Textbook of General Botany. *New York: Wiley, 1939.*]

Stolons are stems that grow horizontally along the ground. A runner is a stolon with long internodes originating at the base or crown of the plant. At some of the nodes, roots and shoots develop. A well-known example of runners is found in the strawberry.

Spurs are stems of woody plants whose longitudinal growth is greatly restricted. They are characterized by greatly shortened internodes, and appear laterally on branches. In mature fruit trees, such as apple, pear and quince, flowering is largely confined to spurs. Spurs are not irrevocably static,

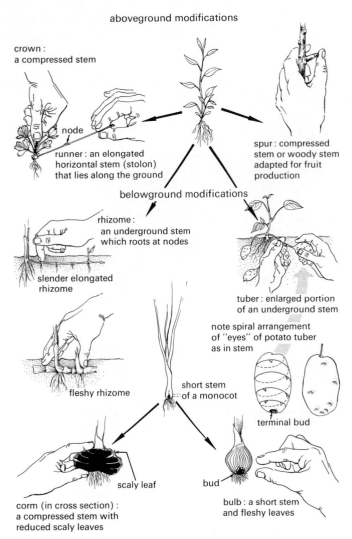

aboveground modifications

crown :
a compressed stem

node

runner : an elongated
horizontal stem (stolon)
that lies along the ground

spur : compressed
stem or woody stem
adapted for fruit
production

belowground modifications

rhizome :
an underground stem
which roots at nodes

slender elongated
rhizome

fleshy rhizome

tuber : enlarged portion
of an underground stem

note spiral arrangement
of "eyes" of potato tuber
as in stem

short stem
of a monocot

terminal bud

scaly leaf

bud

bulb : a short stem
and fleshy leaves

corm (in cross section) :
a compressed stem with
reduced scaly leaves

FIGURE 7-14. *Stem modifications. Note that all have nodes and leaf-like structures.*

and may revert to normal stem growth even after many years of fruiting.

Belowground stem modifications. **Bulbs** appear as compressed modifications of the shoot, and consist of a short, flattened, or disc-shaped stem surrounded by fleshy, leaf-like structures called scales. They may enclose shoot or flower buds. Bulbs and corms are found only in some monocotyledonous plants. The scales, filled with stored food, may be continuous, and form a series of concentric layers surrounding a growing point, as in the onion or tulip (tunicate bulbs), or may form a more or less random attachment to a small portion of stem, as in the Easter lily (scaly bulbs). Bulbs commonly grow under the ground or at ground level, although bulb-like structures (bulbils) may be formed on aerial stems, as they are in some lily species, or even in association with flower parts, as in the onion.

Corms are short, fleshy, underground stems having few nodes. The corm is stem; the few rudimentary leaves are nonfleshy. The gladiolus and crocus are propagated by corms.

Rhizomes are horizontal, underground stems. They may be compressed and fleshy, as in Iris, or slender with elongated internodes, as in turfgrasses (for example, Bermuda, bluegrass). Normally, roots and shoots develop from the nodal regions. Such weeds as quackgrass and Canadian thistle are particularly insidious because they spread so rapidly, owing to their natural propagation by rhizomes.

Tubers are greatly enlarged, fleshy portions of rhizomes. They are typically noncylindrical. (The word "tuber" is derived from a Latin word meaning lump.) The edible portion of the potato is a tuber. The "eyes" arranged in a spiral around the tuber are the buds. Each eye consists of a rudimentary leaf scar and a depressed cluster of buds.

THE LEAF   Leaves are the principal photosynthetic organs of higher plants. The leaf blade is usually a flat appendage of the stem arranged in such a way as to present a large surface for the efficient absorption of light energy, and is usually attached to the stem by a stalk or petiole. Leaf-like outgrowths of the petioles, known as stipules, may be present. Leaves may contain many secretory structures such as glands or ducts. In many plants, particularly those native to the humid tropics, water can be discharged through small openings in the epidermis, called hydathodes; they are often situated at the leaf tip.

The anatomical structure of the leaf is shown in Figure 7-15. Note that its vascular system consists of a branching network of veins. The veining is typically net-like in dicots and parallel in monocots. The leaf blade, although commonly bilaterally symmetrical, is not radially symmetrical, since it has distinct upper and lower sides. Beneath the upper epidermal layer, which is often characterized by heavy deposits of cutin, lie series of elongated, closely packed palisade cells, which are particularly rich in chloroplasts. The irregular, loose arrangement of cells beneath the palisade cells results in a sponge-like region (spongy mesophyll) that provides the air space necessary for gaseous exchange in photosynthesis and transpiration. Depending on the species, stomata occur on the lower, upper, or both upper and lower epidermis.

Leaves of plants vary in shape from the flat, thin blades just described to the needle-like leaves of conifers or the stem-like fleshy structures found in cactus. The tendrils of peas are modified leaflets attached to the leaf. Leaves are the edible portions of many plants, such as lettuce, cabbage, and spinach, and are the most prominent feature of some plants, such as grasses.

THE FLOWER   The flower shows great variety in structure, composition, and size. The principal flower parts (Fig. 7-16) are as follows:

Sepals (collectively, the calyx) enclose the flower when it is in bud. They are usually small, green, leaf-like structures below the petals.

Petals (collectively, the corolla) are the most conspicuous parts of most flowers. They may be highly colored, but are usually not green. The extremely large, showy flowers of many cultivated ornamentals are the result of rigorous selection for this character.

Stamens, the "male" reproductive structures, are composed of pollen-bearing anthers supported by a filament. When the pollen is mature, it is discharged through pores or through the ruptured anther wall. Nectaries, which produce a viscous, sugary substance, and often a perfume, are most commonly found at the base of stamens.

Pistils (made up of one or more carpels), the "female" reproductive structures, are usually modified into an ovule-bearing base (or ovary)

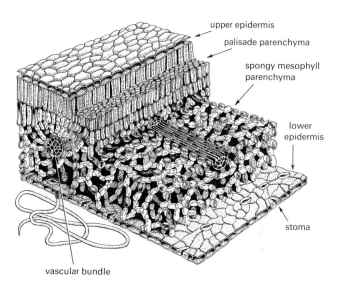

FIGURE 7-15. *Structure of a leaf.* [*After Eames and MacDaniels*, An Introduction to Plant Anatomy, *New York: McGraw-Hill, 1947.*]

supporting an elongated region (or style) whose expanded tip (or surface) is called the stigma. The ovule develops into the seed. The mature ovary becomes the fruit.

The petals and sepals of the flower, as well as the reproductive parts (the stamens and pistils) are essentially modified leaves. That the stamens are modified leaves can be clearly seen in the stamenoids or "extra" petals of the cultivated rose. These flower parts are borne on an enlarged portion of the flower-supporting stem, called the receptacle.

Flowers composed of sepals, petals, stamens, and pistils are referred to as complete. Incomplete flowers lack one or more of these parts. For example, they may lack stamens (pistillate, or "female" flowers) or pistils (staminate, or "male" flowers). Those that contain both stamens and pistils (perfect, bisexual, or hermaphroditic flowers) are referred to as perfect even if they lack calyx or corolla and are thus incomplete.

Similarly, plants are referred to as staminate, pistillate, or perfect on the basis of the type of flowers they bear. When both staminate and pistil-

130

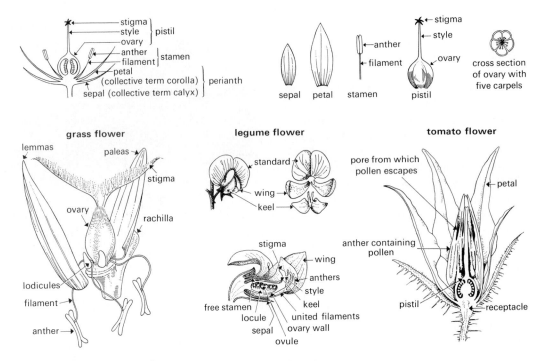

FIGURE 7-16. *Structure of the flower.*

late flowers occur on the same plant, as in maize, the sex type is monoecious. Species in which staminate and pistillate flowers are borne on different plants are dioecious (date palm, papaya, spinach, asparagus, hemp). Other combinations of flower types also occur. For example, muskmelons have perfect and staminate flowers on the same plant; this sex type is referred to as andromonoecious.

There are many ways in which the flowers are arranged. The term **inflorescence** refers to a flower cluster. Some of the more common types of inflorescence are diagrammed in Figure 7-17.

THE FRUIT   The botanical term "fruit" refers to the mature ovary and other flower parts associated with it. Thus, it may include the receptacle as well as withered remnants of the petals, sepals, stamens, and stylar portions of the pistil. It would also include any seeds contained in the ovary.

The structure of the fruit is related to the structure of the flower. Fruits are classified according to the number of ovaries incorporated in the structure (for example, as simple, aggregate, or multiple fruits), the nature and structure of the ovary wall, whether they split when ripe (dehisce), and the way in which the seed is attached to the ovary.

Simple fruits.   The majority of flowering plants have fruits composed of a single ovary, and are referred to as simple fruits. In the mature fruit (when the enclosed seed is fully developed), the ovary wall may be fleshy (composed of large portions of living succulent parenchyma) or dry (made up of nonliving sclerenchyma cells with lignified or suberized walls).

The ovary wall, or pericarp, is composed of three distinct layers. From outer to inner layer, these are the exocarp, mesocarp, and endocarp. When the entire pericarp of simple fruits is fleshy, the fruit is referred to as a berry. The tomato, grape, and pepper are berry fruits. The muskmelon is a berry (specifically, a pepo) with a hard rind made up of exocarp and receptacle tissue. Citrus fruit is also a berry, called hesperidium, in which the rind is made up of exocarp and mesocarp; the

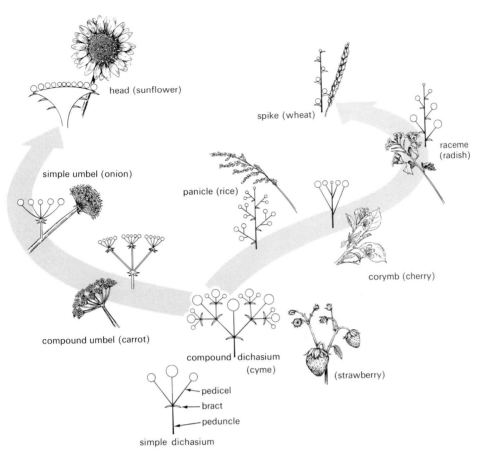

head (sunflower)

spike (wheat)

raceme (radish)

simple umbel (onion)

panicle (rice)

corymb (cherry)

compound umbel (carrot)

compound dichasium (cyme)

(strawberry)

pedicel

bract

peduncle

simple dichasium

FIGURE 7-17. *Common infloresences.*

"edible" juicy portion is endocarp.

Simple fleshy fruits having a stony endocarp are known as drupe, or stone, fruits (peach, cherry, plum, olive). The skin of these fruits is the exocarp; the fleshy, edible portion is the mesocarp. Simple fleshy fruits in which the inner portion of the pericarp forms a dry, paper-like "core" are known as pomes (apple, pear, quince).

Dry fruits may be either **dehiscent** or **indehiscent.** Dehiscent fruits usually have many seeds and are classified by the manner in which dehiscence occurs and the number of carpels from which they develop. For example, the pod (legume) of leguminous plants, such as the pea or bean, develops from a single carpel and separates by both sutures. Among the other types are the silique (crucifers), follicle (larkspur), and capsule (Jimson weed). Indehiscent fruits generally bear one seed. When the pericarp lies close to, but separated from the seed, the fruit is called an achene (sunflower). When the pericarp becomes fused to the seed, the fruit is called a caryopsis. Caryopsis is the type of fruit found in most grasses, including maize and wheat. Nuts are characterized by a hardened pericarp. Other indehiscent fruits include the samara (maple), in which the pericarp forms a winged appendage, and the schizocarp (carrot), in which the fruit divides into two or more one-seeded indehiscent parts. These are diagrammed in Figure 7-18.

Aggregate fruits. Aggregate fruits are derived from flowers that have many pistils on a common receptacle. The individual fruits of the aggregate are drupes in the blackberry and achenes in the

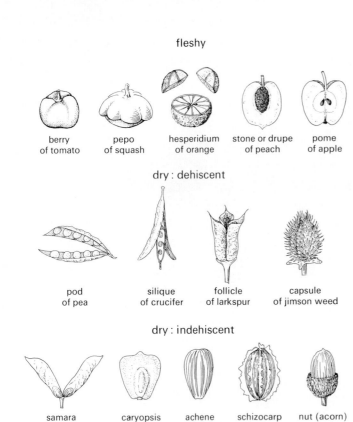

fleshy

berry
of tomato

pepo
of squash

hesperidium
of orange

stone or drupe
of peach

pome
of apple

dry : dehiscent

pod
of pea

silique
of crucifer

follicle
of larkspur

capsule
of jimson weed

dry : indehiscent

samara
of maple

caryopsis
of corn

achene
of sunflower

schizocarp
of carrot

nut (acorn)
of oak

FIGURE 7-18. *Various types of simple fruits.* [*After Holman and Robbins,* A Textbook of General Botany. *New York: Wiley, 1939.*]

strawberry. The fleshy, edible portion of the strawberry is the receptacle.

Multiple fruits. The multiple fruit is derived from many separate but closely clustered flowers. Familiar examples of multiple fruits are the pineapple, fig, and mulberry. The beet "seed" is a multiple fruit.

THE SEED   A seed is in essence a miniature plant (the embryo) in an arrested state of development. Most seeds contain a built-in food supply (the orchid seed is an exception). Structurally, the true, or botanical, seed is a mature ovary. The agricultural definition of a seed refers to the unit that is planted, regardless of its structure (Fig. 7-19), and includes true seeds (alfalfa, cucumber); one-seeded dry indehiscent fruits, such as the caryopsis

in grasses and the achene in lettuce; and multiple-seeded fruits, in which additional accessory structures are present, as in beet and New Zealand spinach.

The seed contains a miniature plant, or embryo, which usually develops from the union of generative cells or gametes. The details of fertilization and seed development will be discussed in Chapter 9, Reproduction and Propagation.

The fully developed embryo commonly consists of a hypocotyl-root axis, a transition zone between the rudimentary shoot and root. On the upper end it contains one or more specialized seed leaves called **cotyledons.** (In grasses the shield-like cotyledon is called the **scutellum.**) The embryonic shoot bud is called the **plumule** (Latin for little feather). On the lower end of the axis there may be an embryonic root called the **radicle.** In grasses a protective sheath, or **coleoptile,** covers the plumule and a similar sheath, the **coleorhiza,** covers the radicle.

The food stored in seeds is present as carbohydrates, fats, and proteins, one of these forms usually predominating. Seeds are thus a rich source of food as well as of fats and oils for industrial purposes. The reserves in cereals and legumes provide a large share of the world's food supply for man and animals. In some seeds this stored food is derived from a tissue called the endosperm, which is formed as a result of the overall fertilization process. In mature seeds of some plants the endosperm produces a specialized region, as in maize; in others it is absorbed by the developing embryo, in which case the cotyledons serve as the food-storage organ, as in beans and walnuts.

Seeds vary greatly in size, form, and shape. Most plants can be identified by their seeds alone. In addition, great variation exists within some species. Examples of such variation are the presence or absence of spines (spinach), color variation (beans), and differences in the chemical composition of stored food (sugary versus starchy maize).

Seed germination refers to the change from arrested development to active growth. The subsequent seedling stage, the interval during which the young plant becomes dependent on its own food-manufacturing structures, is diagrammed in Figure 7-20.

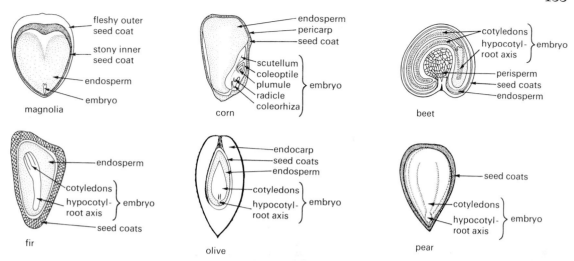

FIGURE 7-19. *Structure of seeds and one-seeded fruits.* [*After Hartmann and Kester,* Plant Propagation. *Englewood Cliffs: Prentice-Hall, 1959.*]

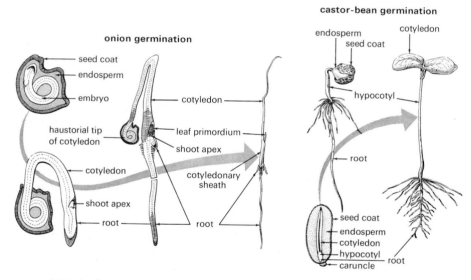

FIGURE 7-20. *Seed germination and seedling morphology in onion (a monocot) and castor-bean (a dicot).* [*From Foster and Gifford,* Comparative Morphology of Vascular Plants. *San Francisco: W. H. Freeman and Company,* © *1959.*]

Selected References

Eames, A. J., and L. H. MacDaniels. *An Introduction to Plant Anatomy.* McGraw-Hill, New York, 1947. (A well-known textbook on the anatomy of vascular plants.)

Esau, K. *Plant Anatomy* (2nd ed.). Wiley, New York, 1965. (An authoritative treatise and advanced textbook on the anatomy of seed plants.)

Heyward, H. E. *The Structure of Economic Plants.* Macmillan, New York, 1938. (This book is espe-

cially noted for its treatment of particular species of important crop families, including maize, wheat, onion, hemp, beet, radish, alfalfa, pea, flax, cotton, celery, sweet-potato, potato, tomato, squash, and lettuce.)

Weier, T. E., C. R. Stocking, and M. G. Barbour. *Botany: An Introduction to Plant Biology* (4th ed.). Wiley, New York, 1970. (An excellent introductory textbook.)

CHAPTER 8

Growth and Development

The living state of organisms is characterized by systematic increases in bulk and complexity. These events can be discussed in terms of the interrelated processes of growth and development. **Growth,** in a restricted sense, refers to an irreversible increase in size, reflecting a net increase in protoplasm. **Development** infers differentiation, a higher order of change that involves anatomical and physiological specialization and organization.

PLANT GROWTH PROCESSES

Growth in plants can be measured in many ways; the most straightforward is by the increase in dry weight. This increase in weight of an organ (such as a fruit) or an entire plant may result from increases in either the number or the size of cells, or both. The growth of multicellular organisms is based on increases in cell number provided by cell division. The mechanical aspects of cell division will be discussed in more detail in Chapter 9.

The increase in protoplasm is brought about through a series of events in which water, carbon dioxide, and inorganic salts are transformed into living material. This involves the production of carbohydrates (**photosynthesis**), the elaboration and degradation of complex molecules (**metabolism**), and the uptake and movement of water and nutrients (**absorption** and **translocation**). The required chemical energy is provided by **respiration.** These physiological processes are functions of individual cells and of multicellular organisms; although these processes are interrelated, it is useful to classify them into separate systems.

Photosynthesis

Photosynthesis—the process by which plants use light energy to convert carbon dioxide and water into energy-rich organic compounds—is one of the most significant of the life processes. All of the organic matter in living things is ultimately provided through photosynthesis. The reactions that constitute this process take place in the chlorophyll-containing plastids—the complete structural and biochemical units of photosynthesis—and are mediated by certain enzymes. The process has been stated in the form of a chemical reaction on page 29. The series of photosynthetic reactions can be grouped into a light phase (the reactions that require light) and a dark phase (the reactions that do not require light).

The first step in this series is independent of temperature, and consists in the trapping of light energy, which accomplishes the cleavage (photolysis) of the water molecule into hydrogen and oxygen. The oxygen is released as gaseous molecular oxygen, and the hydrogen is trapped by a hydrogen acceptor, nicotinamide adenine dinucleotide phosphate (NADP). Thus, the liberation of oxygen in photosynthesis is independent of the synthesis of carbohydrates. This step has been referred to as the **Hill reaction** (the NADP serving as the natural Hill reagent). The trapping of light energy by the conversion of adenosine diphosphate (ADP) to adenosine triphosphate (ATP) occurs in a process known as **photophosphorylation.** The combination of Hill reaction and phosphorylation is known as the **light phase** of photosynthesis.

The conversion of light energy to chemical energy is achieved by the formation of energy carriers such as ATP and NADP. ATP is involved in energy transfers in many vital processes of the cell. It is formed from ADP through the addition of a third phosphate group. The captured light energy stored in the third phosphate bond becomes available following the conversion of ATP back to ADP. NADP accepts electrons and hydrogen atoms produced in photosynthesis and transfers these to other compounds. Indeed the crucial reactions of photosynthesis are the conversion of ADP to ATP and the reduction of NADP to $NADPH_2$.

The **dark phase** of photosynthesis (also called the **Blackman reaction**) is greatly affected by temperature and has been shown to be independent of light. Essentially the hydrogen atoms from water are transferred by the hydrogen-carrying acceptor ($NADPH_2$) to a low-energy organic acid to produce, with the help of the energy of the ATP, a "carbohydrate" of higher energy, from which sugars are formed.

This reduction reaction—that is, the addition of electrons and hydrogen atoms to carbon dioxide—results in the formation of sugar units. Although

the precise pathways by which carbon dioxide is synthesized to sugar are not completely known, the main routes have been worked out. A substance formed early in this synthesis has been identified as a 3-carbon phosphorus-containing compound, phosphoglyceric acid, two molecules of which eventually give rise to a single 6-carbon sugar and, finally, to starch grains in the chloroplasts, as shown in Figure 8-1.

Photosynthetic efficiency (the rate of net photosynthesis) is equal to the rate of "gross" photosynthesis less any photosynthate that may be lost through respiration:

Net photosynthesis
= gross photosynthesis − respiration

Recent studies have indicated that there are two distinct categories of plants in regard to net photosynthetic rate (Table 8-1). One group, the most typical, fixes carbon from ribulose diphosphate into a 3-carbon acid (phosphoglyceric acid) utilizing the enzyme ribulose diphosphate carboxylase. Thus, this type of photosynthesis is referred to as the C_3 type. The second group of plants, which have a much higher rate of net photosynthesis,

Table 8-1 COMPARATIVE RATES OF MAXIMUM PHOTOSYNTHESIS.

TYPE OF PLANT	MAXIMUM PHOTOSYNTHESIS (MILLIGRAMS CO_2 FIXED PER SQUARE DECIMETER OF LEAF SURFACE PER HOUR)
C_3 photosynthesis	
Slow growing perennials (desert spp, orchids)	1–10
Evergreen woody plants	5–15
Deciduous woody plants	15–30
Rapidly growing annuals (wheat, soybeans, sugar beet, sunflower)	20–50
C_4 photosynthesis	
Tropical grasses, sugar cane, corn, *Amaranthus, Atriplex*	50–90

Source: Jarvis and Jarvis, *Plant Physiology* 17:645–666, 1964.

136

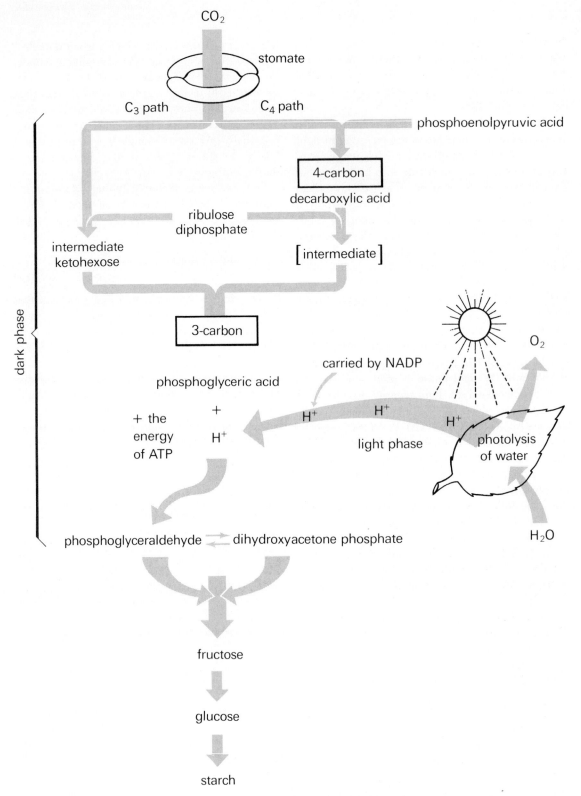

FIGURE 8-1. *Photosynthesis. Note the C_3 and C_4 pathways.*

include tropical grasses, sugar cane, and species of *Atriplex* and *Amaranthus* (these last two genera include many fast-growing weeds). In this group, CO_2 is first fixed into phosphoenolpyruvate to yield 4-carbon acids (such as oxaloacetic, malic, and aspartis acid) and involves the enzyme phosphoenolpyruvate carboxylase; hence it is known as the **C_4** type of photosynthesis. C_4 plants contain enzymes for this transformation that are not present in C_3 plants.

Although the C_4 pathway requires slightly more energy than the C_3 pathway, this requirement is offset by other features. The most important advantage is the apparent absence of a type of light-dependent respiration (hence the term **photorespiration**) which is linked to the photosynthetic cycle in C_3 plants. Photorespiration appears to lower the efficiency of CO_2 assimilation. The process is different from normal ("dark") respiration (see the discussion under *respiration* below), which is independent of light. The occurrence of photorespiration was difficult to detect because it is hard to distinguish the CO_2 fixed in photosynthesis from the CO_2 given off by photorespiration. However, it is now established that the greater efficiency of C_4 plants is due to the absence of photorespiration.

The difference in photorespiration between the C_3 and C_4 systems can be demonstrated by enclosing plants in a sealed, illuminated container. As photosynthesis increases, CO_2 is taken up by the plant and fixed, and the effect is to lower the amount normally present in the air. For C_3 plants, a steady state of CO_2 concentration in the air (known as the **compensation point**) is reached when there remains about 50 ppm at 25°C (77°F). This amount is as large as it is because photorespiration releases CO_2. With a C_4 plant such as corn, which lacks photorespiration, the amount of CO_2 in such a sealed container diminishes to less than 10 ppm. A soybean plant (C_3) enclosed with a corn plant (C_4) will die because the corn exhausts the CO_2.

The rate of photorespiration increases with temperature faster than gross photosynthesis. Thus many C_3 plants are nonproductive at high temperatures (25–35°C; 77–95°F) whereas C_4 plants such as tropical grasses increase in productivity at these higher temperatures.

Other specific adaptations are associated with the C_4 system. These include enhanced translocation of photosynthetic products, specialized leaf anatomy and chloroplast structure, and greater ratio of chlorophyll *a* to chlorophyll *b*.

Respiration

Energy is required to run the machinery of the cell. The energy incorporated in the chemical bonds of the sugars formed from photosynthesis cannot, of course, be harnessed by the cell from high-temperature combustion, but must be provided at low and constant temperatures in delicately controlled reactions. Respiration, the process of obtaining energy from organic material, is accomplished with great efficiency in the cell. It is in a superficial sense the reverse process of photosynthesis:

$$C_6H_{12}O_6 + 6O_2 \rightarrow 6H_2O + 6CO_2 + \text{energy}.$$

The captured energy of light is released from the low-temperature oxidation (removal of hydrogen) of sugars. Although a small part does appear as heat, the useful energy is channeled into chemical work, initially as high-energy phosphates, and later in the synthesis of organic materials required in growth and development (Fig. 8-2).

The biologic combustion of sugar is accomplished through an extremely complicated series of reactions involving many specific enzyme systems and energy carriers. The steps from 6-carbon sugar to carbon dioxide involve the transformation of phosphorylated sugars to 3-carbon pyruvic acid ($CH_3COCOOH$). This step, known as **glycolysis,** is common to many organisms. Organisms that do not require oxygen (**anaerobes**) transform pyruvic acid to alcohol or lactic acid. Organisms that do require oxygen (**aerobes**) transform pyruvic acid to water and carbon dioxide. This involves the participation of a number of plant acids in the cyclic series of steps known as the citric acid cycle, or (in honor of its discoverer) **Krebs cycle.**

The rate of respiration depends on many factors. It is highest in rapidly growing tissues and lowest in dormant tissues. The rate of respiration is greatly influenced by temperature, and approximately doubles for each 10°C (18°F) rise over a range of 4°–36°C (40–97°F). Other factors influence the

138

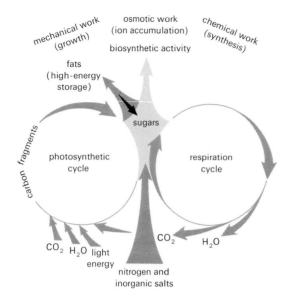

FIGURE 8-2. *Energy cycle in green plants.*

respiration rate, including the availability of oxygen and carbohydrates and the age and condition of cells and tissues. Respiration is a common feature of all living material, and consequently the detection of evolved carbon dioxide is often utilized as a test for life.

Metabolism

All of the various materials produced in the plant are ultimately derived from the carbon fragments produced by photosynthesis and from the inorganic nutrients and water absorbed from the soil (Fig. 8-3). The synthesis (anabolism) and degradation (catabolism) of these organic materials is known as **metabolism.** The degradation of sugars and fats and the release of energy in respiration are examples of catabolic metabolism. The step-by-step elucidation of the pathways in plant metabolism comprises some of the brightest chapters in biochemistry.

Plants are cultivated for the complex molecules they synthesize; carbohydrates, proteins, and lipids are the compounds of most concern because they are the major constituents of food (see Chapter 3). The metabolism of carbohydrates has already been

dealt with briefly in the various discussions of photosynthesis. Proteins are second in abundance to carbohydrates as plant constituents (Table 8-2). An understanding of the interrelationships between carbohydrate metabolism and nitrogen metabolism is necessary to the comprehension of plant development.

Green plants can survive in an inorganic world. They are autotrophic for nitrogen as well as carbon. Except for the nitrogen-fixing legumes and a few other plants that depend on their bacterial partners, plants are dependent on the nitrate and ammonium ions in the soil solution as a source of nitrogen. Since microorganisms in the soil rapidly oxidize the reduced forms of nitrogen, the most common source of nitrogen taken up by plants is the nitrate ion, NO_3^-. Before combining with organic acids to form amino acids, nitrate is reduced through a series of enzymatic reactions. The reduced nitrogen groups NH_2 and NH_3 combine with the carbon frameworks formed during the oxidation of sugars to form amino acids.[*]

Amino acids join to form proteins through a complex series of reactions regulated by the nucleic acids present in the cell. The primary sites of protein synthesis in plants are the tissues in which new cells are being formed, such as tips of stems and roots, buds, cambium, and developing storage organs. The green leaf is also an important site of protein synthesis, since both the carbon frameworks and the inorganic nitrogen necessary for the formation of amino acids are readily available. Leaf protein is in a continuous state of turnover. When leaves are placed in the dark, are excised from the plant, or reach the state of senescence, protein breaks down and soluble nitrogen compounds are formed that can be reincorporated into amino acids in other leaves or organs of the plant.

Because nitrogen metabolism and carbohydrate metabolism of the plant are closely linked, both

[*] The ammonia (NH_3) initially combines with the organic acid α-ketoglutaric acid to form the amino acid glutamic acid, an energy-requiring process. Other amino acids are constructed via the transfer of the amino group (NH_2) from this amino acid to other organic acids from the metabolic pool (a process known as transamination).

processes are markedly affected by gross changes in the environment of the plant, such as heavy nitrogen fertilization, reduced light intensity, or drought; the effects on these processes are reflected in altered patterns of growth and development.

A great number of organic compounds, many of great complexity, are formed in metabolic processes. (Hundreds of different compounds contribute to the flavor of coffee alone!) Some of these metabolites are commonly referred to as essential oils, pigments, vitamins, and so forth. These terms, however, are not specific in the chemical sense and often refer to mixtures of materials, many unrelated. For example, pigments—substances that preferentially absorb certain wavelengths of visible light—differ structurally and chemically. Not all substances produced in the various metabolic pathways are, so far as is known, essential to the

Table 8-2   NUTRIENT CONSTITUENTS OF CERTAIN PLANT PARTS.

		NUTRIENTS AS PERCENT OF DRY MATTER			
PLANT PART	PERCENT DRY MATTER	CRUDE PROTEIN[1]	FAT	CARBO-HYDRATE	MINERAL
Cabbage leaf	15.8	16.5	2.5	62.0	19.0
Beet root	16.4	9.8	0.6	82.9	6.7
Potato tuber	21.2	10.0	0.5	84.0	5.2
Wheat seed	89.5	14.7	2.1	81.0	2.1
Soybean seed	90.0	42.1	20.0	32.8	5.1

[1] Includes protein and nitrogen-containing compounds.

plant. Such substances are characteristically produced only by certain plants or plant groups. Many of them (for example, rubber, menthol) are highly prized and constitute the prime reason for the

FIGURE 8-3. *Metabolic pathways in green plants.*

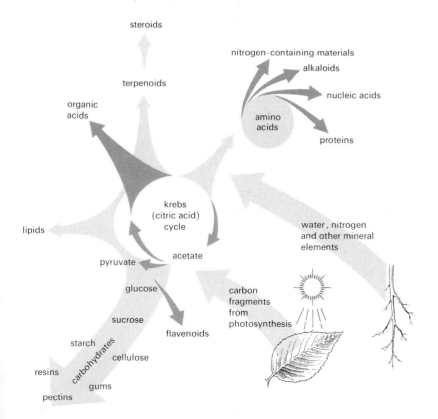

cultivation of the species from which they are derived.

The following brief sections deal with the various categories of metabolic substances other than carbohydrates, proteins, and lipids.

ORGANIC ACIDS AND ALCOHOLS   Plants manufacture a great variety of organic acids. A number participate in metabolic cycles like those involved in the biological combustion of sugar to carbon dioxide. Organic acids are found in the vascular sap and commonly accumulate in certain plant organs, usually the fruit (citric acid in lemons, malic acid in apples).

Alcohols occur only in very low concentrations as free alcohols, and are generally found combined with organic acids as esters. Fats are esters of alcohol and fatty acids. The characteristic fruit odors and flavors are due to a combination of the volatile organic acids, esters, and other compounds such as ketones and aldehydes.

AROMATIC COMPOUNDS   Aromatic compounds, in contrast to aliphatic or "straight-chain" carbon compounds, have structures containing at least one benzene ring. Many, but not all, of these compounds have very characteristic odors, hence the name "aromatic." The simple phenolic compounds (those containing one benzene ring), such as vanillin and methyl salicylate, are responsible for the odors of vanilla and wintergreen.

Vanillin   Methyl salicylate

The volatile phenolic compounds (specifically, the phenylpropane derivatives) are responsible for the

Phenylpropane skeleton

characteristic flavors and odors of cinnamon, clover, and parsley.

Two commercially important plant constituents, lignins and tannins, are mixtures of complex aromatic and carbohydrate materials. Lignins have already been discussed briefly as constituents of the cell wall (p. 115). Tannins have the property of precipitating proteins and are used to transform animal hides into leather. Tannins are extracted from a number of plants—the word "tan" originally referred to the bark of the oak tree. Unripe persimmons are high in tannin, as any person unwise enough to sample one will discover. The extreme astringent taste is due to protein precipitation on the tongue.

Flavonoids are a group of aromatic compounds characterized by two substituted benzene rings connected by a 3-carbon chain and an oxygen bridge.

Flavonoid skeleton

The flavonoids are commonly attached to sugars to form glycosides. Among the flavonoids are many common pigments, including the anthocyanins and flavones. Anthocyanins are responsible for the reds and blues of many fruits and flowers; flavones, for the yellow of lemon. The insecticide rotenone is related to the flavonoids.

TERPENOIDS AND STEROIDS   Such compounds as essential oils, steroids, alkaloids, and various pigments are related in the chemical sense to fused units of the 5-carbon substance isoprene. Two units of isoprene may form a terpene unit, the basic structure of a number of important plant constituents called terpenoids. (See next page.)

The essential oils are mixtures of volatile, highly aromatic substances that have distinctive odors and flavors. They are formed in specialized glands, ducts, or cells in various plant parts. Although called "oils," these compounds are not lipids. An example is menthol, which is the major constituent of both peppermint oil and turpentine. The odors, flavors, and other properties of essential oils make

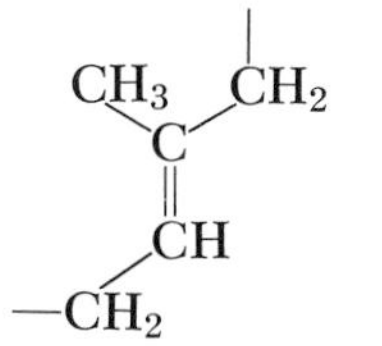

CH₃ CH₂
C
‖
CH
—CH₂

Isoprene skeleton

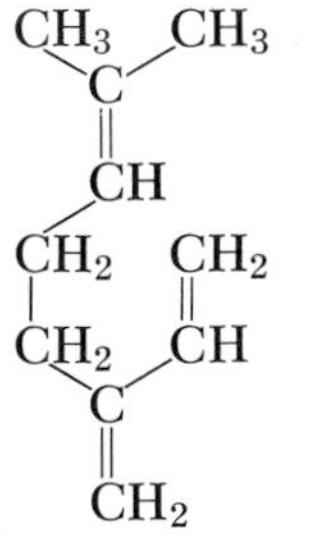

CH₃ CH₃
C
‖
CH
CH₂ CH₂
CH₂ CH
C
‖
CH₂

Myrcene
An open-chain terpene

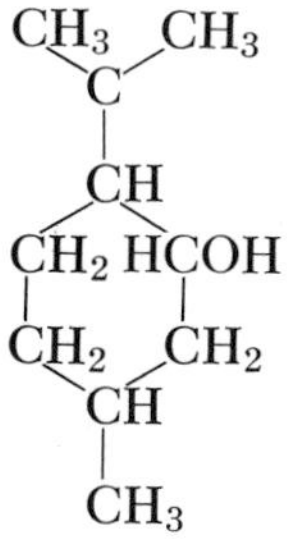

CH₃ CH₃
C
‖
CH
CH₂ HCOH
CH₂ CH₂
CH
CH₃

Menthol
A cyclic terpene

them economically important for a variety of uses.

Resins—gummy exudates of many plants—are terpenoids consisting of three to six isoprene units. Resins are often associated with essential oils (and have the ability to harden as the oils evaporate). Resin synthesis occurs near wounded tissue, and the resins serve to retard water loss and entry of microorganisms. The ability of plants to produce resins is lowered if the general vigor of the plant is poor.

Steroids are complex cyclic terpene compounds composed of eight units of isoprene (tetraterpenoids). Although their function in plants is not well understood, it is known that many of them have important metabolic effects in animals. Cortisone, sex hormones, and vitamin D are steroids.

The carotenoids are complex tetraterpenes— yellow to red pigments that occur in many different kinds of tissue. Common members of this group include carotene (composed exclusively of carbon and hydrogen) and xanthophyll (composed of oxygen in addition to carbon and hydrogen). The most widespread carotenoid is the orange pigment β-carotene, which when split in the digestive tract of animals gives rise to two molecules of vitamin A.

Rubber is a high-molecular-weight terpenoid containing 3,000 to 6,000 isoprene units. Although it is produced in many dicotyledonous plants, only a few produce enough for commercial purposes. The most important is the rubber tree, *Hevea brasiliensis*. The rubber is extracted from latex, a milky, sticky liquid exuded by glandular structures associated with the phloem.

NON-PROTEIN NITROGEN COMPOUNDS Nitrogen is found in a wide variety of compounds in addition to amino acids and proteins. These include the nitrogen bases and a heterogeneous group of compounds called alkaloids. Heterocyclic rings containing both carbon and nitrogen are a common structural form (Fig. 8-4).

The nucleoproteins, the source of the genetic material, are composed of proteins and nucleic acids. Nucleic acids are high-molecular-weight polymers of nucleotides, each formed from three constituents: a sugar (specifically ribose or deoxyribose), phosphoric acid, and a nitrogen base having the structure of either a purine or pyrimidine ring. In deoxyribonucleic acid (DNA), the genetic material in chromosomes, there are

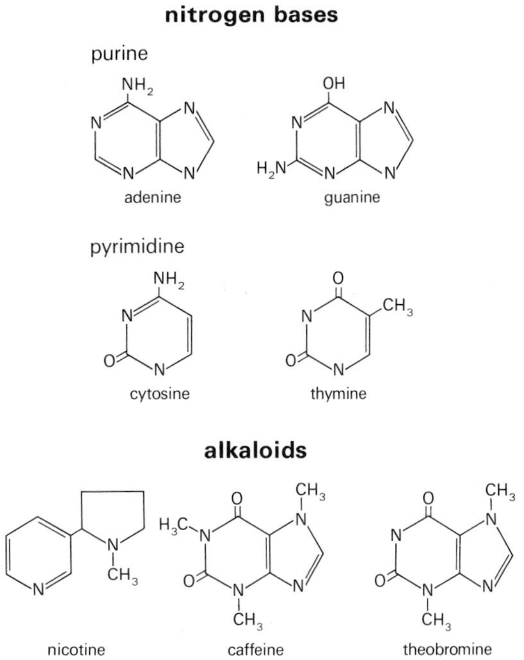

nitrogen bases

purine

adenine

guanine

pyrimidine

cytosine

thymine

alkaloids

nicotine

caffeine

theobromine

FIGURE 8-4. *Cyclic nitrogen compounds.*

two purines (adenine and guanine) and two pyrimidines (cytosine and thymine). Thus there are four nucleotides, depending on the specific base involved.

The sequence of each of the four possible nucleotides spells out a message that is decipherable by the cell. This genetic code specifies the sequence of amino acids in protein. Enzyme specificity is determined by its sequence of amino acids (as well as by its structural configuration); thus, DNA controls the destiny of the cell. The gene, in essence a functional portion of the DNA molecule, is the genetic information passed from generation to generation through the gametes.

Alkaloids, as their name indicates, are organic bases. (Although the nitrogen bases discussed above fit into the broad definition of alkaloids, they are usually treated separately.) The function of many alkaloids in plants is obscure. Nevertheless, the many effects that these substances have on human physiology render them of extreme pharmacological interest. A list of alkaloids derived from plant sources, along with their medicinal properties or other uses, is shown in Table 8-3. It is of interest to note that the popular beverages, coffee, tea, cocoa, and maté all contain stimulating alkaloids.

A number of extremely important pigments are complex alkaloids called porphyrins. Chlorophyll is a magnesium-containing prophyrin. The cytochromes are iron-containing porphyrins that function in energy-transfer. Hemoglobin, the oxygen carrier in animal blood, is also a porphyrin.

Nutrient absorption and translocation

Although chemical analysis of plant cells indicates the presence of many different elements, only 16 have been shown to be essential. The most abundant elements, carbon, hydrogen, and oxygen, are derived largely from carbon dioxide and water. The other 13 (iron, potassium, calcium, magnesium, nitrogen, phosphorus, sulfur, manganese, boron, zinc, copper, molybdenum, and chlorine) are derived ultimately from the soil in the form of inorganic salts. Plant growth is dependent on the availability of the essential nutrients. Since

Table 8-3 ORIGIN AND USES OF SOME PLANT ALKALOIDS.

ALKALOID	COMMON PLANT SOURCE	MEDICINAL AND OTHER USES
Atropine (belladonna)	*Datura stramonium, Atropa belladonna, Duboisia spp.*	Relaxant of gastro-intestinal tract, parasympathetic nervous system
Caffeine	Tea, coffee, maté, cola nuts	Stimulant of central nervous system
Cocaine	*Erythroxylon coca*	Surface anesthetic, minor use
Colchicine	*Colchicum autumnale*	Chromosome doubling, relieves pain of gout
Emetine	*Uragoga (Cephaelis) ipecacuanha*	Emetic, amoebicide
Ephedrine	*Ephedra gerardiana*	Stimulant of central nervous system, relief of nasal congestion
Hydrastine	*Hydrastis canadensis*	Antihemorrhagic
Morphine	*Papaver somniferum* (opium poppy)	Analgesic
Nicotine	*Nicotiana tabacum*	Insecticide
Pelletierine	*Punica granatum*	Vermifuge (tapeworm)
Pilocarpine	*Pilocarpus microphyllus*	Diaphoretic (sweat inducer)
Quinine	*Cinchona* spp.	Antimalarial, cardiac depressant
Reserpine	*Rauwolfia serpentina*	Tranquilizer, sedative
Strychnine	*Strychnos nux-vomica*	Poison, stimulant of central nervous system
Tubocurarine	*Chondodendron tomentosum*	Muscle relaxant
Theobromine	*Theobroma cacao*	Diuretic, stimulant
Yohimbine	*Pausinystalia yohimba*	Aphrodisiac

nutrients and water are ultimately supplied to the cell from the soil, the study of plant nutrition is largely concerned with the biology and chemistry of the soil (see Chapters 12 and 16).

With respect to absorption, the cell can be considered as a mass of protoplasm surrounded by a differentially permeable membrane that permits passage of water and inorganic salts but restrains the passage of most large complex molecules, such as sucrose. Molecules move through a selectively permeable membrane by **diffusion.** The movement of water through such a membrane is referred to as **osmosis,** and involves diffusion as well as bulk flow due to hydrostatic pressure differences. The osmotic movement of molecules can be demonstrated in nonliving closed systems by immersing a differentially permeable membrane that contains sugar water into a solution of pure water (or one with a lesser amount of sugar). The water moves from the solution of high solvent concentration (pure water) to the solution of low solvent concentration (sugar solution). Living cells, however, are able to accumulate certain ions in a manner unaccounted for by diffusion. The cell appears to act as a metabolic pump. This process, known as active uptake, requires energy, which is supplied by respiration. The ability of molecules to move in and out of plant cells is related to the size of the molecules, their lipid solubility, and ionic charge; membrane permeability is affected by the ionic concentration of the nutrient medium. Monovalent ions (K^+, Na^+, Cl^-) appear to increase the permeability of membranes, whereas polyvalent cations $(Ca^{++}$ and $Mg^{++})$ decrease membrane permeability. Furthermore, different ions interact in their effect on membrane permeability.

Translocation may be defined as the movement of inorganic or organic solutes from one part of the plant to another. The transport of water and solutes in and out of single cells and simple multicellular plants is accomplished largely by diffusion. In higher plants, however, this conduction of solutes is carried out largely in distinct tissue systems. Physiological specialization in muticellular plants is made possible because of the rapid, large-scale transport of substances within the plant. This movement is largely a two way stream, in which water and its dissolved contents move up

from the roots through the xylem, and synthesized sugars move out of the leaves to other parts of the plant through the phloem (Fig. 8-5). There is, however, some movement of minerals in the phloem, and the xylem of woody stems functions in the upward movement of organic compounds, especially at certain seasons of the year.

The upward movement of water and solutes in the xylem of higher plants is related in part to **transpiration,** the evaporative loss of water vapor

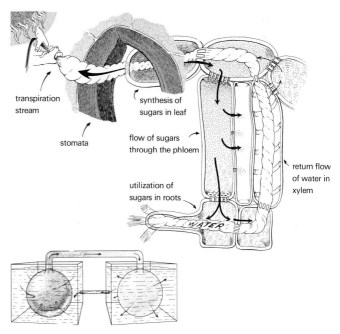

FIGURE 8-5. *A schematic diagram of the translocation of water and elaborated sugars in the plant. The upward movement of water through the xylem is due to tension on the continuous water column in the plant. This tension, produced by the evaporation of water from the leaf (transpiration), is transmitted to the absorbing cells of the root. Sugars synthesized in the leaves move through the sieve tubes of the phloem. Phloem transport is a pressure flow brought about by a high osmotic concentration in the leaf cells and a low concentration in the receiving cells. A model of this system, called the osmometer, is shown at lower left.* [After Bonner and Galston, Principles of Plant Physiology. *San Francisco: W. H. Freeman and Company,* © *1952.*]

from the leaves through the numerous stomatal openings. As water is lost by the cells, a diffusion-pressure deficit draws the water from the xylem elements, which form large numbers of continuous tubes from roots to leaves. Thus, the tension is transmitted through the entire column to the root cells and results in increased water absorption. The rate of transpiration is affected by the degree of stomatal openings and by environmental factors, such as temperature and vapor pressure of water in the atmosphere, that affect the rate of water evaporation. The opening of the stomata is a mechanical process regulated by the turgidity of the guard cells (see p. 196).

The movement of sugars occurs principally in the phloem. Phloem transport appears to be accomplished by increased osmotic concentration in the leaf mesophyll cells brought about by the high concentrations of dissolved photosynthates. These sugars then move into the sieve tubes of the phloem by a process that is not clearly understood. The resultant sugar gradient appears to result in a flow, and other substances are swept along the presumed sieve tube stream. The sugars are utilized in the receiving cells through respiration, growth, or storage processes. There is also evidence of lateral transport between xylem and phloem.

PLANT DEVELOPMENT

The plant is more than the sum of its physiological processes. The orderly cycle of development that the whole plant undergoes involves complex patterns of change in cells, tissues, and organs. This cycle begins with seed germination and progresses through juvenility, maturity, and flowering. Upon fruiting the essential cycle of plant development is completed. In perennials the plant is ready to recycle after a period of quiescence. In annuals and biennials fruiting is a signal to the organism to enter the final phases of plant growth—senescence and death (Fig. 8-6).

Differentiation

The problem of differentiation is one of the great themes in biology. There are two levels of complexity to the problem. The first concerns development of individual cells. Unicellular organisms are capable of undergoing complex transformations with no other apparent internal stimulus than their own genetic makeup. The differentiation of individual cells must involve some systematic turning on and off of the multiplicity of complex instructions potentially receivable from the genetic material. Differentiation in multicellular organisms comes about as a result of differential growth within and among cells. This is accomplished in an orderly and systematic way, with the mitotic process in cell division insuring genetic continuity of all cells! The differentiation of genetically identical cells in multicellular organisms is indeed a mystifying process. The way in which this is accomplished appears to depend on the interaction between the cell's genetically controlled processes and their external environment. In a multicellular organism the environment of one part of a cell may be quite different from that of another part. Investigations of tissue and organ differentiation have shown that many of these differences involve the interaction of substances produced from different parts of the organs. For example, to artificially culture roots in tomato, not only must root meristem be present, but also thiamine and pyridoxine (familiar as vitamins in the B complex), which are normally provided by the leaves.

In higher organisms particular cells take over the control of differentiation through the action of "chemical messengers." Naturally occurring organic substances that affect growth and development in very low concentrations and whose action may be involved in sites far removed from their origin are known as **hormones.** A great many such substances occur in plants. Minute amounts (as low as one part per billion) exert measurable physiological effects. The term "growth regulators" has been coined to include all naturally occurring and synthetically copied, or created, substances that affect growth and development. They may be either inhibitors or promotors of growth, sometimes one and sometimes the other, depending upon their concentration.

Growth regulators

From continuing research in the chemical control systems affecting differentiation in plants, a number of important groups of hormone-like sub-

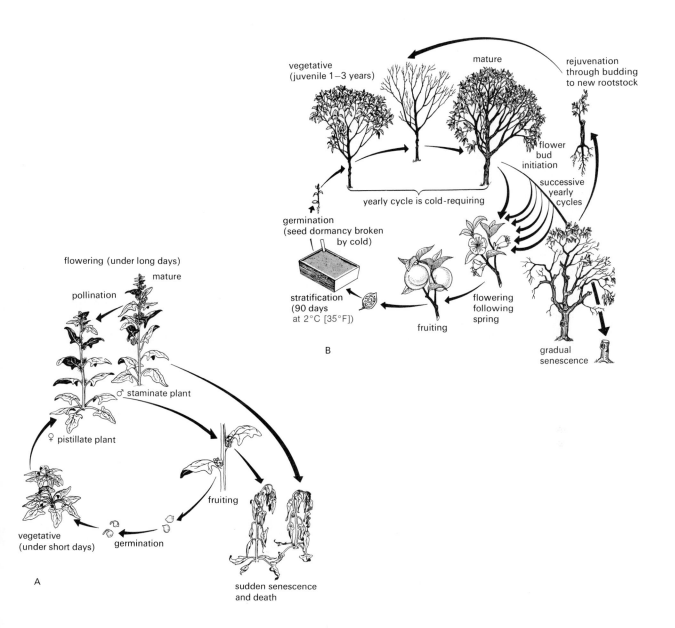

FIGURE 8-6. (A) *Developmental history of a herbaceous annual (spinach). The seed of spinach germinates quickly under proper environmental conditions for growth. If the seedling is grown under short days it forms a distinct vegetative stage (rosette). Under long days the stem elongates to form a seedstalk and initiates flowers. Since spinach is normally dioecious, staminate and pistillate flowers form on separate plants. Soon after the staminate plants flower, and soon after the pistillate plants fruit, they undergo rapid senescence and die. (B) Developmental history of a woody perennial (peach). The peach seed germinates only after a period of exposure to cold. The juvenile stage lasts about 2 to 3 years. The yearly cycles must be interrupted by a period of cold. Usually flower buds are initiated in the third or fourth year and open in the following spring. The plant may live for 40 to 50 years, and undergoes a very gradual deterioration. The plant may be rejuvenated, however, from buds close to the base of the tree. If vegetative growth is continually budded onto new rootstocks the "original" plant remains, in effect, immortal.*

stances have been defined: **auxins, gibberellins, kinins, dormin,** and **ethylene.** These substances are so grouped on the basis of structural and physiological similarities, but their functions overlap and interact. There are also undoubtedly other groups. The basis for the chemical control of many distinct growth patterns, such as flowering, remains unknown. In addition, many natural occurring compounds known to inhibit growth and developmental processes fit no well defined group.

AUXINS   The auxins, the first class of growth regulators to be discovered, have received considerable attention in plant physiology. Auxins are growth-promoting plant hormones. Cell elongation, the simplest example of anatomical differentiation, is directly affected by auxin concentration. The mode of action here appears to involve alterations in the plasticity of the cell wall. This fundamental property of auxin has been used in assaying its activity. The most basic test for activity consists of measuring the elongation of dark-grown oat coleoptile sections. Similar tests assay the rate of curvature of longitudinally halved stems in response to auxin application, as in the split pea test. The most common natural auxin is indoleacetic acid (IAA).

CH₂COOH

Indoleacetic acid (IAA)

Some very fundamental growth responses have been shown to be controlled by auxin. Phototropism, the bending toward light of the stem apex (growing point) can be explained as the result of differential cell elongation caused by the accumulation of auxin on the darkened side of the meristem. The cells on that side elongate and cause the stem to bend toward the light. Similarly, auxin greatly affects growth patterns in the plant. For example, **apical dominance**—the inhibition by the growing point of the growth of dormant buds below it—appears to be a function of auxin distribution. Auxin is produced in greatest abundance

in a vigorously growing stem apex, and high concentrations emanating therefrom have been shown to inhibit bud break. Removal of the auxin-producing stem tip increases lateral bud break and subsequent branching, usually directly below the cut. Thus the form of plants can be changed by the manipulation of apical dominance through pruning.

Auxin is at present the best understood of the many substances affecting plant development. It is formed in the stem and root apices, from where it moves to the rest of the plant. Its distribution, however, is not uniform. The resultant concentration of auxin in various parts of the plant has been correlated with inhibition and stimulation of growth (Fig. 8-7) as well as with differentiation of organs and tissues. Such processes as cell enlargement, leaf and organ abscission, apical dominance, and fruit set and growth have been shown to be influenced by auxin. Auxin has been associated with flowering and sex expression in some plants. Research on auxin has had a deep impact in agriculture. Among the uses are herbicides, rooting promoters, and fruit-set, thinning, and drop control materials.

The role of auxin has been associated with so

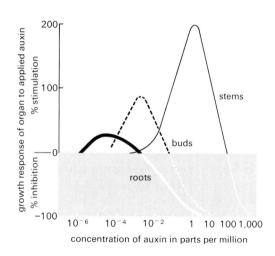

FIGURE 8-7. *The effect of auxin concentration on the growth of roots, buds, and stems.* [*From Machlis and Torrey,* Plants in Action. *San Francisco: W. H. Freeman and Company,* © *1956.*]

many diverse growth systems in plants that it has at times been tempting to suggest that it must behave as a "master" hormone affecting growth and differentiation, but the importance ascribed to any particular developmental material is influenced by the intensity of study that it receives. Growth and development appear to depend on interactions of many factors. The conclusion is now inescapable that auxin, though an important substance, is only one of many significant materials involved in differentiation and growth.

GIBBERELLINS   The gibberellins are a group of at least nine closely related, naturally occurring terpenoid compounds. They were discovered through studies of excessive growth of rice occurring as a response to a fungal disease. The gibberellins affect cell enlargement and cell division in subapical meristems. The most startling effect is the stimulation of growth in many compact plant types. Minute applications transform bush to pole beans or dwarf to normal maize (Fig. 8-8). This effect is utilized as a biological assay. In addition

to the dwarf-reversing response, gibberellins have a wide array of effects in many developmental processes, particularly those controlled by temperature and light (photoperiod), including seed and plant dormancy, germination, seed-stalk and fruit development. Barley is treated with gibberel-

Gibberellic acid (GA₃)

lic acid in the malting stages to increase the enzymatic content of malt. The use of gibberellins has had impressive agricultural applications, particularly in grape production. In Japan, seedlessness is induced in grapes, and in California the 'Thompson Seedless' cultivar ('Sultanina') shows a remarkable increase in berry size after gibberellin treatment (Fig. 8-9).

untreated   gibberellin-treated   untreated   gibberellin-treated
bush bean   bush bean   dwarf corn   dwarf corn

FIGURE 8-8. *Gibberellin treatment stimulates normal growth in dwarf plants. In beans 20 μg changes the bush to a vine habit; in corn 60 μg causes a change from dwarf to normal growth.*

FIGURE 8-9. *Effect of gibberellic acid (GA) on 'Thompson Seedless' grapes: (Above left) control; (Above right) 5 ppm GA; (Below left) 20 ppm GA; (Below right) 50 ppm GA. [Courtesy R. J. Weaver.]*

KININS   The term kinins (or cytokinins) has been applied to the group of chemical substances that have a decisive influence in the stimulation of cell division. Many kinins are purines, such as adenine (see Fig. 8-4). Kinetin, the original cell division stimulant isolated from yeast, has been identified as 6-furfurylamino purine.

Kinetin
(6-furfurylamino purine)

Many kinins have been detected in research involving tissue culture. Vascular strands laid on top of cultured pith tissue stimulate differentiation in otherwise nondividing cells, an effect also induced by adenine and kinetin. Kinin and auxin interact to affect differentiation: high auxin and low kinin gives rise to root development; low auxin and high kinin gives rise to bud development; equal amounts result in undifferentiated growth. Kinins are found in abundance in fruit and seeds (for example, corn endosperm, coconut milk), and are probably important in promoting growth and differentiation of the embryo. Tumorous cells are also rich in kinins.

Kinins affect such diverse physiological mechanisms as leaf growth, light response, and aging. Recently kinins, and other plant hormones, have been found to be active in regulating protein synthesis, possibly by turning gene transcription on and off or acting through transfer RNA to control translation of the gene product.

DORMINS   Dormins have only recently been established as a class of inhibitory compounds that affect bud and seed dormancy and leaf abscission. The only naturally occurring dormin identified is abscisic acid (ABA), also called at one time "abscisin II" and "dormin," but similar synthetic substances have been reported. ABA has a wide range of effects including the promotion of dormancy in buds and seeds, an acceleration of abscission, as well as the promotion of flowering in some short-day plants. The dormancy response may be through an effect of RNA and protein synthesis. Some effects of ABA seem to be reversed by gibberellins.

Abscisic acid

ETHYLENE   Although it has been known for many years that ethylene has a number of striking effects on plant growth and development, only recently has ethylene been considered to be a regulatory hormone. Thus the introduction of ethylene has been a standard practice to achieve ripening in banana and ethylene (as well as closely related acetylene and calcium carbide) has been used to obtain uniform flowering in pineapple. Ethylene has enormous effects on dark-grown seedlings, causing swelling and disorientation; such effects were used as a method of assay before the days of gas chromatography. Ethylene also influences cell division; thus, tomatoes grown in high concentrations of ethylene show extensive rooting up and down the stem. It is wise not to store budwood in apple storages (which produce ethylene as they ripen); ethylene will cause splitting and bark peeling. In the past decade it has been clearly shown that the natural or "endogenous" ethylene produced by plants influences the natural course of development in etiolated seedlings, abscission, floral initiation (in some plants), and fruit ripening. Ethephon, or (2-chloroethyl)phosphonic acid, a material that is transformed into ethylene in the plant, has a wide array of interesting effects including the induction of uniform ripening (pineapple, tomato), sex conversion, induction of staminate flowers in gynoecious (pistillate) plants of cucumbers and squash, and the stimulation of latex flow in rubber trees.

Ethylene

INHIBITORS Natural as well as synthetic inhibitory substances are often placed together as a diverse class of growth regulatory materials (Fig. 8-10). Natural inhibitors help control such processes as seed germination, short growth, and dormancy. Several synthetic inhibitors have found important agricultural applications. Maleic hydrazide has been effective in preventing the sprouting of onions and potatoes. A number of materials that inhibit the natural formation of gibberellins by the plant act to dwarf plants. CBBP and chlormequat show promise for use in reducing the height of many ornamental flowering plants without unduly interfering with flowering time or flower size. A gibberellin suppressor, SADH, in addition to dwarfing, affects fruit maturity, hardiness and many other plant processes.

Vegetative physiology[°]

GERMINATION Germination includes all the sequential steps from the time the seed first imbibes water until the seedling is self sustaining. After the imbibition of water the complex reserve substances are emzymatically converted to simple soluble forms that are readily translocated to the embryonic plant. Here some substances undergo respiratory breakdown and release energy; others are utilized in synthesis. The seed must, of course, be supplied with ample quantities of oxygen and water to satisfy the respiratory requirements and to provide a medium for enzymatic activity and synthesis. The temperature of the environment must be such that the biochemical processes of degradation and synthesis can operate. Although light is not usually required, it can either trigger or inhibit the germination process in some plants (for example, lettuce).

SEED DORMANCY Seed that is viable yet fails to germinate in the presence of normally favorable environmental conditions is said to be dormant. The cause of dormancy may be physical or physiological (Fig. 8-11).

Physical dormancy takes the form of structural limitations to germination, such as hard, impervious seed coats that provide a mechanical barrier to the entrance of either water or, in some plants, oxygen. The legume family provides the greatest number of examples of this type of dormancy. Some hard seeds resist water imbibition by means of a "one-way valve" in the hilum, the scar formed from the attachment of the seed to the pod. Although the seed may lose moisture in a dry environment, a moist environment effectively closes the valve. Under natural conditions these seeds do not germinate until soil microorganisms or weathering have sufficiently weakened the seed coat to permit the entrance of water. Thus setting the stage for germination may require a number of years rather than only a small part of one season. Seeds having an impervious coat may be artifi-

[°] The section on vegetative physiology is based on the chapter Plant Development by C. E. Hess in Janick, *Horticultural Science* (2nd ed.). San Francisco: W. H. Freeman and Company, 1972.

cially worn or weakened in order that germination can take place uniformly and without delay (see Chapter 9).

Physiological dormancy may be due to a number of mechanisms. These commonly involve growth-regulating systems of inhibitors or promotors. Inhibiting substances that block the germination process may be present in the flesh of the fruit, in the seed coat, or even in the endosperm of the seed. Germination of seed within a tomato is a rare occurrence, yet as the flesh of the tomato is removed and the seeds are rinsed, germination takes place without delay. This inhibition of germination is due not to the low pH or high osmotic values of the tomato flesh but to a specific chemical entity. Although the inhibitor in tomato fruit has not been isolated and characterized, there is evidence that it belongs to a group of compounds known as the unsaturated lactones. A naturally occurring member of this group is coumarin (see Fig. 8-10), a substance responsible for the aroma of freshly cut hay. The mode of action of such substances as coumarin consists in blocking or activating enzymes essential to the germination process. For example, it has been shown that the activity of α- and β-amylases is blocked in the presence of a germination inhibitor. The amylases are essential for the hydrolysis (splitting) of starch into simple soluble forms. A number of chemical materials may function as germination inhibitors.

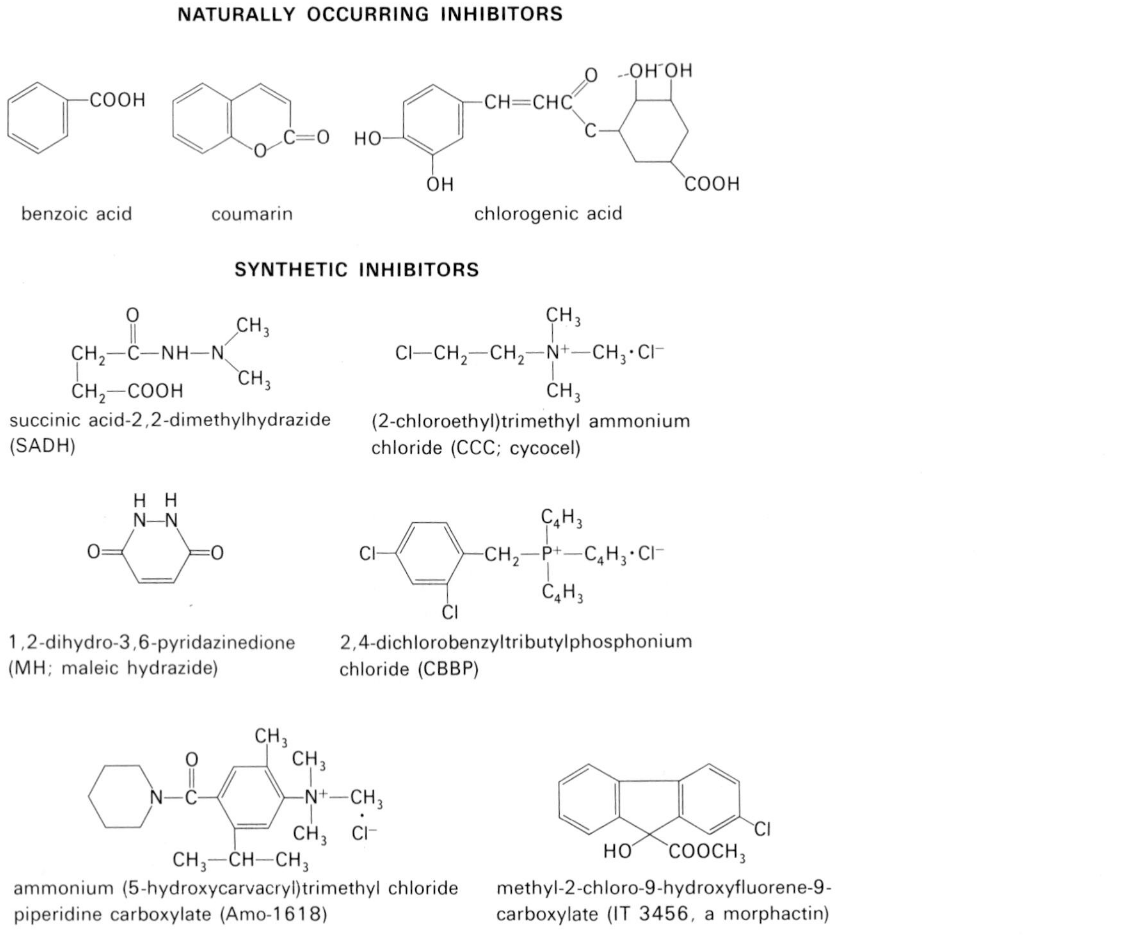

FIGURE 8-10. *Structural formulas of some naturally occuring and synthetic inhibitors.* [*From Weaver*, Plant Growth Substances in Agriculture. *W. H. Freeman and Company. Copyright* © *1972.*]

They are removed naturally from the fruit and seed coats through leaching by rain and degradation by microorganisms.

Physiological dormancy may also be due to internal factors, such as embryo immaturity. In the seeds of some species the embryo consists of only a few undifferentiated cells even when the fruit is ripe; the seed of the American holly (*Ilex opaca*) is an example. Therefore, an **afterripening**, process—maturation of the seed after harvest—must occur during which the embryo differentiates at the expense of the endosperm. In the seeds of other species, particularly woody species of the temperate zone, the causes of internal dormancy are more complex. In addition to the inhibitors that block germination, a germination stimulator is also required. To induce production of the stimulator, the seed must be subjected to a period of **cold stratification**—storage under moist, aerobic conditions at temperatures above freezing but below 10°C (50°F). A period of about 4 to 6 weeks at 5°C (41°F) is usually required.

Germination of seed may be blocked by more than one form of dormancy. Such **double dormancy** is characteristic of several members of the legume family. In the redbud (*Cercis canadensis*) both physical and physiological blocks to germination are present, the former being an impervious seed coat and the latter an internal dormancy that is broken by exposure to cold. For seeds with double dormancy to germinate, the physical barrier must be removed before the physiological barrier. If a seed coat is impervious, water essential for biochemical reactions cannot enter, and cold treatments administered to remove inhibitors or to promote the synthesis of stimulators will not be effective. In nature, seeds with double dormancy often require two years for germination. The first year is required for soil microorganisms and weathering to remove the physical barrier. During the winter of the second year internal dormancy is broken, and germination occurs the following spring. Physiological dormancy in some seeds may be broken by artificial treatments that either interfere with, dilute, or remove inhibitors. In others, a missing growth regulator, such as gibberellin, will break dormancy.

Dormancy of seeds is a biological mechanism

Some seeds proceed without hindrance provided they have adequate moisture, proper temperature and oxygen; some need light as well.

Internal physiological barriers are overcome in many seeds by low temperature.

Inhibitors in fruits or seed coats may block germination. They may be removed by leaching and by microorganisms.

Physical barriers, such as an impervious seed coat, prevent entrance of water or oxygen. They may be removed naturally through weathering and bacterial action or by treatment with caustic substances.

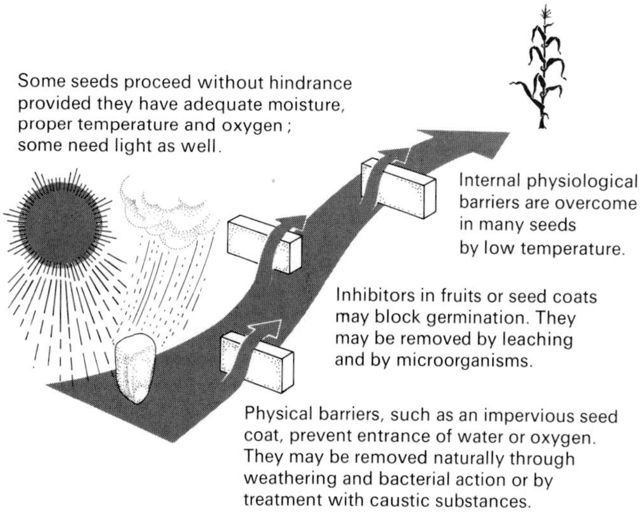

FIGURE 8-11. *Germination is the route from seed to seedling. Physical or physiological dormancy may prevent germination even though favorable environmental conditions for growth are present.*

that provides protection against premature germination when environmental conditions may not be favorable for seedling growth. Seeds from woody plants of the temperate zone will not germinate during the late fall, but are internally delayed until the following spring. Weed species possess, almost without exception, one or more of the major types of dormancy. In contrast, the seeds of most crop plants are not blocked by either physical or physiological forms of dormancy. Dormancy mechanisms have been eliminated through selection for rapid germination.

JUVENILITY A seed is considered germinated when it has produced a plant that, under proper environmental conditions, is potentially capable of continuous and uninterrupted growth. From the time this stage is reached until the first flower primordium is initiated, the plant is considered to be in the vegetative phase of growth. If during the vegetative phase the plant cannot be made to flower, regardless of the environmental conditions imposed, it is said to be **juvenile.** Juvenility and maturity, however, are relative terms. In many species these growth phases blend into each other.

The end of the juvenile period is indicated when the plant responds to flower-inducing stimuli.

The juvenile phase is characterized by the most rapid rate of growth of the overall organism and, in some plants, by distinct morphological and physiological features. The juvenile phase varies in length from one to two months for annuals to a period of a few years for the fruit trees. Some plants, such as bamboo, require scores of years. Among the morphological features that are associated with juvenility, and which are lost or altered at maturity, are the presence of thorns (*Citrus*), leaf lobing (cotton) or the lack of lobes (*Philodendron*), and the angle of the branches with respect to the main axis of the plant (spruce). The geotropic response varies with developmental stage in some plants. In its juvenile phase of growth, English ivy (*Hedera helix*) is a trailing vine, climbing only with support; in the mature stage of development the plant grows upright (Fig. 8-12). The ability of juvenile plants to initiate adventitious roots readily is a common feature of many species. This ability decreases or is lost entirely in mature forms.

In some plants it can be shown that there is a gradual change from juvenility to maturity in the same plant. The basal portions can remain juvenile even though the shoot extremities have reached maturity. This feature is utilized in the propagation of certain apple clones. In order to obtain shoots

that root easily (a characteristic of juvenility), growth is continually forced from the basal portion of the tree by pruning.

Juvenility can be experimentally terminated by chemical or physical treatments, but many species resist the change from the mature to the juvenile stage. Intergrafting juvenile and mature wood has provided some insight into the physiology of this phenomenon. The termination of juvenility may be explained as an aging of the growing point. This may be the result of dilution or exhaustion of some substance responsible for the juvenile condition. If a mature branch of English ivy is grafted onto a juvenile stock, the mature branch will develop juvenile shoots at first, but after a few weeks growth the juvenility gradually disappears and the branch again becomes mature. This sort of result lends support to the assumption that new shoots on the mature scion first utilize some "juvenile factors" in the stock and gradually mature when the juvenile factors become exhausted. No hypothetical "juvenile factor," however, has been isolated.

BUD DORMANCY AND REST Vegetative growth is not a continuous process but one that is interrupted by periods of arrested development. Dormancy may affect the whole plant or be restricted to certain parts, particularly seeds and buds. In the potato, dormancy is typically limited to vegetative buds, whereas in coffee, dormancy is limited to flower buds.

One type of arrested growth is brought about by unfavorable environment. For example, the growth of bluegrass, a common turf species, ceases when water is lacking, and top growth may die under continued drought. When conditions are more favorable, growth resumes from underground rhizomes. Many plants similarly survive periods of temperature extremes by undergoing a period of quiescence or rest. In tropical areas the periodic flushing and resting of vegetation has been correlated with moisture, light, and temperature. In addition, it has been suggested that internal factors, such as stage of development or age of the plant, influence the rhythm of growth. In the banana each stalk or pseudostem stops growing after flowering.

In **bud dormancy,** as in seed dormancy, growth is temporarily suspended even though all the ex-

FIGURE 8-12. *Morphological changes associated with growth phases in English ivy* (Hedera helix). *In the juvenile stage the leaves are lobed, and growth is horizontal. When the plant becomes mature the leaves are entire, and shoots grow upright and bear flowers.*

mature

juvenile

ternal conditions normally required for growth are provided. For example, woody plants of the temperate climates develop vegetative buds at each node throughout the growing season. The lack of growth of these buds after formation is initially an expression of apical dominance, controlled by auxin distribution. At this time, these buds can be induced to grow by removing the growth-inhibiting apical meristem. With the onset of autumn, however, the buds of many species become truly dormant and will not grow even if the plant is pruned and moved to a warm greenhouse unless they have been exposed to a suitable period of cold weather. The degree of dormancy varies not only among species, but also among buds on the same plant. The flower buds of many trees, such as peach, cherry, and apple, require less chilling than do the vegetative buds. This is why the flower emerges before the leaves in the spring. In *Forsythia* the cold requirement is so minimal that it may be seen to flower in the fall after a brief period of cold weather.

The physiological basis of dormancy seems to depend on the accumulation and disappearance of growth inhibitors and promotors. In woody plants of the temperate climates, the onset of dormancy is conditioned by short day length and low temperatures. Bud dormancy in these plants is probably internally regulated by the formation of growth-inhibiting substances. It is broken naturally with cold temperatures. This cold reaction is localized. If an isolated stem of a dormant plant is cold-treated while the remainder of the plant is kept at a warm temperature, only the buds on the treated stem grow.

The period of cold treatment required to break dormancy not only varies with the species but is sensitive to selection within species. For example, peaches have cold requirements that vary from 350 to 1,200 hours below 7°C (45°F). Cultivars having low chill requirements are selected for areas where the periods of cold weather may be brief. If varieties having high chill requirements are grown in mild-winter regions they will leaf out poorly or not at all in the spring. This type of dormancy prevents the production of most temperate fruit crops in subtropical regions.

The survival value of dormancy is clear. A physiological mechanism that prevents growth is a biological check on the perversity of weather. If woody plants lacked internally imposed cold dormancy they might initiate growth under favorable periods in the late fall or winter only to succumb to succeeding severe weather.

The chemical control of dormancy holds promise of important application. A means of extending dormancy would protect the plant against growth during unsuitable periods, such as times of frost. Similarly, a means of breaking dormancy would permit precise control of crop operations. For example, the application of ethylene and related compounds to dormant potato tubers overcomes dormancy and ensures prompt, rapid growth.

Reproductive physiology

A plant is mature when it is potentially capable of reproduction. Maturity is characterized by abundant and general flower development, although other physiological and morphological changes may occur. The mature plant, although capable, may not necessarily flower. The environment to which the plant is exposed at the time of maturity determines whether the plant will reach this ultimate expression of the mature state.

FLOWERING Flowering represents a wide spectrum of physiological and morphological events. The first event, the most critical and perhaps least understood, is the transformation of the vegetable stem primordia into floral primordia. At this time subtle biochemical changes take place that dramatically alter the pattern of differentiation from leaf, bud, and stem tissue to the tissues that make up the reproductive organs—pistil and stamens—and the accessory flower parts—petals and sepals. Among the events that follow initiation are development of the individual floral parts, floral maturation, and anthesis.

Once a meristem has been signaled to change from the vegetative to the reproductive state, microscopic changes in its configuration become apparent. Growth of the central portion is reduced or inhibited, and the meristem becomes flattened in contrast to the conical shape characteristic of the vegetative condition. Next, small protuberances develop in a spiral or whorl arrangement around the meristem. Although this phase of reproductive differentiation is quite similar to vegetative differentiation, a basic difference does exist in that there is no elongation of axis between the successive floral primordia as there usually is between leaf primordia. In most plants the transformation from the vegetative to the reproductive state is irreversible, and the floral parts will continue developing until **anthesis**— the time at which the flower is fully open—even though environmental conditions change. By the time anthesis takes place, the pollen and the embryo sac are completely developed, and the plant is prepared for fruiting, the next major step in its development.

Floral initiation and development have been studied primarily through manipulations of the plant's environment, particularly nutrition, light, and temperature. The rapid advances in the knowledge of the purely nutritive aspects of plant physiology in the latter part of the nineteenth century created an atmosphere in which it was believed that perhaps all aspects of plant growth could be explained or regulated by an alteration or adjustment of a plant's nutrition. In 1918 E. J. Kraus and H. R. Kraybill proposed that the development of flowering was regulated by the **carbohydrate:nitrogen** relationship of the plant. When tomato plants were grown under conditions favoring photosynthesis, and at the same time were supplied with an abundance of nitrogen fertilizers, vegetative growth was lush and flowering was reduced. But when the nitrogen supply was reduced while photosynthesis was maintained at a high level, vegetative growth was reduced, and flowering was abundant. With the combination of low nitrogen and low photosynthesis both vegetative growth and flowering were reduced. This concept of nutritional control of flower development was readily accepted, but the results of extensive investigations indicated that many plants flower over an extremely wide range of carbohydrate:nitrogen ratios. In view of our present appreciation of the tremendous physiological effects of minute quantities of growth-regulating substances, it is not difficult to understand why the gross ratios of total carbon compounds to total nitrogen compounds does not provide a consistent indication of the physiological condition of the plant in relation to flower initiation and development.

It can be demonstrated that the presence of new growth has an inhibitory effect upon floral initiation. Tomato plants flower abundantly with the continued removal of new leaves, even when only the cotyledons are left for photosynthesis. From these results it is tempting to suggest that the reason nitrogen reduces flowering in the tomato is only incidentally associated with the carbohydrate: nitrogen relationship, but rather that the stimulation of new growth by nitrogen inhibits flower initiation. In support of this concept is the observation that almost any means by which growth can be reduced, such as bending a branch from an upright to a downward position, or applying growth retardants such as Alar (N-dimethyl amino succinamic acid), results in increased floral initiation.

PHOTOPERIODIC EFFECT   The discovery of photoperiodism—the growth response to the length of light and dark periods—by W. W. Garner and H. A. Allard, scientists of the United States Department of Agriculture, completely changed the concept of flower initiation. Decreasing the daily exposure to light during the summer induced a fall-flowering variety of tobacco to bloom profusely, a feat unattainable with all other environmental changes. It was subsequently demonstrated that the length of the dark rather than the light period is the critical factor in the photoperiod response. Interrupting the dark period with light reverses the effect of a long night, whereas interrupting the light period with darkness has no effect. Nevertheless, "day length" remains the conventional term for describing the photoperiodic response. As the results of investigations accumulated, it became apparent that a majority of plants fell into one of three categories: short-day, long-day, and day-neutral plants. Short-day plants require a dark period exceeding some critical length to flower, and cannot flower under continual illumination. These include spring- and fall-flowering tobacco, poinsettia, and chrysanthemum. Long-day plants are inhibited from flowering when the dark period exceeds some critical length, and they can flower under continual illumination.

Included in this group are many of the summer-flowering plants of the temperate zones, such as radish, lettuce, spinach, potato, and winter barley. Day-neutral plants apparently can initiate flowers under any night length. These include the dandelion, buckwheat, and many tropical plants that either flower the year round or, if they do not, can be shown to be affected by other environmental conditions. With the continued study of plant response to day length, other categories have been added. For example, some plants are non-obligate; that is, they will flower regardless of the day length but will flower earlier or more profusely when the day is either long or short. In still another category are the plants that flower only after an alteration of day lengths; these are known as long-day, short-day plants. Such plants require first an exposure to long days and then to a period of short days.

Investigations relating to photoperiodism have clearly demonstrated the hormonal control of flower initiation. The mature or newly expanded leaf is the perceptor of changes in day length. In some plants the leaves need only to be exposed to one light-dark cycle of the proper day length to cause flower initiation, although most require several to many cycles. Once the leaves have received the photoperiodic message, they produce a substance, or a precursor of a hypothetical substance, called **florigen.** Unfortunately, however, florigen remains only a name, for it has proven to be one of the most elusive of all plant-growth substances. Its transport from the leaves to the growing point can be demonstrated indirectly up and down stems, across graft unions, and from one plant to another, but the substance has not been isolated. Other evidence from grafting experiments indicates that the flowering stimulus is identical in long- and short-day plants. A major advance that may lead to the identification of florigen is the recent isolation from the leaf of the pigment system called **phytochrome,** which specifically receives the external photoperiodic message. This was discovered by H. A. Borthwick and Sterling B. Hendricks, scientists of the United States Department of Agriculture, and first reported in 1959. Phytochrome is a pigment present in small amounts in all plants. Under short-day conditions

an interruption of the dark period with light (specifically red light) causes an intramolecular shift in phytochrome that brings about flowering in long-day plants and prevents flowering in short-day plants. The reason for this differential response is not known. The effect of red light (ca. 0.67μ) on the structure and thus the activity of phytochrome can be reversed rapidly by exposure to far red light (ca. 0.72μ). (See Fig. 8-13.) In addition, this shift is also influenced by the length of the dark period and by temperature. This may explain the light-temperature relationship in photoperiodism and may explain why the light break is most effective at the middle of the dark period.

The characterization of phytochrome is still being studied. It is a protein with chromophoric

FIGURE 8-13. *The length of the dark period is the critical factor in the photoperiodic control of flowering. Short-day plants flower under periods of long night; long-day plants flower under periods of short nights. Interrupting the dark period with light prevents flowering in short-day plants and permits flowering in long-day plants. The flowering effect can be reversed by substituting red and far red light during night interruption.* [After Galston, The Life of the Green Plant, Englewood Cliffs: Prentice-Hall, 1961.]

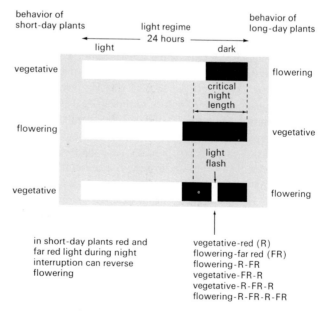

(pigment) groups. The molecular weight of phytochrome is estimated to be 60,000 with three chromophores per molecule. The chromophore is a bile pigment closely similar but not identical to the chromophore of c-phycocyanin, an algal chromoprotein.

Phytochrome appears to have many effects upon growth and development. It has been shown to be responsible for the promotion and inhibition of germination in seeds of some plants. In lettuce, seed germination can be inhibited or promoted by alternating exposure to red and far red light. Although all seeds contain phytochrome, those of many plants do not require this "light signal" to germinate. It probably is involved in other photoperiodically controlled processes such as tuber and bulb formation, and dormancy.

Although the discovery of phytochrome has made a very significant contribution toward explaining the first step in the mechanism of flower initiation, much experimentation must be done before the substances specifically causing the transition of the apical meristems from the vegetative to the flowering condition can be isolated and their mode of action determined.

TEMPERATURE EFFECTS   Photoperiodism does not have exclusive control over flowering. Temperature, for example, has both indirect and direct effects. It can influence flowering indirectly by modifying the plant response to a given photoperiod. Thus, a poinsettia will initiate and develop flowers in 65 days when grown under a short day length at 21.1°C (70°F). When grown under the same day length at 15.6°C (60°F), however, 85 days are required before initiation occurs. A more striking example is that of the strawberry. At temperatures above 19.4°C (67°F), the June-bearing strawberry behaves as a short-day plant and will not initiate flowers in day lengths longer than 12 hours. At temperatures below 19.4°C (67°F), its response is that of a day-neutral plant, and the plant will initiate flowers even in continuous illumination.

The direct effect of low temperature upon flowering was recognized more than a hundred years ago when an American, J. H. Klippart, described a process for "converting" winter wheat to spring wheat. (Winter wheat is planted in the winter and produces a crop the following year; spring wheat produces a crop in the same year it is planted.) The process consisted essentially of reproducing the sequence that occurs in the field. The winter wheat is partially germinated and further development is prevented by maintaining the seed at a temperature near 0°C (32°F). The treated seed, when planted in the spring, produces a crop in the same year. The elimination of the normal biennial habit of the winter wheat has been accomplished by satisfying the cold requirement. Klippart's research received little attention until his work was repeated by the Russian agronomist T. Lysenko,° who in 1928 coined the term **vernalization** to describe the effect upon future flowering behavior of subjecting a plant to low temperatures during one or more of the developmental phases of growth. The low-temperature effect may be obtained by treating the whole plant or part of the plant. In the grains only the moistened seed need be treated. Sometimes a vernalized plant—that is, a plant that has received a cold treatment—can be grafted onto a nonvernalized plant and both will flower. The implication is that some substance has passed from the vernalized plant, across the graft union, to the nonvernalized plant. A similar experiment has been conducted with annual and biennial forms of the same plant, such as henbane. The annual form requires no cold treatment to flower and induces flowering in the biennial form when grafted to it. Apparently the annual form produced the flowering stimulus without a cold treatment. In wheat, this difference between annual (spring) and biennial (winter) forms of wheat has been shown to be controlled by a single gene. Although the flower-inducing substance produced during vernalization has not been isolated, a clue to its identity is found in the fact that the cold requirement of some plants (dicotyledonous biennials, but not the cereal grasses) can be partially or completely replaced with gibberellins.

Further evidence that temperature can have a profound affect upon flowering is provided in the

° Lysenko later became notorious for his unorthodox and discredited views on genetics and plant breeding.

phenomenon of devernalization. Some plants exposed to a low temperature for a period of time sufficient to initiate flowering (usually at least six weeks) will revert to the original nonflowering condition if exposed to a high temperature. Onion growers take advantage of the devernalization on a commercial scale. Onion sets (small bulbs produced by crowding) are stored during the winter at temperatures near freezing to retard spoilage. The onion sets are vernalized at this temperature; and if they were planted directly from cold storage, they would quickly flower and bulbs would not be formed. Therefore, the onion sets are exposed to temperatures above 26.7°C (80°F) for 2 to 3 weeks before planting. The devernalized sets form bulbs and do not flower.

Although photoperiodism and vernalization have many similarities and are definitely interrelated, the stimuli produced in the two responses to environment do not seem to be identical. It is possible to separate the effect of vernalization from those of photoperiodism. For example, some biennial plants when exposed for the required time to low temperature still will not flower unless they are placed under the proper day length, usually a long day length. Similarly, when gibberellins are substituted for the cold treatment, flower initiation will not occur in some plants unless the day length is correct.

In addition to these temperature effects, it has been shown that the alternation of warm and cool temperatures influences flowering (as well as vegetative growth). This phenomenon, called **thermal periodicity,** has been utilized in the culture of greenhouse tomatoes. The tomato initiates the most flowers when grown at 26.7°C (80°F) during the day and 17.2 to 20.0°C (63 to 68°F) during the night.

FRUIT DEVELOPMENT   Fruit development can be divided into four phases: (1) initiation of the fruit tissue, (2) prepollination development, (3) postpollination growth, and (4) ripening, maturation, and senescence. The increase in fruit size in the prepollination phase of development is primarily the result of cell division. After pollination, cell enlargement is responsible for most of the increase in size. However, in some large-fruited plants (for example, watermelon and squash) cell division continues for some time after pollination, the final size being a consequence of an increase in both cell number and cell size. The stimuli and nutrients for prepollination growth are supplied primarily by the main body of the plant. In plants with perfect flowers, the stamen primordia are often differentiated before the ovary primordium, and have been shown to be a source of growth stimulus. Early surgical removal of the immature stamens in the flower bud can adversely affect the growth of the ovary. Extraction of the unripe anthers reveals the presence of large amounts of auxin.

POLLINATION   One of the most critical events in the growth and development of a fruit is pollination. Pollination has at least two separate and independent functions. The first is the initiation of the physiological processes that culminate in **fruit set** (more precisely, the inhibition of fruit or flower abscission); the second is to provide the male gametes for fertilization. The separateness of these two functions can be demonstrated in orchids by the use of dead pollen where fruit set and some growth occurs but not fertilization. The fruit-setting function of pollen can be substituted for by the application of auxin (Fig. 8-14). The effectiveness of water extracts of pollen in inducing

FIGURE 8-14. *The role of pollen in preventing abscission of the flower may be replaced by the application of auxin.* [*From Bonner and Galston,* Principles of Plant Physiology. *San Francisco: W. H. Freeman and Company,* © *1952.*]

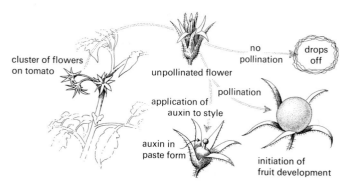

cluster of flowers on tomato

unpollinated flower

no pollination

drops off

pollination

application of auxin to style

auxin in paste form

initiation of fruit development

fruit set has led to the postulation that the pollen contains an auxin. But the minute amount of auxin present in the pollen that lands on a stigma cannot account for the auxin response obtained. Instead, it seems that the pollen contributes either an enzyme that converts precursors present in the stigma to auxin or provides a synergist that renders effective the auxin already present.

Even though pollination takes place, and fruit set is obtained, fertilization is not absolutely assured. Sometimes pollen does not germinate, or if it does, the pollen tube may burst in the style. The germination of pollen is dependent upon the presence of a medium of the proper osmotic concentration (as shown by the effect of various sugar concentrations upon germination) and is stimulated by the presence of certain inorganic substances, such as manganese sulfate, calcium, and boron. In the lily it has been shown that the highest concentration of boron occurs in the style and stigma.

If the growth of the pollen tube is very slow, the style or even the entire flower may absciss. This may be artificially prevented, however, by the application of an auxin, such as α-naphthalene acetamide. When fruit set and growth are obtained without fertilization the fruit is said to be **parthenocarpic.** Although the fruits of some plants have no seeds, they may develop seedlike structures as do some seedless varieties of holly, orange, and grape. The effect of pollination on fruit set and the influence of the resultant seed on fruit growth make pollination a crucial phase in the production of many fruit crops.

POST FERTILIZATION DEVELOPMENT   After fertilization, the plant enters a phase of physiological activity that is second in intensity only to germination. The developing fruit no longer depends primarily upon the parent plant for a source of growth stimuli but instead receives them from developing seed within the fruit. The role of the seed can be empirically demonstrated by observing that misshapen fruits result from uneven distribution of developed seeds. In many fruits, a direct correlation exists between either weight or length and seed number. This effect of the seed on fruit development is mediated through chemical sub-

stances. For example, extracts of immature seeds can stimulate growth of unpollinated tomatoes. Furthermore, it is possible to correlate various physiological events in the development of a fruit with the presence of growth substances. It has been demonstrated that the auxin levels reach a low at the time of flower drop, and particularly during the natural abscission of partially developed fruit that occurs when a heavy crop is obtained in fruit trees. The growth of the strawberry receptacle (the fleshy part) has been shown to be controlled by the achenes—one-seeded fruits scattered over its surface. The achenes contain high concentrations of auxin, whereas the receptacle tissue contains little. The addition of an auxin (indoleacetic acid) dramatically stimulates the growth of an acheneless receptacle. Indoleacetic acid has been demonstrated to be present in very high concentrations in immature corn kernels. More recent studies indicate that the embryo and the endosperm of the developing seeds are the sites of many fruit-growth-promoting substances, such as several auxins, gibberellin-like compounds, and kinins.

Although the control center of fruit growth is located in the seed, the raw materials for fruit development are supplied by the plant. Thus, the nutrition and moisture availability of the plant directly affect fruit size. It has been calculated that at least forty leaves on a mature apple tree are required to support the growth of one apple. If the forty-to-one ratio is substantially reduced by an abnormally high fruit set, the quality and size of the individual fruits is greatly reduced. Therefore, it is a common orchard practice to artificially reduce the number of fruit by thinning either the flowers or the partially developed fruit.

FRUIT RIPENING   A final, dramatic, physiological event, the **climacteric,** marks the end of maturation and the beginning of senescence of fruits in some species. It is characterized by a marked and sudden rise in the respiration of a fruit prior to senescence. The respiration rate then returns to a level equal to or below that which existed prior to the climacteric. The climacteric is characterized not only by the abrupt increase in the production of carbon dioxide, but also by qualitative changes related to ripening, such as changes in pigment. The transition from green to yellow in certain varieties of apple, pear, and banana takes place during or immediately following the climacteric. The peak of acceptability, or "edible ripeness," of pears coincides with the peak of the climacteric; maximum acceptability in the apple, pear, and avocado is reached immediately after the climacteric. A marked increase in the susceptibility of fruits to fungal invasion follows the climacteric.

Temperature has a profound effect upon maturation and the climacteric. For example, in the apple maximum respiratory activity is five to six times as high at 22.8°C (73°F) as at 2.2°C (36°F), and it takes twenty-five times as long to reach the climacteric at the lower temperature. However, the total amount of CO_2 liberated during the time between harvest and the end of storage life is approximately the same for both temperatures, equivalent to 16 to 20 percent of the reserve carbohydrates initially present in the fruit.

It is becoming clear that fruit ripening, like other phases of plant development, is a DNA-controlled process. A rise in RNA occurs in climacteric fruits, followed by increases in enzyme proteins such as the synthetases, hydrolases, and oxidases. Therefore, the climacteric is associated with both degradative and synthetic reactions. Ethylene and perhaps other volatiles appear to have key functions in the ripening process, and ethylene has been referred to as a fruit-ripening hormone. It is true that ethylene can accelerate fruit ripening; treatments to remove ethylene from within tissues of preclimacteric fruit can delay ripening. How ethylene exerts its effect on the initiation of ripening is not yet clear.

Senescence

Senescence is the phase of plant growth that begins at full maturity and ends in death. The erosive processes that accompany aging are the most baffling and least understood aspects of plant life. Senescence in plants may be partial or complete (Fig. 8-15). **Partial senescence** refers to the deterioration and death of plant organs, such as leaves, stems, fruits, and flowers. The death and abscission of cotyledons in bean plants, the death of two-year-old raspberry canes, and the death of the

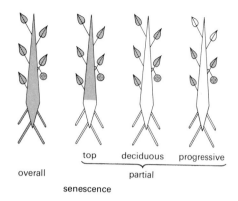

FIGURE 8-15. *Patterns of plant senescence. The dead portions of the plant are indicated by shading. [After Leopold, Plant Growth and Development. New York: McGraw-Hill, 1964.]*

entire shoot of tulip in early summer are all examples of partial senescence. **Overall senescence** refers to the aging and death of the entire plant except for the seeds. The termination of the life cycle of true annuals and biennials is often sudden and dramatic. For example, after fruiting whole fields of wheat die in a synchronized pattern during the growing season. In contrast, the senescence of such perennial plants as the apple or peach appears as a gradual erosion of growth and viability. In addition, perennials can be rejuvenated. Mature apple trees may be revitalized by severe pruning and fertilization or by encouraging the growth of adventitious buds at the base of the tree. The senescence of annual plants after flowering and fruiting is irreversible. Spinach plants kept vegetative under short days will not senesce, but after flower induction, death is a certainty.

The earlier explanations for the senescence of annual plants are associated with a depletion hypothesis. It is suggested that during flowering and fruiting essential metabolites are drained from the main plant and accumulate in the fruit and seed. By the time the fruit is mature the plant is depleted to the extent that further growth is impossible, and death rapidly follows. It can be shown that the removal of fruits can significantly postpone senescence, whereas the greater the number of flowers or fruits left on the plant, the more rapidly senescence occurs. The depletion concept of senescence, however, is an oversimplication of a complex process. Death is delayed, but not averted, by the removal of fruits and flowers. Unpollinated pistillate plants and staminate plants of a dioecious species such as spinach also senesce. Apparently, the biological signal that causes an annual plant to senesce is associated in some way with the flowering process and is merely accentuated through fruiting. The relationship between the senescing signal and flowering is unfortunately not at all clear at present.

Selected References

Annual Reviews, Inc. *Annual Review of Plant Physiology*. Annual Reviews, Inc., Palo Alto, Calif. (Recent advances in plant physiology are brought up to date in timely, technical reviews. Published each year since 1950.)

Leopold, A. Carl. *Plant Growth and Development*. McGraw-Hill, New York, 1964 (An advanced treatment of the physiology of growing plants.)

Rabinowitch, E., and Govindjee. *Photosynthesis*. Wiley, New York, 1969. (A broad introduction to the physiology of photosynthesis and the enzymatic processes associated with it.)

Salisbury, F. B., and Ross, C. *Plant Physiology*. Wadsworth, Belmont, California, 1969. (An excellent modern text in plant physiology.)

Weaver, Robert J. *Plant Growth Substances in Agriculture*. W. H. Freeman, and Company, San Francisco, 1972. (A thorough and comprehensive review of growth regulators from an agricultural point of view.)

Zelitch, Israel. *Photosynthesis, Photorespiration, and Plant Productivity*. Academic Press, New York, 1971. (A modern treatment of photosynthesis and its effects on plant productivity.)

CHAPTER **9**

Reproduction and Propagation

Reproduction refers to the sequence of events involved in the perpetuation and multiplication of cells and organisms. There are two essentially different types of reproduction in plants: **sexual,** the increase of plants through seeds formed from the union of gametes; and **asexual,** the increase of plants through ordinary cell division and differentiation. The basis of asexual reproduction is the ability of many plants to regenerate missing parts. **Propagation,** the utilization of reproductive processes in the controlled perpetuation of plants, is in a real sense the basis for agriculture.

REPRODUCTION

Reproduction involves a controlled self-duplication of the life-controlling mechanism. A self-duplicating machine is an intriguing and complex concept. Although man has devised devices that can duplicate themselves, the living organism certainly remains the most elegant and the most fascinating system. We appear to be on the verge of understanding the mystery of reproduction.

Replication of living systems

The principle involved in the replication of living material is somewhat analogous to that involved in a machine that forms type (for example, a linotype). The machine arranges the type in certain sequence to transmit information—that is, produce the printed page. New type is continually being formed from a mold or template of the original type. In this analogy the reproducing element is type; in living systems the reproducing elements are the nitrogen bases in DNA.

In the English language 26 letters of the alphabet produce a dictionary of tens of thousands of words. The same results can be achieved with the two symbols of the Morse code. The four symbols of the genetic alphabet, or code, are the four nitrogen bases that make up the nucleotides in DNA: adenine, guanine, thymine, and cytosine (see p. 141). If two symbols can express 26 letters, thousands of words, and thus an infinite series of messages, is it not conceivable that four bases can make up the code by which the information for maintaining living matter is transmitted? The way in which this is accomplished is the theme of intense study. The general theory is that the sequence of bases in DNA directs the sequence of amino acids in the synthesis of protein, specifically enzymes. Enzyme specificity is related to its amino acid sequence, and it is not unreasonable to assume that enzyme control can account for the biochemical control of living systems.

FIGURE 9-1. *Portion of the DNA molecule, shown untwisted. The molecule consists of a long double chain of nucleotides made up of phosphate-linked deoxyribose sugar groups and a side group of one of the four nitrogen bases. Hydrogen bonds (broken lines) link pairs of bases to form the double chain. The bases are always paired as shown, although the sequence varies. (Below) The double helix form of the DNA molecule.* [*After Allfrey and Mirsky, "How Cells Make Molecules."* Copyright © 1961 by Scientific American, Inc. All rights reserved.]

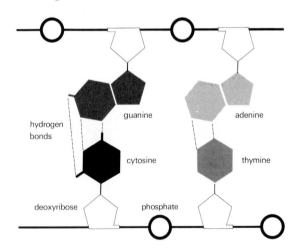

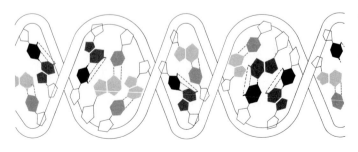

The replication of DNA appears to be accomplished by a system quite analogous to the template system of the type machine. The template model, however, is based on chemical bonding rather than physical impression. Chemical analysis of DNA shows that although the relative quantity of the four nitrogen bases varies, the amount of adenine is always the same as thymine, and the amount of guanine is always the same as cytosine. This suggests that adenine and thymine, and guanine and cytosine, occur in pairs. The significance of this becomes apparent when the structure of DNA is considered (Fig. 9-1). DNA normally occurs in double strands, and each strand is connected by linkages between the nitrogen bases. But because of the configuration of the bases, adenine is always paired with, and opposite to, thymine; and guanine is always paired with, and opposite to, cytosine. Thus if the sequence of one strand is given, the sequence of the opposite strand is fixed. Each strand, then, is the "template" for the other.

The theory of DNA replication follows nicely from its structure. The separation of the strands of DNA is apparently followed by a realignment of complementary bases on each strand (Fig. 9-2).

The structure of DNA accounts for both its ability to replicate and synthesize. The actual synthesis of protein is carried out not in the nucleus but in the cytoplasm. The information from the nucleus must be transferred to the structural cytoplasmic sites of synthesis, called ribosomes. The information is carried by a substance called ribonucleic acid or, RNA. (The components of RNA are almost identical to those of DNA except that the sugar portion of the molecule is slightly different—ribose takes the place of desoxyribose, and the nitrogen base thymine is replaced by an altered form called uracil.) The particular form of RNA that carries the genetic instructions, in the form of a complementary copy of the DNA series of bases, is called messenger RNA. Another form of RNA, transfer RNA, presumably brings the amino acids to the ribosomes to construct protein (Fig. 9-3).

Protein synthesis can be achieved outside of the living organism by placing together the component parts: ribosomes (from bacteria), an energy-generating system, amino acids, and RNA. Very elegant

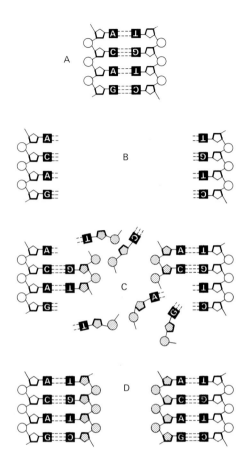

incorporates leucine. The four-letter alphabet and base sequence is shown in Table 9-1. The sequence of amino acids in the polypeptide (protein unit) of the enzyme is related to the three-letter code sequence of DNA. Thus the four-letter alphabet represented by the four nitrogen bases is used to make three-letter "words," each corresponding to one of the approximately twenty amino acids. Each sentence spells out the amino acid sequence for one protein. This "sentence" is the gene!

Genes and chromosomes

The complete set of instructions is separated into chromosomes, linear structures within the nucleus. A chromosome is in essence a sequence of DNA. A specific function-controlling sequence of DNA is the gene. Our understanding of the molecular, or "fine structure," of the gene is derived from studies that began in the late 1940's. A tremendous amount of information on the transmission and action of genes has accumulated from work on a wide spectrum of organisms, including crop plants. These investigations trace back to the beginnings of genetics, conveniently marked by the rediscovery in 1900 of Mendel's 1865 paper on inheritance. The theory that the units of inheritance, the genes, are particulate bodies on chromosomes and that they are transmitted from cell to cell and from generation to generation is the contribution of "classical" genetics and was clearly established by 1925.

Studies of many organisms have indicated that genes function through the regulation of chemical events. Genes control each step in the biochemical pathways of metabolism. Our understanding of the gene's role in metabolism dates back to flower pigmentation studies of the 1920's. It was found, for example, that two anthocyanin pigments that are different shades of red are chemically distinguished by a single hydroxyl group. The addition of this single hydroxyl group changes the color of the plant's petals from scarlet to deep red (see Fig. 9-4). This biochemical step is gene controlled. Thus, the petals of a plant that contain a gene (C) controlling the transformation of pelargonidin to cyanidin will be deep red. A plant having an altered version of this gene (c), one that is incapa-

experiments involving the direction of protein synthesis have been possible through the use of artificial RNA. In this way, the genetic code has been deciphered; that is to say, the sequence of bases in RNA has been associated with the incorporation of a particular amino acid into protein. The message has been shown to be in the form of a three-letter code; that is, a sequence of three bases directs each amino acid. For example, RNA composed only of uracil (UUU) incorporates the amino acid phenylalanine; the sequence UUA

Table 9-1   THE GENETIC CODE DICTIONARY[1].

FIRST BASE	SECOND BASE				THIRD BASE
	U	C	A	G	
U (Uracil)	Phenylalanine	Serine	Tyrosine	Cysteine	U
	Phenylalanine	Serine	Tyrosine	Cysteine	C
	Leucine	Serine	Nonsense	Nonsense	A
	Leucine	Serine	Nonsense	Tryptophan	G
C (Cytosine)	Leucine	Proline	Histidine	Argenine	U
	Leucine	Proline	Histidine	Argenine	C
	Leucine	Proline	Glutamine	Argenine	A
	Leucine	Proline	Glutamine	Argenine	G
A (Adenine)	Isoleucine	Threonine	Asparagine	Serine	U
	Isoleucine	Threonine	Asparagine	Serine	C
	Isoleucine	Threonine	Lysine	Argenine	A
	Methionine	Threonine	Lysine	Argenine	G
G (Guanine)	Valine	Alanine	Asparagine	Glycine	U
	Valine	Alanine	Asparagine	Glycine	C
	Valine	Alanine	Glutamine	Glycine	A
	Valine	Alanine	Glutamine	Glycine	G

Source: This code is derived from the work of M. Nirenberg and others.

[1]The three bases of RNA (remember that uracil replaces thymine of DNA) specify the amino acids indicated. The nonsense combinations act as "periods" to stop polypeptide synthesis.

ble of adding a hydroxyl group, will have scarlet red petals. Alternate forms of a particular gene are referred to as **alleles** (*C* and *c* are alleles). The change in the gene's internal structure, which gives rise to new alleles, is known as mutation. **Gene mutations** are changes in the genetic information (changes in the sequence of the nitrogen bases in DNA), and are ultimately responsible for the inherent variation in all living things.

Only one set of instructions is necessary for the proper functioning of many organisms. That is, only one set of chromosomes is present in each cell, and there is one gene for each function. There are obvious disadvantages to such a system. Malfunctioning of any part of the essential instructions sacrifices the whole organism. Thus, a system has evolved in which the set of genetic instructions occurs in pairs. Higher plants and animals normally have two complete sets of chromosomes in each cell.

The characteristic number of chromosomes in the cells of higher plants is known as the $2n$, the somatic or diploid number (for example, $2n = 12$ in spinach and rice, 14 in barley and red clover, 16 in onion and peach, 18 in cabbage and sugar beet, 20 in corn and soybean, 22 in watermelon and bean, 24 in tomato and tulip). The reproductive cells, or **gametes,** however, receive only one chromosome from each pair, the haploid number (n). Fertilization subsequently restores the diploid number to the zygote, the fertilized egg.

As each pair of genes may be present in more than one form (for example, *C* and *c*), a particular pair of alleles may be present in any one of three combinations: *CC*, *Cc*, or *cc*. A plant containing two identical genes, *CC* or *cc*, is **homozygous** for that allele. When the alleles are different, as with *Cc*, the plant is **heterozygous.** In a plant containing the heterozygous pair *Cc* the gametes may be *C* or *c*.

What is the difference in outward expression (**phenotype**) when the genetic constitution (**genotype**) is *CC*, *Cc*, or *cc*? With reference to our example of petal color, if the allele *c* is completely nonfunctional, as regards hydroxylation of pelargonidin, two alleles (*cc*) should not be any more efficient. That this is so has been shown experimen-

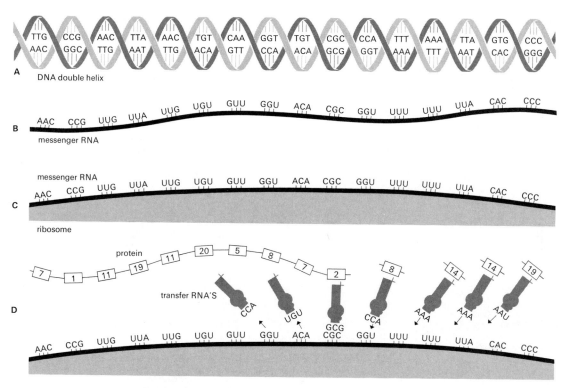

FIGURE 9-3. *A schematic representation of protein synthesis as related to the genetic code. The complementary code, here derived from the dark-lettered strand of DNA (A), is transferred into RNA (B) and moves to the ribosome (C), the site of protein synthesis. (D) Amino acids (numbered rectangles) are presumably carried to proper sites on messenger RNA by transfer RNA and are linked to form protein. [After Nirenberg, "The Genetic Code."* Copyright © 1963 by Scientific American, Inc. All rights reserved.]

tally, for flowers of *cc* plants are scarlet, and chemical analysis of their petals yields no cyanidin. The difference between plants that are homozygous (*CC*) and those that are heterozygous (*Cc*) cannot be predicted. The allele may be efficient enough to produce sufficient enzyme such that plants with

FIGURE 9-4. *The conversion of pelargonadin to cyanidin involves the addition of a single hydroxyl group and is controlled by a single gene.*

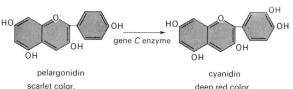

pelargonidin
scarlet color,
as in scarlet asters
or pelargoniums

cyanidin
deep red color,
as in cranberries
or deep red roses

only one functioning allele (*Cc*) cannot be distinguished phenotypically from those having two functioning alleles (*CC*). When this is the case it would appear that the allele *C* dominates *c*. The condition in which heterozygous plants *Cc* are indistinguishable from homozygous *CC* plants is termed **dominance**. In genetic terminology *C* is **dominant,** *c* **recessive.** If the heterozygote *Cc* is intermediate in phenotype between the two homozygous types *CC* and *cc*, dominance is said to be incomplete.

Cell division

The basic difference between sexual and asexual reproduction lies in the way the chromosomes are distributed during cell division—**mitosis** and **meiosis** (Figs. 9-5, 9-6).

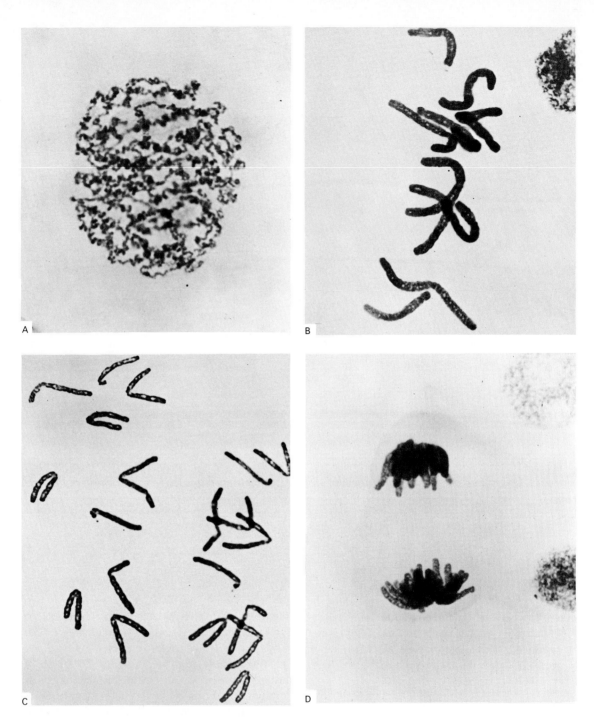

FIGURE 9-5. *Mitosis in the California coastal peony. The vegetative cells of this species have 10 chromosomes* (2n = 10). *Although mitosis is continuous, the process is broken down into a number of stages for descriptive purposes. (A) Prophase. The coiled linear structure of the chromosomes becomes distinct. (B) Metaphase. The chromosomes line up in the equatorial plate. The chromosomes have reduplicated, and appear visibly doubled. (C) Anaphase. The chromosomes separate and approach the poles of the cell. (D) Telophase. The contracted chromosomes are pressed close together at each end of the cell. A wall is subsequently formed across the cell, forming two "daughter" cells with the same number and kind of chromosomes as existed in the original cell. [Courtesy M. S. Walters and S. W. Brown.]*

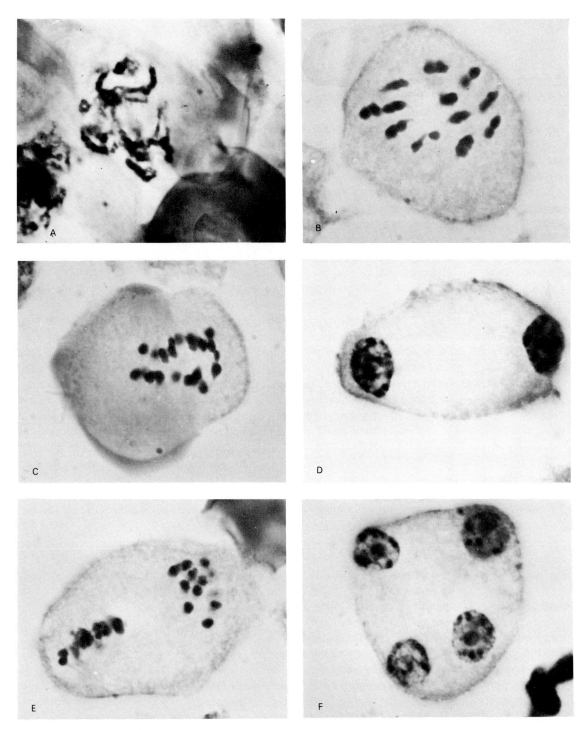

FIGURE 9-6. *Meiosis in pepper. This species has 24 chromosomes in the vegetative cells (2n = 24). Note that there are 2 divisions. (A) In the prophase of the first division, the chromosomes reduplicate and pair. A visibly doubled chromosome pair can be seen in the bottom of the cell. (B) At metaphase of the first division, the 12 chromosome pairs line up on the equatorial plate (face view). Each consists of 2 doubled chromosomes (4 chromatids). (C) Anaphase of the first division. (D) Telophase of the first division. (E) Metaphase of the second division: one plate is a face view; the other is a side view. (F) Telophase of the second division. Walls will form 4 pollen grains, each containing 12 chromosomes, half as many as the original cell.*

MITOSIS   Mitosis is the type of cell division that occurs in growth. It is a synchronized division in which both the chromosomes and the cell divide (Fig. 9-5). The chromosomes duplicate themselves by splitting longitudinally, each half moving to alternate ends of the cell. The two resulting daughter cells thus receive exactly the same number and kind of chromosomes. The division of the cell distributes equally the other constituents of the cell.

The differentiation of cells into tissues and organs ordinarily is not related to any chromosomal difference. That each cell contains all the necessary genetic material implies that any cell holds the potential to give rise to the entire organism. The production of an entire plant from a single carrot parenchyma cell has been achieved. The formation of shoot buds from roots, or of roots from leaves, is a potential result of the genetic continuity made possible by mitosis. The differential patterns of growth would indicate that the instructions carried in the chromosome can be turned on and off. This system itself is no doubt also under the control of the genes.

MEIOSIS   Meiosis is a specialized sequence of two cell divisions that occur prior to the formation of gametes. Meiotic divisions reduce the number of chromosomes by half (Fig. 9-6). In somatic (vegetative) cells the two chromosomes of a pair (homologues) are morphologically similar and contain the same kind of genes, although each member of the gene pair may not be identical because of some mutation. The combination of different alleles, as well as different genes, is responsible for the genetic variability between living things. The sexual process is one mechanism that provides for the reassortment and recombination of genetic factors, a feature that permits organisms to adapt through time to an ever-changing environment. The reassortment of genetic factors between and within chromosomes is accomplished by meiosis; the recombination is accomplished by fertilization.

Meiosis is basically a series of two divisions in which the cells divide twice but the chromosomes only once. This results in four cells, each having the haploid number—that is, half the number of chromosomes present in the somatic cells. Each of these four cells may give rise to a gamete. Fertilization, the fusion of two gametes, subsequently restores the diploid number.

The way in which meiosis differs from mitosis becomes apparent at the first division. The chromosomes divide longitudinally, as in mitosis (into half chromosomes called chromatids), but the homologous visibly doubled chromosomes pair (synapse). During synapsis there may be an actual exchange (crossover) of segments between chromatids of homologous chromosomes. The precise way in which this occurs is still not clearly understood. The attraction between the synapsed chromosomes changes to repulsion and each doubled chromosome moves to opposite poles of the cell without separation of the chromatids. The resultant shuffling of chromosomes occurs at random. In the second division, which immediately follows the first, the chromatids separate.

The most obvious effect of meiosis is the reduction in chromosome number from the diploid to the haploid number necessary to accommodate the fusion of gametes at fertilization and to maintain a uniform chromosome number from generation to generation. This is accomplished by the synchronization of two-cell division and one-chromosome division. (The chromosomes duplicate by the first division and separate in the second.) The less obvious but equally important effect in the long run is the reassortment of the genetic material in the gametes. The reassortment involves not only chromosomes, but chromosome segments that result from crossovers.

Life cycle of plants

The life cycle of plants consists of a series of developmental events involving an alternation of diploid (**sporophytic**) and haploid (**gametophytic**) stages, as is illustrated by the example shown in Figure 9-7.

The haploid stage is produced by the sporophyte through a meiotic division. The resulting haploid cells undergo mitotic divisions and produce distinct entities—gametophytes. The fusion of the male and female gametes restores the diploid condition and gives rise to the embryo from which the sporophyte is produced. In the seed plants, the sporophytic (2n) stage, or generation, is the dominant, independent one, and the gametophytic (n) generation is short and dependent, or parasitic, on the sporophyte. In more primitive plants, such as the ferns, these two generations alternate, each in turn leading an independent existence. In plants that are even more primitive, such as the mosses, the gametophytic generation is independent and of long duration, and the sporophytic generation is transitory and dependent.

In higher plants the male gametophyte, or **pollen spore,** is produced in anthers and usually consists of two sperm nuclei (the gametes) and a tube nucleus. The female gametophyte is produced in a specialized region of the ovary, called the ovule. Although there are various patterns, the developed female gametophyte commonly consists of eight nuclei and is known as the **embryo sac.** One nucleus is the egg; two polar nuclei eventually contribute to the formation of the endosperm, a tissue that supports the young embryo. The function of the five remaining nuclei is obscure.

POLLINATION   Pollination is the transfer of pollen from the anther to the stigma. The transfer of pollen within the same flower (or any flower on the same plant or clone) is known as **self-pollination;** the transfer of pollen to a flower on a plant of different genetic makeup is **cross-pollination.** Self-pollination may be accomplished by gravity or by actual contact of the shedding anther with the sticky surface of the stigma. In cross-pollination wind and insects are the important agents of pollen transfer. Most plants both self- and cross-pollinate naturally, with the proportion of each mode depending upon functional or structural features of the flower or upon genetic incompatibility. Plants may be grouped by their usual method of pollination as **self-pollinated** (cross-pollination less than about 4 to 5 percent), **cross-pollinated** (cross-pollination predominant), and **self- and cross-pollinated.** The mode of pollination has important consequences to the plant's genetic structure (see Chapter 19). The likelihood of reproducing a particular plant exactly depends on its natural method of pollination. Seed propagation duplicates plants that are highly self-pollinated because such plants tend

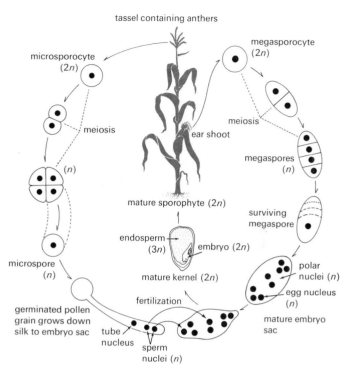

FIGURE 9-7. *The life cycle of corn, illustrating formation of pollen and embryo sac.* [*From Srb, Owen, and Edgar,* General Genetics. *San Francisco: W. H. Freeman and Company,* © *1965.*]

to be homozygous. Since cross-pollinated plants are highly heterozygous, they can be duplicated exactly only by asexual methods. Nevertheless, a high degree of uniformity in some characters may be achieved in the seed propagation of cross-pollinated plants by constant selection.

Natural self-pollination is achieved through functional and structural features of the flower. Perfect flowers—those that contain both stamens and pistils—lend themselves to self-pollination. In the violet, the pollen sheds before the flower is open; in the tomato the pistil grows through a sheath of anthers.

Cross-pollination is brought about in many different ways. Stamens and pistils may occur on separate flowers (**monoecism**), as in corn and cucumbers, or on separate plants (**dioecism**), as in spinach, asparagus, hemp, hops, and date palm. However, many perfect-flowered plants cross-pollinate. This is achieved by anatomical or physiological features of the flower that prevent self-pollination (selfing). For example, the differential maturation of stamens or pistils will prevent natural selfing. The structural features of the flower that insure cross-pollination are often adapted to pollen transfer by insects. Among the special adaptations that aid in insect pollination are petal color, odor, and presence of nectar. In some plants an intimate interdependence exists between structural features of the flower and insect pollination (Fig. 9-8).

Incompatibility (self-sterility) is a physiological mechanism that prevents self-fertilization. A genetic factor (or factors) serves to prevent pollen tubes produced by the plant from growing in the style of the same plant. Incompatibility factors prevent self-pollination in such crop plants as alfalfa, cabbage, tobacco, and apple.

FERTILIZATION   After a pollen grain lands on the stigmal surface of the pistil, it absorbs water and such substances as sugars, and forms a pollen tube, which literally grows down the style to the embryo sac. The pollen tube penetrates the embryo sac, where one male gamete unites with the egg to form the **zygote.** After mitotic division the zygote becomes the **embryo** of the resultant seed. In angiosperms the other male gamete fuses with the two polar nuclei and forms the **endosperm.** This complete process is referred to as **double fertilization.** The endosperm of gymnosperms is formed from mototic divisions in the gametophyte.

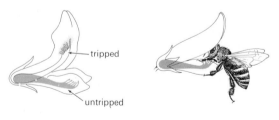

FIGURE 9-8. *Flower structure and pollinating mechanism in alfalfa. Pollen is deposited on the bee when the staminal column (pistil and stamens) is tripped.* [*After Poehlman,* Breeding Field Crops. *New York: Holt, Rinehart & Winston, 1959.*]

SEED MATURATION   Fertilization initiates rapid growth of the ovary and subsequent development of the seed. Usually the ovary will not develop unless it contains viable and growing seed. Common exceptions (**parthenocarpy**) are seedless grapes and oranges, in which the developing seed breaks down at an early stage of development.

In the developing seed the development of the embryo is preceded by endosperm growth. When rapid growth of the embryo begins, it does so at the expense of the endosperm. The amount of endosperm at maturity varies with different plants. The endosperm may be completely absorbed at seed maturity, in which case stored food may be in the embryo itself, as in the cotyledons of bean seeds.

Seed development is completed with the formation of hardened integuments, or seed coats. In one-seeded fruits, these integuments include maternal ovarian tissue.

PLANT PROPAGATION

The basic objectives of plant propagation are to achieve an increase in number and to preserve the essential characteristics of the plant. Propagation may achieved sexually by seed or asexually by utilizing specialized vegetative structures or employing such artificial techniques as grafting, layering, and cutting.

Seed propagation

Seed is the most common means of propagating self-pollinated plants, and is extensively used for many cross-pollinated plants. It is often the only possible or practical method of propagation. There are many advantages in propagating by seed. It is usually the most inexpensive method of propagation. Seeds also offer a convenient means of storing plants over long periods of time. When kept dry and cool, seeds normally remain viable from harvest to the following planting season, and under suitable storage conditions seeds can be kept in viable condition for many years. Seed propagation also makes it possible to start "disease-free" plants. This is especially important with respect to virus diseases, since it is almost impossible to free most virus-infected plants. Most virus diseases are usually not seed transmitted. The disadvantages of seed propagation are the genetic segregation in heterozygous plants and the long time required in some plants from seed to maturity. Potatoes, for example, not only do not breed true from seed but do not produce large tubers the first year. These disadvantages are overcome by vegetative propagation.

SEED TECHNOLOGY   The practice of saving seed to plant the following year has developed into a very specialized technology. Seed technology involves all the steps necessary to assure the production and utilization of high-quality seed (seed with high viability, freedom from disease, purity, and trueness to name)—specialized growing and harvesting techniques, cleaning and separation, moisture regulation, and a number of distinct processes for improving the viability and germination powers of the seed or the subsequent performance of the crop. Examples include delinting of cotton seed, separation of multigerm seed such as beet, stratification and scarification of dormant or hard seeds, inoculation of seeds of nitrogen-fixing legumes with bacterial cultures to improve crop performance, and chemical treatment to control seed- or soil-borne diseases. Finally, the marketing of seeds involves distinct storage, packaging, labeling, and transporting procedures. The seed is a crucial stage in the entire agricultural cycle. The regulation and control of the movement and labeling of seed and the establishment of seed improvement and certification agencies to effect the distribution of improved seed are associated with a highly developed agriculture.

The location of the seed-producing area varies with the crop. When the seed or the associated fruit is the commercial part of the crop, as it is in the grains, seed is usually produced within the crop-producing areas, in which case seed- and crop-production practices differ little. But when the seed is not the usable part of the plant (as in root crops, forage crops, many vegetables, and flowers), seed production tends to concentrate in areas separate from crop-production areas, and develops into a very specialized practice. The use of distinct seed-producing areas may be imposed by specific flowering requirements, such as cold induction or photoperiod. For example, seeds of many biennial vegetables, such as cabbage and carrots, cannot be produced in tropical areas, as they require cold to flower. Among other requirements that must be taken into account for economical seed production are environmental factors such as low moisture at harvest which permits proper seed maturation and the reduction of diseases that attack the seed. Isolation is often necessary to prevent contamination in cross-pollinating species. Relatively little tree and shrub seed is grown commercially. Forest tree seed is usually collected from natural stands. The intensification of forest tree plantings, however, has encouraged the development of seed orchards consisting of groves of selected, genetically superior trees intensively managed to produce a large seed crop. Seed to produce rootstocks for fruit trees is often obtained as a by-product in the fruit-processing industries.

The harvesting of many seed crops includes cutting the plant (reaping) and separating the seed from the plant (threshing). Modern combines reap, thresh, and partially clean seed in one operation. Although many seed crops are mechanically harvested, a considerable amount of hand labor is still used on crops whose seed is dispersed easily (shatters) or is produced so gradually that there may be mature seed pods, flowers, and flower buds at the same time. In some plants, such as carnation, the entire plant is cut and placed on sheets of canvas or in windrows to dry and is then threshed. In many crops the seed must be extracted from the fruits and then cleaned. Seed is removed from dried pods or seed heads by milling. The separation of the seed from fleshy fruits, such as the tomato, is accomplished through fermentation of the macerated pulp.

A great number of special machines have been developed to separate and clean seed through the utilization of differences in the physical characteristics of the seeds.

Size.   Separation of seed by size is usually accomplished by the use of screens of different sizes and shapes. Small, light fragments are removed at the same time by passing an air stream through the seed as it passes from one screen to another.

The air and screen machines are basic processing apparatus, and practically all seeds are first processed in this manner. The machines vary from small two-screen farm models to large industrial types. Seeds whose widths equal their thicknesses can be separated by length with disc separators. A wheel covered with indentations is passed through a mass of seed, and each indentation picks up a seed. The size and shape of the indentation are determined by the seeds being cleaned.

Specific gravity. Seed separation can be achieved by taking advantage of differences in specific gravity. It is thus possible not only to separate different kinds of seed, or remove foreign material, but to achieve a quality separation of the heavy seed of the same crop. The separation of a mixture of grain and chaff by blowing away the light fragments or tossing the mixture in the air is a very ancient technique. One ingenious machine, the specific gravity separator, functions by stratifying a mixture of seed by the use of an air column over a vibrating platform. The different layers can be separated by a combination of deck motion and gravity.

Aspirators are used to separate materials according to their terminal velocity, the speed at which a material falls through an air stream. The terminal velocity is affected by size, shape, and specific gravity.

Surface characteristics. A number of machines have been designed to take advantage of differences in the surface characteristics of seed, such as texture, shape, and color. For example, rollers covered with a velvet-like material lift out rough seeds (such as dodder) and leave smooth seeds behind. Seed with a rough, gelatinous, or granular surface can be separated from smooth seed on the basis of differential retention of a fine iron powder applied to the moistened seed. The actual separation is achieved over a magnetized metal drum. Differences in roundness are exploited by passing seed down a spiral incline. Rounder seed falls faster and is forced out of the channel by centrifugal force; less round and slower rolling seed slides down the spiral. Seed of different colors can be separated by machines equipped with photoelectric cells. These machines are also used to grade such seed crops as coffee.

SEED STORAGE   Seed is living material capable of surviving in a more or less metabolically suspended state. The storage life of seeds varies with the species and with environmental conditions. When storing seed for extended periods it is necessary to avoid conditions favorable for respiration and enzymatic action in general by controlling moisture, temperature, and oxygen availability.

Low moisture content is critical for extended storage life of most seed. Because seeds frequently contain 16 to 20 percent moisture when harvested, drying is necessary to retain maximum viability. There are exceptions, however; sugar maple and citrus species, for example, lose viability when dried. Some seeds dry naturally in the field except under conditions of high humidity. Others may be dried artificially by forced air, which is often heated to increase its drying power. The moisture content of seed fluctuates with the humidity of the surrounding atmosphere, consequently the storage humidity is very important. Most seeds retain their highest viability with a low relative humidity; a relative humidity of 65 percent or more is unsafe for storage.

The optimum temperature for long-term seed storage lies in the range -18 to $0°C$ (0 to $32°F$).* Moisture and temperature are interrelated; if either is high the other must be low to insure seed viability (Table 9-2). For most seeds, however, a temperature between 0 and $10°C$ (32 and $50°F$) and a relative humidity of 50 to 60 percent is adequate to maintain full viability for at least one year.

The storage life of seed can be extended by controlling the storage atmosphere. Reducing the oxygen content and increasing the CO_2 content decreases respiration. Seeds stored in sealed containers build up a controlled atmosphere naturally through respiration. The atmosphere in sealed containers can be replaced with an inert gas such as nitrogen, which prolongs the storage life of some seeds.

The long-term storage of agricultural seeds offers a number of important economic advantages. Most important, perhaps, is that it permits adjustment

---

* Seeds of the arctic tundra lupine, uncovered after being frozen for 10,000 years, not only germinated but produced normal, healthy plants.

of supply and demand. In high-value speciality crops (hybrid cauliflower) it is often most practical to produce more than a year's supply of seed in one season. This also permits the crop to be tested before the seed is sold. Long-term storage is necessary for such purposes as breeding and experimentation. There are also biological advantages to maintaining a supply of old seed. For example, certain seed-transmitted pathogens (bacterial wilt organism in tomato) have a short storage life, and the disease can be controlled by using old seed.

GERMINATION   Germination—the series of events leading from dormant seed to growing seedling—is dependent upon seed viability, suitable environmental conditions, and in some crops, the breaking of dormancy. The germinating seed and young seedlings are vulnerable to certain diseases, and may need protection.

Seed viability.   Seed viability refers to the percentage of seed that will complete germination, the speed of germination, and the resulting vigor of newly germinated seedlings. The viability of seed lots can be determined by standardized testing procedures. Probably the most significant measure of viability is the germination percentage—the percentage of seed of the species tested that produces normal seedlings under optimum germinating conditions (Fig. 9-9). Germinating tests are usually performed on moistened absorbent paper under rigidly controlled environmental conditions. The length of the test varies with the rate at which different species germinate. Perhaps the greatest problem is in distinguishing between dormant and

Table 9-2   EFFECT OF MOISTURE CONTENT AND STORAGE TEMPERATURE ON VIABILITY OF COTTONSEED STORED IN SEALED CONTAINERS.

APPROXIMATE MOISTURE CONTENT (IN PERCENT)	PERCENT GERMINATION AFTER STORAGE FOR FOLLOWING NUMBER YEARS							
	1	2	3	5½	7	13½	15	19
Stored at 0.5°C (33°F)								
7	87	87	87	90	94	90	91	92
9	92	87	93	89	92	93	91	92
11	89	88	91	79	89	88	93	89
13	90	87	86	87	92	91	72	51
14	88	90	85	61	34	10	0	—
Stored at 21°C (70°F)								
7	93	91	90	84	89	85	73	51
9	87	91	82	81	59	0	—	—
11	86	89	68	1	—	—	—	—
13	72	23	3	—	—	—	—	—
14	17	0	—	—	—	—	—	—
Stored at air temperature (uncontrolled)								
7	87	90	83	—	88	—	0	—
9	91	88	84	—	—	—	0	—
11	85	69	18	—	—	—	—	—
13	49	0	—	—	—	—	—	—
14	0	—	—	—	—	—	—	—
Stored at 32.2°C (90°F)								
7	86	86	59	0	—	—	—	—
9	50	33	0	—	—	—	—	—
11	21	0	—	—	—	—	—	—
13	0	—	—	—	—	—	—	—
14	0	—	—	—	—	—	—	—

Source: After D. M. Simpson, *Agronomy Journal* **49**:608, 1957.

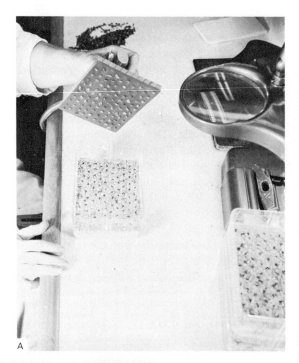

FIGURE 9-9. *Seed testing. (A) Vacuum seed counter picks up 100 seeds and places them on germination medium. (B) Germination rooms with complete environmental control at Indiana State Seed Laboratory. (C) Germination of corn seed is being tested by the "rolled towel" technique. One hundred seeds are placed on moist paper toweling, which is sealed in wax paper, rolled, and stored under controlled temperature. After seven days, the percentage germination is evaluated. [Photographs by J. C. Allen & Son.]*

nonviable seed. Seed dormancy must be overcome to obtain a reliable test. A rapid chemical test with tetrazolium (2,3,5-triphenyltetrazolium chloride) makes possible a rough prediction of viability in nongerminating dormant seed. Living cells turn red, but nonliving cells do not.

Breaking dormancy. The breaking of dormancy and the creation of a suitable environment are necessary to initiate the germination process for some species. The treatments depend on the type of dormancy involved (see Chapter 8). They include scarification, stratification, embryo culture, and various combinations of these treatments, with suitable environmental control. The germination of seed that contains an impervious seed coat may be promoted by **scarification**—the alteration of the seed coat to render it permeable to gases and water. This is accomplished by a number of techniques, mechanical methods involving abrasive action being the most common. The action of hot water, 77 to 100°C (170 to 212°F), is effective in honeylocust seed. Some seed can be scarified by the corrosive action of sulfuric acid.

The seed of some plants requires dry storage for a period of days or months before it will germinate. The physiological basis of this type of dormancy is not clear but seems to depend on the time required for the dissipation of volatile inhibitors. The afterripening of some seeds requires a period of moist, aerobic storage known as **stratification.** Cold-stratification—the afterripening of dormant embryos by storing them at high moisture and low temperature—is a prerequisite for the uniform germination of many temperate-zone species, such as apple, pear, redbud. The cold-stratification of apple and pear seed involves storing the moist, aerated seed at around 2°C (36°F). The germination percentage increases with time in these species until the third month of treatment. The stratification medium consists of moist soil, sand, peat, or substances such as vermiculite. Warm-stratification—moist storage above approximately 7°C (45°F)—promotes germination in some species as a result of microbial decomposition of the seed covering. Seeds that possess double dormancy, such as those of viburnum, are first warm- and then cold-stratified. In some seeds a combination of scarification and cold-stratification is used.

Environmental factors affecting germination. The germination of seed that does not require afterripening, or of seed that has had this requirement satisfied, depends upon external environmental factors: water, favorable temperature, oxygen, and sometimes light.

Water is an essential requirement for germination. The amount of water required varies somewhat with different species. For example, celery requires that soil moisture be near field capacity, whereas tomato will germinate with soil moisture just above the permanent wilting point. For most seed, overwet conditions are harmful, since they prevent aeration and promote conditions favorable for disease.

The effect of temperature upon germination varies with the species, and is related somewhat to the temperature requirements for optimum growth of the mature plant. In general, the germination rate increases with increasing temperature, although the highest germination percentage may be obtained at a relatively low temperature. An alternating temperature is usually more favorable than a constant temperature for grasses.

Because of its critical role in respiration, oxygen is necessary for seed germination in all plants except some water-loving species, such as rice. The use of proper drainage and tilth in seedbeds promotes rapid germination, largely as a result of good aeration. The effect of light in stimulating or inhibiting the germination of some seed, is a reversible red—far-red phenomenon (p. 155). The action of certain salts has been shown to influence germination. At concentrations of 0.1 to 0.2 percent, potassium nitrate will increase germination in a number of plants, and is used in seed testing. Germination is inhibited, however, by high salt concentrations caused by excessive fertilization.

Embryo culture. The artificial culture of the embryo is used to facilitate seed germination in certain species. For example, the embryo in many early-ripening peaches (for example, the Mayflower) is not sufficiently mature to germinate when the fruit is ripe. This problem, a serious impediment to the breeding of early-ripening peaches, can be overcome, however, by excising the embryo from the pit and culturing it under aseptic conditions in media providing certain nutrients.

Tukey's Solution for Culturing Mature and Relatively Immature Embryos

Stock Chemical	Amounts (g/liter H_2O)
KCl	7.5
$CaSO_4$	1.9
$MgSO_4$	1.9
$Ca_3(PO_4)_2$	1.9
$Fe_3(PO_4)_2$	1.9
KNO_3	1.5

The routine germination of orchid seed also involves culture in artificial media. These seeds, which are almost microscopic in size, consist of a simple, undifferentiated embryo with no reserve food.

Disease control. Disease is a constant threat to the germinating seed. This is especially true for seed that must be stratified or which requires an extensive period of time to germinate. The control of these diseases is an integral part of the technology of seed propagation.

Damping-off, the major disease of germinating seeds, is caused by a number of separate fungi, mainly species of *Pythium, Rhizoctonia,* and *Phytophthora.* The disease is expressed either by the failure of the seedling to emerge or by the death of the seedling shortly after emergence. A common symptom is the girdling of young seedling stems at the soil surface. Damping-off usually occurs only in young succulent seedlings during or shortly after germination, but older plants may be affected by severe attacks, which may occur in either field or greenhouse plantings.

Protection from damping-off and other seedling diseases requires both the direct control of the organism and the regulation of environmental conditions such that they favor the rapid growth of the plant rather than the growth of the pathogens. A number of seed treatments are available, either to eliminate the organisms from the seed or to provide protection to the seedling when planted in infested soil. These consist in coating the seed with a suitable fungicide, particularly compounds containing mercury (mercuric chloride, various organic compounds of mercury) or copper (cuprous oxide). A common seed treatment to eliminate surface contaminants consists of a 5-minute dip in a 0.5 percent solution of sodium hypochlorite. The treatment of seeds in hot water, 50°C (122°F), for 15 to 30 minutes has been used for the control of such seed-borne diseases as loose smut of wheat when the organism (*Ustilago tritici*) is present inside the seed. Such treatment must be precise, however, or the seed may be seriously injured.

Soil may be treated by applying fungicides to the upper surface or by applying heat. Raising the soil temperature to 82°C (180°F) for 30 minutes (pasteurization) is always recommended for potting soils to control weeds and nematodes, as well as organisms that cause damping-off. Complete sterilization of soil interferes with nutrient availability and should be avoided. Sphagnum moss has proved satisfactory as a germination and stratification medium for some seeds because inhibitors and low pH prevent the growth of many of the organisms that cause damping-off.

The temperatures most favorable to the fungi responsible for damping-off are 21 to 29°C (70 to 85°F). Thus, damping-off tends to be severe when cool-season crops are germinated at temperatures that are too high. For best control, germination temperatures should be optimum for the crop. Good viable seed and rapid seedling growth are important. Many of the fungi responsible for damping-off are water loving, and are encouraged in wet, poorly drained soils. Cloudy weather and periods of poor drying encourage the damping-off complex.

Vegetative propagation

Asexual reproduction, the basis for vegetative reproduction, enables plants to regenerate tissues and parts. In many plants vegetative propagation is a completely natural process; in others it is more or less artificial. The methods of vegetative propagation are many, and the choice of methods depends on the plant and the objectives of the propagator. The advantages of vegetative propagation are readily apparent. Heterozygous material may be perpetuated without alteration. In addition, vegetative propagation may be easier and faster than seed propagation, as seed dormancy problems may be completely eliminated and the juvenile state reduced. Vegetative propagation also makes it possible to perpetuate seedless clones—for example, Washington Navel orange, Gros Michel banana, Thompson Seedless grape.

The various methods of vegetative propagation are summarized as follows:

1. Utilization of apomictic seed (citrus)
2. Utilization of specialized vegetative structures.
   Runner (strawberry)
   Bulb (tulip)
   Corm (gladiolus)
   Rhizome (iris)
   Offshoot (banana)
   Stem tuber (potato)
   Tuberous root (sweet-potato)

3. Induction of adventitious roots or shoots
Layering: regeneration from vegetative part while still attached to the plant.
Cutting: regeneration from vegetative part detached from the plant.
4. Grafting—the joining of plant parts by means of tissue regeneration.
5. Tissue culture

UTILIZATION OF APOMICTIC SEED **Apomixis** is the development of seeds without the complete sexual process. It is therefore a form of vegetative reproduction. The most significant type of apomixis is that in which the complete meiotic cycle is eliminated. The seed is formed directly from a diploid cell, which may be either the nonreduced megaspore mother cell or some cell from the maternal ovular tissue, such as the nucellus. As a result of apomixis, a heterozygous cross-pollinating plant will appear to breed true.

Although apomixis is widespread within the plant kingdom, it is not a common means of asexual propagation. It is utilized in the propagation of Kentucky bluegrass, citrus, and mango. These species, however, are only partially apomictic, and seed will be derived from both the sexual and asexual processes.

Virus diseases are not readily transmitted through seed, including apomictic seed. This is utilized to free virus-infected clones of citrus. These "nucellar" clones, being asexually produced, will retain all of the characteristics of the parent plant, but will be free of virus infection. Apomictic seedlings, like ordinary seedlings, show juvenile characteristics.

UTILIZATION OF SPECIALIZED VEGETATIVE STRUCTURES The natural increase of many plants is achieved through specialized vegetative structures. These modified roots or stems are often also food storage organs (bulbs, corms, tubers), although some function only for natural vegetative increase (runners). These organs enable the plant to survive adverse conditions or to give the plant a means of spreading. They renew the plant through adventitious roots and shoots and are commonly utilized by man as a means of propagation. The propagating process is referred to as **separation**

when the structures divide naturally, and as **division** when they must be cut.

Stem modifications. **Bulbs** are shortened stems with thick, fleshy leaf scales. In addition to their development at the central growing point, buds develop at the axils of the leaf scales. These buds form miniature bulbs (**bulblets**), which are called **offsets** when grown to full size.

Bulbs are common in the lily family (Liliaceae). The development of bulbs from initiation to flowering size takes a single season in the onion, but most bulbs, like those of the daffodil and hyacinth, continue to grow from the center, becoming larger each year while continually producing new offsets. The asexual propagation of bulb-forming plants is commonly achieved through the development of scale buds. The individual scales, the offsets, or the enlarging mature bulb itself may be used in propagation. The bulb of hyacinth is commonly wounded to encourage adventitious bulblets to form. The bulblets develop into usable size in 2 to 4 years.

Corms, although they resemble bulbs, do not contain fleshy leaves, but are a solid-stem structure containing nodes and internodes. The gladiolus, crocus, and water chestnut (*Eleocharis tuberosa*) are examples of corm-forming plants. In large, mature corms, one or more of the upper buds develop into flowering shoots. The corm is expended in flower production, and the base of the shoot forms a new corm above the old. At season's end, one or more new corms may develop in this manner. **Cormels,** or minature corms, are fleshy buds that develop between the old and new corms. Cormels do not increase in size when planted, but produce larger corms from the base of the new stem axis. These require 1 to 2 years to reach flowering size. This is the usual method of propagating corm-producing plants. Corms may also be increased by division, but this is not commonly practiced because of disease problems.

Runners are specialized aerial stems that develop from the leaf axils at the base or crown of some plants. They provide a means of natural increase and spread. The commercial propagation of strawberries is through runner-plant production. Leaf clusters, which root easily, are formed at the second node of the runner, and may in turn produce new runners. Runnering is photoperiod-sensitive, being commonly initiated when the day length is 12 hours or longer and temperatures are above 10°C (50°F). The yield in plants per mother plant varies with the variety, but averages somewhere between 20 and 30; under optimum conditions it may be as high as 200. Some species of strawberry are nonrunning, and many of the everbearing varieties usually form relatively few runners. Nonrunning plants may be vegetatively propagated by crown divisions, but the increase in plants is much lower.

Horizontally orientated stems growing underground are called **rhizomes.** Rhizomes contain nodes and internodes of various length, and readily produce adventitious roots. Rhizomes may be thick and fleshy (iris) or slender and elongated (Kentucky bluegrass). Growth proceeds from the terminal bud or through lateral shoots. In many plants the older portion of the rhizomes dies out. If new growth proceeds from branching, the new plants eventually separate. Rhizomatous plants are easily propagated by cutting the rhizome into pieces containing a vegetative bud.

Tubers are fleshy portions of rhizomes. The potato, the best known example, is propagated by planting either the whole tuber or pieces containing at least one bud cluster, or "eye." If the whole tuber is planted, the terminal eye commonly inhibits the other buds, but this apical dominance is destroyed when the tuber is cut. Commonly, the "seed pieces" are kept at 1 to 2 ounces to provide sufficient food for the young plant. The seed pieces may be cured (stored in a warm place) to effectively heal the cut surface. This is accomplished by active cell division at the cut surface and the formation of a waxy substance called suberin. Chemical treatments are used to prevent disease.

In many plants lateral shoots (**offshoots**) develop from the stem, which when rooted serve to duplicate the plant. They are called offsets, suckers, or crown divisions, depending on the species. Propagation of plants that produce offshoots is easily made by division.

Root modifications. Roots as well as stems may be structurally modified to propagative and food storage organs. Fleshy, swollen roots that store food materials are known as **tuberous roots.** Shoot buds are readily formed adventitiously. The sweetpotato is commonly propagated from the formation of rooted adventitious shoots called slips. Tuberous roots of some species may contain shoot buds at the "stem end" as part of their structure. In the dahlia, the roots are divided, but each tuberous root must incorporate a bud from the crown; in the tuberous begonia, the primary taproot develops into an enlarged tuberous root that can be propagated by division if each section contains a bud.

INDUCTION OF ADVENTITIOUS ROOTS AND SHOOTS The regeneration of structural parts in the propagation of many plants is accomplished by the artificial induction of adventitious roots and shoots. The process is called **layerage** when induction of new roots or shoots takes place while the original vegetative part is attached to the plant, and **cuttage** when induction takes place after the vegetative part has been detached from the plant of origin. These two processes are part of the same phenomenon—the ability of vegetative plant parts to develop into a complete plant.

Layerage often occurs naturally. In the black raspberry, the drooping stem tips tend to root when in contact with the soil. Because the regenerated stem is still attached and nourished by the parent plant, the timing and techniques of layerage are not as critical as in cuttage. Layerage is thus a very simple and effective means of propagation. Rooting may be facilitated by such practices as wounding, girdling, etiolation, and disorientation of the stem, which affects the movement and accumulation of carbohydrates and auxin.

Cuttage is one of the most important means of vegetative propagation. The term "cutting" refers to any detached vegetative plant part that can be expected to regenerate the missing part (or parts) to form a complete plant. Cuttings are commonly classified by plant part (root, stem, leaf, bud).

ANATOMICAL BASIS OF ADVENTITIOUS ROOTS AND SHOOTS The formation of adventitious roots can be divided into two phases: (1) initiation, characterized by cell division and the differentiation of certain cells into a root initial; and (2) growth, in which the root initial expands by a combination of cell division and elongation. Although the two processes usually occur in rapid sequence, in some plants, such as the willow, the time between initiation and development is well separated.

Root initials are formed adjacent to vascular tissue. In herbaceous plants, which lack a cambium, the root initials are formed near the vascular bundles close to the phloem. Thus, roots appear in rows along the stem, corresponding to the major vascular bundles. In woody plants, initiation commonly occurs in the phloem tissue, usually at a point corresponding to the entrance of a vascular ray. Adventitious roots and shoots from leaf cuttings both commonly originate in secondary meristematic tissues—cells that have differentiated but subsequently resume meristematic activity.

Adventitious roots and shoots may be derived from different tissue; for example, in African violet leaf cuttings the roots are initiated from cells between the vascular bundles, whereas shoots are initiated from cells of the epidermis or cortex.

PHYSIOLOGICAL BASIS OF ROOTING   The ability of a stem to root is a variable character, depending on the plant, its age, its environmental history, and type of treatment. Some insight into the physiological basis of rooting has been developed from studies on closely related **easy-to-root** and **difficult-to-root** plants. The ability of a stem to root has been shown to be due to an interaction of inherent factors present in the stem cells, as well as to transportable substances produced in leaves and buds. Some of these transportable substances are auxin, carbohydrates, nitrogenous compounds, vitamins, and others that have not been identified. Substances that interact with auxin to induce rooting are called **rooting cofactors.** In addition, such environmental factors as light, temperature, humidity, and oxygen play an important role in the process. The physiological factors involved in rooting are only beginning to be understood; it is still not possible to induce rooting easily in many plants—for example, in blue spruce, rubber tree, and oak.

Auxin level is closely associated with adventitious rooting of stem cuttings, although the precise relationships are not clear. Rooting of stems appears to be triggered by the accumulation of auxin at the base of a cutting. The increase in rooting

achieved by applying indoleacetic acid or some other auxin supports this concept. It is certain, however, that auxin is not solely responsible, for rooting of many difficult-to-root cuttings is not improved by auxin alone. Many specific compounds are known that either stimulate (catechol) or inhibit rooting (5-fluorouracil).

The presence of leaves and buds exerts a powerful influence on the rooting of stem cuttings. In many plants the effect of buds is due primarily to their role as a source of auxin, whereas the rooting stimulation provided by leaves can be shown to be due to additional transportable cofactors, which complement both carbohydrate and auxin application (Fig. 9-10). Rooting is also affected by the nutritional status of the plant. In general, carbohydrates stimulate rooting.

The accumulation of auxin as well as carbohydrates explains in part the effectiveness of ringing and wounding in stimulating rooting. Callusing of the wounded surface also increases the efficiency of water absorption.

The effectiveness of stem rooting varies with the stage of development and the age of the plant, the type and location of stem, and the time of year. Owing to the great variation between species, precise conclusions concerning the relationship of these factors to rooting cannot be made. In general, greater rooting ability is associated with the juvenile stage of growth. Such plants as English ivy, apple, and many conifers become very difficult to root when they reach the mature stage. Mature,

FIGURE 9-10. *The rooting of a cutting is dependent on auxin, carbohydrates, and other rooting cofactors. The cofactors interact with auxin to trigger rooting.*

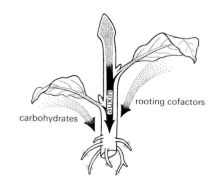

difficult-to-root plants may be made easy-to-root by a reversion to the juvenile stage. Generally, adventitious shoots from the base of the mature plants tend to assume juvenile characteristics. In mature plants that become difficult to root, adventitious shoots may be induced by severe pruning. A form of layering called **stooling** maintains the juvenile stage of growth by repeated pruning to induce the development of adventitious shoots at the base of the plant.

The ability of a stem to root is also affected by its position on the plant; lateral shoots tend to root better than terminal shoots. Vegetative shoots also tend to root better than flowering shoots. These differences are probably related in part to auxin level and amounts of stored food.

Cuttings vary in their ability to root, depending upon the type of stem tissue from which they are derived. Cuttings may be made from succulent, nonlignified growth (**softwood cuttings**) or from wood up to several years old (**hardwood cuttings**). Although almost all types of cuttings of easy-to-root plants root readily, softwood cuttings taken from deciduous, woody plants in the spring or summer generally root more easily than hardwood cuttings taken in the winter. Dormant hardwood cuttings are used when possible, however, because of the ease of shipment and handling. Dormant cuttings must be stored until the shoot's rest period is broken, although rooting is less affected by dormancy than is shoot development. The time at which softwood cuttings should be taken varies greatly with the species. In the azalea, softwood cuttings root best in the early spring; in other broad-leafed evergreens, the optimum time for rooting may be from spring to late fall.

ENVIRONMENTAL FACTORS AFFECTING ROOTING

Humidity. The death of the stem as a result of desiccation before rooting is achieved is one of the primary causes of failure in propagation by cutting. The lack of roots prevents sufficient water intake, although the intact leaves and new shoot growth continue to lose water by transpiration. Leaves or portions of the leaf may be removed to prevent excessive transpiration, but this practice is not desirable because the presence of leaves

encourages rooting. The use of mist maintains high humidity and also reduces leaf temperature by maintaining a water film on the leaf. This allows a high light intensity to be employed in order that photosynthesis will not be reduced. The use of automatic controls to produce an intermittent mist is desirable because excess water may be harmful to many plants and because higher temperatures can be maintained in the rooting media.

Temperature. The use of bottom heat to maintain the temperature of the rooting medium at about 24°C (75°F) facilitates rooting by stimulating cell division in the rooting area. The aerial portion may be kept cool to reduce transpiration and respiration. Daytime air temperatures of 21 to 27°C (70 to 80°F) are optimum for the rooting of most species.

Light. Light in itself appears to inhibit (or conversely, the lack of light encourages) root initiation. The role of light in inducing rooting varies with the plant and with the method of propagation. Softwood and herbaceous cuttings indirectly respond to light because of its role in the synthesis of carbohydrates. Deciduous hardwood cuttings that contain sufficient stored food, and to which artificial auxin can be supplied, root best in the dark. Root promotion may be achieved by the use of opaque coverings that etiolate the stem. The reason the absence of light favors root initiation in stem tissues is not clear. Etiolation probably affects the accumulation of auxins and other substances that are unstable in light.

Rooting medium. The rooting medium must provide sufficient moisture and oxygen and must be relatively disease free. It is not necessary that it be a source of nutrients until a root system has been established. The rooting medium may, however, have an effect on the percentage of cuttings rooted and on the type of roots formed. Various mixes containing soil, sand, peat, vermiculite (expanded mica), and perlite (expanded volcanic lava) have been widely used. Perlite used alone or in combinations with peat moss has proven especially effective because of its good water-holding properties, drainage, and freedom from root-rotting fungi. Sand or water alone may be satisfactory for some easy-to-root cuttings. When water is used alone, improved results are achieved with aeration.

GRAFTING Grafting involves the joining together of plant parts by means of tissue regeneration, in which the resulting combination of parts achieves physical union to grow as a single plant (Fig. 9-11). (Natural grafts may form as a result of the close intertwining of roots or stems.) The part of the combination that provides the root is called the **stock;** the added piece is called the **scion.** When the graft combination is made up of more than two parts, the middle piece is referred to as an **interstock.** When the scion consists of a single bud, the process is referred to as **budding** (Fig. 9-11).

There are two basic kinds of grafts: approach and detached scion. In the **approach graft** the scion and the stock are each connected to a growing root system. In the **detached scion graft,** the most common method, only the stock supplies roots, since the scion is severed from any root connection. Approach grafting is used when it is difficult to obtain a union by the ordinary procedures. The root connection to the scion acts as a "nurse" to the scion until union is achieved, at which time the scion is severed from its own roots.

Grafting has many uses in addition to propagation. The interaction of two or more genetic components in a graft combination may affect growth (dwarfing or invigoration), disease resistance, and hardiness. Other uses include variety change (topworking) and repair (inarching, bridging, and bracing).

The graft union. The **graft union** is the basis of graftage. It is formed from the intermingling and interlocking of callus tissue produced from the stock and scion cambia in response to wounding. The cambium is continuous in perennial woody dicots. Monocots with a diffused cambium apparently cannot be grafted. Callus tissue is composed of parenchymatous cells. Under the influence of the cambium, the callus tissue differentiates to form new cambial tissue. This new cambium differentiates to form xylem and phloem to form a living, growing connection between stock and scion (Fig. 9-12). The basic technique of grafting consists in placing the cambial tissue of stock and scion in intimate association such that the resulting callus tissue produced from stock and scion interlocks to form a continuous connection. A snug fit is often obtained by utilizing the tension of the split stock and/or scion. Tape, rubber, and nails are also used to achieve contact.

The graft union provides vascular continuity, and offers little if any barrier to translocation in xylem and phloem or to the movement of hormones or viruses. Consequently, grafting is used in virus identification. A plant that is suspected of having a virus infection, but does not show obvious morphological symptoms, is grafted to a plant that is sensitive to the virus and will show the symptoms of infection; this process is called **indexing.**

Incompatibility. Because plants have no antibody mechanism they have a greater tolerance to grafting than do animals. The ability of two plants to form a successful graft combination is related in part to their natural relationship and thus to their structural and physiological similarities. The inherent failure of two plants to form a successful union—**graft incompatibility**—may be due to both structural and physiological effects. Incompatibility may be expressed in a high percentage of grafting failures; poor, weak, or abnormal growth of the scion; overgrowths at the graft union; or poor mechanical strength of the union, which in extreme cases results in a clean break at the graft. Incompatibility may be manifested immediately or be delayed for several years. In some cases incompatibility has been traced to a virus contributed by one of the graft components, which was itself virus tolerant. Such incompatibility is due to the virus sensitivity of the other component. If the sensitive component is the rootstock, the entire tree may be adversely affected.

In general, grafting within a species results in compatible unions. Grafts made between species of the same genera or species of closely related genera vary in their degree of compatibility. For example, most pears (*Pyrus communis*) may be successfully grafted to quince stock (*Cydonia oblonga*), but the reverse combination, quince scion on pear stock, is not successful. Incompatibility may be bridged by an interpiece that is compatible to both components.

Factors influencing grafting success. In addition to the inherent compatibility of the plant, there are a number of factors that affect successful "take" in grafting. Skillful grafting or budding

182

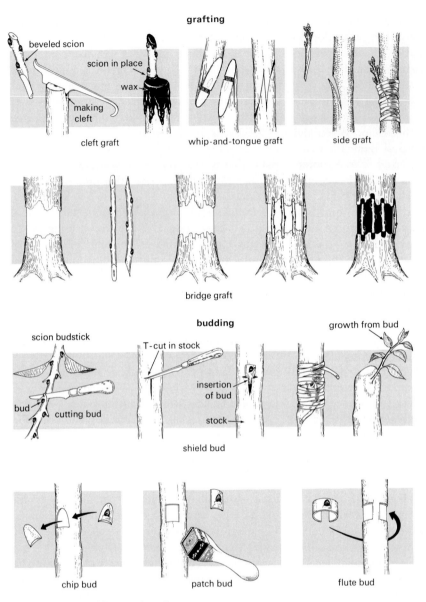

FIGURE 9-11. *Budding and grafting.*

techniques are, of course, necessary. Success is then dependent upon environmental factors that promote callus formation. In general, callus formation is optimum at about 27 to 29°C (80 to 85°F). Grafts of dormant material are often stored in a warm, moist place for a week or two to stimulate callus formation.

It is very important that high moisture be maintained to prevent desiccation of the delicate callus tissue. Special waxes are available that consist of various formulations of resins, beeswax, paraffin, and linseed oil. Bench grafts should be plunged in warm, moist peat to prevent desiccation during the period of healing. The use of plastic films has

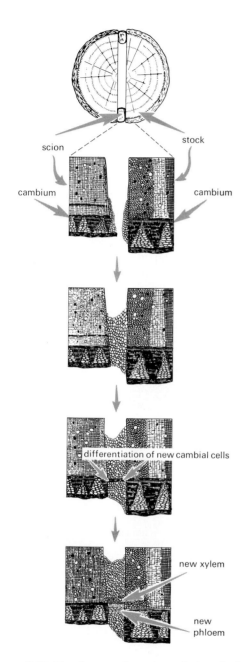

FIGURE 9-12. *Developmental sequence during the healing of a cleft graft union. The graft union is formed from the redifferentiation of the callus tissue under the influence of the stock cambium.* [*After Hartmann and Kester,* Plant Propagation. *Englewood Cliffs: Prentice-Hall, 1959.*]

proved successful in conserving moisture. In some plants (for example, the grape), oxygen has been shown to be required for callus formation, hence waxing should not be used.

The percentage of "take" in grafting is often improved if the stock is in a vigorous state of growth. The scion, however, should be dormant to prevent premature growth and subsequent desiccation. Growing trees are often grafted with dormant scions from refrigerated storage. In summer budding, irrigation should be supplied before budding in order to invigorate the stock. When grafted in late summer, the leaf buds of most woody plants of the temperate zone remain naturally dormant until the following spring.

TISSUE CULTURE **Tissue culture** is the aseptic growth of tissue or organs in artificial media. Although the culture of plant tissue has long been a tool of the physiologist, recently these techniques have become increasingly used as a means of plant propagation. There are two types of plant propagation involving tissue culture (also referred to as **micropropagation**): meristem culture and callus culture.

Meristem culture. **Meristem** or **shoot culture** involves the culture of meristematic tissue directly. Tissue growth is exogenous; that is, the epidermis remains continuous and growth does not erupt adventitiously. The meristematic region may be the embryo or the shoot apex (bud, runner tip, stem tip, and so forth) along with rudimentary leaves.

The meristem is often free of pathogens, including virus. Thus meristem culture is being used to obtain disease-free clones of carnation, chrysanthemum, geranium, and other plants.

Meristem culture has proven very efficient in the clonal increase of some orchid genera. This is potentially very valuable because the usual method of vegetative propagation of orchids by offshoots is very slow. In some genera (*Cymbidium, Cattlea, Dendrobium*) the culture of shoot tips results in the development of bulblets resembling protocorms; a stage between seed germination and full development of seedlings. Protocorm-like bodies may be multiplied by cutting these bodies into quarters. Vegetative increase may be extremely rapid (quadrupling every 10 days). Vegetative propagules of orchid under such a system are called **mericlones,** however, such rapid propagation is not possible with all orchids. Species of *Vanda* and *Phaleonopsis* do not form such protocorm-like bodies.

Callus culture. **Callus culture** involves (1) induction and growth of callus tissue and (2) organ differentiation ultimately leading to a mature plant (Fig. 9-13). In the first step some plant tissue is induced to produce undifferentiated growth or callus. These relatively isodimetric and rapidly dividing cells usually arise from active cell division of the cut surface but may also come from single cells. Recently even pollen grains have been induced to proliferate callus tissue. The change from growth of the callus to organ differentiation can be obtained with changes in the media after some critical mass of tissue is produced. In many cases the critical factor is the concentration and balance of auxin and cytokinin (discussed in Chapter 8). High auxin promotes root initiation, high cytokinin promotes shoot initiation.

Tissue culture techniques require surgical removal of tissue explants under aseptic conditions. Special media have been developed for different plant material. These involve inorganic salts (Table 9-3), sugar, vitamins, growth regulators, and organic complexes such as coconut "water" or yeast extract. Semisolid or solid media is made with the addition of agar; liquid media is often agitated and aeration may improve results. Diffuse light is required in most instances after callus formation.

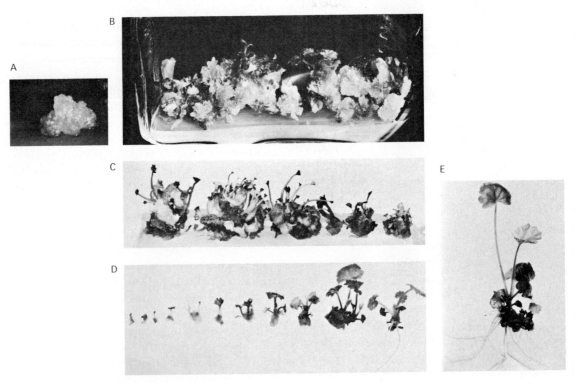

FIGURE 9-13. *Callus culture of geranium.* (A) *Undifferentiated callus;* (B) *Various stages in callus differentiation;* (C) *Shoot development;* (D) *Root development;* (E) *A plantlet two inches long derived from callus culture.* [Courtesy A. C. Hildebrandt.]

Table 9-3 INORGANIC CONSTITUENTS USED IN VARIOUS MEDIA FOR CULTURE OF TISSUES, ORGANS, AND EMBRYOS (IN MG/LITER).

CHEMICAL COMPOUND	KNOP	WHITE	HELLER	MURASHIRE AND SKOOG	
				BASAL	HIGH SALT[1]
KNO_3	200	80		80	1,900
$Ca(NO_3)_2 \cdot 4H_2O$	800	200		144	
$NaNO_3$			600		
NH_4NO_3				400	1,650
KH_2PO_4	200			12.5	170
$NaH_2PO_4 \cdot H_2O$		17	125		
Na_2SO_4		200			
$MgSO_4 \cdot 7H_2O$	200	360	250	72	370
KCl		65	750	65	
$CaCl_2 \cdot 2H_2O$			75		440

Source: Hartmann and Kester, *Plant Propagation* (2nd ed.), Prentice Hall, Englewood Cliffs, New Jersey, 1968.

[1]Also contains $MnSO_4 \cdot 4H_2O$, 22.3, H_3BO_3, 6.2; $ZnSO_4 \cdot 4H_2O$, 8.6 KI, 0.83; $Na_2MoO_4 \cdot 2H_2O$, 0.25; $CuSO_4 \cdot 5H_2O$, 0.025; and $CoCl_2 \cdot 6H_2O$, 0.025; all in mg/liter.

Also add $Na_2 \cdot EDTA$, 37.3 plus $FeSO_4 \cdot 7H_2O$, 27.8 (as 5 ml/liter of stock solution of $FeSO_4$ $7H_2O$, 5.57 g, plus 7.45 g $Na_2 \cdot EDTA$ per liter of water).

Seed and plant certification

Seed or plant certification is part of a distribution system designed to preserve and maintain the identity and quality of crop varieties. A new or improved variety may literally start with a handful of seed or a single vegetatively propagated plant, although many tons of seed or millions of plants may be ultimately required in a single growing season. In the multiplication and subsequent distribution, some control must be exercised to assure that the essential characteristics of the variety are not lost by contamination, mix-up, or faulty production practices. This control is best handled through an organization association, or agency with legal or official status.

The certifying agency obtains a new or improved strain of crop plant identified by a distinct varietal name from the originator. The original seed is known as breeder, or basic, seed. Because in many crops the first multiplication of breeder seed will not be sufficient to supply the needs of the grower, a series of multiplicative generations is necessary. Although each generation removed from breeder's seed increases the chance of alteration, the most damaging alterations are those which occur early. Consequently, standards must

be highest in the early steps of multiplication. The certifying agency assumes the responsibility for each step, although the actual production may be contracted to private growers. The certifying agency provides supervision, inspection, laboratory services, and sets and maintains production standards. It receives a payment for its services through membership and inspection fees. Fees may be based on the number of fields, acres, or bushels, or all three. The certification tag specifies varietal identification and the year for which it is certified. Certified seed, like all other seed in the United States, is subject to seed control laws that require detailed labeling on germination percentage, mechanical purity, amount of weed seed, origin, and moisture. The labeling usually also includes the date the seed was produced and the name or code number of the individual grower involved. The first, second, and third generations removed from the original breeder's seed are identified in various ways, usually as foundation, registered, and certified seed, respectively. Foundation and registered seed is used for further seed production. Certified seed is used for commercial crop production.

Standards of certification for vegetatively propagated crops are concerned primarily with diseases, particularly virus disease. Foundation plants, for example, are checked each year by suitable indexing procedures to insure freedom from virus infection. In the United States the propagation of potatoes, strawberries, and citrus is covered by plant certification programs.

The success of these programs is well documented. Ultimately, everyone—the breeder, the seed producer, the grower, and the eventual consumer,—profits from a well-executed program.

Selected References

Crocker, W., and L. V. Barton. *Physiology of Seeds.* Ronald Press, New York, 1953. (A comprehensive review.)

Hartmann, H. T., and D. E. Kester. *Plant Propagation.* (2nd ed.). Prentice-Hall, Englewood Cliffs, N.J., 1968. (A valuable, practical book; the most authorative in the field.)

Kozlowski, T. T. (editor). *Seed Biology* (3 vols.). Academic Press, New York, 1972. (A series of monographs on the physiology, ecology, and pathology of seeds.)

Srb, A. M., R. C. Owen, and R. S. Edgar. *General Genetics* (2nd ed.). W. H. Freeman and Company, San Francisco, 1965. (A well-recognized genetics textbook. Clear and simple exposition of reproduction and the chemical basis of heredity.)

U.S. Department of Agriculture. *Woody Plant Seed Manual.* USDA Miscellaneous Publication 654, 1948. (A manual on all phases of seed handling of North American forest tree species.)

U.S. Department of Agriculture. *Seeds.* Yearbook of Agriculture, 1961. (The story of seeds.)

PART III

PLANT ENVIRONMENT

CHAPTER 10

Light

SOLAR RADIATION

The sun is the earth's source of energy. The sun's energy is transferred across 93 million miles of space in the form of radiation. Solar radiation has two phases, one electric and one magnetic, and unless deflected travels in a straight line at 186,000 miles/sec. It can gain or lose energy, and its effects on matter can be mathematically predicted. For some purposes it is best construed as bundles of energy called **photons.** For other purposes it is best interpreted as moving in **waves,** and is commonly described in terms of its wavelength, just as sound is described in terms of its pitch. But in spite of all the effort that has been spent in study and experimentation, the true nature of solar radiation eludes science.

The transfer of energy is accomplished by radiation, conduction, and convection. **Radiation** refers to an organized flow of energy through space. It travels in the form of electromagnetic waves. When absorbed by a surface radiant energy can be transformed into heat energy and produces an increase in temperature. In **conduction,** energy flows through material from warmer to cooler places. The transfer of heat through the soil is an example of conduction. The ability of a substance to conduct heat is called its conductivity, a property that varies with the material (Table 10-1). The movement of energy by **convection** involves the mass movement of energized particles. For example, warmed air or water rises because its density changes. Still another phenomenon that concerns us greatly is **reflection.** Heat energy as well as light can be reflected from a surface; a sheet of polished metal reflects both heat and light, but not with equal efficiencies.

The quantity and quality (spectral composition) of radiation depend on the temperature of the radiating body. The higher the temperature, the greater the rate of radiation and the greater the proportion of short-wavelength (high-frequency) radiation. Solar radiant energy received at the edge of the earth's atmosphere has a great frequency range—from about 1 billionth of a micron (μ) to 3 million μ or longer. However most is shortwave radiation in the visible, 0.4 to 0.7 μ, to near visible (ultraviolet, 0.1 to 0.4 μ, and infrared, 0.7 to

190

Table 10-1 HEAT CONDUCTIVITY OF VARIOUS SUBSTANCES.

SUBSTANCE	VALUE (CAL/CM-SEC-°C)
Silver	1.0
Iron	0.1
Water	0.0013
Dry soil	0.0003
Sawdust	0.0001
Air	0.00005

4.0 μ) frequency. The radiant energy occurring as visible light is but a small fraction of the frequency range of the electromagnetic spectrum (Fig. 10-1).

Shortwave solar energy from the sun is absorbed at the earth's surface and transformed into heat. The earth then becomes a radiating body at a lower temperature, which averages 14°C (57°F). The energy radiated by the earth is entirely in the long-wave (low-frequency) range.

Many things happen to solar radiation as it traverses the earth's atmosphere. This is readily apparent in cloudy weather, although not so evident on bright and sunny days. Noticeable changes in the solar spectrum include a shift in peak energies toward the red. Blue light is scattered, producing the blue color of the sky.

The atmosphere in the first 10,000 feet above sea level absorbs about 20 percent of the total radiation received, but on a dark cloudy day only 5 to 10 percent of the total radiation may reach the surface of the earth. A high percentage of solar radiation is deflected by scattering caused by molecular nitrogen, oxygen, water, and other gases.

Ozone absorbs much ultraviolet and cosmic

FIGURE 10-1. *The electromagnetic spectrum and action spectra of certain plant processes.* [*After Machlis and Torrey*, Plants in Action. *San Francisco: W. H. Freeman and Company,* © *1959.*]

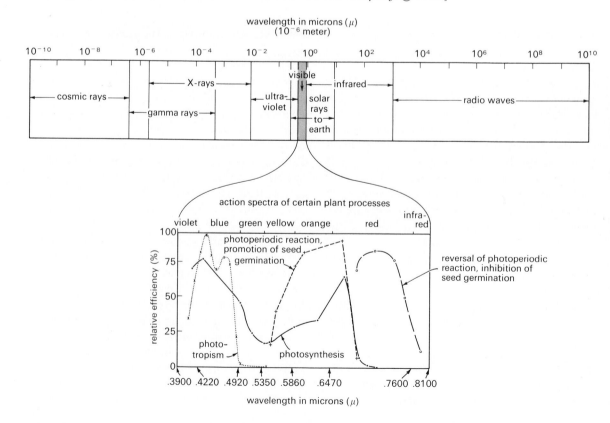

radiation, so that radiation shorter than 2.9 μ seldom reaches the earth. The ozone layer in the atmosphere is one of the most effective filters in nature. It also absorbs and reradiates long-wave radiation from the earth. Water vapor absorbs only about 14 percent of the incoming shortwave radiation, but absorbs 85 percent of the earth's long-wave radiation. This tends to maintain surface temperatures of the earth much higher than they otherwise would be. The earth's atmosphere thus acts as a pane of glass, transparent to the sun's shortwaves, but partially opaque to the earth's long waves; hence the name **greenhouse effect** for this phenomenon (Fig. 10-2).

Radiation absorbed by plants and soil takes one of several routes. Some is conducted into the ground as heat energy. A large part is used in the evaporation of water. A smaller amount is transferred by convection through the layers of air next to the soil. A small but extremely important part, 1 or 2 percent, is captured by photochemical processes. Eventually, however, all is lost back into the atmosphere as long-wave radiation.

THE DISTRIBUTION OF SOLAR ENERGY

Although visible light is but a small fraction of the total range of the electromagnetic spectrum, it is the most important source of energy in today's world, for radiant energy in this special range is converted by plants through photosynthesis to chemical energy. In addition, both shorter and longer wavelengths have a profound influence on plant growth. Some of the solar energy captured by plants in past ages has been converted into coal, oil, and gas, and can be released today by burning.

The amount of light energy that any portion of the earth receives is related to factors that affect the intensity and duration of incoming radiation (Fig. 10-3). The intensity of intercepted radiation depends on the angle at which the solar rays penetrate the earth's atmosphere. Water vapor and, to a smaller extent, air and dust diffuse, reflect, and absorb radiant energy.

The solar radiation received is greatly affected by latitude. Because the earth is spherical, the sun's

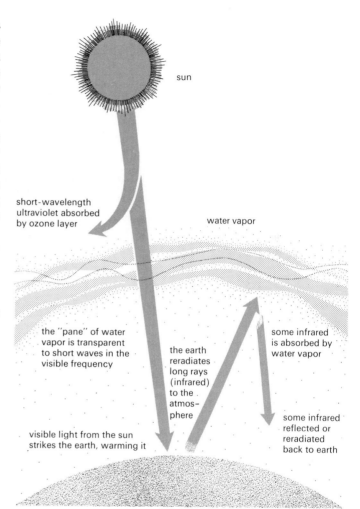

FIGURE 10-2. *The "greenhouse effect" of the earth's atmosphere is confined to the zone within which there are enough particles to interecept a significant proportion of the radiation that passes through it.* [*After Galston,* Life of the Green Plant. *Englewood Cliffs: Prentice-Hall, 1961.*]

rays strike the polar regions at oblique angles. At latitudes near the poles the sun's rays are distributed over a larger surface of the earth and travel longer through the atmosphere than they do at the equator, where the sun's rays strike at right angles to the surface. At the equinox the sun's rays must penetrate an air mass at the poles equivalent to 45 air masses at the equator (Fig. 10-4).

192

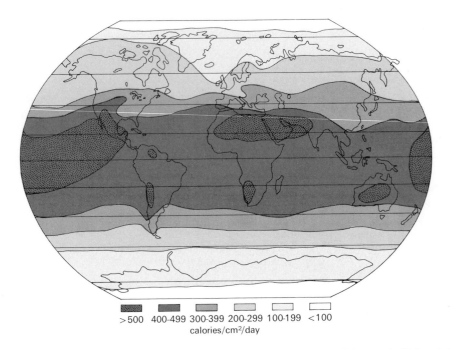

> 500 400-499 300-399 200-299 100-199 <100

calories/cm²/day

FIGURE 10-3. *The mean annual distribution of solar radiation on the surface of the earth.* [*After Ashbel, "Solar Radiation," in* Third International Geophysical Year. *Jerusalem, Israel: The Hebrew University, 1961.*]

We are all familiar with seasonal differences in light. The duration of exposure to radiation changes seasonally as a consequence of variation in day length and cloud cover. Seasonal differences in light intensity occur because radiation passes through a thicker layer of atmosphere in winter than in summer. Since atmospheric gases filter out

FIGURE 10-4. *The effect of latitude on the dispersion of solar energy over the earth's surface during the equinox.* [*After Reifsnyder and Lull, "Radiant Energy in Relation to Forests," in* USDA Tech. Bull. 1344, 1965.]

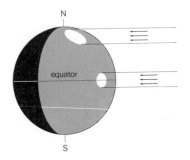

some parts of the spectrum more effectively than others, there are also seasonal differences in the distribution of wavelengths.

The thickness of the air layer through which radiation must pass before reaching the earth changes with altitude. The higher the altitude, the more intense the radiation, particularly of the shorter wavelengths. Some types of radiation, such as cosmic rays, are more abundant at high elevations and may result in higher mutation rates. Mountain climbers must take special precautions to protect exposed parts of their bodies from the intense ultraviolet radiation at higher elevations; the danger of sunburn is much greater at high altitudes than at sea level.

Local topography also determines how much radiation reaches the earth's surface. In the Northern Hemisphere a cool, shady, northern slope may receive only one or two hours of direct radiation during the entire day, and less if it is in a cove or ravine. Slopes that face southward may be droughty as a result of the intense radiation they receive. The patterns of vegetation in mountains develop partly in response to the amount of light the slopes receive.

Vegetation has an influence on the amount of radiation that filters through to the soil surface. The Black Forest in southern Germany, whose dark and mysterious interior was so dreaded by travelers and merchants of the Middle Ages, was so named because of the dense growth of spruce and fir trees, which intercept a high proportion of solar radiation.

Finally, dust and smoke, a special problem in and around cities, may intercept large amounts of radiation. The Los Angeles smog, a prime example, is composed primarily of gases from automobile exhausts, but industries also contribute their share. In addition to being a light screen, smog also has toxic effects on plants and people.

MEASURING RADIANT ENERGY

There are many units in which the energy of solar radiation can be measured, depending on the objective. For engineering purposes energy is often expressed in British Thermal Units (BTU). One BTU is the amount of heat energy required to raise the temperature of one pound of water 1°F. In the metric system the gram calorie is the amount of heat required to raise the temperature of one gram of water 1°C. One BTU is equivalent to 252 gram calories. A langley is equivalent to one gram calorie per square centimeter. Radiant energy can also be expressed in terms of electrical energy (watts) per unit area.

Radiant energy can also be measured in terms of wavelength. The intensity of visible light can be measured in footcandles or lumens. In the English system, a footcandle was originally the amount of illumination shed by a standard candle on a square foot of surface with a curvature radius of 12 inches from the candle. In recent years, however, more reproducible standards have been developed. In the metric system, the lux, equal to one lumen per square meter, is the unit of measurement.

For many purposes, the footcandle has gained wide acceptance. Full sunlight at sea level in the middle of a clear day may reach an intensity of about 10,000 footcandles. On a bright night the

moon provides about one-fiftieth of a footcandle. About 20 footcandles is required for comfortable reading. Although it is convenient to express the light requirements of plants in terms of footcandles, this unit is not quite precise because it is based on the sensitivity of the human eye. When measuring energy required for photosynthesis the quantum is the most appropriate unit. A quantum is the amount of energy delivered by one photon.

Many kinds of instruments have been devised for measuring solar radiation. Most are useful for measuring a particular narrow range of energies. Instruments useful for measuring all wavelengths of solar radiation are called **radiometers.** They are generally sensitive in the range from 0.3 to 100 μ. **Net radiometers** contain two sensitive elements, one facing the sun and the other facing the earth. Long-wave radiation from the earth is subtracted from incoming radiation to give net values.

One of the standard instruments used for measuring direct incoming radiation from the sun is the Eppley normal incidence pyrheliometer. It is designed so that a blackened surface containing 15 thermocouples intercepts a narrow beam of solar radiation and transforms it to heat energy. This heat energy causes the flow of a small current that is measured by a meter.

It is frequently desirable to measure the radiation from a hemispherical surface, for this is the geometry to which a plant is exposed. An Eppley pyrheliometer is one of the instruments used for this purpose. It has a 16-junction thermopile that is exposed to all radiation passing through a hemispherical glass bulb 3 inches in diameter. Other instruments that measure radiation from a hemisphere are based on either the differential bending of metals or on the evaporation of liquids.

The intensity of visible light may be measured by various techniques and devices, such as photocells, exposure meters, photographic paper, and extinction meters. Unfortunately the exact wavelengths measured by most techniques are usually unknown. Spectralradiometers, which measure the spectral composition of radiation, have recently become available and will have many uses. We can expect the future development of meters that will be sensitive to the wavelengths and intensities that plants use in photosynthesis.

EFFECTS OF LIGHT ON PLANTS

Light and plant growth

Light has tremendous effects on the growth processes of plants. In the absence of light plants will grow until their food reserves are exhausted, but their growth in darkness is abnormal and they develop a syndrome known as **etiolation** (Fig. 10-5). They have whitish, spindly stems, and leaves are not fully expanded. Root systems develop poorly, and all tissues are highly succulent. Internodal elongation is extreme, and plants are frequently not strong enough to remain erect.

Normal plant growth requires the entire range of the visible spectrum (white light, or sunlight), because the various light reactions of plants (photosynthesis, phototropism, photoperiodism, and so on) depend on photochemical reactions carried on by specific pigment systems that respond to various wavelengths. The Dutch Committee on Plant Irradiation has described several bands of radiation

FIGURE 10-5. *Etiolated (left) and light-grown (right) bean seedlings.* [*From Bonner and Galston,* Principles of Plant Physiology. *San Francisco: W. H. Freeman and Company,* © *1952.*]

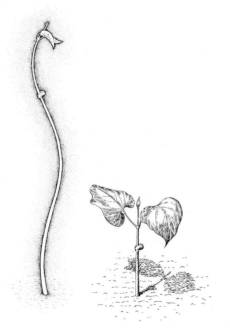

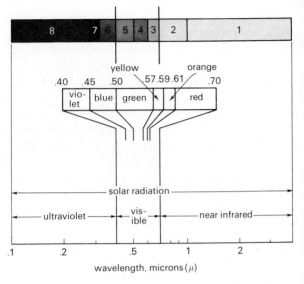

FIGURE 10-6. *Wavelength of solar radiation and spectral bands as defined by the Dutch Committee on Plant Irradiation. Band 1: > 1μ. No specific effects of this radiation on plants are known; to whatever degree it is absorbed by the plant it is transformed into heat without interference of biochemical processes. Band 2: 0.7 to 1 μ. The region of specific elongating effect on plants. Band 3: 0.61 to 0.7 μ. Strongest absorption of chlorophyll and strongest photosynthetic activity. In many cases the region of strongest photoperiodic activity. Band 4: 0.51 to 0.61 μ. Low photosynthetic effectiveness. Band 5: 0.4 to 0.51 μ. Strong chlorophyll absorption and absorption by yellow pigments, and strong photosynthetic activity in the blue-violet. Band 6: 0.315 to 0.4 μ. Fluorescence and strong photographic action. Band 7: 0.28 to 0.315 μ. Antirachitic and germicidal action. Band 8: < 0.28 μ. Below the atmospheric transmittance limit for sunshine.* [*After Reifsnyder and Lull, "Radiant Energy in Relation to Forests," in* USDA Tech. Bull. 1344, 1965.]

according to the biological effectiveness of certain wavelengths, disregarding the vague references to color (Fig. 10-6). Variation in light quality does differentially affect many physiological processes; the light quality over the surface of the earth, however, is relatively uniform. Thus the critical aspects of light as an environmental factor are light intensity and light duration.

In order for a photoreaction to occur, the energy of light must be transferred to the electrons of atoms. When photons of light hit a light-absorbing molecule or atom (chlorophyll), an electron moves away from the nucleus to an orbit further out. When this happens, the atom is said to be "excited." It is then at a higher energy state and capable of delivering its energy for incorporation into chemical bonds. Recent research in photosynthesis has shown that the high-energy electrons in excited chlorophyll molecules actually escape from the molecule. They are accepted momentarily by an iron-containing carrier, such as ferrodoxin, which in turn reduces the nucleotide NADP to $NADPH_2$, which then participates in the photophosphorylation of adenosine diphosphate to adenosine triphosphate. The electron from the chlorophyll molecule is replaced by an electron from another source, presumably an H-ion.

It has been found that corn, sugar cane, sorghum, and some species of tropical grasses are photosynthetically much more efficient than might be expected. Two Australian scientists, M. D. Hatch and C. R. Slack, have found that these species carry on certain aspects of the photosynthetic process in a manner unlike most other plants (refer to p. 135 for a discussion of C_4 photosynthesis). Photorespiration, a deterrent to productivity, is nearly absent in C_4 plants because the plants lack the necessary enzymes! Particularly at high light intensities, photorespiration can result in large depressions in productivity. It is exciting to contemplate the possibilities of increasing crop production by experimentally or genetically repressing photorespiration in economic crops of many kinds.

Light has many direct and indirect influences on plants. Light provides the energy for the synthesis of carbon compounds. It also has a profound effect on many physiological processes. Germination, flowering, dormancy, and plant movements are directly affected by light in various plant species. Light affects differentiation of many tissues and organs. It also is a source of heat, although most of the heat received by the plant is from reradiation of long wavelengths outside of the visible range.

Plant growth depends on the amount of carbon

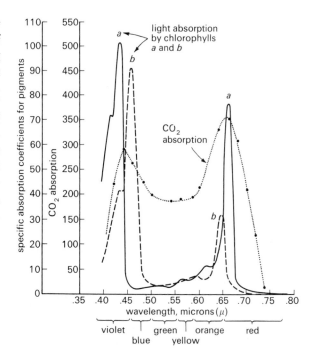

FIGURE 10-7. *The maximum apparent rate of photosynthesis indicated by the rate of CO_2 absorption by plants is the average of light absorption by chlorophylls a and b.* [*After Curtis and Clark,* Introduction to Plant Physiology. *New York: McGraw-Hill, 1950.*]

fixed during photosynthesis. Neither photosynthesis nor chloroplast development can take place without light. Photosynthesis is most effective at 0.655 μ (red), although there is a secondary peak at 0.440 μ (blue). These peaks agree very well with the light absorption spectra of chlorophyll (Fig. 10-7). Chlorophyll formation, however, is greatest at around 0.445 μ (blue) and next greatest at 0.620 to 0.660 μ (orange-red). Thus overall plant growth is usually much less in the green wavelength than in either the blue-violet or orange-red. Photosynthesis and energy have been discussed in Chapter 2; photosynthesis as a growth process has been discussed in Chapter 8.

Carbohydrate accumulation is associated with high light intensity. Therefore many growth processes such as flower, fruit, and seed production are indirectly affected by light. Root growth is also affected, and occurs after the requirements of tops for phosynthate have been met. The resistance of plants to diseases and insects, as well as stress brought on by drought and high temperature, is also influenced by the vigor and reserves of plants.

Light affects plant movement or phototropism. Phototropic curvatures are affected most strongly by light in the blue end of the spectrum. Yellow pigments, possibly carotenoids, probably flavonoids, absorb the wavelengths effective in phototropism. The familiar twisting of sunflower to "follow" the sun is due to the effect of light on auxins (see p. 146). Usually auxins move down uniformly along the stem, but light may penetrate and either result in destruction of or migration of auxin away from the side exposed to light. As a result elongation of the stem takes place at a faster rate on the side away from the light. The reaction requires very low intensity. Even a small amount of weak light is often sufficient to prevent the elongation of etiolated seedlings.

Light indirectly influences the opening of stomates, thereby determining in part the rates of gas exchange between leaves and their environments. Stomatal opening is affected by blue light. Light operates through its effects on the acidity of guard cells, which regulate the conversion of starch to sugar. When sugar concentration is high, water osmotically diffuses into the guard cells, causing

them to swell and bend away from each other, widening the stomatal openings. With darkness, acidity is changed again, sugars are converted to starch, and the process reverses.

Light influences germination (Fig. 10-8) and flowering. In both processes the pigment phytochrome seems to be the seat of activity. Phytochrome is influenced by the red and far red part of the spectrum. Lettuce seeds kept in darkness or exposed to far red wavelengths will not germinate, but germination will proceed if the seeds are exposed to shorter wavelengths in the red end of the spectrum. This is a reversible process. The process requires very little energy. Exposure of loblolly pine seed to light lasting as little as 1/2,000 of a second stimulates germination.

The effect of day length and night length (photoperiodism) on flowering has been discussed in more detail on p. 153. Photoperiodism also affects other developmental processes, such as plant dormancy, tuber and bulb formation, and sex expression.

The chemical composition of plants is affected by light. The brilliant red coloration of deciduous trees in autumn is a function of high light intensities. Cool bright conditions favor the conversion of starch to sugar, which in turn is then available for synthesis of red anthocyanins. During cold nights chlorophyll is decomposed, exposing the more stable yellow carotenoid pigments. The variations in fall coloration of different species is due to differences in pigment composition. In most plants anthocyanin formation is stimulated by light; in some, anthocyanin formation is light-dependent.

Light promotes structural differentiation. The internal structure of plants growing in bright sun is likely to be different from those of plants in shade. Sungrown plants usually have fewer chloroplasts, more layers of palisade cells, smaller cells in the leaves, a thicker cuticle, and pubescence on the lower surfaces of leaves.

Light and the plant community

Not all plants use light with equal efficiency. Many deciduous plants can persist and grow in shade that would be insufficient for conifers because they are more efficient photosynthetically. Some weeds use

FIGURE 10-8. *The relative germination of slash pine* (Pinus caribaea) *seeds in darkness and when exposed to 16 hours of light daily.* [*Courtesy U.S. Forest Service.*]

light with extreme efficiency and grow to maturity in the shade of crop plants. Efficiency also changes with stage of development. For example, the internal structure of the leaves of juvenile pine seedlings is arranged much more effectively for capturing light and carrying on photosynthesis than are the mature leaves of older plants.

Many principles of crop production are based on the relative light requirements of different species. This knowledge is considered part of the professional's expertise for his crop.

Some species show a number of structural adaptations to light. Adaptations that promote light capture include thin cuticles on leaves, several layers of chlorophyllous palisade cells, and high efficiency in using low intensities of reflected light. Trees are well adapted structurally for capturing and fixing light energy because they have large numbers of leaves that are scattered over a wide range of height. In the canopy formed by a stand of trees, both the sides as well as the tops of at least some trees receive direct light, thereby increasing efficiency. Plants that prosper best in bright sun and even require it for survival are called **heliophytes** (sunplants). Most plants that require intense light are photosynthetically inefficient. Several species of plants, including turkey oak, manzanita, and prickly lettuce, have leaves that are vertically rather than horizontally oriented. Vertically oriented leaves are less exposed to the direct rays of the sun, and subject to smaller transpiration stresses than horizontal leaves. Plants with this characteristic are nearly always found in hot, exposed habitats.

In most locations important crops are not part of the natural vegetation but exist only by virtue of special care. Many commercially important species of pines must be cultivated and manipulated vigorously to maintain them as a component of forest stands. If left to propagate themselves, they will be succeeded by native vegetation, which is more tolerant of shade and better adapted to the local environment. In turn, each successive vegetation type will be succeeded by still another, until a shade-tolerant species that reproduces and grows in its own shade predominates. Other factors, water, nutrients, and temperature, also influence successional relationships.

THE CONTROL OF LIGHT

Light control has an important role in agriculture. Techniques presently used to obtain maximum utilization of light involve the choice of location, the modification of plant density and distribution, and the use of supplemental illumination or shade.

Geographic location and light

The world distribution of many plant species is greatly determined by their photoperiodic response. The duration and intensity of light both depend on the climatic or geographic regimes. In the tropics the day length approaches 12 hours throughout the year, whereas in polar regions day length varies from 0 to 24 hours. It is useless to attempt to grow many kinds of plants for flowers or fruit in the tropics, for they will remain vegetative indefinitely. The total amount of light energy received by a plant is also affected by local topographic features and by local atmospheric conditions, which determine the number of sunny versus cloudy days during the growing season. Light greatly affects quality in many crops. The bright, clear summer days in Washington State favor the production of high-quality, attractive apples. Similarly, grape production is intensive on sunny slopes that receive high amounts of reflected light from rivers.

Plant density and distribution

The control of density is a basic method for controlling the amount of light received by plants. Plant density may be modified by plant spacing, pruning, and training.

The control of plant population is one of the simplest ways to control the amount of light received by individual plants. When plants are grown close together mutual shading prevents direct light from reaching all but the tops.

A good example of plant density modification is found in forest management. Approximately 2 to 5 times more seedlings are planted than are expected to grow to full maturity. This practice helps to suppress weed species by fully occupying the site with the desired species while the trees are small and promotes the early natural death and shedding of lower branches (self-pruning). But every 5 to 10 years or so a number of trees must be removed (thinning) to permit full crown development of the remaining trees.

For row crops, spacing within and between rows determines density. Usually the amount of space between rows is determined by the nature of cultivation equipment. The trend toward narrow rows in order to make maximum use of available light has encouraged the manufacture of specialized equipment. The use of chemical weed control may eliminate the need for row cropping in many cases.

Density has a tremendous effect on plant performance, largely because of light competition. In general, high yields per unit area are achieved by high populations because maximum use is made of light early in plant development. Eventually however, the performance of individual plants decreases because of competition for light and other growth factors. Plants respond by decreasing in size and/or in number of parts. Optimum density is dictated by the economic factors involved in optimizing returns. The yield value is a function of yield per plant, number of plants per unit area, yield quality, price, and production costs. When quality is not a factor (as for example in production of cellulose for dissolving pulp) yield per unit area continually increases, although at a decreasing rate with increase in population. But when quality is a factor, as in yield of timber, optimum yield often occurs only at a particular population.

The distribution of plants is also a factor in efficient light utilization. In general, equidistant plant spacing is more efficient than any other spacing because the onset of light competition is delayed (Fig. 10-9).

The orientation of rows can be utilized to make maximum use of light. Crops planted in east-west rows utilize light more efficiently than those planted in north-south rows, because the latter shade each other. In most situations, however, the direction of the row is usually governed by the prevailing slope of the land or by convenience.

Competition for light between parts of the same plant can be modified by pruning and training. **Training** refers to the orientation of the plant in space. Fruit trees may be trained to wires to make maximum use of available light. A new training system for grapes, which results in greater exposure to light and is called the Geneva Double Curtain, has increased yields 40 to 90 percent over those obtained by the standard (Umbrella Kniffen) system. **Pruning** refers to the judicious removal of plant parts. Pruning may be performed for a great number of reasons. For example, forest trees may be pruned to produce high-quality knot-free lumber. Fruit tree crops are pruned for structural strength, and ornamentals for esthetic reasons. The main consideration in pruning most fruit crops is to control competition for light within the plant to affect crop size and quality. High yields of extremely small, poorly colored fruit may be economically useless. Pruning is performed to open up the tree so that light may reach normally shaded parts. Flowers and fruit may also be removed by a number of very sophisticated techniques in order to adjust the fruit load to the light-absorbing capacity of foliage.

Finally weed control can be considered in part a practice to eliminate competition for light between crop and weed. This is discussed further in Chapter 18.

Shading

In some cases it may be necessary to reduce the amount of light. Shading is important in many nursery operations to reduce light and temperature and to lower moisture requirements. In permanent slat houses shade is determined by the spacing between the slats. In recent years plastic screen has been manufactured with various meshes that pass different amounts of light. It is thus possible to change the intensity of light throughout the course of a growing season by changing the screen.

FIGURE 10-9. *Plant distribution patterns.*

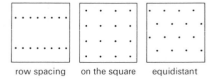

row spacing   on the square   equidistant

Day length may be shortened by shading with black opaque cloth. This is a regular practice to induce or delay flowering in many florist crops.

Tobacco for cigar wrappers is grown under a cover of shade to promote the growth of large, thin, undamaged leaves. This is achieved by placing "tobacco cloth," which is a kind of strong cheese cloth, on frames 8 to 10 feet above the plants. Entire fields of tobacco are grown in this manner, with the cloth suspended high enough to permit cultivation in the enclosure by conventional tractors.

Shading is used to inhibit pigment development in crops in which lack of color is an important factor in "quality." In England celery is "blanched" by mounding the base of the crop with soil to produce pure white (etiolated) stalks.

Supplemental illumination

Supplemental illumination can be used either to satisfy photosynthetic requirement or to extend day length in photoperiod control. Because of the high energy requirements of photosynthesis and the cost of power, it is not economically feasible to use supplemental light to increase photosynthesis except in special situations where plants are especially valuable, such as in seedling production and high-value florist crops. Extension of the photoperiod by illumination is economical on a commercial basis because of the low light intensity required. An increase in photoperiod usually is achieved by extending the day length to about 17 to 18 hours. The same effect can often be achieved by interrupting the dark period for about 3 hours. In terms of power, altering the dark period is more efficient than extending day length. This effect can be made even more efficient by the use of brief light flashes (4-second flashes at 1-minute intervals).

The control of day length by utilizing supplemental illumination and shading has become a standard practice in florist crop production, particularly in greenhouses. In this way plants can be induced to flower "out of season," and production can be synchronized more closely with market demand. The artificial lengthening of short days (as in winter), or interruption of dark period,

promotes flowering of long-day plants and prevents flowering of short-day plants. (Similarly, under naturally long days, shading to reduce day length prevents flowering of long-day plants and promotes flowering of short-day plants.) The manipulation of photoperiod is standard practice in growing chrysanthemums, the most important florist crop in the United States. Alteration of photoperiod is a valuable tool for the breeder who may wish to cross plants that normally do not flower simultaneously.

Artificial light sources differ greatly in their spectral distribution (Fig. 10-10). Tungsten lamps, which emit light from filaments heated to extremely high temperatures (2870°K), produce a continuous spectrum from blue to infrared. The radiation within the visible spectrum lies mainly in the red and far red, although the greater part of the overall emission is in the invisible infrared. Fluorescent lamps emit light from both low-pressure mercury vapor and fluorescent powder. Their emission spectrum contains both the continuous spectrum from the fluorescent material and the line spectrum of the mercury vapor. Light from ordinary fluorescent lamps is very low in red and deficient in far red. This is why fluorescent bulbs are cool. Different types of lamps vary in their spectral distribution. Special bulbs are available that produce light rich in red.

Because of the energy lost in heat (the infrared

FIGURE 10-10. *Spectral emission of tungsten filament and daylight fluorescent lamps as compared to the spectrum of sunlight reaching the earth. The weak line spectra of the mercury discharge from the fluorescent lamps is not shown. The chart offers no quantitative comparison with respect to the energy output of the sources.* [*After General Electric Co.*]

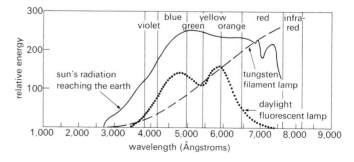

radiation), tungsten lights are rather inefficient. Only about 5 percent of the energy input is transformed into the light range required by plants, as compared to more than 15 percent for fluorescent. Consequently fluorescent lights are better for providing the high energy required for photosynthesis. They have not been widely used in commercial greenhouses, because of the high cost of ballasts and installation. Tungsten lamps have proven very efficient for extending the day length because the red light so important in the photoperiodic response is proportionately high.

Combinations of light sources are used to achieve satisfactory growth with complete artificial light in sophisticated experimental growth chambers called **phytotrons.** Tungsten and fluorescent lights complement each other to produce a spectrum close to that of sunlight. Special cooling must be provided to dispose of the high heat output.

Selected References

Bickford, E. D., and Stuart Dunn. *Lighting for Plant Growth.* The Kent State University Press, Kent, Ohio, 1972. (A reference manual for scientists and commercial producers working with the control of light in greenhouses or plant growth chambers.)

Clayton, R. K. *Light and Living Matter: A Guide to the Study of Photobiology* (2 vols). McGraw-Hill, New York, 1970–1971. (The physical and biological aspects of light are considered.)

Evans, L. T. (editor). *Environmental Control of Plant Growth.* Academic Press, New York, 1963. (A collection of papers dealing with the effects of many environmental factors on plant growth.)

McElroy, W. D. (editor). *A Symposium on Light and Life.* Johns Hopkins Press, Baltimore, 1960. (Comprehensive papers of a technical nature.)

Reifsnyder, W. E. and H. W. Lull. *Radiant Energy in Relation to Forests.* USDA Technical Bulletin No. 1344, 1965. (An introduction to radiant energy.)

Ruechardt, E. *Light, Visible and Invisible.* University of Michigan Press, Ann Arbor, 1960. (A useful reference on the physical nature of light.)

Seliger, H. H. *Light: Physical and Biological Action.* Academic Press, New York, 1965. (A book for advanced students.)

Terrien, J., G. Truffant, and J. Carles. *Light, Vegetation, and Chlorophyll* (translated by M. E. Thompson). Hutchinson, London, 1957. (A short, interesting volume dealing with the effects of light on plants.)

Veen, R. van der. *Light and Plant Growth.* Philips Technical Library, Eindhoven, Holland, 1959. (A brief discussion of light as it affects plant growth.)

CHAPTER 11

Heat

All matter is composed of atoms that are in a state of vibration that depends upon their relative heat. The temperature of a substance is a measure of the relative speed with which its atoms are vibrating. If they are vibrating fast, the temperature is high. Theoretically, at absolute zero ($-273°$C) all vibration ceases and atoms at that temperature are absolutely still. This low temperature is never found in nature, however, and scientists have not been able to produce it in the laboratory, although they have approached it very closely.

EFFECTS OF TEMPERATURE

Physical and chemical processes are controlled by temperature, and these processes in turn control the biological reactions that take place in plants. For example, temperature determines the *diffusion rates* of gases and liquids in plants. As temperature decreases, the viscosity of water increases. The same is true of gases, for the kinetic energy of carbon dioxide, oxygen, and other substances and compounds change with temperature.

The *solubility* of various substances is temperature dependent. Carbon dioxide is perhaps twice as solublé in cold water as it is in warm water. The inverse is true of most solids; sugar is much more soluble in warm water than in cold water.

The *rate of reaction* is affected by temperature; usually, the higher the temperature, the faster the reaction. Thus temperature has a very marked and important effect on respiration. However, the relationship of temperature to the biochemical reactions that occur in plants is seldom directly proportional because of certain complicating factors. For example, the end products produced, such as sugars, may accumulate and block further reactions. In some reactions the availability of raw materials may be a limiting factor.

It is useful to think of the Q_{10} of reaction velocities. The symbol Q_{10} refers to the rate of change in reaction activity that results from a change in temperature. The rates of uncatalyzed chemical reactions increase approximately 2.4 times for each $10°$C ($18°$F) rise in temperature. Of course, this implies a physical system in which limiting processes are minimal. For the overall growth process the Q_{10} of chemical reactions is more likely to be only 1.2 to 1.3, for there are many factors that slow the rates of chemical reactions in cells and living systems.

Temperature affects the *stability* of enzyme systems. At optimal temperatures, enzyme systems function well and remain stable for long periods of time. At colder temperatures, they remain stable but are nonfunctional, while at high temperatures they completely break down. An enzyme system that is permanently stable at 20°C (68°F) may be active for only a half hour at 30°C (86°F) and for only a few seconds at 38°C (100°F).

The *equilibria* of various kinds of systems and compounds is a function of temperature. For example, the balance between sugars, starches, and fats is altered when temperature changes. During the fall season, sugars in some species decrease, whereas starches and fats increase (some plants store food mainly as fats, others as starches). When spring arrives, however, there is a change from both starch and fats to sugars, which are translocated to actively growing parts of plants.

Since temperature has strong effects on physiological and biochemical reactions in plants, it also determines the rates of various plant functions, such as the absorption of mineral elements and water. Not only is the *viscosity* of water greater at low temperatures, but cytoplasmic membranes through which water must pass seem to be less permeable. Photosynthesis is notably slower at low temperatures, and consequently the growth rate is slower. Temperature also affects the rate of cytoplasmic streaming within individual cells.

Plant growth in relation to temperature has been of great interest to many scientists. For example, roots of most plants grow more or less continuously if temperatures are suitable; root growth stops only when temperatures become too cold or too hot. Roots of apple trees start growing at 7.2°C (46.7°F), reach a maximum growth rate at 18.3°C (64.9°F), and stop growing at 32.2°C (89.9°F). Other temperate-zone species that have been investigated are similar in this respect. Near Durham, North Carolina, the roots of loblolly pine will continue to grow all winter if soil temperatures are high enough. Bluegrass roots grow, even if slowly, whenever soil is not frozen.

In view of its profound effects on the biochemical reactions of the plant, the critical effects of temperature on many developmental processes are not unexpected. Among these processes are flowering, sex-expression, seed-stalk formation, dormancy of seed and plant, as well as various maturation processes (see Chapter 8).

TEMPERATURE MEASUREMENT

The temperature of a substance may be expressed in terms of several different scales, Fahrenheit (F) in the United States, but Centigrade (C) in most other parts of the world and in the scientific community. However, because it is based on a system in which 0 is the freezing point and 100 the boiling point, the Centigrade scale is coming into much greater use—as the metric system is being increasingly used in the United States for all types of measurements. For some scientific purposes, the Kelvin (K) scale, a scale whose zero point is absolute zero, is used. The gram calorie is the amount of heat energy required to raise the temperature of one gram of water from 14.5° to 15.5°C.

Temperatures can be measured by various devices. Probably the most commonly used devices are those that are based on the expansion of liquids, solids, and gases when heated. As temperature increases, molecules move faster and require more space. Mercury and alcohol are the most common liquids used in inexpensive household and fever thermometers. Liquid-in-glass thermometers, though relatively simple, can be made very precise.

When a high degree of precision in temperature measurement is required, thermocouples are frequently the first choice. A thermocouple, or thermoelectric couple, consists of two pieces of dissimilar metals in contact. As the point of contact is heated or cooled, the electrical potential between the two metals changes. Since the change in potential is proportional to the change in temperature, a thermocouple can be calibrated to measure temperature. Thermocouples are easily made and comparatively inexpensive.

Another device for measuring temperature is based on the change in electrical conductivity that metals undergo when their temperatures change. In recent years, certain metal alloys have been invented in which the change in conductivity is very great for very small changes in temperature. These metals have been incorporated into devices called **thermistors.** Thermistors, some of which are as small as a pinhead, have found many applications in many kinds of electrical circuits as well as for temperature measurement.

The melting point of solids is sometimes used to measure temperature. By exposing a series of waxes or metallic alloys that melt at different temperatures, it is possible to tell how high a temperature was reached for many purposes. This method has been used also to regulate kiln temperatures. A metallic tablet placed between two electrical contacts inside the furnace will melt and shut off the flow of electrical current when the desired temperature is attained. Specially manufactured tablets have also been used for determining maximum temperatures of forest fires and in other situations where it is not possible to use instruments. The main shortcoming of the tablets is that they do not tell how long the temperature remained at the temperature indicated by the last tablet to melt.

Since the long-wave infrared radiation emitted by objects is a function of their surface temperature, it is possible to use this radiation as a measure of surface temperature. Instruments that measure long-wave radiation, called **infrared radiometers,** operate on a number of principles. They are generally devised so that they "sense" long-wave radiation emitted by an object, compare it with a standard within the instrument, and give a meter reading that indicates the object's temperature. Such instruments, however, divulge nothing about internal temperatures.

Unusual methods for measuring temperature include the use of sonic frequencies and the refractive index of air. However, these techniques are not in wide use nor is it likely that they will be for biological purposes in the near future.

At times it may be desirable to measure the total amount of heat in a particular environment over a period of time. Such a measurement is called a **heat summation.** If a sterile solution of a disaccharide (compound sugar), such as sucrose, is placed for a period of time in a vial with a given amount of a sugar-digesting enzyme, such as invertase, the inversion of disaccharides to monosaccharides (simple sugar) will be proportional to the sum of the heat to which the vial was exposed (provided that temperatures do not rise above the point of diminishing returns). This also gives a measure of potential biological activity, which depends on the amount of heat available.

Other devices for determining temperatures make use of the transmissivity of sound at different temperatures. Such sophisticated techniques have relatively little use in biological research today, but they may become important in the future.

CLIMATE AND TEMPERATURE

In discussing temperature and other climatic elements, it is nearly always necessary to designate the space or area to which they apply. Several expressions for doing this have evolved. The term **macroclimate** refers to the climate of a relatively large part of the earth's surface. It implies a regimen of weather, and is measured by a standardized weather station, which may be 25 to 50 miles distant from some of the locations it represents climatically. Macroclimate is discussed further in Chapter 14.

At the other extreme, **microclimate** refers to the weather conditions around a plant, a leaf, or even a stomate. It expresses the weather of an extremely small volume of space and is not necessarily representative of any place other than the one where measurements are made.

Still another expression, **local climate,** is useful for indicating the climate of a volume of space intermediate in size between that represented by macroclimate and microclimate. Local climate may refer to the weather conditions within a stand of trees, a patch of peas, or a meadow; it is measured by a standard weather instrument shelter placed strategically to give the best representation of gross weather conditions within the crop or other unit of vegetation.

German foresters have an interesting and frequently used expression called **trunk space.** The trunk space, of course, is the space that could be occupied by the trunks of trees, and is usually considered to extend from the soil up to the lower part of the crown. While trunk space is not a scientific term, it is useful insofar as it provides a reference point for estimating local temperatures in forests.

The temperature at any point on the earth's surface depends on the geographic ordinates of latitude and altitude, season and time of day, and the mediating influence of microclimate. The major factor that determines temperature is the amount of solar radiation received, which depends upon both the intensity and the duration of solar radiation. The duration of the solar radiation is affected both by the day length and by the variable effects of cloud cover. As cloudiness increases, the amount of solar radiation decreases.

Variation in environmental temperature

Temperatures over the earth's surface range from −71°C (−91°F) at Verkhoyansk, Siberia, to 58°C (136°F) at Azizia, Libya. Mean annual temperatures range from an estimated −30°C (−22°F) at the South Pole (elevation 8,000 ft) to a record 30°C (86°F) at Massawa, Eritrea, Africa. In annual crops the important climatic elements are the mean and extreme temperatures and the length of the growing season. Perennial plants are affected by temperature throughout the year. Both seasonal variation and average temperature must be considered in relation to plant growth.

As a general rule, variations in environmental temperatures are greatest near the interface between the soil and the air. Solar radiation is generally intercepted by the surface of the earth with a relatively high efficiency, and so one might expect to find the highest temperatures at this point. At the surface of the soil it is not uncommon for night and day temperatures to differ by 38°C (100°F) during the course of a day. Soil temperatures were measured in Florida by covering the bulb of a mercury thermometer with a thin layer of soil a thirty-second of an inch deep. Temperatures varied from 29.3°C (74°F) to 74°C (165°F) during a single 24-hour period. The variation decreases above and below the soil (Fig. 11-1). At elevations of several hundred feet, temperatures are usually neither as high nor as low as at the soil surface. The decrease in temperature with increase in elevation averages 1.4°C (2.6°F) for every 1,000 ft, approximately 1,000 times the latitudinal rate of temperature change. The reason for this is that much of the atmospheric thermal energy is radiated from the earth's surface. Moreover, because the lower tropospheric air contains more water vapor and dust, it is a more efficient absorber of terrestrial radiation; *ergo* the snow-capped mountains on the equator.

The greater the distance from the earth's surface, the more constant the temperature. At great distances from the earth, temperatures are constantly cold because there is very little matter to absorb radiant energy. Similarly, at soil depths of 24 to 30 feet, temperatures may not change more than a degree or two during the course of an entire year. At 10 to 12 feet, there are likely to be seasonal variations in temperature, but soils at these depths do not undergo daily fluctuations.

Soil temperatures vary much less than air temperatures. Soil beneath the surface is not warmed immediately by the radiation incident on the surface, and a temperature lag occurs because heat must be conducted through the soil. In the spring, soil temperatures may not rise enough for germination to occur for several months after air temperatures have risen enough to support plant growth. Conversely, once soils have been warmed by the hot summer sun, they may retain their heat and delay the dormancy of roots even after air temperatures have become quite low.

Factors that regulate environmental temperatures

Of all the factors that determine soil temperatures, meterological events and conditions are the most important. The amount of solar radiation striking the soil surface is a primary consideration because it determines the maximum potential input of heat. Vapor, clouds, dust, the angle of the sun's incidence, and many other factors are important influences on soil temperature.

A snow cover on the ground can reflect a high proportion of radiation back into the atmosphere. But snow also insulates the soil, with the result that soil temperatures under a heavy snow are frequently higher than in an area without a snow cover.

Soil texture and condition influence heat input, heat retention, and heat loss. Soils with a high proportion of fine materials conduct heat faster than coarse-textured soils. In fine-textured soil, particles are in intimate contact with each other and the transfer of heat from one soil particle to another by conduction is relatively efficient. Moisture content is important too, for water has a much greater conductivity than air; the greater moisture content, the greater the heat conductivity.

Organic matter, unless it is wet, has a very low specific heat, and neither absorbs nor retains much heat. Soil with a high content of organic matter is likely to be a very poor conductor of heat. In some soils the organic content of the surface may be so great that it acts as a mulch, thereby stabilizing the subsurface soil temperatures. Surface color of the soil is important. Dark surfaces absorb a much higher proportion of radiation than light surfaces.

Soils with a heavy litter are likely to have more constant temperatures than those with no litter. The litter serves to prevent convection losses, stops incoming radiation, and precludes the loss of a large part of the long-wave radiation.

Vegetative cover has a marked effect on soil temperature. During the early part of the growing season, when crops are small and have relatively little foliage, a high percentage of radiation reaches the soil. This is beneficial in the spring of the year because the soil is cold and must be warmed before crop growth can proceed at a maximum rate. By midsummer, plants are large enough so that their tops form a continuous canopy and shade the soil.

Vegetation of any kind reduces the range of temperatures because it reflects heat back into the atmosphere. At the same time, it absorbs heat from the soil and prevents losses into the atmosphere. As a general rule, vegetation raises the average minimum temperature of the environment while lowering the average maximum temperature. The magnitude of this effect will vary according to the

FIGURE 11-1. *Temperatures around crop plants are subject to very regular variations, which decrease with both altitude and soil depth.* [*After Neuman and Blair,* Crops and Soils Magazine, *June–July 1964.*]

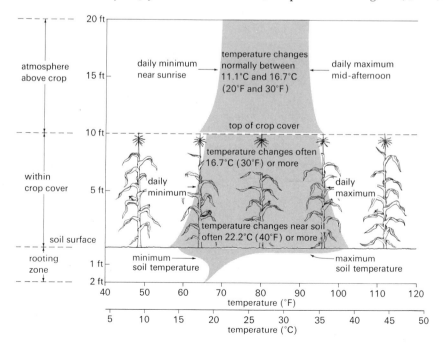

size and density of vegetation. Forest vegetation is usually much more effective in ameliorating the effects of solar radiation than grasses or row crops, but even a very sparse cover of short grasses may modify air temperatures greatly.

Still another way in which forest vegetation can change air temperature is to decrease wind velocity and permit the buildup of heat within the plant cover. This has an opposite effect from some of the factors mentioned previously. It should be obvious that any vegetational ecosystem is a complex mass of interacting forces.

CRITICAL TEMPERATURES

Cardinal temperatures are those at which the changes in rates of processes that take place are critical to the survival, growth, or reproductive capacity of a plant. It has been commonly assumed that the cardinal maximum temperature for plant life is about 54°C (130°F) and that the cardinal minimum temperature for growth is about 5°C (41°F). All this means is that most plant will not live if temperatures exceed 54°C and that they will not grow at temperatures below about 5°C. The sustained maximum cardinal temperature for growth of temperate-zone plants seems to vary from about 32.2 to 33.3°C (90 to 92°F), but the sustained cardinal minimum for life is more obscure. For some tropical plants the cardinal minimum may be a few degrees above freezing, whereas some subarctic plants may survive temperatures of −30 to −50°C (−22 to −58°F) with no apparent ill effects. There is a wide variation in temperature-zone plants, too. Experimentally, stem sections of mulberry (*Morus alba*), a woody plant, have survived temperatures of −196°C (−321°F) for 160 days.

At temperatures of 4.5 to 5.5°C (40 to 42°F) most plants cease to grow. There are exceptions, for some subarctic species continue to function normally at temperatures of 1 to 2°C (33.8 to 35.6°F). It is true that the rates of metabolism are slow at these temperatures, but most plants that live in such cold, rigorous climates are small; some of the higher species may require 150 years or more to reach a height of 1 or 2 feet.

The concept of cardinal temperatures for life processes is, of course, unsatisfactory if not qualified. Does the cardinal maximum for life refer to protoplasm, cells, tissues, organs, or entire plants? Obviously, many plants live and even grow when exposed to occasional temperatures that exceed 71°C (160°F). Individual living cells of the plant would be killed by these temperatures, but they are protected by characteristics that prevent internal temperatures from being lethal, such as heat resistant bark and coloration that reflects radiation. Entire plants, then, are obviously unsatisfactory as indicators of temperature effects on protoplasm, and protoplasmic temperatures are extremely difficult to measure.

The cardinal temperatures of different plant parts differ considerably. Flower buds of temperate-zone plants are much more sensitive to low temperatures than vegetative buds. Different vegetative stages vary in sensitivity, too, for young leaves are more sensitive to both high and low temperatures than mature leaves.

Time is also a factor. A plant that can survive exposure to very high temperatures for a brief interval may die when exposed to lower temperatures for an extended period of time.

Humidity of the environment is important in determining the lethality of temperature. When humidity is low, plants and other organisms can generally tolerate much higher temperatures than when humidity is high. This is partially due to the fact that organisms dissipate heat as long-wave radiation and by convection more efficiently when humidity is low.

High-temperature injury

There are several fundamentally different ways in which heat can injure plants. Perhaps the most readily apparent method is by its effects in directly killing protoplasm. When the temperature of protoplasm reaches about 54°C (130°F), it coagulates in the same manner that the white of an egg solidifies when cooked. Most plants are provided with a measure of protection against lethal internal temperatures by thick bark or large radiating surfaces such as those of leaves.

Direct heat injury is most apparent when particular areas of plants have been injured. **Sunscald,** a condition in which cambium is killed by high temperature, occurs frequently on species of trees with thin bark. If an apple tree is severely pruned, solar radiation may raise bark temperatures and kill cambial tissues beneath bark previously shielded by leaves. The resultant injury may provide a portal of entry for disease-causing microorganisms. Forest species are also subject to sunscald, which results in a deterioration of wood quality. Birch, a thin-barked species, may suffer severe sunscald if trees in a dense stand are thinned heavily.

High temperature is also harmful through its desiccating effect. When temperatures are high, the vapor pressure gradient between the insides of leaves and the outside environment is large, and transpiration takes place at a very fast rate. Wilting is the first indication of desiccation. It is possible for a plant to wilt and still recover if it has not lost too much water. Tomato plants wilt practically every day in the middle of the summer, but at night their water balance is restored, and they become turgid and firm again. Each day this process is repeated, and continues as long as soil water is abundant enough to replace the losses. Such wilting is called **incipient wilting.** When a plant fails to recover after water is made available, it is said to be **permanently wilted.**

Another kind of heat injury occurs when plants get hot and respire at an abnormally high rate. When the rate of respiration is such that the photosynthate is being used faster than it is being made, the plant is in a state of **metabolic imbalance.** When temperature is so high that respiration exceeds photosynthesis over a period of time the plant is "operating in the red." To reverse this and obtain a net increase in photosynthate, temperature must be lowered. Of course, a plant is normally in a state of some kind of imbalance most of the time. During the day, when photosynthesis is proceeding, carbohydrates are manufactured faster than they are used. At night carbohydrates are not made but are used in respiration and various growth processes. When the ratio of carbohydrate synthesis to breakdown is less than 1 over a period of several days the imbalance may become dangerous to the plant. If a plant continues to use carbohydrates faster than they are manufactured over a long period of time, it will eventually deplete its reserves and die. Metabolic imbalance can be brought about by disease organisms, insects that eat plant parts, and other factors that either limit the accumulation of reserve foods or destroy photosynthetic tissues.

High temperatures may have indirect effects on plants, too. When timber stands of birch in the Northeast are thinned, soil temperatures may increase one or two degrees because of the increased amount of radiation reaching the ground. This small increase in temperatures is apparently enough to stimulate a fungus that is always present in the soil to become pathogenic and cause a disease called "birch dieback."

Low-temperature injury

FREEZE DAMAGE Cold temperature injury to plants is often thought of as a simple freezing and death phenomenon. However, there are several mechanisms through which cold temperatures can kill plants, depending upon the nature of the cold weather and the condition of the plants. The simplest kind of cold-temperature injury is freezing of the living matter of the cells. When protoplasm freezes its structure becomes disorganized, proteins precipitate, and death results. But this kind of freezing injury takes place only when temperatures drop to subfreezing levels rather quickly.

If temperatures slowly fall below freezing over a period of eight or ten hours, or gradually over a period of days, freeze injury is of quite a different nature. As temperatures decrease, the intercellular water between cells is the first to freeze. This water is not a part of protoplasm, and its sugar content is usually quite low. Upon freezing, its diffusion pressure decreases to practically zero while the diffusion pressure of the unfrozen water inside cells remain relatively high. As a result, water moves out of the cell protoplasm, where diffusion pressure is high, into the intracellular spaces, where the diffusion pressure is low. Once there, it freezes and its vapor pressure drops. This process continues until much of the water in the protoplasm of cells has moved out of the cells and into the intercellular

FIGURE 11-2. *Glaze damage to hard maple trees. Limbs broken off by the weight of their ice coating provide portals of entry for disease organisms. [Courtesy USDA.]*

spaces, causing the proteins of the protoplasm to clump together and precipitate. Thus, this injury is a form of cell dehydration.

Still another kind of cold-weather injury is desiccation. If the weather has been cold for several months, as in late winter, the soil will have had a chance to reach a temperature equilibrium with the air, so that both air temperatures and soil temperatures are low. Sometimes air temperatures become unseasonably warm, with the result that the tops of evergreen plants may start transpiring and lose large quantities of water, even though growth processes are not taking place. If the weather remains warm even for several days, transpiration will exhaust the supplies of moisture that are immediately available to the plant, and in the absence of continued root growth and water absorption the plant becomes desiccated. When soil moisture becomes frozen and unavailable, this results in **winter burn**—a very serious and costly form of damage to ornamental shrubs that may not be in precisely the ecological niche for which they are best adapted. Winter burn results in the death of the most tender parts of leaves. Only the tips of needles turn brown on conifers, but broad-leaved evergreens may become uniformly brown.

Cold-temperature injury usually occurs in winter, and is seldom of importance in relation to annual crops, which are not usually exposed to winter temperatures. Winter temperatures should not harm woody perennials either, if selections have been made carefully and with full consideration of the extreme lows to which temperatures can fall every 25 to 50 years. In northwest Florida, thousands of acres were planted in oranges, only to be completely destroyed by a short period of unusually cold weather during the winter of 1894.

Glaze damage (Fig. 11-2) can occur when relative humidity (see p. 257) is high and temperatures drop below freezing. The weight of the ice coating on tree limbs causes them to break off.

Other kinds of problems may arise due to low temperatures. Trees may split and crack like a rifle shot when temperatures fall very low. In Canada and Northern Europe, a fungus disease called "snow blight" sometimes attacks coniferous seedlings growing in nurseries while they are covered by snow. In Sweden the fungus does not grow unless 16 to 24 inches of snow remain on the ground throughout the winter. In temperate parts of the world, seedlings planted in the fall may be pushed out of the soil by **frost heaving.** Frost heaving results when water that has accumulated in cracks, or in holes in which the soil is packed loosely, freezes and expands upward in a column and out of the hole.

FROST Cold injury is a common hazard in temperate and subtropical climates as a result of spring or fall frost. **Frost** is the thin layer of ice crystals that is deposited on soil and plant surfaces when temperatures drop below freezing. Two types of weather conditions produce freezing temperatures: (1) rapid radiational cooling, which results in a **radiational frost,** and (2) the introduction of a cold air mass with a temperature below $0°C$ ($32°F$), which produces an **air mass freeze.** The former commonly occurs when skies are clear and calm; the latter, however, usually occurs when skies are overcast and windy. Both occur when temperatures are at or below the freezing point. Because of its local nature, frost may occur when the so-called official temperature (taken nearby and usually at a height of six feet) is above freezing.

The earth radiates energy, as do all bodies. At night, when the earth receives no solar radiation, there is a net loss of heat. Frost occurs when the loss of heat from the ground permits the temperature to drop to or below the freezing point. The **white frost** (hoarfrost) commonly seen in the morning results from frozen dew. Its occurrence depends on the **dewpoint**—the temperature at which relative humidity reaches 100 percent. When the air temperature is below the dewpoint but above $0°C$ ($32°F$), water vapor condenses in the form of dew. White frost occurs when the air temperature is below the dewpoint and below the freezing point. If the humidity is low, frost damage may occur when the air temperature is below freezing but above the dewpoint. This is known as a **black frost,** because the only indication of it comes when the vegetation turns dark due to cold injury

Frost injury is potentially extremely destructive to perennial fruit crops that flower in the early spring because flower parts are extremely sensitive to freezing. It is also a problem in tender vegetable transplants.

Meteorological conditions that are especially favorable for frost formation are those conducive to rapid and prolonged surface cooling. These occur on cloudless, dry, calm nights when long-wave radiation from plant and soil surfaces is likely to be at a maximum. When atmospheric humidity is high or when the sky is cloudy, water particles intercept radiation and reflect it back to the earth. The absence of wind leaves the coldest air undisturbed. Hazardous conditions also form in **frost pockets**—depressions into which cold air "drains" from the surrounding higher ground and accumulates. A field or an orchard may have low areas that require special protection during times of forest danger.

Frost may also occur when there is a **temperature inversion**—a situation in which a cold air mass is trapped below a warmer air mass. Normally, temperature decreases with height. Thermal inversions are sometimes very local in extent, but most frequently take place over large areas on windless, overcast nights. As the sun warms the soil during the early morning, conditions that are conducive to inversions are altered.

The methods used to prevent frost involve either the reduction of heat lost through radiation or the addition of heat. The loss of heat by radiation may be reduced by the use of fogging, hot caps, or cold frames. Heat may be obtained from the earth by improving the conduction characteristics of the soil, or it may be obtained from the air by disturbing the temperature inversion. Heat may be added directly by using heaters, and indirectly by means of sprinkler irrigation or techniques that increase the absorption of insolation. These techniques will be discussed later in this chapter.

CHILLING INJURY Some plants are found to be sensitive to temperatures slightly above freezing. **Chilling injury** affects the peanut, velvet bean, sweet-potato, various cucurbits, and many tropical plants. Chilling injury can also interfere with the ripening of such fruits as tomato and banana.

Chilling injury affects appearance and may result in actual breakdown of plant tissues.

COLD HARDINESS

Cold hardiness refers to the ability of plants to withstand cold injury. Cold hardiness is a variable characteristic that differs greatly with species and with seasonal change. For example, many plants that survive extreme winter freezing may be severely injured by spring frost. The natural change in hardiness in woody plants is related to temperature, water availability, and day length. The onset of cooler weather and short days in the fall brings about dormancy and a physiological toughening in some plants.

Cold resistance of the whole plant develops in a number of steps. As the end of the growing season approaches, the rate of growth slows down, and carbohydrates begin to accumulate. Reserve foods accumulate during this period partly because they are not being utilized in growth and respiration. They are translocated to other plant parts, particularly roots, where they remain over the winter as insoluble fats and starches, and are used at least to some extent in respiration. Also during this period, water in the individual cells moves from the protoplasm to the vacuoles. These changes in water relations and in the availability of certain carbohydrates promote stability of the protoplasmic proteins, which have a higher proportion of bound water during winter than during the summer. There seems to be a surface effect on protoplasmic colloids, so that they do not clump or precipitate so readily. During the onset of winter the concentrations of colloidal and dissolved materials become much greater than during the summer, which may contribute to cold temperature resistance.

Hardening in its broad sense refers to processes that increase the ability of a plant to survive the impact of unfavorable environmental stress. In its more restricted meaning hardening usually refers to processes that enable plants to withstand cold injury, just as the term hardiness usually refers specifically to cold hardiness.

There are several ways to condition plants for cold weather. One of the simplest techniques is to withhold water and fertilizer, especially nitrogen, toward the end of the growing season. Nitrogen stimulates vegetative growth, and if applied late in the season may cause plants to continue to grow until late in the season rather than **harden off** prior to winter. Succulent green shoots, stimulated by the addition of nitrogen, are not frost hardy, but are quickly killed as temperatures approach freezing. On the other hand, additions of phosphorous and potassium seems to increase frost hardiness, but even these must not be applied too late in the growing season.

Promoting plant vigor during the growing season increases the supply of carbohydrates. Plants with high levels of reserve carbohydrates are better able to withstand cold weather than those in poor condition. Respiration continues to some extent throughout the winter, and there must be a supply of carbohydrates to maintain life processes during this period. Some grape varieties overbear, and if the excess fruit is not removed, the plants are subject to winter injury.

As a general rule, resistance to cold injury increases with age until senescence. When exotic ornamental shrubs and other plants are introduced into a region, seeds are not planted directly in the field. Rather, they are planted in nurseries and carefully sheltered and protected for several years to get them past the juvenile stage. After they have ceased to be tender young plants and are "hardened" to some extent, they survive much better than if they are planted directly in the field. Some species of ornamentals must be 6 to 8 years old before they will survive field planting.

The hardening of transplants may be achieved by any treatment that materially checks new growth. This may be accomplished by gradually exposing plants to cold, by withholding moisture, or by a combination of these two treatments. In general, about 10 days are sufficient to harden plants. This treatment is designed to produce a stocky, toughened plant in contrast to a soft, tender, "leggy" plants. Hardened plants are often darker green in color, and hardened crucifers have an increased waxy covering of the leaves. The induced cold resistance brought about by the hardening treatment in such cool-season crops as cabbage or celery may be considerable. Unhardened cabbage plants show injury at $-2.2°C (28°F)$, whereas hardened plants can withstand temperatures as low as $-5.6°C (22°F)$. In warm-season crops, such as tomato, the degree of cold hardiness imposed is slight, but hardening inures the plant to transplanting shock and hastens plant establishment under adverse conditions. The reduced growth rate enables the plant to withstand desiccation until the root system becomes established. Tender plants, which respire rapidly, have little chance of survival under warm, windy, dry conditions following transplanting. The cessation of growth under the hardening treatment, however, may severely interfere with subsequent performance; thus care must be taken to avoid overhardening. Under ideal transplanting conditions hardening may not be necessary.

HEAT UNITS

The harvest date of some annual crops can be changed by planting earlier or later. But a change in the date of planting does not always cause an equal change in the harvest date. For some crops the time required to reach a harvestable stage is a function of the total heat received. When temperatures are below the minimum required for growth, as may be the case if seeds are planted early, an earlier planting date will not necessarily affect the date of maturity.

For some crops the time required to reach the harvestable stage may be expressed in terms of temperature-time values called **heat units** by calculating time in relation to temperature above a certain minimum. If the minimum temperature for growth of a particular crop is $50°F$, then a day with a mean temperature of $60°F$ would provide 10 degree-days of heat units. A day with an average temperature of $40°F$ would provide zero degree-days of heat units. The harvest date can be ascertained by an accounting of accumulated heat units. In Wisconsin it has been shown that 1,200 to 1,250 degree-days are required for the 'Alaska' cultivar of pea to mature if planted in early spring. Under increasing spring temperatures it takes longer to accumulate heat units in the early part of the season

than later. Assuming that all temperatures above a minimum have similar effects on growth, there would be a decreasing interval between planting and harvest dates as the season progresses.

The heat-unit system has a number of limitations. For example, soil temperatures more accurately indicate conditions for early growth than do air temperatures. Differences in day-night temperature shifts, day length effects, as well as the differential effect of temperature on various stages of plant growth also affect results. In addition, all temperatures above a minimum may not have a similar effect on growth, but may act exponentially, approximately doubling many physiological processes with every $10°C (18°F)$ rise in temperature. The precise determination of harvest date by the accumulation of temperature data depends upon a knowledge of the general climate of an area and upon experience with harvest dates based upon planting dates for each crop. The system works rather well for some crops, not so well for others.

TEMPERATURE CONTROL

The control of environmental temperatures around plants can be extremely important in crop production. Environmental temperatures may be controlled in several ways.

Choice of site

The choice of site is often a way of controlling temperature. Sunward slopes intercept more radiation than shady slopes or level ground. Selecting these warm slopes from crop production extends the growing season in cool climates. Site selection is very important for frost protection. In order to escape the damaging effects of frost, orchard sites subject to cold air drainage must be avoided (Fig. 11-3).

Proximity to large bodies of water has a stabilizing effect on air temperature. Frost protection afforded by large bodies of water is due to the high specific heat of water as compared to that of the land; water absorbs and gives up heat slowly. Solar radiation penetrates water more deeply than land, and the continuous internal movement of water

212

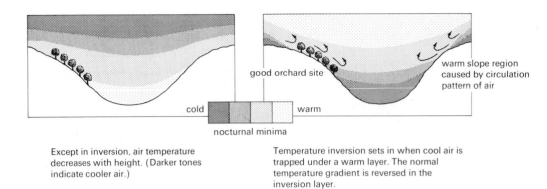

cold | warm

nocturnal minima

good orchard site

warm slope region caused by circulation pattern of air

Except in inversion, air temperature decreases with height. (Darker tones indicate cooler air.)

Temperature inversion sets in when cool air is trapped under a warm layer. The normal temperature gradient is reversed in the inversion layer.

FIGURE 11-3. *(Left) Normal air-temperature gradient. (Right) Thermal belt formed on a hillside as the result of a temperature inversion.* [*After Geiger,* Climate Near the Ground. *Cambridge: Harvard Univ. Press, 1957.*]

results in heat distribution throughout the water mass. Thus large bodies of water become "heat reservoirs" in the fall and "heat sinks" in the spring. In the winter and early spring the influence of large bodies of water keeps temperatures moderately low and prevents premature plant growth. In the late spring they may provide enough heat to prevent frost; in the fall they tend to delay the advent of frost. This temperature lag is felt mainly along the windward side of large bodies of water. The fruit industry of the northeastern United States is localized in counties bordering the Great Lakes to take advantage of the ameliorating effects of these large bodies of water. Crops frequently escape extreme low-temperature winter injury as well as crippling spring frosts in these locations. The effect of warm ocean currents is also important. England and Labrador are in the same latitude, but England is warmer because of the influence of the Gulf Stream. The southern coast of Alaska is warmed by the Japanese currents, and commercial truck gardens near the coast produce many vegetables that cannot be grown inland.

Soil mulch

Soil temperatures are frequently controlled by mulches, which are ground coverings of various kinds that reflect intense solar radiation in summer and prevent loss of heat from the soil in winter. In northern Florida two inches of pinestraw mulch decrease surface soil temperatures in summer by

25°C (45°F). In more extreme but less common situations, mulches may depress surface temperatures as much as 33°C (60°F). Plant litter or such by-products as leaves, straw, pine needles, wood chips, and hay are good insulating materials, and therefore make good mulches. Not only the cost of the mulching material but the cost of application must be considered. Some materials are toxic, and others may considerably lower soil pH excessively; consequently care must be exercised for these reasons too in choosing a mulch. Organic mulches have the added advantage of preventing erosion, conserving moisture, inhibiting weed growth, and contributing plant nutrients after decomposing. Because of the importance of mulches in the production of such crops as pineapples or strawberries, it is economical to use polyethylene or paper.

Soil temperatures can also be controlled by spreading a thin layer of sand on the ground. The sand reflects light and also serves as an insulating soil mulch.

When plant litter is burned the soil is blackened and more radiation is absorbed. Mean surface temperatures in areas that have been burned may be several degrees higher than in unburned areas. As interesting experiment was done in Siberia to determine the effect carbon black might have on hastening the melting of snow by raising the surface temperature. Carbon black spread over the snow at 4,000 pounds per acre resulted in the earlier maturation of cotton.

Irrigation

Irrigation can be used for temperature control as well as for supplying moisture. Water used in irrigation is relatively warm compared to the surface temperature of soils in fall, winter, and spring. When pumped into a field, it has a warming effect. Moreover, after a field has been thoroughly wetted, heat is more readily transferred from lower depths to the surface because saturated soil conducts heat better than dry soil.

Water serves not only as a vehicle for the transfer of heat but functions in other ways to control temperature. The formation of ice is accompanied by the release of large amounts of heat. The heat released helps maintain plants at the freezing temperature of water for as long as water is being frozen. Of course, once all the water is frozen, temperatures will fall again. If this principle is to be used water must be continually applied until the air temperature is above freezing. Spray irrigation is commonly used to protect strawberries from frost.

Structures

Structures of various kinds have long been used to control temperature. **Cold frames**—low, enclosed beds covered with a removable sash of either glass or plastic—are used to extend the growing season so that young plants can be started before the growing season begins. Temperatures are maintained at a relatively high level because radiant energy passes through the transparent top and is absorbed by the soil. It is lost from the soil as long-wave radiation, which is prevented from leaving the glass or plastic cover. As a result, heat builds up in the cold frame during the day and warms the soil, which releases its heat gradually at night and keeps plants warm.

Hotbeds are cold frames with supplemental heat, which may be supplied in a number of different ways. Today, electrical heating cables, steam pipes, and hot water pipes are buried in the soil. In the past, the supplemental heat was supplied by the respiration of microorganisms, which decomposed a layer of organic matter (usually manure) under the plants.

FIGURE 11-4. *Greenhouses are structures covered with glass (above) or plastic (below) for temperature control in plant growth. In the glass greenhouse fan and pad cooling is used as an economical system for lowering summer greenhouse temperatures. Cooling is achieved by the evaporation of circulated water through excelsior or some other material with a high surface area per unit volume. Fans opposite the cooling pads draw the cooled air across the greenhouse. The plastic greenhouse is a relatively inexpensive structure. The polyethylene is removed in the spring when temperatures get too high and is replaced in the fall. The frame can then be covered with shade cloth and the structure converted to a shade house. During the winter, temperature control is achieved with ventilators and steam carried in from the greenhouse range. [Upper photograph courtesy Acme Engineering and Manufacturing Corp., Muskagee, Okla.; bottom photograph courtesy P. H. Massey, Jr.]*

Greenhouses are essentially large hotbeds in which heat is provided usually by steam (Fig. 11-4). Temperatures in winter are controlled automatically. Summer temperatures are regulated by shading, by spraying the tops with water, by pulling air in through wet pads (and thus cooling it by evaporation), and by refrigeration. Glass surfaces are sometimes painted or speckled with an easily removable white paint or whitewash to reflect a part of the solar radiation. Greenhouse maintenance is a highly specialized operation, requiring a knowledge of plant materials and practical mechanics.

Inexpensive plastics, such as polyethylene film are being used for greenhouses in place of glass. Rigid plastics are also used in this manner. They are usually made by constructing a frame of lumber or metal tubing. The flexible plastics used for such structures usually last only a single season before replacement is necessary. They cannot be used in areas that are very windy or subject to frequent hail.

Cones of translucent paper or plastic, called "hot caps," are sometimes placed over the tops of individual plants as they are growing in the field. They are usually used for crops that are very tender and must be transplanted (Fig. 11-5).

The use of a portable, tent-like glass sash (cloche) over individual plants has long been used in European market gardens to facilitate early vegetable production. This technique fell into disuse because of the tremendous labor inputs

required. However the principle has been revived on a large scale, especially in Israel for winter vegetable production, with the innovation of plastic tunnels (Fig. 11-6). Tunnels are made from sheets of polyethylene laid down mechanically, usually over wire hoops. Soil may be used to seal the sides of the plastic. The higher temperature within the tunnel encourages early production. A number of systems ameliorate built-up heat as temperatures increase with the approach of summer. In one system the plastic may be temporarily slipped off the hoops during periods of warm weather. In others, the plastic is first perforated for ventilation and then slit in increasingly more places at intervals as temperatures rise. Plastic tunnel and plastic mulch may be combined; in some systems the plastic tunnel is converted to a plastic mulch.

In England, it is possible to grow peach trees if they are located near a masonry wall, which will absorb radiant heat and release it gradually during the night. Thus, many English "kitchen gardens" have a welcome addition, otherwise unavailable because of the cool climate.

Although most structures are intended to increase the temperature of the plant environment, slat houses, or shade houses, are used to reduce

FIGURE 11-5. *Hot caps protect early tomatoes in California's San Luis Rey district. The hot cap is made of a translucent paper and acts as a miniature greenhouse.* [*Courtesy USDA.*]

temperatures. This kind of structure provides partial shade by admitting a fraction of the radiation that strikes the structure. Evenly spaced wooden or metal slats or boards were originally used for this purpose. Today, however, plastic screening, which allows a known percentage of radiation to pass through it, is commercially available. Thus by changing the screening, the amount of shading may be controlled.

Wind machines

Air itself is a reservoir of heat that can sometimes be used to advantage. A large part of the heat of a plant is normally transferred to the air by convection, or vice versa. On calm, clear nights temperature inversions may occur. The air temperature 50 feet above the ground may be more than 2.5°C (5°F) warmer than that in the vicinity of the crop plants. By using an 85 to 100 horsepower wind machine during the night, a local turbulence can be created to cause air to mix and thereby increase temperatures in the vicinity of crop plants as much as two degrees (Fig. 11-7). Changes as small as this can be critical during the flowering period. The wind machines merely stir up the air and make the heat in the warm upper layer available to plants. On cold, windy nights, wind machines give no protection, but during calm, clear nights, when temperature inversions occur, they can be extremely helpful. Wind machines are usually permanently mounted on towers at a level about 20 to 25 feet above the soil surface. Their elevation is important, because low areas are more subject to the accumulation of cold air than are high ones.

Smudge clouds and heat

In regions where citrus crops are subject to occasional frost, oil smudges and heat may be used. If orange growers have a warning from the U.S. Weather Bureau that there will be a frost on any particular night, they prepare smudge pots by filling them with enough oil to burn throughout the anticipated danger period and place them in orchards in strategic locations (Fig. 11-8). Before the cold front moves in, the heaters are

FIGURE 11-6. *Plastic row covers for winter cucumber production in Southern California.* [*Courtesy Bernarr Hall.*]

lighted. The radiant fraction of the thermal output of heaters provides the most protection, although part of the convection heat is also useful, particularly under conditions of good inversion where wind machines can redirect some of the warm layer overhead back into the orchard. Although the smoke from smoky heaters may be of some value in acting as a radiation "blanket" if atmospheric conditions permit its accumulation in an inversion ceiling during the night, this benefit will be more than lost if more than one night of protection against radiation frost is required, for the smoke will reduce the incoming solar radiation in the succeeding day. "Smudging" is illegal in California and Washington citrus-growing areas because of the air pollution that it causes. It seems probable that smudge will not be allowed any place in the world, in the near future. Environment-conscious people are doing all in their power to outlaw such sources of air pollution.

Various kinds of heaters have been used to ward off cold temperatures in high-value fruit crops. A recent innovation is the use of solid fuel candles. Made of solid petroleum wax, they are about eight inches in diameter and burn with a low flame for

FIGURE 11-7. *Wind machines mix the warm upper air with the cool air around the citrus crop when temperature inversions exist.* [*Courtesy Food Machinery and Chemical Corp.*]

eight hours. Two such heaters beneath a grapefruit tree will raise the average air temperature within the canopy of the tree by about 4°C (7.2°F).

Manipulation of vegetation

Vegetation itself can be manipulated to control temperatures within crops. Anything that prevents the accumulation of radiation during the day will increase frost. Vegetation that shades the soil will reduce the amount of heat stored in the day. Thus a sodded or mulched area is more liable to become frosted than one under "clean cultivation." The necessity for controlling frost is one of the main reasons why peaches are not grown under permanent sod.

If plants are widely spaced, more solar radiation is admitted to the soil. Plants themselves are relatively poor at storing heat, but soil materials and water are good storage "containers." The low vegetation beneath a forest canopy may be extremely effective in reducing the solar radiation reaching the soil, for shrub covers of rhododendron, laurel, and blueberries may form a very tight layer that admits very little radiation. In both forest and field crops, the density of the canopy

FIGURE 11-8. *Heaters placed at regular intervals between the rows of trees in a citrus orchard are now used instead of smudge pots to protect the trees and fruit from frost.* [*Courtesy USDA.*]

can be regulated by careful spacing. If the canopy is too dense after planting, then some plants may be removed to make the spacing more appropriate. In Norway and Sweden, forests are frequently thinned to admit radiation through the canopy to the soil surface in early spring. By doing this, the growing season of conifers can be effectively extended.

Although it seems impractical at this time, it is quite possible that in the future entire fields may be warmed by subsurface systems that rely on nuclear fission as a source of heat. Water-conducting pipes of the heating system could be laid beneath the soil to carry heat in the same manner as it is carried through the floors of some houses.

Selected References

American Society for Horticultural Science. "Cold Hardiness, Dormancy, and Freeze Protection of Fruit Crops" (A Symposium). *HortScience* **5**:401–432, 1970. (A series of papers on the cold hardiness of plants and techniques for protecting plants from low-temperature injury.)

Daubenmire, R. *Plant Communities.* Harper and Row, New York, 1968. (An introductory text.)

Geiger, R. *The Climate Near the Ground.* Harvard University Press, Cambridge, 1957. (A classic work on microclimatic relationships.)

Levitt, J. *The Hardiness of Plants.* Academic Press, New York, 1956. (A short but thorough monograph dealing with the hardiness of plants.)

Platt, R. B., and J. F. Griffiths. *Environmental Measurement and Interpretation.* Reinhold, New York, 1964. (Discussions of techniques for measuring temperature and other climatic elements.)

Weber, R. L. *Temperature Measurement and Control.* Blakiston, Philadelphia, 1941. (The nature of heat and its measurement.)

Weiser, C. J. "Cold resistance and injury in woody plants." *Science* **169**:1269–1278, 1970. (A review of low-temperature injury physiology.)

Wijk, W. R. van (editor). *Physics of Plant Environment.* Interscience (Wiley), New York, 1963. (A collection of technical papers on the reaction of plants to their environment.)

CHAPTER 12

Soil

Soil is different things to different people. To the housewife it is something to be washed from clothes and swept out the door. To the geologist soil is the unconsolidated, weathered part of the earth's mantle, an insignificant, small fraction of its total volume. To the civil engineer it is the medium for the support of structures and a construction material. But to the agriculturist, soil is that part of the earth's crust in which the roots of plants grow. It is a vital, living component of the environment—a component that can be manipulated to affect crop performance. When soil is misused crops become less productive; when handled with due consideration for its biological and physical nature, it can continue to yield crops throughout countless generations of cultivation and use.

Soil has three primary functions in sustaining plant life:

1. It supplies mineral elements, serving both as a medium of exchange and as a place of storage.

2. It supplies water and serves as a storage reservoir.

3. It serves as a medium within which the roots of terrestrial as well as many aquatic plants anchor themselves.

One might ask whether soil is really necessary for plant growth. **Hydroponics**—the soil-less cultivation of plants with nutrient solutions alone or with sand or gravel—is practiced today on a limited scale. The United States Armed Services uses hydroponic gardens to provide fresh vegetables for personnel on isolated islands in the Pacific where soil is not available. Even though it is possible to maintain a limited agriculture without soil, the massive production of plant materials that the world requires cannot at present be accomplished economically without soil under present economic conditions.

SOIL SYSTEMS

For the purposes of crop production, soil must be considered to be a delicate balance of interwoven and interacting systems: (1) inorganic minerals, (2) organic matter, (3) soil organisms, (4) soil atmosphere, and (5) soil water.

Although it is convenient to consider each of these systems individually, they are neither separate nor independent in nature. To change one of them results in a change in all. The nature of a soil is greatly modified when one system dominates. For example, a high water table produces a kind of soil that has particular and definite characteristics, no matter where it is found in the world.

Inorganic minerals

The inorganic minerals of the soil are ultimately derived from the parent materials upon which soil rests. They may be **original**, relatively unchanged from the parent minerals (such as quartz in the form of sand), or **secondary**, formed by the weathering of less resistant minerals (such as clays). The amount of inorganic material varies greatly, from more than 99 percent of the weight of sandy and clayey soils, to as little as 1 percent in organic soils. The inorganic component in soils consists of a mixture of particles that differ in size, in composition, and in physical and chemical properties.

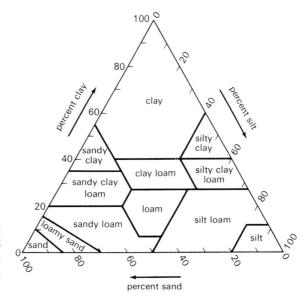

FIGURE 12-1. *The texture triangle shows the relative percentages of sand, silt, and clay in each textural class. (In the United States the term "loam" refers to a soil with more or less equal proportions of sand, silt, and clay. As used in Great Britain the term refers to a soil high in organic matter, a "mellow" soil.)* [Courtesy USDA.]

Table 12-1 CLASSIFICATION OF SOIL PARTICLES.[1]

PARTICLE	DIAMETER (mm)
boulders	>256
cobbles	256–64
pebbles	64–4
gravel	4–2
fine gravel	2–1
coarse sand	1–0.5
medium sand	0.50–0.25
fine sand	0.25–0.10
very fine sand	0.10–0.05
silt	0.05–0.002
clay	<0.002

Source: After USDA and Wentworth.

[1] The International Classification (Atterberg) System refers only to soil particles under 2 mm:

Coarse sand	2.0–0.2mm
Fine sand	0.2–0.02
Silt	0.02–0.002
Clay	<0.002

SOIL TEXTURE The mineral particles of the soil can be arranged according to size from very coarse to very fine. The term **soil texture** refers to the size of the individual mineral particles. Soil particles have been classed according to size, as shown in Table 12-1. Textural designations are also used to describe soils (Fig. 12-1). The main ones are sand, silt, and clay. The nontechnical terms "lightness" and "heaviness" refer to soil texture. "Heavy soils" are high in clay and other fine particles; "light soils" are low in clay and high in sand and other coarse particles.

The coarse materials such as sands and gravels are usually composed of many small particles cemented together either chemically or by a matrix material. These are bound relatively firmly and present only a single outer surface. The physical and chemical properties of these coarse materials do not differ greatly from those of their parent materials.

Silt particles are more or less unweathered, but their surface is coated with a clayey matter. The properties of silt are therefore somewhat intermediate between sand and clay.

The clays, the smallest of the soil particles, show distinct chemical and physical properties. Clays are colloidal°—viscous and gelatinous when moist but hard and cohesive when dry. Their structure can only be seen with an electron microscope. Clays are composed of particles called **micelles,** which are formed from the parent materials by a crystallization process; they are not merely finely divided rock. The micelles are sheetlike (laminar), with internal as well as external surfaces, and are held together by chemical linkages or ions between the plates. The tremendous surface area relative to volume is one of their most significant features.

The structure of clay micelles may be complex (Fig. 12-2). Clay particles are negatively charged and thus attract, retain, and exchange positively charged ions. The adsorbed water on clays acts both as a lubricant and as a binding agent. Clay platelets behave like a stack of wet poker chips. This to a large degree explains the plasticity of clay. Wet clay soils low in organic matter and low in weakly hydrated cations, such as calcium, become sticky or puddled. It is the unique chemical and physical properties of clay that are responsible for most of the important properties of soils.

Soil texture affects the retention of water and the rate of water infiltration. Coarse soils permit the rapid infiltration and percolation of water, so that there is no surface runoff even after a heavy rain. In contrast, clay soils are so finely textured that very little water penetrates to lower levels, especially after the surface clays become wet and expand. Coarse soils, however, are incapable of retaining large amounts of water. The range of capacity for holding moisture available for plant growth among soils of different textures is given in the following tabulation:

° A colloid is a small, insoluble, nondiffusable particle greater than a molecule but small enough to remain suspended in a fluid medium without settling. The soils' colloidal particles may be inorganic or organic.

Soil type	Inches of water per foot of soil
Very coarse texture—very coarse sands	0.40–0.75
Coarse texture—coarse sands, fine sands, and loamy sands	0.75–1.25
Moderately coarse texture—sandy loams and fine sandy loams	1.25–1.75
Medium texture—very fine sandy loams, loams, and silt loams	1.50–2.30
Moderately fine texture—clay loams, silty clay loams, and sandy clay loams	1.75–2.50
Fine texture—sandy clays, silty clays, and clays	1.60–2.50
Peats and mucks	2.00–3.00

SOIL STRUCTURE **Soil structure** refers to the gross arrangement of the soil particles into aggregates. A soil may either have a simple or a compound structure. Sands and gravels, examples of soils with a **simple structure,** have very little cohesion, plasticity, and consistency (the resistance of the particles in the soil to separation). Simple-structured soils are usually composed of materials that are relatively resistant to weathering, such as quartz sand. They are also said to have a **single grain structure.**

Most agricultural soils have a **compound structure;** their aggregates stick together. Several distinct sizes of compound structures are recognized.

Structural type	Aggregate size (diameter in mm)
columnar	>250 mm
blocky	5–25
granular	3–5
crumbly	1–3
massive	completely puddled or compacted

Soil structure develops when small colloidal soil particles clump together or **flocculate** into granules. Granulation is promoted by freezing and thawing, the disruptive action of plant roots, the mixing effects of soil fauna, the expansion and contraction of water films, and the presence of a

A. kaolinite

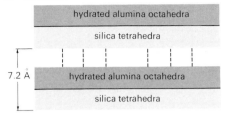

B. montmorillonite

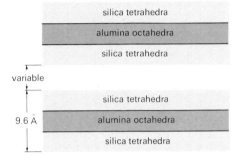

C. illite

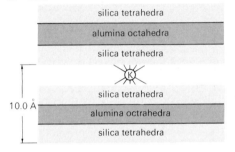

FIGURE 12-2. *The crystalline structure of clay. (A) The simplest clay is called* kaolin. *It is composed of two different layers, one of silica and the other of alumina. Kaolinitic clay miscelles are relatively large and are bound together tightly. The distance between the alumina and silica layers is rigidly fixed and does not increase when water is adsorbed on the surfaces of the individual miscelles. The internal space is not available for surface reactions. Such clays do not shrink greatly when dry or expand when hydrated. (B) Montmorillonitic clays are complex, and are composed of an alumina layer sandwiched between two layers of silica. These clays are not bound together tightly, and swell when wet because the hygroscopic surfaces between layers adsorb water and force the layers apart. All surfaces can adsorb water and minerals. (C) Crystals of illite consist an alumina layer between silica layers. The crystals are held together by potassium bonds. Surface adsorption can take place on the surfaces between crystals to a limited degree. Illite is less abundant than kaolinite or montmorillonite.*

network of fungal hyphae. However, unless the granules are stabilized by coatings of organic matter or their own electrochemical properties, they will coalesce into larger and less active clods.

Good soil structure is very important for agricultural soils. Highly granulated soils are well aerated and have a high water-holding capacity because of the increased size of the soil pore space. The **pore space** of soil is occupied by water and air in varying proportions, the soil acting as a huge sponge. The total pore space of soil, about 50 percent of the total volume, is not as important as the characteristic size of the pore spaces. Clay soils have more total pore space than sandy soils, but because of the small size of the pores in clay soils, air and water move through them slowly. When the small pores of clay soils become filled with water, the lack of aeration so essential to root growth becomes limiting. Large pore spaces become filled and are drained by gravity, whereas small pores absorb and retain water by capillary action. Capillary water is of the utmost importance to the plant: it is the soil solution.

The crumbly nature of good agricultural soils depends on soil texture and on the percentage of humus (decomposed, stable organic matter). Clay soils low in organic matter typically have poor structure. In order to maintain good compound structure in clay soils, they must be carefully managed. If worked when too wet, the structure may be damaged. When clods are exposed they become dry, hard, and difficult to work back into the soil (Fig. 12-3). In heavy soils it is necessary to add organic matter to maintain good structure. In sandy soils, where structure is not as critical, it is necessary to add organic matter to increase the water- and nutrient-holding capacities.

EXCHANGE CAPACITY The capacity of a soil to retain and exchange cations such as H^+, Ca^{++}, Mg^{++}, and K^+, is called **exchange capacity.** Exchange capacity, a measure of the reactivity of the soil, varies inversely with particle size (Fig. 12-4). Because a given volume of small soil particles has much more surface area than an equal volume of large particles, fine soils accumulate and retain many times more cations than do coarse soils. The soil's colloidal particles, clay and humus, are nega-

FIGURE 12-3. *Puddled clay soils may form large cracks when they dry out. This condition often indicates an unproductive soil that has been poorly managed.* [*Courtesy U.S. Department of Interior.*]

tively charged and attract cations. When these particles in a soil are saturated with hydrogen ions, the soil has a strong acid reaction. Soil reaction and pH are discussed more fully in Chapter 16.

The exchange capacity of the soil's colloidal particles is of tremendous importance. Nutrients that would otherwise be lost by leaching are held in reserve and when exchanged become available to the plant.

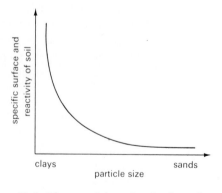

FIGURE 12-4. *The reactivity of soils depends on the total surface area of the soil particles per unit volume of soil. The total surface area varies inversely with particle size.* [*Courtesy USDA.*]

The process of base exchange is not a random one. Cations differ in their ability to replace one another; if present in equal amounts,

$$H^+ \text{ replaces } Ca^{++} \text{ replaces } Mg^{++}$$
$$\text{replaces } K^+ \text{ replaces } Na^+.$$

If one cation is added in large amounts it may replace another by sheer force of number (mass action). This is largely what occurs with the addition of fertilizer.

The release of hydrogen ions in soils tends to promote the exchange of cations, making them available to plants. Hydrogen ions are made available by the dissociation of carbonic acid formed from the carbon dioxide released by roots and from the even larger amount released by the enormous population of microorganisms in respiration and in the decomposition of organic matter:

$$CO_2 + H_2O \rightleftharpoons H_2CO_3 \rightleftharpoons H^+ + HCO_3^-.$$

The cations are replenished by the decomposition of the inorganic fraction of the soil, by the degradation of organic materials, and by the application of fertilizer. The cations in a productive soil are in equilibrium between the soil particle, the soil solution, the soil microorganisms, and the plant.

The cation exchange capacity of a soil is expressed in milliequivalents (meq), of H^+; that is, the number of milligrams of H^+ that will combine with 100 g of dry soil. The exchange capacity of a soil depends on the percentage of humus it contains and on the percentage and composition of its clays. Clays differ markedly in their ability to exchange cations. Kaolinite has an exchange capacity of 10 meq/100 g, whereas montmorillonite clay has an exchange capacity of about 100 meq/100 g. The exchange capacity of humus ranges from 150 to 300 meq/100 g. The ranges of exchange capacity for various soils are given in Table 12-2.

The ability of a soil to supply mineral ions for absorption by plants is a measure of its **fertility.** It is quite possible for soils to contain relatively large amounts of minerals and yet be infertile because the ions are unavailable to plants. In a region with moderate to large amounts of rainfall, simple soils are usually exceedingly low in fertility because minerals are readily lost by leaching. Such

Table 12-2 RANGE OF CATION EXCHANGE
CAPACITY FOR VARIOUS SOIL TYPES.

SOIL TYPE	CATION EXCHANGE CAPACITY (meq/100 g)
Sands	2–4
Sandy loams	2–17
Loams	7–16
Silt loams	9–26
Clay and clay loams	4–60
Organic soils	50–300

Source: Adapted from Lyon, Buckman, and Brady. *Nature and Properties of Soils.* Macmillan, New York, 1952.

crops as peanuts, which are grown in sands, must be fertilized frequently and carefully.

Soils may be fertile without being **productive.** Some desert soils have an exceedingly high natural fertility, but must be irrigated to render them capable of producing crops. A fertile soil can be productive only when moisture, temperature, and other environmental factors are in good balance. The aim of soil management must be to make maximum use of the productive capacity of the soil.

Soil organic matter

Soil organic matter is the fraction that is derived from living organisms. Most apparent is the **litter** on the surface. It is composed of undecayed leaves, branches, reproductive parts, and other residues from the top parts of plants. **Duff** refers to partially decayed litter. It is frequently matted together with fungal mycelium, and in this condition is called **leaf mold.** Duff forms when soil is moist enough to supply the water essential for microbial activity, and when litter is thick enough to prevent evaporative water losses. Leaf mold is an important component of forest soils, but is never found in cultivated agricultural soils. Plant roots and the excreta, sheddings, and bodies of soil organisms, although not as apparent, also contribute to soil organic matter.

The upper layers of the soil are often high in an organic fraction called **humus.** Humus is relatively resistant to further breakdown and decomposition. Humus, unlike mineral colloids, is non-

crystalline. The primary sources of humus in the upper soil layers are surface litter and plant roots. As surface litter is broken down by mechanical action and decomposed into fine particles by micro-organisms, it is washed down into the soil, where it becomes a part of the soil complex. The washing action can only affect the surface of most soils. The decomposition of dead roots provides organic matter throughout the upper soil. Prairie soils, which receive small amounts of precipitation, have a low rate of biological decomposition, so that organic matter from dead grass roots accumulates to make them dark in color, friable, and extremely fertile.

Perhaps one of the most important contributions of organic matter to soil is its water-holding capacity. Organic matter acts like a sponge: it can absorb large amounts of water in relationship to its weight. Organic matter is also a source of mineral elements, which are made available when it decomposes. The decomposition by bacteria, fungi, and other organisms of organic matter to form water and carbon dioxide with the release of minerals is called **mineralization.** This is an important aspect of **chemical cycling** in stands of vegetation. Chemical cycling consists of (1) the absorption of minerals through roots and their incorporation into chemical compounds of various kinds in plants, (2) the death of plants and plant parts, (3) and the decomposition of the plant material and release of minerals into the soil. The minerals are then reabsorbed and recycled. The high adsorptive capacity of organic matter is also important in the retention and exchange of mineral cations. When organic matter decomposes or when fertilizers are added to the soil, the mineral elements made available are subject to leaching. Organic matter may retain large quantities of minerals and thus prevent their losses from the soil. Organic matter may account for as much as 90 percent of the absorptive and adsorptive capacities of sandy soils.

Organic matter helps maintain the structure of cultivated soils. Finely divided organic matter covers mineral particles and keeps them from sticking together. Clay soils with an appropriate quantity of organic matter are less inclined to be sticky and are more readily cultivated. Crop growers say that soils that are friable and easy to cultivate have good **tilth.**

Soils are frequently classified according to their content of organic matter. One such classification is as follows:

Soil name	Percent organic matter
mineral	0–10
muck	10–40
peat	40–100

Although most agricultural soils are mineral soils, mucks and peats also may be rendered extremely productive under proper management. Mucks and peats develop when organic matter from living plants is covered by water and does not decompose. Once these soils are exposed to air by cultivation and drainage, bacteria and fungi begin to decompose them, with the result that the soil surface is actually lowered. In the Florida Everglades, drained and cultivated peat soils have actually subsided six feet over 40 years of cultivation (Fig. 12-5).

Soil organisms

Soil organisms play an important part in soil development. In addition to the roots of higher plants, the soil is inhabited by a wide variety of

Table 12-3 AVERAGE WEIGHT OF ORGANISMS IN THE UPPER FOOT OF SOIL (IN LB/ACRE).

ORGANISM	LOW	HIGH
Bacteria	500	1,000
Fungi	1,500	2,000
Actinomyces	800	1,500
Protozoa	200	400
Algae	200	300
Nematodes	25	50
Other worms and insects	800	1,000
Total	4,025	6,250

Source: Allen. *Experiments in Soil Bacteriology.* Burgess, Minneapolis, Minn. 1957.

plant and animal life (Fig. 12-6). In fact, a soil does not usually develop until the inorganic material is "invaded" by various kinds of organisms. The total weight of soil organisms (excluding higher plants) in the upper foot of fertile agricultural soils is impressive, as much as 6,000 lb per acre (Table 12-3), equivalent to 20 to 30 marketable hogs! This is about 1/1,000 of the weight of the soil they occupy.

Higher plants must be considered the principal soil organisms; the roots of trees and other higher plants penetrate crevices in rock, expand them, and split the rocks by the tremendous forces they exert when growing, thereby performing an important function in the continuing process of soil formation. Roots exude many kinds of organic acids and other substances that hasten the solution of soil minerals and make them available to plants. Living roots give off CO_2, which raises the carbonic acid content of the soil solution and increases the rate at which soil minerals dissolve. When leaves fall, or when plants die, their organic matter becomes incorporated into the soil, contributing to its fertility. Channels remaining after dead roots have decayed serve as pathways for the movement of soil water.

Bacteria are present in all soils. Estimates of the weight of living and dead bacteria in forest soils have been as great as 5,600 lb per acre. Fertile agricultural soils contain 500 to 1,000 lb of bacterial matter per acre.

Bacteria decompose organic matter of all kinds, releasing minerals which become available for

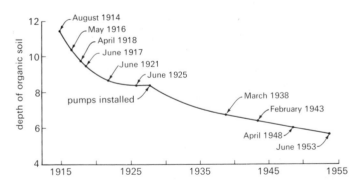

FIGURE 12-5. *The subsidence of drained organic soils in Florida over a period of 40 years. Gravity drainage began in 1914. It was necessary to install pumps for drainage in 1927 in order to keep surface soils dry enough for cultivation.* [Courtesy USDA.]

another cycle of plant growth. Some kinds of bacteria, such as those in the genus *Azotobacter*, convert molecular nitrogen of the atmosphere into nitrogenous compounds that can be used by plants. Other species live in symbiotic association with the roots of certain kinds of plants and perform the same function. **Nitrogen-fixing bacteria** supply nitrate to the plant while the plant supplies carbo- hydrates to the bacteria. Some bacteria play roles in soil formation and degradation but these proc- esses are poorly understood.

Not all soil bacteria are helpful. Many species are **pathogenic**, causing crop diseases and large economic losses. Other harmful types oxidize fer- rous iron to the slightly soluble ferric form, thereby contributing to the formation of soil horizons

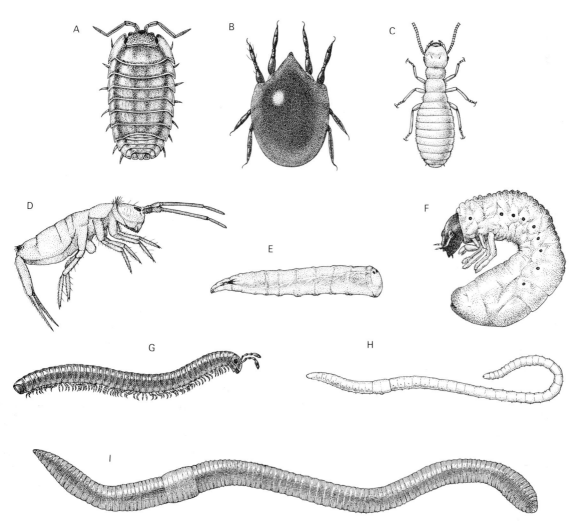

FIGURE 12-6. *Nine small animals are responsible for the fertility of most forest soils. They include (A) the wood louse, (B) the oribatid mite, (C) the termite, (D) the springtail, (E) the fly larva, (f) the beetle larva, (G(the millipod, (H) the enchytraeid worm, and (I) the earthworm. [From Edwards, "Soil Pollutants and Soil Animals."*

called **hardpans,** which are extremely rich in iron and quite hard. Hardpans may prevent the drainage of excess water.

Fungi perform many of the same functions that bacteria carry out in decomposing organic matter and in cycling nutrients. Many soil-borne fungi are also pathogenic. **Saprophytic fungi** decompose the dead organic matter of soils. When temperature, moisture, oxygen, and soil acidity are in the right combination, these fungi decompose protein and cellulose, lignin, and other carbohydrates, changing them to humus. Even the bark of trees, which is highly resistant, is decomposed by fungi.

Many species of fungi live in such close symbiotic association with roots (particularly of woody plants) that organic and inorganic materials pass back and forth. Such an association is known as a **mycorrhiza.** Water and minerals move from the fungi to the roots, and carbohydrates and other organic materials move in the other direction. Some mycorrhizal fungi actually penetrate the interior of the roots; others remain on the outside. The importance of these fungi is becoming increasingly apparent. Most species of pines cannot survive without them.

Most types of algae are aquatic, but several groups, mainly the blue-green (*Cyanophyceae*) and the green (*Chlorophyceae*), are also found in the soil, where they hasten the weathering of soil minerals. Algae normally occur near the soil surface, as they require light for photosynthesis. **Lichens** are made up of an alga and a fungus living in symbiotic relationship. They occupy a special niche in the process of soil formation. They become established on rocks, which they very slowly dissolve, and also serve as traps for dust, which accumulates and may form the inorganic soil in which other plants become established. Organic acids are leached from living and dead lichens, increasing the rate of solution of the rocks to which they are attached. In making possible the growth of other plants, lichens hasten their own demise, for they soon become covered with litter and die.

Arthropods (see Table 18-1) are among the more obvious soil fauna. Into this large classification fall the crayfish (important in poorly drained areas of the southeastern United States) mites, ants, centipedes, millipedes, sow bugs, insects, and numerous other species that physically cultivate the soil and contribute to its organic matter when they die. They are numerous and have a short life span and thus contribute hundreds of pounds of organic matter per acre to soil.

Ants, in particular, cultivate great amounts of soil. In parts of Germany, the red wood ant has been introduced into many forests to compensate for centuries of soil degradation due to timber cutting. Ants also loosen and aerate the soil and destroy many kinds of insects that are harmful to plants. Many ants and termites are severe pests in tropical areas. The large termite mounds in Brazil seriously interfere with cultivation. The leaf-cutting ant is one of the most destructive of tropical pests. Mites and myriapods play an important role in the breakdown and decomposition of forest litter.

The segmented worms (Annelidae), commonly called earthworms, and round nonsegmented eelworms, or nematodes (Nematoda), are both important in agricultural soils. An early study of earthworms by the famous biologist Charles Darwin concluded that $\frac{1}{10}$ to $\frac{2}{10}$ inches of soil may be brought to the surface each year by earthworms and deposited as castings. Nematodes are economically important because of their harmful effects in distributing parasitic fungi and increasing the extent of root-rot diseases. Some species kill bark beetles and other insect pests; some give rise to serious disorders in mammals and birds.

A number of vertebrate animals, including many species of burrowing mammals, must be considered as part of the soil fauna. Their effect is mainly one of cultivation, although their excrement contributes organic matter. Some, however, also destroy crops. Badgers, gophers, moles, voles, foxes, shrews, mice, ground squirrels, woodchucks, and even some species of birds, such as woodcocks, frequent the soil for a part of their lives. When present in great numbers they may become serious pests. Moles, for example, do considerable damage to turf grasses and certain farm crops. Beavers, through their dam building over the millennia, have had a profound impact on the forest soils of boreal areas.

The soil atmosphere

The soil atmosphere exists in the pore spaces not filled with water. These pores contain the same gases as the atmosphere above ground, but in different proportions. The soil atmosphere is not necessarily a continuous system, for there may be isolated, unconnected pore spaces.

Humidity of the soil atmosphere is near 100 percent much of the time. The carbon dioxide content is greater than that of the air above the soil because of the decomposition of organic matter, and increases with depth because of the slow rate of movement into the upper atmosphere. Conversely, the oxygen content of the soil atmosphere is less than that of the air above the soil, and decreases with depth. Oxygen is used in respiration carried on by roots and microorganisms, and is not readily replaced.

As water is added to a soil, the air in the soil is squeezed out. Consequently, plant roots may be deprived of oxygen in flooded or very wet soils. For some plants even a few days of flooding may be disastrous, especially during the growing season.

Perhaps one of the most characteristic features of the soil atmosphere is its variability. Because the distribution of dead roots, living roots, microorganisms, and structure is not uniform, the rates of gas production and rates of exchange with the atmosphere are not equal, and show great seasonal variation. Conditions in cultivated agricultural soils are likely to be much more uniform than in forest soils.

Soil moisture

Soil moisture includes free water as well as capillary water, hygroscopic water, and water vapor. Soil moisture is discussed at length in Chapters 13 and 17.

SOIL ORIGINS

Soil is formed as a result of all the interacting forces that affect the parent rock materials, air and water movement of particles, and the composition and fate of living organisms that inhabit it. Although much has been written about the fascinating subject of soil genesis, only a few of the most important contributing factors can be considered in this brief discussion.

Parent material

The **parent materials** of soils are the rock or unconsolidated materials from which the mineral components of soils have been derived. It is the parent materials that determine the physical and chemical properties of soils. The wide variety of soils found within relatively small areas frequently reflects a diversity of parent materials. **Residual materials** are formed by the weathering of rocks that have not been moved from the place where they were formed. **Transported materials,** often more important than the residual materials, are those that have been moved to the place of soil formation by wind, water, gravity, ice, or a combination of these forces. When the Pleistocene glaciers were growing in size and moving southward in the northern hemisphere, great quantities of rock and other materials were scooped up on the front edge of the ice sheet and were deposited at great distances from their sources when the ice melted. Locally, landslides deposit unconsolidated materials downslope, which later become the **colluvial** (gravity deposited) parent materials of soils. Soils deposited in the bottomlands along the rivers are called **alluvial.**

Three main kinds of rocks are igneous, sedimentary, and metamorphic (see Table 12-4). **Igneous** rocks are formed when molten lava or magma solidifies. **Sedimentary** rocks are formed when sediments that have accumulated in water or have been deposited from the air are compressed and cemented into rocks. **Metamorphic** rocks are formed when igneous, sedimentary, or previously metamorphosed rocks are altered and recrystallized by heat or pressure, or occasionally infiltrated by solutions of minerals that subsequently precipitate.

Table 12-4 REPRESENTATIVE TYPES OF ROCKS OF DIFFERENT ORIGINS.

IGNEOUS	SEDIMENTARY	METAMORPHIC
Basalt	Chalk	Gneiss (from granite)
Granite	Conglomerate	Marble (from limestone or chalk)
Mica	Limestone	Quartzite (from sandstone)
Obsidian	Sandstone	Schist (from micaceous shales)
Pumice	Shale	Slate (from shale)
Rhyolite		

Weathering and soil formation

CHEMICAL PROCESSES   Many changes take place in the weathering of parent material, both on the surface of rocks and within them. The main chemical processes are hydrolysis, carbonation, oxidation, and hydration.

Hydrolysis is a decomposition process in which water is one of the reacting agents. It can be illustrated by the hydrolysis of potassium aluminum silicate to aluminum silicate, an important soil mineral:

$$KAlSi_3O_8 + H_2O \rightarrow HAlSi_3O_8 + KOH.$$

Carbonation may be illustrated by the decomposition of calcite to calcium bicarbonate:

$$CaCO_3 + CO_2 + H_2O \rightarrow Ca(HCO_3)_2.$$

The bicarbonate is highly soluble and readily leached from soils.

Oxidation involves the loss of electrons to a receptor, which is frequently oxygen. Iron is a familiar compound that is easily oxidized. We call this particular iron oxide "rust." Iron pyrite is an important soil mineral that is oxidized to yield ferric hydroxide and sulfuric acid:

$$2FeS_2 + 7H_2O + 15O \rightarrow 2Fe(OH)_3 + 4H_2SO_4.$$

Hydration is the combination of molecular water with a compound. In the hydrated form it may be more subject to weathering processes. Calcium sulfate is hydrated to gypsum by the addition of water:

$$CaSO_4 + 2H_2O \rightarrow CaSO_4 \cdot 2H_2O.$$

TEMPERATURE   Temperature is important in the weathering of parent material because changes in temperature cause differential expansion and contraction of minerals and rocks. Since rocks are aggregates of minerals that each have different coefficients of expansion, changes in temperature result in internal stresses that cause cracks. Even large boulders can be rent asunder by relatively small differences in temperature.

In rocks exposed to solar radiation a very large temperature gradient is established within the first few millimeters beneath the surface. When this happens the hot outer surfaces may expand and flake off (**exfoliate**). American Indians and members of other primitive cultures used the same principle to fashion arrowheads by repeatedly dropping water on heated flint on the spot where a chip was to be removed.

Low temperatures cause water to freeze in the cracks of rocks and split them apart. When water freezes it exerts a force of about 150 tons per square foot. Freezing is probably the main weathering agent at high latitudes and altitudes. The rubble formed by freezing has characteristic sharp edges.

The dissociation of water molecules into H^+ and OH^- ions is much greater at high temperatures than at low temperatures. Since H^+ ions are particularly effective in dissolving minerals, the rate of weathering is faster when temperatures are higher. If soil is frozen or very cold throughout the year, bedrock may be weathered only a few inches deep. On the other hand, weathering in the tropics extends to depths of 160 feet or more!

RAINFALL   Soil moisture is obtained in most areas through rainfall, although fog and dew are important in some parts of the world. The pounding of raindrops over thousands of years plays an important role in the weathering of exposed rocks and in the erosion of weathered materials. Such effects are minor, however, when compared to the effects of soil moisture in dissolving minerals. And these in turn are minor when raindrop joins raindrop to form streams, which erode and grind away mountains of material, redistributing them hither and yon.

Rainfall is frequently the most important climatic element in soil development. In regions of moderate to high rainfall, soil development is likely to be greater than in those where rainfall is sparse. Desert soils are usually poorly developed, yet, in adjacent areas with adequate rainfall to sustain plant growth, soils originating from identical parent materials may be well developed and productive.

TOPOGRAPHY   Topography frequently plays an important part in soil development. If the land is flat, low, and poorly drained, water may accumulate and drain so slowly that biological activity will be arrested for long periods. Under such conditions, soil material, both inorganic and organic, may remain in a state of preservation until drained. When topography is very steep, parent materials may be subject to erosion by water, wind, and ice to such an extent that soil material is removed from the site as soon as it is weathered from the parent material. The most favorable topographic situation for soil development is one in which the slope is steep enough to carry off excess water, but not so steep as to permit the removal of weathered materials that lack the protection of a vegetative cover. In addition, for "normal soil" the water table should be below the weathered parent material so that roots can occupy it to the greatest possible extent.

TIME   The factors that contribute to the development of soil are straightforward. They fall into a discrete pattern of events. Yet, unless the time element in soil formation is recognized, the significance of the events that take place cannot be truly appreciated. A mature soil is not formed in a decade, or even ten decades, but over thousands of years. Soil development and vegetational succession are related processes. When vegetational development reaches a **climax condition,** in which the vegetation type is in balance with its environment and is relatively unchanging, soil will also have reached full development and will be in balance with its environment. The soil is then said to be **mature.** There is a large variation in the absolute length of time required for complete maturation. In the tundra of the far north, cold weather precludes rapid biological and physical weathering; consequently, five to ten thousand years might be required to form a soil only a few inches deep. In the tropics, however, the much faster rates of reaction can lead to the development of soils that may be five to ten feet deep in a much shorter period of time.

The age of a soil is determined by the degree to which it has developed toward maturity. It may be chronologically young (in terms of absolute years) and yet may be mature in terms of its zonal structure—the soil profile.

The soil profile

In excavating through the soil, it is necessary to dig through distinct zones called horizons. The morphology of these horizons is called the **soil profile.** Each soil type has a distinctive profile that is due to the particular combination of factors that produced the soil. By the time most forest and prairie soils are mature they have developed three major horizons, designated (from the top down) as A, B, and C. The underlying bedrock (usually but not always the true parent material) is called the D horizon.

The A horizon is the zone of leaching (eluviation). It is the most abundant in roots, bacteria, fungi, and small animals (for example, worms, nematodes). It is poor in soluble substances and has lost some clay and iron and aluminum oxides.

The B horizon is the zone of accumulation (illuviation). It is less abundant in living matter. It is higher in clay and iron and aluminum oxides, and is thus stickier when wet and harder when dry.

The C horizon consists of weathered rock material, often true parent material. When clay (hard) pans are present (as in prairie, wheat, and corn raising regions) they are a part of this horizon. (In Great Britain the C horizon is not considered a part of the true soil.)

Often it is desirable to subdivide the A or B horizons further. They are then each divided from top to bottom into three strata: A_1, A_2, and A_3; B_1, B_2, and B_3. The A_1 is often dark in color, owing to its high content of organic matter. The middle subdivision (A_2 or B_2) is always the typical one for that horizon; the others tend to grade into their neighbors. The layer of partially decomposed duff immediately above the A horizon can be referred to as the A_0 horizon and the fresh litter above that as the A_{00} horizon. Not all profiles show complete development, and soils that have been used for agricultural purposes for long periods of time are especially likely to be lacking in one or more of the uppermost horizons. Their most prominent feature is the plow horizon—the horizon that has been repeatedly disrupted by plowing, disking, and harrowing.

Soils with well-developed profiles are called **zonal,** or **normal, soils.** Zonal soils are found where the factors of soil formation are conducive to profile development. **Azonal soils** have no distinct profile and are frequently alluvial (water deposited) or colluvial (gravity deposited) soils of recent origin. **Intrazonal soils** are those that are limited in extent and are under the control of a local factor, such as high salt content or poor drainage, that does not permit the development normally associated with zonal soils.

Some soils have been **truncated.** That is, they consist of only a portion of the original profile, the upper portion having been lost through erosion (Fig. 12-7). Most agricultural soils are truncated remains of the original soil that developed under natural vegetation, and some are completely new, having been formed by the addition of soil or organic matter from another location in order to increase productivity. At times, truncated soils are "buried" under soil deposited by wind or water (Fig. 12-8).

Major soil groups

Differences in soil-forming processes lead to different zonal soils. The major soil groups are based on climatic differences. In the humid regions three broad groups have been recognized: tundra, podzolic, and lateritic soils. In the arid and semiarid regions specific soils have developed in response to low rainfall. These include the chernozems, or black soils, and various desert soils.

The **tundra soils** are formed in cold climates, where biological activity is minimal. Organic matter accumulates on the surface, and the deeper horizons remain frozen throughout the year (permafrost). In winter they are completely frozen. Soil horizons are only poorly differentiated. These soils are of little agricultural significance,

In cold, humid climates a particular combination of soil-forming processes, collectively called **podzolization,** produces **podzolic soils. Podzols,** the most extreme type, have a deep accumulation of litter and humus, a strongly acid reaction, and an upper horizon from which iron and aluminum oxides (the so-called sesquioxides) have been leached. (The name "podzol" is derived from the Russian and means "ash beneath," with reference to the color of the A_2 horizon.) Fungi are the main soil-forming organisms. True podzol soils are found at the higher latitudes and coincide with the boreal forest. The mild summers and long, cold winters in these latitudes permit the accumulation of the

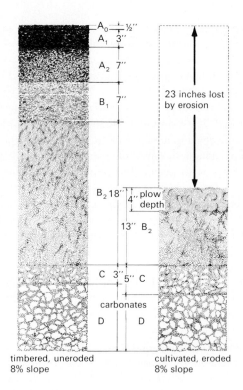

FIGURE 12-7. *Uncontrolled erosion can result in the loss of productive topsoil, as shown by these profiles taken in Miami silt loam on adjoining fields. Profile at left was taken on a field still covered by virgin timber; profile at right was taken on a field cleared and farmed for 50 to 75 years.* [After USDA.]

litter and humus necessary for their formation. Even in temperate climates, however, and sometimes under special conditions in the subtropics, podzolization is a factor. The widespread forests of the temperate zone grow predominantly on podzolic soils. They are potentially good agricultural soils, topography permitting, but are easily degraded by poor cropping practices.

In the humid tropics and subtropics the combination of soil-forming processes is collectively called **laterization** and forms **lateritic soils. Laterites,** the extreme type, have a very shallow accumulation of litter and humus, a neutral soil reaction, and an accumulation of iron and aluminum oxides in the upper horizons. Bacteria are the primary soil-forming organism. Laterites are typical of tropical rain forests, and are characterized by extremely

FIGURE 12-8. *Silt was deposited on a flood-plain soil, and a second soil developed as a result. The soil whose former surface shows at the 8-foot level is called a buried soil.* [*Courtesy USDA.*]

intensive chemical weathering. Even the most resistant compounds, such as silica, are broken down and leached by the heavy precipitation in the tropical rain forests. These soils are so deeply weathered that some geologists regard laterization as a geologic rather than a soil-forming process. The name "laterite" is derived from the Latin word *later*, which means brick; these soils are widely used as building materials. Also, some soils in climates with low precipitation accumulate carbonates on and near the surface. This process, called calcification, tends to produce very alkaline soils, which are characteristic of deserts and steppes.

Podzolization and laterization, the dominant soil-forming processes, are compared in Table 12-5. Figure 12-9 shows the relationship of all three processes.

Table 12-5 COMPARISON OF PODZOLIZATION AND LATERIZATION IN SOIL FORMATION UNDER OPTIMAL CONDITIONS.

VARIABLE	PODZOLIZATION	LATERIZATION
Climate	cold	warm to hot
Natural vegetation	conifers	tropical rain forest
Most active soil-forming microorganisms	fungi	bacteria
Raw humus	abundant	sparse
Soil reaction	acid (pH 4–4.5)	nearly neutral (pH 6.5–7)
Earthworms	scarce	abundant
Solubility of silica	low	high
Iron and aluminum	leached out of A horizon	accumulates in A horizon
Hydrolysis	weak	intense

232

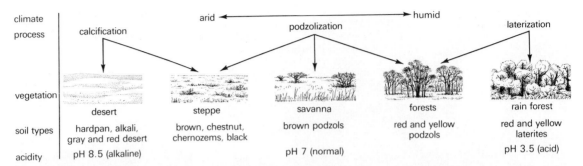

FIGURE 12-9. *Soil-forming processes are not discrete, easily identifiable units. They form a continuum, so that some soils are the result of a combination of processes, while others are more clearly a function of a single process.* [*Courtesy USDA.*]

Chernozems are formed in areas that receive 15 to 25 inches of rain per year. The natural vegetation is grass. The upper horizon is dark due to the accumulation of organic matter. Chernozem means "black soil." In the temperate zones these are among the most fertile agricultural soils, and produce much of the world's grain.

Desert soils are only slightly weathered and leached because of the low rainfall. Nutrients other than nitrogen are moderate or high. As a result, under irrigation and proper management these soils may become very productive.

SOIL CLASSIFICATION

Soils may be classified on the basis of genetic origin—that is, parent materials—or on the basis of morphology. The National Cooperative Soil Survey initiated a new system called the United States Comprehensive System of Soil Classification, which contains 8,000 soil series and about 80,000 soil types and phases. This system, which took 15 years to devise, does not alter the basic content or form of the soil survey report that soil conservationists and land use planners have used in the past. Neither does it require major changes in the technical guides for engineers, biologists, and other users of soil survey information. It does, however, permit a higher degree of accuracy in classifying soils, permit greater efficiency in mapping, and allow the classification of soils that for various reasons could not be included in other systems.

The classification system used previously was developed and published in 1936 by C. F. Marbut, who was chief of the U.S. Soil Survey at that time. A drastic modification, outlined in the 1938 Yearbook of Agriculture (*Soils and Men*) was used until 1965. The Comprehensive System differs from all other systems in two important respects: (1) the nomenclature of the higher categories is entirely new, and (2) definitions of the classes are much more quantitative and specific.

The new system emphasizes the properties of soils as they exist today! Previous classifications were based on assumptions of what the soil was when the country was settled, even though cultivation and erosion might have changed them greatly. It was always difficult to get agreement on the genesis of many soils. Thus, the new classification is much more utilitarian than the older ones.

A new language was created for the higher categories of classification so that there would be no confusion with names or words formerly in use. Soil scientists selected the diagnostic properties for each class, and classicists from the University of Illinois and the University of Ghent proposed Greek and Latin names for them.

In the new system there are 10 **orders** of soils, each defined by a few distinguishing characteristics (Table 12-6; Fig. 12-10). As an example, one of the orders is designated "Oxisol," the *oxi* referring to oxide and *sol* to soil. These are the soils of tropical regions; they contain large amounts of iron and aluminum oxides.

Table 12-6   SOIL ORDERS ACCORDING TO THE COMPREHENSIVE CLASSIFICATION.

ORDER	FORMATIVE SYLLABLE	DERIVATION	MEANING	DIAGNOSTIC FEATURES	OLDER EQUIVALENTS
1. Entisol	ent	Coined syllable	Recent soil	Very weak or no profile development	Regosols, lithosols, alluvial, some low humic gley
2. Vertisol	ert	Latin: *verto*, turn	Inverted soil	Self-mulching; expanding lattice clays; subhumid to arid climates	Grumusol, regur, black cotton, tropical black clays, smonitza, some alluvial
3. Inceptisol	ept	Latin: *inceptum*, beginning	Inception, or young soil	Weak profile development but no strong illuvial horizon; cambic horizon present	Brown forest, subarctic brown forest, tundra, ando, and some lithosols, regosols, and humic gley
4. Aridisol	id	Latin: *aridus*, dry	Arid soil	Soils of arid regions; often have natric, calcic, gypsic, or salic horizons	Desert, red desert, sierozem, reddish brown, solonchak, some regosols, and lithosols
5. Mollisol	oll	Latin: *mollis*, soft	Soft soil	Thick, dark A_1 horizon; usually develops under grassy vegetation	Chernozem, brunizem (prairie), chestnut, reddish prairie, some humic gley, rendzinas, brown, reddish chestnut, and brown forest soils
6. Spodosol	od	Greek: *spodos*, wood ash	Ashy (podzol) soil	Illuvial horizon shows accumulation of iron and organic collodids; weak to strongly cemented pan	Podzols, brown podzolic, groundwater podzols
7. Alfisol	alf	Coined syllable	Pedalfer (alfe) soil	Argillic horizon of relatively high base saturation ($>35\%$); usually under boreal or deciduous broad-leaf forest	Noncalcic brown, gray-wooded; many planosols, some half-bog soils
8. Ultisol	ult	Latin: *ultimus*, last	Ultimate (of leaching)	Argillic horizon of low base saturation ($<35\%$); plinthite often present; humid climate; usually forest or savanna vegetation	Red-yellow podzolic, reddish-brown lateritic, rubrozem, some gley and groundwater laterites
9. Oxisol	ox	French: *oxide*, oxide	Oxide soils	Argillic horizon very high in iron and aluminum oxides	Latosols, and most ground water laterites
10. Histosol	ist	Greek: *histos*, tissue	Tissue (organic) soils	Organic surface horizon ($>30\%$ organic matter) more than 6 inches thick	Bog and some half-bog soils

Source: After *Soil Classification, a Comprehensive System*, Soil Conservation Service, 1960 (revised 1964).

Subcategories include **suborders, great groups, subgroups, families, series,** and **soil type.**

Suborders are named according to the distinguishing features of the horizons that develop. Thus, an "Aquox" is a wet soil, gray or bluish in color, with large amounts of oxides.

Great groups define the soils still more specifically, whereas suborders may make reference to minor properties which are also found in great groups. Families are based on properties important to the growth of plants.

The soil series is a collection of individual soils that are essentially uniform with respect to differentiating characteristics, including color, zonal development, and the depth limit within which each horizon is found. The series is classified ac-

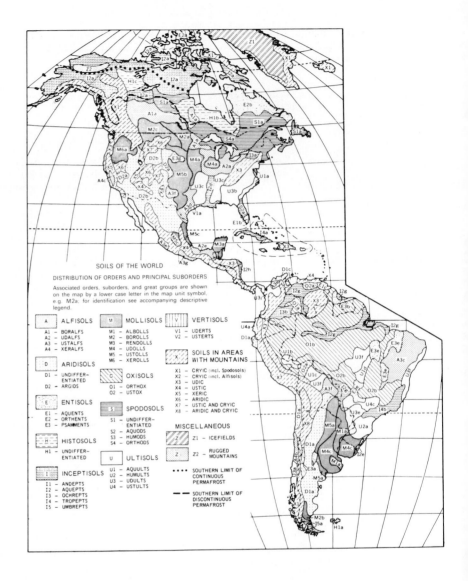

SOILS OF THE WORLD

DISTRIBUTION OF ORDERS AND PRINCIPAL SUBORDERS

Associated orders, suborders, and great groups are shown
on the map by a lower case letter in the map unit symbol.
e.g. M2a. for identification see accompanying descriptive
legend.

	A	ALFISOLS		M	MOLLISOLS		V	VERTISOLS
	A1 – BORALFS			M1 – ALBOLLS			V1 – UDERTS	
	A2 – UDALFS			M2 – BOROLLS			V2 – USTERTS	
	A3 – USTALFS			M3 – RENDOLLS				
	A4 – XERALFS			M4 – UDOLLS				
				M5 – USTOLLS			SOILS IN AREAS	
				M6 – XEROLLS			WITH MOUNTAINS	

	D	ARIDISOLS		G	OXISOLS	X1 – CRYIC (incl. Spodosols)
	D1 – UNDIFFER-			O1 – ORTHOX	X2 – CRYIC (incl. Alfisols)
	ENTIATED			O2 – USTOX	X3 – UDIC
	D2 – ARGIDS				X4 – XERIC
					X5 – USTIC
					X6 – ARIDIC
	E	ENTISOLS		S	SPODOSOLS	X7 – USTIC AND CRYIC
	E1 – AQUENTS			S1 – UNDIFFER-	X8 – ARIDIC AND CRYIC
	E2 – ORTHENTS			ENTIATED	
	E3 – PSAMMENTS			S2 – AQUODS	MISCELLANEOUS
				S3 – HUMODS	
	H	HISTOSOLS		S4 – ORTHODS	Z1 – ICEFIELDS
	H1 – UNDIFFER-				Z2 – RUGGED
	ENTIATED		U	ULTISOLS	MOUNTAINS
				U1 – AQUULTS	
	I	INCEPTISOLS		U2 – HUMULTS	••••• SOUTHERN LIMIT OF
				U3 – UDULTS	CONTINUOUS
	I1 – ANDEPTS			U4 – USTULTS	PERMAFROST
	I2 – AQUEPTS				
	I3 – OCHREPTS				——— SOUTHERN LIMIT OF
	I4 – TROPEPTS				DISCONTINUOUS
	I5 – UMBREPTS				PERMAFROST

cording to type, the lowest category of classification based on texture.

Fortunately, the crop grower will still be able
to call his soil by the same name as used before
the advent of the new system. As a matter of fact,
he need not even know that an old familiar soil
such as "Lackland sand" is also an "ultic quarzop-
samment!"

Land capability classification

The capability classification of land has been
widely used in the United States. In this system,
arable soils are grouped according to their potential productivity and their limitations for crop
production without special treatment. Nonarable
soils are grouped in eight classes according to their
risk of damage and their potential for sustained
production. The first few classes are suitable for
cultivation if good management practices are followed: Classes V through VII are not suited for
cultivation, but can be used for grazing and forest
management. Class VIII land is useful only for
recreation, wildlife, and watersheds. Details of the
classes are as follows:

Class I soils are potentially the most productive

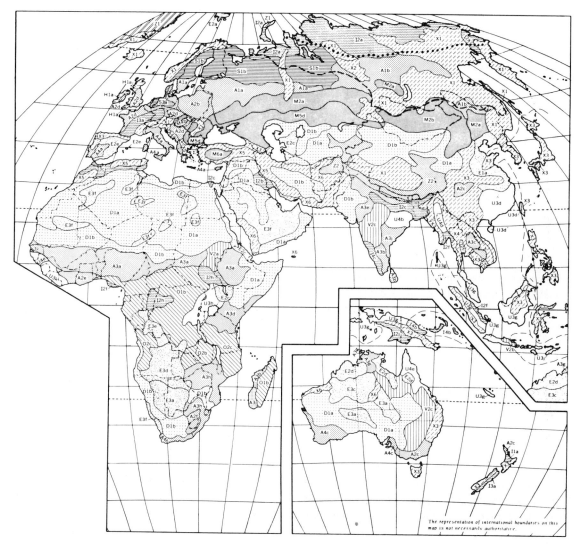

FIGURE 12-10. *Soils of the world: distribution of Orders and principal suborders.* [*Courtesy USDA.*]

of all soils. They are nearly level, and have a low erosion hazard. They are deep and well drained, with a good nutrient supply and water holding capacity, and have all of the other physical and chemical characteristics required for intensive crop production. They can be used for the production of cultivated crops, pasture, range, forests, and wildlife. Irrigated soils may be placed in Class I if water is supplied by a permanent irrigation system.

Class II soils have some limitations. They may have gentle slopes, moderate susceptibility to ero-

sion, a soil depth that is less than ideal, slightly unfavorable soil structure, wetness that can be corrected by drainage, slight climatic limitations, moderate salinity or alkalinity, or other slight limitations. Farm operators have a smaller choice of crops to produce on these soils than they do with Class I soils. Class II soils can be used for crops, pasture, range, forests and wildlife.

Class III soils have severe limitations that reduce the choice of plants, frequently requiring the use of special conservation practices. They may have moderately steep slopes, high susceptibility to wind and/or water erosion, overflows which result in crop damage, only slightly permeable subsoil, a shallow depth to rock or pan layer, which restricts rooting, poor fertility, and moderate salinity or alkalinity, either all in combination or singly. They may be used for cultivated crops, pasture, forest, range, or wildlife. Management of these soils must be good if they are to remain productive.

Class IV soils have very severe limitations. They may be on steep erodable sites, have a history of erosion, be too shallow, be excessively wet, be severely alkaline, have a poor water holding capacity, or be subject to overflows. Although these soils may be used for crops, pasture, forests, range, or wildlife, extremely careful soil conservation practices must be applied. It is not advisable to use them for crops that require extensive cultivation.

Class V soils have little or no erosion hazard, but they have other limitations so that they may be used only for pasture, range, forests, or wildlife production. They may be bottomland soils which are subject to frequent overflow, nearly level soils where the growing season is short and limits the production of cultivated crops, level soils that are rocky, and ponded areas in which drainage is not feasible.

Class VI soils have severe limitations. They may be steep, highly erodible or eroded, stony, shallow, wet, subject to overflow, saline, or alkaline, or they may have a poor water-holding capacity. They can be used for pasture, range, forest, or wildlife. Special practices, such as water control by contour ditches, water spreading, or drainage are usually necessary. Some of these soils are adapted to special crops such as sodded orchards and blueberries.

Class VII soils have such severe limitations that their use is restricted to grazing, forest, or wildlife. They are more severely restricted than Class VI soils in that one or more of their continuing limitations may be be compensated for or corrected. They may be too steep, stony, erodible, wet, alkaline, saline, or deficient in some other manner.

Class VIII soils are useful only for recreation, wildlife habitat, watersheds, or aesthetic purposes.

They cannot be managed for crops, grasses, or trees. Uncorrectable limitations may include erosion or erosion hazard, wetness, stoniness, salinity or alkalinity, or poor water-holding capacity.

Capability subclasses have the same kinds of limitations. Four kinds of limitations are recognized.

Subclass e soils have experienced severe erosion, or have a potential erosion hazard.

Subclass w includes soils for which excess water is the dominant limitation. Poor drainage, overflow, and high water tables are likely to be problems.

Subclass c consists of soils for which temperature and lack of precipitation are likely to be problems. This subclass contains soils for which climate is the only limitation.

Subclass s includes soils having characteristics that restrict rooting. These soils may be shallow, stony, alkaline, or saline, they may have low fertility that is hard to correct, or they may have a low water-holding capacity.

The capability classification of land has been used in the past solely for agricultural practices. However, land-use planners and municipal officials are now beginning to realize that this system can be applied in many other fields as well, such as in planning the location of new subdivisions. For example, a shallow soil with an impermeable layer close to the surface would preclude the use of septic tanks and drainage fields.

SOIL CONSERVATION

Conservation means "careful use." The careful use of our soil resource is an obligation not only to ourselves but to future generations as well. This obligation has not always been acknowledged or acted on. As early as 1894 Assistant Secretary of Agriculture Charles W. Dabney wrote, "Thousands of acres of land in this country are abandoned every year because the surface has been washed and gullied beyond the possibility of profitable cultivation" (Fig. 12-11). Unfortunately, three-quarters of a century later, this statement holds true, on an even larger scale!

Soil conservation has many meanings. For crop producers it implies a careful and intensive series of management practices involving plants and soils. In forests, soil conservation requires the use of logging practices that will not permit erosion, the regulation of grazing, and protection from uncontrolled fire. To others less concerned with growing crops, it may mean the stabilization either of sand dunes along a beach or of steep slopes by planting appropriate plant species.

The basis of agriculture is a layer of topsoil that averages only seven inches in depth over the earth's surface. This soil cannot be recklessly exploited indefinitely; it must be preserved and refurbished. Although the reduction of productive capacity through the loss of fertility and structure is considerable, the most serious problem is erosion. Nutrients can be artificially supplied, but the loss of topsoil cannot be so quickly or easily remedied. The loss of soil due to wind and water are national problems. It contributes to silt-clogged rivers, alternate drought and flood, dust bowls, and poverty.

FIGURE 12-12. *A stroboscopic camera has caught the impact of a raindrop on moist soil, bursting upward and outward and carrying soil with it. [Courtesy USDA.]*

FIGURE 12-11. *Uncontrolled erosion eventually results in the complete loss of the soil resource, accompanied by poverty and misery. This tragedy has been commonplace in many parts of the world. [Courtesy USDA.]*

Erosion of the soil is a natural process influenced by climate, topography, and the nature of the soil itself. Where permanent and undisturbed plant cover exists, erosion is more or less gradual and in equilibrium with soil-forming processes. Accelerated erosion comes about in the absence of plant cover. Areas that are unable to support a permanent plant cover as a result of climate or topography undergo large-scale erosion of the kind that carved the Grand Canyon. The accelerated soil erosion brought about by agricultural cultivation or overgrazing comes about largely through the action of water in humid climates and of wind in arid climates.

The maintenance of vegetative cover is basic to soil management. Vegetative cover retards erosion by cushioning the beating force of the rain (Fig. 12-12), increasing the absorptive capacity of the soil, and holding the soil against both water and wind. The soil cover increases the infiltration of water through the soil by preventing the clogging of the soil pores by fine surface particles. The techniques used for increasing soil cover include the increased use of sod culture, proper rotation, cover cropping, and mulching.

Water erodes the soil by literally carrying it away. The carrying power of water increases with its speed and volume. The volume of excess water depends upon the amount of rainfall and the rate at which it is absorbed by the soil. The speed with which this water moves is directly related to the slope of the land and the amount of cover. Any technique that either increases absorption or reduces the speed of the runoff will prevent soil erosion.

The absorptive capacity of soil may be increased by deep plowing, by increasing the organic matter of the soil, or by increasing drainage. Thus the burning or removal of organic matter is a poor conservation practice. Where natural drainage is poor, tiling may be necessary to remove water and provide air. In some regions clay pans have to be periodically broken up with plows cutting as deep as six feet.

The control of erosion by reducing the speed of runoff may be accomplished in a number of ways. Most basic is contour tillage, in which plowing, cultivation, and the direction of the "row" follows the contour rather than the slope of the land (Fig. 12-13). This affects the speed and power the surface water attains and thus the ability of the tilled soil to absorb water. The use of intertillage or strip cropping, which alternates strips of sod and row crops planted along the contour, helps to slow runoff by interposing barriers with high absorptive capacity. The alternating of sod with row cropping serves to achieve the benefits of rotation. On steep slopes where greater amounts of surface water must be accommodated, the use of **waterways**—permanently sodded areas—facilitates water removal and minimizes erosion.

Where contour cultivation and strip cropping are not sufficient to check erosion, terraces constructed on the contour must be used. **Terracing,** an ancient practice, consists in cutting a slope into a number of level areas. Terraces appear as giant steps on the hillside. Although the steps of ancient terraces were made of stone, modern terraces are made by building low rounded ridges of earth

FIGURE 12-13. *Water was held on this 6 percent slope for 36 hours after a 4-inch rain without washing out the corn. Nearby fields planted in straight rows up and down the slopes had to be replanted.* [*Courtesy USDA.*]

across the sloping hillside. Terracing slows down the speed of surface runoff, and although it is designed primarily to prevent erosion, it facilitates the storage of available water. Thus terracing is an important practice in areas of low rainfall and where the scarcity of arable land necessitates the exploitation of steep hillsides.

Some of the erosion caused by wind, especially in open prairies or plains, may be checked by planting windbreaks, one or more rows of trees or shrubs planted at right angles to the prevailing winds. The effectiveness of the windbreak is local and is related to the thickness and the height of the trees. The maintenance of a permanent plant cover in conjunction with the windbreaks effectively reduces wind erosion where it is a problem. On organic soils small grain rows are used as temporary windbreaks to protect seedlings.

Cropping practices almost always result in the loss of soil, much more so for row crops such as potatoes and corn than for crops that provide a permanent soil cover. Until recently, soil losses from sloping cropland could be estimated only in relation to level cropland losses. Scientists at the University of Tennessee, using data from experimental plots, devised an equation for predicting soil losses for a given field under different management practices, which applies directly to Tennessee agricultural land. The following components make up the equation:

(A) *Average annual soil loss* refers to the annual soil loss, in tons per acre, which occurs in the course of a period of years. It may not be valid for any one particular year.

(R) The *rainfall factor* is the erosion potential in a given locality. This factor expresses the intensity of rainfall and its kinetic energy.

(K) *Soil-erodibility* is a reflection of soil structure, texture, organic matter, permeability, depth, and all of the other characteristics that influence the rate of soil erosion.

(LS) The *length and steepness of slope* factor is extremely important. Longer and steeper slopes have more erosion.

(C) The *cropping-management* factor is the ratio of soil loss under a specific management system to losses under continuous fallow.

(P) The *conservation-practice* factor assigns a value for conservation practices such as contour plowing and terracing, which reduce soil losses.

(T) The *soil-loss-tolerance* value is the estimated average annual soil loss that can be tolerated in tons per acre, if sustained, economical crop production is yet to be achieved.

In the equation, $A = RKLSCP$ is computed, and then compared with the T value. One acre-inch of soil weighs about 150 tons. If the average soil-loss is 5 tons per acre per year, then 30 years would be required to erode 1 acre-inch of soil from a field that is being cropped. If this rate of loss is too high for a particular situation, then either additional conservation must be initiated or cropping practices must be changed.

In the past few years, much attention has been given to developing cropping methods that disturb the soil as little as possible. Aerial seeding and fertilization are now established practices in certain parts of the western United States. **Sod-cropping,** growing with a permanent soil covering of sod, is a new innovation that is presently being tested. **Stubble-cropping,** which is planting seeds directly in the stubble of the previously harvested crop without cultivation, is a form of **minimum-tillage** agriculture. Leaving bands of permanent sod, on the contour, between strips of crops that must be tilled is a method sometimes used. All of these practices are potential causes of decrease in total crop production, and yet they are developments that have taken place, in part, because of the continuing concerns of agriculturists for environmental conservation.

Soil conservation has become a matter of ethics for crop producers, equally as important as those dealing with honest weights and measures and with the humane treatment of livestock. To make conservation a vital factor, it may be necessary to sacrifice large short-term gains for smaller profits that accumulate over a long period. In the end, soil conservation implies acceptance (and perhaps pursuit) of a level of biological productivity that can be sustained indefinitely. Soil conservation often provides immediate benefits in terms of plant performance and must be considered a requisite of sound soil management.

Soil has sometimes been called "our most important heritage." Without a doubt, it is the backbone of agricultural productivity, and the greatness of nations can be measured in terms of their skills and persistence in producing agricultural products. The developing nations will do well to heed the lessons of history, and see that their agricultural base, which is circumscribed by their climates and soils, is carefully developed and protected.

Selected References

Baver, L. D. *Soil Physics* (4th ed.). Wiley, New York, 1972. (A widely used text.)

Black, C. A. *Soil-Plant Relationships*. Wiley, New York, 1968. (An outstanding and comprehensive text which treats soil as a medium for plant growth.)

Buckman, H. O., and N. C. Brady. *The Nature and Properties of Soils* (7th ed.). Macmillan, New York, 1969. (A basic and widely used textbook on agricultural soils.)

Committee on Tropical Soils of the National Research Council. *Soils of the humid tropics*. National Academy of Sciences, Washington, D.C., 1972. (A collection of 12 papers concerning problems in the management of tropical soils.)

Corbett, J. R. *The Living Soil*. Martindale Press, West Como, New South Wales, Australia, 1969. (An elementary text in soil science.)

Forth, H. D., and L. M. Turk. *Fundamentals of Soil Science* (5th ed.). Wiley, New York, 1972. (An introductory text.)

Postgate, J. R. (editor). *The Chemistry and Bio-chemistry of Nitrogen Fixation*, Plenum Press, New York, 1971. (A technical overview of an often overlooked but fundamentally important aspect of biology, comprising a comprehensive survey of what is ordinarily a widely divergent field).

Russell, E. W. *Soil Conditions and Plant Growth* (9th ed.). Longmans, New York, 1961. (The most famous work on agricultural soils. The first seven editions were written by Sir John Russell.)

Taylor, S. A. *Physical Edaphology* W. H. Freeman and Company, San Francisco, 1972. (A textbook on soils that also deals with the practical aspects of irrigation, fertilization, plant-water relationships, and soil behavior under cultivation.)

U.S. Department of Agriculture. *Soil*. Yearbook of Agriculture, 1957. (This is an updated version of the famous 1938 yearbook, *Soils and Men*.)

Wilde, S. A. *Forest Soils, Their Properties and Relation to Silviculture*. Ronald Press, New York, 1958. (An original textbook that contains many unusual and intriguing ideas about soils.)

CHAPTER **13**

Water

Consider, for a moment, a substance that is often taken largely for granted—water. We drink it, breathe its vapors, cook with it, and bathe ourselves with it. In the form of perspiration, it cools our bodies. Water can extinguish fire, dissolve either rocks or sugar, and is transparent. When frozen, unlike most substances, water becomes lighter and floats. In short, this common substance has rather uncommon properties.

Water comes closer than any other liquid to being a universal solvent. Given time, it can dissolve practically anything on earth. This is biologically important, because practically all of the substances of which organisms are composed are either immersed or dissolved in it. In effect, people and plants are columns of water that serve as the media for both passive and active metabolic transfer of materials. Its distribution throughout plants provides pathways of movement from the tips of the roots to the topmost leaves of even the highest redwoods.

PROPERTIES OF WATER

Once the temperature of water has reached the boiling point, it will rise no further until it changes from the liquid to the gaseous state. But before the change in state occurs the water will absorb 539.3 calories per gram. The energy required to change water from the liquid to the gaseous state is called the **latent heat of vaporization** of water (Fig. 13-1). The uniquely high latent heat of vaporization of water adds greatly to its stabilizing effect on surface temperatures of the earth.

A similar absorption of heat at constant temperature takes place when ice melts. The energy required to melt ice is called the **latent heat of fusion.** After the temperature of ice has risen to 0°C, it will absorb 79.7 calories per gram before melting. Conversely, 79.7 calories of heat is released when one gram of water changes to ice. Few substances require this much energy to change from the liquid to the solid state or vice versa.

242

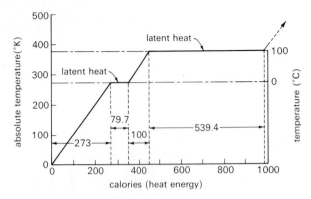

FIGURE 13-1. *The latent heats of water. One gram of water changes from ice to liquid at constant temperature with the absorption of 79.7 calories. It changes from liquid to gas with the absorption of 539.4 calories.* [*After Davis and Day,* Water, the Mirror of Science. *Copyright © 1961 by Educational Services, Inc. and Heinemann Educational Books, Ltd.*]

The **specific heat** of a substance is the amount of heat in calories required to change the temperature of one gram of the substance 1°C. As a consequence of the definition of the gram calorie, the specific heat of water is 1 calorie per gram °C. Water has an exceptionally high specific heat, as is shown in the accompanying table. Few other substances found in nature require so much heat to change its temperature so slightly. It is easy to see why water has a stabilizing effect on the temperature of organisms. Imagine how rapidly the temperature of an organism, either plant or animal, might change if 90 percent of the weight of its living parts were not made up of water.

Specific heats of various substances (cal/gram °C)

Water	1.00
Paraffin	0.69
Wood	0.42
Cellulose	0.37
Chalk	0.21
Glass	0.19
Granite	0.19
Quartz	0.19
Iron	0.11
Copper	0.09
Zinc	0.09

Water is also a good conductor of heat. It is not as good a conductor as some other compounds, but nonetheless it ranks high in this respect.

One of the most remarkable things about water is that, although it contracts as its temperature decreases, just as other substances do when cooled, it undergoes a rapid and abrupt expansion as its temperature decreases from 4° to 0°C. Molecules of crystallized water assume a polar relationship to each other, a condition which requires more space than when they are in the liquid state. This is why ice floats.

Water has the greatest surface tension of all common liquids. Surface tension is the force of cohesion that exists between the molecules in the surface layer of a liquid. It is this cohesive force that causes liquids to pull into a round shape as they drop. Molecules of water cohere to each other so strongly that a force of approximately 210,000 pounds would be required to pull apart a column of perfectly pure water having a diameter of 1 inch. Of course, there is no pure water on earth, and so this is only a theoretical value. Even so, experiments have demonstrated that 2,000 pounds of tension is required to pull apart a column of water. It has a tensile strength approximating that of steel!

Not only is water cohesive, it is also strongly adhesive, meaning that it adheres to other substances with which it comes in contact. Among the few substances that water will not wet are fats and waxes; this is why water forms droplets, rather than a film, on the surfaces of cutinized leaves.

The transparency of water is of immense importance, because visible radiation can pass through it and through the interiors of leaves and other organs. If water were not transparent photosynthesis would be confined to a very narrow layer on the surface on leaves. But because it is transparent, cells deep inside plant tissues can carry on photosynthesis.

Because water is relatively inert it serves as a medium in which practically every chemical reaction in the plant takes place, without itself taking part in the reaction. In this manner it serves as a catalyst. (The photolysis of water in photosynthesis is one important exception.)

Although most people don't realize it, water

decreases in viscosity as its temperature increases. Water is 36 percent as viscous at 40°C (104°F) as it is at 0°C (32°F).

Temperature (°C)	Relative viscosity (%)
0	100
10	73
20	56
30	45
40	36

Pure water is a good electrical insulator. Of course, pure water is never found in nature, and the substances that are dissolved in it frequently make it an excellent conductor.

THE HYDROLOGIC CYCLE

Water circulates through a system that extends from 10 miles up in the atmosphere to about a half mile below the surface of the oceans. Within this vertical range of space, the waters of the earth are constantly moving and circulating. The **hydrologic cycle** (Fig. 13-2) has neither a beginning nor an end, and all the waters of the world are subject to perpetual re-use.

The hydrologic cycle is sustained by solar energy. The sun supplies the energy that evaporates water from the surface films of plants, from the soil, and from the waters of rivers, lakes, and oceans. Although most atmospheric moisture comes from the oceans, transpiration accounts for a large fraction in areas of high vegetation. The combination of evaporation and transpiration is called **evapotranspiration.** Among the various phenomena that cause air to rise are the frontal sliding of masses of warm air over wedges of cold air; the upward squeezing of air between two relatively cold air masses (convective circulation); and orographic lifting, the result of rising air being

FIGURE 13-2. *The hydrologic cycle has neither beginning nor end. It is as old as the earth and as relentless as time.* [*From Gilluly, Waters, and Woodford,* Principles of Geology, *3rd ed., San Francisco: W. H. Freeman and Company,* © 1968.]

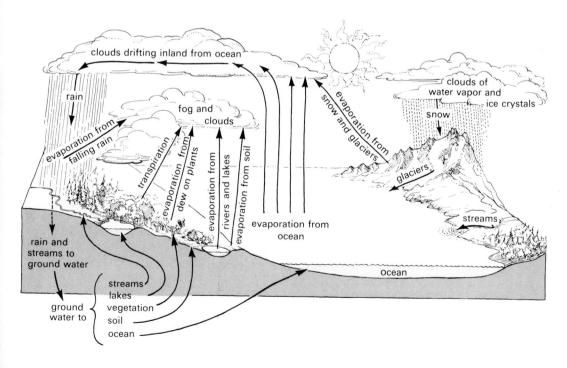

forced upward over mountains. Precipitation completes the hydrologic cycle.

The movement of moisture-laden air masses in the atmosphere is influenced by gravity, centrifugal forces caused by the rotation of the earth, and atmospheric temperature gradients. Differential heating of the earth's surface directs the movement of moisture-laden air. Because solar radiation is not as intense near the poles as it is near the equator, the general circulation of air near the earth is toward the equator. After it is heated at the equator it rises and passes at high altitudes over the cooled air to the poles, where it is cooled, drops, and repeats the cycle. Since the atmosphere always contains moisture, and since water has a very high specific heat, great transfers of energy always accompany the movement of air masses.

The temperature of the oceans and other large bodies of water remains relatively constant because of the high specific heat of water. Land masses, however, have a comparatively low specific heat, and tend to cool rapidly in winter and heat rapidly in summer. Consequently, low-pressure air masses generally form over the land in the summer, and high-pressure masses over the oceans. The hurricanes that originate in the Gulf of Mexico and Atlantic, and sweep through the Central and Southeastern United States in the summer, are good examples of this phenomenon. In winter, the reverse occurs. The resultant pressure differentials cause air to move from areas of high pressure to areas of low pressure.

Movements of air masses are also influenced by seasonal and diurnal variation in radiation. Differences between night and day and between the seasons are eventually averaged out, but in the process they strongly influence the direction and speed of the hydrologic cycle.

We can do relatively little to alter the hydrologic cycle. Building dams, modifying vegetation, tapping groundwater via deep wells, and other practices can produce economically significant local changes, but we cannot make major interruptions, and it is unlikely that we will want to in the near future. Before this can be done with safety, we need to know much more about the effects it might have on the ecology of the **biosphere**—that part of the earth's environment in which organisms live.

Unfortunately, the desire of man to increase the supply of water available for agricultural and industrial use has resulted in the exhaustion of local water supplies, with the resultant "death" of some communities.

ATMOSPHERIC MOISTURE

Forms of atmospheric moisture

There are several forms of atmospheric moisture, including vapor, rain, snow, sleet, hail, and dew. Atmospheric moisture is important to plants in two basically different ways. As precipitation it supplies moisture to soil and permits the uptake of both water and nutrients by plants. As water vapor, it decreases the loss of water from plants (and soil) by transpiration (and evaporation). These two aspects of moisture have consequences that bear directly on the fundamental life processes of plants.

Rain forms when the condensation of atmospheric moisture takes place above the freezing point. **Snow** forms when atmospheric moisture condenses below the freezing point. When rain falls through a layer of subfreezing air, **sleet** is formed, but if air currents carry the frozen particles back up into the sub-freezing layer of air, and accumulate a thicker covering of ice before falling again, **hail** is formed. Hail particles may accumulate enough ice to become as large as baseballs under certain atmospheric conditions, and can cause considerable damage to plants and man-made structures.

Dew forms at night when atmospheric moisture condenses on cold surfaces. It can form only when the sky is clear enough for the soil and plants to lose enough heat such that their temperature falls below the dewpoint. Clouds, haze, and dust prevent the loss of heat at night, when the energy balance of the earth is normally negative. Dew cannot, of course, supply all the moisture for plant growth, but it has great potential as a supplemental source of water for crop plants in semi-arid areas, such as Israel, where the total amount of rainfall amounts to about one inch per year.

A number of plant species are able to use the moisture from fog and mist. Spanish moss (*Tillandsia* sp.) grows as an epiphyte (a plant that grows on other plants but receives no substance from them) on the branches of several tree species and even on telephone lines. It is never found in areas that are not subject to mist and high relative humidity. The Torrey pine and several other plants of California grow on very dry, poor soils near the ocean, where fog from the ocean supplies moisture.

Fogs and mists are low clouds. They do not settle, and so they cannot be measured by conventional rain gauges. Even so, the vegetation that develops in areas subject to frequent fogs may be far richer and more abundant than might be expected from precipitation records. Campers on Mount Mitchell, in North Carolina, will vouch for the fact that when a fog "rolls" in and is intercepted by the foliage of fir trees, the resultant drip of water to the ground beneath is equivalent to a heavy rain.

Measuring atmospheric moisture

There are several ways to measure humidity. Perhaps the most commonly used one is the thermodynamic technique, in which temperatures of dry bulbs and wet bulbs are used. The most commonly used instrument for this purpose is the **psychrometer.** A psychrometer usually consists of two thermometers, one of which has a wettable wick around the bulb. The two thermometers, fastened side by side, are spun so that a stream of air passes over their bulbs. The thermometer without a wick will indicate the air temperature, and the other will indicate a lower temperature because the water evaporating from the wick causes a heat loss. When the temperature of the water in the wick has been lowered until the vapor pressure of the water remaining is equal to the vapor pressure of the atmosphere, no more water will be lost and the temperature of the wet-bulb thermometer will remain constant. Relative humidity can be obtained by readings of temperatures from the two bulbs and the use of appropriate calculations or tables.

Another kind of device for measuring humidity is called a **hygrometer.** Hygrometers contain substances that are affected by the moisture content of the environment. The elements of hygrometers are made from hairs and from thin animal membranes that stretch when they become moist. Both the hairs and the membranes are connected, through a series of levers, to a chart or a needle, calibrated in terms of relative humidity.

Also useful are such chemicals as cobalt chloride, which changes color as humidity changes. One of the first techniques used for estimating transpiration consisted of putting a small filter paper, which had been saturated with cobalt chloride and dried, in contact with the leaf surface. As water diffuses through the stomates, the color of the cobalt chloride changes from blue to pink. The rate of change and the intensity of color was thought to indicate the internal water conditions of the plant.

Optical techniques also have some value. Certain wavelengths of radiation are absorbed by water to a much higher degree than others. By measuring the absorption of radiation in these wavelengths with a special cell, it is possible to determine how much water vapor is in the air at any time.

Humidity can also be measured electronically with the use of **resistance psychrometers.** This is done by coating a nonconducting glass or plastic element with lithium chloride or some other hygroscopic salt, and measuring the electrical resistance of the material between two electrodes with a meter. The salt will adsorb more water at higher humidities so that electrical resistance between electrodes decreases as atmospheric moisture increases.

Liquid precipitation can be measured in a number of different ways, but by far the most common is to use rain gauges. Rain gauges may be simple buckets, fashioned from cans of various sizes, or they may be much more sophisticated devices that record intensity as well as time of rain fall. In the United States, receptables with an opening 8 inches in diameter are most frequently used; this size of opening is practically regarded as standard.

Recording rain gauges, which usually weigh the water that falls into the bucket, are best when intensity and time of precipitation measurements are required. They are nearly always supplemented, however, with simple collecting gauges, which usually give a more precise measurement of total rainfall.

At times, it may be desirable to measure rainfall

in remote locations, as in deserts and mountains, where daily or even weekly access to the gauges is not possible. In such locations as these, a can is placed in an appropriate location with a quarter of an inch of mineral oil in the bottom. The mineral oil being less dense than water, rises to the top of accumulated rainfall and prevents evaporation. This technique can be used for stations visited as infrequently as once every six months. Snow gauges, important at high elevations and in cold climates, are frequently made of rain gauges with slatted wooden baffles added on the outside to still the wind so that it drops its burden of snow.

It is frequently desirable to determine the amount of rainfall that has been intercepted by the canopy of a forest or by the vegetative cover of some other crop. For this purpose, two sets of rain gauges are used, one placed under the canopy at ground level, and the other placed either above or to one side of the crop. The difference in amounts of precipitation measured in the two cans represents the amount of precipitation intercepted by the vegetation. However, one source of error in such measurements lies in the fact that some of the intercepted precipitation finds its way to the soil via stem flow. This source can be counteracted by the construction of troughs around the trees or other plants to catch the water going down the stem, whereupon it is retained and measured.

There are several devices used to estimate evaporation. The simplest consists of nothing more than an open pan of water, frequently about a meter in diameter, placed in an exposed location. Evaporation is measured merely by noting changes in the water level. This device has been used often in various parts of the world. A more sophisticated device is the **atmometer,** which consists of a porous clay bulb connected by a water-filled tube in a reservoir of water. As solar radiation strikes the outside of the clay bulb and heats it, the water in the tube evaporates and is replaced by water that moves up from the reservoir. Still another device, the **evaporimeter,** consists of a wick wetted by water from a reservoir. As water evaporates from the wick, it is replaced by water from the reservoir. The amount of water lost indicates the relative rate of evaporation.

SOIL MOISTURE

Forms of soil moisture

Water is found in several forms in the soil. Let us imagine a bucket full of soil, to which enough water has been added to saturate the soil and form a puddle on the surface. If the bucket were equipped with a removable bottom that could be replaced by a screen the water that would drain out first would be **free water.** It is not bound to the soil but percolates through it and drains through the screen at the bottom under the influence of gravity. Most of the water remaining in the soil is called **capillary water** (Fig. 13-3), because it is held in capillary pores by its own surface tension and adhesive properties. Capillary and free water can be withdrawn and used by plants. Although capillary water cannot move through the soil, it often is the most important component of soil moisture available to plants.

The rate at which plants extract capillary water from soil is largely a function of root abundance, root hairs, mycorrhizae, and, to some extent, of mineral concentration of the soil water. Since there are many more roots in the surface soil, water is extracted at a faster rate, and more com-

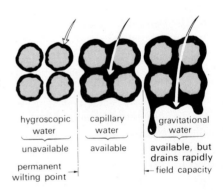

FIGURE 13-3. *Capillary water is the most important in sustaining plant life. Free water (gravitational water) drains away before it can be used, whereas hygroscopic water is bound too tightly to be removed by plants. [Adapted from Bonner and Galston,* Principles of Plant Physiology. *San Francisco: W. H. Freeman and Company,* © *1952.]*

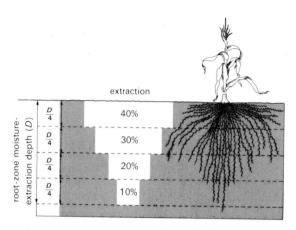

FIGURE 13-4. *Average moisture-extraction pattern for plants growing in a soil without restrictive layers and with an adequate supply of available moisture throughout the root zone.* [*Courtesy Soil Conservation Service, National Engineering Handbook, 1964.*]

pletely, than from soils at lower depths (Fig. 13-4). The rate of extraction from lower soil zones does increase, however, in times of drought. It is generally believed that the lower soil zones supply the moisture that sustains some trees and other deep-rooted crops during these critical times.

Still another form of soil moisture is called **hygroscopic water.** Hygroscopic water adheres tightly to soil particles and is not normally available to plants. During periods of drought, however, it may evaporate and form water vapor, so that some hygroscopic water may become available, even though the amount may be small. Water vapor does not constitute a large fraction of the moisture in most soils, but in some, especially in sandy, porous soils, it may move from one part of the soil to another—for example, from a low horizon to a higher horizon. Vapor transfer of moisture in the soil is poorly understood.

Water is also present in the soil as **chemically combined water.** The chemical compounds with which water is combined are said to be hydrated. Moisture in this form is bound tightly and is unavailable for use by plants.

When all free water has drained from a soil, it is said to be at **field capacity.** The significance of this term is that it takes into account the physical

nature of the soil. In terms of absolute amounts of moisture, field capacity differs with soil type. Field capacity of a heavy clay soil may be 30 percent, whereas it is only 3 percent in a sand. Various direct and indirect methods have been devised for measuring soil moisture in terms of percentage of field capacity.

The water content of soils can also be expressed in terms of availability to plants. After capillary water has been depleted, moisture is said to be at the **wilting point,** because plants lose turgor. Actually, it is more nearly correct to say **wilting range,** because there is a small variation in water tension. The percentage of water depends upon the type of soil, and is relatively independent of the species.

Movement of soil moisture

There are a number of factors which influence **infiltration,** the movement of water into soil, and **percolation,** the movement of water through soil. One of the most important factors is soil moisture content at the time precipitation falls. If the soil is dry, much more water can be absorbed than if it is wet, provided there are no barriers to absorption. In the spring of the year, in temperate regions the soil is usually saturated, and cannot absorb and store additional water so that runoff is likely to cause floods. Not only is there less pore space through which water can move when the soil is wet, but the soil particles of colloidal size are swollen by hygroscopic water, further reducing available pore space for water storage.

Vegetation has an important influence on infiltration of water. Layers of vegetation, even a single leaf, can break the speed of rain drops and slow them down so that they do not erode the soil (Fig. 13-5). Surface litter cushions the impact of rain drops and keeps them from beating directly against the soil and causing raindrop erosion, a significant factor in the deterioration of clay soils. Surface litter also forms myriads of small dams that retain water and permit it to infiltrate the soil. Vegetation also cools the soil, prevents evaporation, and decreases surface runoff by absorbing water. It slows down the wind and prevents the fine soil particles from blowing away, especially

FIGURE 13-5. *Plants protect soil from heavy rain. When a large drop strikes a leaf or stem it bursts into many little drops, some of which evaporate while others coalesce and drop short distances to the soil.* [*Courtesy USDA.*]

the organic particles, which are comparatively light in weight.

A layer of vegetation maintains the humidity and keeps the surface from drying up completely. As soil dries out, complex compounds that are hard to wet are formed on the surface. Vegetation also increases the depth of soils. As litter decays and accumulates on the surface, a part of the partially decomposed humus is washed down into the soil. Most important, roots decay and form channels through which water can move. These channels provide a rapid access to the subsoil for both precipitation and roots of succeeding plants, thereby increasing the available moisture storage capacity of the soil.

The use of heavy equipment and even trampling by animals and people, as on paths and roads, will compact surface soil so tightly that water will not infiltrate. Sandy soils do not compact because their particles do not normally stick together. However, for clay soils, in which the fine soil particles can cling to each other, even rain may cause surface soils to puddle and thereby decrease infiltration.

The inwash of fine material may affect infiltra-tion and percolation. If inwash is composed of sticks, stones, rocks, and large particles, it has little effect on the soil that it covers. But if it is fine material such as clay, bits of organic debris, or fine sand, then it is likely to clog the small pores of the surface and prevent infiltration.

Measuring soil moisture

There are several ways of expressing the amount of water in soil. Experienced persons can approxi-mate moisture content rather closely by "feeling" the soil—that is, by rubbing a small amount of soil from the root zone between the hands to deter-mine its consistency (Table 13-1).

The simplest way to express soil moisture quan-titatively is as a percentage of the dry weight of the soil from which it has been extracted. To determine moisture content gravimetrically, soil samples from the field are usually placed in cans with tight lids, weighed, oven-dried to a constant weight, and reweighed to determine how much water was lost. The data obtained in this fashion can be used to compute the percentage of soil moisture on either a dry-weight or a wet-weight basis. This method does not, of course, take soil structure into account. To do this requires deter-mining the water content for a known volume of soil.

It is also desirable to express soil moisture in terms of the tension with which it is bound to the soil (Fig. 13-6). For example, free water is held with very low tension. Capillary water is bound more tightly, but not extremely so. Hygroscopic water is bound very tightly, and the tension re-quired to remove it is high. Tension may be ex-pressed in atmospheres of pressure or in one of several other units. The greatest advantage to using tension to express soil moisture is that the critical values, wilting coefficient and field capacity, are the same for all soils, even though the total amount of extractable water may be vastly different.

An apparatus called a **tensiometer** measures the tension with which soil water is held in the soil (Fig. 13-7). When the tensiometer is filled with water and placed in moist soil, water is extracted from a porous ceramic cup until it comes into equilibrium with the soil water. The tension in the ceramic cup is read from a manometer or a vacuum gauge.

Moisture content can be estimated by measuring the electrical conductivity of a volume of soil. Conductivity is not reliable, however, because the soil texture varies greatly from one spot to another, soil minerals are not distributed evenly, and it is difficult to repeat or duplicate measurements.

A modern technique for measuring the moisture content of soils is to use the thermalization and scattering of fast neutrons. Fast neutrons have an energy of about 2 million electron volts, but after colliding elastically with the hydrogen in water, the speed of the neutrons comes into thermoequilibrium with the speed of the molecules composing the material that they are in, and become slow, or thermal, neutrons. Thermal neutrons have energies of about 0.025 electron volts. If a source of fast neutrons is placed in the soil, the number of fast neutrons that will be converted to thermal neutrons is a function of the amount of water in the soil, and they can easily be detected by a thermal neutron counter. This technique has been found so useful in recent years that a number of commercial devices have become available. Its major drawback in the field is that it does not distinguish between water in the soil, water in the roots, and hydrogen in soil organic matter.

An interesting but seldom used technique is based on the ability of water to attenuate or absorb gamma radiation. This method can be used very easily with small containers, such as pots. A radiation source is placed on one side of the pot and a detector on the other side. The amount of moisture present at any given time is an inverse function of the amount of radiation that can penetrate the pot.

Many investigators have used Bouyoucos blocks to measure soil moisture in experimental plots. These blocks, invented by a scientist with the same name, are of porous ceramic clay with two electrodes embedded in each, a fixed distance apart. They are "buried" in the soil in locations where soil moisture is to be measured with wires going to the surface from the electrodes. When a moisture measurement is desired, an electrical current is applied by means of the exposed wires, and the amount of current flowing through the moist

Table 13-1 FEEL CHART FOR THE DETERMINATION OF MOISTURE IN MEDIUM- TO FINE-TEXTURED SOILS. WITH SANDY SOIL, THE BALLS ARE MORE FRIABLE AND FRAGILE THROUGHOUT THE WHOLE RANGE.

DEGREE OF MOISTURE	FEEL	PERCENT OF FIELD CAPACITY
Dry	Powder dry	None
Low (critical)	Crumbly, will not form a ball	Less than 25
Fair (usual time to irrigate)	Forms a ball, but will crumble upon being tossed several times	25–50
Good	Forms a ball that will remain intact after being tossed five times; will stick slightly with pressure	50–75
Excellent	Forms a durable ball and is pliable; sticks readily; a sizable chunk will stick to the thumb after soil is squeezed firmly	75–100
Too wet	With firm pressure can squeeze some water from the ball	In excess of field capacity

Source: From Strong, in *Sprinkler Irrigation Manual,* Wright Rain, Ringwood, England, 1956.

blocks is measured. The greater the soil moisture, the greater the flow of current.

Dry soil is a poor conductor of heat. But as water content increases, so does thermal conductivity. Thus thermal conductivity can be used to estimate the amount of moisture present. This technique requires precise calibration, however, and so far has not found wide acceptance.

PLANTS AND WATER USE

There are several classifications of plants on the basis of their water requirements. At one extreme are **hydrophytes,** plants that grow in water. Hydrophytes are frequently characterized by having a low oxygen requirement and poorly developed root systems. **Mesophytes,** intermediate in water requirement, have well-developed root systems, and are characteristic of mature, well developed, soils. At the other extreme of the moisture scale are **xerophytes,** which grow in dry habitats and

250

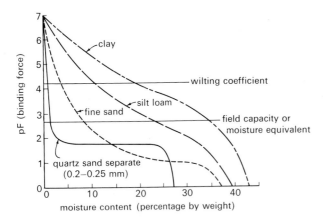

pF (binding force)

moisture content (percentage by weight)

clay

wilting coefficient

silt loam

fine sand

field capacity or
moisture equivalent

quartz sand separate
(0.2–0.25 mm)

FIGURE 13-6. *As the moisture content of soil increases, the soil water is bound with less force. Note that at wilting point and at field capacity a clay soil may contain four times as much water as a sand.* [*Adapted from Baver,* Soil Physics. *New York: Wiley, 1940.*]

have undergone various adaptations that enable them to grow and reproduce in very rigorous situations.

Drought

Drought is usually described as a period in which soil moisture is limiting to the growth of plants. The precise definition of drought is elusive; none of those in use satisfies everyone. Drought has been referred to as "a period in which the soil contains little or no moisture." How much is little or no moisture? Some clay soils may contain very little moisture in relation to the amount they are capable of holding, and yet may have a hundred times more than sandy soils that are at field capacity. Another definition of drought is "a period following 14 consecutive days without rainfall." This definition also has limitations, because although some soils may be deficient in moisture after 14 days, others can go for 50 to 90 days and still provide sufficient moisture for plant growth. "Dry-farming" techniques are based on the supposition that there will be very little rainfall during most of an entire growing season. Drought has also been defined as "a period of 21 days or more during which rainfall

is 30 percent of normal rainfall for the period."

All of these definitions, though useful in various specific, isolated circumstances, are of little use in generalized descriptions of drought. Perhaps the concept of drought is best understood by using the term **drought day**—a 24-hour period during which soil moisture is limiting to the growth of plants. This definition, proposed by the American soil scientist Van Bavel, is useful for a number of crops, for it provides a reproducible criterion if soil moisture tension is used as a parameter of water availability to plants. Obviously, it is not possible to measure moisture content on a 24-hour basis in every area where data are needed, so the drought-day criterion is based on a single measurement made at some time during the day. The time of day that the measurement is made must also be standardized, because there may be substantial variation in moisture content during the course of a single day.

Plants respond to drought in different ways. There are many adaptations that enable plants to survive under drought conditions. Usually drought survival is due to a complex of several factors

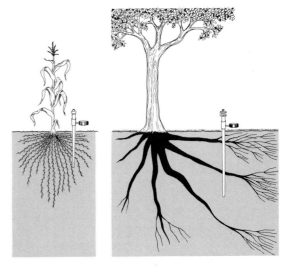

FIGURE 13-7. *Tensiometers are placed in the vicinity of roots to measure the force with which water is held in soil.*

rather than a single one. Desert plants are especially interesting in this regard, because they have several mechanisms for living through extended rainless periods.

Some desert plants persist and reproduce in areas with little rainfall by a drought-escaping mechanism. Many small plants are able to germinate, grow to maturity, flower, and produce seed within a period of several weeks. Such plants are invariably small, and are sometimes called "belly plants" by ecologists because to observe them closely it is necessary to assume the prone position. The spring rains that frequently occur even in the driest deserts are sufficient to provide these plants with enough moisture to complete their life cycle. The rest of the year, they persist as seeds and may remain in this state for years.

Tolerance to desiccation is well demonstrated by creosote bush (*Larrea tridentata*). Leaves of this species are able to withstand desiccation to about 50 percent of their normal weight. Some of its leaves are less tolerant of drought than others, however, and these fall when long periods pass without rainfall. During extended droughts the remaining leaves may turn brown and appear to be dead. But when precipitation occurs they turn bright green and reassume their normal function.

Some plants seem to rely exclusively on deep root systems that are continually in contact with moist subsoil in order to withstand long periods without precipitation. Creosote bush, an obligate deep-rooted plant, never grows in soils with a hardpan, since a hardpan precludes deep root development. Even during droughts alfalfa produces relatively high yields of forage because of its deep root system, whereas the growth of shallow-rooted cereal crops and grasses is greatly retarded. It is always a surprise to find that a plant has roots that extend into the soil to a depth of a hundred feet or more, but this is not uncommon in desert climates. Except when landslides occur, one rarely has the opportunity to observe roots at great depths. Such opportunities are usually the fringe benefits of canal projects, mining operations, or well excavations.

The amount of surface area on roots influences the response of plants to drought. Plants that have large roots with few ramifications have little absorbing surface area and extract less water as compared to those with highly divided, fine root systems. Roots that have a thick suberized surface layer absorb less moisture than those that do not.

The growth potential of roots is a significant feature of drought response. If a plant has ample supplies of food reserves, roots continue to grow through drought periods and invade new supplies of capillary soil moisture that would otherwise be unavailable. This is one of the indirect ways in which light can have an effect on drought resistance.

Drought resistance in certain species is influenced by the presence of mycorrhizal fungi. The fungal hyphae, in intimate association with the root and in contact with soil, absorbs water otherwise unavailable to the plant. In the conifers, mycorrhizae function in lieu of root hairs.

Since most of the water extracted from soil by plants is lost by transpiration, leaf modifications affect drought resistance. Plants whose leaves have fewer and smaller stomatal openings are less subject to water losses. A thick cuticle will prevent the loss of water through epidermal cells. Some plants have a thick covering of hairs on the lower surfaces. When hairs are present, the air movement of air across them is slowed and humidity is more likely to build up around stomatal openings. As a result, the vapor pressure gradient between the inside and the outside of leaves is reduced.

Stomatal action is much more rapid in some species than in others. Extreme sensitivity in responding to mild drought by closing results in the conservation of water that would otherwise be lost. This is not only true of desert plants, but of species of eastern deciduous forest trees. In some species, especially conifers, the great thicknesses of leaf veins per unit of leaf area make the leaves stiff and decrease the susceptibility of cells to shrinking during periods of drought. If cells do not shrink, then protoplasm is less likely to coagulate. Grasses have leaves that curl or fold during periods of drought, causing less surface areas to be exposed directly to wind and radiation. This mechanism will not stop transpiration, but slows it down periods of drought, thereby decreasing transpiration. Finally, some desert plants dispense with leaves altogether.

Excess moisture

Excess soil water affects both plants and soils. Crops with a high requirement for oxygen, such as wheat, cannot be grown on poorly drained soil. Soils that are waterlogged usually have a high CO_2 content. The carbon dioxide released by respiration does not move into the upper atmosphere, because water in the soil pores restricts diffusion. At the same time, oxygen is used but is not replaced. The increased CO_2 content which accompanies flooding decreases the permeability of the cytoplasmic membranes of root cells and inhibits the uptake of water. As a result, it is possible for living plants to undergo **physiological drought** while their roots are completely submerged in water. Winter flooding is usually less serious than summer flooding, because roots neither grow as fast nor absorb as much water during cold weather as during warm weather, and some plants are dormant.

After the oxygen in a waterlogged soil has been used, anaerobic decomposition of organic matter by soil organisms results in the production of marsh gas (methane), methyl compounds, and aldehydes. In addition, several kinds of ions are reduced and may accumulate to toxic levels.

Oxidized state	*Reduced state*
ferric ions	ferrous ions
manganic ions	manganous ions
nitrate ions	ammonia, nitrous oxide, nitrogen
sulfate ions	hydrogen sulfide and sulfide ions

High water tables restrict root development. In forested areas with a water table 18 to 24 inches below the surface, roots of many species fully occupy the surface but penetrate below the water table only a few inches. As a result, root systems may grow in an "inverted umbrella" configuration, which makes trees highly susceptible to wind-throw.

Poor aeration is directly related to the incidence and severity of some diseases. Plants weakened by poor aeration are much more susceptible to diseases caused by soil-borne pathogens.

Wet soils are slow to warm in the spring, with the result that the growing season for plants may actually be shorter than on adjacent dry soils. Five times the amount of heat energy is required to raise the temperature of water one degree than is required to raise the temperature of an equivalent weight of dry mineral soil by the same amount. In one study it was found that the heat input required to evaporate water from a saturated soil would raise the temperature of a soil with optimal moisture content more than 5°C (23°F).

Soil structure is adversely affected by water saturation, with the result that root development is restricted. In constantly flooded land the lack of periodic wetting and drying also prevents granulation.

WATER RESOURCES AND MAN

About five thousand years ago, people living in the Indus Valley of India designed and built systems that supplied water for irrigation, swimming pools, baths, and other uses. At about the same time, the first man-made dam was built in Egypt to store water. In ancient Babylonia, King Hammurabi oversaw the construction of a network of irrigation canals and enforced laws providing for their upkeep. Several thousand years later Mohammed said that "No one can refuse surplus water to a thirsty person without sinning against Allah and against man."

Today, matters have changed only in magnitude. People still build dams, make laws providing for the equitable use of water resources, and wage battles both legal and military to determine who will avail themselves of water supplies, which are limited. Today, the dams and the battles are bigger, and so are the problems. Furthermore, the technology that we employ today for the manipulation of water resources differs from that of years past only in its degree of precision, not much in methodology. New ideas relating to methods of water resource development and allocation have not been put into practice for thousands of years.

Is it possible that radically new methods of entrapping and moving water may be found? Could large amounts of water be moved from one place to another as clouds, rather than through canals and pipes? Can atmospheric moisture be condensed by cloud seeding with regularity and in a magnitude large enough to provide water for cities of millions of people? Can fresh water entrapped as ice at the poles be towed behind barges to places where it is needed? Can sea water be economically desalinated on a large scale? There must be new and startling developments in the wise manipulation of the hydrologic cycle if the water requirements of future generations are to increase, as predicted, and still be met.

Agriculture, too, has its particular problems relating to water availability. In the past, at least in the United States, the issue of sufficient water for agricultural purposes has received relatively little attention. Nevertheless, the problem continues to become more and more urgent, and consequently we can expect that in the future it will be necessary to make use of new concepts in our attempts to meet the increasing demand for water. Perhaps the prayers and pleas of many persons will be answered by the use of slightly saline water, obtained at least in part from the ocean, for irrigation.

Selected References

Crafts, A. S., H. B. Currier, and C. R. Stocking. *Water in the Physiology of Plants*. Ronald Press, New York, 1949. (A monograph on plant and water relations.)

Davis, K. S., and J. A. Day. *Water, and Mirror of Science*. Doubleday, Garden City, N.Y., 1961. (This is a short reference book on the physical and chemical nature of water.)

Fox, C. S. *Water, a Study of Its Properties, Its Constitution, Its Circulation on the Earth, and Its Utilization by Man*. Technical Press, London, 1951. (A popular treatment of the subject.)

Frank, B. *Water, Land, and People*. Knopf, New York, 1950. (A discussion of the results of misuse of water resources.)

Kozlowski, T. T. *Water Metabolism in Plants*. Harper and Row, New York, 1964. (A short book on the function and uses of water in plants.)

Kramer, P. J. *Plant and Soil Water Relations*. McGraw-Hill, New York, 1949. (A textbook on plants and soils, notable for its lucidity.)

Water and Agriculture. A.A.A.S., Washington, D. C., 1960. (An excellent collection of articles easily understood by nonscientists.)

CHAPTER 14

Climate and Crop Geography

Crop geography deals with all of the factors related to the distribution and cultivation of crop plants. In addition to the physical and biotic environment, it must also be concerned with the social, economic, political, technological, and historical forces that shape cropping practices over the entire world. By its very nature crop geography must treat large areas and long periods of time. Even though crop geography deals solely with crop plants, its scope is broad because of the interplay of social customs and historical events with crops and cropping patterns.

Crop ecology is concerned with the relationships between crop plants and their physical and biological environments, frequently with no regard for economic motivation. Its scope ranges from continental land areas to the immediate environment of a single stomate. It deals with the same environmental factors and relationships that concern plant ecology, but the point of focus is on crops. Today, the differences between plant and crop ecology have become small, because many of the exclusive domains of the traditional plant ecologists—virgin forests, tundra, and deserts—are

being critically examined to determine whether they will serve a productive function in our economic structure. It seems probable that in the future there will be little or no distinction between plant and crop ecology, because even the trees of wilderness areas and the plankton of oceans will be considered as crops to be manipulated for perpetuation and production.

Several examples of the complex interplay of factors involved in crop geography are in order. Historical forces have a great effect. During and shortly after the American Civil War (1861–1865) millions of acres of cropland in the southeastern United States were abandoned because of a labor shortage. A high proportion of these fields were naturally seeded by pines, which reached a peak of sawlog merchantability in the period 1920–1940. Cotton never entirely recovered its place in the regional economy in the period after the war, and as a result the agricultural base of the region changed from cotton to forest products. Tobacco production, however, did not shift because tobacco has an exceptionally high value per acre, and because soils of the region are favorable for certain types highly desired for cigarettes.

Cultural factors and personal preference also play a role in the choice of crops. Wheat recently shipped to a nation in the throes of a famine was rejected by all but the starving because they preferred rice. Even though the climate and soils of such a country might be well adapted to wheat production, several generations would be required for its people to develop a taste for wheat.

Technological forces are also important. In the tropics a form of agriculture is practiced in which all merchantable trees in an area are cut and removed and those remaining are felled and burned, the soil being enriched by ashes. Several successive crops, such as hill rice, are grown on the land, until it becomes too unproductive to justify further cultivation. It is then either allowed to revert to forest or is planted with a desirable timber species such as teak (*Tectonia grandis*). The repetition of this process over many years has resulted in vast changes in the vegetative cover of the landscape. This pattern of **shifting agriculture** (see Chapter 15) is considered by some to be one of the major agricultural problems of the tropics.

CLIMATE

Since climate is the principal factor that controls crop distribution, an examination of the major climates of the world will reveal the patterns that help establish the distribution of crops. In order to interpret descriptive data, it is first necessary to understand how daily weather data is compiled.

Temperature

Temperature, the simplest weather element to measure, is probably more frequently used than any other kind of data. Temperature and precipitation records are often the only ones that have been collected at many stations throughout the world, and limit the number of ways weather data can be universally expressed.

The **absolute maximum** and **absolute minimum temperatures** are the highest and lowest temperatures reached, if even for only a second, throughout the course of a year. They are usually accompanied by date and sometimes by duration.

The **yearly mean temperature** is the arithmetic average of the daily mean temperatures (an average of the daily maxima and the minima) on a yearly basis. Mean temperatures of winter months are sometimes contrasted with mean temperatures of summer months for climatic comparisons.

The **mean maximum temperature** is the average of the daily maxima for a given period of time. Its antithesis is the **mean minimum temperature.**

Deviation from normal is used, as it is with reference to precipitation, to express deviation from the average of the mean temperatures on any particular date. The United States Weather Bureau regularly announces this information, deriving it from a "normalized" curve based on averages for a number of years.

In a limited number of locations **soil temperatures** are recorded. However, these data are frequently collected for ecological studies in locations under sods or common field grasses that may be far removed from crop fields.

Precipitation

Precipitation is expressed in many ways, but perhaps most often as **average annual precipitation** (Fig. 14-1). There is a limit, however, to the time period to which an average value may be applied. Even though data for several hundreds of years are available at some stations in England, the entire period is not used because over this long a period of time a true climatic change may have occurred. Therefore, in using a value that is meaningful under today's climatic regime, any average should be based on the past 25 to 35 years of accumulated data. **Mean winter precipitation** and **mean summer precipitation** are frequently used when describing the climate of a region. When precipitation is in the form of snow, the total depth of snow is given along with a factor that can be used to convert the snow to inches of water. A 1:10 ratio of water to snow is sometimes used when actual conversion data are not available.

Growing season rainfall is an expression frequently used for agricultural purposes, but because

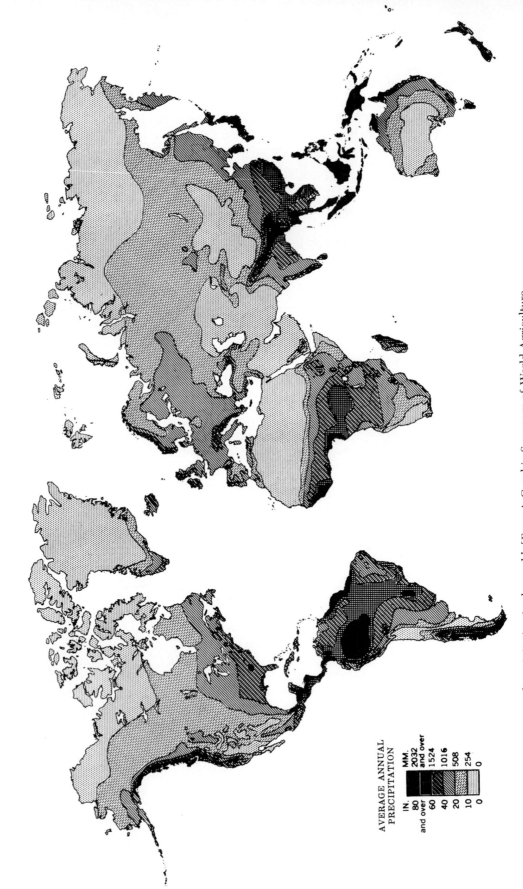

AVERAGE ANNUAL
PRECIPITATION

IN.	MM.
80 and over	2032 and over
60	1524
40	1016
20	508
10	254
0	0

FIGURE 14-1. *Average annual precipitation in the world.* [*From A Graphic Summary of World Agriculture,
USDA Misc. Publ. 705, 1964.*]

the growing season is not of equal length at all latitudes, such data are not always directly comparable. Growing season is considered to be the average length of the period between the last spring frost and the first fall frost, the **frost-free season.**

Graphic techniques, showing **average monthly precipitation** are useful for making rapid surveys. The **total monthly precipitation** is sometimes used as a general index of drought conditions, especially during summer months. When it falls below the necessary minimum determined for a particular area, drought conditions are considered to prevail.

The **deviation from normal precipitation** is an expression frequently used by weather reporters. This value (and its equivalent for temperature) may be useful when studying growth performance of native plants. **Normal precipitation** is defined as the average precipitation for any particular period.

Humidity

Atmospheric moisture may be described in a number of ways. **Relative humidity,** the most frequently used unit of measure, is the actual quantity of water vapor in the atmosphere expressed as a percentage of the amount that can be present at the same temperature. The amount of water vapor that the atmosphere can hold increases with temperature. Thus the **drying power** of air, which is proportional to the water vapor deficit below saturation, is related to relative humidity and temperature. At high temperatures small differences in the relative humidity represent large differences in drying power; at low temperatures the reverse is true. **Absolute humidity** is the mass of water vapor per unit volume of air. It is usually expressed as grains of water (one grain equals 1/7,000 lb) per cubic foot of air. **Specific humidity** is the weight of the water vapor per unit weight of air (including the weight of the water vapor). Thus absolute humidity is a function of volume, and specific humidity is a function of weight.

Light

Unfortunately, light is not generally expressed directly in weather data. Even actual sunshine data are not generally available, and so **potential sunlight** (sunrise to sunset) obtained from tables calculated for different latitudes is generally used. Such tables do not provide a good quantitative estimate of light because of the variable nature of clouds and fog. The total hours of sunshine can be recorded by an instrument called a Stokes-Campbell sunshine duration recorder, which concentrates the sun's rays and burns a line in a strip of paper.

The length of the daily photoperiod varies greatly at different latitudes. Natural vegetation has responded to latitudinal variations by adapting to photoperiod. Thus many plants cannot be successfully moved from one latitude to another even though other environmental factors are compatible.

KÖPPEN CLASSIFICATION OF CLIMATE

Climatic classifications of greatest agricultural value are those based on the interactions of temperature and precipitation. The most widely known and used system was devised by the Austrian geographer W. Köppen. It is based on temperature, precipitation, seasonal characteristics, and the fact that natural vegetation is the best available expression of climate of a region. A distinctive feature of the **Köppen system** is its use of symbolic terms to designate climatic types. The various climates are described by a code consisting of letters, each of which has a precise meaning.

Köppen acknowledges five basic climates: *A,* tropical rainy; *B,* dry; *C,* humid, mild-winter temperate; *D,* humid, severe-winter temperate; *E,* polar. Each is subdivided to describe different subclimes, which are denoted by a combination of capital and small letters. The capital letters *S* (steppe) and *W* (desert) subdivide the *B,* or dry, climates. Similarly *T* (tundra) and *F* (icecap) subdivide the *E,* or polar, climates. Small letters further differentiate climates. Some of these are indicated in the following list, in which the basic climates to which the small-letter codes pertain are given in the column to the right.

A partial listing of the major and minor categories is given in Table 14-1 and in Figure 14-2.

a = Warmest month above 22°C (71.6°F)	C, D	
b = Warmest month below 22°C (71.6°F)	C, D	
c = Warmest month below 22°C (71.6°F); less than 4 months below 10°C (50°F)	C, D	
d = Coldest month below −38°C (−36.4°F)	D	
f = No distinct dry season	A, C, D	
h = Hot, average annual temperature above 18°C (68°F)	B	
k = Cold, average annual temperature below 18°C (68°F)	B	
m = Monsoon (short dry season)	A	
n = Frequent fog	B	
s = Dry season in summer	C	
w = Dry season in winter	A, C, D	

Critics have expressed the opinion that the Köppen classification is based on too few data, and that boundaries between the various climatic regions are too arbitrary. But in spite of these objections this system has gained widespread recognition and use. Its simplicity and general adherence to vegetational zones has made it the basis for many revisions and other classifications.

Tropical rainy (A)

Tropical rainy climates are characterized by the absence of winter; there is no month with an average temperature less than 18°C (64.4°F). This type of climate prevails over about 36 percent of the total surface of the earth and over about 20 percent its land surface. Rainfall is usually abundant, seldom less than 30 inches per year. The principal climatic types in the humid tropics are distinguished by the amount and distribution rainfall.

TROPICAL RAINFOREST (Af)   This type of climate is found in parts of Africa, Central America, Brazil, Madagascar, the Philippines, and elsewhere. It is characterized by uniformly high temperatures and heavy, more or less uniformly distributed precipitation. Even though temperatures may be more moderate than those in higher latitudes, the high

Table 14-1   SELECTED CLIMATIC TYPES ACCORDING TO KÖPPEN.

SYMBOL	MAJOR CLIMATE	CHIEF CHARACTERISTIC	SUBCLIMATES (INCOMPLETE)
A	Tropical rainy	Coolest month above 18°C (64.4°F)	Af, Tropical rainforest Am, Monsoon rainforest Aw, Tropical savannah

B	Dry	Evaporation exceeds precipitation	BS, Steppe
			   BSh, Tropical and subtropical steppe
			   BSk, Middle latitude steppe
			BW, Desert
			   BWh, Tropical and subtropical desert

C	Humid, mild-winter temperate	Coldest month between 18°C (64.4°F) and 0°C (32°F)	Cs, Mediterranean (dry summer subtropical)
			   Csa, Hot-summer Mediterranean
			   Csb, Cool-summer Mediterranean
			Ca, Humid subtropical (warm summer)
			   Caw, with dry winter
			   Caf, with no dry season
			Cb, Cc, Marine west coast (cool summer)

D	Humid, severe-winter temperate	Coldest month below 0°C (32°F); warmest above 10°C (50°F)	Da, Humid continental, warm summer
			   Daw, with dry winter
			   Daf, with no dry season
			Db, Humid continental, cool summer
			Dc, Dd, Subartic

E	Polar	Warmest month below 10°C (50°F)	ET, Tundra EF, Icecap

humidity makes the heat oppressive. Violent thunderstorms occur daily at the same hour and with clocklike precision during certain parts of the year.

Plants that grow in tropical rainforests are called **megatherms,** and require continuous high temperatures with abundant precipitation. The vegetation of tropical rainforests is vigorous and luxuriant. The number of plant species is large. The trees are tall, broad-leaved, held together as a unit by vines and lianas. During World War II, when campsites were cleared in the midst of tropical rainforests, a few trees were nearly always left for shade, but this practice was discontinued because high winds blew many of the unsupported trees down.

Soils of the rainforest are deeply weathered because of high temperatures and chemical action. When cultivated they are quickly exhausted unless fertilized at a high rate. Rainforest soils may also lose their porosity under cultivation, in which case poor aeration becomes a problem.

MONSOON RAINFOREST (*Am*) Another type of *A* climate, the monsoon rainforest (*Am*), is characterized by a long season of fairly evenly distributed rainfall, which supports rainforest vegetation, and a short dry season. The classic monsoon climate is found in India, and has been the subject of many sagas. Indochina, Burma, and the eastern Amazon valley also have this type of climate.

TROPICAL SAVANNAH (*Aw*) Savannah is transitional between tropical rainforest and dry climates. There is less yearly rainfall than in rainforests and it is less evenly distributed throughout the year. Temperatures are much like those of the rainforest climate (*Af*), but yearly ranges are greater. There are distinct cool-dry and hot-wet seasons. Savannah climates are found in South America, Africa, and Australia.

Natural vegetation consists typically of tall, coarse grass punctuated with clumps of trees, giving a parklike appearance. Bunch grass, which may reach heights of 12 feet, predominates. The trees are always widely spaced and are leafless during the dry season.

Savannah soils are lateritic. They are low in organic matter, poor in fertility, but porous and easily cultivated. In dry savannah regions, the soils resemble prairie soils.

Owing primarily to the difficulty of weeding and keeping land clear so that crops can be established, the major agricultural exports of tropical rainy (*A*) climates are products of woody plants—trees, shrubs, and canes. Woody perennial plants, which need be regenerated only periodically, are much less costly to establish. The culture of field crops is limited. Important crops include cacao, banana, sugar cane, coffee, rubber, black pepper, vanilla, and sisal.

Sugar cane (*Saccharum* spp.) requires abundant sunshine, 60 to 65 inches of rain annually, and an ample supply of mineral nutrients. Climates in which there is bright sun and frequent short periods of heavy precipitation are best. Although strictly a tropical plant, sugar cane is cultivated in some parts of the world that are on the margins of regions with tropical climates. In Louisiana, where the frost-free season lasts about 250 days, it is cut in an immature stage.

Cacao (*Theobroma cacao*) beans, used for cocoa and chocolate, grow in pods from 6 to 10 inches in length and 3 to 5 inches in diameter. The trees grow up to 40 feet tall, bear fruit for as long as 50 years, and have two harvests per year (one more abundant than the other) in well-adapted regions. *Af* climates are best, for cacao requires at least 80 inches of rainfall. Average annual temperatures should be from 24°C to 27°C (75 to 81°F).

Bananas (*Musa sapientum*) are adapted to *Af* and *Aw* climates. They require at least 75 inches of water annually, a temperature of 24°C (75.2°F), and fertile, well drained soil.

Coffee (*Coffea* spp.) is indigenous to Africa, but has been introduced throughout the world wherever climate permits. It is better adapted to *Aw* climates than to *Af* climates, and is grown from sea level to about 7,000 feet. Some species, however, such as *Coffea arabica*, can be grown at higher altitudes. The mean annual temperature in most coffee-growing regions is at least 20°C (68°F) or higher; coffee is not grown in regions with a mean minimum of less than 13°C (55.4°F). From 45 to 70 inches of rainfall annually is required, and good soil drainage is a requisite.

Dry (B)

The main features of dry climates are low humidity, scanty and unreliable rainfall, intense radiation, and violent winds. Desert (*BW*) and steppe (*BS*) are the principal subtypes.

Potential evapotranspiration (transpiration and evaporation from land and water surfaces) exceeds precipitation in dry climates. In general, the drier the climate, the more uncertain the rainfall. Since dryness is defined as a function of potential evapotranspiration, it is difficult to associate a particular temperature with dryness. For example, Paris, France, has a cool marine climate but receives only 23 inches of rainfall per year; it is in the midst of forest vegetation typical of humid climates. In the United States temperatures are much higher in areas with such low rainfall, and no forests grow in regions with less than 25 inches of precipitation per year.

DESERT (*BW*) Annual rainfall in most deserts is usually less than 5 inches. In Cairo, Egypt, rainfall

FIGURE 14-2. *Climates of the earth.* [*From Trewartha,* An Introduction to Climate, *New York: McGraw-Hill, 1954.*]

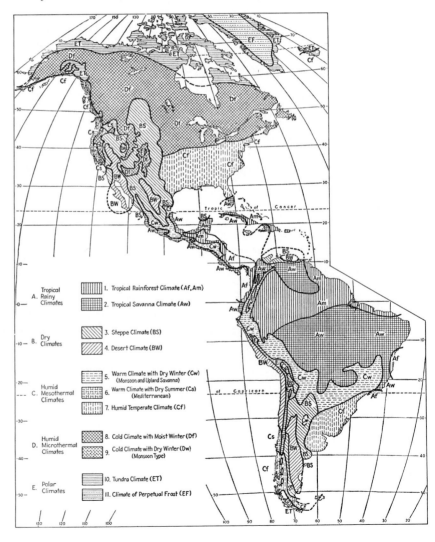

averages less than 2 inches, whereas in some parts of the world, such as Calama, Chile, there is little evidence to show that rain has ever fallen. When rain does fall in desert climates, it is likely to create local torrential floods.

Desert vegetation is notable by its absence, but when present is highly seasonal in character, bursting out after rainfall and rapidly completing its life cycle or yearly growth (if perennial) in a period of about a month, before available soil moisture is depleted. The mechanisms by which plants survive drought are discussed in Chapter 13. Wide spacing is characteristic of desert vegetation, with

bare soil between shrubs and bunch grasses (Fig. 14-3).

Deserts of the middle latitudes tend to be cooler than those of lower latitudes, and frequently have a partial plant cover. The meager vegetation has slight value as food for livestock, but has been exploited for this purpose in practically every part of the world. In contrast to the highly leached soils of wet climates, desert soils are often highly fertile, and may become productive when irrigated, but require careful management. They are highly variable, however, and tend to be alkaline. The accumulation of salts (salinity) is a recurrent problem.

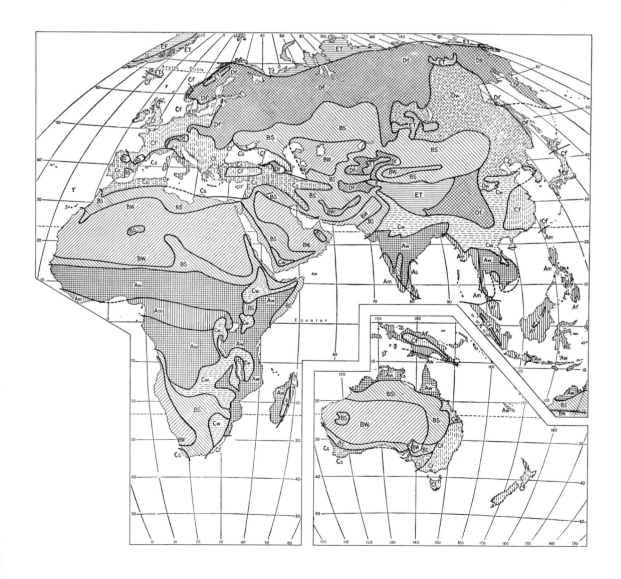

FIGURE 14-3. (*Left*) *Desert vegetation is always widely spaced. In some areas the spacing is so regular that the desert shrubs appear to be planted.* (*Right*) *In contrast, the natural green fescue range in more humid climates forms a dense ground cover.* [*Photograph at left courtesy U.S. Atomic Energy Commission; photograph at right courtesy USDA.*]

STEPPE (*BS*) Steppe climates are semiarid. Where rainfall is greatest, short grasses prevail; elsewhere, bunch grasses dominate the vegetation. The steppes of the South Africa veldt (grassland in which there may also be scattered shrubs) once supported vast herds of wild animals. It has recently been established that the native animals produce more meat for human consumption than the beef cattle that have been introduced.

Steppe soils are dark in color and have no well developed structure. Those of tropical regions are generally less well developed than in higher latitudes. Although cultivation is easy, there is not enough rainfall to support a well-developed agriculture unless irrigation is provided.

Wheat (*Triticum* spp.), one of the world's principal grain crops, is well adapted to steppe climates, although it grows well in many other climates. Climate determines the type of wheat that must be grown. High-protein (hard) wheats are grown in *Bs* climates; the starchy (soft) wheats are most frequently grown in *Db* climates. There is not a month of the year during which wheat is not harvested somewhere in the world.

Humid, mild-winter temperate (C)

These climates are characterized by seasonal variations in temperature. Plant dormancy is usually due to cold rather than drought. Whether conditions in *C* climates are likely to be changeable due to conflicting masses of air from tropical and polar sources. Cyclonic storms are frequent. Regions with *C* climates produce a wide variety of crops, including forage grasses, olives, cotton, grapes, tobacco, cereal grains, and vegetables. There are three principal climatic subtypes: dry-summer subtropical (Mediterranean) (*Cs*), humid subtropical (*Ca*), and marine west coast (*Cb*).

DRY SUMMER SUBTROPICAL (*Cs*) This climate is characterized by hot, dry summers, mild winters with rainfall, and brilliant sunshine throughout the year. It is found on the borders of the Mediterranean Sea, in central Chile, in South Africa, in southern Australia, and in central and coastal California.

Because of its delightfully mild winters and bright sunny weather, this is one of the best-known climates, even though found in only 2 percent of the world's land area.

Winter temperatures typically average between 4.4° and 10°C (40 to 50°F) and summer temperatures between 21.1 and 26.7°C (70 to 80°F). Rainfall averages 15 to 25 inches, less than enough to support a forest vegetation of large trees with a closed canopy. The effective precipitation is high, since the bulk of the rain comes in the winter. Summers are generally hot and dry.

Soils are varied because of the diversity of parent materials. In the Mediterranean region a distinctive soil, known as **terra rosa** (red soil), has developed on the hard limestone bed rock. Soils tend to be easily leached when cultivated. Soils in humid regions with abundant vegetation tend to be poor, whereas those in regions with less rainfall are often very fertile. At both extremes of precipitation, productivity of crop plants may be poor.

Vegetation in this climate is variable and falls into complex patterns. No single type dominates, although trees are always present except in local areas where soil type or fire is the dominant ecological factor. Glades and open woodland comprise the tree vegetation. Shrubs and trees are usually dwarfed and have small, thick, glossy leaves which prevent excessive transpiration. The cork oak (*Quercus suber*) is characteristic, as is the olive. With irrigation, it is possible to grow an extremely wide variety of crop plants.

The fruit crops of *Cs* climates include citrus, fig, date, apricot, olive, grape, and plum. The dry heat of summer is ideal for fruit drying; the fruit is merely laid out in the sun. California vegetables grown under irrigation are shipped throughout the United States and Canada because of their high quality and winter availability.

HUMID SUBTROPICAL (*Ca*) The *Ca* climate has abundant precipitation, which may be concentrated in the summer, and is usually found on the eastern sides of continents. Every continent has regions dominated by this type of climate, which has a high potential for the production of agricultural products. There is nearly always enough precipitation to support crops that mature in late summer or in early fall. Summer temperatures average from 24 to 26.5°C (75 to 80°F), and winter temperatures average from 4.4 to 12.8°C (40 to 50°F). There is a wide difference between day and night temperatures, sometimes as much as 20°C (36°F).

Precipitation ranges from 30 to 65 inches per year. There is abundant snow, but in lower latitudes it may remain on the ground for no more than a day. Violent typhoons sometimes occur near land masses, and late summer rains and hailstorms may destroy nearly ripened crops.

Rice (*Oryza* sp.) is the major cereal crop of this climate. It is an annual grass that germinates and grows in an aquatic environment and has specialized tissues through which oxygen moves to roots. Although grown over a range of temperatures, 20 to 21°C (68° to 70°F) is best for grain formation. Rice can be grown in a wide variety of soils, the main requirement being that the soil must hold water for a considerable period of time.

Cotton (*Gossypium* spp.) is used throughout the world. It originated in both the New and the Old World and grows in a wide range of climates, but does best where there are at least 200 frost-free days, an average growing-season temperature of at least 15° to 17°C (59 to 63°F), and light soils.

Tobacco (*Nicotiana* spp.) makes very rapid growth during a short season and quickly exhausts soils of mineral elements. In some areas it is the main agricultural crop.

MARINE WEST COAST (*Cb* and *Cc*) These climates

are usually found in relatively narrow strips along the west coasts of continents in the middle latitudes. Characteristic air masses move in from the oceans, resulting in cool summers and moderate winters. Mean summer temperatures are seldom higher than 20°C (68°F). Mean winter temperatures are above freezing, although killing frosts are not unusual. Frequently the transition from a marine climate to a severely cold one (as in Scandinavia) or from a marine climate to a very dry one (as in California) may take place within a mile or two.

In some places there is as much as 150 inches of precipitation per year but in others no more than 20 inches, which nevertheless may be adequate for crop growth because of cool temperatures. Snow is usually abundant only at high elevations. Always accompanying the precipitation is a high degree of cloudiness. Fog and mist are common, and the sun may not be seen for several weeks.

Soils are quite variable and generally deep. For the most part they are capable of good agricultural production.

Evergreen forests are predominant in some areas, deciduous ones in others. In North America gigantic redwoods that grow to more than 10 feet in diameter occupy the foggy western slopes of California's coastal ranges. In Chile the forests are of broad-leaved evergreen trees. Evergreen shrubs such as heather, gorse, and juniper take the place of trees in areas that have been subjected to severe glacial scouring, as in Wales and Scotland. In Europe the dominant trees are deciduous.

These are rich climates for such cool-season crops as apple, pear, strawberry, peas, and lettuce, but too cool for growing many warm-season crops outdoors, with the result that greenhouses are used in Holland and England to grow such crops as tomato and cucumber.

Humid, severe-winter temperate (D)

There are two main subdivisions of D climates of agricultural importance, both distinguished by cold, snowy winters and wide annual ranges in temperature. Because their controlling climatic forces orig-

inate over large land masses, they are frequently called continental climates. Only in the northern hemisphere, and there only in North America and in Eurasia, are there land masses large enough for their development. The climates of Canada and the Soviet Union are typical.

Soils of the northern limits of D climates are podzols with a heavy cover of litter. They are poor for agricultural purposes, but support dense stands of conifers. Soils of the deciduous forests have podzolic characteristics, but generally are better than those further north. Grasslands have fertile, dark, prairie soils. Their general use for agricultural production testifies to their desirability. The grasses are tall and originally supported large herds of animals. Today, there are few places where relics of this vegetation in North America have been preserved. The forests were originally maple and mixed oak and hickory in the southern parts, with conifers along the northern edges.

HUMID CONTINENTAL WITH WARM SUMMERS (Da) Conditions in these climates are excellent for the growth of certain crops. Mean temperatures may be as much as 30°C (86°F), and highs of 38.5°C (105°F) have been recorded. There may be as many as 200 frost-free days. The climate of the north-central United States is an excellent example of this type.

Most rainfall occurs in the summer months. In some years rainfall is sparse, resulting in drought. Snows are occasionally extremely deep, but winter precipitation does not usually exceed 5 inches (water or melted snow). The snow cover prevents losses of heat from the earth during the winter, with the result that soil warms faster than might be expected on the basis of air temperature measurements.

Corn (Zea mays) has reached its greatest development in Da climates, but also does well in some of the Db, Ca, and Cb climates. A native of the American tropics, by 1492 it was grown throughout North America by the Indians. In the corn belt of the United States the mean summer temperature is 14.4°C (58°F) or higher. Temperature requirements for germination are 10°C (50°F). The most critical period for soil moisture is shortly before and after the silking time. Corn is a major crop on every continent that has large areas of *Da* climate.

HUMID CONTINENTAL WITH COOL SUMMERS (*Db*) Regions with the *Db* climate lie to the north of those with the *Da* type, and are primarily wheat producers. Although summers are a few degrees cooler than those of the hot-summer type, winters are 5 to 15°C (9 to 27°F) lower. The long summer days in higher latitudes compensate to some extent for the colder temperatures. But in all, the climate is dominated by winter. In the parts of Europe subject to this climate, winter temperatures are tempered somewhat by oceanic air masses. The frost-free season may be no longer than 50 days.

Precipitation is highly variable. There is, however, considerable snow, and the snow cover may last for more than 120 days. Crops grown in *Db* climates include potato, pea, sweet corn, turnip, carrot, cabbage, and other plants that need only a short growing season.

Polar (E)

Polar climates are agriculturally unimportant. The presence of the sun above the horizon for six months, its absence for an equal period, the bitter cold, and the perpetual snows or icecaps (*EF*) signal high adventure for those who wish to test their courage, endurance, and skill.

There are, however, interesting possibilities for storing surplus agricultural produce (if it ever again exists) in snowpacks and ice that are not subject to lateral movement. The only cost would be that of transportation.

Not all polar climates are characterized by permanent snow. In the warmer tundra (*ET*), found over a large area in the northern hemisphere, there is a brief growing season and meager vegetative cover, even though the soil may be permanently frozen to a depth of a few inches to a few feet below the surface, a condition called **permafrost**. Tundra vegetation is basically of two types: one a carpet of moss and lichens; the other a mixture of mosses, sedges, flowering herbs, and low woody shrubs. These shrubs (often trees in warmer climes) are dwarfed and stunted, have distorted and crooked stems, and are known as elfinwood, or

krummholtz. The soil in which tundra vegetation grows may be no more than a few inches deep—an environmental characteristic shared by plants that grow at high elevations.

THORNTHWAITE CLASSIFICATION OF CLIMATE

The classification of climate proposed by C. W. Thornthwaite resembles Köppen's insofar as combinations of letters are used to designate individual climates and plant response is used to integrate the climatic elements. Three factors enter into Thornthwaite's description of climatic types—precipitation effectiveness, seasonal distribution of precipitation, and temperature efficiency.

Precipitation effectiveness is represented by a P-E ratio in which monthly precipitation is divided by monthly evaporation. The P-E index is the summation of monthly P-E ratios. There are five humidity provinces each associated with a particular vegetation type:

LETTER	HUMIDITY PROVINCE	VEGETATION	P-E INDEX
A	Wet	Rainforest	128 and above
B	Humid	Forest	64–127
C	Subhumid	Grassland	32–63
D	Semiarid	Steppe	16–31
E	Arid	Desert	under 16

The humidity provinces are subdivided into subtypes on the basis of seasonal distribution.

LETTER	DISTRIBUTION OF PRECIPITATION
r	abundant all seasons
s	sparse in summer
w	sparse in winter
d	sparse all seasons

The **temperature efficiency** index (T-E) is obtained by summing the monthly mean Fahrenheit temperatures, subtracting 32°F, and dividing by 4.

LETTER	TEMPERATURE PROVINCE	T-E INDEX
A′	Tropical	128 and above
B′	Mesothermal	64–127
C′	Microthermal	32–63
D′	Taiga	16–31
E′	Tundra	1–15
F′	Perpetual frost	0

Of the 120 possible combinations, only 32 are recognized as actual climatic types. Figure 14-4 illustrates the relationships between P-E, T-E indices, and the natural vegetation. The world distribution of climates according to the Thornthwaite classification is shown in Figure 14-5.

CROP GEOGRAPHY

The distribution of natural vegetation depends on (1) environment, (2) plant response, (3) plant migrations, and (4) the evolution of floras and climax vegetation. These all have somewhat different meanings for crops. For example, the purposeful introduction of exotic species by man is quite a different matter from the natural development of floras by migrations that occur over tens of thousands of years. There is, however, a remarkable coincidence between the occurrence of certain types of natural vegetation throughout the world and the occurrence of agricultural regions (Fig. 14-6). The principal factors that influence the geographical distribution of crops are discussed in the following sections.

Climatic control

Plant geographers have always acknowledged the relationship between the climate of a region and its natural vegetation. Climatic zonation coincides in a broad and general way with **climax vegetation,** the ultimate type that can develop under a particular local pattern of temperature, precipitation, soil parent material, topography, and other meterological conditions. As a result, natural vegetation has been used as a kind of meteorological "instrument" to define boundaries in the classification of climate, with the result that savannah, steppe, and rainforest are names for climates as well as vegetation types. Climate is also a primary controlling factor in the growth of crops. The effects of climate on crops can, of course, be altered. Inadequate moisture may be compensated for by irrigation or drainage. Low fertility may be overcome through the application of fertilizers. In greenhouse production practically all climatic influences can be modified. But fertilization, irrigation, and other cultural practices can be used only to the extent that they yield sufficient return to the grower. It is an expensive process to modify natural climatic influences.

In 1840 Justus Liebig formulated a hypothesis that plant growth is limited by the nutrient available to it in least quantity. This concept, now popularly known as the **law of limiting factors,** can be expanded to include all environmental factors that contribute to growth. It implies that growth, like a chain with its proverbial weakest link, is limited by the factor that is least favorable. However, all sorts of interactions and compensations come into play with environmental factors.

Although plant growth is affected by subtle differences in climate, the extremes produce the greatest effects. Thus, the climatic extremes of a particular region may be more important in maintaining a vegetation type than the average. An oc-

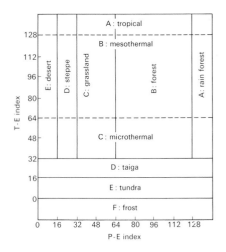

FIGURE 14-4. *Temperature and humidity provinces, based on natural vegetation.* [*After Thornthwaite, Geographical Rev.* **21**:*649, 1931.*]

Table 14-2 CLASSIFICATION OF COMMON FRUIT CROPS BY
TEMPERATURE REQUIREMENTS.[1]

		TEMPERATE	
TROPICAL	SUBTROPICAL	MILD WINTER	SEVERE WINTER

coconut
banana
cacao
mango
pineapple
papaya
	coffee
	date
	fig
	avocado
	
	citrus
	olive
	pomegranate
		almond
		blackberry
		grape (European)
		persimmon (Japanese)
		quince
		    peach
		    cherry
		    apricot (blossoms tender)
		    strawberry ⎱ (very hardy with
		    blueberry ⎰ snow cover)
		    raspberry
		    cranberry
			pear
			plum
			grape (American)
			    currant
			    apple

LOW-TEMPERATURE SENSITIVE	SLIGHTLY FROST TOLERANT	TENDER	WINTER-HARDY
NONCOLD REQUIRING		COLD REQUIRING	

[1] Variation in tolerance depends to a large extent on species, variety, plant part, and stage of growth.

casional severe period of drought may preclude the development of forest vegetation even though annual precipitation is no less than average. Cotton cannot be grown in regions that have heavy rainfall during boll formation. Climatic extremes have great influence on the distribution of perennial crops such as fruit (Table 14-2). Although citrus can be grown in northern Florida and southern Alabama, the hard freeze that can be expected every few years precludes the establishment of an industry in these areas. Some tropical crops, such as cacao and bananas, are injured by low temperatures above freezing. The relation between temperature and the growth of temperate-zone crops may be quite complex. For example, areas of peach culture are bracketed by high and low temperature requirements. Low temperature is required for dormancy, but injury follows extremely low winter temperatures, and the flowering stage is very sensitive to frost. Relatively high temperatures are required during the growing season.

The boundaries of agricultural regions are determined by climatic factors. In the Northern Hemisphere the limits are determined by low

268

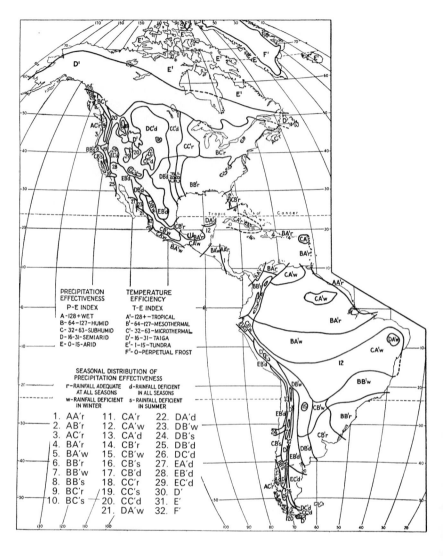

rainfall and low temperature. In Africa low rainfall is limiting on both the northern and southern boundaries. In South America the boundaries are set by low temperatures in the south, the harsh conditions of high altitudes and poor soil near the equator, and low rainfall in the northeast and parts of the Pacific coast. In Australia intensive production is confined to coastal areas that have adequate rainfall.

The action of any single environmental factor is affected by all other environmental factors; that is, the environment is **holocentric.** The physiological responses of plants are conditioned by the collective action of the environment.

The stress that one environmental factor imposes can sometimes be relieved by making another factor more favorable. For example, the debilitating effects of drought may be somewhat lessened as a result of prior fertilization.

Opposite climatic types may be compensatory. In the far north long days may partially compensate for short growing seasons and permit the cultivation of field crops that otherwise could not be grown. In the western United States 45 to 60 inches of precipitation is required for the development of mature forest vegetation, including such species as Douglas-fir. Yet this species grows well in England, where precipitation may be no more

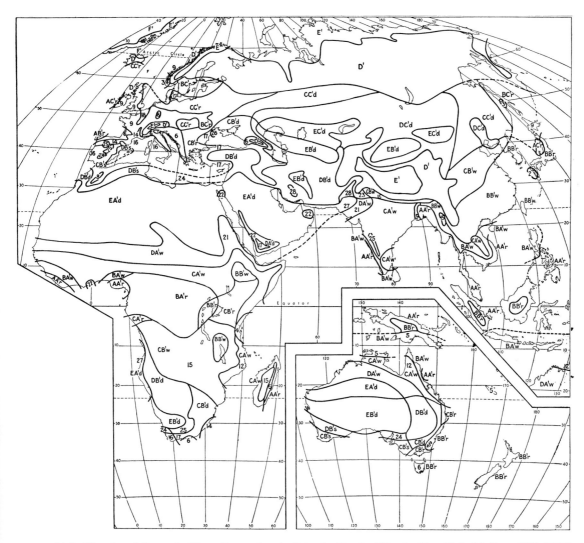

FIGURE 14-5. *Climates of the earth.* [*From Trewartha,* An Introduction to Climate, *New York: McGraw-Hill, 1954.*]

than 25 to 40 inches, but where heavy mists and fogs reduce transpiration requirements.

The compensation of one environmental factor for another is most frequent near the edge of a species' range. For example, spruce (*Picea*) and fir (*Abies*) grow in the cooler climates of high latitudes, but both species grow at high elevations far south of the region of their best development. Thus elevation compensates for latitude. In the southern United States white pine (*Pinus strobus*) is frequently found at low elevations far south of regions in which it might be expected on the basis of its temperature requirements, but it is always found in coves or on slopes that are sheltered by natural topographic features. Topography can therefore compensate, to some extent, for latitude.

Climate exerts a major influence on natural vegetation through its effect on soil development. Natural vegetation progresses through successional stages in concert with soil. If climatic elements or the parental material are adverse to either the physical or the biological processes of soil formation, there can be no progress to more advanced stages.

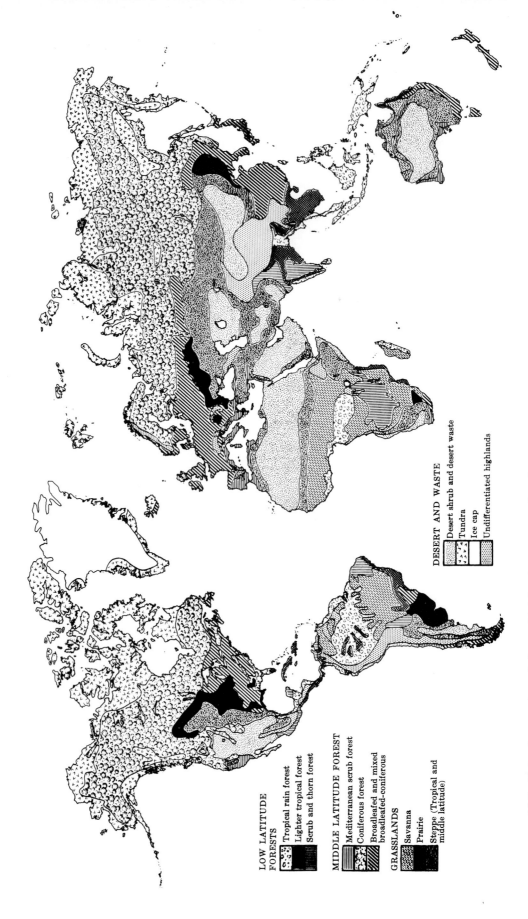

LOW LATITUDE
FORESTS
Tropical rain forest
Lighter tropical forest
Scrub and thorn forest

MIDDLE LATITUDE FOREST
Mediterranean scrub forest
Coniferous forest
Broadleafed and mixed
broadleafed-coniferous

GRASSLANDS
Savanna
Prairie
Steppe (Tropical and
middle latitude)

DESERT AND WASTE
Desert shrub and desert waste
Tundra
Ice cap
Undifferentiated highlands

FIGURE 14-6. *Natural vegetation of the world (above) as compared with approximate cropland area (facing page).* [From A Graphic Summary of World Agriculture, USDA Misc. Publ. 705, 1964.]

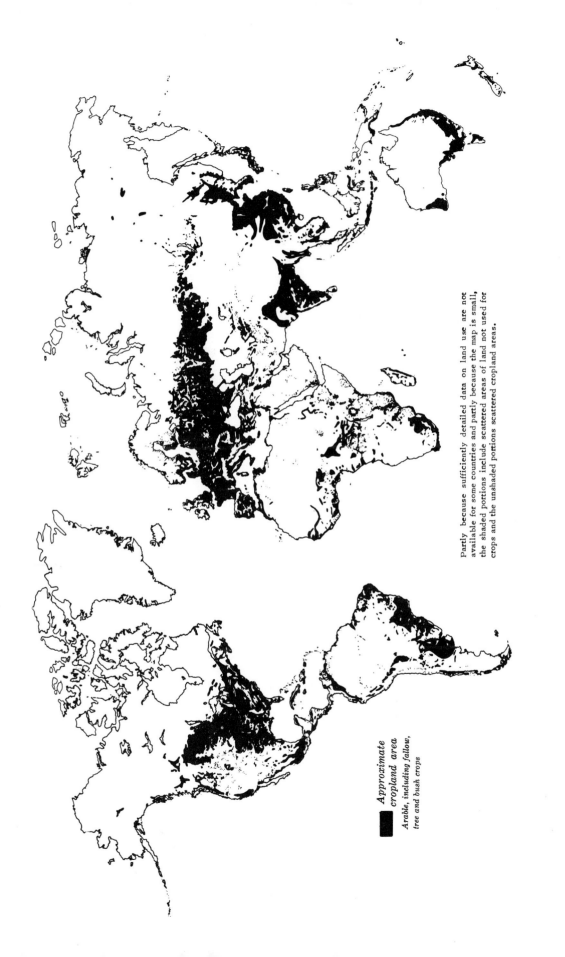

Partly because sufficiently detailed data on land use are not available for some countries and partly because the map is small, the shaded portions include scattered areas of land not used for crops and the unshaded portions scattered cropland areas.

Approximate cropland area
Arable, including fallow, tree and bush crops

Soil may become the limiting factor for some species even when all other climatic factors are favorable. For example, soils composed mainly of quartz sand, which is extremely resistant to weathering, can never develop into deep, rich soils such as those in regions where limestone is the parent material. Limestone is dissolved by carbonic acid, which is a natural component of rainwater and also forms from decomposition of organic matter. Some species of trees require deep soils that allow the development of extensive root systems. Sandy soils may not have enough water-holding capacity for the production of some crops, even though rainfall is abundant. Conversely, if the water table is too high, normal root development may be restricted. Soils that hold water on the surface make good rice paddies but will not produce corn. In some soils a deficiency or excess of minerals can be the limiting factor. Nevertheless, soil must be considered of secondary importance in controlling the distribution of crops. If soil conditions are not extreme they can often be altered to render the soil productive. If the demand for a particular crop warrants the expense, the soil may be physically and chemically altered. Fertilization, irrigation, subsoiling, drainage, and fumigation are the most common means used to alter soils. These will be discussed in greater detail in Chapters 16 and 17.

Environmental gradients

Environmental factors vary along gradients from one area or region to another. Probably the most extreme example of an environmental gradient is the change in temperature from the equator to the poles. Striking temperature gradients commonly occur where topography causes cold air to drain from smooth, gentle slopes and accumulate at the bottom. Temperature, humidity, and precipitation also vary with distance from bodies of water. Precipitation declines from 45 inches or more per year in the deciduous forests of the eastern United States to less than 10 inches on the great plains.

Environmental gradients are usually not steep except where topography or some other factor intensifies them. The temperature gradient from the coast to the Piedmont in the southeastern United States is very gradual. Similarly, the precipitation gradient between the deciduous forest and the great plains is gentle. Soil moisture shows steeper gradients, for even within a few meters (or yards) moisture conditions may change greatly. The sharp changes in environmental conditions where topography is irregular are troublesome for crop producers, who frequently wish to establish large production units that are managed uniformly. Large-scale agriculture is difficult in mountainous areas.

The environment is constantly changing, although the changes occur only very gradually. The Wisconsin glacial stage, the most recent advance of the ice sheet that once covered part of North America, ended about 12 thousand years ago. The climate has since shown a warming trend, and the extent of glacial ice in high mountain valleys has been observed to decrease markedly during the past 50 years. Other major climatic changes have resulted in a change in vegetation from spruce and fir to such forms as oak, hickory, and grasses, which can tolerate drier climates.

Because climatic change is gradual many crop growers would maintain that it is not realistic to take it into account, but within the large climatic changes and cycles there are smaller changes that can be witnessed within a generation. For example, the United States Weather Bureau has abandoned the practice of using the total weather records available for a station to determine mean and average values, and now uses a shorter period of time, 25 to 30 years. However, in the tropics, an even shorter period may be used because of the more nearly constant weather conditions.

Although environmental changes are gradual there is often a lag in the kinds of natural vegetation which occupy an area. Plant migration does not always immediately follow site availability. This means that some species may be successfully introduced into regions in which they are not native because the rate of migration has not kept apace of climatic change. Slash pine (*Pinus eliottii*), native to Florida and the Gulf coast, is a notable example; this species has been planted as far north as North Carolina, where it is sufficiently productive to justify planting for timber.

Species range and tolerance

The **tolerance** of a species is represented by the range of climatic and soil conditions within which it can exist and reproduce. The expression **ecological amplitude** is sometimes used to describe this range of conditions. There is great variation in tolerance. Some species can exist only in a very narrow range whereas others can tolerate a broad spectrum of environmental conditions.

In the ecological literature many theories of tolerance have been proposed. Most are concerned with the historical development of natural vegetation and the function that time has served. V. E. Shelford's generalized law of tolerance is of great interest and value to ecologists and geographers, and is defined in terms applicable to crops. Shelford's concepts can be stated as follows:

1. Species with a wide range of tolerance are likely to be widely distributed.

2. Some species may have a wide range of tolerance for one environmental factor and a narrow range for another.

3. When one environmental factor is limiting, the range of tolerance to other factors is likely to be decreased.

4. The range of tolerance is likely to be narrowest during the period of reproduction.

Tolerance has a genetic base. Environment can influence the growth and development of organisms only within genetically imposed limits. Species with a wide range of genetic diversity are more likely to respond favorably to new and changed environments that to those with a narrow genetic base. The evolutionary forces that determine the range of species involve differentiation and speciation. Long-term forces apply only incidentally to crop plants, because man has imposed evolutionary change through artificial hybridization and other kinds of genetic manipulation. Adaptability becomes one of the main goals of modern plant breeding (see Chapter 19). The relatively slow migration of natural species, hampered in many cases by natural topographical barriers, has given way to widespread and rapid introduction of exotic species.

Different stages of growth may have different tolerances. Plants, like other organisms, change in structure as they develop. The developmental phases respond differently to environmental factors. The limits of tolerance are often narrow in the seedling and juvenile stage. For example, the stem of a newly generated long-leaf pine seedling has no protective bark and may be killed by temperatures of 60°C (140°F) or less. The same tree, when 10 years old or more, may tolerate temperatures of several hundred degrees or more for a few minutes, because by that time a thick layer of bark will have developed. Tobacco plants must be protected when young, but after a few weeks grow well under field conditions. As a seedling, baldcypress (*Taxodium distichum*) must not be submerged, but later in its life history the water level of the habitat in which it grows is not nearly so critical.

In practically all crops, the range of tolerance during the flowering period is critical. Excessively hot dry weather during flowering of maize can cause crop failure if pollen is released before silking occurs, and can even kill pollen. Similarly, cold wet weather during apple bloom adversely affects pollination because bee activity is reduced.

Because the range of species is tolerance-limited much effort has gone into a worldwide search for areas with similar climates in order to weigh the prospects of successfully introducing new crop species. Areas with similar climates are called **homoclimes,** or with reference to agricultural production, **agro-climatic analogues.** The crop ecologist M. Y. Nuttonson has studied agro-climatic analogues of the United States using mean temperatures, dates of killing frost, average precipitation, precipitation effectiveness, and other types of climatological data. Beer Menuha, in Israel, and Calexico, in California, were found to have comparable climates, and are nearly thermal equivalents. Irrigated barley is the most important cereal crop in Calexico because it tolerates soil salinity and grows in a wide variety of soil conditions. Since the same conditions prevail in Beer Menuah, barley might be expected to grow profitably there if equivalent cultural techniques are used.

Biotic factors

The members of the biologic community, in their roles as parasite, predator, symbiont, competitor, or vector of pollination and propagation, strongly influence the distribution of natural plant communities. Consider the role of the vast herds of animals in helping to maintain the great South African veldt.

Biological factors also have great significance on the distribution of crops. The crop producer works in situations where the natural biologic equilibria are disturbed. Under these conditions the biotic factors of greatest concern are populations of crop pests. The inability to exert sufficient control over a predator or disease can virtually eliminate crop species. Thus Panama disease has eliminated commercial banana production from large areas of Central America. A number of diseases of the *Hevea* rubber tree have had a great effect in moving the industry from its native habitat in Brazil to Southeast Asia. The bacterial disease fire-blight has eliminated commercial pear production in most of the warm humid areas of the United States.

Crop pests are frequently controlled directly by chemical or other cultural means. Control through genetic manipulation of the crop plants is also widely used to produce new varieties with genetic resistance. The adversaries are often biologically resourceful, however, and the occurrence of new races of some pests has made the incorporation of genetic resistance only a temporary stopgap technique. **Biological control,** the manipulation of biotic factors in the control of plant and animal pests, has recently received a great deal of attention (see Chapter 18).

Other biotic factors have influenced crop distribution. Many crops depend on biological vectors for pollination. The production of smyrna fig in California was unsuccessful until the *Blastophaga* wasp was introduced for pollination. The value of many legumes as crops depends on their association with particular nitrogen-fixing bacteria. Many legumes proved ineffectual in tropical areas until specific forms of bacterial symbionts were obtained.

Introduction and adaptation

The term "natural migration" infers that a species has been transported to a new area, become established, and by successfully reproducing has become a permanent part of the natural flora. This process is slow and evolutionary. In contrast, the successful migration of crop plants, which leads to the permanent incorporation of a particular species into the agriculture of an area, can be a rapid, explosive, revolutionary process.

The migration of crop plants typically results from deliberate importation. The widespread dispersal of plant species from the Old and New Worlds during the Age of Exploration took place within a century. At present, plant expeditions to uncover new crop plants and new forms of old ones is a continuing endeavor.

Introduced crop plants do not necessarily have to be able to reproduce in the new area. Many crop industries depend on seed produced outside the area of production. This is often true of plants grown for some part other than the fruit.

Low tolerance in critical periods may be compensated for by cultural practices. The adaptability of crop plants depends heavily on man; most of our crops could not survive on their own in most locations. Maize is not native anywhere except as a crop plant.

Finally, adaptibility is encouraged through genetic manipulation. For example, for a long time cabbage would not reproduce itself in the subtropical climate of Brazil, and seed had to be imported each year. But recently, in an abandoned field in the state of Sao Paulo, some variants responded to slight chilling and flowered. Selection soon produced types adapted to local seed production, called *repolho loco*, or "crazy cabbage." The tomato is constantly being altered by breeding to maintain adaptability to new locations (it is grown from the equator to as far north as Fort Norman, Canada, 65° N lat.), to resist local pests, and to adapt to new systems of management such as mechanical harvesting. New strains of wheat must be continually produced to compensate for mutations that occur in the destructive stem rust fungus (*Puccinia graminis*).

In some crop plants it is desirable to prevent change in genetic structure. When crops are particularly adapted to the needs of man, and where asexual propagation is possible, particular genotypes may be kept intact for long periods of time. The Bartlett pear, the main variety grown in the

United States, Canada, and France, originated in England before 1770.

The accidental introduction of destructive weeds has been costly to crop growers. Russian thistle (*Salsola kali*), Japanese honeysuckle (*Lonicera japonica*), and witchweed (*Striga asiatica*) are examples. The migration of fungi and bacteria also causes serious problems. Fifty years after the fungus *Endothia parasitica*, which causes chestnut blight, was introduced to the New York Botanical Garden on an imported chestnut, it had virtually eliminated the native chestnut from the eastern United States. Governmental agencies maintain large and costly systems of quarantine and inspection in an attempt to prevent the introduction of unwanted species. In the United States even individual states have also found it necessary to maintain inspection stations at their borders.

Economic and cultural factors

Crops are a part of the economic and social life of man. The distribution of crop plants is thus affected by many factors in addition to biologic factors. The agriculture of any nation strongly reflects the level of economic development and the standard of living. On a local level, such factors as land costs, availability of labor, distance to market, and transportation facilities frequently override biological considerations of climate and soil suitability. Finally, governmental policy in the form of taxes, subsidy, and tariff barriers become important considerations. Even though a crop may be well adapted to a particular climatic regime, there must be a demand for it by consumers and a profit incentive for producers. Countries in which agriculture is organized in such a manner that production incentives are dampened by collective farming or by absentee landlord systems, for example, have found it extremely difficult to achieve rapid growth in agricultural production. The economics of crop production are discussed in Part VI.

The preferences of people in particular countries and regions for certain kinds of agricultural products has developed over countless generations.

Rice is the staple carbohydrate in the Far East; wheat occupies this niche in the West. It would be difficult to reverse the use of these products even if such a change were technologically desirable. Food habits often reflect religious attitudes. Observant Hindus refrain from beef, even though India probably has the largest population of bovines of any country in the world. Although preferences are frequently dictated by local productive potential and capacity, they develop into habits or biases that are difficult to change rapidly. Mobility has increased greatly in the past several decades, resulting in the introduction of new foods and other plants products, but for most of the world's peoples significant dietary changes are unlikely.

THE FUTURE

Agriculture is the world's largest occupation. Even so, only 8 to 9 percent of the total land surface of the earth is under cultivation—approximately $2\frac{1}{2}$ to 3 billion acres. Most of this cultivated land is in the middle latitudes, between 30° and 60° north and south of the equator. Four countries—the United States, the Soviet Union, China, and India—have a fourth of this land. The distribution of land that is suitable for cultivation is primarily determined by the nature of the physical environment—climate, soil, and land forms.

Vast portions of the earth do not now meet the requirements for crop production. In addition, probably no more than one-half to two-thirds of the available land is used for agriculture during any one year. Although no one knows how much more cropland can be added to that which exists, there is agreement that we have not reached the limit. But practically all of the best land is now under cultivation. Advances in agricultural technology are required to make profitable use of the great areas in the tropics and in the deserts that are now agriculturally unproductive. The expansion of the frontiers of present-day agriculture will be more than a challenge for future scientific pioneers.

Selected References

Cain, S. A. *Foundations of Plant Geography*. Harper, New York, 1944. (A widely used textbook on plant geography.)

Clements, F. E., and V. E. Shelford. *Bio-ecology*. Wiley, New York, 1939. (A treatise which deals, in part, with the broader aspects of the interrelations between climate and organisms.)

Gleason, H. A., and A. Cronquist. *The Natural Geography of Plants*. Columbia University Press, New York, 1964. (An interesting and highly readable elementary book on plant geography.)

Klages, K. H. W. *Ecological Crop Geography*. Macmillan, New York, 1947. (An excellent textbook.)

Trewartha, G. T. *An Introduction to Climate*. McGraw-Hill, New York, 1954. (An elementary text of climatology. The appendix contains a detailed presentation of the Köppen system of climate classification.)

U.S. Department of Agriculture. *Climate and Man*. Yearbook of Agriculture, 1941. (A collection of articles dealing with many aspects of climate and crop production.)

Williams, C. N., and K. T. Joseph. *Climate, Soil and Crop Production in the Humid Tropics*. Oxford University Press, New York, 1970. (A text dealing with many aspects of crop production in the tropics, including moisture, heat, light, soils and cultural practices.)

Wilsie, C. P. *Crop Adaptation and Distribution*. W. H. Freeman and Company, San Francisco, 1962. (An excellent textbook or reference dealing with the environmental biology of crop plants.)

PART IV

STRATEGY OF
CROP PRODUCTION

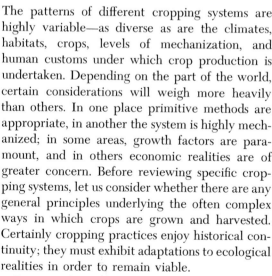

CHAPTER **15**

Cropping Systems and Practices

The patterns of different cropping systems are highly variable—as diverse as are the climates, habitats, crops, levels of mechanization, and human customs under which crop production is undertaken. Depending on the part of the world, certain considerations will weigh more heavily than others. In one place primitive methods are appropriate, in another the system is highly mechanized; in some areas, growth factors are paramount, and in others economic realities are of greater concern. Before reviewing specific cropping systems, let us consider whether there are any general principles underlying the often complex ways in which crops are grown and harvested. Certainly cropping practices enjoy historical continuity; they must exhibit adaptations to ecological realities in order to remain viable.

Primitive harvesting undoubtedly consisted of little more than casual collecting of wild seeds, fruits and other plant products. The hunting of game supported nomadic tribesmen, but was incidental in increasingly sedentary societies on the highroad towards civilization. Sedentary tribes came to rely chiefly upon the nearby abundance of food that domestication and cropping provided. Thus since mankind's earliest days two general "cropping systems" have been in operation: one centering on grazing and browsing animals (the animals harvest the vegetation, and man, secondarily, the milk and meat—a system still basic in some parts of the world, such as East Africa where it is practiced by the Masai); the alternative system is the direct harvest by man of the crops he sows and tends (though sharing some of his produce with livestock), and forms the basis for most of today's agriculture and horticulture.

Forages for livestock are discussed in Chapter 20. The rumen or "second stomach" of cattle enables them to digest cellulosic materials indigestible by man and carnivores. The rumen is populated by a microflora of flagellates, which symbiotically make possible the digestion of forages and fodders useless to other animals.

Grasses and legumes are the major products used for livestock feed and various management procedures conforming to climate, soil and other environmental considerations are followed. Grazing is an especially effective way of utilizing "secondary" land, which is too steep, rocky, dry, or otherwise unsuitable for economical growing of direct-harvest crops. Even today the system of grazing varies from merely turning out half-domesticated livestock onto open range, to cultivating pastures that are as meticulously organized for maximum production as is any agricultural pursuit (with appropriate combinations of grasses and legumes, properly fertilized, irrigated, and otherwise managed). Open-range herding is really not very different in principle than was the taking of wild game, whereas advanced animal husbandry is systematized to yield as much as 400 kilograms or more of high-quality beef to the hectare annually. In practice, of course, livestock raising is usually combined with direct crop production; in primitive societies the pigs may be allowed to root in the garden on occasion, whereas in technically advanced operations the grain, hay and silage may be delivered to the livestock.

A cropping system must relate to the climate or biome (Figure 15-1) in a general way. Mean annual temperature and mean annual precipi-

tation are major determinants of climate. Notice in the Figure 15-1 that certain types of natural vegetation result from the interaction of temperature and rainfall, and these in turn are indicative of the cropping practices suited to the region. We will see in this chapter that swidden (slash-and-burn) and plantation agriculture are especially suitable for tropical forest habitat, diversified farming for the deciduous forest and moist grasslands, ranching and dry-land cropping for the drier grasslands, irrigation agriculture for the deserts, tree cropping for the coniferous forests, and grazing for the alpine or arctic meadows.

Whatever the cropping systems that evolved in each of these dissimilar habitats, undoubtedly each underwent long evolution by trial and error, and depended upon domestications made from the wild there or nearby.

During such long evolutionary periods even climate may change (as almost surely it seems to have done in northern Africa), leading to still newer cropping innovations. Or man may impose change on older cropping systems, as occurred with the development of the Mexican wheats by Norman Borlaug of the Rockefeller Foundation, and of the new rice cultivars by the International Rice Research Institute. Such new cultivars usually call for greater environmental control by man (such as implementation with fertilizer), and the growing season is likely to be different than that for the traditional cultivars (and thus upsetting to established habits). Although changes like these seem trivial on the surface, the influence can be profound on delicately balanced economic and political systems geared to other considerations. For example the new cultivar may favor the well-to-do and knowledgeable grower (who is able to finance fertilizer purchase and use it intelligently) over the peasant farmer (who may have to sacrifice his independence as a landholder for economic reasons, with consequent, and possibly widespread, political dissatisfaction).

On the whole, the ecosystems (the ecological relationships between all elements of the environment) of the tropics are more complex and more diverse, and allow for many more alternative paths of energy flow, than those of temperate and boreal climates. A tropical ecosystem is therefore usually

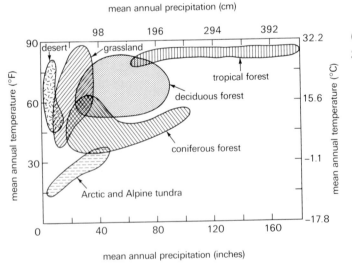

FIGURE 15-1. *Six biomes of the ecosystem characterized by mean annual temperature and precipitation.* [*Courtesy National Science Foundation; from* Science 175:47, 1972; © 1972 by the American Association for the Advancement of Science.*]

Table 15-1 AREAS AND NATIONS RANKED IN ORDER OF PESTICIDE USAGE PER ACRE AND IN ORDER OF YIELDS OF MAJOR CROPS.

AREA OR NATION	PESTICIDE USE		YIELDS	
	OUNCES/ACRE	RANK	LB/ACRE	RANK
Japan	154.0	1	4,890	1
Europe	26.7	2	3,060	2
United States	21.3	3	2,320	3
Latin America	3.1	4	1,760	4
Oceania	2.8	5	1,400	5
India	2.1	6	730	7
Africa	1.8	7	1,080	6

Source: Adapted from W. B. Ennis, Jr., *Weed Science* **19**:632, 1971.

more stable than a nontropical one, and able to absorb more abuse without serious alteration. Lateritic tropical soils, naturally deficient in organic content, withstand the drastic imposition of cultivation poorly, however. Conversely, of course, the ecosystems of colder, harsher environments are less complex, contain fewer species (in larger populations), and show less stability and resistance to the impact of a cropping system imposed upon it.

Any cropping system implies some manipulation of the ecosystem. This ranges from mere care not to exhaust a resource completely (as do still the Australian bushmen or remote tribes in Africa and South America), to nearly complete environmental control, in which moisture, fertility, pests and even temperature are carefully managed (as in the intensive growing of strawberries and fresh vegetables in California, in which even the soil is sterilized before planting).

Considering the relative richness of tropical ecosystems, it is perhaps inevitable that tropical cropping would have gravitated more towards multiple-species than has the agriculture of such climates as the United States and Europe, taking advantage of the abundance of ecological niches. Simpler systems are appropriate to the harsher environments of temperate climates, more rigidly controlled by technological means (examples of such systems being the extensive monoculture of corn, wheat, annual cotton, and so on). Attempts

to impose a simplified cropping system on tropical habitat (say, the plantation system, which will be discussed), have, on the whole, not met with persistent success, and today there seems to be a trend towards smaller, individual plantings supplemented by multi-crop door-yard gardens.

Yet it is impossible to dispute the productivity of cropping systems that utilize the maximum density of highly selected individual plants, even though such a system may be naturally unstable and require close technological supervision. Notice in Table 15-1, in which designation of pesticide usage indicates intensive cropping, that total yields per unit of land area are greatly favored by this highly technical approach. However, it must be recognized that such intensive cropping, if it is to be effective, is heavily dependent on outside resources—capital (which provides the many forms of labor-saving equipment), energy (especially electricity and gasoline to power operations), and knowledge (the technological background gained from an advanced educational system, as well as a familiarity with supplementary aids such as pesticides).

Simplification of the ecosystem by means of monoculture presents a constantly changing pattern. Advantage is taken of new regulative practices to improve yields and especially man-hour efficiency. Production at a 1920 rate on even the finest Corn Belt soils would be unremunerative today. At that time, plowing was done with a single moldboard pulled by a mule or span of horses, the farmer handling the plow. Readying the seedbed and planting were similarly tedious. With open-pollinated corn or any other crop, seed for the next planting was usually from a previous harvest on the farm; even under ideal conditions such strains were incapable of yields half that of modern hybrids. Weed control was solely by tillage, and required additional passes over the field, which were very costly in man-hours (and doubtless caused some damage to crop roots). Because of the need for row space along which a horse could pass, plant populations per acre were considerably lower than at present. Harvest required much hand labor, field shocking, and multiple gathering or threshing operations.

Yet, this was a beginning for mechanized monoculture. In America farming has gradually evolved into the highly mechanized systems discussed later in this chapter. Multi-plow tractor-drawn equipment has replaced the independent moldboard, and, with some models, one pass completes seedbed preparation, planting, fertilizing and pesticide application. Herbicides eliminate the need for subsequent weed-control cultivation. Large plant populations in narrow rows are harvested by a field combine, and on-the-farm cribs with grain-on-the-cob are mostly a thing of the past. Yields per acre have increased four-fold, and yields per man-hour ten-fold. As we will see, the end is not in sight, as more and more of the environmental variables are brought under control.

Most efficient, in terms of low input per unit of yield, would be "primitive" forms of gardening, still practiced in remote locations such as New Guinea, parts of central Africa, and the Amazon basin. In these regions all ecological niches are utilized, and the result is a mixed planting that looks quite untidy to someone from the temperate zone who is accustomed to neatly kept fields. In taking maximum advantage of sunlight, space, and the generally abundant rainfall, growers in these areas use a thin canopy of trees as a crop overstory, which shelters an understory of bush and herbaceous plants, many of which avail themselves of underground space, too, thereby yielding comestible rootcrops and tubers as well as the fruits and greens found aboveground. Such a cropping system is quite balanced, and requires minimum maintenance once it is established. The shade and crop competition help restrain weeds, and regrowth occurs rapidly to compensate for portions harvested. Moderate attention to weeding and the maintenance of improvised fences to keep pigs or other livestock out (until appropriate occasions when they might undertake harvest for themselves) are the chief demands upon the custodians.

Tropical systems utilize vegetatively propagated crops much more than do temperate systems. Most root and tree crops are propagated by division or from cuttings, and even herbaceous plants, if not perennial in the tropical climate, are more conveniently started from vegetative slips rather than from seed. Quite the opposite is true for temperate zone agriculture, especially where agriculture is based primarily upon grains and legumes massively planted by mechanized means.

A diversified subsistence garden in the tropics may continue to be productive for a number of years in favorable locations. Less diversified plantings exhaust soil fertility within a few years, and must be re-established on "virgin" ground elsewhere—the so-called shifting agriculture, also called the slash-and-burn, or swidden, system of cropping. A forest that has been felled and burned for manioc or maize is a typical example. Once the protective canopy of trees is destroyed, ecological change is drastic; soil fertility is quickly exhausted, and sun-loving weeds often become impossible to control. Although conservative of inputs, the swidden system is extravagant of land use, and suited only to lightly populated regions, which are now fast vanishing from the face of the earth. When population pressure necessitates continuous cropping of the same soil, and use of seeded plants to increase the yield, problems quickly arise in the tropics. Many lateritic soils become brick hard and unusable when subjected long to cultivation.

The greatest advancements in tropical cropping have not come in the dense rainforest regions, however, but in habitats where rainy and "dry" seasons alternate. Apparently the contrasting ecological regimens that result provide a rich range of habitat, well suited to origination and domestication of special crops and cropping systems. It appears that many (especially the starchy root crops) originated in climates of this sort, where the monotony of rainforest is interrupted by a dry season that opens new opportunities.

Intensive cropping is quite demanding, and best confined to good soil and level land. A crop of 2.5 tons of grain (100 bushels) per acre will deplete the land of 95 pounds of nitrogen, 17 pounds of phosphorus, 22 pounds of potassium, 19 pounds of magnesium, and 8 pounds of sulfur. If crop fields are to remain productive for intensive use, obviously nutrients must be replaced continuously, and to do so in adequate amounts is possible only in parts of the world where fertilizer is readily available. Although in some places food needs could no longer be met were cropping systems to be limited to the "natural" methods characteristic of subsistence agriculture, it appears that intensive cropping can be continuous if measures are undertaken to replace nutrients, and if organic content is maintained at adequate levels by the incorporation of plant residues. Admittedly, however, intensive cropping is too recent to provide the experience that can only be gained in many decades (and even centuries) and is needed to evaluate permanence of the system, unlike some other cropping systems that have been carried on in certain areas since the dawn of civilization (such as the paddy farming in parts of the Orient, or the terrace agriculture and irrigation systems of the Near East and the South American Andes).

It is quite evident that in many parts of the world misuse of the land has caused the destruction of civilizations that once existed there. The tendency of lateritic soils to become unworkable when exposed to cultivation has been mentioned. Perhaps an even more significant cause of soil depletion is erosion in hilly country, and on overgrazed or overbrowsed lands (like those where feral goats range in the Near East and Oceania). Sardis, in southern Turkey, is an example of a once wealthy community now virtually abandoned. Lands of the eastern Mediterranean, now denuded and bare (and unfit for any cropping unless accorded expensive renovation), were once covered with magnificent forests (such as the cedars of Lebanon). Although soils respond differently in susceptibility to erosion, depending upon such factors as cultivation, slope, rainfall, and so on, a study in Puerto Rico shows how serious erosion can be: nearly 48 metric tons per hectare of Pandura sandy loam were lost in a single simulated storm when exposed by clean cultivation. If a cropping system is to be enduring, soil protection must certainly be afforded. Unfortunately, scant attention is given to soil conservation in many parts of the world, especially in some of the remote areas where the inhabitants have little if any modern agricultural knowledge and more primitive cropping practices prevail.

THE MODERN CONTEXT

In general, crop producers aspire to the greatest possible yield commensurate with reasonable cost. The balance struck varies with location and occasion. In well-to-do parts of the world where population pressure is not yet excessive, as in the United States, the cost of labor is generally the prime consideration, more important than total production and even yields. It is not economical to farm types of land, kinds of crops, or inconvenient locations where extra man-hours of effort are required for limited additional production. In other, more crowded parts of the world, such as the heavily populated Orient, labor is far less important than is total production. There each square foot of productive land will be farmed, even at the expense of hours of tedious hand labor. Given aids to increase growth, such as fertilizer, per-acre yields may be greater than under highly mechanized, massive farming in America, where detailed attention to the land cannot be practiced because it is too costly.

Goals and methods will continue to change, depending upon economic conditions of the future and upon the ability of man to limit the size of the world population. The day is already at hand when American agriculture is concerned with increasing production rather than managing surpluses that have plagued the agricultural economy for some decades. In Japan, where trained labor is becoming more costly, a trend toward greater mechanization in agriculture can be expected. The cultural system best adapted to the various parts of the world depends in part upon the overall human ecology there.

One thing can be foreseen. There will come a time when a ceiling on production is imposed beyond which it is impossible to go, practically. No matter how rich the soil is made, and no matter how well crops are tended, space, sunlight, and crop potentiality limit further increase. Maximum possible yields will have been achieved, because the crops will have been grown on the richest possible soil under the most suitable conditions obtainable. Somewhat before this ceiling is reached, the law of diminishing returns will have made impractical further increments of possible increase. The cost of practices needed for this last measure of return will be unjustifiable. Most of the world, however, is a long way from such a ceiling, and the increased use of fertilizers alone should greatly extend the world's capacity to produce crops. Fortunately, this gives civilization some margin of time to institute population control and reduce rapidly mounting demands on natural resources. It is sobering to note that even with all the improved cultural practices developed since World War II, and the consequent increased yields, the world is no better fed today than it was decades ago.

Quality of production must be assessed, too. How long can the United States, for example, continue to enjoy a basic meat diet—an extravagant source of protein in terms of the resources needed to produce it? Will not the time arrive when a much greater share of protein must be gained directly from plants? And what cropping practice, what type of fertilization will yield suitable chemical quality in each foodstuff, even if at the expense of some yield? Cropping practices of the future must take this into consideration. Moreover, man will have to determine not only the best means of growing a crop, but which kind of crop best fits the economic situation of the moment!

Complicated interrelationships are involved in agricultural policy decisions. Should such crops as soybean, which is relatively low in yield but high in protein, be grown in preference to crops that are higher in yield but lower in protein, such as grains and root crops? What about select sugar canes, in which the yield of dry substance is powerfully increased at the expense of nitrogen content?

What combination of such crops provides the best balance? Maximum use is made of photosynthesis with the sugar canes, but certainly the protein is deficient. Broad cultural systems under maximum population pressure will have to provide a daily intake of about 2,500 kilocalories. The ratio of protein to carbohydrates plus fat should run about 1 to 11. Rice is one staple crop that matches good yield with this protein quality. Various analyses made to evalute such factors as yield, protein content, and utilization of straw show that rice is capable of sustaining 25 percent more persons per square mile than wheat. One can hypothesize sustaining even greater populations by using a high-yielding sugar to ferment proteinaceous yeast. Of course some help will come from the sea; already there is talk of "farming" the continental shelves with marine plant species. Let us hope, however, that man learns to regulate his numbers before turning to yeast and algae just to stay alive. The world will not be a happy place in which to live if it becomes necessary to sustain upwards of 20,000 persons per square mile, though this is theoretically possible with good crop land.

TECHNIQUES AND SYSTEMS

There is no end to the schemes that might be employed for raising and harvesting a crop. But certain systems adapt better than others to certain circumstances. Although ill-defined and with many variations, the following procedures cover most of the cropping mankind undertakes to gain his livelihood from the land in all parts of the world.

Shifting agriculture (swidden cropping)

Shifting agriculture is a slash-burn-and-abandon process employed mainly by primitive cultures in sparsely populated regions. The **milpa** of the American tropics and the **ladang** system of the Orient are examples. The storehouse for vegetational wealth in the tropics is lush topgrowth, not soil. Shifting agriculture endeavors to release these stores for down-to-earth cropping, making but temporary use of the stubborn lateritic soils.

Forest trees are either girdled or felled. The dead vegetation is burned during the dry season (or sometimes left to decay and deteriorate through termite action). The resulting open ground is automatically fertilized with wood ashes. After only a few crops, fertility is typically exhausted. The clearing is then abandoned, the process repeated elsewhere. On unfavorable terrain appreciable soil is eroded.

Cropping entails the planting of seed or more commonly vegetative parts as soon after burning as the soil softens with onset of the rainy season. Under primitive circumstances this is done simply by dibbling holes in the soft soil with a planting stick. Some weeds will volunteer in the crop, and hand pulling many be needed to aid the crop. Brush or grasses typically overrun the clearing within a very few years.

Obviously, such a system makes extravagant use of land and is wasteful of trees. Societies depending upon it must either keep on the move or extend operations ever farther from the central city as nearby lands become exhausted. Frequent clearing of the same land degrades the vegetation into an unmanageable tangle of brush. Maybe this is why empires based upon shifting agriculture have not endured as well as those established on valley flood plains.

Tree horticulture

Fruits, derived mainly from woody plants, are important crops. They were gathered from the wild by primitive man, and today support the economy in many parts of the world. Tree crops, mostly fruits and nuts, are particularly suited to tropical cultural systems, and, as was noted earlier in this chapter, allow for a more stable cropping system there than does swidden cropping. In addition to fruits, woody plants are grown for tea, maté, wax, rubber, quinine, and cork, often found in plantations (which will be discussed subsequently). In temperate climates, orchards—counterpart to the tropical plantation—yield a variety of fruits and nuts and even decorative materials such as holly.

The tropical climate is generally better suited to tree growth than to the production of annual crops, which peoples living in temperate climates have come to regard as agricultural mainstays. The lateritic soils of the tropics become hard and refractive; powered equipment for cultivation is expensive, hard to procure and maintain; efficient means of mass handling and storage are generally lacking. But tree groves accompanied by family gardens support individual families and small villages well. They are enduring and harmonize with the tropical forest environment. Thus, as settlements become too mature for burn-and-abandon agriculture, villagers turn to tending groves of coconut and other palm fruits, breadfruit, banana, citrus, avocado, and so on. This kind of tree horticulture is supplemented by diversified home garden patches, and usually some domesticated animals are used to help keep down prolific tropical growth and supply manure for the plantings. Fishing and hunting often yield additional protein.

In some systems trees are cultivated as a source of fertilizer. The tops are constantly cut back to the trunk, the vegetation being used to fertilize cleared ground, usually by burning but sometimes by composting.

Plantation agriculture

A successful agricultural method where forest has been cleared is the **plantation system,** hallmark of tropical agriculture in many regions. A simple example might be plantain growing in Ghana. Locally selected cultivars are perpetuated by vegetative "suckers," after felling of forest and burning of brush in the dry season. Reasonable attention is given to mulching, irrigation, propping of fruit bunches, intercropping with vegetables, spraying for disease and pests, timing of harvest, and so on. Only where an enlightened supervision practices sanitation, selection of disease-free suckers, and fertilization, can a planting endure for many years and thereby circumvent the swidden system of moving on to new ground after only a few crops.

A good planatation combines the better features of shifting agriculture with those of horticulture. Under trained managers, often sent from agriculturally advanced parts of the world, plantations generally utilize the land efficiently. The plantation system requires responsible management and the capital resources to supply the necessary machinery and facilities and to open up marketing opportunities. Although there is no doubt that many foreign-controlled plantations have exploited native populations unfairly, it is regrettable that the recent rise of nationalism and anticolonialism has tended "to throw out the baby with the bath water." The rubber plantations of Sumatra and Malaya, the cacao plantations of Africa and Central America, the nutmeg orchards of Grenada, and the sugar and coffee plantations throughout the tropics are probably the soundest way of utilizing the land under prevailing circumstances. The native populations of these parts of the world often lack agricultural training, and must be supervised—a situation that must be changed if world food demands are to be met. There should be some means of combining social enlightenment with the efficiences of large-scale operations, especially in the less-advanced tropical countries where small holdings are not well managed.

Orcharding

Although the Orient and the tropics still rely a great deal upon backyard plantings of a few fruit trees, in technologically advanced countries commercial orcharding has evolved into a large-scale specialization (Fig. 15-2). Blemished fruits typical of home-yard growing no longer find a market in the United States. Select varieties with proper keeping qualities, exact spray schedules, and harvesting standards, typical of professional orcharding are needed. It is not unexpected that production centers have shifted to particularly favorable climates—in the United States mostly to the coastal Northwest, and Great Lakes region. Europe has seen similar specialization—for example, the heavy plantings of pears in the Rhone Valley, and of both apples and pears in the Po and other valleys of northern Italy. Southern Australia, South Africa, and southern Argentina also produce considerable fruit crops.

Eastern Asia is particularly well suited to fruit production, having climatic conditions equal to other fruit-producing parts of the world, plus the blessing (for fruit production) of unlimited manpower. Skillful gardening rather than extensive agriculture has an ancient tradition in Asia. It is said that there are over 500 species of native

FIGURE 15-2. *A portion of land owned by the Dixie Orchard Company, Vincennes, Indiana. This large enterprise manages more than 1,000 acres of apples and peaches. [Courtesy U.S. Soil Conservation Service.]*

chinese plants that produce edible fruits, of which 120 species are in cultivation. These include apple, pear, persimmon, jujube, peach, grape, mandarin orange, sweet orange, pineapple, banana, longans, and litchi nut. Peach varieties such as the Chinese Cling, reknowned as a parent of the Belle of Georgia and Elberta varieties, have come from China, and this area with its long history of civilization (during which selections and adaptations were made) promises a wealth of untapped breeding potential still.

In the United States and Europe apple, citrus, pear, peach, plum, apricot, cherry, pecan, walnut, and grape are among the principle orchard species. Throughout the tropics, coconut and banana are extremely important; and in special locations such as Hawaii, pineapple is a major crop. Lesser fruits grown in the tropics include mango, papaya, date, fig, avocado, olive, and pomegranate; in temperate climates, almond, persimmon, blueberry, chestnut and various bramble species.

Each type of orchard, each soil and climatic zone, demands its own cropping practices. In general, select planting stock is usually grown in a nursery, through budding, grafting, and rooting of cuttings (see Chapter 9). Saplings are transplanted to the orchard, where they are spaced geometrically to accommodate eventual tree size and the equipment used to tend the orchard. In the United States trees are generally spaced 20 to 45 feet apart. In Europe, where land space is at a premium, dwarfed "hedgerow" orcharding is very common, especially with apple and pear. There are obvious efficiencies in harvesting fruit that grows low enough to be reached from the ground or from a moving platform. In northern Europe trees are often trained to wire trellises or are maintained as dwarfs either by pruning or by using size-controlling rootstocks. In Italy and France plantings range from spaced trees to narrow "fruiting walls." The Golden Delicious, an apple variety popular in the United States, responds well to hedgerow growing and is now extensively planted in Europe under this system.

Few orchard trees demand as fertile a soil as do annual crops, but, instead, thrive best in deep, well-drained locations. Their root system, more extensive than that of annuals, is more continuously active but does not cross-transfer nutrients well from one side of the plant to the other. Fruit orchards often do well on light soils, and tolerate wide ranges in pH. But they may be susceptible to shortages of minor elements.

Location is important; blossom loss from freezing can be prevented by planting orchard trees in equable climates leeward from bodies of water (which warm later in spring and cool more slowly in autumn, allowing late blooming and ripening), and on rises rather than depressions (where cold air settles, causing frost).

Various soil management schemes are practiced in the orchard, ranging from clean cultivation (particularly when the planting is young) to maintenance of sod or legume covers to hold and build the soil. With modern herbicides sod and weeds can be restrained without cultivation, resulting in less competition for water and fertilizer and the elimination of cultivation injury to surface feeder roots.

The modern orchard requires elaborate maintenance to produce fruit of marketable quality. As many as a dozen pesticidal sprayings may be required during the growing season. Seasonal pruning is needed to regulate shape and contain the trees for convenient harvesting and maximum fruit production. Fertilization, and perhaps irrigation, must match local soil requirements.

Cultivars must be properly selected to insure pollination and fruit set (although this can be implemented with chemical sprays). Many fruits are self-sterile and must cross-pollinate, sometimes with an entirely different variety. This entails appropriately spaced interplantings and, possibly, introduction of pollinating insects (such as honeybees). Moreover, some fruits set well only in alternate years. Cross-pollination is generally necessary for a good crop of apples; many pears, sweet cherries, and plums are likewise self-unfruitful. Peaches, nectarines, apricots, and sour cherries self-pollinate adequately.

Growth-regulating materials have found great application in modern orcharding. In temperate fruit crops such as apple, pear, and peaches, **chemical thinning,** is commonly practiced to adjust crop load. This is necessary in order to control fruit size and prevent alternate bearing and winter injury—problems that arise from overbearing. Chemical thinning materials may be either caustic substances (such as dinitro compounds) that burn off the blossoms or auxin derivitives (particularly naphthalene-actic acid) that bring about embryo abortion. Auxins are also used to prevent premature fruit drop in the fall. In grape production gibberellins have been used to induce seedlessness and particularly to increase fruit set and fruit size in seedless cultivars. The chemical SADH (succinic-2,2-dimethylhydrazide) has interesting effects—dwarfing, the induction of earlier production, increased fruit color, and improved storage behavior.

Added to the orchardist's tasks may be such things as nematode control and protection against freezing (citrus orchards in Florida), timing of harvest (bananas for keeping during shipment, rubber for yield), and perhaps above all the securing of seasonal hand labor for harvesting. Fruits that can be shaken from a tree to fall on nets or canvas without damage are increasingly being mechanically harvested (with padded devices that vibrate the tree or bush).

Floodplain irrigation

Most remunerative of all cultural systems in the tropics and subtropics is floodplain cropping with irrigation, a system that had its beginnings in Tehuacan and Mesopotamia. Floodplain irrigation supports dense, sedentary populations along the floodplain of the Nile and throughout the Orient, where individualized paddies are created by terracing and intensive land management. Practical skills and much hand labor are needed, with the result that by and large floodplain irrigation has been most successful where the people are docile, intelligent, and dexterous. The system has been singularly successful in Southeast Asia.

Burning-grazing

Many parts of the world are in grass, and its utilization for grazing (originally by wild animals, later by livestock) can support a moderately concentrated population. Where climate is seasonally too dry for forest or parkland, prairie persists, but even there it can be argued that periodic burning is required to keep brush and grass in balance. This is particularly true in tropical savannas or in man-made clearings, which are burned annually to prevent the incursion of forest. Often savannas are burned merely to get rid of lignified old leafage, which is unpalatable to stock and contains little protein; the animals prefer tender new sprouts. Dispute is endless over whether such burning causes long-term deterioration, which is likely if burning is accompanied by overgrazing. But certainly burning is essential in some regions to prevent reversion to forest. Tough grasses that have invaded tropical clearings are often intentionally destroyed by controlled burning just as new growth starts; food reserves in underground parts are low then, and the grasses are easily weakened.

Irrigated desert

Although stream borders in desert valleys have always been among the world's most productive agricultural locales, it remained for modern technology to make practicable the massive, intensive, highly remunerative farming seen today in such naturally unlikely spots as the interior valleys of California, southern Arizona, and the high plains from Colorado into Texas. Irrigation projects that transport water for many miles are the prime means for turning "worthless" desert into productive agricultural land. Well-managed and scientifically irrigated holdings in California and Arizona can produce twice as much cotton as the famed cotton lands of the Southeast. The same is true with forages such as alfalfa, in comparison to yields in the Northeast under natural rainfall (and shorter season). Where irrigation is not possible through diversion from streams, increased drilling capabilities and huge pumps enable well-capitalized farmers to draw groundwater from as much as 2,000 feet below the surface (the water table, unfortunately, is thereby being lowered in many parts of the country at an alarming rate). **Trickle irrigation,** in which water emission is restricted to a driplike application, is finding increasing approval as a practice that is more conservative of water than is flood irrigation or sprinkling (apparently it also permits use of somewhat more saline water, which is typically found in arid regions).

In arid regions, naturally inhospitable to man, irrigation permits very remunerative yields under the highly technical and intensive farming carried on. Some of the world's most specialized large-scale agricultural operations have been established on level desert. Desert soils may not be as good as prairie soils, but their deficiencies can be compensated for by fertilization and proper irrigation. How involved these operations can become is exemplified by the efforts of the Kern County

Land Company in California to trap seasonal runoff for underground storage, to be pumped for irrigation later. Soil treatments that will maximize surface infiltration are assiduously sought, and ponding fields are built merely as insoak reservoirs to get mountain stream runoff into the water table. Almost no water runs "uselessly" to the sea any more in southern California.

The huge outlays generally required for large-scale irrigation farming are reminiscent of the management practices of plantation agriculture in the tropics. Small, self-sufficient home farms are seldom seen. Many of the large land holdings are owned by absentees and managed by hired professionals. Even with private operations the "farmer" acts more like a "businessman" (not uncommonly commuting hither and yon in his private airplane) than the overalled cultivator of the soil so dear to the American political ethic. These owners and managers are just as interested in credit, national money policies, the stock market, and the Washington political scene as are the executives of industrial corporations. In groups or associations they may even control marketing outlets and have connections overseas. This is probably but a foretaste of what most agriculture will be like eventually in advanced countries.

Diversified general farming

The farming tradition familiar to western civilization of temperate climates—the system of diversified general farming—embraces most of the temperate-zone crops, planted where well adapted and economically renumerative, by the landholders themselves. This sort of cropping developed in climates where rainfall is sufficient to maintain forest. Until quite recently Europe and North America have been particularly dependent upon farming of what was formerly forest land. By and large a very flexible program is possible. Though the soils are not fertile by nature, they are quite responsive to fertilization practices and proper handling. Much of the best cotton, soybean, corn and dairy-pasture lands, so important in the eastern United States, are of this type.

The colonist regarded the virgin temperate forest as somewhat inimical to his economic well-being.

Trees were to be utilized, of course, for timber, fuel, furniture, and so on. But the forest seemed limitless, and the main task at hand was to cut down the trees and grub out the stumps, so that land could be cultivated and cropped. In North America even the Indians had made a few clearings in the forest, where they planted corn and other crops in hills. Gradually the hand felling of trees and manual hoeing of soil, which were characteristic of the early Virginia colonies, changed to soil cultivation with moldboard plows drawn by oxen or horses, and then to today's use of elaborate farm machinery. This is at the heart of American economic progress, and will be discussed in more detail under "Mechanization."

General farming allows great self-sufficiency. A New World colonist or a European peasant could provide himself with life's chief essentials, and perhaps a bit extra for barter. It was a system that encouraged individualism, so dear to the American political image. Gradually farming of the humid temperate lands has given way to greater specialization, although diversification and crop rotation are not yet things of the past. Some once-forested lands prove particularly suitable for vineyards or orchards —especially in equable climates near bodies of water, such as in western Michigan and from northern Ohio to upstate New York. Cotton underwrote a one-crop economy in the upper South a century ago, where, after a good many fields were nearly ruined, diversification eventually returned (tobacco, beans, peanuts, corn, small grains, orchards, hay and forage). Once-forested lands in Ohio and Indiana are today among the best in the Corn Belt, where corn is rotated with soybeans and occasionally small grain or pasture. It is hard to imagine that the state of Ohio, once almost solid forest, has within a century and a half been transformed into a man-made "prairie" with scarcely a virgin stand of trees to be found anywhere in its tens-of-thousands of square miles.

Prairie

The temperate prairies, the modern world's breadbasket, were among the last of the major vegetational complexes to be harnessed. The harsh, windy environment provided little in the way of fuel and shelter. Until sturdy, powered implements became available the plowing down of thick sod was a back-breaking chore. It is not surprising that in the colonization of America, cultural systems patterned after those of the Old World carried settlement easily to the boundaries of the prairies, where migration paused until new techniques and attitudes (built around wheat cropping) could replace the diversified farming tradition long familiar in the East.

Prairie life breeds cooperation rather than individualism. The very nature of the environment calls for large acreages devoted to but one or few crops adapted there. Neighbors, though more distant, are also more cherished—the sole social outlet in a culturally barren environment. And the lifeline for existence itself is a tenuous road or railroad leading to industrialized parts of the world, from which come the tools and supplies not locally procurable. With an existence that is anything but self-sufficient, being dependent upon distant markets, there is real concern about other parts of the world, and a cooperative rather than isolationist political outlook develops.

Modern transportation and powered equipment have completed destruction of the native prairie even more quickly than the eastern forests were felled. Only fits and snatches of original prairie remain, although it once covered hundreds of thousands of square miles in the United States, from Illinois to the Rocky Mountains and from Texas to Canada. Much of this has been replanted to domesticated grasses and legumes for hay and forage, and even more to cultivated crops such as wheat and grain sorghum. Unfortunately, at times prairie lands that were too dry for cropping have been stripped of their protective sod cover, losing soil to dust storms. Other prairie has been so overgrazed that grassland has become degraded to brush and desert.

Dryland farming

The western reaches of grassland in the United States embrace the drier short-grass prairie, where a system of alternating crop with fallow permits acceptable yields under minimum rainfall. Alternate strips, or fields, are sowed (usually to wheat) and left unplanted (but cultivated to control weeds). Such rain as does fall soaks into the fallow ground and is not dissipated by plant growth. Moisture "stored" in this fashion for one year is usually sufficient to sustain a crop the second, after which the process is repeated. Dryland farming is man's adaptation to the severest part of the prairie environment, and requires a hardy breed of agriculurist to stand up to the adversity of dry cycles for the sake of bounty in relatively rainy ones. Soils are good, and dryland farming is increasingly giving way to irrigation as facilities become available.

Forest management

Forests have only recently become regarded as crop rather than as natural bounty to be freely taken. With much of the world's virgin timber cut, there is now concern not only for quantity but quality also. Save for a few still wild parts of the world, such as the Amazon valley and the remote subarctic regions, forests are today more or less managed. In some cases their management is just as intensive as that of cultivated crops, and includes specialized techniques for planting and replanting as well as harvesting.

Compared to the systems used for annual crops, however, and even orchards, forest cropping systems are still in their developmental stages. Over most of the tropics and much of the temperate zone, management is minimal. Shifting agriculture continues, and in many places trees are cut for timber, pulp, and firewood much as by early settlers. But technically advanced societies have come to realize that forest is not unlimited. There is little hope that improved forestry practices will be imposed over most of the tropical woodlands any time soon; rather, high-grading (removal of only the best trees of the best species, leaving inferior types to reproduce) will continue, as will wanton deforestation for fuel and new cropland. But more intelligent use of temperate forests is at hand, both with regard to their use as a source of raw materials (Fig. 15-3) and as an esthetic and recreational resource.

Forest management has many problems that are peculiar to it. Most obvious is the long time required for tree growth. Up to the present, forestry has depended upon old growth timber, some of it thousands of years old. New growth generally takes 15 to 150 years to reach merchantable size for timber and significantly less time for pulp, depending upon the part of the world. It is apparent that any mistakes made in selecting new planting stock, or in getting it started, will be perpetuated for decades. Besides causing a loss of time, mistakes of this kind cannot be corrected without the additional cost of cutting useless stock and replanting again.

In many parts of the world, forestry relies for its seed mainly upon native trees. It is assumed that natural selection will have provided a suitable genotype for any given area. A specialized forest seed industry has now been developed, whereby seed from years of abundance is properly stored for use in leaner times. In most temperate countries where seedling culture is practical, a 5- to 7-year supply of seeds is always on hand.

Ecological studies are identifying more clearly the environmental factors that favor individual forest species. The modern forester has some information on the silvicultural practices favoring pine, mixed conifers, hardwoods, and so on. It is apparent that temperature, thermoperiodicity, light, moisture, and soil condition all interact to determine site suitability. Of course, these are in turn affected by elevation, slope, and so on. Temperate-forest yields (stem wood volume) are about 10,000 to 13,000 cubic feet per acre per year, the yield decreasing with increasing dryness or elevation. Evergreen forest is more productive at elevations of 5,000 feet or higher or at latitudes north of about 45°. An example of the change in (and overlap of) dominant forest species as the environment alters is shown in Figure 15-4. Ponderosa pine and Douglas-fir, for example, are adapted to the warmer and drier forest belts of the Pacific Northwest, and are gradually replaced by other species as the climate becomes colder and wetter with increasing elevation.

Armed with this sort of information, the forest managers can base their decisions on measurements rather than guess work. Still, there are many aspects of tree growth about which little information has accumulated. Compared to annual crops, trees are peculiar in the tremendous accumulation of

FIGURE 15-3. *A carefully managed pine planting, characteristic of sustained-yield operations in the southeastern United States.* [*Courtesy U.S. Forest Service.*]

materials (energy) in their bulk. Just how might the trapped chemical elements be most efficiently replaced? Recirculation through leaf fall is very much related to temperature. Tests in the great Smoky

292

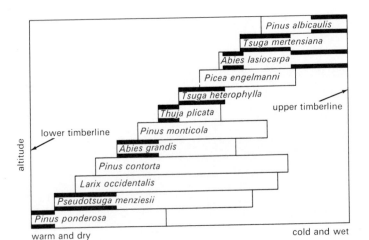

block cutting
Trees like the Douglas-fir need full sunlight for proper growth. Also, their seeds are carried a good distance by the wind.

FIGURE 15-4. *The usual order in which coniferous trees are encountered with increasing altitude in eastern Washington and northern Idaho. The horizontal bars designate upper and lower limits of the species relative to the climatic gradient. The heavy lines indicate that portion of a species' altitudinal range in which it can maintain a self-reproducing population in the face of intense competition.* [After R. Daubenmire, "Vegetation: Identification of Typal Communities," Science **151**:291–298, 1966. Copyright © by the American Association for the Advancement of Science.]

seed-tree cutting
Southern pines also need sunlight, and their seeds are carried by the wind. Isolated trees are relatively windfirm.

Mountains show that fallen leaves decompose only about 35 percent in the first year at 5,000 feet, but 46 percent at 1,000 feet. In the tropics recirculation may take but a few weeks or months. Measurements of energy exchange for individual species and stands are very scanty. Whether one species can trap more energy than another or whether uneven-aged stands can trap more energy than even-aged stands is not known. Forestry is still quite a way from completely scientific cropping.

Although the ultimate aim in cropping forests is the same as with annual crops—profitable return on investment—the business outlook has to be adjusted to long-term objectives (Fig. 15-5). It is impossible to see fifty years into the future, which is the time needed in some areas to mature a new tree crop. A logical approach under these circumstances is to cut one-fiftieth of the holdings yearly; after fifty years the first parcel would be ready for

selective cutting
Young red spruce will grow in the shade of larger trees. Thus trees of all ages and sizes are found growing side by side.

For these reasons the accepted way of harvesting Douglas-fir is to cut all marketable trees in a block of 100 acres or less. Ample standing timber is left between blocks.

Seeds blow in from the surrounding forest and new trees appear. The new stand is thinned periodically to provide growing space and to remove undesirable trees.

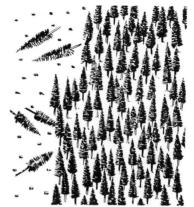

In about 40 years the first cut-over block is mature enough to provide seed for the neighboring block. After a few more decades the first block can be harvested again.

It is therefore possible to remove all trees from a large area except for 4 or 5 per acre, which act as seed trees. These remain until the new seedlings are established.

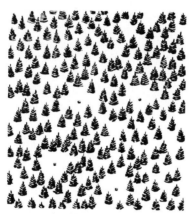

When the seedlings have grown enough so that they are reasonably safe from fire—5 to 10 years—the seed trees can be removed in their turn.

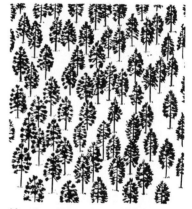

After about 20 years the new stand of pine should be thinned to prevent crowding. Under favorable conditions the pines can be harvested 30 years after original cutting.

In harvesting red spruce the most mature and marketable trees are selected for cutting. (In this picture they are the trees shown slightly darker than the others.)

After the marketable trees have been removed their neighbors provide seed for the new trees that will replace them. In the meantime other trees are maturing.

In 10 to 20 years these trees are ready to be harvested in their turn. If carefully harvested, a forest of red spruce can provide a continuous supply of marketable trees.

FIGURE 15-5. *Harvesting systems in forest management.* [*After St. Regis Paper Co.*]

harvesting again. This is the principle behind **sustained yield.** It is also possible to manage forest made up of trees of diverse ages on a sustained-yield basis by cutting only a portion of the volume during each cutting cycle. In practice, 100 percent sustained yield may be difficult to achieve in any but the largest holdings, although a practicable ideal to strive for in temperate forests, where essentially solid stands of one or a few species occur. In the tropics, where there is a tremendous diversity of species, ecological complications are more likely to occur. For one thing, tropical forests do not lend themselves to clear-cutting. Rather, selective cutting is necessary—one tree here, another there—in order to procure a uniform product for economical processing, handling, and marketing.

In many temperate forests clear-cutting of timber over extensive areas has given way to sequential harvesting in smaller blocks. This practice leaves adjacent stands of seed trees, which under favorable circumstances will recolonize the cut area within a few years. The size and treatment for cleared blocks vary with local conditions; whether woodland should be clear-cut or selectively cut involves such considerations as whether regenerating seedlings demand full sunlight or are of a type that prosper under partial shade, and whether slope is such as to cause material soil erosion when all plant cover is removed. In areas where insufficient seed trees persist or are of types that do not recolonize naturally, replanting of the forest is necessary. Various degrees of success have been obtained in reforesting. Where huge areas have been devastated, such as in the Tillamook burn of Oregon, the least advantageous sites have regenerated poorly over a period of nearly 40 years. Neither direct seeding nor planting has been very successful in many parts of the burn, perhaps because soils exposed by the forest fire have been altered and eroded, and perhaps for other ecological causes.

Replanting cut-over forest is now commonplace, and in the future will largely replace the hit-or-miss and time-consuming practice of natural regeneration. Tree cropping takes advantage of mechanization, too, although not to the same extent as in field cropping. Some seed is scattered by airplane and by helicopter, usually about one-half pound per acre with Douglas-fir in the Northwest. Mechanically pulled devices are used for planting nursery seedlings on fairly level land with few rocks. In rough, rocky country hand planting is still necessary, although expensive. Seedlings such as Douglas-fir can be grown in the nursery for a penny or two each, but hand planting them in forest sites may cost as much as $160 per acre, quite a sizable investment.

Tree seed is still collected mostly from the wild. The difficulties are formidable, what with most pine and fir cones growing at the tips of branches 50 to 250 feet above ground. Much seed is gathered from felled trees after logging. Sometimes cones cut by squirrels are appropriated; a bushel or more may be cached by a squirrel in one hideaway. Such a system allows little directed genetic selection (although weak seedlings are eliminated in the nursery) and no planned cross-breeding. Tree-breeding possibilities should be tremendous once suitable techniques are more widely developed.

The tree seed industry is just beginning to come into its own. Cones of coniferous trees are oven-dried and tumbled to free the seed. A temperature of about 49°C (120°F) is best for some pines, whereas 32°C (90°F) is better for fir. Seed is de-winged by brushing or friction, and the residues are blown away. It is typically stored at subfreezing temperatures. At these low temperatures viability is retained for 5 to 10 years. Most tree seeds must be cold-stratified for good germination. Seed planted into the nursery in autumn is cold-stratified naturally by winter weather. Seed for spring planting is typically soaked in water most of a day and then held at temperatures slightly above freezing for a number of weeks. It will then germinate in 7 to 10 days. Nursery seedlings are tended in much the same way as orchard trees; they are watered, fertilized, weeded, and usually cultivated for one to four years (Fig. 15-6). In the southern United States seedlings are sufficiently mature for field planting in one year.

Of course, direct seeding is much simpler and more economical than nursery operations. It is likely, too, that the root system of a seedling sprouting in place is more adequate than one that has been transplanted once or twice. Direct seeding is generally the only practicable means of replanting large and inaccessible areas; it is also the only

FIGURE 15-6. *Slash pine seed beds in Georgia. This nursery grows about 30 million seedlings per year. Increased selection of germplasm has a tremendous impact upon eventual production.* [*Courtesy U.S. Forest Service.*]

means of replanting species that do not transplant well, such as longleaf pine. The comparatively heavy seeds of hardwoods are often more successfully seeded directly than are softwoods (conifers). Potential loss is great with direct seeding. Birds and small mammals may eat the seed, and weeds may smother the seedling. Rodents are sometimes poisoned with treated grain prior to seeding. Where terrain permits the use of tillage equipment, seed is liable to be buried. Coatings are applied to the seed (pelleted seed) to repel animals, insects, and birds.

A helicopter or aircraft can seed as much as 3,000 acres in a single day. Broadcast seeding from the ground, with hand or tractor-mounted seeders, is tedious but generally more precise and quite satisfactory for small tracts. Seeding from air or ground is more effective if the soil has been mechanically prepared beforehand to the extent that the terrain allows. Mechanized row drilling is even possible on favorable sites.

Large forest products companies, today accounting for most of the pulp and lumber produced in the United States, are as elaborately organized as are the gigantic field crop farms discussed earlier. Virtually all of the large lumber and plywood producers now use computers for linear programming in operations analysis to streamline their planning, production, and sales. Lumber itself may be graded electronically. Raw material—logs—is thus more efficiently turned into wood products. Logs are no longer inexpensive; they must be fully utilized for profitable operations. The trend is to use all types of timber cut, and programming can indicate quickly whether a certain log of a certain type at a certain time should be used for lumber or plywood. Operations research may suggest that lower-grade logs are satisfactory for plywood sheeting, thus freeing higher-grade logs for premium sale for other uses; or that spruce can be substituted for white fir in interior plys, thus saving costs (largely because spruce has a shorter drying time than fir and because its use relieves a bottleneck at the kiln and speeds the product to market). Or it may take into account glue-spreading capacity and market demand to determine whether plywood should be

3-ply or 5-ply. Computerized predictions can be made for markets months and years ahead, to better adjust product mix to needs. One of these days, no doubt, the processing of forest products will be scientifically controlled from raw material to finished product, as in the processing of farm products. The small, inefficient forestry operation is already as obsolete as the small, inefficient farm.

MODERN CROPPING INNOVATIONS

Mechanization

Mechanization has brought significant new opportunities to plant cropping, especially for seeded crops in temperate climates. Mechanical harvesting is now widespread, and is employed not only for grains but for tree and bush fruits, root crops, leafy vegetables and even crops as perishable as ripe tomatoes. The efficiency in terms of human labor that mechanization brings is nowhere more evident than in North America, where today one farmer produces sufficient food for nearly 50 non-farming people—a situation in striking contrast to the nearly universal participation in farming found in primitive societies (and in the United States during the colonial period). Mechanization, the biological engineering that provides tailormade cultivars, and all of the accouterments for crop care, not only have had a quantitative effect but have on the whole improved quality as well.

The trend of industrialized societies is to "use more capital, less labor." The pressure of labor costs encourages extensive and often rapid mechanization. Since it often proves impossible to hire efficient hand labor in the United States at any wage, much less low wages, it is no wonder that in two years the picking of canning tomatoes in California changed from 100 percent hand picking to 85 percent mechanized picking. Even though a picking machine may cost upwards of $20,000 with its accessories, it still pays for itself in a reasonably short time.

The increase in number of certain farm machines in the two decades between 1940 and 1960 is phenomenal (see Fig. 28-5). There are now nearly a million corn-picking units in operation, most of them put into use since World War II. Sprinkler irrigation has risen fiftyfold in the same interval, with only 0.1 million acres irrigated in 1945 compared to more than 5 million acres in 1965. Twelve million horses and mules gave way to 5 million tractors within one generation in the United States, and the trend is for tractors to get ever bigger and more powerful. Some units are essentially a self-propelled chassis with power take-offs to which special attachments are hitched for a wide variety of operations. There are cultivators that barely scuff the topsoil, others that chisel deep into the subsoil; rotary hoes that chop weeds, and tractors that pull four or more plows; spray machines that arch over tall crops, and others that carry spray booms a hundred feet wide.

Harvesting machines are able to determine the readiness of head lettuce for picking, to cut asparagus under photoelectric control, to shake tree fruit onto cushioned catchers, to vacuum-sweep pecan and tung fruits from the ground, to dig potatoes and sort the tubers while traveling at speeds of more than 3 kilometers per hour, to collect pineapples on conveyers that are many meters wide, and so on. A multitude of devices for separation, cleaning, and sorting of agricultural products minimize the drudgery of yesteryear. Many crops are prepared for harvest with chemical sprays to control ripeness and abscission. Indeed, it seems that some piece of equipment has been invented or adapted for almost every once-laborious agricultural operation. The situation is nicely summarized in Figure 15-7.

The effects wrought by this mechanical revolution are truly astounding. The greater output per man-hour was accomplished by substituting tractors for horses and by putting into use nearly a million each of pickup balers, corn picker-shellers, and grain combines (Fig. 15-8). Cotton is now picked mechanically throughout most of the cotton belt. Airplanes are used to sow rice and to spread pesticides. There are mechanical tobacco harvesters, motors to lift hay and move grain. Machines limb, top, cut, and handle entire trees. Gadgets shake fruit and nuts from trees, prevent frost, grind and blend feeds, dry grain and hay. There is equipment to place just the right amount of fertilizer in just the right soil location, to spray

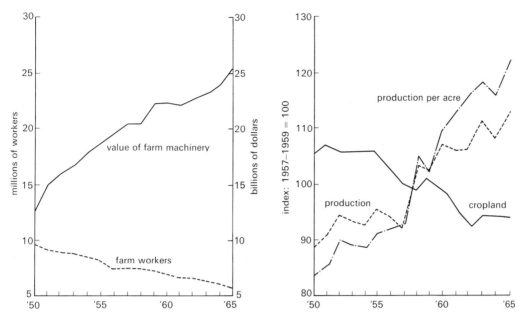

FIGURE 15-7. *Increases in the ability to produce in the United States have been achieved by fewer farm workers and more equipment.*

the exact row width with herbicide, to meter seed exactly 12 rows at a pass. For a long time, metering was accomplished by slotted planter plates that accepted seed of a particular size and shape, thus necessitating uniform seed. Now plateless planters can utilize the more economical seed not standardized for size, and are less likely to crack the seed coats. Furrow openers can work through crop residues left on the soil surface to provide an organic soil mulch, with precise depth control even at a speed of 8 kilometers per hour. No wonder corn yields have risen from about 34 bushels per acre in 1945 to more than 100 bushels (on well-run farms), with yields of 200 bushels in sight (Fig. 15-9). It took about 350 man-hours to produce 100 bushels of corn in 1800, only 34 hours by 1950, $3\frac{1}{2}$ hours in 1960, and less than 1 hour in 1966. Tobacco yields have doubled since 1945, wheat yields gone up 80 percent, soybeans 30 percent. Today's agribusiness uses capital and brains to increase production and profit.

Such a degree of mechanization is possible only under a fortunate combination of industrial ca-pacity, capital, and technical knowledge. It could hardly be operative today in much of Latin America, Africa, or the heavily populated parts of the Orient (where human labor is often much less expensive than machines). It is ironic that acceptance of agricultural innovation is greatest and fastest where agriculture is already profitable and progressive, though the world will be in sore need of whatever increases in productivity that mechanization might be capable of bringing in less advanced parts of the world. Mechanization has tended to accentuate the already appreciable disparity between advanced and primitive crop production practices. It is most adaptable to large farms in sophisticated societies. Even within the United States it is more easily adapted to the prosperous farms of Iowa than to the hills of Appalachia (for a variety of reasons). The fruits of mechanization have been little enjoyed in most tropical countries; except to a small degree where plantation agriculture is practiced. Even today much of the world's crop land is tediously scratched by oxen pulling a forked log as a plow.

A

B

Agribusiness—the new system

In agriculturally advanced parts of the world, such as Japan, the United States, Canada, and parts of Europe, the simple farmer of yesteryear is fast becoming a manager of crop production, drawing upon chemistry, biology, engineering, economics, and many other disciplines. He runs a rural "factory" producing proteins, fats, and carbohydrates. His goal is the same as that of any businessman—maximize return. Thus in these parts of the world the trend in farm operations will be toward a high-yield system, drawing upon many technical resources. If soybeans are planted, they will be a select variety, for which a special fertilizer is compounded in keeping with local soil needs (as determined by test). Pre-emergence weed control will be provided. Planting machinery will adjust spacing and row width for maximum utilization of sunlight. Chemical growth regulators may even control shape of the plants, prevent lodging, and defoliate plants before harvest. Irrigation will probably be used, perhaps automatically applied as soil moisture reaches a predetermined critical level. Pesticides will be accurately and efficiently applied at exactly the correct season (before diseases or insects get out of hand). All of these activities may very well have to be integrated by business machine data processing, programed by specialists who never work in the fields. As noted previously, this is already almost the case in advanced southwestern agriculture.

Specialized services make such systems click. Much of American agriculture is already beyond general servicing by a county agent, trying to cover all facets with demonstrations, tours, and bulletins. Experts in many fields are needed, operating as a team. Diverse inputs are demanded that must be correct, timely, sophisticated. The system is concerned with the whole sequence of steps dealing with the organism from growth to distribution and marketing. The agribusiness manager must be aware of government programs and financial trends, as well as of new varieties, cultural techniques, and pest hazards. Eventually computers may relate these inputs, and perhaps be used to help anticipate and relate expected rainfall, storage facilities, processing capacity, and to help determine the most favorable distributional pattern.

The less well developed parts of the world cannot support such a sophisticated agribusiness, and in heavily populated lands mechanization to relieve labor (as opposed to increasing production) is hardly the answer. What many parts of Africa, Asia, and Latin America need at this stage are simple, practical agricultural techniques developed through local research, not involved investigations or computer controls. Agriculturalists from advanced countries may provide guidance, but should not assume that techniques developed for

FIGURE 15-8. *Mechanized grain harvesting. The invention of the combine, a unitized mechanism for harvesting grain and freeing it of chaff and detritus, has had a tremendous impact on agriculture. A single machine substitutes for hand-cutting, shocking, flailing (threshing), winnowing, and associated procedures of primitive agriculture. (A) In the self-propelled wheat combine an adjustable reel holds grain-bearing stalks against the basal cutter bar; the auger guides cut stems to a central beater and eventually to the rasp-bar cylinder that rotates within the open-grate concave (just above front wheels). Ninety percent of the grain separation occurs at the concave; the grain is abraded free from the seed head by the cylinder, and falls through open slots in the concave to conveyer belts below. Additional mechanisms in the upper rear part of the combine further agitate stems and chaff to retrieve remaining grain, the spent straw being issued from the rear of the machine. Screening mechanisms in the lower rear part of the machine clean the grain further by siftings and air blasts before it is elevated to the grain tank (behind operator). The clean grain can be unloaded directly into trucks drawn alongside the combine, through the unloading auger (behind operator). (B) The corn combine, similar in principle, has a gathering mechanism designed to accommodate rows of corn. Ears are freed from stems by a snap-bar mechanism (instead of cutting the whole stalk). The rasp-bar cylinder and concave are designed to shell the corn from the cob. Grain is retrieved and cleaned as in the wheat combine. [Courtesy John Deere Co.]*

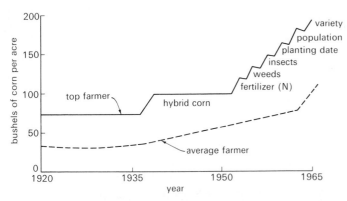

FIGURE 15-9. *Corn-yield trends in Illinois. By successively incorporating improved cultural practices in his cropping system the farmer achieves dramatic increases in yield in relatively short periods of time. He can be expected to adopt any technological improvement coming to his attention if it provides even small economic return or increased efficiency.* [Courtesy J. W. Pendelton.]

the temperatue zone can be applied elsewhere. Programs for underdeveloped areas are often impractical, as was the ambitious attempt to grow peanuts in what were once British colonies in Africa.

In agriculturally advanced regions the trend is toward fewer but larger farms. Most production may eventually come from incorporated agribusinesses rather than individual holdings too poorly capitalized to operate efficiently. It has been predicted that as few as 500,000 farmers could provide the necessary production for the United States. The trend is unmistakable: from 1954 to 1964 the size of the average farm increased from 251 acres to 356, each now representing an average investment of more than $50,000. The number of farms decreased from 5 million to 3 million, of which $1\frac{1}{2}$ million provide 90 percent of the total output. During that decade $1\frac{1}{4}$ million farmers left the farm, and total farm acreage decreased about 3 percent; yet crop production increased remarkably. The farmers themselves now work less strenuously and put in fewer hours.

Agribusiness has immense social overtones. For example, migrant workers work seasonally and may be unemployed in the off-season, causing grave social problems. And what of the political consequences, as fewer and fewer votes come from the farm, more and more from the city? Farm programs will no doubt change, and the interests of the non-farmer will be given greater consideration.

Perhaps as the world-wide demand for food increases, the "parity" and "support prices" of various mid-century agricultural acts in the United States will become a thing of the past. Labor laws, involving such questions as minimum wage, are almost sure to be extended to agribusiness. Promotional efforts by farm groups to solicit support of city dwellers will be intensified, as the 50 billion dollar agribusiness adopts public relations approaches similar to those used by other industries. Yet agriculture remains as fundamental to the well-being of man as ever; indeed, it is likely to be appreciated more in a world that is fast turning hungry. The feeding of dense populations in underdeveloped regions—the "foreign aid" of the future—has political and practical ramifications for all advanced countries.

Present techniques and trends

Improvements in efficiency are eagerly sought in agriculturally advanced parts of the world. Midst the plethora of improvements in product and practice, several trends seem to be shaping up.

Minimum tillage is increasingly practiced, its usefulness varying from soil to soil and from crop to crop (Fig. 15-10). It is a marked departure from the traditional idea that fields are best thoroughly cultivated. Soil is better protected from erosion and is less compacted by machinery. Crop yield is usually equally as good as under conventional cultivation, yet with an economy of operation (at least one plowing or disking is avoided). Equipment now in use can prepare the soil, apply the fertilizer, and plant the seed in one pass over the field. Extreme examples of minimum tillage involve relatively little soil disturbance of any kind. For example, a sod blanket may be killed chemically, left as a mulch, and a row crop planted in slits cut into the dead sod.

FIGURE 15-10. *"Minimum tillage" implies reduced soil cultivation and separate planting operations. Here wheel-track planting creates two distinct soil zones, the row zone and the interrow zone. The row zone is smooth and firm, providing good soil-to-seed contact for fast germination. The interrow zone is left rough to catch and absorb rain faster and reduce weed-seed germination. The planting machine shown here is equipped with dry-fertilizer, insecticide, and liquid-herbicide attachments. [Courtesy John Deere Co.]*

Vegetables and small fruits can sometimes be grown with no postplanting tillage at all. Weeds are controlled entirely by herbicides. Solid plantings of course take much better advantage of space, since cultivating equipment need not pass between rows. With some crops this may facilitate harvesting, and certainly surface feeder roots are not physically disturbed. Furthermore, no reserve weed seed is brought to the surface where it would sprout. Feasibility, of course, depends upon local conditions—for example, whether the soil is of a type that allows adequate water infiltration without cultivation. In any event, several root crops, including potatoes, have been grown in Britain without postplanting tillage. Yields of potatoes, parsnips, carrots, gooseberries, broad beans, and Brussels sprouts have all been close to or slightly better than yields where cultivation is practiced. The cost of herbicide treatment must be weighed against the cost of cultivation for a particular area.

Another trend is toward higher plant populations, to take maximum advantage of sunlight. Increased fertilization and irrigation are needed to accommodate the more densely planted fields. Dense plant populations often shade weeds and protect soil somewhat more adequately from ero-

sion. In some crops nowadays seeds are even being oriented during planting so that leaf position (as in corn, which has leaves in a single plane) is crosswise, not overlapping and partly shading other plants in the row. Dense populations are being used effectively with cotton, grain sorghum, sugar beets, soybeans, and corn just about as fast as new equipment to fit the narrower rows becomes available. The narrower rows may give an additional 20 percent yield if fertility and moisture are adequate.

Corn populations denser than 24,000 plants to the acre are generally impractical; excessive crowding can cause lodging and undeveloped ears. Simple arithmetic shows that more symmetrical spacing is had with more closely spaced rows. On traditional 40-inch rows (the old standard, devised to accommodate the mule), the distance between seeds in the row must be about 6 to 7 inches to achieve a population of a 24,000 plants per acre. But a 30-inch row allows about 8 to 9 inches between plants, with less crowding in the direction of the row. Perhaps even larger yields would be obtained if approximately 16 inches were allowed between rows and plants, possible if herbicides were made to substitute for cultivation.

Where cropland is abundant, and high yield with fewer plants is more important than yield per acre, such as those areas in which acreage controls are in force, plant yield can often be increased by skip-row planting. A row or two of crop is sowed, then a row or two left fallow. The plants then receive light from the sides as well as from above, resulting in more photosynthesis and higher yields per plant. Even greater growth can sometimes be stimulated by placing reflecting boards in the vacant rows, which, like mirrors, direct more sunlight to the sides and base of the plants.

There is increasing interest in **field processing** of many crops to avoid unnecessary haulage and storing. Preliminary treatment of hay crops in the field ("conditioning," a partial chopping to facilitate drying) is increasingly practiced, as is automatic heat-drying of forage (often as mechanically formed pellets) before storage and marketing. Combines pretty well clean up grain in the field, casting straw and other residues back on the land. Corn is increasingly shelled in the field, too. Inexpensive procedures for oxygen-free storage of silage are being developed, as are simplified feed-lot operations.

The following are a few examples of present trends according to crop.

HAY AND FORAGE Field pelleting or wafering is increasing as more dependable and economical machinery becomes available. It is common in dry climates, such as the alfalfa fields of central California. Field conditioning of hay before baling is widespread everywhere. Improvements have been made too in mechanized bale handling and in gathering green chop (see Box 15-1).

CORN The trend toward denser plant populations has already been mentioned. The distance between rows is increasingly being reduced from the conventional 40 inches, and corn is increasingly being harvested with a four-row corn-head on a combine, which can handle 10 bushels per minute. Field shelling is well accepted, now that drying systems and high-moisture storage facilities are more available.

COTTON Narrower rows, and planting to assure an even stand without thinning, are on the increase. Herbicidal weed control, chemical defoliation prior to picking, and machine harvesting are

---

Box 15-1 PLANNING HIGH TDN FOR LIVESTOCK

TDN, "total disgestible nutrients" available to live-stock from feed and forage, is the key to profitable animal husbandry in technically advanced lands. A high TDN is best gained from select cultivars, of course. Yet there are options, and systems must be chosen in view of local conditions such as moisture availability, length of growing season, soil type, slope, and so on. On good land that has been progressively cropped, 5 tons of TDN to the acre can be obtained either from 25 tons of corn silage, 225 bushels of corn, or 10 tons of alfalfa hay. Which route is best? Or if a meadow can be harvested for silage instead of for hay, to yield an additional 25 percent of TDN, does the added expense justify this choice?

On poorer lands, especially on slopes unsuited to cultivation, yields may amount to only $1\frac{1}{2}$ tons of TDN per acre. This type of land may be most efficiently exploited by letting livestock graze rather than by trying to produce the 8 tons of silage or 67 bushels of grain needed to match this TDN yield. Cost of inputs (fertilizer and capitalization of technology) for high yields on poor land may be so great that intensive cropping may not be the most efficient use! Urging a highly responsive forage such as alfalfa (each ton of which removes about 100 pounds of lime, 12 pounds of phosphate, and 45 pounds of potash from the soil) to maximum production on poor soil might require doubling of inputs for an increase of only 50 percent in output! Adding grass to alfalfa may reduce TDN

yields, but may still be the wisest course for protection of soil on slopes!

Obviously, management is a vital factor. Proper timing of fertilization is important, not only to provide maximum yields when needed, but to control the botanical composition of the pasture. Choice of cultivar is significant. For example, white clover may prove to be better on a shallow soil (if moisture is not limiting) than deeper-rooted alfalfa or red clover, which are ordinarily more productive. Where moisture is insufficient for top corn yields, sorghum may be preferable, or, on stony hills perhaps tall fescue forage! In the southern United States, Coastal bermudagrass, which provides a high TDN on good soil, would be inferior to bahiagrass on poor land.

Maximum advantage is not taken of TDN opportunities if a pasture is undergrazed. Yet rotational grazing to allow recovery from overgrazing can increase TDN 50 percent. If corn is harvested in the milk stage it may yield only two-thirds as much energy as silage feed, made from the same corn allowed to mature to the dent stage. Yet, delaying alfalfa harvest can impair nutritive quality so much that 50 percent additional feed grain may be required for equivalent animal gain. But too early (or too frequent) cutting of alfalfa can hurt the stand, increase the weeds!

Obviously, managing modern forage production intelligently requires no little technical knowledge. Profitable farming is not for the untrained.

today standard in efficient cotton farming. Hybrid cotton is under development.

VEGETABLES Root crops such as beets and leaf crops such as lettuce typically are densely planted in rows and then thinned. Not only is there machinery for automatic thinning now, but techniques are being rapidly developed for properly metering seed during planting so that thinning is not necessary. Plastic or resinous mulches are often used to assure a better stand. Systems for mechanical harvesting are used in practically all vegetable crops.

In tomato fields it is now possible to fumigate the soil (with ethylene dibromide) and prefertilize with anhydrous ammonia at the same time. Raised beds are formed and planted, and a pre-emergence herbicide and starter fertilizer drilled in by the same machine. Seeding is metered for a plant population as high as 50,000 plants per acre. During growth insecticide and fungicide sprays control insects, mites, and disease. A mechanical harvester cuts spreading vines at root level, lifts them, and shakes the tomatoes into a bin. The old vine is discarded, while a small crew sorts out green tomatoes and tosses them back onto the field. The harvested fruit is handled in bulk for dispatch to processing facilities.

To some extent high-value greenhouse crops are being fertilized with carbon dioxide (vital for photosynthesis but sometimes limiting because of its scant supply—0.07 percent or less is present in the air) as well as with optimum mineral nutrients.

CONCLUSION

Cropping systems vary with crop, time and location, and level of technology. They have evolved with the development of man and society. This development can be considered to have proceeded in three stages. The first, the most primitive, consisted in **gathering from the wild.** This was the basis of pre-agricultural societies and still exists in many primitive parts of the world. Many of the plant products used in advanced countries are still gathered from the wild: chicle, Brazil nuts, blueberries, gumarabic, and occasionally rubber. The next stage, the **management of natural stands,** was the basis of early orcharding. Much of the world's forest and range land is still handled in this manner, as is the herding of livestock on open range. Finally, the most advanced stage, **cultivation,** involves management of the plant throughout its life cycle. It usually depends on the introduction of exotic species and complex techniques of soil manipulation, environmental control, and alteration of plant growth processes. The intensity of management varies greatly. The scale of operation ranges from extensive outdoor farming to concentrated cropping in small garden-like areas. It extends to the artificial environments found in greenhouse production, and to synthetic factor systems of the pharmaceutical and fermentation industries.

Outdoor cultivation systems usually start with various levels of land and soil preparation. In areas new to agriculture this may involve land clearing, adjusting land slope to prevent erosion and washouts, and the creation of irrigation and drainage systems. It usually involves adding soil amendments (lime and fertilizer) and making various physical changes in soil to create conditions favorable to tender seedlings.

The procedures to increase yield involve a wide spectrum of processes to modify the plant environment and to control plant growth processes. Growth control can be achieved by such ancient techniques as training and pruning, graft combinations, genetic change, or today with growth regulating materials. Regulation of plant physiology is limited only by the extent of our understanding of growth and development—the more complete our knowledge, the more sophisticated the control. It becomes possible to affect not only growth itself but such developmental processes as flowering and fruiting. Advances in this area have been a by-product of basic physiological studies. The field was given great impetus with the discovery of substances that kill plants in very small amounts, modern herbicides. A number of materials are known that dwarf plants, set or remove fruit, and induce rooting and flowering. Many are routinely used in intensive horticultural operations. It is very likely that future changes in cropping patterns will involve chemical manipulation of plant growth processes, and may well usher in revolutionary methods of crop production.

Cropping practices also extend to market preparation of the plant product and to storage and distribution. The path of agricultural products from the growing plant to the ultimate user is complex. **Marketing,** the activities that direct the flow of goods from producer to consumer, often becomes an integral part of cropping systems. This will be discussed further in Part VI.

Selected References

Childers, N. F. *Modern Fruit Science* (5th ed.). Somerset Press, Somerville, N.J., 1973. (A popular text on orchard management with special emphasis on temperate tree fruits.)

Duckham, A. N., and G. B. Masefield. *Farming Systems of the World,* Chatto and Windus, London, 1970. (A worthwhile "textbook" on comparative agriculture: it relates farming systems to ecological, economic and sociological factors, utilizing case studies in various parts of the world).

Eddowes, M. *Crop Technology.* Hutchinson Educational, London, 1971. (A crisp, to-the-point presentation of cropping fundamentals and practices in the United Kingdom.)

The Farm Quarterly. F & W Publishing Corporation, Cincinnati, Ohio. (A prestigious American quarterly on agriculture with articles highlighting new technology and modern practices.)

Hillel, D. (editor). *Optimizing the Soil Physical Environment Towards Greater Crop Yields.* Academic Press, New York, 1972. (The Proceedings from a special invitational symposium held in Israel, dealing with a wide range of topics that influence crop growth and production yields).

Hughes, Harold D., and Edwin R. Henson (revised by Harold D. Hughes, Darrel S. Metcalfe, and Iver J. Johnson). *Crop Production.* Macmillan, New York, 1957. (An elementary text on agronomic crop production in the United States.)

Janick, Jules. *Horticultural Science* (2nd ed.). W. H. Freeman and Company, San Francisco, 1972. (An introduction to the scientific and technological bases of modern horticulture.)

Linton, Ralph. *The Tree of Culture.* Knopf, New York, 1955. (Early civilizations and the rise and fall of empires; agriculture in the context of history.)

Martin, John H., and Warren H. Leonard. *Principles of Field Crop Production.* Macmillan, New York, 1949. (A valuable reference on field crops. The cereal crops are covered more extensively in a more recent book; see Leonard and Martin reference in Chapter 20.)

Meggers, B. J., *Amazonia: Man and Culture in a Counterfeit Paradise.* Aldine-Atherton, Chicago, 1971. (Discusses the relationship between tropical forest environments and human societies, suggesting that the tropical rainforest is an unsuitable environment for development and persistence of complex societies, intense agriculture, and appreciable population density).

Ochse, J. J., M. J. Soule, Jr., M. J. Dijkman, and C. Wehlburg. *Tropical and Subtropical Agriculture* (2 vols.). Macmillan, New York, 1961. (A voluminous work dealing with tropical agriculture. The first volume deals with general cultural practices, the second with crops.)

Smith, David Martyn. *The Practice of Silviculture* (7th ed.). Wiley, New York, 1962. (A widely used textbook on practices and techniques of North American forestry.)

Thomas, William L., Jr. (editor). *Man's Role in Changing the Face of the Earth.* University of Chicago Press, Chicago, 1956. A collection of analyses by eminent authorities relating to the drastic alteration in world ecology as a result of man's progression to a modern state of civilization.)

CHAPTER 16

Crop Nutrition

The control of plant nutrition is one of the foundations of modern agriculture. Although the influence of soil on plant growth is recorded in ancient writings, the role of inorganic minerals was unknown until relatively recently. The first experimental studies on plant nutrition were conducted by a Belgian chemist, J. B. Van Helmont (1577–1644). He planted a willow branch in a known weight of soil and found that after five years growth, the tree increased in weight by 164 pounds while the weight of the soil decreased only two ounces! Van Helmont erroneously (but understandably) assumed that the increase in the plant's weight was all derived from the water added to the soil. Today we know that the two ounces of material removed from the soil were mostly various inorganic minerals that are essential to plant life and growth. This concept, however, was not established until the early nineteenth century, when the Swiss physicist Theodore de Saussure showed that the ashes of plants contained various mineral elements and that plant nutrition involves the uptake of nitrates and mineral elements from the soil. In 1960 Sachs and Knop determined those mineral elements—calcium, magnesium, potassium, and nitrogen, which are required by plants in relatively large quantities. The nutrition of crops is heavily dependent upon the chemical and physical properties of soil, which influence its ability to hold and provide water and nutrients for plant growth.

FACTORS INFLUENCING SOIL FERTILITY

The fertility of a soil is only indirectly related to the chemical composition of its primary inorganic minerals. Often the most important factor is the form in which the nutrients exist in the soil. The availability of nutrients depends upon many factors, among which are the solubility of the nutrients, soil pH, cation-exchange capacity of the soil, soil texture, and the amount of organic matter present.

Plant nutrients

About 90 percent of the entire weight of a living herbaceous plant is water; the remaining 10 percent, the dry matter, consists mainly of three elements: carbon, hydrogen, and oxygen. A small but important fraction of the dry matter consists of other elements that are indispensible for growth. Although soil may supply a large number of minerals, only 13 (in addition to carbon, hydrogen, and oxygen) have been proven to be absolutely necessary—that is, **essential** for higher plant life and growth.

These 13 essential elements are divided into two categories on the basis of the abundance with which they are required by plants. **Major elements** are required in relatively large amounts, usually expressed as parts per hundred (percent) per unit of dry matter. These include nitrogen, phosphorus, potassium, calcium, magnesium, and sulfur. **Minor elements** are required in very small quantities, usually expressed in parts per million (ppm) per unit of dry matter. These include iron, boron, manganese, zinc, molybdenum, copper, and chlorine. The essential elements will be discussed individually later in this chapter.

Some elements are not required by higher plants but are nearly always present. These may be necessary to animals and certain microorganisms. Similarly, some algae and fungi do not require calcium, an essential element for higher plants.

Cobalt is found in higher plants, but apparently serves no function. Yet animals that do not have a supply of cobalt suffer from a vitamin B_{12} deficiency. Blue-green algae have been shown to require cobalt. The addition of one-billionth gram of cobalt to a culture solution in which *Nostoc* was grown increased the yield of dry weight.

Vanadium is apparently necessary for the growth of certain bacteria and algae but is not required by higher plants.

Sodium is not absolutely required but is found in all plant materials in quantities large enough to detect. It is, however, required by animals. Sodium can partially substitute for plant needs otherwise satisfied by potassium.

Iodine is required by animals and has been found either in or associated with organic compounds in plants. It has not, however, been found essential for higher plants.

Fluorine is frequently found in plants, even though it has not been proven to be essential. But it is absolutely necessary for the development of animal skeletons.

Silicon and **aluminum** are found in most plants in comparatively large quantities, yet have never proven to be necessary to sustain life. Silicon, however, apparently imparts structural strength to a great many plants. Scouring rushes are so-called because their silicon content is so great that they were used by early settlers in America for scrubbing pots and pans. The leaves of many tropical plants also have an extremely high content of silicon.

The qualitative determination of which elements plants require has been achieved by growing plants without soil in solutions of inorganic salts. The combination of inorganic salts that supports normal growth makes up a **complete nutrient solution.** Plants may be grown in the solutions alone, if aeration is provided, or in quartz sand "watered" by the solution. Whether a particular element is essential is determined by comparing long-term growth in complete solutions with growth in a solution lacking that element. Special precautions are required, because some elements are required in minute quantities. The container, the sand, the chemicals, the distilled water, and even the atmosphere may contain traces of the element under test.

The compositions of two standard nutrient solutions are shown in Table 16-1. Either solution must be supplemented by other essential elements, such

Table 16-1   COMPOSITION OF TWO NUTRIENT SOLUTIONS.

CONSTITUENTS	APPROXIMATE AMOUNT IN OUNCES PER 25 GALLONS OF WATER
Solution 1	
Potassium phosphate (monobasic)	0.5
Potassium nitrate	2
Calcium nitrate	3
Magnesium sulfate	1.5
Solution 2	
Ammonium phosphate	0.5
Potassium nitrate	2.5
Calcium nitrate	2.5
Magnesium sulfate	1.5

Source: From Hoagland and Arnon, *Calif. Agr. Exp. Sta. Circ.* 347, Berkeley, 1950.

as boron and manganese. Frequently these will be supplied in sufficient quantities by contamination when chemicals that are not of high-grade analytical quality are used.

Foliar analyses are frequently useful in determining nutrient requirements from the mineral composition of plants. Leaf samples collected from plants growing under a wide variety of conditions are analyzed chemically and spectrographically. If other environmental factors are not limiting, growth rate can be related to the content of each element and response curves developed. Assumptions based on foliar analysis must be made with great care until standards have been developed for each crop and each element.

Soil reaction

Soil reaction, the degree of acidity or alkalinity of a soil is determined by the relative number of hydrogen ions (H^+) and hydroxyl ions (OH^-) in the soil solution. A predominance of hydrogen ions makes the soil acid; a predominance of hydroxyl ions makes it alkaline.

Expressing the concentration of these ions in the usual chemical terms results in decimal fractions that are difficult to use. A Danish biochemist,

S. P. L. Sorenson, devised a system that is extensively used today for expressing acidity and alkalinity in terms of pH, the logarithm of the reciprocal of the hydrogen ion concentration in moles per liter.

$$pH = \log \frac{1}{H^+}$$

When pH is below 7, soil reaction is acid; above 7, soil is alkaline; at 7 soil is neutral. Table 16-2 gives the concentration in moles per liter of H^+ and OH^- for pH values of 0 to 14. Note that the molar concentration of H^+ times the molar concentration of OH^- equals a constant of 10^{-14}. Since the scale is logarithmic, a soil with pH 5 has 10 times as many H^+ as soil with a pH of 6. Likewise, a soil of pH 10 has 100 times as many OH^- as a soil with pH 8.

Hydrogen ions exist in soils in many chemical combinations. Those in the soil solution are in balance with those adsorbed on the surfaces of soil particles. The combination of these two sources of hydrogen ions is called the **total acidity** of a soil. Comparatively few hydrogen ions are in solution, as compared with those adsorbed on soil surfaces. Hydrogen ions are released from these surfaces as fast as they are removed from the soil solution, so that the acidity of the solution is not changed greatly. This resistance to change of acidity

Table 16-2 THE CONCENTRATION OF H^+ AND OH^- WITH VARYING pH.

pH	SOIL REACTION			H^+ CONCENTRATION (MOLES/LITER)°	OH^- CONCENTRATION (MOLES/LITER)†	REACTION OF COMMON SUBSTANCES
0				10^{-0}	10^{-14}	
1				10^{-1}	10^{-13}	
2				10^{-2}	10^{-12}	
3		very strong		10^{-3}	10^{-11}	
4		strong		10^{-4}	10^{-10}	lemon juice
5	Acidity	moderate		10^{-5}	10^{-9}	orange juice
6		slight		10^{-6}	10^{-8}	
7		neutral	soil range	10^{-7}	10^{-7}	milk
8		slight		10^{-8}	10^{-6}	pure water
9		moderate		10^{-9}	10^{-5}	sea water
10	Alkalinity	strong		10^{-10}	10^{-4}	soap solution
11		very strong		10^{-11}	10^{-3}	
12				10^{-12}	10^{-2}	
13				10^{-13}	10^{-1}	
14				10^{-14}	10^{-0}	

° 1 mole of H = 1 g
† 1 mole of OH = 17 g

is called **buffering.** Most soil solutions are highly buffered.

Either an abnormally high soil pH (above 9) or a low pH (below 4) is toxic to plant roots, but within this range the direct effect of acidity on most plants is not great. However, the indirect effects of acidity and alkalinity on soils are tremendous. Soil pH affects nutrient availability (Fig. 16-1). For example, iron, manganese, and copper become less available to plants as alkalinity increases. The chlorotic condition (chlorosis) found in some plants grown in high pH is often a result of iron deficiency resulting from the formation of insoluble iron compounds. Soil organisms, especially bacteria are also affected by pH. For example, highly acidic soils are not favorable for the growth of the bacteria that fix atmospheric nitrogen and help rot plant residues.

Crop plants vary in their response to pH (Fig. 16-2). Most crops perform well with a pH between 6.0 and 6.5. Acid-loving plants, many of them in the Ericaceae family (rhododendrons, gardenias, azaleas, camelias, cranberries, blueberries) perform best at low pH (4.5 to 6.0). Soil reaction may be used to control soil disease in crops that prove less sensitive than the pathogen involved. Thus potatoes may be grown at pH 5.2 to control scab, a disease caused by a fungus not adaptive to acid soils. But potatoes perform equally well at a higher pH in "disease-free" soil.

MEASURING SOIL pH   Soil reaction may be determined approximately by the use of chemically treated papers such as litmus papers, which change color in response to different pH. The pH of dilute soil-water systems provides an acceptable approximation of soil acidity. The device usually used for measuring acidity is called a glass electrode pH meter. It has two electrodes, usually glass and calomel, which are placed in a soil and water mixture. The glass electrode consists of a thin glass bulb filled with a KCl solution. An electrical potential, which is proportional to the hydrogen ion concentration of the solution being tested, develops across the glass membrane and is registered on a meter that is read directly in units from 0 to 14. The calomel electrode is a reference for measuring the potential of the glass electrode.

REGULATION OF SOIL REACTION   The reaction of soils is due to the interaction of climate with the soil components. Soil pH depends on whether the clay particles are charged with hydrogen ions or mineral cations (Fig. 16-3). In general, acid soils are common where the precipitation is high enough to leach appreciable amounts of exchangeable bases from the surface layers. The ions may be removed through absorption by plants, leaching, and other processes. The soil colloids become charged with hydrogen ions made continually available by the dissociation of carbonic acid formed from the dissolved carbon dioxide released by living roots in respiration, and from the biological decay of carbohydrates (see p. 52). Alkaline soils generally occur in arid regions where there is a higher degree of cation accumulation.

Soil reaction can be modified. Soil can be made more alkaline; that is, the soil pH can be increased by adding basic cations such as calcium, magnesium, sodium, or potassium. Calcium is the most economical cation for increasing pH and in addition has other beneficial effects. The addition of calcium or compounds of calcium and magnesium to reduce the acidity of the soil is known as **liming.** Although the term "lime" refers specifically to calcium oxide (CaO), it is used in the agricultural sense to include such limestone-derived materials as oxides, hydroxides, carbonates, and silicates of calcium or both calcium and magnesium. Liming

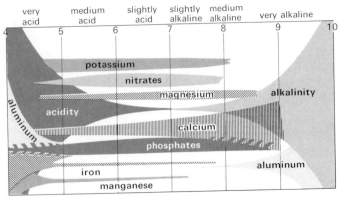

FIGURE 16-1. *Nutrient availability at different pH ranges.* [*Courtesy Virginia Polytechnic Inst.*]

may significantly improve the performance of crops grown in acid soils. The amount of lime required depends upon the degree of pH change desired, the exchange capacity of the soil, the amount of precipitation, and the liming material and its physical form. When adding lime enough must be used to neutralize the hydrogen ions that are released from absorbed colloid surfaces as well as those that are in solution; that is, the buffering effect of the soil must be overcome. Figure 16-4 illustrates the reactions that take place when an acid soil is limed.

Soil may be made more acid by placing hydrogen ions on the soil colloid. This is done by adding substances that tend to produce strong acids in the soil. Some nitrogen fertilizers increase soil acidity, but sulfur is by far the most effective substance. In warm, moist, well-aerated soils, bacterial action converts sulfur to sulfuric acid.

Soil texture has an influence on the ease with which pH can be changed. Clay soils are more difficult to neutralize than silt or sandy soils because they have more surface area for adsorbing, holding, and supplying hydrogen ions.

Organic matter

Organic matter, formed by the decomposition of plant and animal residues, affects the physical condition and fertility of the soil. Once organic matter becomes a part of the soil, it begins to decay and is soon converted to water, carbon dioxide, and a few mineral elements (Fig. 16-5). Some fractions such as lignins, waxes, fats, and some proteinaceous materials resist decomposition and through complex biochemical processes form a dark, noncrystalline, colloidal substance called **humus.** Humus has adsorptive and absorptive properties for nutrients and moisture that are even higher than those of clay. Thus small amounts of humus greatly affect the structural and nutritive properties of soil.

The accumulation of organic matter reaches an equilibrium in undisturbed soil. Because the organic matter of the soil is largely under the control of biologic and climatic forces, one must differentiate between a temporary and a long-term increase in organic matter. Unfortunately, it is easy

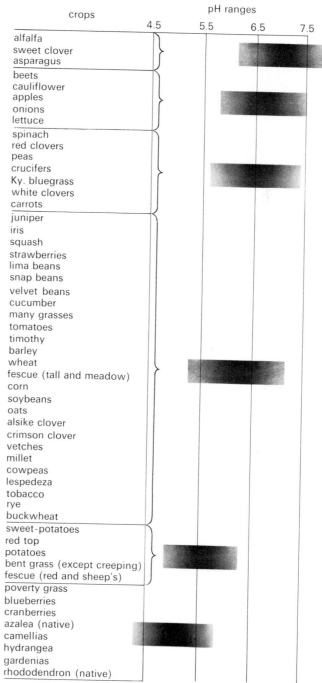

FIGURE 16-2. *Suitable pH range for certain crop plants.* [*Adapted with permission of the publisher from* The Nature and Properties of Soils *by Lyon, Buckman, and Brady, 5th ed.,* © *1952 by the Macmillan Co.*]

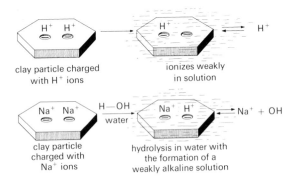

clay particle charged
with H⁺ ions

ionizes weakly
in solution

clay particle
charged with
Na⁺ ions

hydrolysis in water with
the formation of a
weakly alkaline solution

FIGURE 16-3. *The soil reaction depends on whether the clay particles are charged with hydrogen ions or mineral cations.* [From Bonner and Galston, Principles of Plant Physiology. *San Francisco: W. H. Freeman and Company,* © *1952.*]

to decrease the organic matter of a soil, but relatively difficult to increase it. The major loss of organic matter comes about from increased oxidation as a result of cultivation and from crop removal. The rate of decomposition is affected by aeration, temperature, nitrogen availability, soil moisture, soil reaction, and the nature of the decomposing organisms.

Organic matter affects the physical condition of the soil in a number of ways. It improves the water-holding capacity of the soil because it acts as a sponge to absorb and hold water that would otherwise percolate through the soil and become unavailable for plant growth. Organic matter also increases soil aeration in this manner. Organic matter coats the mineral particles and gives a crumbly structure to soils, which makes them easier to cultivate.

Organic matter also supplies nutrients, which are released and made available for plant growth as the organic matter decomposes. The decomposition of plant and animal material is accomplished by enzymatic digestion carried out by soil microorganisms. The decomposition of simple carbohydrates (starches and sugars) is a fairly rapid process and results in the release of carbon dioxide in the soil. Water-soluble proteins are decomposed readily to amino acids and then to available ammonium compounds. Ammonium compounds under the action of certain "nitrifying" bacteria are transformed to nitrates, in which form they are available to plants. In addition to nitrogen, organic matter supplies other essential elements.

Because of its high exchange capacity humus also acts as a cation retainer. Organic matter is negatively charged and attracts positively charged ions to its surface. Thus it adsorbs and retains cations, such as calcium and magnesium, which would be subject to loss by leaching. Organic matter can absorb far more cations per unit weight

FIGURE 16-4. *Cation exchange relations when an acid soil is limed.* [After USDA, *Yearbook of Agriculture, 1957.*]

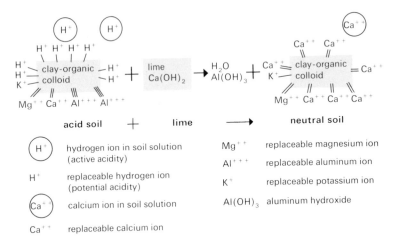

acid soil + lime ⟶ neutral soil

H⁺ hydrogen ion in soil solution (active acidity)

H⁺ replaceable hydrogen ion (potential acidity)

Ca⁺ calcium ion in soil solution

Ca⁺⁺ replaceable calcium ion

Mg⁺⁺ replaceable magnesium ion

Al⁺⁺⁺ replaceable aluminum ion

K⁺ replaceable potassium ion

Al(OH)₃ aluminum hydroxide

than clay (see Table 12-2). Because organic matter absorbs large amounts of cations it greatly increases the buffering capacity of soil.

The content of organic matter in soil may be increased through the extended use of a legume or grass sod. The organic matter produced as a result of root disintegration is protected from excessive oxidation by the constant cover and lack of cultivation. Organic matter can be built up by rotating row crops with legume sods. The plowing under of a growing crop as a **green manure** temporarily increases the organic content of the soil (Fig. 16-6), but this practice cannot be expected to increase the organic matter content for long. The buildup of high populations of microorganisms may actually reduce the net organic matter content of the soil through an unexplained breakdown of the more resistant organic fraction. The rapid breakdown of green manure releases bound nutrients, but minerals that have a high carbon content in relation to their nitrogen content, such as millet or sudan grass, may create a temporary nitrogen shortage because available nitrogen is incorporated into the bodies of decomposing microorganisms. This is called **biological absorption.** It can be

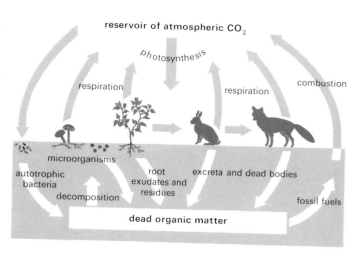

FIGURE 16-5. *The cycle of carbon between organic and inorganic forms can be regulated by cultural manipulations. In natural ecosystems, such as the one illustrated, an imbalance will occur if one of the factors is disturbed.*

avoided either by waiting for the carbohydrates and other organic matter to decompose thoroughly before planting a crop or by adding enough nitro-

FIGURE 16-6. *This buckwheat crop, planted between the orchard trees, will be plowed under at the end of the growing season to provide "green manure" for the trees. [Courtesy USDA.]*

gen to compensate for that which is "absorbed" by the bacteria. Sweet clover and other legumes that have a high nitrogen content in proportion to carbon compounds are rapidly decomposed by microorganisms and release a steady supply of nitrogen. Farm manures are used to supplement soil organic matter, but this is often an expensive procedure. Manure is also valuable for its nutrients.

Fertilizer and fertilization

A fertilizer is a material that supplies nutrients to plants. Fertilizers are usually applied to the soil, but may also be applied through the leaves or stems. Carbon dioxide applied to the air in greenhouses may also be considered as fertilizer.

Fertilizers that supply nitrogen, phosphorus, and potassium are referred to as **complete.** The grade or analysis of fertilizers is the percent by weight of nitrogen (expressed as elemental N), phosphorus expressed as P_2O_5), and potassium (expressed as K_2O). Phosphorus and potassium are not expressed in their elemental form for historical reasons. This may be changed.° An 80-pound bag of a 5-15-30 fertilizer will contain 4 pounds of N, the amount of phosphorus in 12 pounds of P_2O_5, and the amount of potassium in 24 pounds of K_2O. The fertilizer ratio is the analysis expressed in terms of the lowest common denominator. A 10-10-10 analysis has a 1:1:1 ratio. Fertilizers are referred to as "low analysis" when the amount of available nutrients is below 30 percent and "high analysis" when 30 percent or above. Because of lower weight and less handling, high-analysis fertilizers tend to be less expensive per unit of nutrient than low-analysis fertilizers but require more precision in their application. In the United States the analysis of nitrogen, phosphorus, and potassium must be stated on the container, although many fertil-

izers contain, or consist entirely of, other plant nutrients.

Fertilizers may be classified as **natural organics** or **chemical.** In times past, most fertilizers were waste organic materials, such as manure, crop residues, blood and tankage, or fish scraps. Chemical fertilizers, such as ammonium nitrate and superphosphate are synthesized from inorganic minerals. Nitrogen fertilizers can be synthesized using nitrogen directly from the air. Recently, a number of forms of nitrogen-containing organic compounds (urea, cyanamid) have been synthesized. Although they are organic in the chemical sense they are not necessarily derived from living systems. The nutrients in natural organic fertilizers and some synthetic organics undergo gradual chemical transformations to available forms to provide a means of extending the period of nutrient availability. "Urea-forms" are combinations of urea and formaldehyde—the greater the proportion of formaldehyde the lower the solubility of urea and hence the slower the "release." Other types of slow-release materials are composed according to certain specifications in particle size, chemical composition, or solubility. A special slow-release system has recently been developed by encapsulating fertilizer pellets with a plastic coat. The coat acts as a membrane to release the fertilizer salts gradually. These materials are comparatively expensive on a per nutrient basis but have special uses, such as for container-grown nursery stock, because they reduce labor costs.

In order to minimize the cost of fertilization, agricultural practice strives to provide only enough fertilizer to supplement available soil nutrients and raise the level of nutrients to that actually required by the crop. The nutrient requirements of crop plants are determined by correlating plant response with the mineral content of plant tissue and soil. Unfortunately, the total nutrient content of a soil does not usually give a true picture of nutrient availability. The availability of nutrients is related to the cation-exchange capacity of the soil, soil reaction, and organic cycles. Some biological assays of soil have been made by utilizing the response of sensitive plants or microorganisms. Quick chemical tests have been developed, al-

°To change oxide analyses to elements and vice versa, multiply the quantity of the oxide or element by the following conversion factors:

ELEMENT	OXIDE TO ELEMENT	ELEMENT TO OXIDE
Phosphorus	.43	2.33
Potassium	.83	1.20

though these are often not very precise. In many plants severe shortages of certain nutrients produce characteristic **deficiency symptoms,** which can often be used as a diagnostic technique. The good plantsman will not permit nutrient shortages to become this severe.

The relationship between nutrient level and plant performance varies with the species and the nutrient in question. Nevertheless, there is a general response to nutrient level. **Deficiency levels** produce definite symptoms of nutrient starvation. At somewhat higher levels, although deficiency symptoms may not be present, plants may respond by increases in yield and performance. At levels that produce no obvious response to fertilizer the plant may continue to show an increasing level of nutrient absorption termed **luxury consumption.** At abnormally high or **toxic levels** growth is reduced and death may even occur. These levels of course differ with each nutrient. Boron may be toxic to certain plants if tissues contain as little as one part per million.

The level of crop response to fertilizer is related in part to the **productive capacity** of the soil. Crops grown on soils of low productive capacity show a maximum response at a lower level of fertility than those grown on soils of high productive capacity. Productive capacity depends on long-term availability of nutrients and soil condition. Owing to the forces that establish an equilibrium between the soil and the soil solution, optimum fertility is usually not achieved in one quick step. When large amounts of fertilizer are placed on soils of low productive capacity, much of it is wasted by leaching, being tied up in unavailable forms, or by unequal distribution throughout the soil in relation to plant needs. Continued application of fertilizer at the level of optimum plant response appears to increase the productive capacity of the soil, ultimately raising its yield potential.

Fertilizer may be solid, liquid, or gaseous. Most are solids and are applied to the soil. Fertilizer may be dissolved in irrigation water (Fig. 16-7) or be applied to the foliage. Nitrogen may be added to the soil in the form of ammonia gas (NH_3), since it is heavier than air and dissolves quickly in the soil moisture.

FIGURE 16-7. *Metering liquid fertilizer into irrigation water. The flow of water and fertilizer are regulated to get the proper amount of each on each acre. [Courtesy USDA.]*

Proper placement and timing are important factors in fertilization. Plant response, the prevention of injury, and convenience and economy of application must be considered. To be effective, fertilizer must be applied where and when the plant needs it. Single yearly applications may be insufficient for some nutrients and unnecessary for others. Concentrated, highly soluble fertilizers cannot be applied to growing plants, especially when young, because of salt injury. In perennials, or in long-season annuals, nutrient availability may be more efficiently controlled by making repeated applications throughout the season. This is especially true with nitrogen fertilization, as excess amounts may be wasted because of leaching, and time of application is critical to plant response.

There are various methods of placement. **Broadcast application** refers to the scattering of material uniformly over the soil surface, usually before the crop is planted. Surface applications of this kind may not be as effective as a **top dressing**—placement of the fertilizer directly over the growing crop. When plants are subject to injury (fertilizer burn) fertilizer can be put alongside the plants as a **side dressing.** Side-dressed applications of fertilizer are often made along with a cultivation and are thus mixed into the soil. Fertilizer can also be placed in continuous strips between rows (**band placement**) or it can be dropped behind the plow in the bottom of furrows (**plow-sole placement**).

Fertilizer may be applied along with mechanical transplanting either as a band under the plant or dissolved in the supplemental transplanting water as **starter solution.** In **foliar application** the fertilizer is absorbed immediately and plant response may be evident within a day or two, but because there is little residual effect, applications must be more frequent than soil application. Foliar application with nitrogen and minor elements such as manganese and boron has proved practical.

The timing of fertilizer placement may be critical. In perennial fruit crops, such as apple and peach, excess nitrogen is responsible for poor fruit color and soft fruit as well as for undesirable vegetative growth that occurs late in the season and leaves the plant vulnerable to winter injury. Consequently, nitrogen is usually applied only once, early in the spring, in order that any excess will have been used up by late summer. Many plants show a response to spring application of nitrogen because the concentration in cold or wet, poorly drained soils is low due to a lack of nitrification by aerobic bacteria and to excess **denitrification** (the conversion of nitrogen in the form of nitrate to gaseous forms) by anaerobic bacteria. In many crops, nitrogen is applied several times during the growing season because it is subject to leaching and conversion to gaseous or other unavailable forms. Some nutrients that are relatively immobile in the soil, such as phosphorus, are best applied before the crop is sown so that they can be worked into the soil and be evenly distributed by plowing.

ESSENTIAL ELEMENTS IN PLANT NUTRITION

In the ensuing discussion, all the essential elements are treated in turn. Each element is discussed in relation to its function in the plant, deficiency symptoms, soil relationships, and fertilization.

Nitrogen

Nitrogen (N_2) is an inert, odorless, tasteless, colorless gas that makes up about 77.5 percent of the air by weight and about 78 percent by volume. An acre of soil may contain from about 1,000 to 10,000 pounds of nitrogen, whereas the air over it holds about 70 million pounds. Because atmospheric nitrogen cannot be used by plants, it must be "fixed" in a soluble inorganic form before it can be absorbed and assimilated by plant tissues.

FUNCTIONS IN PLANTS Amino acids (building blocks of protein) and many other plant substances, including purines, vitamins, and alkaloids are composed in part of nitrogen. Chlorophyll molecules have one magnesium atom and four nitrogen atoms. Nitrogen accounts for only 1 or 2 percent by dry weight of a plant, but nitrogen-containing compounds make up perhaps 25 percent of the dry weight.

Nitrogen excesses delay maturity and fruiting of some crops by promoting vegetative growth. Severe nitrogen deficiencies can also delay crop maturity and reduce yields. High-nitrogen plants may be less fruitful than low-nitrogen plants.

Nitrogen causes plants to grow rapidly, resulting in a high proportion of succulent, fleshy plant tissue in contrast to stiff fibrous tissue. As a result, herbaceous plants with a high nitrogen content are often blown down (lodged). Freezing injury is also associated with a high nitrogen content of plant material.

DEFICIENCY SYMPTOMS Leaves of nitrogen-deficient plants are usually very light green, but may be yellow or red because when chlorophyll is deficient the color of other pigments shows through. Leaves are also small. Lower leaves are usually the first to show discoloration, and may turn yellow and drop off before the topmost leaves have lost their intense green color. Individual branches may die, and the entire plant is stunted.

SOIL RELATIONSHIPS Nitrogen is found in both organic and inorganic compounds in the soil. Soil nitrogen is most abundant in climatic regions that favor the accumulation of organic matter, such as those of grasslands. Mineralization of nitrogen—the change from an organic to an inorganic form by the decomposition of organic matter—must take place before nitrogen can be absorbed and used by plants.

The forms of nitrogen most usable by plants are the ammonium (NH_4^+) and nitrate (NO_3^-) ions. Ammonium nitrate is widely used as a fertilizer because it is composed of both ions. The nitrite ion (NO_2^-) can be utilized by the plant, but it tends to be unstable and toxic in high amounts. The transformation of nitrogen-containing compounds to available forms is a part of the **nitrogen cycle** (Fig. 16-8). Nitrogen changes from gaseous molecule to protein and back largely through biological pathways.

Nitrogen fixation—the transformation of atmospheric nitrogen into forms available to plants—is accomplished largely by certain species of bacteria (Table 16-3). The most efficient of these are symbiotic; that is, they convert atmospheric nitrogen to combined forms only in association with the roots of certain plants. Symbiotic nitrogen-fixing bacteria are frequently associated with crop plants of the legume family in nodular deformations on their roots (once thought to be a disease condition) (Fig. 16-9). Some nonlegumes, such as alder (*Alnus*), also have nodules. *Rhizobium*, an important genus of bacteria associated with legumes, will continue to fix nitrogen for a time even on severed roots. Seeds of leguminous crops are often "innoculated" prior to planting by coating them with a slurry or a powder containing spores of *Rhizobium*. Nonsymbiotic fixation involves the incorporation of atmospheric nitrogen into organic compounds by free-living bacteria and algae. A genus of bacteria called *Azotobacter* is especially important in this process. Environmental conditions, such as pH, soil moisture content, carbohydrate supply, oxygen, and soil minerals, influence this process.

Atmospheric fixation of nitrogen results when lightning discharges cause nitrogen and oxygen to unite. As much as 50 pounds of elemental nitrogen per acre per year may be added to soil in this manner in regions with high thunderstorm activity, but 2 or 3 pounds is the usual average.

The breakdown of the complex proteins of organic material into simpler forms, such as amino

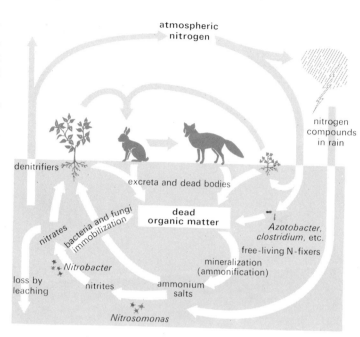

FIGURE 16-8. *The nitrogen cycle.*

acids, is accomplished largely by bacterial action. The nitrogen from this process is only available after the death and disintegration of the bacteria involved in this decaying process. Soil microorganisms have the first call on nutrients. This is especially true for material with a carbon to nitrogen ratio (by weight) greater than 10:1. The subsequent breakdown of amino acids and similar nitrogenous materials to forms of nitrogen available to plants takes place by transformations referred to as **ammonification** and **nitrification**.

The bacteria involved in nitrification are autotrophic and aerobic; that is, they do not require

Table 16-3 NITROGEN-FIXING BACTERIA.

TYPES	GENERA	REQUIREMENTS
Symbiotic	*Rhizobium*	Carbohydrates, inhibited by nitrates and ammonium
Nonsymbiotic		
Anaerobic	*Clostridium*	Carbohydrates, inhibited by nitrates
Aerobic	*Azotobacter*	Calcium, traces of molybdenum

organic nutrition, but do require oxygen. Thus they are greatly affected by soil aeration, temperature, and moisture.

$$\text{Amino acids from protein degradation} \xrightarrow{\text{(many bacteria and other soil organisms)}}$$

$$NH_4^+ \text{ (ammonium ion)}$$

$$NH_4^+ \xrightarrow{\substack{\textit{Nitrosococcus} \\ \textit{Nitrosomonas}}}$$

$$NO_2^- \text{ (nitrite ion)} \xrightarrow{\textit{Nitrobacter}}$$

$$NO_3^- \text{ (nitrate ion)}$$

LOSSES FROM SOIL   Nitrogen is lost through soil erosion, leaching, crop removal, and denitrification. Moreover, nitrate is extremely water soluble, and thus large amounts are lost because of leaching. When rainwater percolates down through the soil, the highly soluble nitrates are carried with it. Losses are greater in coarse soils than in fine soils. Losses are also greater from fallow soils than from those with a vegetative cover; consequently, a winter

"cover crop" may greatly reduce losses due to leaching in humid regions.

Losses due to crop removal are unavoidable. Most agricultural crops remove 25 to 50 pounds of nitrogen per acre during a single growing season, which is 1 to 2 percent of the amount present in various forms.

Denitrification can account for significant losses. Gases are evolved through the bacterial reduction of nitrate to nitrogen gas (N_2), nitrous oxide (N_2O), and nitric oxide (NO). Denitrification is carried out under anaerobic conditions; thus improper aeration can result in the loss of available nitrogen. In compost piles losses of nitrate may be large. For crops grown in wet soils, such as rice, denitrification may be critical. Still other losses are due to the escape of ammonia gas into the atmosphere.

FERTILIZATION   Nitrogen fertilizers may be either organic or chemical. Organic matter has been used for fertilizer for thousands of years. Legume crops have long been used to add nitrogen to soil, often supplemented by tankage and other organic wastes.

Sodium nitrate, imported from Chile, was the first widely used source of inorganic nitrogen. It is still used, but to a smaller extent than in the years before nitrogen could be extracted from air. Nitrogen was first removed from the air for conversion to forms usable as fertilizers on a commercial basis in 1898.

Ammonium sulfate contains 20.5 percent nitrogen. It is called an **acid-forming** fertilizer because the sulfate can be transformed to sulfuric acid. It is usually obtained as a by-product of coke ovens but is also made from ammonia. **Anhydrous ammonia** (dry ammonia) is gaseous in the range of temperatures at which plants grow, but becomes a liquid when cooled and put under pressure. Kept in refrigerated and pressurized tanks until used, it is injected 5 to 6 inches into the soil as a gas and covered immediately to prevent its escape before it becomes fixed. It is also available in solution, in which form it is injected into the soil or added to irrigation water. **Ammonium nitrate** contains 33.5 percent nitrogen and is usually used in the form of pellets. It is one of the most widely used of nitrogen fertilizers. **Ammonium phos-**

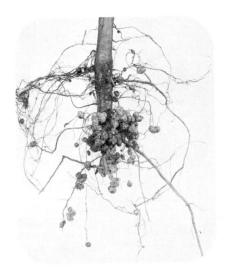

FIGURE 16-9. *Nitrogen-fixing bacteria inhabit the nodules of soybean roots in a symbiotic relationship.* [*Photograph by J. C. Allen & Son.*]

phates, available in both the mono- and diammonium forms, are used for making mixed fertilizers. **Magnesium ammonium phosphate** is convenient to use because it is relatively insoluble and will not "burn" plants, even when used in fairly heavy applications. **Urea,** which is synthesized on a massive scale, is an excellent fertilizer because of its high content of nitrogen (45 percent). Urea can be applied either to the soil or to the foliage.

Phosphorus

Phosphorus, a member of the nitrogen group, reacts vigorously with oxygen in the atmosphere and must be stored under water. Never found free in nature, it was long ago thought to have mystical properties.

FUNCTIONS IN PLANTS Phosphorus is found in plants in many forms. In seeds it is stored as phytin. During germination phosphorus is "mobilized" (converted to a form in which it can be translocated) and becomes a part of the new proteins formed during growth. Nucleoproteins are composed in part of phosphorus. There are many phosphorous compounds involved in metabolic transfer processes; one of the most important is **adenosine triphosphate** (ATP). Surprisingly, only about 0.2 percent of the total dry weight of a plant is phosphorus.

DEFICIENCY SYMPTOMS Phosphorus-deficient plants frequently have purplish leaves, stems, and branches. Maturity is retarded and growth is generally slow. Yields of fruits and seeds are usually poor; fruit often drops prematurely.

Phosphorous deficiency in tobacco results in stunted growth and small, very dark green leaves forming a rosette. Legumes have a high requirement for phosphorus, and seed yields may be greatly reduced under low phosphorus. Oranges from phosphorus-deficient trees have a coarse skin, sour juice, and poor shape.

SOIL RELATIONSHIPS Phosphorus is released into soils mainly by the weathering of rocks that contain the mineral apatite. It occurs naturally as calcium phosphate, iron and aluminum phosphates and as organic phosphorus. A soil may have an abundance of phosphorus and yet have little available for plant nutrition if soil conditions are unfavorable.

The availability of phosphorus to plants is partially a function of soil reaction. In acid soils (pH 4 and less) insoluble iron and aluminum phosphates form, whereas in highly alkaline soils (pH 8.5 and higher) calcium phosphates are equally insoluble.

Drought usually causes a decrease in phosphorus availability that lasts long after the immediate effects of soil desiccation. During the process of dehydration the concentration of solutes in the soil solution increases, and large, complex, mineral crystals form that have less surface area per unit volume and are less readily dissolved than small crystals.

When phosphorus is applied to soil, it seldom moves very far from the point of application because of its numerous complex reactions with clay and organic matter. Consequently, it is usually placed close to the crop row rather than broadcast, because band application reduces surface contact with soil particles. Even though applied in a water-soluble form, phosphorus usually diffuses no more than an inch or so from the point of application in the same form, but is converted to an insoluble compound within 2 or 3 days.

LOSSES FROM SOIL Losses of phosphorus are usually due to the removal of crop plants and to its becoming unavailable in the soil. Since it is a metal, it is not volatile; neither is it readily leached. The phosphorous cycle is illustrated in Figure 16-10.

FERTILIZATION Phosphate rock is treated with sulfuric acid to form **superphosphate** according to the reaction

$$Ca_{10}(PO_4)_6F_2 + 7H_2SO_4 + 3H_2O \rightarrow$$
$$3CaH_4(PO_4)_2H_2O + 7CaSO_4 + 2HF.$$

Superphosphate contains about 8.6 percent available phosphorus. **Triple superphosphate,** which is made by treating phosphatic rock with phosphoric acid, contains about 20 percent elemental phosphorus. Other phosphatic fertilizers are **ammonium phosphates, calcium metaphosphate,** and

318

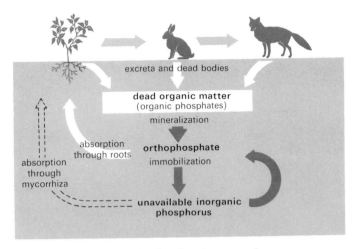

FIGURE 16-10. *The phosphorous cycle.*

finely ground **phosphate rock.** Phosphate rock contains a high percentage of phosphorus, but it is not always readily available to plants.

Because of the high phosphorus requirement of seedlings it is important that adequate levels be made available close to the seed or transplant (Fig. 16-11). Thus phosphorus applications are often banded under the seed. The use of starter solution, water containing a liberal amount of phosphorus (about 1500 ppm), is recommended for many transplants. The phosphorus is supplied in the form of soluble phosphate salts, such as monoammonium or diammonium phosphate and monopotassium phosphate. A popular starter solution contains three pounds of a mixture of diammonium phosphate and monopotassium phosphate (10–52–17 analysis) per 50 gallons of water. When an extensive root system is established plant requirements are met by lower concentrations of phosphorus.

Potassium

Potassium is a soft white metal that can be cut with a knife. It is so reactive that it must be stored under oil or in sealed tubes because it will react with oxygen, water, and carbon dioxide. Potassium in the form of potassium carbonate is commonly called **potash.**

FUNCTIONS IN PLANTS   Potassium was first discovered to be essential for plant growth in 1866.

At that time it was found that oats would not produce flowers unless potassium was present.

Potassium is necessary for the formation of sugars and starches, for the synthesis of proteins, and for cell division. It also neutralizes organic acids and regulates the activity of other mineral nutrients in plants. It activates certain enzymes, helps to adjust water relationships, and promotes the growth of young plants. It improves the rigidity of straw and helps prevent lodging, increases the oil content of fruits grown for oil, contributes to cold hardiness, and is associated with enhanced flavor and color of some fruit and vegetable crops. Even though it is thought to be essential for the formation of carbohydrates and is somehow involved in many other processes, it is never found as a part of organic compounds. About 1 percent of a plant by dry weight is potassium.

DEFICIENCY SYMPTOMS   Plants deficient in potassium produce low yields of crops even before abnormal external colorations or other symptoms can be observed. Potassium-deficient leaves are usually mottled, spotted, or curled, the older leaves showing these symptoms first. Leaves may appear "scorched" or "burned" along the edges and on the tips, with dead areas falling out and leaving ragged edges. Corn plants with insufficient potassium are usually streaked with yellow and yellowish green. Tobacco leaves become mottled and chlorotic, developing dead spots in the centers of mottled areas.

Potassium-deficient plants nearly always have poorly developed root systems and are easily blown over. Carbohydrate synthesis is so greatly impaired that there is not enough to supply both tops and roots. Tops usually have greater priority, partly because they are nearer to the place of synthesis, and the roots get less than they need for optimum growth.

SOIL RELATIONSHIPS   Most potassium in soils is derived from the minerals muscovite, biotite, orthoclase, and microcline. The first two are micas and the second two are feldspars. These minerals are only slightly soluble and are usually found as particles of large size. A small amount of the potassium in soil solutions comes from soluble salts, such as KCl, and is highly available to plants.

Under normal conditions there is a balance between these different forms, and when potassium is added to the soil in the form of soluble salts it is least partially "fixed" and rendered unavailable to plants.

LOSSES FROM SOILS   Much of the potassium applied to soils is removed by crops. Since most of the potassium utilized by plants is in the leaves, crops grown for their foliage, such as forage crops, remove much more than those grown for other parts, such as tubers or fruits. Potassium is not readily leached from most soils. In one study it was found that potassium moved downward through the soil profile no more than 36 inches over a 16-year period. Because of its great mobility in plants, potassium can be leached from living leaves by rain.

Potassium is more likely to be deficient in the upper soil layers than in the lower layers, mainly because it is removed from the upper layers by plants. With increasing depth it becomes more abundant and more uniformly distributed. But in soils that have been highly leached, the surface may contain more than lower depths due to re-entry following leaching from organic matter on the surface. Potassium in the form of potassium carbonate (potash) leaches readily out of wood ashes.

FERTILIZATION   About 95 percent of all potassium is applied in the form of **potassium chloride** (KCl). **Potassium sulfate** (K_2SO_4) and **potash magnesia** ($K_2SO_2 \cdot MgSO_4$) account for most of the remaining 5 percent. Potassium is usually mined as potassium chloride, ground, purified by flotation in a slurry, dried, and screened to the desired size. Cow manure contains about 8 pounds of potassium per ton (wet weight).

Since there is a dynamic balance between the amounts of potassium present in different forms in soils, the choice of fertilizer must be based on the natural ability of soil to supply potassium to crop plants. Soils with a large supply of unweathered potassium minerals usually need little or no additional potassium. Its rate of withdrawal from the exchangeable phase is determined largely by the rate at which it is absorbed by plant roots.

For some plants (oats, wheat, cotton, tomatoes,

FIGURE 16-11. *Proper placement of "starter" fertilizer for corn is made close to the seed.* [*Courtesy Howard Knaus.*]

and alfalfa) sodium seems to serve as a semisubstitute for potassium, producing a crop response when the potassium supply is deficient. Some plants (beets, turnips, and cabbage) respond to sodium even if potassium is abundant. The replacement of potassium by sodium in water-culture experiments has been as great as 80 percent (sugar beets). Most other plants show no response to sodium.

Calcium

Calcium is the most abundant alkaline earth metal. When unexposed to the atmosphere it is white and silvery but rapidly tarnishes when allowed to combine with the oxygen and nitrogen of the air. In the pure state it is malleable and tends to have a crystalline structure.

FUNCTIONS IN PLANTS   Calcium is fixed in cell walls as a calcium salt of the pectic compounds of the middle lamella. It is necessary for cell growth and division in apical meristems and elsewhere. Calcium has complex relationships with other elements. Without calcium some species are unable to assimilate nitrogen.

Calcium is extremely immobile in plants. Thus the older leaves may have large reserves while younger leaves are deficient, although calcium oxalate crystals have been observed to disappear from older leaves during periods of deficiency,

indicating some degree of translocation. Plants contain 0.2 to 0.3 percent calcium by dry weight.

DEFICIENCY SYMPTOMS   Plants deficient in calcium may be green, but have greatly deformed terminal leaves and branches. Although not easily detectable, root growth is impaired. Beans (*Phaseolus* spp.) deficient in calcium turn black. Calcium deficiencies in corn are manifested by gelatinous tips of the innermost leaves, which stick together when they dry. Beet leaves turn pale green around the margins. Dead spots develop near the midrib of potato leaves and gradually form a circular pattern in the middle of the leaf. Roots of many species turn black and die. A number of fruit disorders in apple and pear that often appear in storage have been associated with calcium deficiency. See Figure 16-12 for other symptoms.

SOIL RELATIONSHIPS   Calcium is derived from many minerals, but calcite and dolomite, are the main sources. Calcite and dolomite, the principal mineral components of limestone, readily decompose and are quickly leached from the surface soils of humid regions. In areas where rainfall is less abundant, they accumulate in lower horizons and may form a hard, white calcareous horizon.

Calcium is not only necessary to plants, but has a strong influence on the absorption of ions by soil particles and on the relationships of other elements to plant growth. Lime (calcium oxide, CaO) is

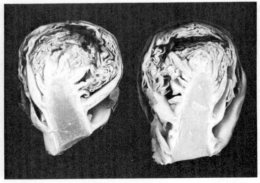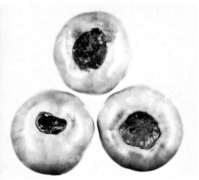

FIGURE 16-12. *Calcium deficiency symptoms: (A) Cavity spot in carrot; (B) internal browning in Brussels sprouts; (C) blossom end rot in tomato fruits.* [*Courtesy D. N. Maynard, University of Massachussets.*]

frequently applied to soil before application of fertilizers containing other elements in order to reduce soil acidity and to reduce phosphorus fixation. When soil acidity is reduced magnesium and molybdenum are changed to forms that are more available to plants, with the result that additions of these elements may be unnecessary. Excess liming, however, can cause problems, because manganese and iron are less soluble at higher pH values. Calcium promotes good soil structure by increasing aggregation, and encourages certain microorganisms, especially the nitrifying and nitrogen-fixing bacteria.

LOSSES FROM SOILS Calcium is lost from soils mainly by leaching but also by crop removal and erosion. Soils derived principally from limestone materials may actually be deficient in calcium if leaching has carried the calcium below the root zone. Carbon dioxide in rainwater (carbonic acid) reacts with calcium carbonate and other clay minerals to form calcium bicarbonate, which is highly soluble. Losses of calcium sulfate and calcium nitrate by leaching are also significant. Losses by crop removal may be significant in crops such as alfalfa, which requires 125 to 150 pounds per acre of elemental calcium for the production of 5 tons of alfalfa hay.

FERTILIZATION Most calcium is added to soil by using limestone with a high calcium content. Calcium and magnesium carbonates are found over the entire United States and are easily quarried and crushed. The finer it is ground, the more readily it is available to plants. **Agricultural meal,** 95 percent of which will pass through an 8-mesh screen sieve, is a form of limestone commonly used. **Marl,** a sedimentary rock composed of precipitated calcium carbonate, is common in some regions and is used because it is immediately available and inexpensive. As a general rule, the calcium in two cubic yards of marl is equal to a ton of limestone. **Oyster shells** and **shell rock** are used in areas where they are readily available. The shell rock is usually pulverized, but oyster shells are sometimes used without being crushed. **Basic slag,** a by-product of iron smelting, contains calcium silicate, calcium oxide, magnesium oxide, and phosphoric oxide, and

is frequently used for liming. It must be ground before application. **Burnt lime** (CaO), or unslaked lime, that has been treated with water is called **slaked lime** ($Ca(OH)_2$). Other names for this material are hydrated lime, caustic lime, and agricultural hydrate. It has some objectionable characteristics and is hard to handle, but reacts very rapidly with acid soils. **Gypsum** (calcium sulfate) is used where it is readily available.

Magnesium

Magnesium is a shiny silvery metal closely related to calcium. It reacts very slowly with water, and its oxide has a melting point of 2800°C (5072°F). The pure metal burns rapidly once ignited, and is a good conductor of heat.

FUNCTIONS IN PLANTS Magnesium is an essential part of the chlorophyll molecule and is also necessary for the formation of amino acids and vitamins. Another important function of magnesium is to neutralize organic acids in plants. It is essential to the formation of fats, the germination of seed, and the synthesis of sugar. Magnesium functions largely as a coenzyme in the above processes. On the average plants contain about 0.2 percent magnesium by dry weight.

DEFICIENCY SYMPTOMS Plants deficient in magnesium are usually chlorotic and may develop brilliant colors. In severe cases, leaves become yellow. Chlorosis is frequently most evident between veins of older leaves, starting at the tips and moving inward (Fig. 16-13). Leaves may also droop.

SOIL RELATIONSHIPS Magnesium is found in soils that include the minerals biotite, chlorite, dolomite, magnesite, olivine, and serpentine as well as in lake brines. After being weathered from mineral sources it is adsorbed onto the surfaces of clay and organic particles. Magnesium is leached much more readily than calcium in lower soil horizons. Magnesium and calcium are similar in their soil relationships. Both cations are found in many of the same rocks and minerals, and behave similarly in response to soil reaction and water content.

FIGURE 16-13. *Progressive effects of magnesium deficiency in potatoes; severity of deficiency increases to right.* [*Courtesy USDA.*]

LOSSES FROM SOILS Magnesium is lost by leaching, plant removal, and erosion. Deficiencies frequently occur in sandy soils exposed to moderate or heavy rainfall, where leaching is intense. Losses of magnesium through transformations to unavailable forms take place in acidic peats and mucks as well as in poorly drained soils. Fertilization with other cations may displace magnesium and can result in deficiencies, because magnesium moves readily to the lower soil horizons.

FERTILIZATION Magnesium is generally added to soils when lime is added. Finely ground **dolomitic limestone** is frequently used to correct deficiencies. When magnesium deficiency is acute foliar sprays of epsom salts ($MgSO_4$), magnesium nitrate, and magnesium chloride are applied. Sprays produce an immediate response, frequently within 24 hours, but their residual effect is of short duration; consequently sprays cannot be considered a complete substitute for soil fertilization.

Sulfur

Sulfur was used medicinally as early as 1000 B.C. It is found in regions where volcanoes have been active and in association with salt domes along the Gulf Coast of North America. Although far less abundant than oxygen, it is in the same group of elements.

FUNCTIONS IN PLANTS Sulfur is an important component of several amino acids, such as methionine and cystine. The vitamins thiamin and biotin also contain sulfur. Sulfur compounds impart flavor to cruciferous plants and pungency to the onion. In general, plants contain perhaps 0.2 percent sulfur by dry weight.

DEFICIENCY SYMPTOMS Sulfur deficiency symptoms are somewhat similar to those of nitrogen. Plants deficient in sulfur tend to be light green because sulfur, although not an essential part of

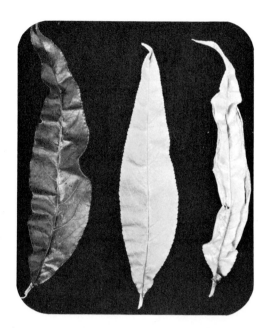

FIGURE 16-14. *Sulfur deficiency in peach; severity of deficiency increases to right. [Courtesy National Plant Food Inst.]*

the chlorophyll molecule, is necessary for its synthesis (Fig. 16-14).

SOIL RELATIONSHIPS   Sulfur occurs in rocks mainly as iron sulfide. Upon weathering it rapidly changes to a soluble sulfate and subsequently becomes incorporated into organic compounds. Small amounts are adsorbed by clay particles. Microorganisms break down the organic compounds and convert them to sulfates. Calcium sulfate seems to accumulate near the surfaces of arid soils; in humid regions it is leached to lower horizons. The sulfur cycle is illustrated in Figure 16-15.

In some regions a large part of the sulfur content of soils originates as sulfur dioxide (SO_2), which is washed from coal smoke in the air by rain. Sulfur dioxide toxicity to plants is an increasing problem in urban and industrial areas, particularly in the vicinity of smelters.

LOSSES FROM SOILS   Sulfur is lost by erosion, leaching, and crop removal. Erosion causes the greatest loss because a large part of the sulfur in

soils is in the organic fraction of the upper horizons. Leaching may account for losses of up to 60 pounds of sulfur per acre per year.

FERTILIZATION   Native sulfur is most frequently mined by drilling into underground deposits and pumping hot steam into them to melt the sulfur and force it to the surface under pressure. It is then shipped, still in the liquid stage, to plants where sulfuric acid is made. (Sulfuric acid is also made by burning iron pyrites, which are by-products of mining operations.) Sulfuric acid is used to make the many sulfur compounds (such as ammonium sulfate, calcium sulfate, and superphosphate) used in fertilizers.

Iron

Iron, the most widely used metal in the world, belongs to the group of elements that includes cobalt and nickel, and the three metals have many similar properties.

FUNCTIONS IN PLANTS   Iron functions as a catalyst in the synthesis of chlorophyll; much of the iron in plants is found in the chloroplasts. It is involved

FIGURE 16-15. *The sulfur cycle.*

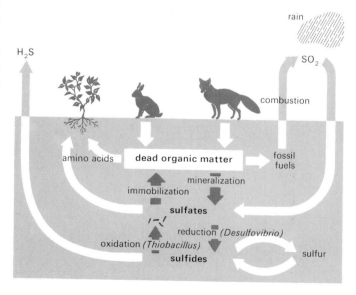

in the formation of many compounds, even though it is not present in their structure. Proteinaceous iron-porphyrin compounds serve as oxygen carriers and "oxygen activators." Iron is a component of certain important enzymes—for example, the oxidases and peroxidases. It is essential to the normal development of the young growing parts of plants, and accumulates in nodes. On the average, plants contain about 0.1 percent iron by dry weight.

DEFICIENCY SYMPTOMS   Iron deficiency in its early stages is manifested by a patternless paling in leaf color, and in its later stages by yellowing of leaves (Fig. 16-16). Veins are not affected at first, but eventually even the larger veins turn yellow. Grasses develop leaf patterns with alternate rows of green and white.

SOIL RELATIONSHIPS   Among the most common iron-bearing minerals are hornblende, chlorite, and biotite. During weathering the iron in such minerals is changed from the ferrous (Fe^{++}) to the ferric (Fe^{+++}) state. In regions with ample rainfall, low temperatures, and an accumulation of litter, iron is dissolved and leached to lower horizons, where it accumulates. In regions where rainfall is abundant, temperatures high, and organic cover

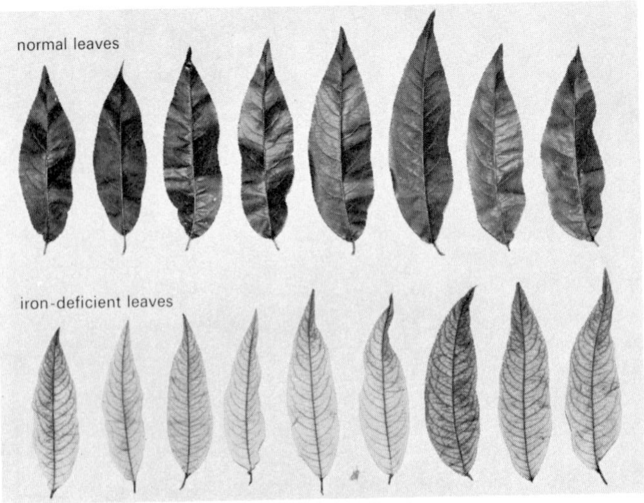

FIGURE 16-16. *Iron deficiency in peach.* [*Courtesy National Plant Food Inst.*]

sparse, iron is converted to a hydrated form and is retained in the horizon where it weathered from rock. The degree of hydration also affects the soil color, which varies from yellow to red.

Under highly alkaline conditions, iron is rendered unavailable to plants by the formation of insoluble ferrous compounds. As a result the total amount of iron is soils may be a very poor indicator of available iron. When soils are alkaline it may be abundant but unavailable to plants.

LOSSES FROM SOILS   Iron may be lost by crop removal, leaching, and erosion. But because iron is required only in minute amounts, losses are likely to be less significant than those of other elements. Availability is of prime concern.

FERTILIZATION   Most soils contain sufficient iron for plant growth. The main problem is to make it available by proper soil management. This may sometimes be done by controlling soil moisture, for ferrous iron accumulates in wet or waterlogged soils. If soil reaction is too high, acidity may be increased by applications of sulfuric acid or ammonium sulfate.

Iron may be applied in several different ways. It is most frequently applied in inorganic form directly to the soil, but in this form it often becomes bound in heavily limed or alkaline soils. Iron can be applied to fruit trees by drilling a hole into the trunk and inserting a gelatin capsule filled with an iron salt.

Perhaps the most effective way to remedy iron chlorosis is to use iron **chelates.** Iron chelates are compounds that bind iron at two or more positions within their structures. The iron ion is held in such a way that it stays in solution and cannot precipitate unless ion exchange takes place or unless the chelate is precipitated by phosphate. Although chelates are expensive to produce, they are widely used in foliar sprays for ornamental shrubs and other high-value crops. For nursery crops they are frequently applied directly to soils. Citric acid is an example of a compound that acts as a chelate. Chelates are not toxic to plants, 1,000 pounds per acre having been tried without ill effects. Chelates are decomposed by microorganisms, and hence the iron they make available again becomes bound.

The effect of small applications of certain chelates on the mineral metabolism of soybeans suggests that they have favorable physiological effects in addition to their capacity to chelate iron in the soil.

Boron

Boron is not an abundant element, but large deposits of the hydrate of sodium tetraborate, called **borax,** exist in the Mojave Desert of California. Boron is used in glass, soap, enamel, pottery, as an antiseptic, and in machinery where high resistance to abrasion is necessary.

FUNCTIONS IN PLANTS Boron affects flowering, pollen germination, fruiting, cell division, nitrogen metabolism, salt absorption, water relationships, and the movement of hormones. It prevents the excessive polymerization of sugars at sites of sugar synthesis. Altogether, at least sixteen functions of boron in plants have been described. Plants contain only a few parts per million of boron, and their requirements for it are small. They must absorb boron throughout their lives because it is not readily transported in the vascular system. Boron uptake seems to be closely related to that of calcium. Plants with an inadequate supply of calcium have a low tolerance for boron; when there is an excessive supply of calcium, boron requirements are high.

DEFICIENCY SYMPTOMS In the early stages of boron deficiency, terminal buds die and lateral branches begin to grow. Lateral buds then die, to be superseded by the growth of new ones, with the result that a rosette of branches may form at the ends of twigs. This seems to be due to a disintegration of meristematic tissue. Leaves may become thickened, curled, and brittle. Fruits, tubers, and roots may become discolored, cracked (Fig. 16-17), and flecked with brown spots.

SOIL RELATIONSHIPS Borates of calcium, magnesium, and sodium occur naturally in soils. Once released from rocks by weathering processes, they are utilized by plants and various microorganisms. Upon death of the organisms the borates are oxidized to inorganic forms. Borates may be leached from soils at any time while they are in an inorganic form. Leaching of borates from clay soils takes place much more slowly than in light-textured soils.

Since most boron compounds are soluble only when the pH is low, soil acidity affects the availability of boron. Liming may therefore result in boron deficiencies due to the fixation or binding of boron on soil colloids at higher pH ranges. Clay and organic soils fix boron in relatively large amounts, whereas sandy soils fix very little. Drying of organic matter also results in fixation. Thus, irrigation increases boron availability.

LOSSES FROM SOILS Boron is most frequently deficient in humid regions where soils are acidic. Every state east of the Mississippi River has areas that are deficient in this element to various degrees. Alfalfa has an unusually high requirement for boron. Vegetable crops also deplete soil boron rapidly. When forest trees are fertilized their accelerated growth causes a rapid depletion of boron.

FERTILIZATION The boron used in fertilizers comes primarily from deposits in the Mojave Desert and from the brine of Searles Lake in California. Borax (sodium tetraborate) is the form that has most often been used for agricultural purposes. Colemanite (a form of calcium borate) is used on sandy soils because it does not leach as rapidly as sodium borate.

Boron is frequently a contaminant in various kinds of fertilizers. Lime contains about one ounce of boron per ton. Barnyard manure contains some boron, and superphosphate also contains appreciable amounts. Thus, for crops with small boron requirements, such materials may provide adequate amounts for crop growth.

Boron compounds have frequently been used as a weed killer. Liberal application of borax to the soil usually results in the death of plants whose roots extend into the treated area. Borate slurrys were used for the suppression of forest fires by aerial "bombing" from tanker planes until it was found that plants would not grow in these areas for many years.

Manganese

Manganese is a reddish-gray metal somewhat like iron, which it resembles in many ways. It is found in many minerals that also contain iron. The pure metal has few commercial uses, but is used in many alloys.

FUNCTIONS IN PLANTS Manganese is required for the synthesis of chlorophyll. It also helps to control some of the complex oxidation-reduction systems that involve iron, glutathione, and ascorbic acid, where it functions as a coenzyme. Plants contain about 300 ppm of manganese.

DEFICIENCY SYMPTOMS Manganese deficiencies in plants are readily apparent unless obscured by other symptoms. Young leaves may show a network of green veins on a light green background of interveinous tissue, a patter similar to that caused by iron deficiency (Fig. 16-18). In more advanced stages light green parts become white, and leaves are shed. Brownish, black, or grayish spots appear next to veins in the leaves of some plants, giving rise to such names as gray speck, white streak, dry spot, and marsh spot.

SOIL RELATIONSHIPS Among the minerals that contain manganese are manganite, hausmannite, olivine, biotite, and garnet. In acid soils manganese is likely to be abundantly available, whereas in neutral or alkaline soils plants often show deficiency symptoms even though the total amount present is great. Manganous manganese (Mn^{++}) is readily available, but the manganic form (Mn^{+++}) is not. In highly acid soils manganese may be available to the extent that it results in toxicity.

FERTILIZATION Plants require very small amounts of manganese, and soil acidification generally cures deficiency problems. When the manganese content of a soil is too low, manganese sulfate is usually

FIGURE 16-17. *Boron deficiency in pear fruits (left) and broccoli plants (right). [Courtesy National Plant Food Inst., and D. B. Maynard, University of Massachusetts.]*

FIGURE 16-18. *Five-month-old tung seedling showing symptoms of manganese deficiency. Lower leaves are normal because they developed before manganese in the seed was exhausted.* [Courtesy USDA.]

applied either to the soil or to the foliage of the crop.

Zinc

Zinc was one of the first metals to be produced from ores. Pure zinc has a white metallic luster when polished but soon tarnishes. It becomes malleable at 100°C (212°F).

FUNCTIONS IN PLANTS Zinc is required in small amounts for auxin formation, internodal elongation, chloroplast formation, and starch formation. Zinc-deficient plants have abnormal roots with enlarged cells and unusual arrangements of root hairs. Tannin, fats, and calcium oxalate crystals tend to accumulate in the root tissues of zinc-deficient plants. Legumes require zinc for seed formation. Monterey pine trees (*Pinus radiata*) planted in zinc-deficient soils in Australia grew poorly except along fence rows, where rainwater washed enough zinc oxide from the galvanized wire fencing to supply their needs. In general, plants contain only a few ppm of zinc.

DEFICIENCY SYMPTOMS Zinc-deficient plants may have mottled leaves with irregular chlorotic areas (*Citrus*), bronze and rosetted leaves (pecan), dwarfed and rosetted leaves (apple, cherry, and other fruits), chlorotic interveinal leaf tissue (various field crops), or rusty brown flecks on leaves (barley and other grasses). Because zinc deficiency leads to iron deficiency, the usual symptoms of both deficiencies are similar. Zinc-deficient corn is illustrated in Figure 16-19.

SOIL RELATIONSHIPS Zinc is found as a trace constituent in soils that contain biotite, magnetite, and hornblende. After being weathered from minerals it is adsorbed by particles of clay and organic matter. It accumulates in surface soils after having been released by decomposing leaves and other plant materials, and deficiencies are likely to occur when the surface soil is eroded or otherwise removed.

Zinc is least available in the pH range from 5.5

FIGURE 16-19. *Leaf of zinc-deficient corn plant.* [Courtesy A. Ohlrogge.]

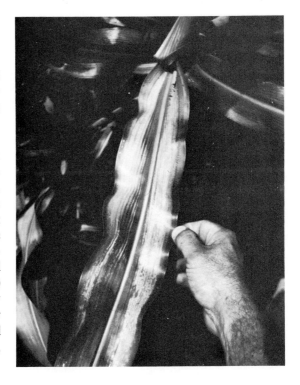

to 7.0. At lower pH values it becomes more soluble, but at higher values, the formation of insoluble calcium zincate is favored, and it may be less available.

Zinc toxicity can result when soils are acidified to increase the availability of other elements and when zinc fertilization is continued over a long period.

FERTILIZATION   Zinc deficiencies can usually be corrected by foliar applications of zinc sulfate. Zinc sulfate is also mixed with fertilizers applied to soils. Galvanized nails are sometimes driven directly into the trunks and branches of deficient trees. In recent years the use of chelated zinc compounds has become common. These compounds are applied to both foliage and soil.

Molybdenum

Molybdenum is a silvery white metal. Most large commercial deposits of this element are in Norway, Colorado, and New Mexico.

FUNCTIONS IN PLANTS   Molybdenum is a component of the enzyme system that reduces nitrates to ammonia. When absent, the synthesis of proteins is blocked and plant growth ceases. Root nodule bacteria of legumes also require it, and when molybdenum is absent nodules fail to develop. It also seems to be required for vitamin synthesis. Seeds are frequently not "filled" when molybdenum is lacking. Plants generally contain about 1 ppm or less of molybdenum

DEFICIENCY SYMPTOMS   Plants lacking sufficient molybdenum may become nitrogen-deficient. Leaves are frequently pale green and have rolled or cupped margins. Yellow spots appear on leaves of some species. Potato tubers may fail to develop, and the glumes of grain do not fill out. Leaves of crucifers become narrow, hence the name "whiptail" for this symptom of deficiency.

SOIL RELATIONSHIPS   Two minerals that contain molybdenum are molybdenite (MoS_2) and powellite ($CaMoO_4$). Molybdenum is sometimes an accessory element in olivine and in hydrous iron and aluminum axides and other clay minerals. After it

weathers from minerals, molybdenum is only slightly soluble. It is fixed much more strongly in acid soils than in alkaline soils. Liming, rather than fertilization, is usually the most suitable corrective measure.

LOSSES FROM SOIL   Molybdenum deficiencies may result either because the element is present only in small amounts in the soil or because it is unavailable. Because it is required only in extremely small amounts, removal by plants is not likely to be a problem.

FERTILIZATION   Molybdenum is usually added to soil at the rate of about two ounces per acre. If even this small amount is added for a number of consecutive years it may accumulate to toxic proportions! It is usually applied either by coating seeds with sodium molybdate powder or by spraying plants with a solution.

Copper

Pure copper was produced by man as early as 5000 B.C. It is malleable, flexible, and a very good conductor of electricity. Almost all countries produce copper, but the largest deposits are in the United States. Rich deposits occur in Peru and Rhodesia.

FUNCTIONS IN PLANTS   Copper is highly toxic when too much is used. It is concentrated in the small absorbing roots of plants and plays a part in nitrogen metabolism. It is a component of several enzymes and may be a part of the enzyme systems that utilize carbohydrates and proteins. The functions of copper in plants are not entirely understood. In general, plants contain a few ppm of copper.

DEFICIENCY SYMPTOMS   The ends of copper-deficient shoots are first to be affected: they die back from the tips, and terminal leaves develop dead spots and brown areas. Rosettes of leaves may precede death of the twigs. Plants that have no branches, such as cabbage, become chlorotic, and leaves fail to form. Deficient corn may become gray at the tips or develop striped leaves (Fig. 16-20). Unlike the symptoms caused by deficiencies of iron, zinc, and many other elements, the

FIGURE 16-20. *Copper deficiency in corn.* [*Courtesy National Plant Food Inst.*]

symptoms caused by copper deficiency are neither clear-cut nor consistent.

SOIL RELATIONSHIPS Clayey and loamy mineral soils usually have a copper content of 10 to 200 ppm; since copper is required by plants only in small quantities, this is generally sufficient. Copper is weathered from cuprite, malachite, azurite, and other minerals, after which it is present in the soil as salts of phosphates, hydroxides, and carbonates. It is bound tightly in organic matter complexes, with the result that plants in organic soils are likely to be deficient in copper. Most copper is held in the latices of soil colloids with such tenacity that it cannot be displaced easily by other soil cations in the pH range 7 to 8. But some is held in the same manner as calcium and is more readily available for plant growth.

LOSSES FROM SOILS Copper is not readily lost from the soil. It is rather commonly present in a form in which it is unavailable to plants. Removal by crops is negligible, since plants usually contain no more than 25 ppm, usually less.

FERTILIZATION Copper is usually added to soils as copper sulfate ($CuSO_4 \cdot 5H_2O$) and copper oxide (CuO) at the rate of 1 or 2 pounds per acre. It should be added only when deficiencies are evident, because yearly additions may accumulate to produce toxicity.

Chlorine

Chlorine is a heavy, greenish-yellow gas that is highly toxic to plants and animals. It is highly reactive, however, and does not occur naturally in the free state.

FUNCTIONS IN PLANTS Chlorine is the most recent element to be included in the "essential" category (1954). It is usually present in plants in small quantities (perhaps 0.1 percent by dry weight), and takes part in the photosynthetic reactions in which oxygen is liberated. At one time it was thought to be necessary for the translocation of carbohydrates, a role that has not been substantiated. Experimental evidence on chlorine behavior is extremely difficult to obtain because of its high solubility and its ubiquitous distribution in soil, water, and atmosphere. An environment absolutely free of chlorine is extremely difficult to produce.

DEFICIENCY SYMPTOMS Deficiency symptoms of chlorine include wilting, stubby roots with excessive branching, chlorosis, and bronzing. Even the odors of some plants are decreased. These conditions only develop, however, when chlorine is deliberately withheld, with great effort, from nutrient solutions in which plants are grown.

LOSSES FROM SOILS Chlorine is usually found in highly soluble forms and is subject to losses by leaching. Removal by plants can account for the loss of a certain amount of chlorine, but in nature it is probably never a serious factor.

FERTILIZER MATERIALS Chlorine is added to the soil by rainfall in amounts sufficient to satisfy plant needs. It is never added intentionally by fertilization, although all commercial fertilizers are "contaminated" by chlorine. Some plants obtain enough chlorine from the air to prevent deficiency symptoms.

WORLD PLANT NUTRITION

The "agricultural revolution" resulted in more sophisticated methods of cultivation, water management, and fertilization, making possible a

330

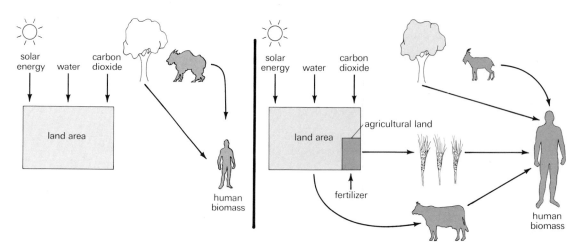

FIGURE 16-21. *Before the "agricultural revolution," hunting, fishing, and food gathering provided sustenance for about ten million persons. The ten percent of the land now being cropped provides the primary support for a population that approaches four billion.* [From Brown, "Human Food Production as a Process in the Biosphere." Copyright © 1970 by Scientific American, Inc. All rights reserved.]

greatly increased production of food, necessary to support the world's expanded population (Fig. 16-21). Fertilization may be the primary key to the problem of feeding even larger future populations; the world is certainly far from having reached its capacity to produce food. It has been estimated that the food requirements of the increasing world population can be met for the next hundred years if the productivity of all existing arable lands can be increased to equal the current rate of production in Europe. The productivity of European agriculture is approximately twice that of the world average, due in part to the greater use of chemical fertilizers. The productivity of arable lands in Japan is about twice that of the European average, which indicates that doubling the productivity of arable lands is by no means an impossible task. One ton of plant nutrients (nitrogen, phosphorus, and potassium) can produce an average increase of about 10 tons of basic food. Ten tons of basic food will supply 50,000 Calories per day for a year, enough to feed 20 more people if the average world caloric intake is considered to be about 2,500 Calories per day.

Countries that are well fed (3,000 Calories per day) are advanced technologically. Asia has 54 percent of the world's population but only 16 percent of the earth's useful land resources available, and the average diet consist of no more than 1,800 Calories per day. In India, even if the acreage of arable land were increased by 25 percent, there would still not be enough land to feed the population expected in 1970 at present productivity levels. But if the rate of productivity per unit land area can be increased to equal that of Europe, then the crisis can be delayed for at least 30 years.

The demand for fertilizers is growing at such a rate that by the year 2000, an estimated 180 million tons of the major fertilizer nutrients N, P, and K, will be needed annually to replace those removed by cropping. This is 15 times greater than the present rate. The increase must come from chemical rather than natural organic sources. Organic materials are of more value for secondary products than fertilizer, and their bulk makes them expensive and precludes widespread use by large crop-producing units.

But the problem should not be oversimplified. Projecting fertilizer needs is not the same as producing the materials. Even though the world has tremendous resources and facilities for producing nitrogenous, phosphatic, and potassic fertilizers, there must be an economic incentive to invite the necessary investment of capital for manufacturing facilities. There is no doubt that these materials can be produced; what is necessary is the motivation, technology, and capital. Furthermore, farmers need special incentives to use fertilizers. Unfortunately, a tax on fertilizer is used to produce revenue in some countries that can ill-afford to discourage good farming practices. Much of the agricultural output is the result of subsistence farming on small holdings. Since their products have no commercial value, farmers have but little incentive to produce more. The wise use of fertilizer is a key factor in agricultural development. It would seem that great social and economic changes must accompany technological change if a balanced, self-sustaining world agriculture is to be attained.

Selected References

Black, C. A. *Soil-Plant Relationships*. Wiley, New York, 1968. (An outstanding and comprehensive text which treats soil as a medium for plant growth.)

Dinauer, R. D. (editor). *Changing Patterns in Fertilizer Use*. Soil Science Society of America, Madison, Wisconsin, 1968. (The proceedings of an important symposium dealing with many aspects of fertilizers and their uses for different crops.)

Hillel, Daniel (editor). *Optimizing the Soil Physical Environment Toward Greater Crop Yields*. Academic Press, New York, 1972. (Proceedings of an international panel.)

Olson, R. A. (editor). *Fertilizer Technology and Use* (2nd ed.). Soil Science Society of America, Madison, Wisconsin, 1971. (A collection of advanced monographs on the economics, production, and use of fertilizer materials.)

Pearson, R. W. (editor). *Soil Acidity and Liming*. Agronomy Monograph No. 12, American Society of Agronomy, Madison, Wisconsin, 1967. (A collection of papers dealing with the soil chemistry of liming and the attendant crop responses.)

Shaw, B. T. *Soil Physical Conditions and Plant Growth*. Academic Press, New York, 1952. (A comprehensive and technical volume on soil and its relationships to plant nutrition and growth.)

Sprague, H. B. (editor). *Hunger Signs in Crops*. National Fertilizer Association, Washington, D.C., 1964. (A volume dealing with deficiency symptoms in selected crops.)

Tisdale, S. L., and W. L. Nelson. *Soil Fertility and Fertilizers* (2nd ed.). Macmillan, New York, 1966. (This college text covers the fundamental concepts of soil fertility and fertilizer manufacture.)

CHAPTER 17

Water Management

The chapters of Genesis, which describe the life of Israel's forefathers 4,000 years ago in Canaan, abound in references to water and wells. To survive, these people had to practice good water conservation. Also, even a modern engineer with his best tools would find it difficult, if not impossible, to improve on the methods of water conservation that were used 2500 years ago by the Nabataean inhabitants of Israel's Negev Desert. The Nabataeans collected the runoff from uncultivated slopes and routed it to cultivated areas. In order to do so, they built elaborate diversion and drop structures, terraces, guide walls, storage impoundments, and irrigation systems for basins. They also practiced dry-farming. Practically all of the systems and structures they developed fell into disrepair because of war.

Today, water management still involves interrupting and manipulating various stages in the hydrologic cycle to make more water available, to remove it when there is an excess, and to improve the quality when necessary. Although water is also managed for industrial, recreational, and other purposes this discussion will be principally concerned with water management in relation to crop production (Fig. 17-1). This involves **irrigation**, the addition of supplemental water, **drainage**, the removal of excess water, and **conservation**, the protection and wise use of our water resources.

IRRIGATION

Irrigation to provide supplemental water for crop growth was a highly developed art even 4,000 years ago. Paintings and sculptures in the tombs of the ancient pharaohs depict slaves lifting water to pour on crop fields. One of the earliest aids to irrigation was the **shaduf**—a long pole balanced on an upright, with a bucket on one end and a counterweight on the other (Fig. 1-2). Water could be raised 6 feet with this device, and even more if several were used in series. The **sakieh**, or chain of pots, was developed later and utilized animal power. It consists of a beam attached to a toothed wheel that engages a vertical wheel which drives an endless rope fitted with buckets. The water in the buckets is spilled into a small flume as they tilt when passing over the vertical wheel. The **water screw,** invented by Archimedes and used about 200 B.C., lifted water about 15 feet. Many of the ancient devices are still used today in some part of the world. It is interesting to see the modern touch that has been given the sakieh, with tin cans replacing leather buckets as bails. The suction pump, driven by wind, steam, or electricity, is the modern contribution to irrigation. An interesting pump called the **hydraulic ram** operates by water power (Fig. 17-2).

In present irrigation practice, supplemental water may be applied in three ways. In **surface irrigation,** water is distributed over the surface of the soil. **Sprinkler irrigation** is the application of water under pressure as simulated rain. **Subirrigation** is the distribution of water to soil below the surface and provides moisture to crops by upward capillary action. Each system is adapted to a particular cropping situation.

For agricultural purposes water is measured in terms of volume and in rates of flow. Volumes are given as gallons, cubic feet, acre-inches, and acre-feet. An **acre-inch** of water is the amount of water that will cover one acre to a depth of one inch, and is approximately 27,000 gallons; an **acre-foot** will cover one acre to a depth of one foot, and amounts to 43,560 cubic feet or about 326,000 gallons. Flow rates are usually given in terms of cubic feet per second, gallons or liters per minute, acre-inches per hour, or acre-feet per day.

Surface irrigation

Water is applied by surface irrigation in arid and semiarid regions. When the surface of a field cannot be contoured to prevent ponding, provisions must be made in advance to drain the lower places. Conversely, when fields are extremely flat, low dikes help to retain water.

Water is usually carried to the place of use in open ditches at a slow velocity in order to prevent erosion. Pipelines and flumes are sometimes used when water is scarce, since they eliminate losses due to seepage and evaporation when water is moved long distances. When ditches are used for this purpose they are frequently lined with asphalt or concrete to eliminate seepage loss. Earth ce-

FIGURE 17-1. *The efficient use of water necessitates multiple use. Storage impoundments can provide water for recreation, crop irrigation, human consumption, industrial use, and many other purposes. [Courtesy USDA.]*

ment, made of cement and soil, is inexpensive and can be mixed "in place" but is not durable. In recent years, plastic has been used for lining ditches, but plastic linings have a comparatively short life. If possible, it is useful to cover ditches with a permanent grass sod. Small portable gasoline pumps and light, easily moved surface pipes are used to carry water from ponds and ditches to fields.

The distribution of water from ditches is accomplished by various control structures or siphons. The flow of water must be carefully controlled to prevent erosion. **Division boxes** are used to divide a stream into two or more smaller streams. **Checks,** dams of adjustable height for controlling the depth of water at turnouts, are installed at places where water is delivered to the area to be irrigated. When water is transported by pipeline it is sometimes helpful to use a **soaker,** a canvas hose or perforated pipe that distributes the water over a large surface area and prevents erosion.

Surface irrigation may be accomplished in a number of ways. In **flood irrigation,** water is

334

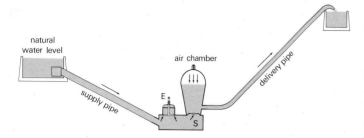

FIGURE 17-2. *The hydraulic ram is an interesting pump that uses water power to move water to a greater height. The principle of operation is as follows: Water moving down a supply pipe closes an escape valve (E) controlled by a spring. The momentum of the moving water causes a ramming action that forces water through the supply valve (S) at the base of an air chamber. The force of the water is captured in a chamber where air is compressed. When the air pressure equals the pressure of the moving water in the supply pipe the supply valve closes and the trapped water is forced out of the delivery pipe to a higher elevation. Immediately after the supply valve is closed there is a rebound or backward flow of water that causes the escape valve to open. At this point the water starts to move in the supply pipe, builds up momentum, and the cycle repeats. The efficiency of this device may be as high as 75 percent under ideal conditions. The hydraulic ram, with only two moving parts, can operate for years without attention. It is especially valuable for providing irrigation where power is unavailable and should have wider use in underdeveloped areas.*

allowed to cover the entire surface of the field in a continuous sheet (Fig. 17-3). It must be deep enough and remain long enough to replenish the root zone with water. Provisions must be made to carry away excess water (water that remains on the surface after the soil has reached field capacity) through drainage ditches. Flood irrigation is frequently used for crops that form a complete ground cover, such as permanent pasture and rice.

Contour flooding is practiced where slopes are steep. A series of dikes are formed along the contour of the land at regular intervals. Each

section of the field is then flooded, resulting in a series of shallow impoundments, each 6 to 12 inches above the other.

Furrow irrigation is one of the most common methods used for crops (Fig. 17-4). Water runs down the furrows between plant rows (but not over them) and moves to all parts of the soil by capillary action or gravity. It is an expensive technique, because constant supervision is required to prevent furrow streams from uniting and forming large channels, but can be justified for truck crops, orchards, and vineyards if soils are not so coarse that seepage is excessive.

Corrugation irrigation is the application of water in small furrows that run downslope from a **head ditch.** The furrows guide rather than carry water. It is adapted to crops on sloping soils that absorb water slowly.

The main advantage of surface irrigation is that under ideal conditions no source of energy is required for the final distribution of water. But since this system depends on gravity flow, it is inefficient in that more water than necessary is supplied close to the source. The chief disadvantage lies in the possible degradation of soil structure. Heavy soils become puddled under the heavy load of water, and are poorly aerated, tending to bake and crack when they dry, and form hard clods when broken by a plow.

Sprinkler irrigation

Sprinkler irrigation was first used in the United States in about 1900. Sprinkler irrigation systems formerly consisted of permanent overhead sprinklers that were used only in intensive cropping operations such as market gardens. Today, the use of lightweight, easily movable aluminum pipe with rapid coupling devices has cut the labor cost and brought the practice into general use.

Water may be applied through sprinkler heads or perforated pipe. **Sprinkler heads,** the most popular, may be used with permanent buried lines or temporary portable lines (Fig. 17-5). Water is delivered through sprinkler heads mounted on riser pipies. Because each sprinkler head applies water to a circular area, distribution is not uniform. There should be an overlap of a fourth to a half

FIGURE 17-3. *Rice culture in Nepal. Terracing permits the re-use of water for flood irrigation at successively lower levels.* [*Courtesy USDA.*]

of the wetted circles, and the distances between lines of sprayers should not exceed 7/10 of the diameter of the area covered by one sprinkler.

Several types of sprinkler heads are available. **Stationary heads** apply water to small areas and at low rates; usually no attempt is made to obtain overlap. **Revolving heads** may have either one or two nozzles. A single nozzle that projects a large water stream is used when wind is likely to deflect a small stream. When double nozzles are used, one nozzle throws water to the outer part of the circle while a smaller one delivers water close to the sprinkler head.

Perforated pipe sprinklers deliver water through small holes drilled along the length of pipes mounted on permanent supports. A water-driven motor causes the pipe to oscillate and throw water alternately on each side. Strips about 40 feet wide can be watered in this way. This kind of system requires a water pressure of approximately 25 pounds per square inch, whereas sprinkler head systems require as little as 15 pounds per square inch.

There are many advantages to sprinkler irrigation. Water can be applied precisely and uniformly to all kinds of soils and crops in the exact amount required. As a result, drainage problems in sprinkler irrigation are usually small. Erosion need not be a problem even on steep slopes. Land preparation is not necessary, and labor costs are smaller than for other methods. Small sources of water can be used efficiently from ponds and catchment areas and applied to higher areas. Sprinklers can sometimes be used to reduce crop damage during periods of frost. Water-soluble fertilizers can be applied by sprinklers, giving very uniform distribution.

There are certain disadvantages to sprinkler

FIGURE 17-4. *Furrow irrigation, using syphons to carry water to the furrows.* [*Courtesy USDA.*]

systems. The initial costs and power requirements are high. Water must be clean, otherwise it will clog the spray equipment. In hot, dry regions evaporation may consume a significant amount of the water while it is in the air. Tight soils with slow rates of infiltration are difficult to irrigate in hot, windy weather. But in spite of these drawbacks, sprinkler irrigation can be adapted to practically any kind of crop.

Irrigation has been a traditional practice in semiarid regions. Within recent years, however, portable sprinkler irrigation systems have become a part of farm equipment even in regions that receive as much as 50 or more inches of precipitation annually. Because occasional periods of drought in such regions can cause the loss of a high-value crop, the cost of owning an irrigation system is often justified even though it may not be used every year.

Trickle irrigation

Trickle irrigation, also called drip or daily-flow irrigation, is a promising new system developed during the 1960's to conserve water by restricting application to the immediate root zone. Water is applied very slowly and in small amounts so that little is lost by evaporation or to portions of the field not being utilized. Soluble fertilizers can also be applied by means of this system. Water is generally applied at a rate of 4 to 8 liters (1 to 2 gallons) per hour under low pressure (15 psi or less). Since moisture stresses tend to equalize throughout the plant, as little as 25 percent of the root system of an individual plant need be wetted.

The delivery system consists of a network of water-conducting plastic tubes, with lines terminating in an "emitter" or leak at each plant. Trickle irrigation first proved useful for watering individual containers in greenhouses where each tube ends in a valve that may be turned on and off and is weighted to keep it in place (Fig. 17-6). The system is flexible because couplings allow fast and simple connections. Many types of emitters can be individually regulated. A popular type of "emitter" is a very thin tube (about .9 mm ID) whose length regulates the pressure—the longer the tube, the slower the rate of delivery. In this way the delivery of water to plants, which are located at different distances from the water source, can be equalized.

Trickle irrigation has great potential for use in arid regions where water conservation is essential. Also the concentrated flow leaches salts from the vicinity of individual plants, thereby reducing the salinity problem.

Subirrigation

Subirrigation is the application of water beneath the surface of the soil. The objective is to create an artificial water table from which water can move upward by capillary action. Usually the artificial water table is maintained between 12 to 30 inches beneath the soil surface.

Ditches are most frequently used to distribute water because they are the least expensive. They have an additional advantage in that water in them can be seen and the water table regulated with relative ease. **Tile drains** can also be used but usually cost several hundred dollars an acre and are the most expensive. **Mole drains** (discussed

FIGURE 17-5. *Sprinkler irrigation. The sections of pipe are lightweight and are easily uncoupled for installation in other fields.* [*Photograph by J. C. Allen & Son.*]

under drainage) are useful only in organic soils and have a relatively short life.

Subirrigation is possible only where the topography is absolutely level. Furthermore, there must be a barrier beneath the irrigation pipe to prevent the loss of water down through the soil by percolation. The barrier may be a water table that is not high enough to provide adequate moisture to the root zone or it may be an impervious soil horizon.

Successful subirrigation in the United States has been carried out in such diverse locations as the flatwoods of Florida, the San Luis valley of Colorado, the Sacramento–San Joaquin Delta in central California, and in the organic soils of Michigan.

Determining irrigation requirements

Irrigation, capable of yielding enormous benefits, may be wasteful and even harmful if applied incorrectly. The determination of when to irrigate and how much water to supply in relation to economic crop production is a crucial part of irrigation technology. In a sense this requires that

an accounting system be established that will determine whether the available soil moisture will meet crop needs in spite of losses due to evapora-

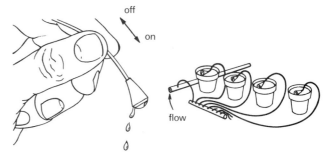

FIGURE 17-6. *Trickle irrigation system for greenhouse watering. (Left) Weighted valve. (Below) Delivery system.* [*From Janick,* Horticultural Science, *Second Edition. San Francisco: W. H. Freeman and Company. Copyright © 1972.*]

tion, transpiration, runoff, and percolation. The net deficiency not compensated for by precipitation may be made up by irrigation. The timing of application in relation to growth stages of the plant can be critical. In beans, for example, moisture stress during flowering and pod formation seriously depresses yield as a result of flower abscission and ovule abortion.

There are two approaches to determining irrigation requirements. One consists in the actual measurement of soil moisture, and subsequent calculation of available moisture. The other consists in calculating the status of water availability from meteorological data.

Irrigation is most effective if applied before soil moisture becomes limiting to plant growth. As a rule of thumb, water should be applied when 60 percent of the available water (that is, 60 per cent of the water between field capacity and the wilting point) in the root zone has been depleted or when the water tension in the zone of rapid water removal exceeds 4 atmospheres. To determine when irrigation is required, it is first necessary to determine how much water is already in the soil. The most commonly used techniques are discussed in Chapter 13. When these measurements are converted in terms of water availability to plants, the amount required in the root zone can be determined.

Meteorological and climatological data offer powerful aids for measuring the status of available water; the procedure involves the calculation of **consumptive use**—the water lost by evaporation and transpiration,—probably the best index of irrigation requirements. Consumptive use varies with a great number of factors: temperature, hours of sunshine, humidity, wind, amount of plant cover, the stage of plant development, and available moisture.

High rates of water consumption are associated with a high percentage of plant cover and with hot, dry, windy conditions. Crops differ in their water requirement largely in relation to their ground covering ability and to some extent to the depth of rooting, even though deep-rooted plants obtain their greatest percentage of moisture from the upper part of the root zone. Because optimum plant performance depends upon an adequate, uninterrupted supply of water, the peak requirements must be considered. Crops have the highest water requirements in the fruiting or seed-forming periods.

The amount of available moisture may be much less than the total precipitation. Owing to soil evaporation and the slow rate of infiltration, showers that provide less than 1/4 inch during the hot summer days may contribute very little to available soil moisture. On the other hand, a high proportion of water from heavy precipitation may be lost by runoff. The effectiveness of precipitation therefore depends upon the intensity, as well as the amount, in relation to temperature and absorptive capacity of the soil.

It has been possible to determine a satisfactory index of consumptive use for a particular area by using monthly averages of mean temperature and hours of sunshine. Empirically derived constants are available for adjusting these values for different crops. Irrigation requirements can then be estimated by a bookkeeping procedure. The moisture content of the soil at the end of any time period is computed by subtracting the amount of moisture lost by evapotranspiration from the amount present at the beginning, and adding the amount of available precipitation during this period. The generalized equation for this relationship is:

$$\text{Final soil moisture} = \text{soil moisture (at start)} - \text{evapotranspiration} + \text{precipitation} - \text{runoff}$$

If the maximum possible soil moisture storage capacity is known, it is a simple matter to compute the soil moisture content at any given time and thus determine whether irrigation is necessary. Computerized programs, which require no more than the accumulation of daily temperature and rainfall data, have been developed to make daily estimates of the status of soil moisture.

Because all of the irrigation water applied is not available to the crop, the amount applied must be based on **irrigation efficiency,** the percentage of irrigation water applied in relation to that which becomes available for consumptive use. Water should be applied to bring the soil up to field capacity at a depth commensurate with the bulk of the feeder root system. The rate must be consistent with the absorptive properties of the soil. The frequency of irrigation is determined in part by soil texture—the coarser the soil, the more frequent must be the irrigation (Fig. 17-7). The amount of water that can be efficiently utilized is also related to the fertility of the soil. The maximum benefits of irrigation depend upon a readily available nitrogen supply.

DRAINAGE

In 1823, Professor W. H. Keating accompanied the Long Expedition to the source of the Minnesota River and described the land south and west of what is now Chicago.

"From Chicago to a place where we forded the Des Plaines River the country presents a low, flat, and swampy prairie, very thickly covered with high grass, aquatic plants, and among others the wild rice. The latter occurs principally in places which are under water; its blades floating on the surface of the fluid like those of the young domestic plant. The whole of this tract is overflowed during the spring, and canoes pass in every direction across the prairie."

Today, there are more miles of drainage outlet ditches in this region than of public highways. The productive fields, numerous good roads, and excellent farm buildings in this area testify to the advantages of drainage when soil and climatic conditions are favorable for crop production. Drainage has made possible the tillage of approximately 30 million acres of land in the United States, and has increased the productivity of about 40 million additional acres.

Excessively high water tables for crop production may exist for more than one reason. Impervious layers may prevent percolation of the water down through the soil to the extent that they become waterlogged. Soils along rivers may be flooded yearly during seasons that are critical to crop production. Soils near the ocean, which may have an elevation of no more than two or three feet above sea level, may be within inches of the groundwater level. Even some irrigated lands must be drained periodically to flush out salts that have accumulated in the surface soil layers. The valley of the Tigris and Euphrates rivers in ancient Mesopotamia has returned to desert, largely because of the accumulation of salts in the **plow layer,** the zone of surface soil that is subject to disturbance by cultivating equipment.

The artificial removal of excess free water from soil may be achieved in two ways: **surface drainage** and **subsurface drainage.**

Surface drainage

Several kinds of ditches are used to drain the surface of waterlogged soil. **Field drains,** which are merely ditches as much as 18 inches deep, collect water from the row crops and land surfaces and carry it to a larger drain. They are usually about 400 feet apart on heavy soils and may be as much as 1,000 feet apart on lighter soils. **Lateral drains,** also called **collecting ditches,** receive water from field drains. They are usually about a foot deeper than the field drains and may have side slopes with a low angle so that they can be easily crossed with motorized farm equipment. **Main drains** collect water from laterals and carry it to drainage outlets.

Topography usually dictates the type of drainage system used. When topography is uniform, it is usually desirable to use a **parallel system** of surface drains. Field drains are usually spaced uniformly, and are at right angles to collecting drains.

So-called **Random systems** are used when topog-

FIGURE 17-7. *Irrigation intervals as influenced by soil texture and depth of root zone at which water is used at the rate of 1 inch per week.* [*Courtesy USDA.*]

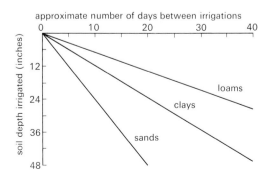

raphy is not uniform, and when only some areas require drainage. Each problem area must be considered separately and enough field drains installed to provide the necessary outlets; collecting drains must be so placed that they do not interfere with the overall farming plan.

A number of surface drainage systems have been adopted in various parts of the world. In Louisiana, one known simply as the **sugar cane system** has been used for sugar cane for over a century. Split, parallel ditches are dug in the direction of the slope at intervals of 100 to 250 feet apart. Quarter drains carry water from crop rows to split ditches and then to cross ditches that empty into canals. **Rice drainage systems** function to remove water in excess of that required to flood paddies.

Bedding, one of the earliest forms of drainage, consists of turning furrows to form high ridges that are flanked by ditches to carry away excess water. Early English systems used beds 3 to 4 feet higher than the bottoms of drainage furrows. In the southeastern United States, beds formed in coastal plain soils are frequently no more than 6 to 8 inches high.

Interception drains are sometimes used to catch water from a higher slope and route it to the place of disposal. Interception ditches are also used to prevent the buildup of water flowing across fields during periods of heavy precipitation.

Subsurface drainage

Tile drains installed below ground are used to remove excess water from high-value agricultural land. Buried below the surface, they occupy no arable area, and thus avoid cutting up the field with ditches. Tile drains also eliminate open-ditch maintenance problems such as weed control. Soil conditions must be suitable, however. Water enters through cracks between the individual tiles and flows by gravity from the root zone down to the water table. Although expensive to install, tile drains have proven their value in many places.

Tile drains can be installed in many patterns, including **parallel line, herringbone,** or **random** arrangements. If necessary, tile can be laid to intercept seepage water that follows the upper surface of an impervious subsoil. **Lateral drains**

remove free water; **submains** collect water from a group of laterals, and **main drains** carry water to the outlets.

Tile drains must be installed at depths great enough to lower the water table by the desired amount. They should be large enough and laid in rows close enough to remove excess water in a specified length of time and give the required amount of protection. Four-inch tile is satisfactory in mineral soils with small amounts of fine silt and sand. Five-inch tiles are the minimum generally recommended for heavy clay soils. Six-inch tiles must be used for peat and muck soils.

It may be necessary to lay the tile on boards if soil in the trench bottom is not stable. Long lengths of drain tile are sometimes the solution to this problem. In one location, better alignment and easier placement were achieved by connecting tiles with short sheet metal tubes slipped over their ends.

Mole drains are temporary drains made by pulling a bullet-shaped tube through the soil by tractor at the desired drainage depth, usually no less than 30 inches. Since they are no more than holes in the ground, they must be renewed every two or three years, although in Great Britain and New Zealand mole drains have lasted as long as 10 to 15 years. They can be installed at less than one-tenth the cost of tile drainage.

Drainage of forest soils

Forest drainage presents problems because of the magnitude of individual projects, sometimes involving as much as 45,000 acres. Fortunately, tree species grown in drained areas are usually fairly resistant to flooding, and there is usually no need for extremely rapid rates of water removal.

Ditches for forest drainage are usually large (Fig. 17-8). Even the **shallow drains** are 2 to 3 feet deep, and the **lateral drains** that they empty into must be several feet deeper. **Main drains,** which carry the water to **drainage canals,** are sometimes 10 feet deep and require the use of draglines for their construction. In dense forest lands, however, rows of dynamite are sometimes exploded to instantly open up as much as a quarter of a mile of ditch.

Many problems are likely to accompany the

FIGURE 17-8. *Drainage can greatly increase forest productivity. The cleared strip adjacent to the drainage ditch can be used as a logging road and for access to fight fires.*

drainage of large areas of organic forest soils. When drained, they frequently become dry and are highly subject to fire. Subsurface fires have been known to smolder and burn in the organic soils of the southeastern coastal plain for many months, and occasionally for more than a year. Such fires are extremely harmful because they are almost impossible to extinguish. They burn out the root systems of plants and make regeneration by sprouting impossible. Because of the rapid rate of decomposition of dried organic soils, it is usually necessary to maintain water table control with a series of **water gates** rather than to permit complete, free gravitational drainage. Roads built on spoil banks dug from drainage ditches may subside because of the action of microorganisms in destroying the drained organic matter.

WATER CONSERVATION

Water is of national concern. In order for a nation to prosper, an abundant source of high quality water must be available for agricultural and industrial use as well as for human consumption, sanitation, and recreation. The misuse of water resources leads to alternate flood and drought, and to pollution—problems that affect us all.

Water conservation implies the best possible use of water. It may involve large programs to control flooding, develop hydroelectric power, and facilitate navigation. Now that cloud seeding seems to be approaching a practical reality when atmospheric conditions are favorable, every phase of the hydrologic cycle is subject to manipulation. Such programs require national effort. Water conservation also involves the control of water resources on a smaller scale. It must be part of the water management of every individual enterprise.

Watersheds

Water supply is closely related to economic development, because water resources must keep pace with other resources. The basic unit of water supply for any particular area is the **watershed.** A watershed is merely a geographic unit from which all precipitation finds its way to a single outlet, called its confluence. A watershed may be no more than a few acres in size, or many minor watersheds may be considered together as part of a vast overall watershed that may occupy an area of many thousands of square miles. Consider the watershed for the Mississippi River. This comprises the entire central part of the United States from the Rockies to the Appalachians. On the other hand, very small watersheds may be of extreme importance for local communities. The problems of maintaining a constant flow of water through the confluence of the watershed are nearly the same, whether they are large or small. They differ enormously, however, in magnitude, and the problems are compounded by political and jurisdictional disputes.

Water leaves the watershed as a liquid in several different ways. After a period of precipitation, **surface runoff** is the first to be lost. Surface runoff is water that never enters the soil but runs directly over the surface into streams, as it would over a rock. Surface runoff is greatest immediately following storms of high intensity and after soil has become saturated with water to the extent that additional water cannot infiltrate the soil (Fig. 17-9).

Water may also leave the watershed as **subsur-**

342

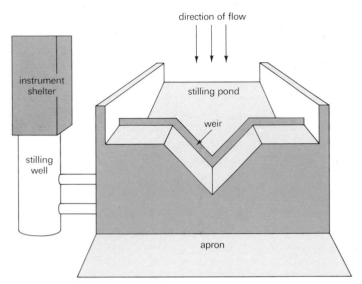

FIGURE 17-9. *A dam-like structure permits collection and measurement of surface water that issues from a watershed as it passes over the V-shaped wier. The well stills the water, the level of which is continuously recorded.* [*From Bormann and Likens, "The Nutrient Cycles of an Ecosystem." Copyright* © *1970 by Scientific American, Inc. All rights reserved.*]

face runoff, which infiltrates and percolates through the soil pores. Subsurface runoff is largely free (gravitational) water and does not take as long to flow into steams as **retained water.** Water in **retention storage** is poorly understood, but a number of studies indicate that it is held in capillary pores any may be "pulled" downward through soil pores due to tension from below. Water in retention determines the regularity of flow of springs, which in turn allow small tributary creeks to flow the year round.

The plant life of a watershed is extremely important in maintaining the flow of water from streams, for the rate of transpiration largely determines how much water is lost into the atmosphere. There is a common belief that forests increase the water yields of watersheds. This is far from the truth, however, because trees have extensive and deep root systems that can withdraw great amounts of water from the soil that cannot be subsequently used for consumption or industrial purposes. For-

ests are beneficial, however, because they not only provide a more even year-round supply but stabilize soil, provide such products as wood, and make good recreation areas. Most important is the fact that the quality of water from forested watersheds is much better than water that flows from denuded areas. Water from forested areas is less difficult to purify for human consumption. In many parts of the world, there must be two water systems: one to supply highly purified water for consumption, and the other to supply less purified water for washing, irrigation, and industrial purposes.

Forests also improve the seasonal distribution of water from the watershed. If one wishes to obtain as much water as possible from a particular area of land, he can do no better than pave it with asphalt or concrete. By doing this, he will stop all losses due to transpiration, and a very high percent of the water from precipitation will flow through the mouth of the watershed. Of course, this is frequently a highly undesirable condition, because under these circumstances the water must be stored in impoundments and reservoirs until it is needed. In some parts of the world there is no alternative to impoundment. Forests, on the other hand, permit the gradual infiltration of water through the soil and allow it to seep through the subterranean channels at a slow rate of speed, so that it is released in the streams for long periods of time. It is apparent, therefore, that watersheds are not only areas for trapping water, but are also important for filtration and as storage repositories.

Conserving precipitation

Preventing immediate runoff and subsequent loss of precipitation is a major means of increasing available water. In agricultural practice precipitation is conserved by terraces constructed on the contour (Fig. 17-10). Terracing slows down surface runoff and thereby increases the time for infiltration. Although terracing may be designed primarily to prevent erosion, it also facilitates water storage (Fig. 17-11).

Since precipitation does not infiltrate compacted soil, rapid runoff may result in flash floods. The regulation of grazing in rangelands, where the trampling of soil by animals causes compaction,

343

is truly a water conservation practice in addition to its function in conserving soil and vegetation.

Rotation cropping, the practice of alternating crops, conserves water as well as soil. Usually a row crop grown for one year is followed by one or more years of grasses and legumes, during which more organic matter is incorporated into the soil and infiltration of water increased. In a typical study carried out in Wisconsin, continuous cropping of corn permitted 19 percent of rainfall to run off, whereas when clover was rotated with corn, only 4 percent of the rainfall was lost.

Spectacular increases in watershed yields have been obtained by appropriate timber harvesting practices. In North Carolina, a 33-acre watershed was completely cut over, and in the first year water yield at the confluence was increased 65 percent. In another watershed, where timber was cut only along the stream channels, the water yield increased by 10 percent. In Colorado, streamflow was increased 31 percent by the complete harvest (clearcutting) of merchantable lodgepole pines, and 19 percent by harvesting methods in which one-third of the trees were left to provide seed for future crops. When trees are removed, not only are transpirational losses decreased but less snow is intercepted by leaves and evaporated or sublimed back into the atmosphere. Snow storage on the forest floor is increased and more water infiltrates the soil when the snow melts. Snow accumulation can also be increased along slopes by the strategic placement of fences or vegetation.

Surface impoundments

The first dam of record was built in Egypt 5,000 years ago. It was about 355 feet long and 40 feet high and stored water for drinking and irrigation. Today, dams are commonplace throughout the world (Fig. 17-12), and there are thousands of dams in the United States alone. Some impound water surfaces of less than an acre and may supply water to irrigate only a single field. But Lake Mead, in Nevada and Arizona, has a surface area of 228.8 square miles, stores 32,300,000 acre-feet of water, and supplies irrigation water throughout an area of several states.

Evaporation from the surfaces of bodies of water

FIGURE 17-10. *Contour plowing and terracing for planting citrus to conserve moisture and prevent erosion.* [*Photograph by W. A. Garnett.*]

can result in great losses of water (particularly serious when the reservoir is in an arid region with low atmospheric humidity), and various materials are used to prevent evaporation. These compounds, called **surfactants,** form a monomolecular layer over the surface of water and can prevent a large part of the evaporative loss. Cetyl alcohol was found to reduce evaporation from a pond as much as 50 percent on calm days, although savings are smaller in windy weather because the monomolecular layer is constantly being broken. Hexadecanol, docosanol, and other long-chain fatty alcohols have also been found useful but are subject to bacterial decomposition and must be replenished frequently. Surfactants may also be helpful when applied to soil surfaces. They increase the efficiency with which plants use water, decrease evaporation losses from soil, and increase the rate of water infiltration into soil. Hexadecanol, when used as a soil treatment, has been found to remain effective for as long as 14 months.

Refilling underground storage reservoirs

A large part of the water used for irrigation comes from natural underground storage reservoirs. In

FIGURE 17-11. *These steep backslope terraces are coupled with underground tile to provide a maximum reduction in soil and water loss. On nonterraced, straight-row field, 10 times as much water runoff and 100 times as much soil loss occurred.* [*Courtesy Cal J. Ward, Concord, Nebraska.*]

California, for example, 60 percent of the irrigation water is from underground sources. There are several methods by which groundwater supplies may be replenished.

Water spreading is the flooding of long narrow basins that surround productive farm land (Fig. 17-13). The artificially perched water tables spread out under the unflooded crop land, giving the water a better chance to infiltrate. This technique can also be used to recharge groundwater supplies for pumping.

Flooding allows water to move slowly over the land in a thin sheet and percolate down to the water-bearing strata (Fig. 17-14). Dikes are some-

times used to ratain water until it moves into the soil.

In **shaft recharge,** an appropriate number of shafts or wells are drilled to permit the water to move directly to the natural underground reservoir, bypassing obstacles. Deep trenches are also used in this manner, but they must be cut through all the soil layers to a good water-conducting stratum.

Replenishment irrigation is the application of excess water to crop land during winter by means of the normal irrigation system. An extra benefit of this system is the control of weeds, which do not survive flooding.

Using dew

Again, the Nabataeans in the center of the Negev Desert, centuries before Christ, made agricultural use of dew. Flint stones were stacked in such a manner that deposited dew dripped to the ground to help support a single plant, such as a grapevine. Piles of stones can still be seen on the hillsides.

At least some dew is directly absorbed by leaf surfaces. Young leaves are less cutinized and absorb more water than older leaves. Dew water runs off leaves and increases soil moisture. It may be possible to increase the deposition of this moisture by breeding for modified plant forms that permit the accumulated droplets of dew to run down the stem.

In an Ohio experiment, an amount of moisture equivalent to 9.1 inches of rain was gained by the

FIGURE 17-12. *Water impoundment for surface irrigation in the mountainous area of Minas Gerais, Brazil.* [*Courtesy Hans Mann.*]

FIGURE 17-13. *By using canals and ditches to "rearrange" the flow of water and make it available to as large an area as possible, erosion is avoided and the best use made of water resources. This technique is called water-spreading.* [*U.S. Dept. of the Interior, Bureau of Land Management.*]

346

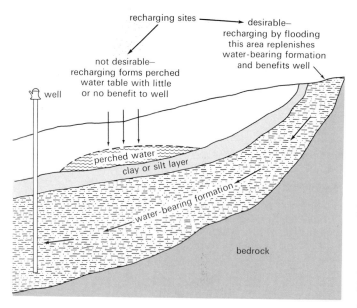

recharging sites → desirable—
recharging by flooding
this area replenishes
water-bearing formation
and benefits well

not desirable—
recharging forms perched
water table with little
or no benefit to well

well

perched water
clay or silt layer
water-bearing formation
bedrock

FIGURE 17-14. *Stratigraphy governs the replenishment of groundwater supplies by flooding. The water-bearing formation (acquifer) can be recharged only if there is not an impermeable layer above it.* [*Courtesy USDA.*]

formation of dew over a 6 month period. This was 20 percent of the total water available to the plants! In California, ponderosa pine can survive on soils below the wilting point, apparently because of the ability of the plants to absorb and utilize dew.

Cultural practices

CHOICE OF CROPS   The amount of water used by a crop during a growing season may determine where it is grown (Table 17-1). If fruit trees require 21 inches of water during the growing season, then they obviously cannot be grown in a region that normally receives only 11 inches of rainfall unless artificial irrigation can provide the other 10 inches.

Differences in the sensitivity of crops to moisture stress can be utilized to make maximum use of available moisture. When subjected to identical drought conditions, corn may die while sorghum survives by "waiting" for the drought to be re-

lieved. Rows of specialized cells in the leaves of many grasses lose turgor as a result of a reduction in moisture and respond by rolling up and curling inward, thus concealing stomates and reducing transpiration. Corn has relatively rigid leaf tissues and is less able to withstand extreme moisture stresses than sorghum. Sorghum is thus a logical choice over corn as a cash crop in a semiarid area.

FALLOW   The term "fallow" refers to the practice of keeping the land bare of vegetation by cultivation, and to the land kept bare by this practice. Its main benefit is the conservation of soil moisture. Fallow, more than any other single practice, has made intensive agriculture possible on the Great Plains of the United States and on the Canadian prairie. This ancient practice is most advantageously used in regions with winter rainfall and dry summers. Fallow is recommended where total annual precipitation is less than 16 inches in the northern United States and in Canada and where it is less than 20 inches in more southerly and warmer regions.

The fallow period between crops is generally 21 months. After harvesting in the late summer or fall, weeds are destroyed by a light surface cultivation. Winter precipitation recharges the soil. A crop is not grown the following year, but the soil is lightly worked, to kill weeds during the

Table 17-1   WATER REQUIREMENTS OF IRRIGATED CROPS IN RELATION TO THE REQUIREMENTS OF ALFALFA.

CROP	RELATIVE SEASONAL CONSUMPTIVE USE IN RELATION TO ALFALFA
Alfalfa	1.00
Pasture	0.90
Clover	0.90
Sugar beet	0.82
Citrus (fallow)	0.70
Corn	0.65
Potato (winter)	0.50
Tomato	0.48
Bean	0.42
Sorghum	0.38
Potato (spring)	0.30

347

growing season. After the second winter's precipitation, soil moisture is abundant enough to produce a crop. Winter cultivation by breaking the surface crust, prevents runoff of water and wind erosion on the smooth surface.

Fallow also permits the accumulation of nitrate. In the fall after a crop is harvested, the stubble, roots, and other organic residues are plowed into the soil. Decomposition of the organic matter releases nitrates and other minerals that remain in the soil if they are not utilized by weeds. Although bacterial denitrification accounts for some nitrogen loss, it is much less than the loss due to weeds.

Herbicides are sometimes used to keep fields in a fallow condition, but supplemental cultivation is often necessary. Chemical weed control is especially useful when erosion is a problem, and it is desirable to leave as much stubble and organic matter on the surface as possible. Stubble standing upright permits snow to drop to the soil rather than be swept away or accumulate at wind breaks (Fig. 17-15).

The excessive use of fallow as a dry-farming technique has led to many problems: wind and water erosion, the eventual depletion of soil minerals, and a decrease in the content of organic matter. Nevertheless, the technique permits the growth of crops in many areas that would otherwise be unproductive. In the future, the increased use of fertilizers and herbicides will offset some of the disadvantages of fallow.

WEED CONTROL   Although weeds have many undesirable effects (see Chapter 18), perhaps the most harmful is their use of soil moisture that would otherwise be available for crop plants. It has been estimated that natural vegetation may consume twice as much water as field crops. Cattails growing in the hot sun along a mile of irrigation canal may use as much water as 8 acres of alfalfa. Weed control becomes an integral part of water conservation; especially where water may be limited.

STRUCTURAL MODIFICATIONS AND MULCHES   A variety of artifical measures for increasing agricultural yield are presently in use in different parts of the world. One example of a structural modi-

FIGURE 17-15. *Stubble mulching is a common practice in dry-farming areas. Stubble helps snow to form an even blanket rather than pile up in drifts. Consequently, when the snow melts, the water is distributed evenly over the field. [Courtesy USDA.]*

fication is an integrated system of both rigid and air-supported greenhouses, in which the environments are strictly controlled, and which provide power, water, and food for the people of Abu Dhabi, an Arabian Gulf shiekdom (Fig. 17-16). Fresh water is produced from sea water by desalting facilities, which harness the heat from small engine-driven generators. Heating and cooling is achieved with sea water. Humidity is very high in the greenhouses, so that moisture is conserved. The yield per unit area is often higher than that from field production. Plant-growing structures are frequently shaded or cooled with a water spray to decrease transpirational losses.

Soil mulches also have a great effect in conserving soil moisture (see Chapter 11). Experimental studies in Alabama have shown that the use of black plastic mulch increases yields of corn, partly as a result of the increased availability of soil moisture.

LIMITING TRANSPIRATION   Crop growers have long sought a method of reducing water losses by plants. Some losses due to transpiration seem to be essen-

FIGURE 17-16. *Abu Dhabians and their camels stroll by the controlled-environment greenhouses, which use sea water for heating, cooling, and irrigation.* [*Courtesy Carle O. Hodge.*]

tial in helping to cool leaves, but at least a part of the water transpired merely moves through a relatively open system at rates dictated by solar radiation. One method of reducing such losses would be to apply a substance to leaves that would cause stomates to close. Most chemicals that do this also kill the leaves, but compounds have recently been discovered that temporarily close stomatal apertures without harming the plant. For example, the glyceryl ester of decenylsuccinic acid and related substances, have proven effective, but field methods of application must still be developed. In the future it may be possible to decrease irrigation requirements considerably by adding such chemicals via sprinkler irrigation systems or by applying them aerially to large acreages of unirrigated crops.

WATER MANAGEMENT AND WATER RESOURCES

All life depends upon water. It usually comes as a shock to find that plants require water in such large quantities, mainly because they lose so much in transpiration. An herbaceous plant will transpire perhaps twice its weight in water each day. A single corn plant in full leaf may require 8 gallons of water each week. A stand of hardwoods may transpire 3 to 4 thousand gallons per acre per day. A single tree requires perhaps 120 gallons of water to produce one pound of wood. The figures required to compute the water requirements of world crops are astronomical.

Water must also be regarded as an economic material. Costs vary from about $25 to $50 for 30 cm of depth per hectare ($10 to $20 per acre foot) in the United States, and this is cheap for the volume acquired; however it is expensive for our crop production because so much is required. For example, if precipitation for crop production had to be obtained at this price, the water alone that was needed to produce an economic forest crop would cost 87 to 100 dollars per hectare (35 to 40 dollars per acre) each year! In terms of value received from the crop, it would be impossible to grow forest crops. Many field crops would be uneconomic for the same reason.

Human biological processes require approximately a gallon and a half of water per person per day. People in primitive cultures can satisfy their needs for minimum comfort with 3 to 5 gallons per day, for such uses as washing and cooking, and may have trouble obtaining even this

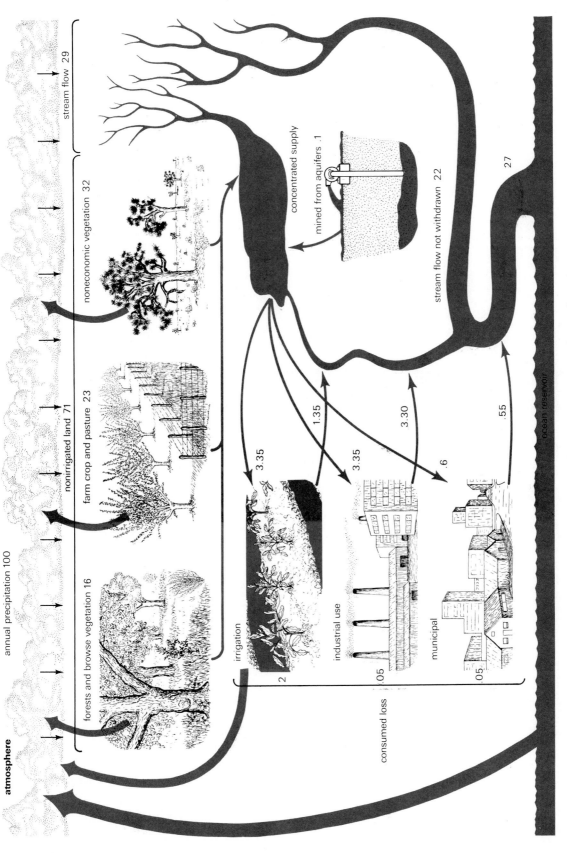

FIGURE 17-17. *The hydrologic cycle of the U.S. is typical of highly developed nations. Irrigation and manufacturing uses require about the same amount of water. However, most of the water applied through irrigation returns to the atmosphere through evapotranspiration. The numbers in the figure refer to the relative quantities of water involved in their hydrologic cycle. [From Revelle, "Water." Copyright © 1963 by Scientific American, Inc. All rights reserved.]*

350

FIGURE 17-18. *Tims Ford Dam and Reservoir in middle Tennessee. This is the first major TVA project planned specifically to provide opportunities for recreation development, hunting, fishing, water supply, water quality control, and shoreline development, as well as to serve the traditional purposes of flood control and power generation. [TVA photograph.]*

small amount. But people living in a modern technological society require tremendous amounts of water. In 1960 the average daily use of water for all purposes in the United States was 1,500 gallons per capita. It has been estimated that it will be 2,300 gallons a day for every man, woman, and child in 1980. Truly, the degree to which water is consumed is one measure of the technological development of a nation (Fig. 17-17).

The power of nations is, in part, reflected in their ability to supply water to their peoples. Wherever the Roman empire flourished, it saw to the building and maintainance of vast systems of aqueducts for irrigation and human use. Ruins of these structures can be seen today in Spain, North Africa, and elsewhere. The water distribution systems failed, or were destroyed, along with the empire that built them.

Water is becoming increasingly precious. Cities, states, and even nations are contesting each other, militarily and legally, for the privilege of using water from particular watersheds, lakes, rivers, and even wells. Water management is of urgent necessity to world health and world peace.

In this chapter, we have dealt with some of the contemporary water management techniques used principally by crop growers. The technical aspects of managing water for crop growth differ from those of providing water for human and industrial use largely in cost.

There is a fantastically large amount of water in the world. Much of it is potable without further treatment, and a large proportion of it is immediately usable for industrial purposes. The problem lies in the fact that it is frequently not present where it is most needed. Transportation costs depend on the size of the operation. It has been suggested that on a large scale, it may be cheaper to bring water from Baffin Bay to Tierra del Fuego than to convert sea water at the tip of South America. The same applies to waste disposal.

A partial answer to the problem of obtaining water for crop irrigation lies in multiple use (Fig. 17-18). Australia provides an excellent example. The great inland desert of that continent is bordered on the east by the Snowy Mountains. The rivers discharge water from these mountains to the southeast into the Bass Strait. The Snowy Mountain-Tumut project has reversed three of the rivers, sending them west through tunnels into man-made lakes. The water then drops, in several stages, through power-houses. At the end of its route, the water is used to stabilize the flow of the Murray and Murrumbidgee Rivers and supply water to their arid valleys.

The Arctic Ocean is probably the largest body of unused "fresh" water in the northern part of North America. It might not be an unsurmountable task to build tunnels and aqueducts to take a part of its water as far south as Mexico.

Cloud seeding is in its infancy. The technological possibilities in this area are limited only by the political, social, and legal consequences that will result when the technique is completely developed.

Solar stills, desalinization of sea water, direct use of brackish water, steam stills using nuclear energy, electrodialysis, compression distillation, freezing, high-pressure diffusion, and other techniques can be used to convert "impure" water to usable water. Crop growers of the future may need to use all of these techniques.

Selected References

Colman, E. A. *Vegetation and Watershed Management.* Ronald Press, New York, 1953. (A widely used reference on watershed management.)

Hagan, R. M., H. R. Haise, and T. W. Edminster (editors). *Irrigation of Agricultural Lands.* Agronomy Series No. 11, American Society of Agronomy, Madison, Wisconsin, 1967. (A monograph that treats irrigation from the standpoints of both engineering and crop production.)

Israelsen, O. W., and V. E. Hansen. *Irrigation Principles and Practices.* Wiley, New York, 1962. (A good reference book on irrigation practices.

La Mer, V. K. *Retardation of Evaporation by Monolayers.* Academic Press, New York, 1962. (Papers on the use of surfactants to reduce evaporation.)

Luthin, J. N. (editor). *Drainage of Agricultural Lands.* American Society of Agronomy, Madison, Wis., 1957. (This is an excellent series of essays on practically all of the many aspects of agricultural drainage.)

Roe, H. B. *Moisture Requirements in Agriculture.* McGraw-Hill, New York, 1950. (A comprehensive manual of agricultural water management both theoretical and practical.)

Sprinkler Irrigation Manual. Wright Rain Ltd. Ringwood, Hampshire, England, 1956. (A compilation by English workers.)

U.S. Department of Agriculture. *Water.* Yearbook of Agriculture, 1955. (A thorough coverage of water in relation to agriculture.)

CHAPTER 18

Crop Hazards

Worms have destroyed half the wheat, and the hippopotami have eaten the rest; there are swarms of rats in the fields, the grasshoppers alight there, the cattle devour, the little birds pilfer; and if the farmer loses sight for an instant of what remains on the ground, it is carried off by robbers . . . . (From an ancient Egyptian scribe.)

The successful cultivation of crops involves more than the proper combination of environmental factors. There are many natural shocks to which plants, like men, are heir. These include pests and various nonbiological factors. Of the latter, heat, frost, drought, flood, and nutritional disorders have already been discussed, and only fire, hail, and wind damage will be reviewed in the present chapter.

PLANT PESTS

Plant pests include all life forms destructive to plants—the whole biological spectrum is represented. Pests may be grouped under the broad terms of pathogen, predator, and weed; their respective injurious effects are disease, damage, and competition. The toll exacted from world agriculture by plant pests is heavy; field and storage losses are estimated to be as high as 35 percent. In the United States the yearly loss in farm crops alone is about $7 billion. Moreover, the annual cost of pest control to farmers in the United States is $430 million for insects, $230 million for diseases, and $2\frac{1}{2}$ billion for weeds! Some pests have such awesome destructive power that they are capable of decimating entire crops. Others, though less spectacular, are equally damaging in the long run; they continually erode our abundance while their damage passes more or less unnoticed. Everyone eventually shares the cost of these losses.

Injury

Broadly defined, the term **disease**—literally, "not at ease"—refers to any injurious abnormality. As generally used, however, the term refers to the injurious effects caused by some infectious agent (the **pathogen**) living in intimate association with its host plant for extended periods of time. This restrictive definition of disease applies nicely to the detrimental effects caused by many microorganisms. The relationship in which one living organism derives nourishment at the expense of the other is known as **parasitism.** Some parasites have evolved to such an extent as to be **obligate:** they can survive only on the living tissue of their host, often a specific one. Pathogens, with rare exceptions, are injurious parasites; all parasites, however, are not necessarily injurious. Organisms that live on dead organic matter or on inorganic material are called **saprophytes.** Many pathogens exist largely as saprophytes. The host range of some pathogens may be extremely large and include many different species; that of others may be very specific, limited even to particular cultivars within a species.

The harmful effects that insects and larger animals such as rodents or birds have on plants are often referred to as damage rather than disease because of the kind of association and the time factor, and the term **predator** is used rather than pathogen. However, the distinctions between predator and pathogen, disease and damage, are arbitrary. Many insects, for example, live in intimate association with certain plants for extended periods and are pathogens under the strictest definition of disease. The distinction between predation and parasitism is somewhat clearer when confined to the animal kingdom. Predators are usually larger than their prey, which they destroy outright. Parasites are usually smaller than their host, at whose expense they obtain room and board. Predators live on capital, parasites on income. In the United States, study of the adverse effects of viruses, fungi, and bacteria on plants has been pre-empted by the field of plant pathology, and the effects are considered diseases. The harmful effects of insects and mites are a specialty of entomology and are referred to as damage. (Jurisdiction over nematodes is claimed by both, rodents and birds by neither.) Any malaise of unknown origin is also the concern of the pathologist. Disorders shown to be caused by climatic or environmental conditions, the so-called physiological disorders, become the concern of physiologists. In the ensuing discussion no special distinction will be made between pathogen and predator or between disease and damage.

The injurious effects of weeds are usually due to direct competition for essential growth factors. A few, however, are also parasitic—for example, dodder, mistletoe, witchweed—and some secrete or leave a harmful residue in the soil.

The association between plants and pathogens and predators is part of the interrelationship and interdependence of living things. The resulting injury should not be regarded as intrinsically evil or particularly unusual. To do so would be to give an egocentric and anthropomorphic interpretation of natural phenomenona—an interpretation that is patently false.

Symptoms

The visible responses of plants to injury are known as **symptoms,** which together with evidence of the particular agent (**signs**) permit diagnosis (Fig. 18-1). Although the basic plant responses to injury are limited, the many changes that take place in different tissues and organs often permit accurate diagnosis from symptoms alone. Death (necrosis) may affect the entire plant, or it may be limited to specific organs (leaves, branches, flowers, fruits) or to small areas that appear as spots or holes. Decline may be gradual and incomplete. For example, chloroplast breakdown appears as yellowing or mottling but does not necessarily result in the death of the plant. Another plant response is reduction in growth rate, which may be general or restricted, and may result in stunting, dwarfing, or incomplete differentiation. In contrast, injury may increase growth to abnormal and morbid types such as enlargements of organs, tissues, or cells, or tumor-like protuberances called galls. Wilting or loss of turgidity may be due to mechanical obstruction of water flow in the vascular tissue, root injury, or increased water loss.

Disease cycle

The **disease cycle** consists of the sequential changes that take place in both pathogen and host. It involves the life history, the morphological forms and stages, of host and pathogen. There is considerable variation in life histories. For example, some lower organisms do not have a sexual stage but

354

FIGURE 18-1. *Symptoms of fungal diseases in apple.* [*Courtesy Purdue Univ.*]

exist as a single continuously repeating form (many bacteria); others reproduce asexually be means of spores (some fungi). Complex life histories involve diverse developmental stages, including a number of asexual cycles and even alternations of hosts in different stages.

The life history is often adjusted to the seasonal cycle. In temperate areas, nonmigratory organisms either die during the winter or survive winter in a stage of inactivity or dormancy. The life history of an organism may involve a number of different morphological stages. The sequence of stages initiated with the advent of the growing season are

called primary cycles, and those initiated subsequently are called secondary cycles. In areas where the change in seasons is not distinct, a secondary cycle may continue uninterrupted.

The disease cycle can be divided into a **pathogenic** phase, in which the organism remains associated with the living tissues of the plant, and an **independent** phase, in which it may be saprophytic, dormant, or pathogenic to another plant. The relative lengths of these two phases varies greatly. For example, the independent phase of plant viruses usually lasts only while they are being transported from plant to plant via insects, mites, or fungi. Some fungi, such as *Fusarium*, live sapro-phytically as a part of the soil flora and become pathogenic with the introduction of a suitable host plant.

The pathogenic phase may be subdivided into a number of stages. **Inoculation** consists of the transfer of some form of the pathogen (the inoculum) to the plant. Pathogens may travel short distances under their own power (for example, as free-swimming spores) or may be transported long distances by some vector, principally wind, water, insects, and man. The **incubation** period begins when the pathogen enters the plant and ends when the plant reacts—the infection stage. Inoculation and incubation are often important stages during which to attempt control, for by the time **infection** is manifested the pathogen is least vulnerable to control measures, and serious damage has been done.

It is impossible to generalize on the relationship between the disease cycle and the life history of a particular pathogen or predator. For example, the primary cycle of the fungus *Venturia inaequalis*, which causes the apple scab disease, begins with sexual spores (ascospores) that overwinter saprophytically on dead leaves. The secondary cycles begin when asexual spores (conidia) are produced during primary infection. The life history of this pathogen takes a whole year, and includes several asexual cycles (Fig. 18-2). In contrast, the life history of the codling moth, a serious insect pest of several fruits, particularly apple, may recur two or three times within a single year, depending upon the latitude. The larval stage is the damaging one.

Pathogen and predator

VIRUSES Viruses are small infectious particles made of a core of nucleic acid surrounded by a

FIGURE 18-2. *Disease cycle of apple scab, caused by* Venturia inaequalis. [*From Pyenson,* Elements of Plant Protection. *New York: Wiley, 1951.*]

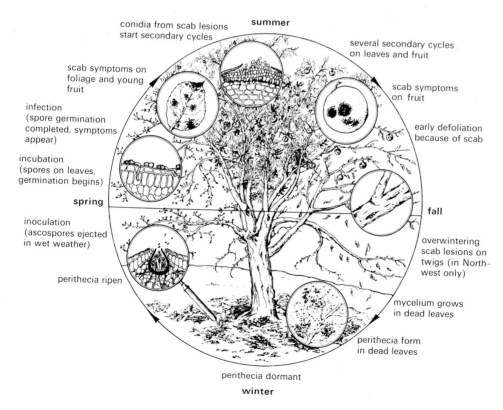

356

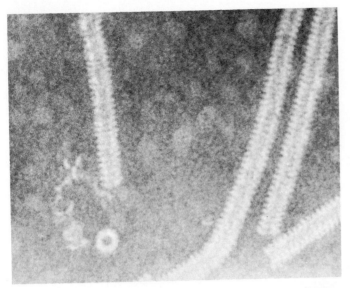

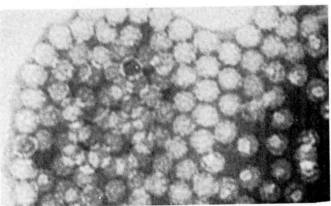

FIGURE 18-3. *Plant viruses. (Above) Portions of tobacco mosaic virus particles. The spiral arrangement of the protein sheath is visible in this electron photomicrograph. The complete virus particle measures 150 × 3,000 Å units (1 Å = 10⁻⁸ cm). (Below) The protein subunits are visible in this photomicrograph of the turnip mosaic virus particles (3,000 Å in diameter).* [*Courtesy R. W. Horne, Cambridge, England.*]

protein sheath (Fig. 18-3). Their small size (approximately one-millionth of an inch) makes them visible only with the electron microscope. Particles may be short or long rigid rods, flexible threads, or "spherical" (actually 20-sided) forms. Viruses are obligate parasites; they reproduce only in living cells of the host, but they may be removed from the organism and remain capable of causing infection. Some viruses remain active in extracted plant juices for many months; the tobacco mosaic virus remains active for years in dried material. Some can be crystallized and still cause the disease when reintroduced into the plant.

The question of whether viruses are living is a matter of terminology or philosophy. If the living system is defined as a self-duplicating entity capable of reconstructing itself from different component parts, then viruses can be considered alive. But viruses are not complete, living systems, because they are unable to generate the energy required for their multiplication and must exploit the enzyme system of infected, living cells.

The genetic structure of a number of viruses that infect bacteria (bacteriophages) has been determined. These viruses mutate and show genetic recombination that is akin to sexual reproduction. Thus various strains are possible. Their genetic material is arranged in linear sequence, and linkage maps (see Chapter 19) can be constructed on the basis of the recombination of characteristics that affect the expression of symptoms.

Viruses cause diseases of many organisms, including many important crop plants. Symptoms are usually systemic; generally the overall effect is a mild stunting with reduced quality and yield. Symptoms can be variable, and include small lesions, a spotted or mosaic pattern of green and yellowish areas, yellowing, stunting, leaf-curling and edge-crinkling, excessive branching, and color-striping or total disappearance of color. Occasionally local lesions are found on certain hosts. The virus is usually named by the description of the major symptom coupled with the name of the host plant where first described. Thus tobacco mosaic virus retains its name when found on tomato. Viruses are identified by a combination of symptoms, serological tests (with the use of antibodies from warm-blooded animals, usually rabbits), electron microscopy, and chemical analysis.

The effect of viruses on the plant depends upon the sensitivity of the host variety and the strain of the virus. Curly top, a virus disease that affects tomato, results in the quick death of the plant. Some viruses produce no obvious symptoms but still cause considerable economic loss by reducing yields and performance. A combination of viruses generally complicates and increases the severity of symptoms.

Viruses are commonly insect-transmitted, many by aphids. A few viruses are soil borne and some of these are transmitted by nematodes or by fungi. Some viruses are transmitted through the seed. But even when a virus is seed-transmitted only a portion of the seedlings become infected. Mosaic viruses may be transmitted by touching healthy leaves after having touched infected leaves or, more commonly, by rubbing infectious sap on healthy susceptible leaves. Viruses are commonly transmitted through a graft union, and some pass from plant to plant through dodder, a parasitic plant.

The vegetative propagation of virus-infected plants also propagates the virus. The only satisfactory way of maintaining virus-free stocks of vegetatively propagated plants is by the perpetuation of plants that are found to be free of the virus. Maintenance of virus-free plant stock is achieved by isolation, roguing of infected plants, and control of insect vectors. As a rule, once a plant contains virus little can be done to free it, although heat inactivation is effective against a few viruses, and excised shoot tips (particularly if rapidly growing) may be free of virus.

VIROIDS A number of virus-like particles implicated as the cause of plant diseases have resisted numerous attempts at identification by conventional techniques. Recently the infectious materials causing potato spindle tuber disease and citrus exocortis disease have been shown to be low molecular weight RNA (about $\frac{1}{10}$ the size of RNA in the smallest known virus). These infectious agents appear to represent a new class of pathogens now referred to as **viroids, infectious** or **pathogenic RNA.** Viroids also appear to be implicated as causal agents in some animal diseases.

MYCOPLASMA For many years a strange group of organisms have been known to exist that have the property of passing through filters that trap bacteria (as do viruses), but that live on inorganic media (as do bacteria). In the late 1960's the discovery was made that these organisms, now called **mycoplasma,** are also responsible for inciting plant diseases. Mycoplasmas are now implicated in a number of plant diseases formerly thought to be caused by viruses, particularly in a group called "yellows," which are principally spread by leafhopper insects. Examples of plant diseases known to be caused by mycoplasma include aster yellows and mulberry dwarf disease.

The size of mycoplasmas is variable. When grown in culture they produce long filaments resembling fungal hyphae, hence their name, which means "fungal-form." However the hyphae may break up into very small, round cells known as **elementary bodies,** ranging in size from 125 to 250 millimicrons. These are the particles that pass through bacterial filters. From the elementary bodies mycoplasma may form either branching filaments or "large cells" (500 mμ) that approach the dimensions of bacteria.

Mycoplasmas, unlike viruses, are definitely lifelike in that they contain enzyme catalysts, energy systems, as well as genetic information in the form of DNA and RNA. Although surrounded by a membrane, they have no exterior cell wall as do most bacteria. This explains their apparent resistance to penicillin, which works on the bacterial cell wall. However mycoplasmas are inhibited by tetracycline compounds (such as chlorotetracycline), and many suspected viral diseases have been proved to be caused by mycoplasmas by the application of these chemicals.

BACTERIA Many crop diseases are caused by bacteria, one-celled "plants" and the smallest of living "organisms." Seven genera of bacteria, none of which form spores, are plant pathogens. Bacteria are able to enter plants only through natural openings, such as the stomata and lenticels, or through wounds. Insects are an important agent in the transmission of bacteria.

One of the most serious bacterial diseases—and the one first shown to be caused by bacteria—is fireblight, a disease of apple and pear caused by *Erwinia amylovora.* Insects may disseminate this bacteria, which penetrates the plant either through the nectar-producing glands in the flower, causing death of the blossom (blossom blight), or through shoot terminals (shoot blight). The bacterial pathogen is also spread from infected parts to other parts of the tree by rain. The organism survives over winter in older bark lesions called cankers. The disease is extremely serious to pear and has confined commercial production in the United States to the Pacific states and the Great Lakes area. Even in these locations, however, careful control measures must be used, such as the constant removal of blighted wood, the use of antibiotic sprays, and the avoidance of rapid, succulent growth.

Symptoms of bacterial diseases include the death of tissues (angular leaf spot of cotton) and the formation of galls (crown gall of apples). The soft rots common in storage diseases are associated with pectin-dissolving enzymes produced by the bacteria. The wilting symptom of some bacterial diseases (Stewart's wilt of corn) are a result of vascular disturbances, specifically, a "plugging" of the vascular system by masses of the bacteria.

FUNGI Fungi cause most plant diseases. Except for certain primitive types, they are characterized by a branching, thread-like (mycelial) growth. Fungi do not have chlorophyll and hence depend ultimately on green plants for their food. They may be saprophytic or parasitic, and many are both. Fungi reproduce by spores, which may be mitotically (asexually) or meiotically (sexually) produced. The life cycle of most fungi is quite involved, comprising many different stages and forms.

Fungi fall into three, large, well-defined classes: **Phycomycetes, Ascomycetes,** and **Basidiomycetes.** Those whose sexual stage is not known (and many apparently never form a sexual stage) are combined in an arbitrary group as **Imperfects,** or **Fungi imperfecti.**

The Phycomycetes are primitive fungi whose characteristic feature is the absence of mycelial cross walls. Because many of the Phycomycetes are aquatic and rather like algae, they are popularly known as water molds. The spores of aquatic forms are motile. Examples of common crop diseases caused by species of Phycomycetes are downy mildews (many host plants including grape, grasses), root rots and damping-off (caused by a number of species in the genera *Pythium* and *Phytophthora*) and the infamous late blight of potato and tomato, caused by *Phytophthora infestans.*

Ascomycetes are distinguished by their specialized sac (*ascus*), which contains the sexual spores. Comprising over 25,000 species of great variety, the ascomycetes range from pure saprophytes to obligate parasites. They are responsible for the largest number of plant diseases: powdery mildews on cereals, fruits, and many other crops; Dutch elm disease; ergot of grains; brown rot of peaches and plums; black spot of roses; and apple scab. Also among the Ascomycetes, however, are truffles, which, being prized by man and pig alike, are imported into the United States from Europe to grace the table of the gourmet.

The Basidiomycetes, the "higher-fungi," produce a specialized sexual spore-forming structure called a **basidium.** They include the smuts, rusts, and fleshy fungi, such as mushrooms, and are almost as numerous as the Ascomycetes. Rusts (red) and smuts (black) are so named because of the color of spores produced on the host plant. They are among the most destructive scourges of grasses, especially cereal crops. The rust fungi are all obligately parasitic. Some require two unrelated plants for the completion of their extremely complex life history. For example, *Puccinia graminis,* which causes wheat rust, requires barberry for its sexual stage; *Cronartium ribicola,* which causes the white pine blister rust, alternates between the pine and species of *Ribes* (wild and cultivated currant and gooseberry); stages of the cedar-apple rust organism, *Gymnosporangium juniperi-virginianae,* infect both hosts. The smut fungi are facultative saprophytes; they may subsist as saprophytes but do not ordinarily complete their life history away from the host. Smuts cause diseases of corn, oats, barley, wheat, and onions. The fleshy fungi are important as wood rotters, and some species are cultivated. A strain of the common field mushroom, *Agaricus campestris,* is the favorite of the United States' mushroom industry.

Fungi spread in a number of ways. Spores are commonly produced in enormous numbers and are dispersed by air currents, water, and insects. Mycelia growing saprophytically spread soil-borne fungi. Most fungi penetrate directly into plant tissue, although wounds and natural openings greatly facilitate entry for some.

Symptoms of fungal diseases are extremely varied. The diseases may affect all of the plant or just parts of it. Plant-pathogenic fungi generally cause localized injury, although some are responsible for vascular wilts. The presence of visible (to the unaided eye) forms of the fungus, such as mycelial growth, masses of spores, or fruiting bodies, are usually an integral part of disease expression and often serve as an invaluable aid in identification (Fig. 18-4).

NEMATODES   Nematodes are roundworms of the phylum Nemata (not to be confused with the segmented earthworms of the phylum Annelida) (Fig. 18-5). Although the majority of nematodes are saprophytes and represent a significant portion of the soil fauna, many species are parasitic on plants and animals, including man. The importance of nematodes in causing plant injury is becoming increasingly apparent.

Nematodes generally are about $\frac{1}{64}$ to $\frac{1}{8}$ inch in length and are thus inconspicuous. The majority of nematodes that attack plants generally feed on the roots. They may feed superficially or be partially or completely embedded in the root tissues. A few species feed on aerial portions (for example, the seed-gall nematode of bentgrass). Some are quite specialized and attack only a few species of plants; others have a wide host range.

There are two common types of plant-attacking nematodes. Those belonging to the genus *Meloidogyne* (root knot nematodes) produce gall-like overgrowths on roots. These readily observable symptoms allow positive identification of nematode injury and are used as a criterion in the inspection of planting stock. The infected roots afford access to other bacterial and fungal rots. The root tissues, especially the vascular, become disorganized; giant multinucleate cells are often formed. The above-ground symptoms appear as excessive wilting and weak yellow growth similar to that resulting from drought injury. Other plant-parasitic nematodes, such as those of the genus *Pratylenchus* (meadow nematodes), do not form galls. Both the root damage and the above-ground symptoms resemble those of root rot. Owing to the diagnostic difficulty, meadow nematode may be easily transported in infected planting stock.

INSECTS AND MITES   Insects and mites are members of the phylum Arthropoda. The arthropods (literally "jointed legged") include the invertebrate animals having an external skeleton; paired, jointed limbs; and a segmented, bilaterally symmetrical body. The Arthropoda is a very large group, containing about three-quarters of all known animal species, of which 90 percent belong to one class, Hexapoda, the true insects. The hexapods have three pairs of legs, and almost all are winged at a certain stage (although the wings of some are rudimentary or degenerate). They have three body regions—head, thorax, and abdomen. Almost 700,000 species of hexapods are known, and the estimated total world number ranges from 2 to 10 million. Mites, the other important group of plant pests, belong to the class Arachnida. Arachnida have four pairs of legs, are wingless, and, unlike the hexapods, do not have a separate head. A brief classification of the Arthropoda, showing the main orders of hexapods that attack plants, is given in Table 18-1. In the following discussion the term "insects," when used nontechnically, will include mites.

Man's struggle with crop-damaging insects is a continuous one. The battle lines are not clearly drawn, for we are aided by the intense competition between insect species; predatory insects must be regarded as beneficial. Furthermore, man rid of all insects might be worse off, since many are necessary for the pollination of crop plants.

Insects have many built-in advantages. These include their small size, which makes them difficult to find; their ability to fly; their extremely rapid rate of reproduction; and their specialized structural adaptions, which enable different species to exist in practically any location and damage any plant species. The division of the life cycle of insects into separate stages (**complete metamorphosis**) is of tremendous advantage to them in that it permits specialized structural adaptation for feeding and reproduction. The insect's life cycle consists of four stages: (1) egg, (2) larva, a feeding stage; (3) pupa, a quiescent stage in which the larva is transformed into the adult form; and (4) adult, the reproductive stage. Some insects, such as grasshoppers, have a gradual or **incomplete metamorphosis,** in which the physical changes from egg to adult are gradual. The intermediate stages are known as nymphs.

The larval stage of insects is often the most injurious to plants; thus, many of the descriptive names given to insect pests (tomato hornworm, apple maggot) refer to this stage. There is generally no obvious resemblance to the adult stage (for example, caterpillar and butterfly). The ordinary terms for the larval stages of the insect orders are not specific. The term **maggot** generally refers to the larval stage of Diptera that tunnel within leaves. **Caterpillar** is the larval stage of Lepidoptera (butterflies and moths). The name **grub** is used with reference to some of the soil-borne large larvae of the Coleoptera (beetles), although the name is loosely applied to any soil-borne larvae. The larvae of "click" beetles are called **wireworms.** The larval and adult forms of beetles that infest grains and seeds are called **weevils.** The name **slug** is applied to any slimy larvae, specifically to larvae of the Hymenoptera (bees, ants, and wasps). (Shell-less snails are also known as slugs.) Moth or beetle larvae that tunnel within roots or stems are known as **borers.** The term "worm" is only incorrectly used to refer to insect larvae.

Insects and mites injure plants directly in their attempts to secure food and shelter, and indirectly by spreading plant pathogens. Moreover, their

FIGURE 18-4. *Symptoms of fungal diseases of corn.* (A) *Stalk rot caused by* Diplodia maydis. (B) *Stalk rot caused by* Pythium aphanidermatum. (C) *Crazy top caused by* Sclerophthora macrospora. (D) *Smut caused by* Ustilago maydis. [*Courtesy of A. J. Ullstrup.*]

B

A

decomposed bodies or excreta contaminate plant products. They exhibit two basic feeding patterns, each associated with specialized mouth parts. **Chewing insects** tear, bite, or "lap" portions of the plants; **sucking insects** pierce or rasp the plant and suck or sponge up the sap, or often passively allow the pressure of the sap to fill them up.

Chewing insects, adults or larvae, may eat their way through the plant, riddling it with holes and tunnels. Leaves may be skeletonized by insects that do not eat the tougher, vascular portions; the entire plant may be devoured by the less selective. Injury due to chewing insects that feed externally is seldom confused with anything else (Fig. 18-6). It is more difficult to diagnose damage caused by chewing insects that girdle the plant or those that feed internally or on roots. The internal feeders

gain entrance to the plant from eggs deposited in the plant tissue or by eating their way in soon after being hatched on or near the surface of the plant. The symptoms of the peach tree borer are typical of the symptoms caused by internal feeders— namely, weakened, devitalized growth of the tree and yellowing of the foliage. A gummy exudate may be observed where the borer has wounded the trunk. Peach or plum trees infested with borers often die in a few years.

Injury by sucking insects (aphids, scale insects, leafhoppers, and the plant bugs) results in distinctly different symptoms, the most common being curling, stunting, and deforming of plant parts, usually stem terminals. Spotting, yellowing, and a glazed appearance of the leaves are also common symptoms. Red spider mites have sucking mouth parts

C

D

FIGURE 18-5. *Nematodes of two genera representing different morphological forms. (A) The swollen female of* Meloidogyne *(root knot nematode). (B) The wormlike female of* Tylenchorhynchus *(stunt nematode).* [*Courtesy W. F. Mai; photograph by H. H. Lyon.*]

and produce similar symptoms as well as webs. Mites are difficult to see without a magnifying glass, and by the time webs are formed damage may be extremely severe.

Overgrowths and galls, the result of abnormal growth of plant tissue, are symptoms of either the feeding of chewing or sucking insects or merely the presence of eggs deposited in the tissue. Such galls, suggestive of cancerous growth, may be quite elaborate and structured. Some appear to do little harm; others obviously are quite injurious.

BIRDS AND RODENTS   Among the chordates, or backboned animals, birds and rodents are the greatest crop pests. Rats and mice are serious field pests and often are responsible for large storage losses. In India the diet of an estimated 2.4 billion rats, five times the human population, has been calculated at 26 million tons of grain, almost one-third of India's present production. The prevention of this loss would transform India from a nation with a food deficit to a one with a food surplus. Mice can be destructive orchard pests. They feed on living trees, which may be completely girdled in winter and early spring. Apple trees are particularly susceptible to injury, and unless prompt action is taken by repair grafting, large trees may be killed outright.

Birds rival rodents as plant pests. Billions of weaver birds (red-billed quelea), one of the most numerous and destructive birds in the world, threaten grain production in Africa. Birds may be quite troublesome to fruit crops, especially cherries and grapes; even a few pecked grapes permit the introduction of diseases and other pests, and in effect spoil the whole cluster.

WEEDS   A weed may be defined as any undesirable plant. More typically the term refers to certain naturally occurring aggressive plants injurious to agriculture. Crop losses are usually a direct result of the competition with weeds for light, water, and mineral nutrients, as well as the lowering of quality. Unchecked noxious weeds tend to dominate crop plants. Weeds also cause crop losses indirectly by harboring insect pests. They present a safety hazard on roadsides and railroad rights-of-way, and may clog irrigation ditches and streams. Some weeds adversely affect the health and comfort of man and livestock. A few are poisonous or irritating (poison ivy, white snake root), and the pollen of some (such as ragweed) is a source of misery to millions of allergy sufferers. Moreover, weeds are unsightly, as any homeowner will testify. The annual cost of weed control in the United States exceeds that of all other types of pest control combined.

The fierce competition offered by certain weeds is due to a combination of their high reproductive capacity and vigorous, exuberant growth. Their

Table 18-1   PARTIAL CLASSIFICATION OF ARTHROPODA, INCLUDING
MAJOR ORDERS OF INSECTS THAT ATTACK PLANTS.

| CLASS AND ORDER | EXAMPLES | TYPICAL MOUTH PARTS | |
		LARVAE	ADULTS
CHILOPODA	Centipedes		
DIPLOPODA	Millipedes		
CRUSTACEA	Crayfish, lobster		
ARACHNIDA	Scorpions, mites, ticks, spiders		
HEXAPODA (INSECTA)	True insects		
Orders with gradual meta-morphosis			
Orthoptera	Grasshoppers, crickets		chewing
Thysanoptera	Thrips		rasping-sucking
Hemiptera	True bugs		piercing-sucking
Orders with complete meta-morphosis			
Hymenoptera	Bees, ants, wasps	chewing	chewing-lapping
Coleoptera	Beetles	chewing	chewing
Lepidoptera	Butterflies, moths	chewing	syphoning
Diptera	Two-winged flies and mosquitos	chewing	piercing-sucking sponging

FIGURE 18-6. *Insect damage. (Left) Cabbage looper on cabbage. (Right) Grasshopper on corn. [Photographs by J. C. Allen & Son.]*

destructive power is a funtion of sheer number. An average of 20,000 seeds per plant—varying from 250 (wild oat) to several millions (tumbling pigweed)—is reported in a study of 181 annual, biennial, and perennial weed species. The seeds of many are structurally adapted for dispersal by wind, water, or animals. For example, many weeds have hard seed coats and remain viable when passed through the digestive tract of animals. Man has unwittingly become one of the chief disseminators through the shipment of crop seeds and plants. Seed numbers alone do not account for their effectiveness. Many crops are prolific seed producers, yet do not become really weed-like. The high reproductive capacity of weeds is due to their extended seed viability and delayed germination brought about by dormancy. The failure of weed seed to germinate may be due to natural (see Chapter 8) or induced dormancy. Induced dormancy is brought about by environmental conditions that prevent germination. Seeds buried deep in the soil by tillage may lack either sufficient oxygen or, in some plants, the light stimulus necessary for germination. The viability of weed seed buried in the soil is usually much greater than that of the seeds of most crop plants. In one famous "buried seed" experiment, the seeds of more than half of 20 species germinated after 25 years and showed some viability after 90 years. The combination of dormancy, extended seed viability, and high seed production complicates weed control. If weed seeds would all germinate at once, control might be accomplished by a rigorous and intensive eradication program. The high weed seed populations of agricultural soils make weed control a continuing and integral part of crop culture.

Many weeds reproduce vegetatively as well as by seed. Some of the most pernicious weeds reproduce in this way: johnsongrass and quack grass by rhizomes; nutsedge by corms; wild morning glory by roots; and wild garlic by bulblets. Vegetative reproduction by roots or underground stem modifications makes control by cultivation particularly difficult.

The competition between weed and crop adversely affects both, but weeds often win out. The exact physiological basis of the growth advantages of weeds is not clear but includes rapid germination and seedling growth and a deeply penetrating yet fibrous root system. Furthermore, weeds often possess a natural resistance to many of the pests that plague crops. Hardiness is undoubtedly another factor that benefits weeds.

The luxuriant growth that is characteristic of weeds might be a trait that has been sacrificed in crop plants during the process of domestication, just as seed dormancy has been lost in most seed-propagated crops. A more reasonable explanation is that weeds must be uniquely adapted to their surroundings to survive, whereas crop plants may be grown in locations far removed from their area of specific adaptation. Weed species change far more dramatically than crop species across the United States.

Plant protection methods

The control of plant pests can be considered as an agricultural discipline called **plant protection.** A highly specialized and rapidly changing technology has been developed around the control of particular pests (microorganisms, nematodes, insects, weeds), and those who have participated in this development have organized themselves into groups, complete with society affiliations, special nomenclatures, and scholarly journals. Although the techniques of control vary with pest and crop, the basic approach is to either interfere with some stage in the life cycle of the particular pest or protect the host plant. The most successful treatments are preventative rather than curative; they aim to deter injury rather than cure already affected plants. All methods depend upon an intimate knowledge of the biological agents, their natural history and ecology, and, especially, their specialized relationship to the crop. Successful control often depends on the proper use of many different methods.

CULTURAL METHODS Cultural practices used to reduce the effective pest population include the elimination of diseased or infested plants or seeds (roguing), the cutting out of infected plant parts (surgery), or the removal of plant debris that may harbor the pest (sanitation). The reduction of pest populations may be achieved by alternately planting crops unacceptable to the pest (rotation). Weed populations can be reduced by rotations involving well-adapted crops like alfalfa and silage corn that may outcompete weeds. Often it is the kind of tillage used in the rotation that brings weeds under control. For example, cultivation of row crops may reduce the grassy weed populations in subsequent small-grain plantings.

Plants, unlike animals, do not have an antibody mechanism that can be utilized to resist disease and thus cannot be made immune by vaccines. Nevertheless, the physiology of the plant can be altered to affect the plant's ability either to resist invasion by the pathogen or to overcome its deleterious effects. For example, the vascular wilt of maple, caused by the fungus *Verticillium albo atrum*, can be compensated for by applications of fertilizer, which will cause the plant to literally outgrow the pathogen. The reverse technique is utilized in fireblight of pears, where infection and growth of the bacterium causing the disease is extremely rapid in fast-growing, succulent shoots. Slowing down growth of the tree by eliminating excessive nitrogen fertilization or by avoiding extensive pruning is one means of control. Application of inorganic nutrients gives protection in some instances. For example, clubroot of cabbage appears to diminish in severity when the ratio of calcium to potassium in the soil is decreased.

Pest populations may be reduced by methods that render the environment unfavorable, including such diverse practices as draining land, flooding, pruning to reduce foliage density and to increase the rate of drying, and varying the temperature and humidity. Potato scab is commonly controlled by lowering soil pH.

Many serious plant pests are introduced from other areas. Although separation of the pathogen and plant cannot be legislated, safeguards against the spread of pests between neighboring countries and states may be enforced by **quarantines**—restrictions governing the importation of certain plants or products. Many countries prohibit the shipment of uninspected plants.

PHYSICAL CONTROL Physical methods may be used to protect the plant against intrusion or to eliminate the pest entirely. The erection of physical barriers to protect crops is among the oldest of treatments. These range from the traditional garden fence to selective screening, such as guards to protect tree trunks against mice, and screenhouses to eliminate insects or birds. In Japan, individual fruits of many types are enclosed in bags to protect them from insect injury.

Traps have been used to control insects. Attractants of various types are used to lure them into lethal solutions. These include "black light" (light of wavelengths between 3,400 and 3,800 Å), various sounds, and chemicals that mimic sex attractants (pheromes). Trapping is recommended in some areas to reduce the population of such birds as the starling.

Physical techniques are obviously difficult to use against microorganisms. When their presence becomes known through symptoms it is difficult to destroy them physically without also damaging the crop. Perhaps the most successful method is the use of heat. Hot water treatment is used for destroying seed- or plant-borne pathogens—for example, loose smut, a fungal disease of wheat; black rot, a bacterial disease of crucifers; and nematodes on dormant strawberry plants. Heat treatment has also been effective in inactivating some viruses. Steaming of potting soils is a common greenhouse practice for eliminating many soil pests, including fungi, nematodes, and weed seed.

Physcial control is still widely used for weed control. The pulling or grubbing of weeds is the simplest and oldest method. The hoe is still the basic farm implement in many parts of the world. Various mechanical devices have been developed to automate this process. The principle is to cut out, chop up, or cover up weeds and thereby destroy them. Weeds propagated by roots are difficult to control by cultivation and may actually be dispersed in this way.

Cultivation and tillage, the loosening or breaking up of the soil, are such widespread agricultural practices that benefits in addition to the control of weeds have been ascribed to them. Many experiments indicate, however, that the primary benefit of cultivation is weed control. The other advantages, such as increased soil aeration, or conservation of soil moisture through the formation of a soil or dust mulch, may be counteracted by the destructive effects of inadvertent root pruning in the upper and most productive soil layer. In addition, extensive cultivation also contributes to considerable soil erosion.

Other physical means of destroying weeds are mulching and fire. Black polyethylene film has been useful as a mulch to control weeds, but cost limits its use to high-value crops. Flame throwers have been adapted to control weeds in such crops as cotton and onions.

CHEMICAL CONTROL   **Pesticide** is the generic term for all chemicals used to control pests. As the name infers they are usually toxic to some stage of the pest. Included under the legal definition of pesticide are **repellents**—compounds that may not be actively poisonous but make the crop plant unattractive to animal predators by virtue of their odor, taste, or other physical properties.

Selectivity.   A selective pesticide is one that kills certain organisms and does not harm others. Selectivity, however, is a relative concept and depends to a large extent on the interaction of a number of factors: dosage, timing, method of application, chemical and physical properties of the material applied, and the genetic and physiological state of the organisms involved. Nonselective pesticides kill indiscriminately. In general, most pesticides are selective to some degree. Thus pesticides are commonly classed according to the organism they affect—for example, bacteriocide, fungicide, nematocide, miticide, insecticide, herbicide, rodenticide, etc.

Many pesticides are selective within a broad group of organisms. For example, insecticides may be classified by their action as stomach or contact poisons. **Stomach poisons** are usually used against chewing insects. They are ingested along with the chewed plant material and cause death when they reach the stomach. Since sucking insects feed on the sap and ingest no external plant material, they are seldom affected by stomach poisons. They must be controlled by **contact poisons,** which kill by penetrating the insect body directly or entering through breathing or sensory pores.

Herbicides, chemicals that kill plants, are absorbed by the plant and cause death by producing some toxic reaction. Those that kill only the area of the plant actually covered (for example, dinitro-compounds, oils, and many arsenicals) are called **contact herbicides;** others that are translocated within the plant (for example, 2,4-D) are called **noncontact herbicides, or translocated herbicides.** Selectivity may be brought about by directing contact herbicides away from the crop plant (positional tolerance) or by taking advantage of inherent morphological differences between the crop plant and the weed that will prevent the crop plant from being killed along with the weed. These include the amount and type of waxy cuticle, the orientation and shape of the leaves, and the location of the growing point. Selectivity may also be achieved by controlling the dose. In high concentrations 2,4-D will seriously damage or kill all plants, whereas in low doses it can be effectively used to kill selectively (Fig. 18-7). Selectivity may be achieved in still another way—by exploiting physiological differences between plants. The mechanisms of physiological selectivity are largely unknown; in fact, the precise way in which most herbicides kill is obscure. Some interfere with enzyme systems; others disturb the metabolism of the plant. Physiological selectivity thus reflects metabolic differences between plant species. Selectivity has been related in some plants to differences

FIGURE 18-7. *Selectivity of 2,4-D is a function of dose. Arrows on mortality curves of two species indicate relative concentration required to kill 50 percent of the plants.* [*From Leopold,* Auxin and Plant Growth. *Berkeley: Univ. of California Press, 1955.*]

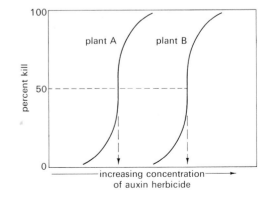

in cell membrane permeability and in others to differences in the ability to be translocated. Physiological selectivity may change drastically with stage of growth (Fig. 18-8). Many herbicides are most effective during seed germination and early seedling growth.

Chemical and physical properties. A great number of compounds, inorganic and organic, natural and synthetic, are pesticidal (Table 18-2). At present, pesticides become outmoded very quickly, as new or improved substances are continually being released by a vigorous and expanding industry.

The earliest fungicides and insecticides were inorganic salts of such metals as copper, mercury, lead, and arsenic; many of these are still useful. Natural organic compounds came into prominence in the nineteenth century. These include the insecticides derived from plant sources—nicotine, pyrethrum, and rotenone. The modern pesticide industry is based on synthetic organic materials. This industry was given great impetus by two widely successful materials whose names are today common terms—DDT and 2,4-D.

The insecticidal value of DDT (dichloro-diphenyl-trichlorethane) was discovered in 1939, although the material had been known 65 years previously. It acts on the sensory nerves of insects and is effective at very low concentrations. Introduced by the Geigy Co., a Swiss dye firm, it was manufactured in great quantities in the United States during World War II and was credited with control of the malaria mosquito in many parts of the world. The phenomenal results attending its use led to the investigation of a great number of similarly constructed compounds or analogs, which led in turn to many other insecticides—methoxychlor, BHC (benzne hexachloride), chlordane, aldrin, dieldrin, and other materials.

DDT is an extremely persistent and fat-soluble molecule that magnifies as it passes along food chains (refer to Fig. 5-17). This magnification aroused great concern and resulted in the application of strong pressure to reduce the indiscriminate use of DDT and other chlorinated hydrocarbons. In the United States, DDT was restricted for most uses in 1972. Many agriculturists have felt that the "banning" of DDT was an emotional issue, because a careful analysis of its action had demonstrated that, when properly used, this chemical posed no threat to environment, wildlife, or man. Despite the ban on most agricultural uses, DDT is still available for public health and quarantine uses.

The origin of the "broadleafed weed killer" 2,4-D can be traced back to research on the herbicidal action of auxin in the early 1940's. The research effort was stimulated by the war, and most of the early studies were shrouded in military secrecy. The precise way in which minute quantities of phenoxy, auxin-like, materials cause death is still not fully known, but it is believed that these materials cause a rapid depletion of energy carriers and the accumulation of toxic metabolites. The phenoxy materials have a close structural relationship to each other and to indoleacetic acid, as is shown in Figure 18-9. However, small alterations in the molecular structure of these materials greatly change their herbicidal and physical properties. For example, the amine form of 2,4-D is

FIGURE 18-8. *Changes of susceptibility of grains to auxin herbicides at various stages of development.* [*From Leopold,* Auxin and Plant Growth. *Berkeley: Univ. of California Press, 1955.*]

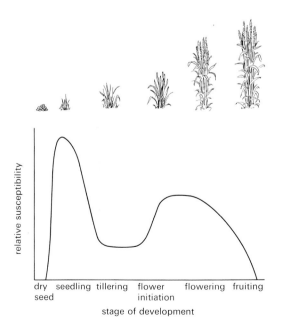

368

Table 18-2 CHEMICAL CLASSIFICATION OF SOME IMPORTANT PESTICIDES.

CLASS OF MATERIAL	SPECIFIC EXAMPLE	MAJOR TYPE OF ACTION
Inorganic		
	lead arsenate	insecticide
	copper sulfate	fungicide, bacteriocide
	mercuric chloride	fungicide, bacteriocide
	lime sulfur	fungicide
	ammonium sulfamate	herbicide
	potassium cyanante	herbicide
	sodium arsenite	herbicide
	sulfuric acid	herbicide
Organic		
Natural compounds and derivatives		
antibiotics	streptomycin, cycloheximide	bacteriocide, fungicide
nicotine	nicotine sulfate	insecticide
pyrethrins		insecticide
rotenone		insecticide
oils	stoddard solvent, diesel oil	herbicide
		insecticide
Synthetics		
Amide-like compounds		
carbamates	benomyl	fungicide (systemic)
	zineb, maneb	fungicide
	Sevin (carbaryl)	insecticide
	IPC, CIPC, EPTC	herbicide
acetamides	propanil, CDAA	herbicide
ureas	monuron, diuron, linuron	herbicide
triazines	simazine, atrazine	herbicide
Chlorinated hydrocarbons	DD, EDB	nematocide
	DDT aldrin, dieldrin	insecticide
Chlorinated aliphatic acids	TCA, dalapon	herbicide
Mercuries	Ceresan, Semesan	bacteriocide, fungicide
Organophosphate	malathion, parathion	insecticide, miticide
Phenoxy acids	2,4-D, 2,4,5-T, MCPA, Sesone	herbicide
Phthalimides	captan, phaltan	fungicide
Substituted phenols	DNBP	herbicide
Miscellaneous		
	DCPA (Dacthal)	herbicide
	amino triazol (amitrole)	herbicide
	trifluralin	herbicide
	Warfarin	rodenticide
	methyl bromide	general pesticide (soil sterilant)

much less volatile than the ester form, and is thus less subject to drifting. They have a short residual life in the soil and are apparently not generally toxic to animals. The selectivity of 2,4-D at low dosages is responsible for its main agricultural use—killing broadleaved plants without affecting grasses (Fig. 18-10). Since it was discovered that 2,4-D is a successful selective herbicide, research has uncovered many others. For example, the chlorinated acetic acids such as TCA (trichloroacetic acid) and dalapon (2,2-dichloropropionic acid)

selectively kill monocots without harming dicots, and thus control grassy weeds in broadleaved crops.

Pesticides may be applied in various forms and ways. Efficient application requires uniform coverage at a controlled rate. As it is seldom practical to apply chemicals in their pure form, they are distributed in dilute form in an inert carrier. If the carrier is a solid (talc, clay, or diatomaceous earth), the pesticide is applied as a dust. Dusts have the advantages of lightness and are particularly suitable for application by airplane or helicopter,

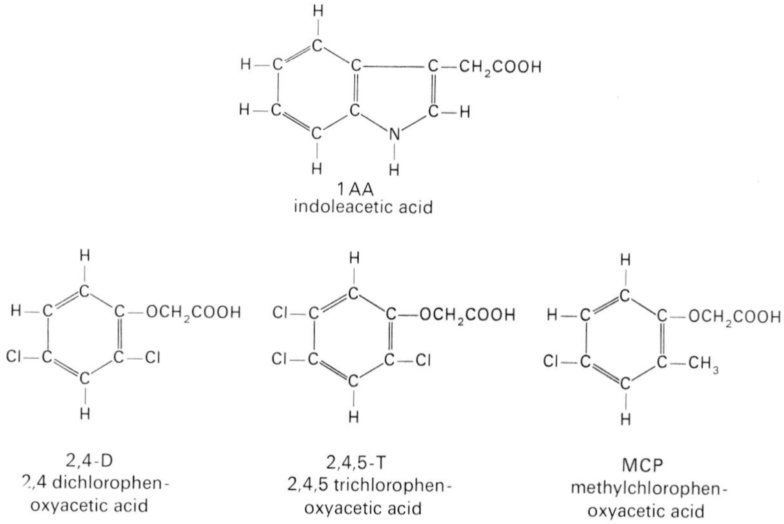

IAA
indoleacetic acid

2,4-D
2,4 dichlorophen-
oxyacetic acid

2,4,5-T
2,4,5 trichlorophen-
oxyacetic acid

MCP
methylchlorophen-
oxyacetic acid

FIGURE 18-9. *Structures of auxin-like herbicides in comparison with structure of indoleacetic acid.*

but they are difficult to apply uniformly, and do not persist for long. Moreover, some materials (for example, oils) cannot be applied in this manner. These disadvantages may be overcome by the use of a liquid carrier, usually water, in which the chemical is dissolved, suspended, or emulsified. The material is applied by pressure in droplets of various sizes but usually in the form of a spray (Fig. 18-11). The disadvantages of the weight and bulk of the water carrier are being overcome by using high concentrations of the active material and achieving dispersion by a blast of air (Fig. 18-12). On heavily cutinized plant surfaces sprays form droplets instead of a continuous film. This irregular dispersion may be overcome by adding wetting agents—chemicals that break the surface tension. Other additives, known as "stickers," are used to improve the adherence and persistence of pesticides.

Volatile pesticides active as gases are known as fumigants. Fumigants are usually stored under pressure as liquids; in aerosol bombs the active ingredient is dispersed in a gaseous carrier. Some fumigants are solid, and volatilize when heated or moistened. For maximum efficiency fumigants must be applied in an enclosed system, such as a storage bin or a greenhouse. Fumigants are finding

increasing usefulness in soil treatment. To prevent the fumigant from escaping, the surface of the soil may be sealed by covering it with plastic film or a layer of water, or merely by compacting the upper layer.

Pesticides may be **systemic** or **nonsystemic** in their action. Systemic pesticides are absorbed by the plant and translocated within it, rendering the plant itself toxic or repellant to the pest. Their use is limited in edible crops unless they break down within the plant before consumption. The action of nonsystemic pesticides takes place on the surface of the plant, although they may be absorbed to some degree.

Timing of application. Effectiveness of chemical control usually depends on timing of applica-

FIGURE 18-10. *The drop-size spectrum.*

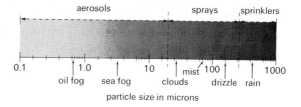

370

FIGURE 18-11. *Orchard sprayer. A blast of air is used as a carrier for high-concentrate sprays [Courtesy Food Machinery and Chemical Corp.]*

FIGURE 18-12. *Results of chemical weed control in peanut. (Left) Untreated. (Right) EPTC applied as a preplanting soil-incorporated treatment has given excellent control of nut sedge and all other broad-leafed weeds and weedy grasses. [Courtesy USDA.]*

tion. A certain pesticide may not be equally toxic to all forms of a particular pest, nor is this necessary; usually one stage or another is especially vulnerable, which is why timing is important. The weak link may be a germinating fungal spore, the young larval stage of insects, or the insect vector of a virus disease. For most diseases, chemical control must be applied before symptom expression. For example, it is difficult to kill fungi after they have entered the plant, but control may be achieved by materials that either kill or prevent spores from germinating on the plant.

Timing is especially important in weed control. Selectivity may be achieved for fumigants and other nonselective herbicides when applied before the crop is planted (**preplanting treatment**). Selective preplanting herbicides are also available that attack germinating weed seeds without harming crop seeds or transplants. Applications made after the crop is planted but before it has emerged from the soil are known as **pre-emergence treatments** (Fig. 18-12). The action of many of these materials is restricted to the soil surface. To be effective, pre-emergence materials must have good coverage and be relatively resistant to leaching. Soil moisture is often critical, for water may be required to activate the herbicide, although an excess may cause leaching. Pre-emergent herbicides are physiologically selective between weed and crops. Selectivity may be enhanced by the early germination of weed seed. The germination of crop seed is usually slightly delayed because of the depth of planting and the time required to imbibe water. Applications of herbicides to the growing crop are referred to as **post-emergence** treatments. Selectivity must be positional (from directed application) or physiological.

Spray injury and residues. The beneficial effects of chemicals must be balanced against potential hazards. To provide for the safe and proper use of these materials restrictions are imposed, especially when these substances are used in connection with foodstuffs. Care must be taken to ensure that the cure is less injurious than the disease! Some chemicals leave soil residues that may interfere with plant growth and future land use. Their eventual disappearance may result from vaporization, chemical breakdown, biological de-

composition, leaching, or adsorption onto soil colloids. Some pesticides are phytotoxic and can reduce yields and quality. Of greatest concern, however, is the possibility that the pesticide or harmful breakdown products may persist in the plant product or in the animal that feeds on the plants.

The potential hazard that pesticidal residues represent to consumers is a topic that has been widely discussed. Every substance, whether it be food, drug, cosmetic, or pesticides, has a finite hazard. All chemicals have some dose at which they are toxic, and some dose at which they are nontoxic. Poison is a relative term: *dosis sola facit venenum*—the dose makes the poison. Many common "poisons"—arsenic, for example—are stimulatory at very low dosages and, in fact, have medicinal properties. As dosage increases they become injurious and ultimately toxic. Some substances have no noticeable effect at low concentrations but accumulate in certain portions of the body. The problems associated with these substances are real. Any undue risks to the consumer as well as applicator cannot be tolerated.

The Environmental Protection Agency of the United States, the governmental agency that regulates the use of pesticides and other chemicals, sets tolerance levels. The tolerance level, often conditioned by a safety factor of a hundredfold, sets maximum detectable amount, zero or otherwise, that is safe. Tolerance levels are determined through toxicity studies on animals. Pesticides are an essential ingredient in agriculture, and, like many substances, they can be used or misused. When used with good judgment they are a blessing to mankind.

BIOLOGICAL CONTROL Plant pests may be also controlled through the manipulation of biological factors. Biological control may be achieved by directing the natural competition between organisms—for example, by introducing natural parasites or predators of the pest (Fig. 18-13) or incorporating natural resistance to the host crop. There are other subtle ways in which biological factors may be manipulated to affect the development of the pest. For example, the screwworm, a severe pest of livestock, has been controlled by releasing

372

FIGURE 18-13. *Insects parasitize other insects. (Above) Cocoons of the Braconid parasites emerging from the body of a tomato hornworm. The cocoons contain the pupal state of a parasitic wasp that lays its eggs on the body of the hornworm. (Below) The potato beetle killer attacking larva of the Mexican bean beetle.* [*Photographs by J. C. Allen & Son.*]

into the population large numbers of males rendered sterile by irradiation. Biological control is an attractive measure, because once put into effect it takes place without the influence of man, and chemical hazards are eliminated.

Natural predators. An example of successful biological control by the introduction of natural predators was the elimination of the threat posed by the cottony-cushion scale, a serious insect pest of citrus in California. This was accomplished by Albert Koebele, who in 1888 introduced from Australia the vedalia beetle, which feeds upon the eggs and larvae of the scale insect. The scale was kept under control until the widespread use of DDT in the late 1940's injured the beetle and upset the biological balance. A more recent example of biological control is the use of the spores of *Bacillus thuringiensis* to control caterpillars; the spores are applied as a spray.

Biological control has also been used to control weeds. Prickly pear (*Opuntia* species) was introduced as an ornamental into Australia prior to 1839 by the early colonists. By 1925 due to the absence of natural predators 60 million acres were infested with the "weed," and the infestation was increasing at the rate of one million acres annually. Ten years after the introduction of the moth borer *Cactoblastic cactorum* from Argentina, control was almost complete, and although there were successive waves of regrowth, they were of diminishing proportions.

The introduction of predators is best suited to the control of pests that lack natural predators, pathogens, or other enemies. For introduced predators to be successful they must thrive in the new habitat and yet not become destructive to agricultural crops. Thus only predators that are highly selective in their feeding habits can be imported. Because of the large-scale areas involved and the inherent problems, this method of control must be placed under the auspices of some national agency rather than left to the discretion of individual growers.

Genetic resistance. The incorporation of genetic resistance, the innate ability of the plant to resist the injurious effects of pathogens or predators, is an ideal method of control (Fig. 18-14). Resistance varies from nil (**susceptibility**), to partial, to complete (**immunity**). **Tolerance** is a type of resistance in which the plant suffers infection and some injury, but is able to live with it.

Many plants are known to have a resistance to viruses, bacteria, fungi, nematodes. These plants owe their resistance to structural features or biochemical effects that either prevent or discourage intrusion and persistence of the particular pest.

Resistance may be complex or simple. Some plants have resistance to whole groups of pathogens; others have only a specific resistance to a particular species or race. The combination of resistance and quality is one of the main objects of the plant breeder (see Chapter 19). Where pathogen and plant are closely adapted to each other (as with obligate parasites), a close relationship exists between the genetic resistance of the plant and the genetic ability of the pathogen to violate or overcome this resistance. The occurrence of a new race of the pathogen by mutation or introduction is one of the major problems the plant breeder faces, and the seeming impossibility of incorporating long-term genetic resistance to some diseases has inspired the search for tolerance or broad kinds of resistance.

Biological balance. The biological balance that exists in the natural environment cannot be ignored in plant protection. The upsetting of this equilibrium by environmental fluctuations is a perfectly natural phenomenon. Agriculture, by its very nature, disturbs the natural biological balance, but does not operate independently of it. For example, often when one fungicide supercedes another,

fungi that were once minor pathogens may begin to cause large-scale damage. Furthermore, the pathogen is not a static factor. New, resistant varieties and new pesticides may act as screening devices for the selection of resistant types. Thus many insects—the common housefly and many others—have developed resistance to DDT since its introduction. Similarly, the development of rust-resistant plant varieties soon becomes negated by the formation of new, virulent races of the rust fungus.

FIRE

The combustion, or burning, of wood is the reaction between carbohydrate fuels and oxygen; the reaction produces heat energy, carbon dioxide, and water (the whitish color of smoke is due to the condensed vapor). The chemical equation for the reaction is the same as that for respiration, but combustion takes place more rapidly and at a higher temperature. Both combustion and respiration are forms of oxidation. The energy accumulated through many years of photosynthesis may be rapidly dissipated by fire. Combustion requires three elements: heat, fuel, and oxygen. The release of stored energy in combustion starts a chain reaction that continues as long as the fuel and oxygen are in adequate supply. Fire suppression techniques strive to break one link in this chain of events.

Forest fire

Fire, especially in forests, is one of the destructive forces that affect crops. A **forest fire** is defined as an unenclosed, freely spreading fire in a forest environment. It is a wildfire, in contrast to the **controlled fire** used as a forest management tool.

Probably the greatest factor governing the severity of forest fire is weather: hot, dry weather lowers the moisture content of fuels. Once a fire has started, wind is extremely critical because it influences the oxygen supply and the rate of spread. Forest type and conditions are also factors. For example, fire is more likely to occur in conifers than in hardwoods; slash-covered or bushy areas

FIGURE 18-14. *Susceptibility and resistance to beta race of anthracnose, caused by* Collectotrichum lindemuthianum, *in beans. (Left) Red Kidney is susceptible (Right) Michelite is resistant. [Courtesy USDA.]*

374

FIGURE 18-15. (*Left*) *Large blow-up fires extend far into the sky.* [*Courtesy* Los Angeles Herald Examiner.] (*Right*) *These extremely dangerous fires can cause total destruction* [*Courtesy U.S. Forest Service.*]

Table 18-3 CAUSES OF FOREST FIRE, AREA BURNED, AND ACRES PROTECTED IN THE UNITED STATES BY FOREST REGIONS, 1935–1954.

REGION AND PERIOD	PERCENTAGE OF TOTAL NUMBER OF FIRES BY CAUSE								AV. ANNUAL NO. FIRES	AV. ANNUAL AREA PROTECTED, M ACRES	AV. ANNUAL AREA BURNED, ACRES	AV. NO. FIRES PER MILLION ACRES PROTECTED	AV. ANNUAL ACRES BURNED PER MILLION ACRES PROTECTED
	LIGHT-NING	RAIL-ROADS	CAMPERS	SMOKERS	DEBRIS BURNING	INCEN-DIARY	LUM-BERING	MISC.					
Rocky Mountain (10 states)													
1935–1939	67.8	2.0	7.7	14.1	3.4	1.2	0.7	3.1	5,554	97,612	105,452	57	1,080
1940–1944	71.6	4.8	4.3	10.0	3.0	0.7	0.6	5.0	6,118	121,289	326,029	50	2,688
1945–1949	70.0	3.7	5.6	10.3	3.3	0.8	0.9	5.4	5,232	159,275	220,485	33	1,384
1950–1954	68.7	2.0	7.2	10.0	4.0	1.0	1.0	6.1	5,336	163,566	160,692	33	982
Pacific (3 states)													
1935–1939	30.2	3.1	7.1	25.8	9.5	11.6	2.7	10.0	8,783	78,615	658,863	112	8,381
1940–1944	34.6	5.4	4.0	23.9	10.3	5.9	2.3	13.6	7,747	94,517	519,028	82	5,491
1945–1949	30.2	3.7	5.3	24.0	11.2	6.0	3.7	15.9	7,287	103,470	417,780	70	4,038
1950–1954	33.7	3.2	6.5	20.8	10.3	4.2	5.2	16.1	7,129	100,024	342,458	71	3,424
North Central (9 states)													
1935–1939	2.9	6.7	5.5	32.6	26.1	14.6	1.2	10.4	7,326	64,097	273,881	114	4,273
1940–1944	1.3	8.4	3.4	29.3	25.7	19.5	1.5	10.9	6,647	69,699	263,175	93	3,776
1945–1949	1.5	13.7	3.4	24.7	25.3	18.0	0.9	12.5	8,374	80,228	287,445	104	3,583
1950–1954	1.0	6.3	2.7	19.2	36.7	20.1	1.0	13.6	9,422	83,278	276,320	113	3,318
South (11 states)													
1935–1939	1.2	2.5	5.8	22.7	12.8	43.0	2.8	9.2	36,698	80,314	1,594,933	457	19,859
1940–1944	1.1	3.7	4.4	20.8	15.1	44.3	2.4	8.2	42,299	90,717	2,137,341	466	23,561
1945–1949	0.8	3.5	3.2	18.9	15.0	45.0	2.3	11.3	39,859	107,999	1,320,583	369	12,228
1950–1954	1.5	1.9	3.4	16.1	19.6	44.9	2.5	10.1	77,162	140,740	2,599,998	548	18,474
Eastern (14 states)													
1935–1939	0.7	9.3	10.2	39.6	19.4	10.5	0.8	9.5	16,161	80,374	332,346	201	4,136
1940–1944	1.0	12.3	7.5	35.0	22.0	9.4	0.8	12.0	17,618	86,079	488,978	205	5,681
1945–1949	1.0	11.4	7.1	33.1	21.4	8.5	1.0	16.5	12,946	93,115	257,353	139	2,764
1950–1954	1.4	5.6	5.8	33.2	25.9	11.3	0.9	15.9	15,096	95,734	594,485	158	6,210
Total U.S.													
1935–1939	9.7	4.4	7.0	27.0	14.5	26.4	2.0	9.0	74,522	400,985	2,965,475	186	7,395
1940–1944	9.7	6.3	4.9	24.1	16.1	27.6	1.8	9.6	80,427	462,301	3,533,551	174	7,643
1945–1949	8.8	6.1	4.3	21.9	16.1	28.5	1.9	12.4	73,775	544,089	2,703,715	136	4,969
1950–1954	6.6	2.8	4.0	18.6	20.5	33.8	2.3	11.4	114,146	583,354	3,953,411	196	6,777

Source: From Davis, Forest Fire, New York, McGraw-Hill, 1959.

are especially hazardous because the rate at which combustion consumes fuel is proportional to fuel volume and surface area.

Small fires, with a rate of energy release of approximately 1,000 BTU per second per foot of fire front, react as a two-dimensional surface. High intensity, or blowup, fires releasing 20,000 to 30,000 BTU per second per foot of front also have a vertical dimension (Fig. 18-15). These "three-dimensional" fires are comparable to thunderstorms.

Some fires are caused by lightning; others are caused by man. Those caused by man may be accidental or wilful (incendiary). The causes of fire in the United States show large regional differences, (Table 18-3). In the Rocky Mountains more than 70 percent of the fires are caused by lightning, whereas incendiary fires amount to only about 1 percent. In contrast, more than 40 percent of the fires in the South are incendiary, whereas lightning causes only 1 percent. Obviously, fire prevention techniques must vary from region to region.

Fire control

Of all the potential hazards to forests, fire poses the greatest threat. Consequently, fire control must be an integral part of forest management. Fire control has several dimensions:

Fire prevention is largely a sociological problem. True, many fires are started by sparks from faulty equipment, lightning, and other causes, but 90 percent of them are started by man. More education is probably needed to reduce the number of fires caused by man. The prevention of fires caused by lightning seems to be out of our hands.

Hazard reduction is the elimination of situations in which fire is likely to occur. Each year tens of thousands of acres are burned lightly to prevent litter fuel from accumulating to dangerous proportions. Firebreaks and fire lines are constructed to prevent the spread of fire in case one starts. Safety devices around logging operations and railroads help reduce the fire hazard.

Preparation for suppression includes all the activities that must be carried out in advance of a fire to be ready to fight it. This includes organizing fire crews, maintaining equipment in a state of readiness, and ensuring the availability of extra manpower for emergencies.

Detection has become a specialized activity, for the earlier a fire is spotted, the sooner suppression activities can begin. In the past, detection was accomplished visually—flame at night, smoke by day—using fixed lookouts or moving air or ground patrols, but these methods are hampered by weather conditions that decrease visibility. Infrared sensing offers a partial solution to this problem, and, coupled with remote television stations, may provide a much more sensitive detection system in all kinds of weather than has ever been available before.

Suppression involves all the work of extinguishing a fire after it has been discovered. Stopping the spread of fire may be achieved by smothering it with water, chemicals, or other materials or by providing a break in fuel supply. Firebreaks can be made by plowing a fire line near a strip free of combustible materials or by backfiring—setting a fire some distance from the wildfire. Fighting fire with fire is spectacular and sometimes dangerous. The use of chemicals or water (when available) to make the fuels unsuitable for combustion accomplishes the same effect. Airplanes are beginning to be used routinely to drop chemicals on fires.

Particular fire fighting tactics that depend on the wind, moisture, topography, and special local conditions are beyond the scope of this text. It is axiomatic that fires are best controlled at an early stage. Blowups, large "three-dimensional" conflagrations that extend far up into the sky, are most difficult to control, and present great danger to timber, wildlife resources, and to human life and limb.

HAIL

The heaviest and largest unit of precipitation existing in solid form is hail. It is a product of vigorous convection currents that occur in the warm season. If a hailstone is cut in half, one can see concentric layers of differing density and opacity that were formed by successive ascents and descents.

Hail begins as rain, but rather than falling to the ground, the drops are carried upward by rapidly ascending updrafts of warm, moist air, into regions where temperatures are below freezing.

FIGURE 18-16. *Hail damage in corn.* [*Photograph by J. C. Allen & Son.*]

They freeze, acquire a coat of frost and snow, and become hailstones. Eventually they enter a weaker convection current and fall, accumulating a coat of water, which also freezes. This process may be repeated many times, with the hailstones accumulating an additional coating of ice each time. Ultimately, they will become so large that convection currents are no longer strong enough to force them upward. Hailstones five inches in diameter and weighing a pound and a half have been recorded in Nebraska.

Millions of dollars are lost yearly due to hail damage, principally through the shredding of leaves (Fig. 18-16) or damage to fruit. For crops in which leaf quality is a factor, such as tobacco, a single hailstorm may result in the loss of an entire crop. Losses in fruit crops can be ruinous. For corn, productivity seems to be reduced by approximately the percentage of damage to leaves.

Efforts have been made to prevent the formation of hail by seeding clouds with silver iodide, which causes moisture to be precipitated at an early stage, thus preventing hail formation. Farmers in South Africa and in the Bayonne region of France have used rockets and ground generators in attempts to seed clouds and prevent hail.

FIGURE 18-17. *Tree plantings act as windbreaks to reduce wind velocity and thus protect crops as well as homes and livestock from cold and hot winds. Shelterbelts consisting of 5- to 10-row strips of trees help to control wind erosion, contribute substantially to the saving of moisture, and protect crops from damage by high winds.* [*Courtesy USDA.*]

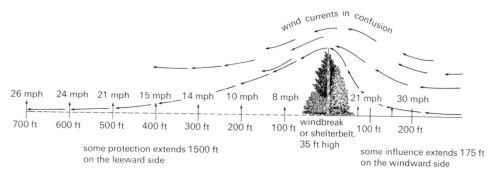

WIND DAMAGE

The harmful effects of wind are most apparent through its mechanical effects. Although wind damage is usually local, some regions suffer more than others. For example, tornadoes occur in a belt from Texas to Michigan, and seasonal hurricanes are a hazard to the Gulf and Atlantic coasts of eastern North America. Even winds of 30 to 40 miles per hour may cause extensive damage when accompanied by hail, sleet, and rain. Winter winds may cause dessication and browning of evergreen ornamental shrubs.

Wind can cause serious erosion problems, as it did in the "dust bowl" during the 1930's, where dry farming practices left soil without the protection of a vegetative cover. Another harmful effect is the transport of propagules of weed plants, insect pests, and spores of pathogenic fungi.

Hot, dry winds can have a debilitating effect of plants by upsetting their internal water balance. When this happens photosynthesis is decreased and crop production is lowered. Excessive wind at flowering time can evaporate the secretions from stigmas and result in reduced seed set. Windbreaks are an important and necessary adjunct to farming practices in the central United States and elsewhere (Fig. 18-17). The benefits of windbreaks in increasing crop yields have been firmly established.

In areas where wind is a hazard sensitive crop plants, such as tobacco, should not be grown. Plants native to regions subject to intense, high winds frequently have structural modifications that protect them against wind damage. Hurricanes seldom uproot mangroves growing in the Everglades of Florida, for their stiltlike roots anchor them firmly. In general, close spacing provides protection against wind damage because the plants then form a barrier against the wind and are "forced" to provide mutual support for each other. In forest management it is sometimes necessary to use a cropping system that does not leave trees vulnerable to wind damage. Thus complete rather than partial harvesting is practiced in many parts of the world where trees are especially liable to such injury.

Selected References

Ashton, Floyd M., and Alden S. Crafts. *Mode of Action of Herbicides.* Interscience (Wiley), New York, 1973. (An advanced treatise.)

Chapman, R. F. *The Insects: Structure and Function.* American Elsevier Publishing Company, New York, 1969. (A text on morphology and physiology.)

Crafts, A. S., and W. W. Robbins. *Weed Control* (3rd ed.). McGraw-Hill, New York, 1962. (An excellent reference and manual.)

Davis, K. P. *Forest Fire: Control and Use.* McGraw-Hill, New York, 1959. (An authoritative text, with much technical information.)

Horsfall, J. G., and A. E. Dimond, (editors). *Plant Pathology, An Advanced Treatise* (3 vols.) Academic Press, New York, 1959, 1960. (A good, detailed work on the science of plant pathology.)

Kasasian, L. *Weed Control in the Tropics.* Leonard Hill (Crown Publishers), New York, 1971. (A guide and elementary text.)

Kearney, P. C., and D. D. Kaufman. *Degradation of Herbicides.* Marcel Dekker, New York, 1969. (The fate of herbicides in soils and plants.)

Kenaga, Clare B. *Principles of Phytopathology.* Balt Publishers, Lafayette, Indiana, 1970. (An excellent elementary text.)

Klingman, G. C. *Weed Control As a Science.* Wiley, New York, 1961. (A general textbook.)

Metcalf, C. L., and W. P. Flint (revised by R. L. Metcalf). *Destructive and Useful Insects, Their Habits and Control* (4th ed). McGraw-Hill, New York, 1962. (A famous work on entomology.)

National Research Council. *Pest Control Strategies for the Future.* National Academy of Science, Washington, D.C., 1972. (Papers on integrated pest control presented at a symposium sponsored by the Agricultural Board.)

National Research Council. *Principles of Plant and Animal Pest Control* (6 vols.) National Academy of Science, Washington, D.C., 1968–1970. (The six volumes concern, respectively, plant disease, weed control, insect-pest management, nematodes, vertebrate pests, and the effects of pesticides on fruit and vegetable physiology. The technical reports are especially useful in offering guidance in resource management).

Pfadt, R. E. (editor). *Fundamentals of Applied Ento-*

mology (2nd ed.) Macmillan, New York, 1971. (Each chapter is contributed by a specialist.)

Roberts, D. A., and C. W. Boothroyd. *Fundamentals of Plant Pathology.* W. H. Freeman and Company, San Francisco, 1972. (An introductory text.)

U. S. Department of Agriculture. *Insects.* Yearbook of Agriculture, 1952. (This reference, together with the next two, gives popular coverage of crop protection.)

U.S. Department of Agriculture. *Plant Diseases.* Yearbook of Agriculture, 1953.

U.S. Department of Agriculture. *Protecting Our Food.* Yearbook of Agriculture, 1966.

Walker, J. C. *Plant Pathology* (3rd ed.). McGraw-Hill, New York, 1969. (An excellent introduction, with an organization based on causal agents. There are many other textbooks whose organization is based on crops or groups of crops.)

Weed Society of America. *Herbicide Handbook* (2nd ed.). Department of Agronomy, University of Illinois, Urbana, 1970. (The nomenclature, structure, and characteristics of herbicides.)

CHAPTER 19

Crop Improvement

Within a given environment, crop improvement may be achieved through superior heredity. **Plant breeding,** the systematic improvement of plants, is an innovation of this century. Genetics, the science of heredity, has placed it on a firm theoretical basis. Plant breeding has become a specialized technology and is responsible for a large share of the current progress in crop production.

Since the needs and standards of man change, the job of the plant breeder is a continuous one. For example, efficient mechanical harvesting becomes possible only with plants that bear the great bulk of their crop at a given time; it further requires that plants be structurally adapted to the machine. The raw materials for the necessary changes may be found in the tremendous variation that exists in cultivated plants and related wild species. The incorporation of alterations into crop plants adapted to specific geographical areas demands not only a knowledge of the theoretical basis of heredity, but also the arts and skills necessary for the discovery and perpetuation of small but fundamentally important differences in plant material.

THE GENETIC BASIS OF PLANT BREEDING

Variation

Variety is more than the spice of life; it is the very essence of it. The variation found in crop plants, as in all living things, ranges from the obvious (sugar beet versus sugar maple) to the almost imperceptible differences between two wheat plants growing side by side.

The variation encountered in biological material is measurable. Just as the general performance of a group can be described by a single statistic called the arithmetic mean, the amount of variation in an array of observations can be expressed by a comparable statistic called the standard deviation, a measure of the average deviation from the mean.

The variation found in biological material (for example, among individual plants in a cotton field) is the result of genetic and environmental components. The genetic component is the result of differences in the inherited makeup of the organism. It is due largely to differences in genes, the main units of inheritance that control development. The sum total of the genetic endowment of an individual is referred to as the **genotype.** Differences between plants grown in identical environments must be due to genotypic differences.

The environmental component of variation is the result of nongenetic factors that influence the expression of the genotype. In our hypothetical cotton field, these factors may include differences in soil fertility and in shading, or any of a number of random events such as insect injury or hail damage. Environmental variation can be observed directly with organisms of identical makeup.

The outward appearance of an organism is known as its **phenotype.** The phenotype may be thought of as the interaction product of genotype with environment. The range of environmental variation is enormous in living things, but its boundaries are set by the genotype. It is difficult, for example, to conceive of any environmental condition that could transform sugar beets to sugar maples.

The variation found in a population of living organisms may be influenced in varying proportions by environmental and genetic factors. the percentage of the total variation due to genetic factors is known as **heritability.** In corn, the heritability of yield and plant height has been estimated as 25 percent and 75 percent, respectively. This means that in segregating populations, 25 percent of the variation in yield and 75 percent of the variation in height is attributable to genetic factors. Consequently, it is much easier to select for plant height than for yield. When heritability is low, then the relative influence of environment is high, and the quickest way to maximize efficiency is to improve the environment.

The question of whether genetic or environmental variation is the more important has little meaning. The pertinent problems are to find out (1) which genotype is best suited to a particular set of environmental conditions; and (2) what combination of environmental variables will permit the optimum expression of a given genotype.

The fundamental discovery that the genotype is inherited in the form of discrete units belongs to the Austrian Abbot Gregor Mendel, whose experiments were done on the transmission of seven traits in the garden pea (1865). Environmental variation is not transmitted in this manner. This was firmly established by the Danish geneticist and breeder Wilhelm Johannsen in 1903, whose studies dealt with the transmission of seed size in beans. Johannsen demonstrated that the inheritance of phenotypic variation is only possible when it is a result of genetic differences.

Genetic factors form the basis for plant improvement through breeding. The unique problem of the breeder is to make the plant better adapted to the needs of man by bringing about a favorable combination of genes.

Mutation

The genetic variation in living things is ultimately due to differences in the kind, structure, and amount of genic material (see pp. 161–165). The source of this variation is found in sudden heritable changes called **mutations.** Changes that affect the gene itself are called **point mutations,** and they result from alterations that affect the internal makeup or arrangement of the functional gene. The rate of these mutations varies with the gene involved. True point mutations may also revert to the original condition (back mutation).

Extragenic changes are also included under the broad category of mutation. These may be due to chromosomal aberrations that alter the structural arrangements of the gene-carrying chromosomes. Chromosomal aberrations are of several types:

Aberrations involving one chromosome
    deficiency: segment deleted
    duplication: segment repeated
    inversion: segment reversed
Aberration involving more than one chromosome
    translocation: exchange of segments

Another type of mutation results from a change in chromosome number (**heteroploidy**). The chromosomes in the somatic cells of the higher plants are composed of sets called **genomes.** Each genome contains a particular number of chromosomes. Although most plants have two genomes, the diploid number, a wide variation in number is possible. Variation in chromosome sets is referred to as euploidy—that is, true ploidy. The following terms indicate the number of chromosome sets a plant contains:

No. of chromosome sets	Name of type
1	monoploid
2	diploid
3	triploid
4	tetraploid
5	pentaploid
6	hexaploid
7	septaploid
8	octaploid

In addition to variation of whole sets, there is variation in the number of chromosomes within a set (aneuploidy). Plants in which such variation occurs are usually low in fertility.

Polyploids, plants in which the number of chromosome sets is higher than the diploid number, may arise by the duplication of sets from a single species (autoploidy) or by the combination of sets from two or more species (alloploidy). Many of our crop plants are alloploid—for example, wheat, cotton, tobacco, and members of the cabbage family. The origin of their genomes can be traced by special genetic and cytological techniques. The complex genomic constitution of the wheats is shown in Table 19-1. Knowledge of the genomic constitution in polyploid crops enables the plant breeder to accomplish the transfer of genes from related wild species. This ability to juggle genes and chromosomes enormously enlarges the potential range of genetic variation available to the breeder. For example, it has been possible to reconstitute certain crop plants from their component species.

In referring to polyploid crops it is important to distinguish between the numbers of chromosomes in the somatic cells (the $2n$ number) and the basic number of chromosomes that make up each genome (the x number). The chromosome number in the somatic cells of the cultivated strawberry is 56 ($2n = 56$). The cultivated strawberry, however, is octaploid, having eight sets of chromosomes. The number of chromosomes in each genome is seven ($x = 7$). The chromosome

Table 19-1   GENOMIC CONSTITUTION OF WHEATS AND THEIR RELATIVES.

SPECIES	GENOME FORMULA	COMMON NAME
Diploid ($2n = 14$)		
Triticum aegilopoides	AA	einkorn (wild)
Triticum monococcum	AA	einkorn (cultivated)
Aegilops speltoides	BB	wild
Aegilops caudata	CC	wild
Aegilops squarrosa	DD	wild
Secale cereale	EE	rye (cultivated)
Tetraploid ($2n = 28$)		
Triticum dicoccoides	AABB	emmer (wild)
Triticum dicoccum	AABB	emmer (cultivated)
Triticum durum	AABB	durum wheat (cultivated)
Triticum persicum	AABB	Persian wheat (cultivated)
Triticum polonicum	AABB	Polish wheat (cultivated)
Triticum turgidum	AABB	(solid stem) wheat
Triticum timopheevi	AAGG	timopheevi (wild)
Aegilops cylindrica	CCDD	goat grass (wild)
Hexaploid ($2n = 42$)		
Triticum compactum	AABBDD	club wheat (cultivated)
Triticum spelta	AABBDD	spelt (cultivated)
Triticum aestivum	AABBDD	common wheat (cultivated)

number in the gametes—the haploid number, or *n* number—is 28. The concept of *n* is halfness, whereas the concept of *x* is oneness. In diploid species *n* = *x*.

Mutation has provided the variation that, through evolutionary processes involving recombination and selection, has given rise to the diversity of living things, each adapted to particular niches of existence. These cataclysmic events occur at very low rates and are usually deleterious. Any random change, however, that increases the fitness of an organism tends to increase through natural selection. Although mutations can at present be induced by a number of techniques, only random mutations can be accomplished.

Inheritance

The transmission of genes is a consequence of the events that occur in meiosis (see p. 165). The transmission of genes from generation to generation can be followed in a cross between two individuals that differ in one trait conditioned by a single gene.

Assume gene *A* with its allele *a*. The heterozygous plant *Aa* will produce two kinds of gametes (*A* and *a*) in equal proportions. Crossing *Aa* and *Aa* will produce three kinds of progeny in a predictable ratio if all gametic and zygotic types are equally viable. It can readily be seen that the possible types of zygotes resulting from all combinations of the two kinds of gametes *A* and *a* will be *AA*, *Aa*, *aA*, and *aa* or 1*AA*, 2*Aa* and 1*aa* (Fig. 19-1). If the *AA* plants are indistinguishable from *Aa* plants, the gene *A* is said to be **dominant** and *a* **recessive.** The phenotypic ratio becomes 3*A*- to 1*aa*.

The genotypes *AA* and *Aa* may be distinguished by a genetic test. The genotype *AA* will produce a single type of gamete (*A*). The genotype *Aa* will produce two kinds of gametes (*A* and *a*). Thus by crossing plants of the dominant *A* phenotype with themselves, or with the double recessive *aa*, these two possible genotypes may be separated on the basis of their progeny:

AA selfed (*AA* × *AA*) →
   all *AA* (*AA* plants "breed true")
in contrast with
Aa selfed (*Aa* × *Aa*) →
   3*A*-:1*aa* (*Aa* plants segregate)
      or
AA × *aa* → all *Aa*
in contrast with
Aa × *aa* → 1*Aa*:1*aa*

Generations are designated by specialized terminology. The first cross (usually referring to homozygous genotypes that differ with respect to a particular character) is the **P₁**, or the **parental generation.** The progeny of such a cross is the first filial generation (called the **F₁**). Homozygous but genetically distinct parents will produce an F₁ that is heterozygous and does not segregate. The **F₂** (second filial generation) results from intercrossing or selfing F₁ plants:

P₁   *AA* × *aa*
            ↓
F₁   *Aa*
            ↓
F₂   1*AA*:2*Aa*:1*aa*.

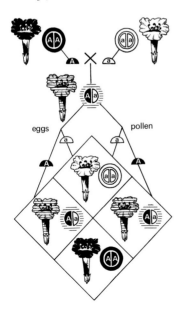

FIGURE 19-1. *Mendelian segregation, illustrating incomplete dominance in the four o'clock (Mirabilis jalapa). Note that the phenotypic ratio of 1 red:2 pink:1 white is also 3 colored:1 noncolored. [Adapted from Dobzhansky.]*

eggs   pollen

384

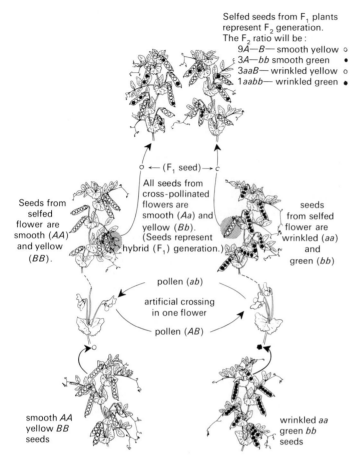

Selfed seeds from F₁ plants represent F₂ generation. The F₂ ratio will be:

9A—B— smooth yellow ○
3A—bb smooth green ●
3aaB— wrinkled yellow ○
1aabb— wrinkled green ●

○ ← (F₁ seed) → ●

All seeds from cross-pollinated flowers are smooth (Aa) and yellow (Bb). (Seeds represent hybrid (F₁) generation.)

Seeds from selfed flower are smooth (AA) and yellow (BB).

seeds from selfed flower are wrinkled (aa) and green (bb)

pollen (ab)

artificial crossing in one flower

pollen (AB)

smooth AA yellow BB seeds

wrinkled aa green bb seeds

FIGURE 19-2. *Independent assortment of two genes in the pea.* [*After Horowitz.*]

The F₂ is typically the segrating generation, which, if large enough, could theoretically include every possible genotype. Further generations are known as F₃, F₄, F₅, and so on. A cross of the F₁ with one of its parents is known as a **backcross**. (Similarly, generations produced by selfing only are denoted as S₀, S₁, S₂, etc.)

The inheritance of characters controlled by two or more genes follows from an expansion of the single gene model. The assortment of one pair of chromosomes at meiosis has no effect on the assortment of the other pairs; consequently, genes on separate chromosomes segregate independently of each other. Thus a plant heterozygous for two gene pairs on different chromosomes (*Aa* and

Bb) will produce four different types of gametes (*AB, Ab, aB,* and *ab*) in equal proportions. The self progeny of the double heterozygote *AaBb* should segregate into a 3:1 phenotypic ratio for each factor, assuming complete dominance. The combined phenotypic ratio would then be 9*A-B*-:3*A-bb*:3*aaB*-:1*aabb*. This type of cross is diagrammed in Figure 19-2.

When two genes affect the same biochemical pathway, the phenotypic ratios give some idea of the gene action involved. For example, the F₂ ratio, which involves a dihybrid in which two dominant genes interact to effect a single character, segregates into two classes in a ratio of 9:7 instead of four classes (in a 9:3:3:1 ratio) when two contrasting characters are involved. Either gene, when homozygous recessive, blocks some essential step, as shown in Figure 19-3. This phenomenon of gene interaction is called **epistasis**. Different types of epistasis can produce strange genetic ratios.

LINKAGE Since a plant contains thousands of genes but only a limited number of chromosomes, each chromosome must contain many genes. In corn, for example, more than 500 genes have been identified, but only 10 pairs of chromosomes. Because genes located on the same chromosome are not randomly assorted, they tend to be inherited together. This condition is referred to as **linkage**. There is a relationship between the physical closeness of genes and the intensity of strength of linkage. The assortment of genes on the same chromosome is affected by crossing over, the actual exchange of chromosome material in meiosis.

By special techniques it is possible to designate on which chromosome a particular gene is located. Since the genes are in a linear sequence on the chromosomes, it is possible to determine their relationship to each other by their linkage strength. In this way "road maps" of the chromosomes may be made, as shown in Figure 19-4 for the ten chromosomes of corn. The chromosome map is constructed only for recognizable genes—that is, genes for which alternate forms are known. The distance between the genes is related to the relative frequency of crossover events.

Linkages may help or handicap the plant

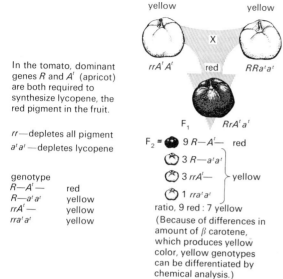

In the tomato, dominant genes R and A' (apricot) are both required to synthesize lycopene, the red pigment in the fruit.

rr—depletes all pigment

$a'a'$—depletes lycopene

genotype	
R—A'—	red
R—$a'a'$	yellow
rrA'—	yellow
$rra'a'$	yellow

ratio, 9 red : 7 yellow

(Because of differences in amount of β carotene, which produces yellow color, yellow genotypes can be differentiated by chemical analysis.)

FIGURE 19-3. *Gene interaction (epistasis) in the tomato.*

breeder. The linkage of two desirable characters makes it easy to keep these characters together. In barley, for example, the genes for stem rust resistance and loose smut resistance are linked. But in order to free the plant of an undesirable gene, linkages between desirable and undesirable factors must be broken. If the genes are very close, extremely large populations of progeny may be required to obtain the exceptional but desired recombinants.

QUANTITATIVE INHERITANCE The plant characters, or traits, that have been discussed may be clearly separated into distinct, discontinuous classes: colored versus noncolored, normal versus dwarf. The **qualitative characters** lend themselves to analysis because of the relative ease with which they can be classified. The greatest advances in genetic theory (beginning with Mendel) have been made using qualitative characters. Although many qualitative characters are of economic importance (for example, the red of tomatoes and dwarfing in beans), most important characters, such as productivity, size, and quality, are continuous and range from one extreme to the other. These are known as **quantitative characters.** For example, there is an almost imperceptible blending of color in wheat seed from very dark red to colorless. Similarly, corn plants vary continuously from very tall to almost dwarf-like. Such quantitative characters are difficult to investigate because of the problems encountered in excluding the effects of environment. Although the transmission of these characters was studied before the rediscovery of Mendel's paper, no general theory of inheritance was developed. It was clearly established, however, that this type of variation is inherited.

The present genetic theory holds that the inheritance of quantitative characters is essentially the same as for qualitative ones. The assumption is that there are many genes that affect these traits and that each contributes a small increment of effect to modify the character. These have been referred to as quantitative genes, polygenes, multiple factors, and modifiers.

The individual effect of each gene involved in a quantitative trait may be different, and the effect of environment may be large in relation to the small contribution of individual genes. As a result, it is difficult to determine the precise pattern of individual gene action. The special techniques used to analyze the inheritance of quantitative traits involve mathematical and statistical analysis of the variation encountered in crosses involving these characters. Some of these methods make it possible to estimate the number of genes involved and to assess their average contribution and average dominance. But because these methods rely on particular assumptions, the conclusions are difficult to prove precisely. For example, the genetic variability of quantitative characters can be divided into **additive, dominance,** and **epistatic components** that reflect the average action of the genes involved. It is assumed that the additive component is responsible for the phenotypic difference between homozygotes of any single gene. All the genetic variability is additive when there are no intragenic (dominance) or intergenic (epistatic) interactions. The amount of additive genetic variance provides the best indication of the degree to which segregating populations will resemble their parents. An understanding of the components of variation that reflect gene action allows the breeder to predict results and thus make accurate decisions concerning various breeding procedures.

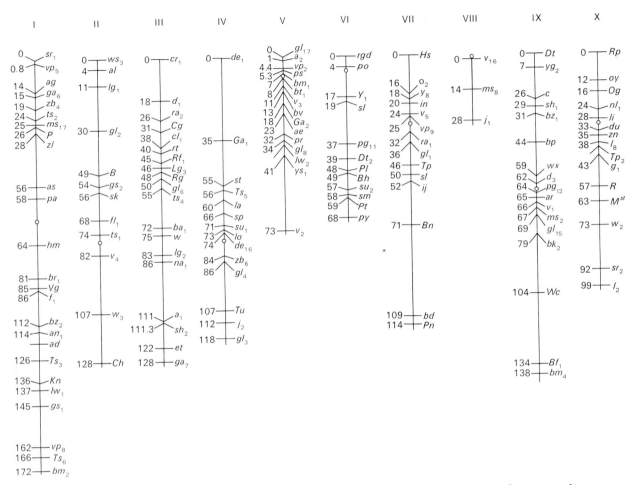

FIGURE 19-4. *Linkage map of the ten chromosomes of maize. The numbers represent distance in the crossover units. The letters represent mutant genes. [Courtesy Dekalb Agricultural Association, Inc., after Neuffer.]*

An example of the inheritance pattern for quantitative traits is shown in Figure 19-5. A cross between two homozygous lines of corn, one short-eared and the other long-eared, produced an F_1 population with a mean ear length intermediate between the parental lines. Presumably the parental and F_1 variation is entirely environmental. The distribution of the F_2 population includes the homozygous parental types. The variation of the F_2 can be shown to be greater than that of either the F_1 or the parental lines. This increase in genetic variation is due to gene segregation and recombination.

These results are similar to those that would be expected if a single partially dominant gene differentiated the parents, and if enough environmental variation was present to obscure the differences between the three expected F_2 classes—for example, AA, Aa, aa. But if a single gene is involved, or if many are involved, markedly different consequences follow when F_2 plants are selfed to produce F_3 lines. If a single gene difference is assumed, only three F_2 genotypes in the ratio of $1AA : 2Aa : 1aa$ would be expected. Therefore, one-fourth of the F_3 lines would resemble each parental type, and one-half would resemble the segregating

F_2 distribution. In actuality, however, there are many more than three F_2 genotypes. It is possible to select F_2 genotypes that give an almost continuous range of ear length from the short to the long parent. This indicates that there are many genes controlling this character.

As the number of genes differentiating the parents with respect to any given trait (such as ear length) increases from one to n, the number of F_2 genotypes increases from 3 to 3^n. In contrast, the number of genotypes resembling each parental type decreases from one-fourth of the F_2 population to $1/4^n$. If only 10 genes differentiate a character, there are 3^{10}, or 58,049, different possible genotypes expected in the F_2. To recover every genotype, a minimum population of 4^{10}, or 1,048,576 plants, would have to be grown. In this population only one plant would be expected to duplicate the male parental genotype, and only one the female.

The genetic value of a seed-propagated plant is the average performance of its progeny. Thus the basic problem in improving quantitatively inherited characters lies in distinguishing genetic from environmental effects. Those plants with the best appearance may not always be the most desirable genotypically. Because of the difficulties inherent in recognizing the effects of individual genes that govern quantitative characters, genetic analysis is heavily oriented toward the statistical approach. The great economic importance of quantitative traits demands that the breeder of crop plants be thoroughly familiar with the basic principles governing their inheritance.

Hybrid vigor

Hybrid vigor, or **heterosis,** is the increase in vigor, usually as compared to either parent, that occurs in certain crosses. The classic example of hybrid vigor is the mule, an interspecific cross of a mare and a jackass. The effect of heterosis in plants has been compared to the effects achieved by the addition of a balanced fertilizer. Hybrid vigor is often associated with an increased number of parts in indeterminate plants (whose main axes remain vegetative and whose flowers form on axillary buds; for example, cucumber) and with increased

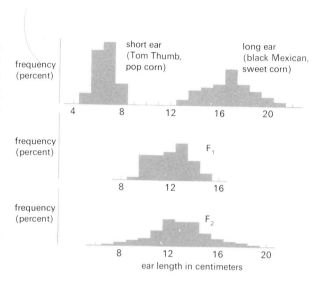

FIGURE 19-5. *Distribution of ear length in a popcorn–sweet-corn cross.* [*Adapted from Srb and Owen,* General Genetics. *San Francisco: W. H. Freeman and Company,* © *1953.*]

size of parts in determinate plants (whose main axes terminate in a floral bud; for example, maize). In perennial plants the vigor persists year after year.

Any organism is a hybrid if it results from a combination of genetically dissimilar gametes. (Although botanists often use the term hybrid to refer specifically to crosses made between species, geneticists use the term to refer to individuals that are heterozygous even for a single gene.) As used in the expression "hybrid corn" the term refers to a particular combination of inbred lines. To avoid confusion the term hybrid should be qualified, such as single gene hybrid, interspecific hybrid, single cross hybrid.

Heterosis is a phenomenon that is intimately assiciated with the decline in vigor brought about in some plants by continued crossing of closely related individuals (**inbreeding**). Selfing (crossing a bisexual plant with itself), is an extreme example of inbreeding. Selfing has no effect on homozygous loci:

$$AA \times AA \rightarrow AA$$
$$aa \times aa \rightarrow aa$$

The selfing of a plant that is heterozygous for a

388

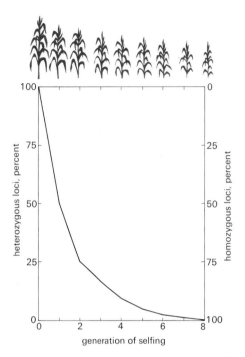

FIGURE 19-6. *Self-pollination reduces the number of heterozygous genes by half in each generation. By the fifth generation more than 95 percent of these genes are expected to be homozygous. Most naturally cross-pollinated plants show a decrease in vigor as homozygosity is reached.*

single gene (*Aa*) produces progeny of which half are homozygous (*AA* and *aa*):

$$Aa \times Aa \rightarrow 1AA : 2Aa : 1aa$$

This increase in homozygosity may be generalized for any number of genes. With each generation of selfing, 50 percent of the heterozygous genes become homozygous (Fig. 19-6). Continued selfing in plants heterozygous for *n* genes will ultimately produce 2^n different lines, each of which is homozygous for all genes. Such plants are known as inbreds, or pure lines. Inbreeding can be defined as any breeding system that increases homozygosity.

The effect of inbreeding depends on the degree of heterozygosity and on the natural crossing mechanism in the plant. The self progeny of homozygous plants, of course, are unaffected. The self progeny of heterozygous plants increase in homozygosity. In some plants inbreeding often leads to a decline in vigor. This decline in vigor associated with an increase of homozygosity, or **inbreeding depression,** is usually high in those plants that are naturally cross-pollinated and therefore in a highly heterozygous state. Heterozygous individuals of self-pollinated plants can be obtained by crossing different homozygous lines. When these heterozygotes are selfed, the sharp characteristic decline in vigor associated with inbreeding in cross-pollinated plants is usually not found or is of a lower degree.

The crossing of unrelated inbred lines of a cross-pollinated crop restores the vigor lost by inbreeding (Fig. 19-7). The progeny of such a cross are referred to as **F₁ hybrids.** In maize, some F₁ hybrids have shown yields as much as 15 to 25 percent higher than those of the original open-pollinated varieties from which the inbred lines were derived. The increased vigor is due to the genotypic control made possible by careful selection of inbreds. An individual F₁ hybrid plant is not necessarily any better than the best plants of an open-pollinated variety. The fact that a vigorous, adapted F₁ hybrid line is composed of genetically homogeneous plants often guarantees increased performance over segregating open-pollinated varieties. This lack of variation in F₁ hybrids, although disadvantageous under certain conditions, is a desirable feature in this age of mechanization.

This increase in vigor associated with the cross-

FIGURE 19-7. *Hybrid vigor in the onion. The F₁ hybrid in the center crate shows increased size and productivity as compared to the parental inbreds on either side.*

ing of diverse selected (usually inbred) lines has been exploited by the commercial production of hybrids in such crops as petunia, zinnia, cabbage, cucumber, sugar beet, and sorghum, in addition to the well-known example of maize.

Crossing various varieties of a self-pollinated crop (varieties that in reality are inbred lines) increases vigor in isolated cases, but the increase is very much less than in cross-pollinated plants. An example of commercially produced hybrids in self-pollinated crops is the tomato. Research is in progress to produce hybrid wheat.

GENETIC INTERPRETATION OF HETEROSIS   Although the incorporation of heterosis is an important feature in plant improvement, its genetic basis is not completely established. Inbreeding depression and heterosis are related phenomena; when one occurs the other can be expected with certainty. Consequently, any genetic explanation must reconcile both phenomena. The numerous explanations proposed for heterosis may be conveniently grouped under two general theories: dominance and overdominance.

The **dominance hypothesis,** first clearly stated by D. F. Jones in 1917, explains heterosis in terms of dominant or partially dominant growth factors. It is based on the assumption that cross-pollinated plants contain a large number of recessive genes conceled in the heterozygous condition. The recessive condition is due largely to the absence of some essential gene function, and homozygous recessives are therefore largely deleterious. As a result of inbreeding, concealed recessives become homozygous and cause loss of vigor. The restoration of vigor achieved by crossing inbred lines from diverse sources is due to the recessives of each inbred being mutually "covered up" by the other's dominant or partially dominant alleles, much as two moth-eaten bathing suits might cover up each other's holes. Thus, in successful combinations the two (or more) inbreds that make up a hybrid mutually complement each other's deficiencies. In its simplest case this situation may be expressed by the cross

$$aaBB \times AAbb \rightarrow AaBb.$$

If each parent lacks vigor because of a deficiency

in either dominant A or B, the hybrid $AaBb$, would be more vigorous than either of its homozygous parents because of the presence of both dominant alleles.

According to the dominance hypothesis, then, the vigor of a hybrid is due to the large number of dominant genes that it contains or to the lack of particular homozygous recessive combinations that, like weak links, ultimately determine the strength of the chain of successful physiological reactions. The high heterozygosity of hybrids is merely a concomitant and incidental part of hybrid vigor. According to the theory, if it were possible to obtain each heterozygous gene in the homozygous dominant condition, the vigor of the resulting genotype would be as great as or greater than that of the heterozygous hybrid. The failure to obtain true-breeding homozygous lines that are as vigorous as F_1 hybrids is due to the large number of genes involved in traits such as yield. Thus the probability of obtaining each dominant gene homozygous in one individual is extremely low. In addition, linkages between deleterious and beneficial genes further reduce the expected frequency of recovery of homozygous dominant types.

The rate of decline of vigor caused by inbreeding depends upon the number of genes affecting a particular character and the frequency of deleterious genes in the population. The lack of a significant inbreeding depression and of heterosis in most naturally self-pollinating crops is explained as being due to the absence of deleterious recessives. Any deleterious recessives quickly become exposed by the natural inbreeding mechanism in self-pollinating crops, and are rapidly eliminated by natural or artificial selection.

The dominance hypothesis has not been an entirely satisfactory explanation of heterosis to some geneticists. This is due in part to the high estimates of recessive alleles required for the phenomenon and in part to other statistical considerations. The **overdominance hypothesis** equates heterosis with heterozygosity. It assumes that heterozygosity, by itself, may be advantageous. If this is true, then the reduction in vigor upon inbreeding is a direct result of the attainment of homozygosity. According to the hypothesis, the heterozygous condition Aa is superior to either homozygote (*AA* or *aa*). This superiority has been termed overdominance.

There are a number of interpretations of overdominance at the gene level. If it is assumed that both alleles of a gene pair are physiologically active, then the combination of these alleles in the heterozygote may be superior to either alone. This may be due to supplementary action, alternative pathways, or the production of some hybrid substance. The combined activity of both alleles in the heterozygous condition produces the heterotic effect. Another explanation proposes incomplete dominance and assumes that the amount of gene product produced by the heterozygote is optimum; the amount produced by the homozygous type is either deficient or inhibitory. There are experimental verifications, although admittedly rare, of these types of gene action.

It is difficult to distinguish in practice between the dominance and overdominance hypotheses, and, furthermore, they are not mutually exclusive. The essential features of the hybrid effect are explained by both. The correct explanation lies in determining the primary action of individual genes affecting heterotic traits. In general, however, the dominance hypothesis is thought to be the most likely explanation of hybrid vigor. This does not, of course, rule out the existence of overdominance, but it is felt by many that overdominance is at most only a contributory factor to hybrid vigor.

BREEDING METHODS

The goal of the plant breeder is to create superior crop cultivars. The **agricultural variety**, or **cultivar**, may be defined as a group of crop plants having similar but distinguishable characteristics. The terms "variety" and "cultivar" have various meanings, however, depending on the mode of reproduction of the crop. With reference to asexually propagated crops, the term "cultivar" is used to mean any particular genotype that is considered of sufficient value to be graced with a name. With reference to sexually propagated crops the concept of cultivar depends on the method of pollination. The cultivar in self-pollinated crops is essentially a particular homozygous genotype—a pure line.

In cross-pollinated crops the cultivar is not necessarily typified on the basis of any one plant but sometimes by a particular plant population, which at any one time is composed of genetically distinct individuals.

Methods of improvement have been developed for cultivars of all kinds. The improvement of vegetatively propagated crops (many fruit crops, woody ornamentals, potato, sugar cane) consists in producing a single desirable genotype. This particular genotype may be of any degree of heterozygosity because elimination of the sexual process by vegetative propagation eliminates genetic segregation. There is no problem of genetic maintenance. Asexually propagated crops are usually cross-pollinating and consequently highly heterozygous; thus in developing populations from which to select a desirable genotype, inbreeding should be avoided. Unrelated plants are usually hybridized in order to obtain a vigorous population from which the most desirable individuals may be selected.

Breeding methods for the improvement of sexually propagated crops depend on the genetic structure of the cultivar, which is governed by the natural method of pollination. The amount of cross-pollination ranges from essentially none in such plants as soybean to 100 percent in dioecious and self-incompatible plants. For breeding purposes, however, two main groups are recognized: naturally self-pollinated plants, in which cross-pollination is less than 4 percent, and naturally cross-pollinated plants, such as maize, in which cross-pollination exceeds 40 percent. The intermediate types (cotton, sorghums) are usually considered with the cross-pollinated group.

Self-pollinated plants are ordinarily homozygous for practically all genes. The exceptions are the result of chance cross-pollination and mutation. Heterozygosity usually is quickly eliminated as a consequence of natural inbreeding. The basic problem in improving self-pollinated plants lies in producing and selecting the best homozygous genotype. Once this is accomplished the problem of genetic maintenance is small as compared to cross-pollinated species.

The genes in naturally cross-pollinated, seed-propagated plants are recombined constantly from generation to generation. The problems involved in improving cross-pollinated plants include maintaining uniformity while avoiding the decline in vigor associated with homozygosity. Once a desirable population is achieved, there is still the perpetual problem of maintenance.

One method of maintaining both uniformity and heterozygosity is to produce hybrids through the crossing of selected diverse inbred lines. At present only technical problems prevent this from becoming a standard breeding technique for cross-pollinated, seed-propagated crops.

Introductions

An abundant source of variation is available to the plant breeder from cultivars or related wild species (Fig. 19-8). The first step in any crop improvement program, therefore, is to assemble the natural variants available. From a study of their performance it may be possible to make an immediate improvement merely by choosing some cultivar not being grown. Because of the wide distribution of most crops, a representative collection should be worldwide in scope. It should include, if possible, old as well as new cultivars. In the United States the Plant Genetics and Germplasm Institute of the Agricultural Research Service (United States Department of Agriculture) carries out the exploration and introduction of genetic variability. Collections of "germplasm banks" for many crops are currently available.

There are many examples of crop improvement by introduction from other parts of the world. 'Bartlett,' the main pear cultivar in the United States, was introduced (about 1797) from England and was originally called 'Williams' Bon Chrétien.' Similarly, the cultivar of pineapple grown in Hawaii is the 'Cayenne,' which was introduced from South America. In Brazil one of the valuable dry bean varieties, 'Rico-23,' was introduced from Costa Rica. Some new selections turn out to be better adapted to areas other than the one in which they were developed. The Indiana wheat cultivar 'Redcoat' was slated for discard, but was subsequently named and released because it performed well in Pennsylvania. The Monterey pine, insignificant in its native California habitat, was introduced into Australia, where it is now one of the most important timber species grown.

With established crops and active breeding programs it becomes more and more difficult to make improvement through direct introduction. Usually, in spite of individually desirable attributes, no one introduction contains the proper combination to be useful directly. Often one or two desirable characters are found in plants that are otherwise unsuitable or unadaptable. Resistance to the apple scab disease is found in species with almost inedible fruit less than 1 inch in diameter. In such plants the approach is to recombine the characters by sexual crossing (**hybridization**) in the hope of obtaining plants that have a more favorable combination of desirable features. When these characters are distributed over many cultivars, or the desirable characters are associated with many undesirable ones, this procedure requires many crosses and many generations.

Genetic variability is the raw material of the plant breeder. The richest source of genetic variability for a particular cultivated species has been shown to be its geographical area of origin. The search for these areas of origin is not only intriguing and fascinating, but is important to the present-day improvement of crop plants. Many types of evidence—genetic, botanical, chemical, archeological, historical—can be brought to bear on individual crops.

Modern study in this field derives from Alphonse de Candolle's pioneering *Origin of Cultivated Plants* (1822). Drawing on many sources, de Candolle surmised that suitable climate and presence of wild relatives were good indications of original habitat, especially if supported by historical evidence. The Soviet plant scientist Nikoli Ivanovich Vavilov, with the benefit of new techniques and widespread collections, worked on the origins of cultivated plants for twenty years, from 1916 to 1936, and succeeded in expanding and revising de Candolle's conclusions. Vavilov was convinced that many crop plants do not occur outside of cultivation, and that wild relatives are often represented by narrow, isolated groups. He reasoned that centers of origin are marked by considerable variability and that as cultivars spread from such centers certain distinctive genotypes become successful

FIGURE 19-8. *Variation in maize. A collection of types grown in Guatemala.* [*Courtesy USDA.*]

and dominate the population. Dominant genes characterize dispersal centers, with recessive forms more likely to gain opportunity at the periphery of a cultivar's range.

Vavilov emphasized that the distribution of plant species is not uniform over the vegetated areas of the earth. For example, two small Central American countries, Costa Rica and Salvador, possess as many native species as the United States and Canada combined, yet have only one percent of the land area. A number of regions especially rich in the concentration of different plants have been characterized as **centers of diversity** and are thought to coincide with the primary and secondary **centers of origin** of present-day cultivated plants (Fig. 19-9).

Centers of origin of important cultivated plants
(after Vavilov)

Old World

I *China*—The mountainous and adjacent lowlands of central and western China represent the earliest and largest center for the origin of cultivated plants and world agriculture. Native plants include millet, buckwheat, soybean, many le-

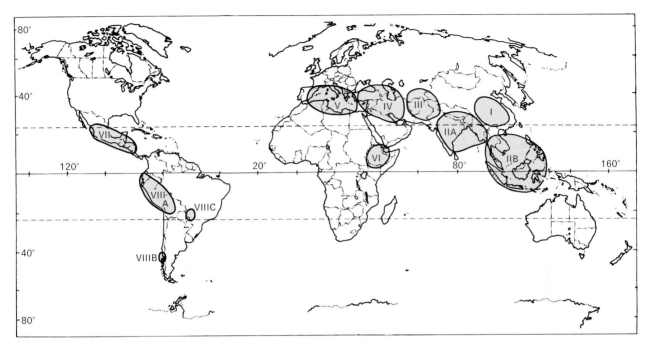

FIGURE 19-9. *Centers of origin of cultivated plants.* [*After N. I. Vavilov,* The Origin, Variation, Immunity and Breeding of Cultivated Plants. *Translated by K. Starr Chester. New York: The Ronald Press Company, copyright 1951.*]

gumes, bamboo, crucifers, onion, lettuce, eggplant, cucurbits, pear, cherry, quince, citrus, persimmon, sugar cane, cinnamon, and tea.

II (A) *South Asia* (Hindustan)—This area is considered the center for rice, sugar cane, many legumes, and such tropical fruits as mango, orange, lemon, and tangerine. An *Indo-Malayan* subcenter (B) includes banana, coconut, sugar cane, clove, nutmeg, black pepper, and manila hemp.

III *Central Asia*—This region is most important as the center of origin of common wheat. Native plants include pea, bean, lentil, hemp, cotton, carrot, radish, garlic, spinach, pistachio, apricot, pear, and apple.

IV *Asia Minor*—At least nine species of wheat as well as rye are native. This is the center for many subtropical and temperate fruits (cherry, pomegranate, walnut, quince, almond, fig) and forages (alfalfa, Persian clover, vetch).

V *Mediterranean*—This is the home of the olive and many cultivated vegetables and forages. The effect of early and continuous civilizations is indicated in the improvement of crops that originated in Asian centers.

VI *Abyssinia*—Wheats and barleys are especially rich; sesame, castor bean, coffee, okra are indigenous.

New World

VII *Central America*—Extremely varied native plants, including maize, bean, squash, chayote, sweet-potato, pepper, upland cotton, prickly pear, papaya, agave, and cacao.

VIII *South America*—A number of South American centers are noted. The West Central area (A) (Ecuador-Peru-Bolivia) is the center of origin of many potato species, tomato, lima bean, pumpkin, pepper, coca, Egyptian cotton, and tobacco. The island of Chiloe (B) in southern Chile is thought to be the source of the potato. The peanut and the pineapple originated in the semiarid region of Brazil (C), and manioc and the rubber tree in the tropical Amazon region.

These areas, comprising only a small portion (2–3 percent) of the land area of the earth, are geographically distinct, being isolated by deserts or mountain ranges. The interaction of the rich local flora and the primitive human populations in these locations appears to have brought about independent pockets or agricultural development. Five-sixths of the 640 species listed by Vavilov are derived from the Old World and one-sixth from the New.

Selection

Selection, as a breeding term, refers to the differential reproduction of individuals in a population. Selection is achieved by preserving favorable variants and eliminating undesirable ones (Fig. 19-10). It is often a natural process, because in the absence of interference by man, those plants most adapted to survive and reproduce leave the most descendants. Almost all of our crops in use today were domesticated before the advent of written history. Improvement of these crops has been continuous as a result of selection by man. This process, carried out both consciously and unconsciously over the years, has been extremely effective; many of our cultivated plants are very different from their wild ancestors; some are known only in cultivation—for example, cabbage, maize. As a result of man's selection over the centuries, they have changed so much that their wild ancestors have become obscure.

The efficient use of the selection process is one of the principal tools of the breeder. It is important to keep in mind, however, that selection does not create genetic variability but merely acts on the genetic variability already available. Thus the breeder must first create a variable population from which to select and then recognize and propagate those individuals with superior characteristics. In order to obtain superior populations from which to select, a wise choice of parental varieties must be made. The efficient evaluation of potential parents is one of the main problems in modern plant breeding. The simplest technique used in choosing parent plants is to look for plants that possess complementing characters. More sophisticated methods involve the sampling of hybrid progeny to predict "prepotency," the ability of an individual to produce desirable offspring. The proper choice of parental materials is one of the crucial decisions in any breeding program.

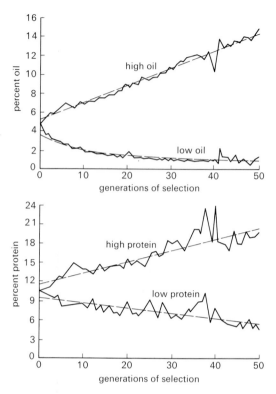

FIGURE 19-10. *Effect of fifty generations of selection for high and low oil and protein content of corn. Solid lines indicate actual data; broken lines indicate trend.* [After Woodworth, Long, and Jugenheimer, Agron. Jour. **44**:60–65, 1952.]

The two general methods of selection used to sort desirable genotypes from segregating progeny are called **pedigree selection** and **mass selection** (Fig. 19-11). In pedigree selection individual plants are selected on the basis of their ancestry or pedigree as well as their phenotype. It is assumed that the best genotypes will be derived from superior plants that occur in superior progeny. Therefore, the progeny of each plant are handled separately. Straight pedigree selection entails laborious and time-consuming effort. It involves keeping extensive records, and unless rigid selection is maintained the program soom becomes unwieldy. As a result, this method limits the amount of material that can be handled, but the method does allow detailed genetic studies to

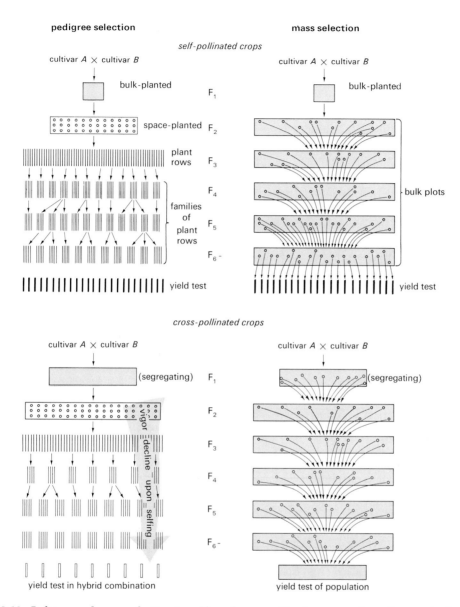

FIGURE 19-11. *Pedigree and mass selection in self- and cross-pollinating crops. In pedigree selection superior plants are selected from best rows, and their progeny are kept separate. In mass selection progeny of selected individuals are grown in mass. The consequences of the selection method depend on the natural method of pollination, as described in the text. Various modifications of these selection methods are possible.*

parallel the breeding program. In addition, self-pollinated crops are extremely responsive to efficient pedigree selection, and progress reflects to a great extent the individual breeder's skill.

In mass selection individual desirable plants are selected solely on the basis of phenotype, and the resulting seed is composited and progenies grown in mass. Individual progeny are not tested separately. Selection may be accomplished either by saving the best plants or by eliminating (**roguing**

out) undesirable ones. Roguing out can be done mechanically for some crops, for example, by screening for seed size or harvesting at a particular date to select for season of maturity. Mass selection is based on the premise that the forces of natural selection will aid the breeder in eliminating undesirable types. The method permits a large pool of germplasm to be manipulated at one time and overcomes some of the limitations of pedigree selection. It is particularly suited to those crops that show decline in vigor under the forced inbreeding inherent in straight pedigree selection.

In actual practice, the processes of mass selection and pedigree selection may be combined. Pedigree selection may be utilized in the early segregating generations to exploit the major genetic differences and eliminate obvious undesirable types. Following this, mass selection techniques may be used. Most modern breeding techniques involve a combination of these systems.

The particular method of selection used depends to a great extent on the mode of reproduction and method of pollination of the crop involved. In general, pedigree selection is most useful for self-pollinated crops and mass selection for cross-pollinated crops.

SELECTION IN SELF-POLLINATED CROPS   Two fundamentally different types of populations of self-pollinated crops exist. One is a mixture of different homozygous lines, as found in a collection of cultivars. Here selection consists of determining the best genotype by testing. The best genotype can be duplicated from its selfed seed. The other type of population is a mixture of different heterogeneous genotypes, as found in the F_2 generation of a cross between two homozygous cultivars. This population consists of many different genotypes with varying degrees of heterozygosity. In order to obtain an improved type, the best genotype must be selected and then transformed into a homozygous, true-breeding line without losing the essential characteristics of the selected individual.

Pedigree selection is widely used for self-pollinating crops. Individual plants are generally selected from a segregating population (typically, the F_2 generation of a cross between cultivars), and selection proceeds between and within progenies in each subsequent generation until genetic purity is reached. Since each selfing brings about homozygosity in a 50 percent increase each generation, the variation between lines is halved each generation. By choosing between lines rather than between plants, there is less chance of confusing genetic and environmental effects. This process is repeated each generation. By the fifth generation the lines derived from single plant selections will be homozygous for more than 95 percent of their genes. Such lines are for practical purposes considered to be true breeding. If testing shows any one of these lines to be obviously superior to existing varieties, it may be named to identify it as a new variety.

Mass selection may also be used in self-pollinated crops by planting segregating populations in large plots and harvesting in bulk. Selection may be practiced each generation by eliminating undesirable plants. After 5 to 10 generations the population will consist of a heterogeneous mixture of somewhat selected homozygous genotypes. The progeny of any plant can be expected to form the basis of a new cultivar. The problem at this stage is to determine the best genotype by testing. The method is successful if the better genotypes are retained. Mass selection by virtue of its economy, permits a large pool of germplasm to be manipulated and carried along.

The combination of mass and pedigree selection is well suited to self-pollinated crops. For example, F_2 plants can be pedigree selected on the basis of F_3 line performance. After the principal agriculturally desirable types are selected, the best F_3 (or F_4) lines may be bulked and carried on by mass selection until homozygosity is reached. At this time pedigree selection is resumed to isolate the superior genotypes. Some varieties may intentionally consist of a mixture of pure lines to provide for some variability.

SELECTION IN CROSS-POLLINATED CROPS   Pedigree selection depends on the evaluation of genotypes by inbreeding. Straight pedigree selection is therefore undesirable as a method of improving cross-pollinated plants when inbreeding leads to less vigor, unless some procedure is set up to restore vigor subsequently. Inbreeding is avoided in the mass selection method by natural interpollination of selected individuals.

Mass selection often leads to an initial rapid improvement of cross-pollinated plants. Sometimes, however, it becomes difficult to increase genetic gain beyond a certain level. To achieve greater control of the genotypes making up the cross-pollinating population, pedigree and mass selection may be combined. This is accomplished by selfing individual plants and "pedigree selecting" them on the basis of their progeny. The selected lines are then allowed to interpollinate to restore "heterozygosity" and vigor. The process may be repeated for a number of cycles. Such a system is called **recurrent selection.** There are a number of variations of this system. The benefit of the technique is that it allows steady progress without eliminating the variability brought about by inbreeding. When the character evaluated is a fruit or a seed, selection must be made after pollination, in which case both pedigree and mass selection are carried out in the same generation. For example, the "female" parent, or seed-bearing plant, is pedigree selected by planting its progeny separately, but the "male," or pollen parent, is mass selected. This technique was first used on maize, in which connection it is known as "ear-to-row breeding."

In a very real sense, F_1 hybrids are an extreme product of recurrent selection. Here the inbreeding process is carried on to homozygosity, usually under a pedigree selection system, and vigor is restored by combining two (or more) inbreds. The selection of inbreds and the choice of inbred combinations results in uniform, genetically controlled hybrids.

In the past mass selection techniques have worked very well for highly heritable characters but have been disappointing for characters that have low heritability, such as yield. As a result mass selection was generally abandoned in such crops as maize, and breeders switched to selection within and between inbred lines—that is, hybrid breeding. As more was learned about the genetics of populations, however, it became clear that what was lacking in the mass selection technique were refined methods for distinguishing genetic from environmental variation. By using special selection techniques with careful replication and statistical control, steady gain in yield has been possible with mass selection. These techniques offer great promise for improving the yield potential of cultivars for use in locations where the uniformity of hybrids may not be desirable (as in mountainous areas) and for providing better populations from which to extract inbreds.

SELECTION IN ASEXUALLY PROPAGATED CROPS Selection is straightforward in asexually-propagated crops, since any genotype may be perpetuated intact. The problem is one of testing to determine the most desirable genotype. Obtaining segregating populations from which superior genotypes may be found is the chief problem in breeding asexually propagated material. Unfortunately, the most desirable selections do not always make the best parents; consequently, potential parents are best selected on the basis of progeny performance.

Somatic mutations of spontaneous origin within vegetatively propagated material are commonly referred to as **sports.** Desirable mutations occurring in adapted, asexually propagated plants may result in an immediate improvement, such as the color sports in many apple varieties and superior tree types in coffee.

Mutations that occur in somatic tissue may involve only a sector of tissue, resulting in a **chimera** (named after a mythical beast with the head of a lion, the body of a goat, and the tail of a dragon.) Such chimeras, when vegetatively propagated tend to be unstable because buds may be formed from tissue with or without the mutation (Fig. 19-12). For example, the 'White Sim' carnation, a sport of the 'Red Sim' cultivar, is a chimera. Its white flowers tend to have islands of red tissue in them, and buds from internal tissue will produce red flowers. The selection of desirable sports is an important means of improving asexually propagated crops. This process is known as **bud selection.**

The backcross method

The backcross has been defined as a cross of an F_1 plant with one of its parents. The backcross method is a breeding technique for transferring single, readily identifiable characters from one cultivar or line to another. Characters controlled by single genes, such as growth habit and certain types of disease resistance, are easily transferred by this method. By repeated backcrossing of the hybrid with the parent variety that carries the most desirable characters (the **recurrent parent**), but selecting in each generation for a single character from the other (**nonrecurrent**) parent, a genotype will eventually be obtained that has all the genes of the recurrent parent plus the selected gene of the nonrecurrent parent.

Two sets of genes must be considered in the backcross method. One set consists of the particular gene (or genes) from the nonrecurrent parent that is to be transferred, and the other is the large group of genes from the recurrent parent to be retained. The genes of the recurrent parent are transferred by an increment of 50 percent in each generation of backcrossing. By selecting plants at random in each generation, the genes of the recurrent parent not only become incorporated but become homozygous. This can be shown easily with respect to a single gene.

Assume a cross involving a single gene pair, $AA \times aa$. The two types of backcrosses are:

$$Aa \times AA \rightarrow 1Aa:1AA$$
$$Aa \times aa \rightarrow 1Aa:1aa$$

Compare these with the self:

$$Aa \times Aa \rightarrow 1AA:2Aa:1aa$$

In backcrossing, as in selfing, half of the genes become homozygous. This can be generalized for any number of genes. The rate of return to homozygosity in backcrossing is the same as in selfing, but all of the homozygous genes resemble the recurrent parent! If backcrossing to the same homozygous type is continued for five or more generations, more than 95 percent of the genes of the hybrid will be identical to those of the recurrent parent and will be homozygous.

If the gene transferred is dominant, the procedure is straightforward. After the last backcross, the gene may be made homozygous by selfing. If the character transferred is recessive, special testing must be carried on in each generation to make sure that the plant selected for backcrossing is heterozygous and contains the desired recessive.

The backcross method has found its greatest usefulness in self-pollinated crops and in improving inbred lines of cross-pollinated crops. The reason for its limited usefulness in cross-pollinated crops is that in such crops backcrossing is equivalent to selfing and thus results in a decline in vigor. This may be overcome by using a large number of selections of the recurrent cultivar to avoid inbreeding.

The main advantage of the backcross method is its predictability. The improved cultivar is often indistinguishable from the old, with the exception of the added character. As a result, extensive testing may be avoided. The backcross method cannot, however, be expected to improve the cultivar more than the increment achieved by the addition of the single selected character. Moreover, unless this character is inherited relatively simply, the method is difficult to use.

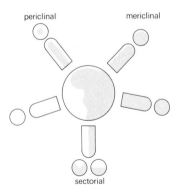

FIGURE 19-12. *Plant chimeras contain genetically different tissues. Various types may be derived from buds arising from a sectored chimeral stem.* [After Jones, Botan. Rev. 3:545–562, 1937.]

Hybrid breeding

Hybrid breeding refers to the production of heterozygous populations from crosses of homozygous lines. Hybrid breeding first came into prominence with maize. Although the feasibility of hybrid corn was suggested in 1909, the first commercial hybrids did not appear until the 1920's. From 1936 to 1945 the total acreage of hybrids in the Corn Belt increased from less than 5 to more than 90 percent of the corn planted. The spectacular success of hybrid corn has given great impetus to this method of breeding. Because of the genetic nature of heterosis, hybrid breeding is most applicable for cross-pollinated crops, but has limited usefulness for some self-pollinated crops.

Seed of F_1 hybrids must be remade each year. Consequently, for the success of this breeding method the extra advantage resulting from the use of hybrids must more than compensate for the additional cost of their production. Thus whether hybrid breeding is used in a particular crop depends on balance between the specific technical problems involved in producing and combining homozygous lines and the amount of the heterotic effect.

There are two technical steps in the production of hybrid seed for cross-pollinated crops: (1) the production of desirable homozygous lines; and (2) crossing of these lines to obtain hybrid seed. Self-pollinated crops are already naturally "inbred," and therefore need only be crossed to produce hybrids.

PRODUCTION OF HOMOZYGOUS LINES The most common method of producing homozygous lines is through continued inbreeding. In self-pollinated crops homozygous lines occur naturally, but in cross-pollinated crops they must be produced by the breeder. The ease of producing inbred lines within a cross-pollinated species varies with the crop species. In perfect-flowered species the inflorescence merely needs to be bagged to exclude foreign pollen. In monoecious crops that have separate staminate and pistillate flowers, the pistillate flower must be protected to prevent random cross-pollination, and pollen must be collected and applied to the stigmal surface of the pistil. In maize this process is relatively simple because all the pollen is produced in one inflorescence, the tassel, and all the pistillate flowers are congregated in the ear.

The production of inbred lines is not straightforward in all crops. Dioecious species, such as spinach, consist of staminate and pistillate plants. Inbreeding in these crops is accomplished by sib (brother-sister) pollination. The rate of return to homozygosity is much lower than under selfing.

A number of crops resist self-fertilization because of incompatibility mechanisms. Self-incompatibility is usually controlled by alleles at a single gene locus. When pollen contains the same incompatibility allele that is present in the style, tube growth is arrested. In a few crops, such as cabbage, incompatibility may be circumvented by pollinating before the flower opens (bud pollination). Self-incompatible lines may be increased asexually; this is being done with alfalfa.

Homozygous lines may be produced in a single step, avoiding the long process of inbreeding by some quite interesting and sophisticated techniques. Monoploids are produced in certain crops, such as corn, in a very low but predictable frequency. They apparently result from a disruption in the double fertilization process. One male gamete combines with the two polar nuclei to form the triploid endosperm, but the fertilization of the egg nucleus does not occur. Monoploids may be recognized in the seedling stage by the use of special marker genes. Many of these monoploid lines double spontaneously to produce "instant" homozygous diploid types. Monoploids can also be produced by tissue culture techniques from pollen. This has been achieved with tobacco, rice, tomato and others.

Another method of producing homozygous lines is through a complex series to translocations. If a line can be synthesized with a portion of each chromosome attached to another, it is possible to create inbred lines in one step. Such a series of translocations occurs naturally in *Oenothera*. A similar complex series of translocations has been synthesized in maize and barley, but the sterility was too high for the method to be practical.

COMBINING INBRED LINES The problem of producing crossed seed is the reverse of inbreeding. For maximum gain selfing must be eliminated. The basic approach consists of transforming one parent of the cross into a "female" line (Fig. 19-13). The simplest method is to eliminate the pollen-bearing organ (emasculation). This is accomplished in corn by removing the emerging tassels. In perfect-flowered species (such as tomato), each flower must

detasseling system

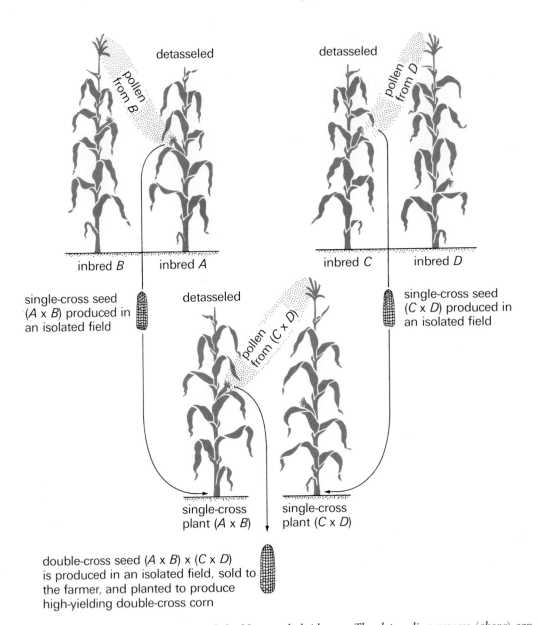

detasseled

pollen from B

detasseled

pollen from D

inbred B   inbred A   inbred C   inbred D

single-cross seed
(A x B) produced in
an isolated field

detasseled

single-cross seed
(C x D) produced in
an isolated field

pollen from (C x D)

single-cross
plant (A x B)

single-cross
plant (C x D)

double-cross seed (A x B) x (C x D)
is produced in an isolated field, sold to
the farmer, and planted to produce
high-yielding double-cross corn

FIGURE 19-13. *The production of single- and double-cross hybrid corn. The detasseling process (above) can be eliminated by manipulating male sterility (opposite page) determined by an interaction between a cytoplasmic factor* $\boxed{S}$ *and a double recessive nuclear gene* rf. *Only the combination* $\boxed{S}$ rf rf *is male-sterile. Plants with* $\boxed{N}$ *(normal) cytoplasm or with the pollen restoring allele* Rf, *either homozygous or heterozygous, produce viable pollen. The cytoplasmic factor passes only through the egg. Appropriate genotypes are shown in shaded rectangles.*

cytoplasmic male-sterile system

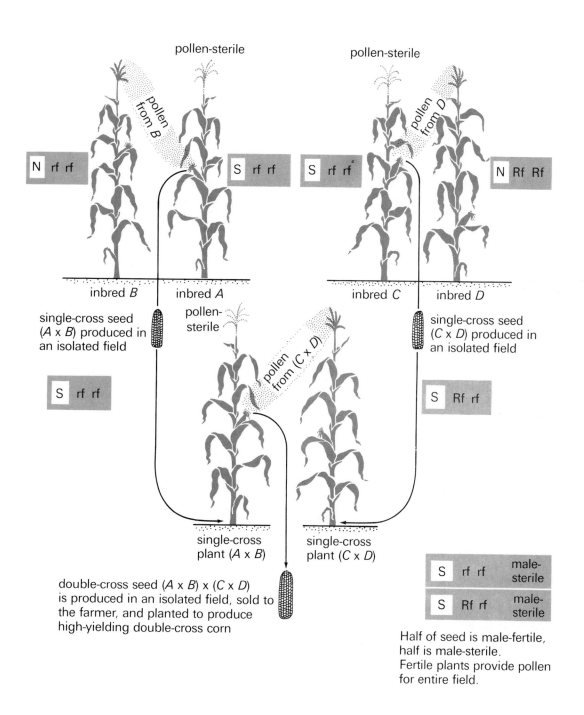

pollen-sterile

N rf rf

pollen from B

S rf rf

inbred B   inbred A

single-cross seed (A x B) produced in an isolated field

S rf rf

pollen-sterile

S rf rf

pollen from D

N Rf Rf

inbred C   inbred D

single-cross seed (C x D) produced in an isolated field

S Rf rf

pollen-sterile

pollen from (C x D)

single-cross plant (A x B)

single-cross plant (C x D)

double-cross seed (A x B) x (C x D) is produced in an isolated field, sold to the farmer, and planted to produce high-yielding double-cross corn

| S | rf rf | male-sterile |
| S | Rf rf | male-sterile |

Half of seed is male-fertile, half is male-sterile. Fertile plants provide pollen for entire field.

be emasculated. In dioecious species, staminate plants must be eliminated completely.

In perfect-flowered crops that produce small flowers or a small number of seeds per flower, the emasculation process is not feasible. In these crops the production of hybrids depends on genetic methods of eliminating selfing. One way to accomplish this is to induce male sterility. Chemicals that cause male sterility are known, but their use is not yet practical. At present, the most efficient way to eliminate selfing is to make use of the genetically determined male sterility that results from the interaction between a cytoplasmic factor and nuclear genes. The cytoplasmic factor is transmitted only through the pistillate parent. Cytoplasmic male sterility was first described in onions and subsequently found in many other crops. Hybrids have been produced commercially by means of cytoplasmic male sterility in onions, corn, and sorghum. This male sterility results from the interaction of the cytoplasmic factor with one or more recessive genes. In the dominant condition these genes are known as "restorers" of fertility. Cytoplasmic male sterility can be incorporated into inbreds by backcrossing.

The use of cytoplasmic male sterility to eliminate detasseling in corn was so successful that by 1970 practically all hybrids had been produced by that method. The system worked so well that most hybrids contained sterile cytoplasm derived from a single plant discovered in Texas (T cytoplasm). However in the summer of 1970 an epidemic of Southern corn leaf blight caused ruinous losses (valued at about a billion dollars, the greatest single plant disease loss in history!) as a result of the appearance and spread of a new (T) strain of the causal fungus, *Helminthosporium maydis*, that attacked *only* plants of the sterile (T) cytoplasm. The situation was totally unexpected because a linkage between susceptibility to a fungus and a cytoplasm was all but unknown before this outbreak. As a result of the 1970 epidemic, a shift to normal cytoplasm and detasseling took place in hybrid corn seed production. The epidemic was a striking demonstration of the way in which uniformity can result in the vulnerability of a major crop. Studies to evaluate new cytoplasms for male sterility are in progress, but it is unlikely that

agriculture will ever again rely on a single source.

In cucumbers, hybrid seed can be produced by a gene that converts a normally monoecious species into a gynodioecious one, which gives equal numbers of monoecious and completely pistillate plants upon selfing. When eliminated (rogued) the monoecious plants leave a completely pistillate line. It is also possible to produce a gynoecious, or "female," line. Such a line may be maintained by self-pollination after the induction of staminate flowers with gibberellins.

Self-incompatibility factors may be utilized to prevent selfing. This is done in the production of hybrid cabbage and alfalfa.

Where it is not possible to completely eliminate selfing, genetic markers may be utilized to distinguish cross- from self-fertilized plants. The "female" line is made homozygous for some recessive seedling character; the male line is homozygous dominant. All hybrid plants will carry the dominant marker gene; plants produced from selfing in the female line will show the recessive phenotype, and can be eliminated in seedbeds or in the field. Hybrid cotton has been produced by this method. A number of methods are used to increase the proportion of natural crossing and reduce selfing, including chemically inducing male sterility and increasing insect activity.

The transfer of pollen in cross-pollinated crops is usually accomplished naturally by wind or insects. In self-pollinated crops the pollen must be transferred artificially. Nevertheless, even in naturally self-pollinated crops, cross-pollination can be often increased by encouraging insect activity. In male-sterile wheat sufficient natural cross-pollination occurs by wind to make hybrids a distinct possibility.

SELECTION WITHIN AND BETWEEN INBRED LINES
The success of hybrid breeding depends on the efficient selection within and between inbred lines. Selection within lines refers to selection in the inbreeding process; selection between lines refers to the choice of combinations in the crossing of inbreds.

The success of an inbred is measured not by its phenotype, or self progeny, but by its cross progeny. The ability of an inbred to produce successful hybrids is known as **combining ability.** Combining ability may be partitioned into general and specific effects. **General combining ability** refers to the average performance of a particular inbred in a series of hybrid combinations. **Specific combining ability** refers to the performance of a specific combination of inbreds.

General combining ability is an inherited character. It can be estimated by crosses with a test variety, usually an open-pollinated cultivar with a broad genetic base. Combining ability may be selected for early in the inbreeding process. Selected plants are crossed with the test cultivar at the same time they are selfed. The selfed seed is used only if the results of the general test of combining ability are favorable.

The genetic improvement of inbreds is a specialized type of breeding. In general, inbreds are handled as self-pollinated crops. Inbred lines can be improved, one character at a time (for example, disease resistance, male sterility), by the backcross method.

One of the most difficult problems is to improve the combining ability of inbreds. Present-day corn hybrids do not show great yielding superiority over the best of the early hybrids. Improvement has come by the incorporation of better agronomic characters, such as disease resistance and specific adaptation. In present-day corn breeding, emphasis is on improving populations in order to produce superior inbreds. This is achieved by recurrent selection, with combining ability a principal criterion of selection.

Although maximum heterosis is obtained by crossing two diverse inbreds, a number of other combinations produce hybrid vigor. The various kinds of crosses that are referred to as hybrids in the trade are designated as follows:

Type of cross	Name of hybrid
inbred × inbred	single-cross
F_1 hybrid × inbred	3-way-cross
F_1 hybrid × F_1 hybrid	double-cross
inbred × variety	top-cross
F_1 hybrid selfed	F_2 hybrid

The double cross, first suggested by D. F. Jones in 1918, was responsible for the commercial feasibility of hybrid corn. Inbred lines of corn are

relatively weak, and their seed yields are low. By crossing certain single crosses, the cost of the seed was greatly reduced with little loss of heterosis. Double-cross yields are predicted from the yields of single crosses. Techniques have also been established for producing double-cross seed by means of cytoplasmic male sterility, which eliminates manual detasseling (Fig. 19-13).

When the production of F_1 hybrids is impractical the utilization of heterosis may be obtained by compositing a number of inbred lines selected for their ability to combine well with each other. The subsequent population is referred to as a **synthetic variety.** Synthetics are not made anew each year but maintained from open-pollinated seed. Synthetic varieties have been most popular in forage crops whose floral structure makes the control of pollination difficult.

The success of hybrid breeding is due in large part to the inherent commercial possibilities. Control of the unique sets of inbreds effectively controls the hybrid, for hybrids do not breed true from seed. In hybrids between cultivars of self-pollinated crops, it is only necessary to know the two cultivars that are used to make the hybrid.

Polyploid breeding

Polyploidy is used in a number of important ways in breeding. The level of ploidy may in itself be important in conferring desirable characteristics. In addition, the manipulation of ploidy greatly increases the amount of variability available to the breeder, because many desirable characters are found in related species with different chromosome numbers.

Polyploids may occur spontaneously or be artificially induced by special treatment. For example, the formation of callus tissue by wounding is effective in inducing tetraploidy in the tomato. Colchicine, an alkaloid derived from the autumn crocus, has the remarkable property of doubling the chromosome number in a wide variety of plants when applied in concentrations of 0.1 to 0.3 percent. This drug interferes with spindle formation in mitosis; the chromosomes divide, but the cell does not. Tetraploids are produced routinely with the use of colchicine in many plants.

Treatment of seeds with colchicine (usually 24 hours) produces tetraploidy in about 5 percent of the surviving seedlings. This drug may also be applied to the growing points of seedlings or other vegetative parts by a variety of methods. Colchicine has not proven efficient with many grasses. Tetraploids may be produced in corn by applying heat or by genetic methods involving the effects of *elongate*, a chromosome-doubling gene.

An artificially induced "autotetraploid" may usually be distinguished from a normal diploid by its larger, thicker leaves and organs (Fig. 19-14) and somewhat slower growth. These **gigas** characteristics are a result of increased cell size. Larger pollen size is thus a reliable indicator of tetraploidy, although actual chromosome counts are necessary for positive identification. Reduced fertility is another indication. Autotetraploids have four, rather than two, chromosomes of each type. Chromosomes that pair two by two (bivalent pairing) disjoin normally at meiosis, but if all four chromosomes synapse (quadrivalent association), there is a possibility of a 3 to 1 distribution, which leads to reduced fertility because gametophytes having

unbalanced numbers of chromosomes are nonviable or at a great selective disadvantage. Because tetraploidy is so often accompanied by a reduction in fertility, it is used to greatest advantage in plants grown for vegetative or floral parts (forages, ornamentals, nonseed food crops).

Inheritance in polyploids is more complex than in diploids. The single-gene ratios in diploid organisms are a result of the assortment of two alleles per gene. In tetraploids, however, there are four alleles per gene. There are two types of homozygotes ($AAAA$ and $aaaa$) and three kinds of heterozygotes ($AAAa$, $AAaa$, $Aaaa$). The genetic ratios of tetraploids also differ from those of diploids. For example, the $AAaa$ heterozygote will produce gametes in a ratio of $1AA:4Aa:1aa$. If A is completely dominant ($Aaaa$ producing the A- effect), the progeny of $AAaa$ selfed will segregate in a phenotypic ratio of $35A—:1aaaa$. This ratio, however, is affected by the location of the gene on the chromosome. Thus, tetraploidy tends to decrease the frequency of recessive genotypes. Consequently, the rate of return to homozygosity upon selfing is very much slower in tetraploids than in diploids.

The promiscuous induction of tetraploidy seldom leads to changes of immediate value. Improvement, if obtained, is usually found only as a result of breeding tetraploids. Examples of economic improvements include the development of large-flowered ornamentals (orchids, snapdragons, lilies, and daylilies) and higher yielding rye and forages (alsike and red clover).

In some plants increases in vigor and in fruit size are associated with the triploid condition. Triploids can be produced by crossing tetraploids with diploids:

$$4n \ (2n \ \text{gamete}) \times 2n \ (n \ \text{gamete}) \rightarrow 3n$$

Triploids also occur spontaneously as a result of the formation of unreduced egg cells.

Sterility in triploids is very high. In the assortment of three homologous chromosomes at meiosis, two move to one pole and the remaining chromosome moves to the other. Since the assortment of each group of three chromosomes is independent, gametes with chromosome numbers from n to $2n$ may be formed. As the chromosome number goes

FIGURE 19-14. *Increased cluster size as a result of colchicine-induced polyploidy in the Portland grape. (Left) Diploid. (Right) Tetraploid.* [*Courtesy Haig Dermen.*]

up, the frequency of balanced n and $2n$ gametes decreases to an insignificant number. The high sterility associated with triploidy has been exploited in watermelon. The seedless watermelon is a triploid produced by crossing a tetraploid and a diploid. Triploids can be propagated asexually (banana, apple), or produced anew each year from tetraploid-diploid crosses.

Almost a quarter of the apple and pear cultivars have turned out to be triploid, originating apparently from relatively rare unreduced gametes. They were selected because of larger fruit size and other desirable features. Triploid apples do not produce viable pollen. (An orchard with a triploid cultivar must have two additional diploid pollinators for all cultivars to set fruit because of self-incompatibility factors.) Recently a number of promising triploid apple and pear cultivars have been produced from crosses involving diploids with naturally occurring tetraploids. In sugar beets triploidy is the optimum ploidy level. they have larger roots and more sugar than do either diploids or tetraploids.

There has been great interest in the use of polyploidy as a means of extending variability by creating new species or by transferring genes from other species. Induced polyploidy makes it possible to overcome sterility associated with interspecific hybrids. This sterility is a result of the inability of chromosomes to pair at meiosis. Fertility is restored by doubling the chromosome number of the sterile hybrid or by crossing autotetraploids of each species. Using capital letters to designate genomes, we can represent this as follows:

AA
$\searrow$
$\times$ → AB — chromosome
BB   sterile   doubling → AABB
       hybrid   fertile
                               amphidiploid,

AA — chromosome
     doubling → AAAA
                    $\searrow$
                    $\times$ → AABB
BB — chromosome   fertile
     doubling → BBBB   amphidiploid.

The resulting "amphidiploid" is an allopolyploid made up of the entire somatic complement of each

species. Fertility results from bivalent pairing within each genome.

These techniques have produced an interesting new crop called triticale, a hybrid of wheat and rye (*Triticum-Secale*). Selection within this amphidiploid is being carried out to combine hardiness of rye with the quality of wheat. Other examples of natural and synthetic amphidiploid are found in cole crops (Fig. 19-15).

The transfer of genes between species offers perhaps the greatest potential value of polyploid breeding. Tobacco mosaic resistance from *Nicotiana glutinosa* ($n = 12$) has been transferred to the cultivated form of tobacco, *N. tabaccum* ($n = 24$), a natural amphidiploid between *N. glutinosa* and *N. tomentosa*. Backcrossing and selection have resulted in the introduction of a whole chromosome from *N. glutinosa*, which contains the resistance factors.

The addition of whole chromosomes has certain disadvantages because of the possibility of the transfer of undesirable features. The transfer of small chromosome segments has been accomplished by irradiation techniques in wheat ($2n = 42$). A small chromosome segment containing resistance to rust has been transferred from a wild Mediterranean grass, *Aegilops umbellulata* ($2n = 14$), to wheat chromosomes by using amphidiploids with a related wheat, *Triticum dicoccoides* ($2n = 28$), as a genetic bridge.

Polyploidy has not proved to be a panacea in plant breeding. It is, however, another in the arsenal of weapons that the breeder can use to change plants to better suit his needs. Polyploid manipulation has greatly enlarged the genetic base from which the breeder may draw. These methods undoubtedly will contribute much to future crop improvement.

Mutation breeding

One of the limitations to crop improvement is the dependence upon naturally occurring variation. Mutations, the source of genetic variation, can be induced by radiation (X-rays, gamma rays, thermal neutrons) and by chemical means (ethyl dimethyl sulfate), and can be increased by special genetic mechanisms. Mutation breeding refers to the use

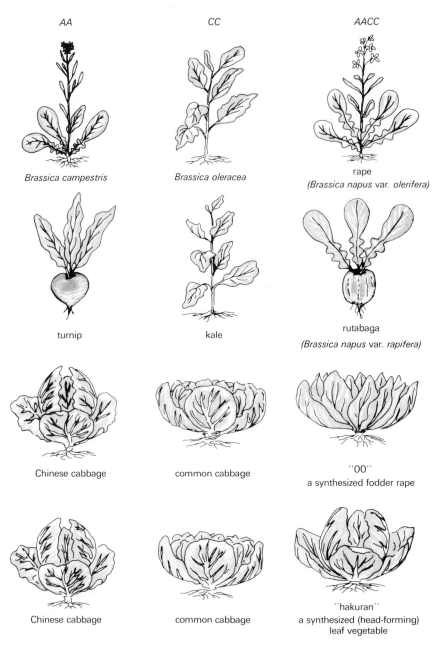

AA   CC   AACC

Brassica campestris   *Brassica oleracea*   rape
(*Brassica napus* var. *olerifera*)

turnip   kale   rutabaga
(*Brassica napus* var. *rapifera*)

Chinese cabbage   common cabbage   ''OO''
a synthesized fodder rape

Chinese cabbage   common cabbage   ''hakuran''
a synthesized (head-forming)
leaf vegetable

FIGURE 19-15. *Natural and artificial amphidiploids in* Brassica. [*From Kihara*, Seiken Ziho, **20:**1–.4, 1968.]

of artificially induced mutations in crop improvement. Present mutagenic agents induce random changes, most of which are harmful, as are naturally occurring mutations. Since such changes are nondirectable and largely deleterious, the problem is one of detection and isolation of the rare desirable alteration. In order for a technique to be practical, an efficient screening procedure must be developed. For example, large populations of oats can be screened in the seedling stage for resistance to the Victoria blight disease, because susceptible plants die soon after inoculation. Mutants that change self-incompatibility genes can be detected by using the stigma as a sorting device, for only mutated pollen grains carrying the desired mutation will grow.

Asexually propagated crops have been suggested as promising material for mutation breeding because adapted material may be screened in the search for a single desirable induced change. But because mutations in such material are largely chimeral, it is often as laborious to isolate and test for these mutations as it is to select for segregants from ordinary breeding programs. The saving in time has been more apparent than real.

A number of useful alterations have been induced. These include beneficial mutants induced in barley and higher-yielding strains of the mold *Penicillium* (for penicillin production) by ultraviolet irradiation. A grape cultivar with too tight a cluster has been improved through the induction of reduced fertility. In general, however, progress with induced mutations has not been proportional to the effort expended. Because of the unpredictability of mutation breeding in a practical improvement program, it can be considered only as a last resort after other sources of variation have been exploited and found wanting.

Breeding techniques

HYBRIDIZATION   Plant breeding relies on the control of the sexual process. Before two plants can be crossed they must be induced to flower. Various environmental factors such as photoperiod, temperature, and nutrition may have to be manipulated to achieve synchronization in flowering. For example, many cultivars of sweet-potato must be grafted to species of morning glory and then subjected to special photoperiods and temperatures to induce flowering. As a result, in many large breeding programs, crossing is often performed under controlled conditions in the greenhouse.

The basic rule in making artificial crosses is to avoid contamination before and after artificial pollination. In monoecious plants the pistillate flower must be protected before pollination from wind or insect pollination by such devices as paper or glassine bags or cheesecloth nets. In perfect-flowered plants, contamination by selfing is avoided by emasculating the flower before the anthers have begun to shed pollen. Selfing may also be circumvented by selective killing of the pollen. Immersion of the flower in hot water 45 to 48°C (113 to 118°F) has been effective for sorghum and grasses, and immersion in 57 percent ethyl alcohol for ten minutes has been effective for alfalfa. The manipulation of genetic male sterility has been discussed.

Pollination can be achieved with fresh or stored pollen. Often the anthers are collected the day before pollination and dried. Pollen is quite variable in its ability to remain viable. Wheat pollen remains viable only minutes at high temperatures, whereas apple pollen stored under low humidity and cool temperatures, −1 to 2°C (30 to 36°F), may retain viability for years. Storage by freeze-dehydration techniques appears promising.

Pollen may be applied with a camel's-hair brush, the fingers, a blackened matchstick, or the dried flower itself. The pollen should be applied in large amounts. The pollination of onions is carried out by enclosing their flower heads in mesh cages containing pollinating insects (blowflies).

After pollination the flower should be protected from contamination by cross-pollination. This is usually accomplished by enclosing the flower in paper or glassine bags; if the petals have been left on, as is customary with many cucurbits and legumes, they may serve as protection if held closed with cellophane tape. When it is only necessary to keep away insects, cheesecloth is usually sufficient. Care must be used in selecting protecting devices that do not act as a heat trap and thus prevent successful fruit set. In citrus and apples, insects will not visit the flowers if the petals are removed during emasculation. Techniques used in hybridizing white pine are shown in Figure 19-16.

TESTING   The testing of genetically different material is a major undertaking in plant breeding. In order to avoid confusing genetic and environmental effects, testing must be systematic. This

FIGURE 19-16. *Techniques used in hybridizing western white pine. (A) Male cones are collected and washed (B) for pollen extraction. (C) Extracting pollen. (D) Pollen is squirted onto enclosed conelets from hypodermic syringe equipped with bulb. [U.S. Forest Service.]*

requires that all plants be treated alike, or at least in an unbiased fashion. Soil differences are difficult to control; two plants cannot be grown in the same place at the same time. In order to minimize experimental error, or at least to be able to test its presence and quantify it, various field plot procedures have been established. These include randomization, replication, the use of controls, and the elimination of border effects. Wherever possible, quantitative measurements rather than qualitative evaluations are made, and the data subjected to statistical analyses to determine the probability that differences are due to the variable being tested or to chance effects. Good field plot technique and subsequent analysis require sound judgment and careful execution. It is axiomatic that accurate records be kept.

A number of artificial treatments are imposed to facilitate the selection process. These include extremes of heat and cold to test for genetic hardiness and the creation of disease epidemics and insect infestations for the evaluation of genetic pest resistance. Techniques for determining quality are often complex. The evaluation of wheat quality involves the testing of baking property. Determining forage quality may require measuring the content of the toxic substance coumarin or checking grazing trails to gauge palatability.

Once a genetic improvement has been made, vigilance is still required to maintain improvement. Mutation, natural crossing, contamination, and, in vegetatively propagated material, diseases, especially those due to viruses, tend to cause the deterioration or "running out" of cultivars. Genetic deterioration may be reduced to a minimum by continued selection and careful propagation. The plant breeder's responsibilities do not end with the release of a superior cultivar.

Selected References

Allard, R. W. *Principles of Plant Breeding.* Wiley, New York, 1960. (An outstanding theoretical presentation of the genetic foundation of plant breeding.)

Elliott, F. C. *Plant Breeding and Cytogenetics.* McGraw-Hill, New York, 1958. (Theory and methodology in plant breeding.)

Frankel, O. H., and E. Bennett (editors). *Genetic Resources in Plants, Their Exploration and Conservation.* F. A. Davis Company, Philadelphia, 1970. (An outline of methods for exploration and introduction in plant improvement.)

Frey, F. J. (editor). *Plant Breeding.* Iowa State University Press, Ames, 1966. (A collection of articles on plant breeding with the emphasis on general methods rather than specific crops.)

Isaac, E. *Geography of Domestication.* Prentice-Hall, Englewood Cliffs, N.J., 1971. (Reviews the traditional concept of diffusion of domesticates from the Near East, suggesting religious motives for the choices of many of the plants).

Poehlman, J. M. *Breeding Field Crops.* Holt, New York, 1959. (Breeding of major field crops are discussed in detail.)

U.S. Department of Agriculture. *Better Plants and Animals* I, II. Yearbook of Agriculture, 1936, 1937. (A valuable reference and a historical landmark of genetics and breeding of agricultural crops in the United States.)

Vavilov, N. I. *The Origin, Variation, Immunity and Breeding of Cultivated Plants.* Chronica Botanica, Waltham, Mass., 1951. (Contains the most important writings of the great Russian plant scientist.)

Wright, J. W. *Genetics of Forest Tree Improvement.* Food and Agriculture Organization, 1962. (A textbook of forest genetics and breeding with good coverage of world literature.)

PART V

INDUSTRY
OF PLANT
AGRICULTURE

CHAPTER 20

Food Crops:
Cereals, Legumes, Forages

There is no more fundamental a problem confronting the human race than that of feeding its expanding populations. A world in which the majority of the inhabitants go to bed hungry each night is inevitably rife with political and social unrest as well as human misery. Other chapters (especially 3, 4 and 15) have discussed this fundamental concern in some detail. This chapter focuses attention on the crop plants that supply the modern world with the major part of its food. It has been shown in Chapter 19 that crop improvement is a never-ending process. Seldom do the cultivars of today satisfy the needs of tomorrow. It is the intent of this discussion to review the overall importance of these species as represented by their various cultivars, not to offer agronomic recommendations or to compare varieties.

The greater proportion of man's food is derived from relatively few plant species. Most of the staple crops are members of the grass family that yield grains. Recent statistics show rice, wheat, and corn closely grouped as the "big three" in world cereal production. Each accounts for essentially one-quarter of the world's total cereal supply of more than a billion metric tons annually. Barley, oats, rye, millets, and sorghum together make up the remaining quarter.

Only one other food plant, the potato, a root crop, gives the cereals much of a run so far as world importance is concerned. Total tonnage of the potato is approximately equal to that of the leading grains but since the potato is fleshy and consists of only 22 percent dry matter, its total food value is considerably less. Sweet-potatoes and yams, various seed legumes (beans, peas, soybean, peanut, etc.), and sugar cane are the other major noncereal food crops. For all practical purposes man relies upon this handful of crop plants for his existence.

Chapter 19 mentioned the centers of origin of most of the major crop plants. Lengthy lists of wild plants utilized by primitive cultures could be recited. The wild plants that made up part of the diet of the American Indian tribes were fairly well documented by the Old World settlers who invaded North America. Among the Indians of the Pacific Northwest, as Lewis and Clark found them, plant cultivation was unknown, except possibly for tobacco. Wild seeds and nuts made up a good part of their diet, but various roots, greens, and berries were also eaten. Piñon nuts, *Pinus monophylla*, were something of a staple from southern areas, but seeds of wild grasses, Chenopodiaceae and Cruciferae, were also gathered in sizable quantities, as were acorns (*Quercus* spp.), hazelnuts (*Corylus cornuta*), and the seeds of the yellow pond lily (*Nuphar polysepalum*). Many species of *Vaccinium* (blueberries, huckleberries, cranberries) were eaten both fresh and dried, as were several related Ericaceous plants. A number of fruits and berries of the rose family were gathered. A great many "roots" were dug, particularly those growing in marshy grounds, such as *Sagittaria latifolia*. Scores of other species could be named.

Perhaps it might be even more interesting to speculate on the wild plants that sustained the primordial civilizations, such as that of the Tehuacan Valley in Mexico (Chapter 1). Archaeological findings, particularly human coprolites (desiccated feces), give definite evidence of primitive diet and the gradual changes that came about as the domestication of plants proceeded. A diet that depended at first upon wild game supplemented with wild beans, agaves, and cactus gradually changed to include foxtail millet (*Setaria*), squashes (*Cucurbita*), peppers (*Capsicum*), and eventually amaranths (*Amaranthus*) and maize (corn, *Zea mays*). Archeological evidence shows that much the same sort of change took place in South America. Almost all important food plants had beginnings of this sort in prehistorical times, the ancestral type being so remote as to be extinct or unrecognized today.

The modern world would grind to a standstill if it could not depend upon the gradual improvement of varieties of these domesticates, which lend themselves so well to modern mechanized agriculture. Many thousands of wild plants were utilized by primitive cultures; some of them, of course, were used only in times of dire need. Table 20-1 categorizes a number of important crops according to whether they originated in the Old World or the New World. Since the fifteenth century these species have been spread widely over the globe, wherever they proved adaptable to local climate. Area of origin and major zone of today's production do not often coincide!

World food production almost doubled between 1950 and 1970. However, it is distressing to note that there was very little increase in per capita food production during this period; in fact, slight

Table 20-1 NEW WORLD AND OLD WORLD FOOD PLANTS BEFORE 1492.

NEW WORLD		OLD WORLD		
maize	pumpkin	wheat	onion	wine grape
manioc	peanut	rye	garlic	apricot
lima bean	chayote	barley	spinach	peach
potato	papaya	oat	eggplant	olive
sweet-potato	avocado	millet	lettuce	fig
kidney and other beans	pineapple	sorghum	endive	almond
tomato	custard apple	rice	celery	quince
green pepper	soursop	buckwheat	asparagus	pomegranate
Jerusalem artichoke	cherimoya	turnip	peas, various	watermelon
sunflower	guava	cabbage	soybean	cucumber
quinoa	chocolate	rutabaga	broad bean	banana
squash	cashew	chard	dasheen	orange
	sapote	mustard	yam	lemon
		radish	apple	lime
		beet	pear	date
		parsnip	plum	mango
		carrot	cherry	clover
		breadfruit	alfalfa	
		mangosteen	bluegrass	

decreases occurred in both Africa and Latin America. Furthermore, it is unlikely that the present overall rate of increase can be maintained for long, since most of the world's best lands are already under cultivation; thus man will increasingly have to depend upon population control (Chapter 4).

Asia, with its huge land area, is the leading food-producing continent, of which the chief food plant is overwhelmingly rice (93 percent of world production). But because of its immense populations, insufficient food is grown for its needs, especially in India. Africa, in large part peopled by primitive societies, is in great measure self-sufficient on a hand-to-mouth basis, though not self-sufficient according to the standards of more sophisticated cultures. Mainstay crops—millets and yams, for example—are often rather quaint to the rest of the world. In Central and South America food production is, by and large, inadequate (although in certain areas this may be more a reflection of poor dispersal than lack of capability). Maize, wheat, and potatoes (Andes mountains) are the mainstay crops.

On the whole, Europe, in spite of its dense population, feeds itself adequately according to modern standards. Because of its industrialization, it is well equipped to "borrow" land elsewhere for food production, and imports foods that might be in short supply locally. Potatoes are perhaps the most important food plant in Europe, followed by wheat, barley, maize, and other grains. North America, with its high degree of mechanization and still relatively sparse population, has great potential for agricultural increase and export. The extent to which food production and export is carried out depends largely upon world economic conditions and the current political posture regarding foreign aid and foreign trade. More than military might, the responsive soils of the United States and Canada may be their real strength. Maize, wheat, and soybeans are the chief North American crops.

THE CEREALS

Cereals are the staff of life to civilization. Around 70 percent of the world's harvested acreage of about a billion hectares, is devoted to growing them. They are the direct source of half the food needs of the world, and the secondary source of much additional food when converted to meat, milk, eggs, and other animal products.

The true cereals are all members of the grass family, the Gramineae. The fruit they yield is a **grain**, a type of fruit in which the ovary wall turns hard and durable, fusing with the single seed (Fig. 20-1). Buckwheat, amaranth, and a few similar seeds of other families ("pseudo-cereals") are sometimes regarded as grains because their seeds are quite similar to those of grasses. This section will be devoted to the true cereals, of which the major crops are rice (*Oryza*), wheat (*Triticum*), maize (*Zea*), barley (*Hordeum*), oats (*Avena*), rye (*Secale*), sorghum (*Sorghum*), various millets (*Panicum, Setaria*, etc.), and a few other species of lesser world importance.

There are a number of reasons why cereals have evolved into man's leading food source. Most of them are annuals, or are at least adapted to cultivation as annuals, permitting facility in cropping, whether on the midden of an early domestication center or as a component of today's highly mechanized farming. Grasses in general are quite versatile, too; they adapt well to a variety of soils, climates, and ways of handling. They are also relatively efficient in garnering the sun's energy, transforming it into usuable food substance. They are generally hardy and recuperative, and are plagued by no more than their share of diseases and pests. But above all, the grain is a neat package of stored energy, conveniently harvested, easily cleaned and handled, amenable to storage without special drying. Is it any wonder that well-filled granaries have been the mark of security, even opulence, throughout the ages! The word cereal itself stems from the legendary Greek goddess Ceres, "giver-of-grain." She and other gods were often propitiated with gifts of grain.

Rice, wheat, and corn are the world's three major cereals, all about equally important in terms of world production. Rice is the prime source of sustenance for tropical populations; it is grown mostly on flood plains or where the land can be seasonally ponded. Wheat is chiefly grown on lands that were naturally prairies, too harsh, cold,

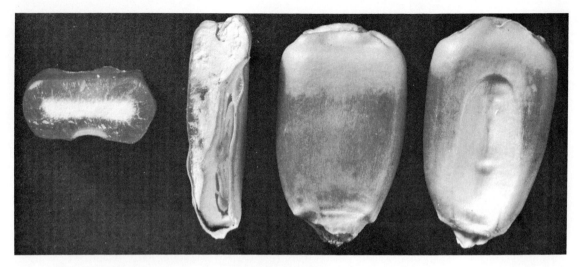

FIGURE 20-1. *Corn grains. Note in the longitudinal section that the embryo occupies only a small part of the seed; the rest consists mainly of endospermic food reserves* [*Photograph by J. C. Allen & Son.*]

and windy for maize. Maize is a crop that performs best with ample warmth and moisture; it is widely utilized as a summer annual in regions where general farming is practiced.

Rice, Oryza sativa

Rice, or paddy, widely grown in Asia from Afghanistan to the East Indies and north to Japan is the principal food for 60 percent of mankind. It is also a staple in parts of Africa, South America, and to some extent in the southern United States (Fig. 20-2). World production exceeds 300 million metric tons annually, mostly from the Far East (China, India–Pakistan, Japan, Indonesia, and Thailand are the major producing countries).

The research on rice genetics has been considerable, but details of domestication and discernment of species still remain quite unclear. The cultivated rices are generally regarded as *O. sativa*, a sort of catch-all complex. Asiatic forms are often considered as *fatua* or *rufipogon* subspecies, American forms as *perennis* (a name of doubtful validity), and African forms as *barthii*. Some systematists have grouped rices into a *fatua* or *rufipogen* series, and a *perennis* complex (of which the American forms are designated *cubensis*, the Asiatic forms *bal-*

unga, and the African ones *barthii*). The wild forms of *fatua* generally grow in shallow swamps as annuals; the *perennis* forms (at least the Asiatic *balunga*), in deep swamps. African rice may spread by rhizomes. All rices are customarily grown as annuals, although under tropical conditions they may be to a degree perennial, perpetuated by new tillers (and perhaps rhizomes).

Rice has been cultivated in China for more than 5,000 years. It is thought likely to have originated from wild *perennis* forms in Southeast Asia. But so widely has rice been spread that it is entirely possible that wild rices from other tropical parts of the world have contributed to *O. sativa*. The diploid chromosome number of most rices is 24, but several American varieties are tetraploid ($2n = 48$). Triploid hybrids are often fertile. A conference sponsored by the International Rice Institute concluded that there are between 13 and 23 validly recognizable rice species, any of which may have made a contribution to *O. sativa*.

For practical purposes rices are generally classified as long-grain, medium-grain, or short-grain. The long-grain types need a longer season to mature, and hence are grown mostly in tropical regions, such as Southeast Asia. Connoisseurs prefer the long-grain type, which is nonglutenous and

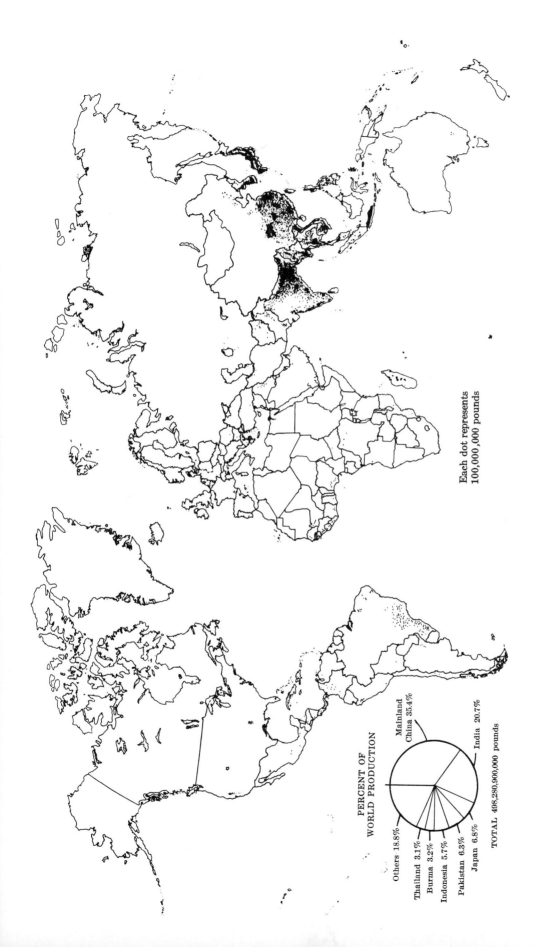

PERCENT OF
WORLD PRODUCTION

Mainland
China 35.4%

India 20.7%

Others 18.8%

Thailand 3.1%
Burma 3.2%
Indonesia 5.7%
Pakistan 6.3%
Japan 6.8%

TOTAL 498,280,900,000 pounds

Each dot represents
100,000,000 pounds

FIGURE 20-2. *World rice production, average 1957–1961.* [*From* A Graphic Summary of World Agriculture, *USDA Misc. Publ. 705.*]

does not turn so soft and sticky when cooked. Medium-grain rices, somewhat softer, are most frequently grown in America. Short-grain types, even more starchy when cooked, are well adapted to more northerly climates, such as that of Japan.

Most rice is grown in paddy fields of standing water. There are, however, various forms of upland rice, cultivated in much the same way as any small grain. Upland rice yields tend to be lower, and its importance is not great; for example, in Indonesia, more than 90 percent of the rice is grown in irrigated paddy land. The best rice land is level, conveniently diked for alternating irrigation and drainage.

Only 1 percent of the world's rice is grown in the United States, but in a technologically advanced fashion. Rice was first introduced into the Carolinas about 1685. Major production has since shifted to the lower Mississippi valley (principally Arkansas and Louisiana) and to California. Immense fields are precisely flooded and drained to permit large equipment to prepare the land, plant the crop, and later harvest the grain in the field. Sometimes rice is sown by aircraft, especially in California. Japan practices the most progressive rice growing in the Orient. Yields average nearly 4 tons per hectare and can be as high as 10 tons. Fertilization in Japan is as intensive as on North American corn land (Fig. 20-3). A complete fertilizer is generally "plowed down" prior to flooding and hand planting of the seedling rice plants. Another "top-dressing" is applied just before heading to increase yields. The fields are drained for harvesting and threshing, typically accomplished with small portable machines on the site. A green manure (vetch) is sometimes planted and plowed down. The paddy fields may also be used for a winter crop, such as barley, wheat, or rape.

Under more primitive circumstances, such as those that prevail in many of the tropical lands, rice growing is an exceedingly laborious and time-consuming hand operation. As noted in Figure 20-3, even fertilization is quite limited in many places, although compost, and sewage have been used since ancient times to fertilize the paddy. After the paddy is flooded to soften the soil, the ground is worked, often by water buffalo pulling a forked stick (small hand tractors may be used in more advanced countries, such as Japan). The rice seedlings are started elsewhere, sometimes under government auspices, and are transplanted by hand into flooded paddies, usually between November and January in southeast Asia, to take advantage of the winter monsoon. About four months later the paddies are drained and the stand harvested— another laborious job, accomplished chiefly by women. The rice stalks are cut individually with a small sharp sickle, banded together, and laid out in the sun to dry. If there is sufficient water, either a second rice crop or some other crop may then be planted to the same land.

Various means are employed to clean and dehull newly harvested rice. In some countries the sun-dried heads are threshed in the field by treading, either by man or live-stock (of course, with considerable loss of grain). In Afghanistan green rice is mixed with heated sand to harden and "crisp" the hulls, which are then separated by crude milling on water-powered devices that pound the grain against the soil. Screening and winnowing follow. In Ecuador much of the rice is fermented by massing it on the floor while damp and covering it with a tarpaulin for a few days. Upon subsequent drying it cooks more quickly than unfermented rice in the high altitudes of the Andes. Fermentation may also provide healthful by-products.

The processing that rice undergoes in the so-

FIGURE 20-3. *Rice yields related to fertilizer usage in several Asiatic production areas.* [After Kemmler, World Crops 14:117, 1962.]

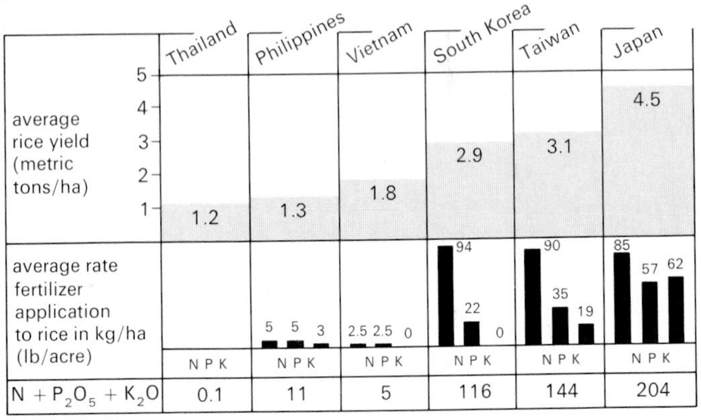

	Thailand	Philippines	Vietnam	South Korea	Taiwan	Japan
average rice yield (metric tons/ha)	1.2	1.3	1.8	2.9	3.1	4.5
average rate fertilizer application to rice in kg/ha (lb/acre)	N P K	N 5 P 5 K 3	N 2.5 P 2.5 K 0	N 94 P 22 K 0	N 90 P 35 K 19	N 85 P 57 K 62
$N + P_2O_5 + K_2O$	0.1	11	5	116	144	204

called advanced countries removes no small part of its nutrients. After the hulls are removed the grain is polished to remove the brown outer layers, which are rich in protein and vitamin B. The insistence of a pearly white grain carries with it the penalty of reduced nutritional value.

An outstanding crop breakthrough has occurred in rice cultivation, thereby extending the "green revolution", begun by Borlaug's work with Mexican wheats (see p. 424), into the Oriental tropics. An extensive breeding program by the International Rice Institute in the Philippines has yielded improved cultivars, notably IR-8 and its pest-resistant successors, capable of markedly higher yields that can help to feed the dense populations of the Orient more adequately. IR-8 is a short, stiff-stem cultivar with an ample panicle, which responds well to nitrogen fertilization. Traditional rice cultivars would lodge if similarly fertilized for high yield.

The introduction of high-yielding cultivars, however, has not eliminated all problems attendant to rice-growing. Their use has even brought about a few new complications! In the Philippines, for example, where the peasant landholder has seldom been able to afford fertilizer, the introduction of high-yielding rice has proved a boon to the advantaged but not necessarily to the masses who soon become caught up in a web of economic bondage. Also, the growing season of the traditional cultivars, different from that of the new cultivars, was more suitable for the established custom of cropping the land twice. In less developed parts of the world, then, methods that work efficiently in technically advanced countries (where research is readily accepted by a well-informed farming population) may not be a panacea.

Indeed, imposition of "improved" growing techniques in technologically less advanced countries is fraught with difficulties and hazards. In 1969 and subsequent years an extensive government-sponsored rice-spraying program was undertaken in Indonesia. Insecticide was applied from the air over hundreds of thousands of hectares, for control of a stem borer, said to cost the nation a quarter of its crop. Aerial spraying was decided upon because it was deemed unlikely that peasant growers could be induced to practice preventive spraying for an "invisible" affliction (the borer is susceptible soon after egg laying on the leaf surface, before it enters the stem and causes damage). A spot check of moth abundance was used to determine the best time for the spraying, which could be done only by trained personnel available through the government-contracted service with a European firm. Spraying was correlated with an "extension" program that included provision of the fertilizer in exchange for a sixth of the crop. Apparently rice yields were increased 25 percent or more, but it is difficult to assess whether the savings (primarily manifest as reduced need for rice imports, although acquainting the peasant populus with improved technology is an intangible gain) justify the drain on the public purse that results from contracting for this tremendous program. The influence on the local ecosystem from massive spraying such as this is seldom considered, and after the foreign technologists leave, most peasants slip back to the traditional way of doing things.

Wheat, Triticum aestivum

World production of wheat, *T. aestivum* (*T. sativum* and *T. vulgare* are synonyms) led that of rice by a narrow margin in 1970 (312 to 307 million metric tons). Wheat is well adapted to harsher environments, and is grown mostly on wind-swept prairies that are too dry and too cold for tropically inclined maize and rice. The U.S.S.R. is the leading producer, followed by the United States, China, India, Canada, France, Turkey, Australia, and Italy, in approximately that order (Fig. 20-4). Wheat can be grown wherever the climate provides moderate moisture and cool weather for early growth, then bright summery months that gradually turn dry for harvesting. It is marginal in humid climates toward the tropics (where disease and debilitation are a hazard), and in the arctic North (where winterkill and short summers are risks).

Wheat is a gift of the Old World, a fair exchange for New World maize. Wheat was apparently domesticated in the Near East. There seem to have been a number of natural crossings with wild

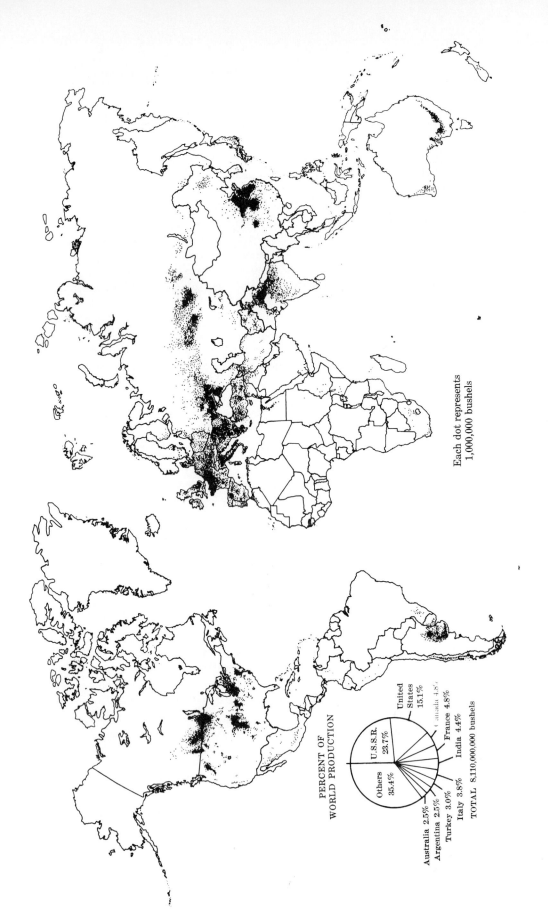

PERCENT OF
WORLD PRODUCTION

U.S.S.R. 23.7%
United States 15.1%
Canada 4.8%
France 4.8%
India 4.4%
Italy 3.8%
Turkey 3.0%
Argentina 2.5%
Australia 2.5%
Others 35.4%
TOTAL 8,110,000,000 bushels

Each dot represents
1,000,000 bushels

FIGURE 20-4. *World wheat production, average 1957–1961. [From A Graphic Summary of World Agriculture, USDA Misc. Publ. 705.]*

species during its early evolvement, as indicated in Table 19-1. As is true of many domesticated plants, a number of uncertainly related species form the genus, many growing in close association near the presumed center of origin (presumably Turkey). Actually, a number of "domestications" of wheat undoubtedly took place, depending upon which species is being considered. Wild wheat, *T. aegilopoides*, and einkorn wheat, *T. monococcum*, possess 14 chromosomes. A number of other wheats are tetraploid, including emmer, *T. dicoccum*, durum, or macaroni wheat, *T. durum*, and *T. timopheevi*. Bread wheat, *T. aestivum*, is hexaploid (2n = 42), as is club wheat, *T. compactum*, and spelt, *T. spelta*. Evidently, there has been natural crossing between several of these and wild species of *Aegilops* and *Agropyron*.

Primitive einkorn and emmer wheats have been discovered in the Jarmo excavations of eastern Iraq, which date back 9 millenia. Other archaeological findings show bread wheats to have been grown in the Nile valley about 5000 B.C., and in India, China, and even England, almost equally long ago. Apparently the species was widely spread by the civilizations that spread from the Near East, where Vavilov believed soft wheats to have originated in the mountains of Asia Minor, durum wheats in North Africa. Wheat was first grown in America in 1602, on an island off the Massachusetts coast.

At first wheat was probably used hulls and all. The grain may have been boiled to make a porridge, or perhaps even fermented for preliminary digestion. It was soon discovered that the wheat grains could be cracked and ground, wetted to make a dough, and the dough baked or fried over an open fire. Thus originated the early unleavened flatbread, still a staple in the diet of many Near Eastern people. There is evidence that raised or leavened bread was used in Egypt as early as 2500 B.C. Leavened bread results from the fermentation of dough by yeast, during which the gaseous carbon dioxide "bubbles" emitted are trapped in the glutenous dough as it bakes. Of the various cereals, only wheat is well suited to the making of leavened bread (rye is somewhat similar, although most rye bread is made from mixed rye and wheat flour).

There are two basic classes of bread wheat, hard and soft, a distinction recognized as early as Roman times. **Hard wheats** are more proteinaceous and make the best formed and longest keeping breads. **Soft wheats** are starchy and are primarily used for pastries or quickly consumed breads, such as "French bread." Hard wheats come from growing areas having limited rainfall, such as the prairie zones of the United States and Canada, most of the U.S.S.R., China, and Argentina. Soft wheats are grown in the more humid general-farming areas —east of the Mississippi in the United States, and throughout Europe. Europe imports hard wheats from the United States to blend them with the soft wheats produced there in order to make better bread. Durum is a type of hard wheat particularly in demand for making macaroni and spaghetti. In the United States it is grown especially in eastern North Dakota, and there is considerable production in southern Europe and central-western India.

Wheats can also be grouped into winter and spring types. Ancestral wheats were probably of the "winter" type, and genetic research in Japan indicates that the winter habit was probably most powerfully reinforced by genes obtained from an *Aegilops squarrosa* parent. **Winter wheat** is planted in autumn, gaining a head start from rooting and tillering before freezing temperatures set in. It is remarkably winter-hardy, standing temperatures well below freezing, especially with the presence of a snow blanket. The crop matures in early summer. Winter wheat is particularly well suited to the southern parts of the wheat belt in the United States, from Texas into South Dakota. North of this zone **spring wheats** are usually preferred, and are planted as far north as the Arctic Circle (where winter wheats would be doubtfully hardy and autumn weather insufficiently protracted to enable a good start then). Spring planting of varieties adapted to a short season (as short as 90 days) still yields an adequate crop under the long summer day lengths of these more northerly regions.

Wheat suffers seriously from the wheat rust disease, *Puccinia graminis*. Rust is especially prevalent in the more southerly winter wheat zones, where certain types of infective spores overwinter and rapidly spread northward on spring winds. Wheat rust can cut yields markedly, and is a serious factor in discouraging wheat growing in sub-tropical latitudes or humid regions. Actually, the best wheat yields are obtained not in the famous wheat belts, but in the intensively managed general-farming areas, such as those of northern Europe and the northeastern United States. But since other, more remunerative crops can usually be grown in these regions, the emphasis on breeding better cultivars is usually reserved for drier regions. The wheat breeder is especially concerned about rust-resistance. Developing rust-resistant wheat has been, and continues to be, a never-ending process, because mutation and natural selection of the rust seem to occur almost as fast as new wheat varieties are produced, rendering new varieties susceptible to attack within a few years of introduction.

Some of the specialty wheats that have probably figured in the origination of bread wheat are still grown locally. Einkorn, *T. monococcum*, yields a scanty harvest in rocky, mountainous areas of southeastern Europe. It is cultivated nowadays chiefly in Turkish Thrace. Fossil evidence of Einkorn has been found in the remains of Swiss lake dwellings that date back to the Stone Age. Emmer, *T. dicoccum*, which also has an ancient lineage in the Mediterranean area, is still cultivated to some extent in mountainous Near Eastern habitats (Yugoslavia, the U.S.S.R.), in southern India, and in Ethiopia. Spelt, Polish wheat, poulard (rivet or English) wheat, and club wheat are still grown in isolated areas, but in general cannot compete economically with modern bread wheat varieties. The one exception is durum wheat, *T. durum*.

As noted in Chapter 15, wheat is a mainstay in areas where dry-land farming is practiced, and is planted in alternate years. In more humid prairie habitats wheat follows wheat year after year, often with relatively little fertilization on the naturally rich prairie soils. In humid climates, where diversified farming is practiced, wheat is often used in crop rotations. If soils are adequately fertilized, 50 bushels per acre° is not out of the question, and is about double the average yield per acre in drier prairie environments where inbred varieties are

° One bushel of wheat amounts to about 60 pounds.

grown. Acre yields of 100 bushels and better are becoming rather commonplace in the Palouse wheatlands of eastern Washington. Yields of the Gaines variety, which is resistant to lodging under heavy fertilization, depend chiefly on the amount of accumulated soil moisture. Four inches of water in the top 6 feet of soil is needed to grow the wheat plant, and every additional inch is worth 7 bushels of yield. Heavy fertilization is practiced, since 2.7 pounds of nitrogen is needed for each bushel of grain, plus phosphorus, potassium, and sulfur.

Wheat lends itself especially well to mechanization, particularly on the prairie lands of the United States, where fields may run on for miles. Plowed soil is typically drill planted in the winter wheat belt late enough in autumn to escape severe attacks by the Hessian fly. Autumn rains are usually favorable for sprouting and consolidative tillering, laying the groundwork for good growth and yields the following spring. If weeds are a pest, there are new herbicides to control most of them, although wheat growing is less able to sustain the expense of weed sprayings than other crops that bring higher returns per acre. As the golden heads of wheat mature—quite a sight across the wind-swept prairies—gigantic self-propelled combines complete the harvest in dawn-to-dust operations (Fig. 20-5). In the southerly parts of North America harvesting begins in late spring and progresses with change in season northward into Canada, where spring wheat may be harvested as late as September. In good wheat years there are not enough local elevators to store all the wheat at harvest time, and loose wheat is mounded in the open until shipping becomes available to transport it to the urban grain centers.

The wheat grain (Fig. 20-6) is relatively proteinaceous; hard wheats usually contain 13 to 16 percent protein, soft wheats 8 to 11 percent protein. Glutenin and gliaden proteins make wheat flours "sticky," suitable for bread. As with rice, the proteinaceous aleurone cells are situated toward the outer part of the grain, and may be removed with the bran during milling. Consequently, whole wheat is apt to be more nutritious than that further processed for white bread. Much of the hard wheat is used for human consumption, much of the soft wheat for livestock feeds. In flour milling, during which the grains are crushed between corrugated rollers, the oily embryos and the bran are screened off in shakers for livestock feed. The starchy endosperm is further mashed between smooth rollers to make flour. In primitive societies wheat is hand crushed in such implements as hollowed-out stones to yield a coarse flour. A landmark of the American Colonial period was the water-powered flour mill, which utilized burred stone wheels to grind wheat.

There is a great deal of excitement about the potential use of hybrid wheat. The breeding problems have essentially been solved, but the economics remain uncertain. Hybrid wheat seed will certainly cost a good deal more than the conventional inbred cultivars, perhaps four or five times as much. If seeded at 50 pounds per acre, average wheat yields (in Kansas about 25 bushels per acre) would have to increase at least 25 percent to justify the use of hybrid seed. Since Kansas accounts for more than one-fifth of the wheat production in the United States, the potential impact of hybrid wheat there may be worth a moment's notice.

Turkish wheat was first introduced into Kansas in 1873, though the state did not become a significant producing area until after 1900. In the last half century, well-adapted cultivars have been unceasingly bred in Kansas; the results include such improved varieties as Kanrad, Tenmarq, Pawnee, Ponca, Bison, Kaw, and Ottawa.

In 1951 the Japanese scientist H. Kihara produced cytoplasmic male-sterility in a cross with wild *Aegilops*. But because no fertility restorer could be found in American wheat cultivars for the Japanese

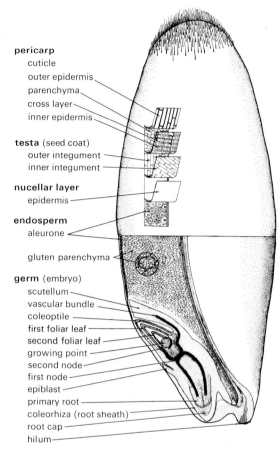

pericarp
  cuticle
  outer epidermis
  parenchyma
  cross layer
  inner epidermis

testa (seed coat)
  outer integument
  inner integument

nucellar layer
  epidermis

endosperm
  aleurone

  gluten parenchyma

germ (embryo)
  scutellum
  vascular bundle
  coleoptile
  first foliar leaf
  second foliar leaf
  growing point
  second node
  first node
  epiblast
  primary root
  coleorhiza (root sheath)
  root cap
  hilum

FIGURE 20-6. *The wheat grain.* [*Courtesy USDA.*]

FIGURE 20-5. *Combining wheat in Nebraska.* [*Courtesy USDA.*]

male-sterile stock, similar crosses were attempted in America with *Aegilops* and primitive wheats to develop a male-sterile line. Most promising was a cross between *Triticum timopheevi* and Bison bread wheat. Subsequently, fertility-restoring factors were also transferred from *T. timopheevi* to bread wheat, setting the stage for successful large-scale production of hybrid seed. A chromosomal sterility factor involving a recessive gene for male sterility has also been discovered; it is restorable through use of an alien chromosome bred into the line. Other research has indicated the possible usefulness of growth regulators, such as ethephon°, which often cause a planting upon which they are sprayed to become self-sterile.

°(2-chloroethyl)phosphonic acid.

A Bison × Scout cross has yielded 57 bushels per acre in a test planting, compared to an average yield of the two parent varieties of around 40 bushels. Generally, wheat hybrids have been running 20 to 37 percent better than the parent varieties, considering western Kansas trials as a whole. If hybrid wheat seed can be produced economically enough to merit its planting to the ten million or so acres sown annually to wheat in Kansas, a half-million acres may be required to produce the needed hybrid seed alone. A whole new facet of the seed industry could arise, of a magnitude exceeding that for hybrid corn.

Wheat is the principal crop underlying the "green revolution"—the development of major higher-yielding crop cultivars to nurture more adequately the burgeoning world populations in the later decades of the 20th century. Norman Borlaug received the Nobel Peace Prize in 1971 for his participation in the wheat breeding program in Mexico under Rockefeller Foundation auspices. Borlaug, utilizing some of the "Norin" germ plasm brought from Japan at the end of World War II, created high-yielding varieties, well suited not only to Mexico, but also to areas of similar climate in other parts of the world, notably India. Norin heredity has also been instrumental in the breeding of high-yielding wheats for the Pacific Northwest of the United States (Gaines and subsequent cultivars). Borlaug's cultivars, having strong, stiff stems, can accept heavy fertilization without lodging. Fortunately, in both Mexico and India, the new wheats were rather quickly accepted, and fertilizer was made available through various programs.

Even so successful an innovation as the Mexican wheats, which have eased the food crisis in many parts of the world, is not without its side effects, the seriousness of which remains to be determined. Like the new rices, cultivation of these high-yielding wheat crops requires more technically advanced operations, thereby placing small landholders and unskilled tenants at such a disadvantage that they may be forced to yield their own lands to the wealthier and more competent, with inevitable social consequences. Perhaps of even greater concern is the likelihood that the new cultivars will so swamp wheat grown even in re- mote areas, that old strains will be abandoned and lost; yet, experience has shown that new germ plasm is needed continuously for the breeding of "wheats of the future," able to resist rust and other afflictions. The only ready sources of such germ plasm have been isolated ecotypes in remote corners of the world unsullied by modern cultivars. If genetic diversity, so essential for long-term crop improvement, is overwhelmed by the widespread success and acceptance enjoyed by present-day cultivars, whence comes the next "Norin" germ-plasm for a future breakthrough?

Maize, Zea mays

Maize (Fig. 20-7), a more valuable find from New World exploration than all the golden tribute, now makes its mark world-wide. But in no part of the world is maize cultivated so intensively as in the United States, where half of the annual quarter-billion tons of corn grain is grown. Nor have many crops attained the degree of sophistication that has come to hybrid corn production in North America. A distant second to the United States in maize production is usually Brazil, with the U.S.S.R., Mexico, Romania, Yugoslavia, Argentina, India, and South Africa following more or less in that order (Fig. 20-8).

Maize appears to have been domesticated in the Tehuacan Valley of Mexico. Maize pollen 80,000 years old has been collected from cores taken from old lake beds near Mexico City, which certainly proves it indigenous to the region. The original wild form has probably long been extinct. Evidence suggests that cultivated maize arose through natural crossing, perhaps first with gamagrass, *Tripsacum dactyloides* (which itself may have originated from a cross of primitive *Zea* with *Manisurus*), to yield Teosinte, *Euchlaena mexicana*, and then possibly with backcrossing of Teosinte to primitive maize to produce modern races. The place of origin has long been debated; if it truly were southern Mexico, which is an excellent possibility, maize soon spread throughout the Americas to become a mainstay food. It might easily have been carried southward into the foothills of the Andes (where it may have picked up flint corn genes), eastward along the northern coast of South America, and thence northward from one Caribbean island to another and on to eastern North America. At the same time it most certainly spread overland northward from Mexico into the western United States, and adapted forms spread northeastward as far as New England, where its famed rendezvous with the early English colonists took place.

Modern cultivars of maize resemble the primordial form only remotely. Indeed, modern maize is wholly a ward of man, unable to survive and perpetuate itself without his care in harvesting and planting. Man's selection through the millenia has yielded a specialized and very queer looking crop, with the female (pistillate) flowers concentrated on one or a few lower cobs, or ears, and the male (staminate) flowers on separate, terminal tassels. The seeds (grains on the cob) shatter little, and because they are enclosed in husks they have scant chance for natural dispersal. Growing of maize today depends upon servicing this odd creation with a variety of contrivances to handle its soil preparation (cultivation, fertilization), planting (planter drills), care (pest control), harvesting (picker-shellers, combines), and cleaning and storing (drying equipment, grain elevators). Add to this all the special accoutrement for hybrid seed production—inbreds, male-sterile lines, detasseling equipment, and so on. Here is a remarkable marriage of biology with technology—interlocking responses and mutual dependence between plant and man.

Early planting of maize by the American Indians was, of course, much more casual. But even by then selections had been made, and this subtropical species had been adapted to a wide variety of habitats, yielding such modern types as popcorn, northern flints, and southeastern dents, ancestral to most of today's Corn Belt cultivars. The Indians planted corn either in slash-and-burn forest clearings, or on drier lands that were seasonally moist or irrigated. Planting was accomplished by such simple techniques as poking holes into the soil and inserting seeds. At least with the Mohawk Iroquois, the women soaked the corn seed for a few days before planting a few to a hill. In certain locations primitive fertilization was practiced, such as burying fish parts at planting. Other crops, such as beans and squash, were often interplanted with the

FIGURE 20-7. *Well-filled ears of maize at the harvest stage. [Photograph by J. C. Allen & Son.]*

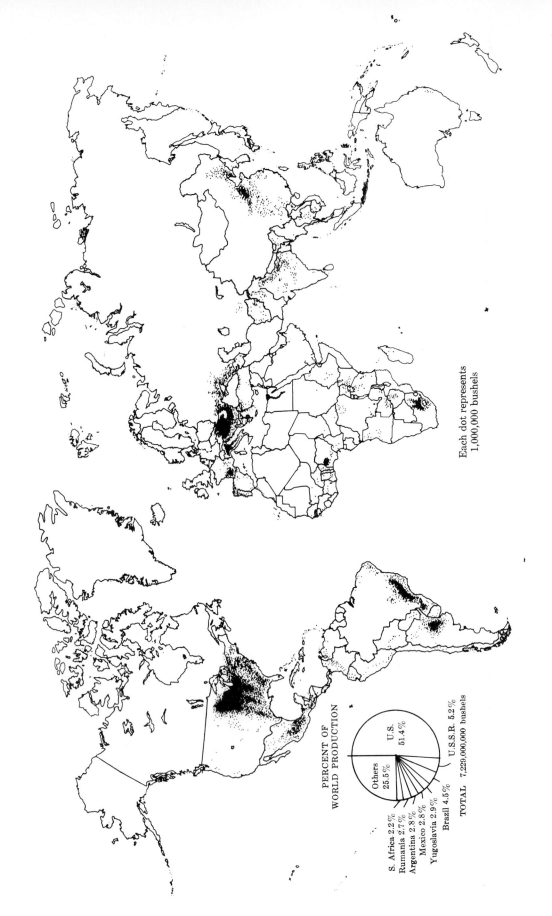

PERCENT OF
WORLD PRODUCTION

Others
25.5%

U.S.
51.4%

U.S.S.R. 5.2%

Brazil 4.5%

Yugoslavia 2.9%

Mexico 2.8%

Argentina 2.8%

Rumania 2.7%

S. Africa 2.2%

TOTAL 7,229,000,000 bushels

Each dot represents
1,000,000 bushels

FIGURE 20-8. *World corn production, average 1957–1961.* [*From A Graphic Summary of World Agriculture, USDA Misc. Publ. 705.*]

corn, and the weeds checked by hand pulling or crude hoeing. Vigil was kept to ward off freebooting birds and marauding animals.

In northern New Zealand, maize was introduced originally in 1772. The Maoris there have adopted only three introduced food plants, one of which is maize. Though they often prepare maize for eating in much the same way as Europeans do, the Maoris have a custom of rotting ears in water for six months or more prior to consumption. This is also done in parts of Peru! The slimy, odiferous kernels are crushed, boiled to a gruel, and eaten warm with milk and sugar, or are fried.

In Indonesia maize is grown only on land unsuited to rice, the preferred cereal. The grain is ground to make puddings, as well as being consumed green as "roasting ears." It is also used as a feed grain. Stalks are utilized for fodder, and the leaves are even employed as cigarette wrappers (a practice common in South America as well). In India flint corns are grown quite a bit in the northeast, usually sown during the period May to June for autumn harvest. With unselected varieties, and primitive cultural techniques, yields are low. Dried grain is ground to a meal used like flour; green ears are roasted or else the kernels are removed from the ear and parched on beds of hot sand (thus enhancing its keeping qualities). The American Indian frequently parched or popped corn in much this same fashion for preservation in storage pits dug into the ground or into the sides of canyons. Corn was also preserved as hominy, by soaking it in lye made from wood ashes. But in dry climates less flinty corns have usually been grown; they keep well if merely allowed to mature naturally, and are consumed as meal made by grinding the kernels between stones, with mortar and pestle, or in recent times in mills. Human consumption in the United States today is chiefly as corn meal, or as grits, ubiquitous breakfast fare in the South prepared from ground hominy.

Crop breeding and mechanization have increased yields of corn fantastically in the United States. Moreover, now that corn borers and other pests can be controlled with pesticides, the trend is toward earlier planting. Corn is basically a long-season, moisture-loving crop. Starting corn with the first warm weather, rather than waiting until late spring, generally results in slightly higher yields from shorter plants that don't lodge so easily and which shade weeds better because of advanced growth (and thus conserve moisture too). They reach the pollinating and ear-forming stages earlier, at which time the extra sunlight of longer days helps make full, high-yielding ears. There is time enough then, too, to let the corn stand in the field for natural drying before harvest.

Noting the tremendous genetic flexibility of maize, which allowed transformation of a primitive ancestor quite unlike modern corn into the plethora of races that exist today, one wonders what can be expected next. Chapter 15 mentioned the trend toward denser plant populations in the field. Will this encourage breeders to develop still newer cultivars that are perhaps squatter, yield even more ears, and are better adapted to field combining? Will combining dictate changes in ear form from that developed for easy picking by hand, with grain attachment secure? Will kernels have to be developed that are less subject to damage by the combine? Of course, breeding for improved standing ability, disease and insect resistance, season-span to match climate, and yield rather than appearance (once important because of rural corn contests) can be taken for granted.

Probably the greatest changes to be wrought by breeding will come from an increase in the attention given to the chemical composition of the grain. It is possible through selection to produce corn high in protein or high in oil (see Fig. 19-10). More sophisticated qualitative changes are possible. For example, in normal corn the starch is composed of 27 percent amylose (straight chain) and 73 percent amylopectin (branching chain). Separation is expensive, and each kind of starch has particular industrial uses; for example, amylose is used for adhesives and thickenings in the food industry for such products as instant puddings, and amylopectin is used in the paper industry and as coatings or sizing in the fabrication of woven fiber glass. The *waxy* mutant produces 100 percent amylopectin; the *ae* gene (*amylose extender*) increases amylose to 60 percent of the total, and selection in these types has increased this to 80 percent. The discovery that certain mutants alter the amino acid composition or corn proteins has far reaching consequences. Corn, like other grains, is not a good sole protein source for monogastric animals (pigs or humans) because it is low in lysine and tryptophan (see Chapter 3). By increasing the proportion of lysine and tryptophan the *opaque-2* gene greatly increases the feeding value of corn. The combination of high protein and improved amino acid balance in corn and other grains offers tremendous potential in alleviating malnutrition in many parts of the world.

Hybrid seed corn is worthy of some special mention, since it has sparked the concept and been the proving ground for heterosis. Because modern corn is monoecious (ears and tassels separate on the same plant) it lends itself well to regulated crossing, which require nothing more than clipping off the tassel to prevent self-pollination. Planting adjacent rows of a different inbred from which the tassels have not been removed assures a hybrid cross for seed collected from detasseled rows. In recent years cytoplasmic male sterility had been introduced into inbred lines, which in effect accomplished the expensive detasseling procedure genetically. The Southern corn blight epidemic of 1970 in the United States—caused by the susceptibility of the sterile cytoplasmic factor to the "T" strain of the fungus responsible for the disease (see Chapter 19, page 402)—has resulted in a reversion to manual detasseling for the production of hybrid seed.

The mechanism of heterosis—the increase in yield due to hybrid vigor—was discussed in Chapter 19. It is widely recognized that select inbred lines can be hybridized to produce superior yields from first-generation progeny. Inbred suitability for such crosses has been determined by trial and error, and more recently through sophisticated mathematical techniques involving a cross to a tester line with a wide genetic base. The discoverer, having control of his suitable inbred lines, has not only a valuable source but an exclusive source of that particular hybrid seed. The hybrid corn seed industry began with a single cross of inbred strains, say A with B. But more remunerative yields of hybrid seed could be had if a seed were produced on a plant that exhibited hybrid vigor. Thus double-cross hybrids became commonplace, with $(A \times C)$ crossed with $(B \times D)$, or occasionally

with $(A \times C)$ merely crossed with B as a pollinating parent (triple cross). The heterozygous seed so formed does not breed true in subsequent generations because of genetic reassortment; thus the farmer must buy new seed the next year from reliable sources of supply.

In the United States much of the corn crop never leaves the farm. Nearly 90 percent is used to feed cattle and hogs. There are three different field corns: (1) the **flint corns,** which have kernels of hard starch, were being used by the northeastern Indians when the pilgrims arrived, and are probably related to certain South American genotypes (by way of dispersal through Central America and Mexico into the Southwest, and then northeastward); (2) the **dent corns,** which have kernels of hard starch capped by soft starch, were developed from stocks originally found in the southeastern United States, and are especially high yielding and much used in the Corn Belt; and (3) **flour corns,** which have kernels of soft starch and are amenable to hand grinding, a method much employed by Indians of the Southwest. In addition to field corns there are the **sweet corns,** the sugary kernels of which are much consumed in the green stage as human food (Golden Bantam, Country Gentleman), and the **popcorns,** whose kernels burst open upon heating, rendering them useful for primitive people lacking grinding mills. In addition to these standard types grown in the United States, there are hordes of other distinctive races grown throughout Latin America, often in isolated regions. Many are now being introduced into the United States as sources of germplasm for the breeding of future maize cultivars.

Corn is an important source of industrial products. Perhaps the chief industrial derivative is corn starch, usually obtained by wet-milling. Starch may in turn be hydrolyzed or fermented to yield corn sugars, syrup, and special gum-like polymers such as phosphomannan, which is used for adhesives. The grain is nearly 10 percent protein some of which is removed by initial steeping in weak acid. This valuable corn steep liquor may be employed as a nutrient broth in the production of antibiotics and in animal feeds. The oily embryo, which makes up 6 to 13 percent of the grain, is next removed, usually by partly crushing the steeped grain, screening away the coarse hulls, and floating off the released "germ" (embryo). Expression or solvent extraction yields corn oil, which constitutes about 50 percent of the embryo. The residues are utilized along with hulls and other by-products in animal feeds. The corn oil is further refined and utilized principally as cooking and salad oils.

After removal of the embryos the remaining mass, chiefly endosperm, is crushed in burr mills, and any hulls are removed by screenings and siftings. Gluten is separated from starch either by differential sedimentation on long, inclined tables, over which the slurry is made to flow slowly, or by centrifugation. The gluten goes back into cattle feeds, and the starch is further washed, dried and pulverized. The starch is either consumed directly as human food or is used by industry to make sizing, laundry starch, urethane plastics, and other products. Some is used as a raw material for further processing by digestion or hydrolysis with enzymes and weak acid catalysts to yield a glucose syrup or various mixtures of simple sugars and dextrins.

Barley, Hordeum vulgare

The annual world production of barley, *H. vulgare* (*H. sativum*), the world's fourth most important cereal, exceeds 120 million tons (Fig. 20-9). Worldwide, the Soviet Union heads the list. France and the United Kingdom are the leading Western European producers. Canada, the United States, Turkey, India, and Morocco also produce significant amounts. The grain is much used for animal feed, little for direct human consumption. About a third of the grain goes for making malt, essential for brewing of beer and liquor and used also in a number of confections and "health" foods.

A number of wild barleys have a diploid chromosome number of 14, including *H. spontaneum* and *H. maritimum*. The cultivated species all have 14 chromosomes, and are considered to be descended from *H. spontaneum*, which crosses readily with them. Tetraploid barleys include *H. jubatum*, *H. marinum*, and *H. bulbosum*. All cultivated barleys have been relegated to *H. vulgare* by many authorities, but others have reserved this appellation for the six-rowed types only, the two-rowed barleys

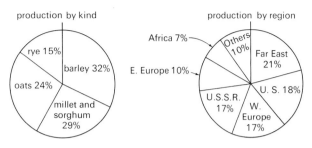

FIGURE 20-9. *Relative world production of some small grains.* [*From* A Graphic Summary of World Agriculture, *USDA Misc. Publ. 705.*]

being considered as *H. distichon* and the irregular in-between types as *H. irregulare.* Six-rowed barley, believed to be more than 8,500 years old, has been excavated in the Near East. The species was probably domesticated at least 9,000 years ago, and, along with emmer wheat, is probably one of the first cultivated cereals. Both Abyssinia and southern Tibet have also been suggested as centers of origin for domesticated barley.

As with wheat, there are spring and winter varieties of barley. **Spring barleys** mature in a short season of 2 to 3 months. In the United States they are planted north and west from middle Texas to southern New England. From the Ohio Valley southward **winter barleys** are generally planted, although spring barleys are winter-seeded near the Gulf for winter grazing, as well as whatever grain eventuates when cattle are removed in spring. As its wide distribution would imply, barley is tolerant of many soils and climates, but like wheat it is not well adapted to hot humid conditions.

Barley is grown in much the same fashion as was described for wheat. It responds well to fertilization, and is generally planted in rotation rather than year after year to the same fields. Although yield tests have been contradictory, the trend toward denser plant populations through closer spacing of rows is being recommended. Attention has been given to production of hybrid barley seed; chemical inducement of male sterility has proved feasible in the greenhouse (with a potassium gibberellate), but has not been practical in the field. Male-sterile lines carrying an extra chromosome (trisomic) derived by the crossing of several species have led to some commercial hybrid seed production. Cultivar yields have exceeded 70 bushels to the acre in Denmark (nearly 4 tons/ha), although the average United States yield has not greatly exceeded 32 bushels (about 1.8 tons/ha). Yields in the rest of the world are considerably less.

Harvesting is done by field combining in the United States. In many of the less developed parts of the world barley is still gathered and processed by hand. Where humidity is relatively high, the grain may be windrowed for drying before threshing or combining. Moisture content of the grain should be below 14 percent for storage.

Barley grain is usually crushed for incorporation into livestock feeds. Some is pearled for use in soups, baby gruels, and other foods. Pearling consists of rubbing the grain against abrasive disks to grind off the hulls and the outer layer of the kernel, much as in the polishing of rice. A minor portion of barley is milled into flour, particularly in Asiatic countries. It is used in India, for example, to make "chapatis."

Much attention has been given malting varieties of barley. Good malting qualities can be determined in the laboratory by protein and diastatic potential analysis. In contrast with wheat, where proteinaceous varieties are most sought for bread, starchy varieties of barley are generally chosen for malt. Brewers demand uniform, carefully cleaned grain, for which a premium is offered. The making of malt is a relatively simple process. Grain is steeped and then germinated under controlled humidity, temperature, and atmosphere. Incipient sprouts develop a high content of amylase and other enzymes, valued for later digestion of starch adjuncts in brewing. Germination is stopped when the enzyme content is maximal by heating and drying the sprouting grain.

In brewing, malt is mixed with other grains to provide the mash; the enzymes of the malt reduce starch to simple sugars in the resulting wort, which is later fermented by yeast to make the beer. A limited supply of malt is also used for flavorings, breakfast cereals, icings, coffee substitutes, malted milk, infant food, malt flours, medicinal syrups, candies, and various industrial fermentations.

FIGURE 20-10. *An oat panicle with the grains separated.* [*Photograph by J. C. Allen & Son.*]

Oats, Avena sativa

In Samuel Johnson's Dictionary, oats are defined as eaten by people in Scotland, but fit only for horses in England. The Scotman's retort to this is: That's why England has such good horses and Scotland such fine men. In spite of Samuel Johnson's scorn, oats are one of the world's important behind-the-scenes cereals, coming to market mainly as livestock. The grain is nutritious and of high feed value.

Oats (Fig. 20-10) are chiefly a European and North American crop. World production is now little more than 50 million tons annually, on the decline ever since tractors replaced horses. The United States, Russia, Canada, Poland, Germany and France are the leading producing countries. As with barley, oats are widely adapted.

The origin of oats is obscure. Classical Greek writers seemed to view oats as a weed, implying cognizance but not domestication in that part of the world several centuries B.C. Some authorities believe that *A. sativa* developed as a mutation from wild oats, *A. fatua*, in Asia Minor or southeastern Europe not long before the Christian era. Most regard it as having entered domestication as an unavoidable weed in barley. Other experts suggest that *A. byzantina,* itself perhaps a subspecies of the wild red oat, *A. sterilis*, is more likely ancestral to at least certain forms of *A. sativa*. All of these are hexaploid (2n = 42). There are also a number of tetraploid and diploid species of little or no commercial importance.

Oats were brought to North America with other grain in 1602 to be planted on the Elizabeth Islands off the coast of Massachusetts. Today they are planted both as spring and winter varieties, the latter sown mostly in the South. Disease prevents oats from really prospering in the humid Southeast, however. In the United States most oats are grown in the Corn Belt, in rotation with corn. Planting and handling are as has been described for other small grains. Oats are harvested in summer (Fig. 20-11), and can be safely stored when moisture content is below 14 percent. There is some processing of oats to flour, to rolled oats and oatmeal. Scotland, true to Dr. Johnson's allegation, is a heavy consumer of oats for human food. Several breakfast cereals made from oat flour are popular in the United States. Oat hulls, a by-product of milling, have served as raw material for fermentation to furfural, a chemical solvent used in refining minerals and for making rosins.

Rye, Secale cereale

Rye is chiefly a European grain, with the Soviet Union the leading producer, followed by Poland and Germany. Total world production is approximately 30 million tons.

Rye is a diploid species (2n = 14). *Secale cereale* has not been authenticated in the wild, although it is familiar as a naturalized escape. It may have been domesticated from wild ryes, *S. montanum* or *S. anatolicum*, of the Mediterranean region and Asia Minor. It is thought that domestication of rye may have come about by inclusion as a weed in barley or wheat crops. As with oats, the domestication of rye seems relatively recent, the domesticated plant not being known in the older archaeological excavations.

Rye prefers cool, dryish climates, and can stand winter weather well. It is widely adapted to temperate climates, and can often be successfully cultivated on soils too poor for remunerative growing of other grains. It is planted and harvested in the same way as other small grains.

The grain, usually about 13 percent protein, is somewhat glutenous like wheat. Thus rye can be made into a bread flour, and is much used in Germany, Poland, and Russia for the making of a dark, bitter, "black" bread; in the United States "rye" bread is made largely of wheat flour with only a small proportion of rye flour. Most rye, however, is produced for hay and pasturage or for stock feed. Yields under intensive management have exceeded 1.8 tons per acre (4 tons per ha). The straw is often employed for stuffing, thatching, and paper making, as well as animal bedding.

FIGURE 20-11. *Harvesting oats with a self-propelled field combine. [Courtesy International Harvester Co.]*

Sorghum, Sorghum bicolor

In recent years sorghums have enjoyed a tremendous increase in popularity, especially in the United States, the world's largest producer. The advent of hybrid seed is largely responsible. India, China, Argentina, Nigeria, the U.S.S.R., Ethiopia, and the Sudan are also important contributors to a world supply that probably exceeds 60 million tons a year.

Sorghum bicolor (*S. vulgare*) is something of a catch-all species. It is important for the grain sorghums, of which there are innumerable cultivars, grown as annuals; and for the botanical variety *S. bicolor sudanensis* (*S. arundinaceum sudanense*), sudangrass, a forage species of some merit from the Sudan. It has been used recently in sorghum-sudan hybrids (Fig. 20-12). Weedy johnsongrass, S. *halepense*, is also closely related. The cultivated sorghums have a diploid chromosome complement of 20, johnsongrass 40. It is believed that most cultivated sorghums originated in Africa, but some are possibly from Asia as well. Sorghum was grown in Egypt before the time of Christ, but was introduced into the United States only during the middle 1800's.

Grain sorghums, like maize, respond well to heterosis. Yield increases of as much as 40 percent are obtained. But because the flowers are perfect, with male and female parts together in a compact terminal head, production of hybrids had to await development of male sterility. Only within recent years have practical male-sterile lines been developed, giving rise to a hybrid sorghum seed industry that is especially important in the plains states from Texas northward into Nebraska. Since the climate there is too dry for successful maize growing, sorghum is an important alternative to wheat. Sorghum-sudan hybrids are also enjoying favor in the northeastern states as forage plants, especially where alfalfa has been hard hit by dry years and the alfalfa weevil. Grazing and green chop are the main uses in such normally humid areas.

Sorghum grain, although nutritious, is small in size and difficult to process. The planting of sorghum for grain rather than forage and silage is especially important in Africa and Asia, where it serves as human food. In India the whole grain is eaten like rice; sorghum flour makes an unleavened bread. Some sorghum is used as fodder. In the Mediterranean region grain sorghum is very important as a dry-land summer crop. Hybrid seed from the United States is used to advantage in Israel. To some extent it is grown under mechanized agriculture there, but in most of the Mediterranean region planting and harvesting are still done by primitive means.

Sorghum has an extensive root system for a modest leaf area, equipping it well for survival in semiarid climates. Since rainfall is light where most sorghum is grown, yields are more dependent upon moisture than fertility. Yields under mechanized agriculture in North America approach an average of 1.3 tons per acre (3 tons per ha), but are only half this in the Near East, and half again less in the Far East. In the United States shortstalked combine cultivars are planted almost exclusively. The old **milo** and **kafir** varieties have almost disappeared. The grain is utilized mostly as a livestock and poultry feed, and to a limited extent for industrial processes, where it can substitute for corn. In Africa sorghum grain is fermented to make a "beer," which is a valuable dietary supplement in the hinterland. About half the acreage is not harvested for grain, but is used instead for forage and silage. If sudangrass used for hay and summer pasture is included, the forage value of sorghum outranks its grain value by a considerable margin.

In some localities **sorgo,** or **sweet sorghum** ("cane"), varieties are still grown, planted much like maize. Although these varieties serve as forage, sorghum syrup can be extracted from the juicy stems, much as with sugar cane. With **broomcorn** varieties the panicle is cut, dried, and bound to make brooms.

Millet

A number of different genera are termed millet, grown for grain principally in the Far East and in Africa. World production is estimated to be around 30 million tons; India, Pakistan, and a number of central African countries are the leading producers.

Archeological evidence dating back 7,000 years indicates that a *Setaria* foxtail millet was one of the earliest domesticated crops in the primitive culture of the Tehuacan Valley of Mexico. By and large the small-seeded millets have remained a local grain of less developed areas. None have been able to compete with the major cereals in world commerce. In agriculturally advanced nations millets are usually used as a component of feeds, especially chick and bird seed, or to make quick soil cover and pasture.

The major millet species of the world are foxtail millet, *Setaria italica;* pearl, or cattail millet, *Pennisetum glaucum* (*P. typhoides*); finger millet, or ragi, *Eleusine coracana;* teff, *Eragrostis abyssinica;* several *Panicums,* such as bread and proso millet, *Panicum miliaceum,* and browntop millet, *Pani-*

FIGURE 20-12. (Above) Harvesting grain sorghum. (Below left) Grain sorghum. (Below right) "Sordan," one of the sorghum-sudan grass hybrids. [Above photograph courtesy Allis Chalmers; lower photographs by J. C. Allen & Son.]

cum ramosum; several fonios or "finger millets," *Digitaria spp.,* especially in Africa; and Japanese barnyard millet, *Echinochloa frumentacea.* They are generally grown as mixed crops in regions of light rainfall, in association with legumes. Pearl and finger millet are important crops in India, where they have been used in unleavened bread since prehistoric times. Teff is an important grain and forage in Ethiopia, and the flour is used in baking the ubiquitous *injara* "pancakes."

Minor cereals

Other seeds of grasses, of course, can serve as grain, and some are locally important as specialty items. Perhaps the most prominent in this respect is wild rice, *Zizania aquatica.* Wild rice was an important Indian food when the Northwest Territory of the United States was first explored. It grows in ponds and sluggish streams, and was usually harvested by flailing it into a canoe paddled into the wild stands. Today the wild stands are much decimated, but more than a thousand tons still comes to market as a luxury food sold at prices more than tenfold that of other cereals. The chief harvest is from northern Minnesota, where the Chippewa Indians have a regulated "monopoly" on wild rice harvested on their reservations. It is still harvested there by flailing it into canoes, although on private lands in Canada there has been some mechanization. As collected the rice is full of wet leafage and lake debris, and shrinks to about 40 percent of its green weight during processing. Of late wild rice has come increasingly to be planted in artificial paddies, which, as in *Oryza* rice growing, can be flooded and drained at will. Seeding may be by plane or boat, and yields are several times larger than those from wild stands.

The most recent taxonomic revision of *Zizania aquatica* recognizes three botanical varieties, of which the variety *interior* is that growing in northern Minnesota and southern Canada. Other varieties occur above tide level along the Atlantic Coast from Canada to Florida, where they are important as a wildlife food but not directly utilized by man. Attempts at cultivating wild rice have been less than a resounding success; for unknown reasons plantings to seemingly adequate habitats do not "take." The species is an annual, and depends upon spontaneous reseeding each year. Doubtless if sufficient research were directed to the species, cultivars suitable to domestication could in time be developed.

In nutritive quality wild rice is on a par with unpolished rice or whole wheat. Although wild rice is still occasionally hulled by tredding and hand winnowing, even the Indians have begun to use such things as small fanning mills for processing the crop. In Canada, where the Indian methods are not protected as they are in Minnesota, there is some harvest by scows fitted with paddle wheels; centralized processing includes drying, parching, mechanical hulling, and removal of chaff by suction.

A number of other grass species yield grain, especially in the tropics. Job's tears, *Coix lacryma-jobi,* is a striking species native to southeastern Asia but widely introduced throughout the tropics. The large, shiny white grains are often used ornamentally, but may be parched, boiled, or milled into flour for food. Seed of mannagrass, *Glyceria fluitans;* guineagrass, *Panicum maximum;* fonio, *Paspalum longiflorum;* bergu, *Panicum vergii* and *Echinochloa stagnina;* reedgrass or canegrass, *Phragmites communis;* and many others may be similarly utilized.

LEGUMES

If plant families can be rated according to importance, the Leguminosae would rank not far behind the grasses. Indeed, the legumes may be more important fundamentally, since the family is noted for having evolved symbiotic associations with certain bacteria, such as *Rhizobium.* The symbionts, usually housed in nodules on the roots of the legume, fix gaseous nitrogen. The legumes are, by and large, the least dependent upon nitrogenous fertilizer of all crop plants, and thus are particularly an asset to primitive agricultural systems. By the same token, this degree of self-sufficiency in nitrogen makes legumes especially rich sources of protein, so often deficient in meager diets. Growing more legumes for food in less advanced parts of the world would materially reduce the prevalence of the dreaded kwashiorkor (protein deficiency) disease. For forages to feed man's livestock, legumes have no equal, especially in partnership with grass. When the agronomist speaks of "grasslands," he is thinking of legumes as well as grasses.

Table 20-2 compares the protein content of several food products. Note how much more protein legumes contain than cereals and major root crops. Neither whole dried milk nor dried egg is richer in protein than are these legumes. The protein content of dried meat or fish (generally 70 to 90 percent protein) is much greater, of course, but plant protein costs far less per pound. No wonder legumes are so important a supplement to the diet of peoples whose subsistence depends primarily upon starchy cereals or root crops.

Legume seeds may be as important for their oil as for their protein. Soybeans are about 20 percent oil, peanuts around 50 percent. Indeed, in technologically advanced parts of the world they may be grown chiefly for their oils, with the protein a "by-product." In the United States 90 percent of the soybean oil is utilized for derived foods, such as margarine, and the remainder mostly for indus-

Table 20-2 PROTEIN CONTENT OF FAMILIAR FOODS AND FEEDS.

FOOD OR FEED	PERCENT PROTEIN
Cereals	
whole rice	7.5–9.0
polished rice	5.2–7.6
wheat flour	9.8–13.5
corn meal	7.0–9.4
Legume seeds	
chick-pea	22–28
soybean	33–42
peanut	25–30
Root crops	
potato (dry weight)	10–13
cassava	1.3
Miscellaneous	
walnut	15–21
dried alfalfa meal	18–23
dried *Chlorella* alga	23–44
dried *Torula* yeast	38–55

trial applications. Ninety-five percent of the protein becomes animal feed, especially preferred for poultry.

Box 20-1 SYMBIOTIC NITROGEN FIXATION TO EXTEND MAN'S FOOD SUPPLY

The advancements in satisfying world food needs, made possible by the "green revolution" (see the discussions on wheat and rice), depend heavily on fertilizer. However, in primitive situations fertilizer may not be available, and it is too costly for general use with typical subsistence farming. In some localities its distribution may be erratic, and means for its application inadequate. Even in technically advanced societies problems exist: in particular, there is concern that fertilization may overload the runoff and ground water with salts, causing a pollution problem.

None of these problems are caused by nitrogen secured through *Rhizobium*-type symbiosis, notably occurring in legume root nodules. Scientists have wondered whether strains of free-living bacteria, algae or fungi that are capable of fixing nitrogen, can be made to thrive in the rhizosphere of monocotyledonous plants such as the cereal grain species, as well as in legume roots. Apparently some phenomenon of this nature occurs in the ages-old rice paddies of the Orient, which remain reasonably productive without appreciable additions of nitrogenous fertilizer. Significant symbiotic nitrogen fixation associated with higher plants other than legumes occurs with *Alnus* in the Arctic, *Myrica* in the temperate coastal plains, *Casuarina* in the tropics, and other genera.

The enzyme, nitrogenase, has been shown to catalyze a conversion of nitrogen to ammonia in those biological systems in which it has been investigated. If nitrogenase is provided the essential protein ambience, a high-energy source (ATP), and a strong electron donor, theoretically an effective nitrogen-fixing system might be developed with any number of unicellular organisms. Nitrogenase activity has been achieved in vitro by combining bacteria and cultured host tissues; why not, then, the "breeding" of broadly adapted microorganisms to live symbiotically in a wide range of host root tissues? Even partial success could do much to help delay worldwide famine.

FIGURE 20-13. *The soybean exemplifies legumes utilized for their seeds. (Left) The growing plant, showing trifoliate, alternate leaves, and seeds borne in pods (legumes), a characteristic of the Leguminosae. (Right) The mature plant ready for combining. [Photograph at left courtesy USDA; photograph at right by J. C. Allen & Son.]*

Various beans and peas have been eaten directly by man since time immemorial. But soybeans (Fig. 20-13)—the world's most abundantly grown seed legume today—have been esteemed only in the Orient. Even as an animal feed, the protein content of the soybean is not tops, presumably because certain minor proteins act as trypsin inhibitors and as mild hemagglutinins. The protein cake that remains after the oil has been extracted from soybeans is improved as a livestock feed by mild heating, which differentially alters these sensitive minor components. Soybean flour is used in human foods largely as an extender for other, more expensive forms of protein. In Israel, for example, government edict requires its inclusion as an "enrichment" in bread flour. Some soybean protein ("nitrogen") is still returned to the soil as organic fertilizer, just as it was in days of yore when its food value was not well recognized. But modern agriculture can hardly justify so expensive and otherwise useful a food being used merely as a source of nitrogen.

Although the legumes are only a little less important in world commerce than the grains, the relative importance of individual legume species is not easily determined. World-wide production of edible beans and peas exceeds 40 million tons annually. Soybeans, serving principally as an oil source, account for equally as much tonnage, with North America the leading producer. Peanuts, chiefly from India, afford nearly half as much tonnage. Dried beans come mainly from China, India, Latin America (Mexico especially), the United States, Japan, Italy, and Turkey. The leading producers of dried peas are China and India, although a fairly large crop is produced in southern Europe too. Broad beans are grown chiefly in Italy and other parts of southern Europe, and in North Africa. Chick-peas are overwhelmingly an oriental crop. Lentils are much grown in India and the Near East.

Soybean, Glycine max

The soybean, indigenous to Southeast Asia, is one of the oldest of cultivated crops. Its cultivation in China is mentioned in a document written by Emperor Sheng Nung in 2838 B.C., at which time the soybean was regarded as one of the five "sacred grains" vital to Chinese civilization. Presumably the cultivated *G. max* was derived from *G. ussuriensis* of central China, with which it crosses readily. Both species have 40 chromosomes, in comparison to other species, such as *G. javanica*, which has 20.

Although the soybean was introduced into Japan

by way of Korea before 200 B.C., it was little known outside of the Orient until the seventeenth century, when Kaempfer, a German botanist, brought the soybean to Europe. According to early records it was first introduced into North America in 1804, but was little grown in the United States until the twentieth century, and only recently has it been grown extensively. So fast has it been adopted in the Corn Belt in recent decades that it now rivals corn as an income crop in many locations. Breeding of improved varieties, and their inherent adaptability to mechanization, have made soybeans a remunerative crop to grow. Moreover, industry has found many ways to use soybean oil, creating a steady and increasing demand for the beans, an inducement to farmers of the Midwest.

Soybeans are well adapted to warm temperate climates as a summer annual. They are grown throughout the world—in middle and southern Europe, southern Russia, South Africa, Egypt, and somewhat in South America. But the United States and China account for more than 90 percent of world production, which has reached more than 45 million tons annually in recent years (Fig. 20-14). The United States alone produces about two-thirds of this, and China most of the remainder. Chapter 15 reviews mechanized soybean growing in the United States; it is presumed that in China soybeans are still cultivated by comparatively unmechanized techniques that involve much hand labor.

Climates with warm moist summers and at least a five-month growing season are ideal for most varieties of soybeans. Soybeans are broadly tolerant of soils, but do best on well-drained heavier types. In America soybeans are generally row planted, and often rotated with corn. Some varieties are erratic in response to fertilizer. Strongly nodulating varieties make better use of nutrients, but for top yields nitrogen fertilizer is used even with these varieties to supplement the *Rhizobium* activity. Soybean cultivars are self-fertile, exhibiting little outcrossing. Pre-emergence herbicides are utilized to control weeds in mechanized growing, especially with narrowly spaced rows (10 inches) for high yields. Weed control is a big help for efficient combine harvesting.

Most of the soybean crop in the United States is processed for oil, and accounts for about one-third of the nation's oils and fats. Oil used to be expressed from the seed, but most is now extracted with hydrocarbon solvents (which cuts oil wastage from 4 or 5 percent to about 1 percent). The extracted oil is made into shortenings, margarine, and salad oils; it is also used industrially in paints, varnishes, inks, caulking compounds, linoleum, and other products. The protein cake remaining after oil extraction may be processed into a soybean flour or be incorporated into animal feeds. Soybean protein balances some of the nutritional deficiencies of such grains as corn and wheat, which are low in the amino acids lysine and tryptophan. Lysine, generally present only in small amounts in feedstuffs, is present in reasonably adequate amounts in soybean meal. Lecithin, an important industrial soybean extract with antioxidant, emulsifying, and wetting properties, is used in foods, pharmaceuticals, printing inks, rubber, leather, textiles, cosmetics, and petroleum products.

As a garden plant soybeans can be eaten green-shelled (though they are not easy to separate from the pods), salted and roasted, or dried. More commonly soybeans are used in human food as flour, incorporated with wheat flour to round out the protein content in bread, cakes, cookies, and crackers; it is also used in soups and as an extender in potted meats, confections, and nut "butters."

Peanut, *Arachis hypogaea*

The peanut, also called groundnut, pinder, and goober, is the strange fruit of *Arachis hypogaea*. The fruit begins as a fertilized flower above ground, but when the seed pod forms, it is gradually pushed into the ground by elongation of the pedicel; the pod and seed mature in the ground (Fig. 20-15). The species is probably indigenous to Brazil, but today is cultivated the world over in the tropics and subtropics. World production amounts to about 18 million tons, most of it in the Far East and Africa (Fig. 20-16). India, China, Nigeria, the United States, Brazil, and Senegal, in about that order, are the other leading producers. The peanut, like the pineapple, the potato and the rubber tree, is a crop that originated in South America but has gained its greatest utilization on other continents!

As with the soybean, peanuts are much utilized

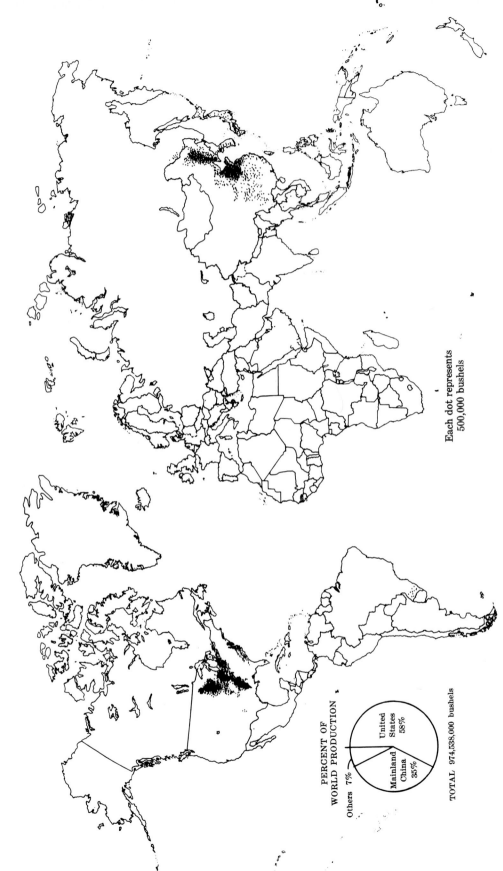

Each dot represents
500,000 bushels

PERCENT OF
WORLD PRODUCTION

Others 7%

Mainland
China
35%

United
States
58%

TOTAL 974,538,000 bushels

FIGURE 20-14. *World production of soybeans, average 1957–1961.* [*From A Graphic Summary of World Agriculture,* USDA Misc. Publ. 705.]

FIGURE 20-15. *Peanuts as they appear on a newly unearthed plant.* [*Courtesy USDA.*]

for oil extraction. As we have seen, peanuts contain about 50 percent oil, 25 to 30 percent protein. Much of the peanut crop becomes stock feed, particularly for hogs, with the hogs themselves rooting out the underground nuts when turned into the peanut fields. Unlike the soybean, however, peanuts are highly esteemed for human food, and are standard fare at American sporting events and at cocktail parties. In the form of peanut butter, the peanut has perhaps nourished more youthful Americans than one would care to count.

Peanuts have two contrasting growth habits— the bunch type, and the runner type. The latter spread widely on the ground, and are generally less densely planted. Most modern varieties, in North America at least, are intermediate between these extremes of habit. Testing indicates that dense plant populations give higher yields. In some experiments peanuts have been profitably planted only three inches apart in the row, in rows 10 inches apart, although such dense spacing in the row is indeed extreme. In the United States two general types of peanut are grown, the small "Spanish" type and the larger "Virginia."

Peanuts do well in light soils, and in the United States are much grown on the southeastern coastal plain from Virginia through the Carolinas, Georgia, and Alabama, and into Texas. The large kernel peanuts come chiefly from the Virginia-Carolina area, the "Spanish" nuts more from the deep South. In keeping with agricultural trends in the United States, cropping is tending more to mechanization. Mechanized harvesting brings in nuts of varying maturity and creates the need for artificial drying, in contrast to the picturesque former method of hanging the vines on stackpoles to dry. One would suppose that inter-row cultivation would provide a looser, more receptive soil for penetration of the maturing pedicels; but research in Israel has proven that this does not increase yields, and that chemical herbicidal control of weeds between the rows is often better for the plants because they are less disturbed than when cultivated. Peanuts are subject to a number of diseases, which are controlled by fungicides or by selecting resistant cultivars. Internal damage—malformation within the nut itself—has been traced in parts of the United States to deficiency of boron, and occasionally calcium.

In most of the Old World, planting, cultivation, and harvesting are still done by hand. Considerable portions of the African and Indian peanut crops are exported to Europe for extraction of the oil. In the United States only a limited portion of the crop is grown for oil; extraction of peanut oil is often done in the same factories that crush and express cottonseed oil. Most of the peanut crop goes into confections, such as peanut butter and candy bars. Peanut oil finds much the same uses as soybean oil, principally in margarine, shortenings, and salad or cooking oils. The oil is high in oleic constituents, modest in linoleic, and low in certain other fatty acid components. It finds several industrial applications, and the protein-rich press cake is a valuable livestock feed.

Pulses—Beans and Peas

Many different genera and species of the *Leguminosae* provide seeds that are directly consumed for food; these are the "pulses" of the Old World, or "peas and beans" (broadly considered) in America (Table 20-3). Included are dry beans, dry peas,

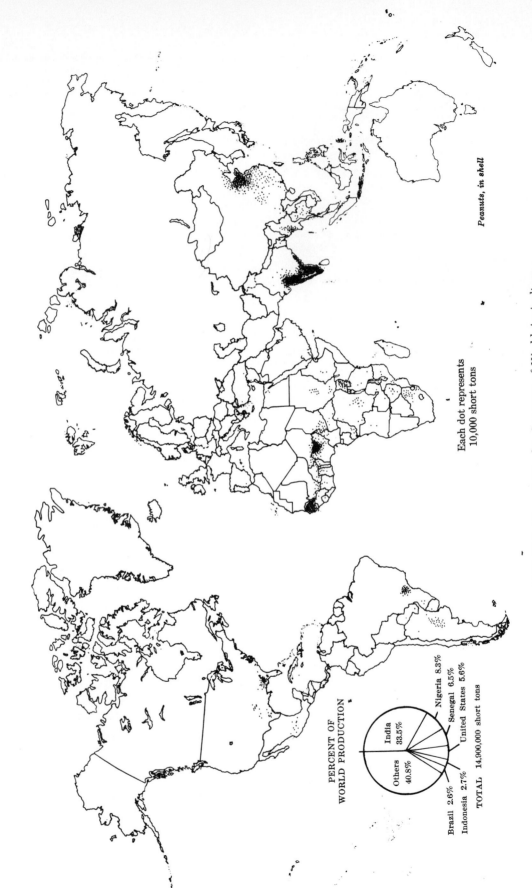

PERCENT OF WORLD PRODUCTION

India 33.5%
Others 40.8%
Nigeria 8.3%
Senegal 6.5%
United States 5.6%
Brazil 2.6%
Indonesia 2.7%

TOTAL 14,900,000 short tons

Each dot represents 10,000 short tons

Peanuts, in shell

FIGURE 20-16. *World peanut production, average 1957–1961.* [*From A Graphic Summary of World Agriculture, USDA Misc. Pubbubl. 705.*]

dry broad beans, chick-peas and lentils, the combined total for which comes to more than 40 million tons annually. The pulses are especially important in the Far East, although much grown in Latin America, too. World production of dry beans is nearly 12 million tons annually, of which better than a fourth are grown in Latin America and about a fourth in the Far East, with North America, Europe, and Africa following more distantly. India, China, Brazil, and Mexico are the leading producing countries. Most dry peas are produced in the Far East, with Europe often a close second. Total annual world production is around 10 million tons, with China and India again leading, followed by the Balkan countries and the United States. Only about 5 million tons of dry broad beans is grown annually, mostly in China, Europe, and northern Africa. Chick-pea production is also relatively small, seldom over 7 million tons annually, with India the leading source. Little more than a million tons of lentils are grown annually, these mostly in the Far East and Near East.

PHASEOLUS The genus *Phaseolus*, especially as represented by *P. vulgaris*, is the most important source of beans. Included in the genus are lima beans, scarlet runner beans, string beans, shell beans, white beans, black beans, pea beans, black-eyed beans and kidney beans. The species now important commercially are widely spread throughout the New World, and have been cultivated since time immemorial by the Indians, though of course not in modern cultivars. The early explorers found such beans an excellent protein-rich food for the long sea voyage between America and the home continent, and quickly introduced beans into most inhabited parts of the world. Today they are the main source of protein for the less affluent societies.

All *Phaseolus* beans are warm-season annuals or short-lived perennials that thrive on hot weather and ample moisture. They are a favorite garden plant, both in bush and climbing types, and are often interplanted with other crops in small subsistence gardens. Indeed, most of the world's bean crop is probably home tended and locally consumed. In the United States commercial production is completely mechanized. Even green beans,

Table 20-3 CONTRIBUTIONS OF DRIED SOYBEANS, OTHER DRIED BEANS, AND MEATS (AVERAGE PORTIONS[1]) IN PERCENT OF DAILY REQUIREMENTS.

FOOD	CALORIES	PROTEIN	CALCIUM	IRON	A	B$_1$	B$_2$
					VITAMINS		
Yellow soybean	10	49.3	26	48.8	1.8	72	86
Kidney bean	10	25	17	72.5	—	24	11
Navy bean	10	28	16	72.5	—	24	11
Roast lamb	7.4	32.4	1.67	15.38	—	9.64	12
Roast veal	5.8	44.25	2.61	34.8	—	20	11

Source: A. Williams-Heller and S. McCarthy, *Soybeans from Soup to Nuts.* New York: Vanguard Press, 1944.
[1]An average portion of dried beans is 3 ounces, or about 1 cup of cooked beans; an average portion of meat equals 4 ounces.

marketed as "vegetables," can be mechanically harvested. The same species that yield green beans as a fresh vegetable also yield dry beans when allowed to mature. Beans are quite responsive to phosphatic fertilizer, and are stimulated by mild dosages of 2,4-D weed killer (especially if accompanied by trace minerals).

The chief American beans, in addition to *P. vulgaris*, are *P. lunatus*, the lima or butter bean; *P. acutifolius*, and *P. multiflorus*. Asiatic species include *P. aconitifolius*, *P. angularis*, *P. aureus*, *P. calcaratus*, and *P. mungo*, all of which have slender cylindrical pods and small seeds. The majority of these species cross partially with each other, and are thus potential sources of improved germplasm for bean breeding.

P. vulgaris is variously known as the kidney, field, garden, or haricot bean. It provides the familiar green, string, snap, or wax beans when the immature pod is eaten entire; and the kidney, pea, pinto, great northern, marrow, and other types of dry bean when matured and shelled. There are both self-supporting "bush" varieties and trailing "vine" or "pole" types. The exact origins and relationships within the species have never been thoroughly worked out.

P. lunatus is the lima or butter bean, another

confusing assemblage of uncertain ancestry. From the culinary standpoint, lima beans are highly prized, eaten both fresh and dried. Both pole and bush forms are grown in a wide selection of seed colors and shapes. Lima beans have had a long history of cultivation in tropical America, being found in some of the most ancient excavations of Peru and Brazil. They are a bit more temperamental to grow than *P. vulgaris*, but require essentially the same conditions.

P. acutifolius is the tepary bean, often grown in northern Mexico and in the southwestern United States. Tepary beans were much utilized by the Indians of that area. *P. multiflorus* is the scarlet runner bean, perhaps more used nowadays as an ornamental than as a source of food seed. *P. mungo* is the black gram, or urd, bean of the Orient, probably native to India. Beans of this type are especially favored in India, where much of the population is vegetarian. *P. aureus*, the mung, golden gram, or green gram bean, is also extensively cultivated in India and China, and the germinating seeds are used as "bean sprouts" in oriental dishes. *P. aconitifolius*, the moth, or mat, bean, is another species cultivated in India and the Orient, both for its small bean seeds and its green pods. *P. angularis* is the adzuki bean, mostly grown in China and Japan, as is the rice bean, *P. calcaratus*.

VICIA *V. faba* is the broad, Windsor, horse, or Scotch bean, probably native to northern Africa or the Near East, where it has been under domestication for many millennia. This is a vigorous annual, well known to ancient Greeks and Egyptians. The diploid chromosome number is 12, and the species crosses with a number of wild vetches, most of which are used for forage rather than seed beans. Historically, the broad bean is important, because it was the only widely used edible bean in the Old World before *Phaseolus* was introduced from America.

PISUM *P. sativum* ($2n = 14$) is the familiar green pea, probably domesticated in central or western Asia, where there are similar wild species with which it can be crossed. Pea remains have been found in ancient Swiss lake dwellings, and the

plant was abundantly cultivated in Greek and Roman times. Originally peas seem to have been consumed exclusively in the dry or mature stage, but in recent centuries varieties grown as sweet or green peas have achieved great culinary status. Nonetheless, there is great world-wide production of dried peas, which in the United States are much used for "split pea" soup. Most of the pea crop, however, is produced commercially for fresh or green peas, which are canned or frozen. The plants are well adapted to cool spring weather or climates with reasonably cool summers, and much of the crop is grown in such northerly locations as Wisconsin. There are both bush and vine forms of *Pisum*. In the United States planting and harvesting are highly mechanized; even special "vining" machines are used to cut the plants and strip them of pods. A threshing device removes the green peas from the pods.

ICER *C. arietinum* is the chick-pea, also called gram pea or garbanzo. It is native to western Asia, and the genus embraces a number of species. The pods are short and contain only one or a few rather large seeds. This is the most important legume grown in India, ranking fourth among food sources. In addition to consumption of the whole seed, or seed in young pods, chick-peas are sometimes ground to flour and used in preparing unleavened bread or sweets. The seeds may be eaten raw, roasted, parched or boiled. The chick-pea is also an important source of livestock feed.

VIGNA *V. sinensis* (often merged into *V. unguiculata*) is the cowpea or black-eyed pea of the southern United States. It is really more a "bean" than a "pea," and like most food legumes is available in either vine or bush varieties. It is related to the "yard-long bean" of the tropics, which has a pod several decimeters in length. The species probably originated in central Africa, but was introduced into colonial America by the early 1700's. Its tropical background is reflected in its inability to withstand cold, though it grows passably on almost any soil.

LENS *L. esculenta* (*L. culinaris*) is the lentil, another of the legumes utilized since ancient times.

FIGURE 20-17. *Forage. (Left) The grass family furnishes many excellent forage species. Jersey cows graze sudan grass on an Alabama pasture. (Right) Fully as important as grass for forage are species of the Leguminosae. Cattle graze on the Hubam cultivar of white clover in southern Florida. [Courtesy USDA.]*

It is believed to be native to southwestern Asia, but was introduced into the Mediterranean area before historical times. Pods are short, and contain only a couple of lens-shaped pea-like seeds. these make a very tasty soup or porridge, the form in which they are usually consumed.

CAJANUS   *C. indicus*, is the pigeon pea, cajan bean, or congo bean. It is widely grown in the Orient and in equatorial Africa. In India it is the second-most important pulse, and is much used as a cattle feed as well as for human food. In Africa the congo bean is often interplanted with cereals and root crops, and the foliage may be used as a forage.

OTHER LEGUMES   The seed of almost any legume can be adapted to the needs of humans or livestock. A few legumes that are somewhat less popular than the species discussed so far are the jack bean, *Canavalia ensiformis;* the sword bean, *Canavalia gladiatus;* guar, *Cyamopsis tetragonaloba;* lablab, *Dolichos lablab;* the geocarpa groundnut, *Kerstingiella geocarpa;* the kasari, *Lathyrus sativus;* the winged pea, *Lotus tetragonolobus;* the kudzu, *Pueraria thunbergiana;* the velvet beans, *Stizolobium* in several species; the bambarra groundnut, *Voandzeia subterranea;* and the pods of many tree

legumes such as the carob, *Ceratonia siliqua;* the honeylocust, *Gleditsia triacanthos;* locust beans *Parkia* spp.; rain tree, *Pithecolobium saman;* mesquite, *Prosopis* spp.; and many others.

FORAGES

Literally, **forages** are any plants consumed by livestock. Most generally the term is restricted to pasture and browse plants, hay, silage, and immature cereals or residues from grain harvest such as straw. Livestock feed grain is normally not considered forage. There is no reason why a species from any plant family might not be considered a forage plant, but practically speaking, only the grasses (Gramineae) and the legumes (Leguminosae) contribute much in the way of forage. Just as cereal grain and legume seeds make excellent complementary foods for human use, so does grass and legume vegetation provide excellent dietary complement for herbivorous animals (Fig. 20-17). In modern animal husbandry cattle are generally fattened on grain before being sent to market in order to provide superior "finished" meat. But there is no reason why a grazing animal could not subsist indefinitely on grass-legume graze so far as its own well-being is concerned.

Thus forages supply additional human food through the intermediary of livestock. Cattle, sheep, hogs, poultry, and other domesticated animals become the instrument for transforming the proteins, carbohydrates, and fats of inedible (by humans) plant foliage into meat, eggs, and dairy products. At the same time that the forage plants sustain this important livestock industry, they also protect the land and build the soil, which cultivated row crops do not. Forages are especially suited to rolling or steeper lands that would erode readily when clean-cultivated, and to other lands that for one reason or another cannot be adapted to the growing of familiar cash crops. Because of this relegation to lands of secondary quality, forages have often seemed the neglected facet of agriculture. However, cumulatively, their extent and value are as great as those of nonforage crops, and their response to management (fertilization, weed control, proper soil handling) is equally as rewarding. Forages are especially useful to expanding civilization, because they provide a means of making good use of land that is not well suited to the conventional cultivated crops that are so much a part of the mechanized agriculture reviewed in Chapter 15.

World-wide, nearly 3.7 billion acres ($1\frac{1}{2}$ billion hectares) of arable land are cropped; 7.5 billion acres (3 billion hectares) are permanent pasture and meadow. In North America cultivated land and permanent pasture are about equal in extent, but in Latin America, Africa, and Oceania pasture land far exceeds cultivated land. In more heavily populated Europe and most of Asia cultivated cropland quite exceeds meadow and pasture. Of the livestock sustained on meadow and pasture, sheep, cattle, and pigs are most important in Europe; cattle, pigs, and sheep in North America; cattle, sheep, pigs, goats, and horses in South America; cattle, sheep, goats, buffalo, and pigs in Asia; cattle, goats, asses, and camels in Africa; and sheep in Oceania. By world total, sheep and cattle run about equal at nearly a billion of each, with about half as many pigs and a third as many goats.

It is impossible to mention more than a few examples of the world's prominent forage plants. Almost all grasses (there are about 5,000 species

in the family) and most herbaceous legumes (there are about 11,000 species estimated in the family) are useful as forages. The feed value of the species used for forage varies greatly with soil fertility, stage of harvest, and handling. Forages make better feed if grown according to advanced agricultural methods. Fertilizer, for example, increases both yields and nutrient value. Protein content of grasses is highest (on a percentage basis) in lush, fresh foliage before strawy seedheading stalks form or the grass starts to turn dormant. Kentucky bluegrass in spring may be almost 20 percent protein, although this percentage decreases to little more than half that amount by summer. One mineral content analysis of bromegrass, a much used forage in the Midwest, shows 2.2 percent nitrogen, 0.32 percent phosphorus and 0.4 percent calcium. This analysis is for grass grown on good soil; on poor soil the mineral content is appreciably less. Alfalfa, the chief forage in the United States, showed 3.5 percent nitrogen, 0.35 percent phosphorus, and 1.6 percent calcium—somewhat superior to the grass as far as protein and minerals are concerned. Mineral and protein analysis is, of course, only a rough indication of feed value. Balance in nutritional factors, including minor or trace elements, is essential to the health of livestock. Sometimes too much leguminous protein (as from heavy fertilization) along with insufficient grass carbohydrate leads to nitrogen poisoning, or livestock bloat. Contaminants sometimes poison livestock (for example, species of *Astragalus* may cause selenium poisoning on the western ranges).

In the United States forages are estimated to have an annual worth of nearly 10 billion dollars. They provide more than half the nutritional requirements of all livestock. About 165 pounds of meat is consumed yearly per person, and nearly 700 pounds of milk; it is said of these that about 25 percent derives from grain (concentrates), and 75 percent from forage (roughage). About 23 cents of the agricultural dollar in the United States is said to come from beef. Where animal products are consumed in such quantities, obviously there is a need for improved forage plants, such as Coastal bermudagrass, capable of yielding more than a ton of beef per acre annually (fed dried and pelleted).

Box 20-2   WHY LEGUMES WITH FORAGE GRASSES?

Combinations of grasses and legumes are planted in most pastures, even though high yields of grass alone could be obtained with fertilization. Some of the reasons are:

Legumes, owing to symbiotic nitrogen fixation, supply a good bit of the grass' nitrogen needs, an obvious economy.

The deeper-rooted legumes aid summer production at a time of year when grass growth often declines.

Proteinaceous legumes improve forage feeding value and micronutrient balance (especially magnesium, the lack of which may cause grass tetany).

Competition for space and survival occurs in the pasture ecosystem, as it does in nature. Grasses tend gradually to dominate legumes. Practices helping to favor the legume include:

Fairly close grazing (or mowing).

Rotational grazing, which provides recovery periods.

Mowing and weed control of grass patches that become undergrazed.

Making hay if growth exceeds livestock capacity to consume.

Using fertilizers rich in phosphorus and potassium (legumes provide much of their own nitrogen).

In most climates the legume component may have to be replanted every few years, although if liming and fertilization are ample many clover stands remain adequate for a decade or two. As legumes disappear, pasture fertilization must be increased if the animal-carrying capacity is to be maintained.

Most grass-legume combinations will provide not only better animal nutrition than would grass alone, but greater daily gains per pound of TDN and more profitable cropping.

Though forage grasses have only recently begun to receive the attention that has been accorded the major cereal grasses, researchers are now exploring all facets of forage production. One need only scan agricultural advertising to note the tremendous diversity and widespread innovation in forage harvesting machinery—cutters, choppers, pickups, blowers, balers, self-loaders, and so on. Almost everything can be done mechanically by a single operator. The plant breeder has been busy, too, and the future of forage plants lies particularly with select cultivars and hybrids. Already benefiting from heterosis are hybrid sudangrasses (crosses made possible by de-riving male sterility from sorghum), sorghum-sudan crosses, hybrid bahiagrasses, a sterile hybrid bermudagrass (Coastal, so excellent a yielder as to justify expensive vegetative propagation), hybrid pearl millet (*Pennisetum glaucum*), and good prospects with white clover, trefoil, and other legumes.

Alfalfa, Medicago sativa

Alfalfa, or lucerne as it is called in Europe, is generally regarded as the "queen of forages." No forage is a better source of protein. Throughout the world more than 60 million acres are planted to alfalfa, something close to 33 million in the United States alone, followed by Argentina, France, southern Europe and parts of Asia. In the United States the annual value of the alfalfa crop is estimated at $1\frac{1}{2}$ billion dollars. About half of the domestic crop comes from the Corn Belt, with Wisconsin, Iowa, Minnesota, Nebraska, Illinois, Michigan, Kansas, and South Dakota the leading producing states there; California, the largest producer outside of this area, produces about the same amount as Wisconsin. Some 70,000 tons of alfalfa seed are produced annually in the United States. It is not uncommon for alfalfa or alfalfa-grass forage plantings to yield 6 tons of hay per acre per year (which amounts to a dry nutrient content of 4 to 5 tons, of which 1.5 tons is protein). On good lands giving such yields alfalfa can be as rewarding as any other crop.

The origin of alfalfa is obscure, and its taxonomic relationships uncertain. Some of its relatives, such as black medic, *Medicago lupulina*, are weeds in lawns. Wild alfalfa is scattered over central Asia, and the cultivated species is believed to have become domesticated in southwestern Asia, possibly Iran. The chromosome number is $2n = 32$, the same as with several other species (*M. arborea, M. gaetula, M. glutinosa, M. hemicycla*), indicating a tetraploid state as compared to black medic and other burclover species ($2n = 16$).

Roman records indicate that the ancient Medes introduced alfalfa into Greece from Persia about five centuries B.C. Even at this early date it was recognized as a forage without equal for chariot horses. The Spanish took alfalfa along with them as feed for their horses when they explored Central and South America. According to early records, the first alfalfa grown in the United States was in Georgia, in 1736. Its use spread slowly, however, until the Gold Rush of 1851, during which it was taken to California and dispersed rapidly throughout the nation. An introduction from Germany by one Wendelin Grimm paved the way for hardy cultivars that could be grown in the northerly states; the Grimm variety was developed in Minnesota through continued selection of this introduction.

Alfalfa is well adapted to a wide range of climatic and soil conditions. It performs best on deep loam soils that are well drained, in climates where rainfall is moderate, winters cool, and summers warm. Alfalfa does well in limed soils with ample potassium. In the northeastern United States it is very responsive to potassium and phosphorus fertilization, but on western lands where soils are rich in potassium, phosphorus is the main nutrient needed. In keeping with legumes in general, alfalfa is symbiotically more or less self-sufficient in nitrogen. When alfalfa and other legumes are planted in association with forage grasses, nitrogenous fertilizer favors the grass and may encourage it eventually to dominate the legume. Similarly, close clipping favors the alfalfa, high clipping the grass. Meadow management is usually adjusted to favor the alfalfa, in order to hold it for a maximum number of years, after which fertilizer nitrogen can maintain grass yields as the alfalfa thins out. Favorably situated stands of alfalfa may persist a decade or longer.

Because alfalfa has a deep taproot that can penetrate many feet into the subsoil on favorable ground, it is rather drought resistant. In the South it is generally autumn-seeded, and in the North spring-seeded, so that stands can become well established before cold winter weather. It is usually drill-planted, often with a companion crop (such as a small grain, which may be harvested the following spring). It is a good plant for improving the soil when used in rotation. Alfalfa gives highest yields on the irrigated lands of the Southwest (Chapter 15), where as many as six cuttings a year can be made, compared to the more conventional three cuttings in the Corn Belt. Formerly alfalfa was swathed and windrowed for the making of hay, but artificial drying and compression into pellets are becoming increasingly popular.

Alfalfa is subject to a number of diseases, and in some of the northeastern states is attacked by the alfalfa weevil. Where weevils are onerous and control costly alfalfa is being abandoned in favor of other forages, such as sorghum-sudan. Needless to say, with so important a crop as alfalfa, there is constant endeavor to develop pest-resistant cultivars that are hardier and give better yields.

New breeding breakthroughs promise eventually to yield hybrid alfalfas. The genetics of alfalfa is somewhat confusing; some strains are self-pollinated, some self-fertile, and others almost obligately outcrossing. Because the flower must be pollinated by insects (leaf-cutting bees are usually used) and because it is perfect (with both male and female parts), practical techniques for producing hybrid seed have been slow to come. The technique that seems most promising is to breed self-incompatability (producing little or no seed from its own pollen) into desirable strains, and propagate these by vegetative cuttings. If another self-sterile strain, also propagated vegetatively, is interplanted in alternate rows, this ensures that all seed produced will be hybrid. Although such a planting is expensive to make, it can be amortized over a period of several years. Hybrid seed generally increases yields 20 to 35 percent over yields from pure lines, and thus justifies a premium price. Whenever hybrid seed does become economically available, it can be expected to supersede present leading cultivars, such as Ranger (already on the decline) and Vernal (on the increase).

Frequently interplanted with alfalfa for forage in the northcentral and northeastern United States are orchardgrass, *Dactylis glomerata;* bromegrass, *Bromus inermis;* and timothy, *Phleum pratense.* About 10 pounds per acre of alfalfa seed, and 6 to 10 pounds of grass seed are usually used, sometimes with oats as a nurse crop (actually a hindrance to establishment of the forage). The alfalfa seed is usually inoculated with an appropriate strain of nitrifying bacteria. Herbicides are being used increasingly to eliminate the less-resistant annual grasses and weeds from alfalfa plantings. Yields from alfalfa-grass mixtures in the Midwest are given in Table 20-4. Yields from these mixed forage plantings are at least as high in protein, and

often quite superior to, those obtainable from corn.

Grazing or feeding of green alfalfa sometimes results in cattle bloat, often serious enough to be lethal. The exact cause of the trouble is not certain, but it may be due to saponins or other minor constituents of "rich" alfalfa fare. Water-soluble phytotoxic substances have also been derived from alfalfa leaf; these substances are at least partially repressive to grasses. Alfalfa foliage has served as a source of certain extractives, particularly the green pigment chlorophyll (not too long ago all the rage as an air purifier and deodorant). Some experimental extraction of protein has been done in Great Britain in the hope of creating food suitable for direct human consumption (Table 20-5). The alfalfa is pulped, the juice expressed, the chloroplasts centrifuged out, and the protein precipitated by acidification and heating. So far, with only about half the protein recovered, this process has not proved economically useful. The most efficient use of poorly digestable alfalfa protein still seems to be conversion into animal products, notably chick ration. The most efficient conversion is probably to milk, which on well-run dairies provides protein yields about equivalent to those derived from soybeans. Channeling alfalfa protein into meat is less efficient.

The clovers, Trifolium

There are a number of important forage species in the "true" clover genus, *Trifolium.* Most prominent are red clover, *T. pratense;* white clover, *T. repens;* alsike clover, *T. hybridum;* and crimson clover, *T. incarnatum.* Hop clovers, Persian clover,

Table 20-4 YIELDS OF ALFALFA-GRASS MIXTURES.

MIXTURE	TONS/ACRE 1963	1964	DIGESTIBLE DRY MATTER (LB/ACRE) 1963	1964	PROTEIN (LB/ACRE) 1963	1964
Alfalfa—Orchardgrass	6.2	5.3	6,925	5,683	1,647	1,500
Alfalfa—Bromegrass	6.6	6.0	6,701	6,262	1,879	1,962
Alfalfa—Timothy	6.6	7.1	6,020	6,899	1,494	1,939

Source: Purdue University.

Table 20-5 THE ESSENTIAL AMINO ACIDS IN LEAVES OF ALFALFA AND OTHER FOODS.

FOOD	ISALEUCINE	LEUCINE	LYSINE	METHIANINE	PHENYL-ALANINE	THREONINE	TRYPTOPHAN	VALINE
				(MG AMINO ACID PER G NITROGEN)				
Alfalfa leaf	357	596	441	91	400	360	100	460
Soybean leaf	362	723	522	82	421	381	120	485
Soybean	284	486	399	79	309	241	80	300
Yeast	239	475	300	69	538	338	150	238
Milk	399	782	450	156	434	278	—	463
Egg	393	551	436	210	358	320	—	428
Beef	301	507	556	170	275	287	—	313

Source: Adapted from J. E. Kinsella and A. Betschart, *New York's Food and Life Sciences Quarterly* 5:16, 1972.

sub clover, strawberry clover, and others are somewhat less used.

Compared with alfalfa, the clovers are smallish plants, generally yielding a bit less forage, but they are excellent sources of protein and are very palatable to livestock. Clovers nodulate well and are perfect companions for grasses in mixed pasture. Indeed, Kentucky bluegrass (*Poa pratensis*) and white clover (*Trifolium repens*) are partners in world conquest, volunteering and supporting one another even where not planted. Together they were known as "English grass" in the American colonies. Though most clovers are perennial, the majority are not long lasting in managed pastures and meadows. The tap root generally rots away by the second year (probably due to a complex of insects and fungi), and the plant must depend upon adventitious roots thereafter. Sometimes it is able to do so, but often it peters out after two or three years.

RED CLOVER, *T. pratense* Red clover probably originated in southwestern Asia, and is a more recent domesticate than alfalfa. It was much used throughout Europe in the late Middle Ages, where it helped initiate crop rotation, which eventually superseded fallowing. Today it is planted and naturalized in almost all temperate climates. It thrives best in humid sections, and is probably the most important hay legume in the northeastern part of the United States. Red clover suffers from a number of diseases, and virus-resistant cultivars are especially sought. The species is largely self-sterile, so that outcrossing results, making it difficult to maintain a pure line for hybrid seed. Cultivars segregate wildly in the second year and thereafter. However, tetraploid red clovers are becoming available. Red clover seed production, although less than alfalfa seed production, amounts to a significant 40,000 tons annually in the United States. Red clover responds well to fertile, limed, phosphorus-rich, well-drained soils.

WHITE CLOVER, *T. repens* White clover, which includes the robust ladino variety from Italy, behaves much like red clover, with which it may cross. Also like red clover, the species is generally self-incompatible. Genetic mechanisms have been discovered that permit self-pollination, giving hope for the development of hybrid white clover in the future. The species is probably a native of southern Europe or the Near East, and undoubtedly served as a forage for the earliest European livestock. Today it is planted or found volunteering in temperate climates around the world, responding well to ample moisture, lime, and generous phosphatic fertilization. White clover generally contains between 20 and 30 percent crude protein (on a dry-weight basis), depending upon the season. Unfortunately, plantings are frequently short-lived. But the plant regenerates and volunteers well, and is undoubtedly one of the world's most important pasture plants (although it can't match alfalfa and red clover for hay and silage).

ALSIKE, *T. hybridum* Alsike clover is much grown and well adapted in northern Europe and in high-altitude wet meadows in the United States. There are both diploid and tetraploid strains. The species perisits poorly, and is often handled as a biennial. Its origin is obscure.

CRIMSON CLOVER, *T. incarnatum* Sometimes called scarlet, Italian, or incarnate clover, crimson clover is much utilized as a winter annual throughout the southeastern United States. It, too, seems to have originated in southeastern Europe or in the Near East, and was widespread in Europe by the Middle Ages. The species tolerates some acidity and some variation in soil types. It usually self-seeds sufficiently to re-establish a subsequent year. It is a good soil improver and excellent green manure. Crimson clover is often sown with annual ryegrass for winter pasture, but can make a stand in established bermudagrass (semidormant through winter); its growth cycle and that of Coastal bermudagrass complement one another nicely. It is also useful for seeding between cotton rows to protect the soil while providing some winter forage.

Other prominent forage legumes

BIRD'S-FOOT TREFOIL, *Lotus corniculatus* Bird's-foot trefoil is the most important of four *Lotus* species used for forage, two of which are annuals. It is a trailing perennial, probably first domesticated in the Mediterranean area. Both diploid $(2n = 12)$ and tetraploid forms occur. The species is much used in Italy, for both hay and pasture, but only in recent years has it become prominent in North America, where it is planted principally east of the Great Plains (Ohio–New York). Bird's-foot trefoil is tolerant of poor soils, but is slow to develop a stand. Like other legumes, it is quite responsive to phosphorus. Its protein and mineral composition is not much different from that of alfalfa or clover. Once well established, it is said to persist in Kentucky bluegrass pasture better than do most legumes, and is then the equal of alfalfa-timothy or ladino-brome in beef carrying capacity.

THE VETCHES, *Vicia* There are about 150 species of *Vicia*, most of them native to Europe and the Near East, of which *V. villosa*, hairy vetch; *V. sativa*, common vetch; *V. atropurpurea*, purple vetch; and *V. pannonica*, Hungarian vetch, are the most important. Other species may be useful as a source of germplasm (for example, *V. sativa* with 12 chromosomes can be crossed with *V. angustifolia* with 10 chromosomes, to give partially fertile F_1). Most vetches are trailing vines with tendrils. Hairy vetch is winter-hardy in the North, but finds greatest use in the United States in the Cotton Belt, where it is widely naturalized. Common vetch is less winter-hardy, and purple vetch the least winter-hardy of all. The vetches are much used as winter-annual cover crops in the United States, especially in the Southeast and on the Pacific coast. They may be drill-planted or broadcast, and are often mixed with rye or oats, especially when used for hay or silage.

SWEET CLOVER, *Melilotus* This genus is native to western Asia and temperate Europe, and was introduced into North America in the early 1700's. Sweet clover has a deep taproot, and is prized as a drought-resistant forage plant and soil builder. Its preference is for well-limed soils. Two species, the yellow (*M. officinalis*) and the white (*M. alba*), are commonly grown in America, both of them as biennials. They are especially planted for forage from Texas northward in the eastern plains, and for soil improvement generally. Both are noteworthy honey plants. A relatively high coumarin content in sweet clover tends to make it less palatable to livestock than many other legumes.

LESPEDEZA The genus *Lespedeza* offers numerous useful forage species, of which the annual Korean lespedeza (*L. stipulacea*) and common, or "Japanese," clover, (*L. striata*), and the perennial sericea (*L. cuneata*) are most important. All are native to eastern Asia. they were not introduced into the United States until the late nineteenth and early twentieth centries, although they are now widely naturalized. Korean lespedeza makes an excellent catch crop for warm weather; it is adaptable to poor soils and even gravel, but is slow to start in spring and is thus a poor competitor against weeds or companion crops. Korean and common lespedeza are most serviceable as a temporary cover while the land is out of other use (such as following early summer grain harvest). Sericea lespedeza is much used as a roadside cover in the South, as well as for hay and pasture.

Other forage legumes

A goodly number of other legumes are sometimes used for forage: kudzu, *Pueraria thunbergiana*, a rank vine from Japan that quickly covers gullied lands in the southern United States; the velvet-bean, *Stizolobium*, adaptable to sandy soils in subtropical climates; the burclovers, *Medicago*, relatives of alfalfa; lupines, *Lupinus* in several species, adapted to cooler climates and considerably used in Europe; *Crotalaria*, in several tropical species; the roughpea, *Lathyrus hirsutus*, native to the Mediterranean region, growing as an escape in the southern United States; alyceclover, *Alysicarpus vaginalis*, a summer annual from tropical Asia; various indigos, *Indigofera*, tropical plants from Africa and Asia; many of the seed legumes in immature stages, such as soybean, tepary bean, cowpea, field pea, and peanut; several native legumes, such as the beggarweed, *Desmodium;* and exotics often used for other purposes, such as Guar (*Cyamopsis tetragonoloba*), Fenugreek (*Trigonella foenum-graecum*), and Sainfoin (*Onobrychis viciaefolia*).

Forage grasses

Almost all grasses can serve as animal graze, for wildlife and for livestock. A few widely cultivated ones are cited here. Some, such as sorghum-sudan crosses and pearl millets, have already been mentioned under the section on Cereals. Others are native American grasses, such as blue grama (*Bouteloua gracilis*), a shortgrass prairie species now being planted on southwestern pastures; several lovegrasses (*Eragrostis*), such as the sand lovegrass from the southern Great Plains; and the famed bluestems (*Andropogon*), long a source of prairie hay in the cool dry plains regions. Rhodesgrass (*Chloris gayana*) is much used in Australia.

As has been pointed out several times, grasses go well with legumes in forage plantings. The legumes, which trap gaseous nitrogen, are the richer source of protein; but grass yields abundantly and persists under grazing or cutting. Where legumes falter, fertilizer nitrogen can substitute to a degree, not only in making forage grasses more productive but also more proteinaceous. By and large, grasslands are very responsive to fertilization.

SORGHUM-SUDAN   Sorghum-sudan crosses, bred for productivity, are especially high yielding. The forage may grow as tall as 15 feet without lodging, and under favorable conditions can produce as much as 18 tons of dry matter per acre. Although first looked upon as a catch crop in case of crop failure, sorghum-sudan is now much planted as a planned pasture, hay, and haylage crop competitive with alfalfa. Sorghum-sudan is not classed as "feed grain," and thus allowable acreage is not restricted under recent United States farm programs. Some 40 to 65 percent of sorghum-sudan is grazed, 15 to 30 percent serves for green chop (hauled to the animals), 5 to 10 percent for hay, 3.5 percent for silage, 1.2 percent for haylage, and the remainder for cover and green manure. Harvesting is often possible within six weeks of planting. Kansas leads in production.

KENTUCKY BLUEGRASS, *Poa pratensis* Bluegrass probably originated in southeastern Europe, and was spread throughout that continent by the Middle Ages. It was introduced into America with early colonization, and adapted so well that it soon became the prime pasture and forage species for the first civilization west of the mountains, near Lexington, Kentucky (Fig. 20-18). There in the early 1800's *Poa pratensis* gained the name "Kentucky" bluegrass. Though not a high yielder, especially during hot summer weather, Kentucky bluegrass is nonetheless one of the richest and most palatable of grass forages. Because it spreads by rhizomes it is also one of the world's finest sod formers and soil protectors. Kentucky bluegrass grows best on limy, phosphatic, well-drained loams, but is adaptable and quite tenacious in almost any cool, humid climate. The nation's foremost lawngrass cultivars are Kentucky bluegrass.

THE BROMEGRASSES, *Bromus* There are about 60 brome species, of which smooth bromegrass (*B. inermis*) is perhaps the most widely utilized.

FIGURE 20-18. *A pasture near Winchester, Kentucky, being harvested for seed before returning it to grazing. Kentucky bluegrass is still an important forage species on the rich phosphatic soils of this area.* [*Courtesy The Lawn Institute.*]

It is apparently native to central Europe, and was introduced into the United States in the late 1800's. It does well on the northern plains, where it withstands moderate drought. Brome is much grown with alfalfa from Kansas to North Dakota and eastward to Michigan.

TIMOTHY, *Phleum pratense* Native to northern Europe and Asia, timothy was first grown in New England in about 1747, where it still holds sway as one of the major forage grasses. It is perhaps the best-yielding hay grass in Ohio and other parts of the eastern Corn Belt, where it is often inter-planted with red clover. With the passing of the farm horse, timothy hay is less grown today than it was a half century ago.

ORCHARDGRASS, *Dactylis glomerata* Orchardgrass is native to western and central Europe, and was introduced into the United States in colonial times. It has proved well adapted to cool, humid regions, where it frequently naturalizes, and is today found world-wide in such climates. In the United States orchardgrass has found greatest use slightly south of the timothy belt, from Virginia to Iowa. It is not as winter-hardy as Kentucky bluegrass, brome, or timothy, but its soil requirements are probably less exacting. In mixed plantings with legumes it is a long-lived perennial, much used for hay or silage. Under close grazing it generally gives way to Kentucky bluegrass or other volunteer growth.

THE FESCUES, *Festuca* The fine or red fescues (*F. rubra*) are most esteemed for lawns (Fig. 20-19); the coarser meadow fescue (*F. elatior*) and tall fescue (*F. arundinacea*) are used as forages. The latter is especially important in the middle latitudes where summers turn too hot for Kentucky bluegrass, timothy, and other more northerly species. By the same token these fescues are doubtfully winter-hardy in northern states. Tall fescue, a vigorous grass with deep roots, adapts well and establishes easily on many types of soil, even in hot and difficult situations. It is much planted to waterways and roadsides as well as to pastures and meadows. In pastures it is usually sown with clover or lespedeza, which improves its nutrient value and palatability. Red fescue, naturalized widely in western Canada, serves as a forage there.

BERMUDAGRASS, *Cynodon* Common bermudagrass, *C. dactylon*, is circumtropical, probably originating in Africa but possibly also in southern Asia. It was introduced into Georgia by at least

1751, and has since spread entirely across the southern United States, where it is often derogatorily spoken of as "wiregrass" or "devilgrass" because of its proclivity for invading gardens, lawns, and fields with its vigorous runners. In many

FIGURE 20-19. *Fine fescue being row-cropped in Oregon; seed of most cultivars comes from that state.* [*Courtesy Union Pacific R. R.*]

states it is classed a noxious weed. Bermudagrass has proved especially well adapted to the area around Yuma, Arizona, where most of the commercial seed is produced today. Mention was made of the cross between a local bermuda and an African introduction, to create 'Coastal.' This high-yielding hybrid must be vegetatively propagated through sprigs or stolons (live stems) because it is sterile.

PASPALUMS   Two species of Paspalum in particular are grown for forage in the southern United States, dallisgrass, *Paspalum dilatatum;* and bahiagrass, *Paspalum notatum,* both native to South America. Dallisgrass is well adapted to the Cotton Belt, where it has volunteered widely and is often a pest in lawns; bahiagrass, in several strains, is much used as a seeded lawngrass in the deep Southeast, where it is also an important pasture species. Both are used more for pasture than for hay, and dallisgrass is often interplanted with lespedeza.

OTHER FORAGE GRASSES   Reed canarygrass, *Phalaris arundinacea,* was introduced from northern Europe in the late 1800's and is especially planted to lowlands and stream banks across the northern United States; the ryegrasses, *Lolium multiflorum* (annual) and *L. perenne* (perennial), native to Asia Minor and the Mediterranean area, are used principally for winter pasture and cover crops in the southern United States and in New Zealand; redtop and the bentgrasses, *Agrostis,* are primarily adapted to damp locations, redtop often being interplanted with alsike clover, and volunteering on poorer soils in the southern Midwest; the wheatgrasses, *Agropyron,* are well adapted to regions of limited rainfall, with both introduced and native species being of service in the northern Great Plains of North America; johnsongrass, *Sorghum halapense,* native to southern Asia and North Africa, is valuable for hay pasture in the upper South, but is so aggressive as to become a serious weed in middle latitudes; carpetgrass, *Axonopus,* is adapted to sandy, boggy soils near the Gulf Coast of the United States, where it may serve as pasture; pangolagrass, *Digitaria decumbens,* is a relative of crabgrass from South Africa, used for pasture particularly in Florida. Additionally, many cereal species serve as forages; included would be oats, barley, and rye for hay or graze; corn and sorghums primarily for silage. In the tropics, too, species of *Eragrostis, Echinochloa, Pennisetum, Setaria, Brachiaria, Cenchrus, Chloris, Hyparrhenia,* and several other genera may be utilized for pasture or fodder.

Selected References

Aldrich, S. R., and E. R. Leng. *Modern Corn Production.* F & W Publishing Co., Cincinnati, 1966. (A practical treatment with technological emphasis on all aspects of production from seed selection through marketing.)

American Society of Agronomy. *Alfalfa Science and Technology.* ASA Monograph 15, Madison, Wisc., 1972. (Consisting of 35 chapters. By 75 authors, this work constitutes a complete review of the "queen of the forages" from its evolutionary background to modern alfalfa production).

Barnard, C. (editor). *Grasses and Grasslands,* Macmillan, London, 1964. (A series of documented essays on the basic biology of grasses and pastures, by Australian authors adopting a world-wide approach; an authoritative over-all review.)

Bolton, J. L. *Alfalfa: Botany, Cultivation, and Utilization.* Interscience (Wiley), New York, 1962. (An excellent monograph.)

Coffman, F. A. (editor). *Oats and Oat Improvement,* American Society of Agronomy, Madison, Wis., 1961. (A technical monograph.)

Doggett, H. *Sorghum.* Longmans, Green and Co., London, 1970. (A thorough discussion of the subject typical of the "world crops" series).

Food and Agriculture Organization. *Production Yearbook,* Rome. (World crop statistics compiled annually.)

Grampian Press, *World Crops,* London. (A journal of international agriculture, published since 1949.)

Grist, D. H. *Rice* (4th ed.). Longmans, London, 1965. (Detailed monograph on rice, its genetics, and its cultivation by an author especially experienced in Malaya.)

Harrison, C. M. (editor). *Forage Economics—Quality.* American Society of Agronomy, Special Publication 13, Madison, Wisc., 1968. (Monographic treatment by a series of authorities having established credentials).

Hitchcock, A. S. *Manual of the Grasses of the United States* (2nd ed., revised by Agnes Chase). USDA Misc. Publ. No. 200, 1950. (A manual of agrostology, the identification and life cycle of grasses.)

Hughes, H. D., M. E. Heath, and D. S. Metcalfe (editors). *Forages.* Iowa State College Press, Ames, 1951. (Sixty authoritative chapters on grassland agriculture, with a chapter devoted to each major forage plant.)

Inglett, G. E. (editor). *Corn: Culture, Processing, Products.* AVI Publishing Co., Westport, Conn., 1970. (Corn production, utilization, breeding, culture, harvesting, storage, and marketing of the corn crop, popularly presented for understanding by a lay audience. Lengthy discussion of industrial processing and fermentation uses).

Leonard, W. H., and J. H. Martin, *Cereal Crops.* Macmillan, New York, 1963. (The production, use, and history of grain crops.)

Matz, S. A. (editor). *Cereal Science.* AVI Publishing Co., Westport, Conn., 1969. (The emphasis is on technology, but general information on cereal crops is also provided.)

Matz, S. A. (editor). *Cereal Technology.* AVI Publishing Co., Westport, Conn., 1970. (Technology and general information.)

Milthorpe, F. L., and J. D. Ivins (editors). *The Growth of Cereals and Grasses.* Butterworths, London, 1966. (A series of technical presentations for the "Twelfth Easter School In Agricultural Science" at the University of Nottingham, embracing germination, morphology, reproduction, ecological response, biochemical quality, and generalized agronomic practices.)

Mueller, K. E. *Field Problems of Tropical Rice.* International Rice Research Institute, The Philippines, 1970. (A book of more restricted scope than the title would indicate, devoted mostly to pest problems and nutritional disorders; pests of rice are described and illustrated. The book would seem of most use as a pocket-size reference for field use).

Myers, R. *Forage Plants.* No. 5 in the Biological Sciences Series, Western Illinois University., Macomb, Ill., 1967. (A concise review designed for introductory economic botany reading, discussing briefly the history, characteristics, and adaptation of the major forage species of the United States, with useful bibliography.)

Pierre, W. H., S. A. Aldrich, and W. P. Martin (editors).

Advances in Corn Production. Iowa State University Press, Ames, 1964. (Recent principles and developments.)

Pomeranz, Y., and J. A. Shellenberger. *Bread Science and Technology.* AVI Publishing Co., Westport, Conn., 1971. (The book is concerned mainly with the art of bread making, but there is some discussion on such subjects as making flour from grain and quality requirements for cereal grains; bread recipes from around the world are included.)

Purseglove, J. W. *Tropical Crops, Dicotyledons* (vols. 1 and 2). Longman Group, Ltd., London, and Wiley, New York, 1968–1972. (A thorough coverage of tropical crops in dicotyledon families.)

Purseglove, J. W. *Tropical Crops, Monocotyledons* (vols. 1 and 2). Longman Group, Ltd., London, and Wiley, New York, 1972. (A thorough discussion of tropical crops in monocotyledon families, rounding out excellent earlier volumes on dicotyledons.)

Quisenberry, K. S., and L. P. Reitz (editors). *Wheat and Wheat Improvement.* American Society of Agronomy, Agronomy Monograph 13, Madison, Wisc., 1967. (A comprehensive review of this particular crop, by a number of authorities).

Scott, W. O., and S. R. Aldrich. *Modern Soybean Production.* The Farm Quarterly, Cincinnati, Ohio, 1970. (Growth pattern, variety selection, planting and caring for the soybean; the soybean's future for food and feed).

Sprague, G. F. (editor). *Corn and Corn Improvement.* Academic Press, New York, 1955. (A technical monograph under the auspices of the American Society of Agronomy.)

U.S. Department of Agriculture. *Agricultural Statistics.* (A yearly compilation of U.S. agricultural production.)

U.S. Department of Agriculture. *Grass.* Yearbook of Agriculture, 1948. (A comprehensive treatment of grass, especially in relation to agriculture.)

Wall, J. S. and W. M. Ross. (editors). *Sorghum: Production and Utilization,* AVI Publishing Co., Westport, Conn., 1970. (A major effort to compile in one volume information on the production and uses of sorghums; a worldwide overview by many authorities).

Whyte, R. O., G. Milsson-Leissner, and H. C. Trumble. *Legumes in Agriculture.* Food and Agriculture Organization, Rome, 1953. (A general monograph with information on specific countries and individual legume species.)

Woodroof, J. G. *Peanuts—Production, Processing, Products.* AVI Publishing Co., Westport, Conn., 1966. (Cultivars, planting, care, and processing are given technical coverage.)

CHAPTER 21

Food Crops:
Roots, Stems, Fruits, Nuts

The chief food staples, mainstays for feeding the world, have been reviewed in Chapter 20. This is not to say that the crop plants to be discussed in this chapter are minor, but they are in general less widely distributed, less easily preserved, and less conveniently transported than are cereal grains and legume pulses. First discussed are the root crops. As was noted in Chapter 15, lush tropical ecosystems have tended to spawn vegetatively propagated crops rather than seeded annuals. The majority of the root crops are of tropical origin, even though some, such as the potato and sweet potato, are now grown as annuals in temperate climates. Biennial root crops such as beets and onions, started from seed, are also adapted to annual cropping.

Root crops, along with the cereals (and pulses), are the world's chief source of energy (and protein); but the "vegetables" and fruits included in this chapter supply most of the nutritional supplements—the vitamins, the various trace substances, and—not least in importance—the appealing flavors dear to the gourmet. Meals would be far more dreary to most palates were it not for the wide choice of vegetables and fruits that advanced cul-

tures, in particular, have come to take for granted as part of a balanced diet. In most parts of the world today preservation techniques—drying, sugaring, canning, freezing—make these foods widely available and procurable at all times of the year. Except for the more massive root crops, then, the foods discussed in this chapter are "delicacies."

ROOT CROPS

"Root crops" is a convenient catch-all for a wide assortment of species scarcely less important as food plants than the groups discussed in the previous chapter. The common feature of root crops is their fleshy underground storage organ, often a true root, but sometimes a rhizome (tuber), corm, or bulb. Compared to cereal grains and legume seeds, food stored in fleshy roots and tubers is watery and less concentrated. Thus root crops are not generally as easily kept, transported, and marketed. Most are abundant in starch or sugar and low in protein and oil, and hence are useful "energy foods" but do not by themselves supply a balanced diet. Some consumed as fresh vegetables (such as carrots, garlic, parsnips, radish, and onions) are prized more for their flavor, vitamins, and subtle nutrient qualities than as sources of energy. Just the opposite is true of potatoes, sweet-potatoes, cassava, and sugar beets.

World gross tonnage of the major root crops is not much less than for the major cereals, although, as noted, dry nutrient content must be discounted (a potato, for example, may be 80 percent water). Annual world production of potatoes approaches 300 million tons, sweet potatoes and yams about 135 million tons, manioc or cassava almost 100 million tons. More than 200 million tons of sugar beets are harvested annually. There are no esti-mates of world production and consumption of fresh vegetables, except for onions, which total about 11 million tons.

Numerous roots and rhizomes of wild species can be consumed for food in times of emergency, as many an early explorer learned perforce. Wild-life and primitive peoples have always made use of indigenous bulbs, roots, and tubers. Some spe-cies, changed little if at all from the ancestral form, are collected or planted in limited areas today. In the Peruvian Andes, for example, a great many root plants little known in other parts of the world (*Arracacia*, *Lepidium*, *Oxalis*, *Pachyrhizus*, *Trop-aeolum*, *Ullucus*) are regularly eaten and marketed. The same is true of certain sedges (*Cyperus*), aroids (*Araceae*), and *Coleus* ("fra-fra potatoes") in Africa. But no other root crop promises to rival the potato as a staple food in temperate climates, or the cassava in tropical climates.

Potato, Solanum tuberosum, *Solanaceae*

The common potato is a member of another large and important plant family, the Solanaceae. Among its close relatives are eggplant and tomato, and in the same family is tobacco. The genus *Solanum* is a large one, comprising about 2,000 species, many of them poorly described and incom-pletely understood. Various wild *Solanum* species produce small potato-like tubers, which from time immemorial have been grubbed from the ground by the Indians. One or more of these was domesti-cated in the Andes Mountains of Bolivia and Peru, certainly prior to 200 A.D., to become the culti-vated *S. tuberosum*. The diploid chromosome com-plement in *Solanum* is 24, with *S. tuberosum* ($2n = 48$) being an autotetraploid. Triploid and pentaploid relatives are known as well. *S. tubero-sum andigena* is hypothesized as the ancestral subspecies of the cultivated potato, itself perhaps descended (at least in part) from *S. sparsipilum* or *S. stenotomum*. In typical Indian plantings in the American tropics, many "wild" potatoes flour-ish about the fields and between the planted rows. Doubtless some hybridization and introgression continues to take place today, as must have oc-curred abundantly in the evolution of *S. tubero-sum*. Archaeologists have found representations of the potato incorporated in designs on early Andean pottery. The potato is still cultivated as a staple in its Andean homeland, where, to better preserve its food value and build reserves against a poor harvest, potatoes are made into "chuño." Chuño is prepared by trampling and drying the potato during alternate freezing and thawing.

The potato was presumably first seen by a Euro-pean in 1537, when the Spanish landed in what is now Colombia. New World explorers and monks became familiar with it in the decades that fol-lowed, and it was brought back to Europe by 1570. It was cultivated throughout the Continent before 1600, and in Ireland by 1663. The cultivated potato is said to have been first introduced into North America in 1621 (presumably via Bermuda). Not until 1700, however, did the potato become extensively planted. One reason for its sudden prominence in Europe during the 1700's was that reigning sovereigns, recognizing its food potential, compelled the people by royal edict to plant it (Germany, 1744; Sweden, 1764). Especially in Ireland was the potato adopted as a mainstay food, and when the late blight disease (caused by *Phyto-phthora infestans*) wiped out the crop two years in a row in the 1840's, famine forced large-scale emigration to America. One might speculate that the Irish introduction stemmed from one narrow genetic source, and that—had the great genetic variability of the potato in its South American homeland been available—famine might have been avoided and the course of history materially changed. Even before the Irish immigrants, intro-ductions of the potato into New England from Ireland had given it the appellation "Irish" potato.

It is sometime referred to as "white potato" to distinguish it from sweet-potato (a different family entirely), but the use of this adjective ignores the many colored varieties.

It may appear strange that a plant from the New World tropics should have become so important in temperate Europe. Keep in mind, however, that in the New World the potato was a highland crop, grown at elevations in the Andes that were too cold for maize. Basically it is adapted to a cool environment, and it will not yield well when temperatures average above 21°C (70°F). Today the potato is grown chiefly in Europe (Fig. 21-1), about 225 million tons annually if Soviet production is included. A poor second is North America, with around 15 million tons, and a distant third is the potato's ancestral homeland (Latin America all together produces only about 9 million tons). The Soviet Union, Poland, and Germany are the leading potato countries. In the United States 7 to 8 million tons of potatoes is produced annually for the table, and another 5 million tons is processed (mostly for potato chips, frozen french fries, dehydrated mashed potatoes, and starch). Nearly a million tons each may be utilized for animal feed and propagation.

The food value of the potato varies, depending on variety, growth conditions, storage, and handling. Analyses have indicated its composition to be 70 to 81 percent water, 8 to 28 percent starch, 1 to 4 percent protein, with traces of minerals and other food elements. In comparison with other familiar foods, the potato is fairly economical, the cost per calorie of nutrient being roughly in the same range as bread and margarine, much cheaper than meat, but half again to three times as costly as sugar. In Europe much of the potato crop is fed to livestock, and an appreciable portion is used for fermentation or for starch for industrial purposes. In the United States cull potatoes—those not used as food because of poor appearance— are utilized for animal feed and for industrial purposes. Potato starch is nowadays less important than corn starch, but potatoes mill more easily than corn, as they do not require the preliminary soakings and separations, although final concentration by centrifugation or in starch tables is similar to the handling of corn starch. Potato starch goes chiefly into sizings for textiles and paper, and to some extent into confections and adhesives.

The potato tuber is anatomically stem, with external buds (eyes) that are able to sprout into new growth (see Chapter 7). Sections of potato with a bud (eye) are conventionally used to vegetatively propagate and maintain potato varieties, since sexual reproduction through true seed would be slower to yield and would risk changes due to genetic segregation. Potatoes saved for propagation, some 7 percent of the crop, are unfortunately termed "seed" potatoes, but of course are not true seed.

Only since 1850 have there been serious attempts to improve the potato. A century ago plantings started "running out," gradually decreasing in yield, probably because of accumulation of tuber-transmitted virus diseases. Since then new introductions have been secured from throughout the Americas and entered into a breeding program. A large germplasm collection is maintained for breeding at the USDA Experimental Station at Sturgeon Bay, Wisconsin.

Potatoes are grown in all states of the United States, but Aroostook County, Maine, is responsible for about three times as much production as the next most productive counties (Kern County, California; Bingham County, Idaho; and Suffolk County, New York). Yields under irrigation in Idaho have been phenomenal—averaging 15 tons to the acre, and some fields reaching 35 tons (some agronomists predict 50-ton, even 100-ton, yields eventually). Few food plants can boast this productivity potential! Winter and early spring potatoes come from Florida and southern California, late spring and early summer potatoes from the southern and border states, and late summer and fall potatoes from the northern half of the nation, where nearly three-fourths of the domestic crop is grown.

Sandy, well-drained loams, and soils high in organic content, are generally best for potato growing. In Idaho, the leading producing state in the United States, vital soil moisture is generally provided by irrigation. The potato is responsive to daylength; in general, long days stimulate stem elongation, and short days stimulate tuberization. Under short days with low light intensity, foliage

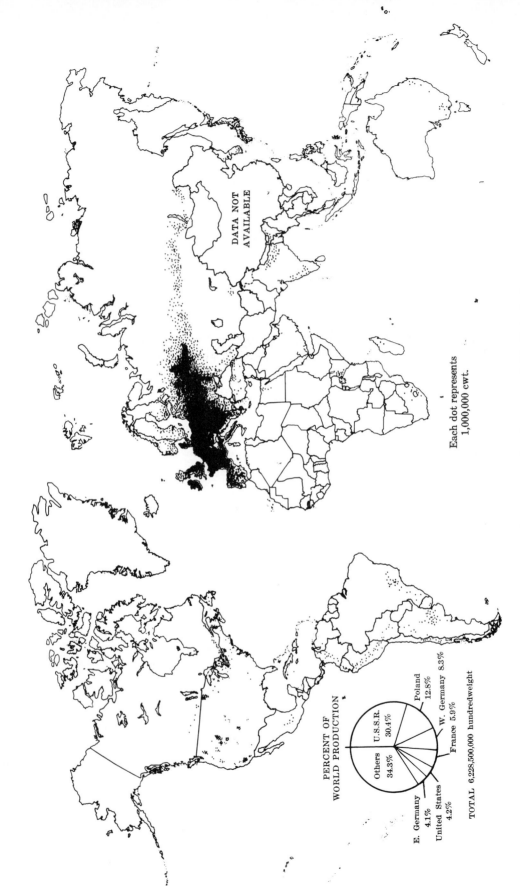

PERCENT OF
WORLD PRODUCTION

Others
34.3%

U.S.S.R.
30.4%

Poland
12.8%

W. Germany 8.3%

France 5.9%

United States
4.2%

E. Germany
4.1%

TOTAL 6,228,500,000 hundredweight

DATA NOT
AVAILABLE

Each dot represents
1,000,000 cwt.

FIGURE 21-1. *World potato production, average 1957–1961.* [*From* A Graphic Summary of World Agriculture, *USDA Misc. Publ.* 705.]

FIGURE 21-2. *Machine harvesting of potatoes from muck soil.* [*Courtesy Purdue Univ.*]

tends to be soft and more susceptible to late blight, whereas under long days and higher light intensities foliage tends to be more resistant to this blight. Thus a variety grown in Florida in winter may have considerable blight trouble, but may be relatively immune when grown in Maine in summer.

Potatoes are planted in well cultivated land, often after plowing down a legume or green manure. Ample organic matter keeps soils crumbly, beneficial to potato growth, and facilitates digging. Moderately acid soils favor scab-free tubers. As much as a ton of fertilizer per acre is used in commercial potato growing, bringing the average yields per acre to above 10 tons in the United States (triple the yield of only 30 years ago).

Seed tubers certified free of disease are the preferred planting stock. Fusarium wilt, wart, corky ringspot, and powdery scab are examples of diseases that might permanently infest the soil if uncertified tubers are used. It has been shown that the geographical source of seed, as well as storage conditions, can have an influence on yields of subsequent plantings. Moreover, varieties differ in dormancy requirements, ability to sustain repeated desprouting in storage, and response to chemical dormancy treatment. It was once common practice to disinfect tubers before sectioning for planting, but with improved seed stock this is not so critical. In some areas whole (small size) potatoes may be planted as seed, both to avoid spread of disease (on the sectioning knife) and the cost of sectioning. In the United States both sectioning and planting are mostly done mechanically.

Choice of spacing, depth of planting, and mode of cultivation are determined partly by variety and partly by local conditions. Spraying and dusting to control insects and diseases are common practices today, in keeping with intensified modern agriculture. Dozens of afflictions can affect tuber quality, and equally as many preventatives can be advocated. Weeds are controlled chiefly by cultivation, but pre-emergence herbicides are increasingly being used.

Potato tops must be eliminated to facilitate mechanical digging. This can be done chemically or physically. The digging machines then "sift" the soil from slightly below tuber depth, screening the tubers and depositing them on elevator lifts that carry them to adjacent vehicles for transportation from the field (Fig. 21-2). More than 90 percent of the crop is now harvested mechanically in the United States.

Considerable loss can arise from bruising potatoes during harvest. Care must be taken in handling and storage to avoid this injury as much as possible. Most potatoes are washed (with spray jets) to remove soil before marketing, and this, too, may encourage injury and rot. In northern producing areas a goodly portion of the crop is stored on the farm in such facilities as pit houses, where low temperatures discourage sprouting. In larger storage facilities special air circulation and ventilation may be needed. Potatoes are preferably kept at about 4°C (39°F). Above 10°C (50°F), sprouting is stimulated within a few months, although this can be inhibited by chemical treatment with maleic hydrazide. The processing quality of potatoes intended for french fries and potato chips is improved if they are held for some time (from a few days to a couple of weeks) at 15 to 21°C (60 to 70°F) before processing to avoid an undesirable brown discoloration caused by reducing sugars. This treatment is called reconditioning. During this period the concentration of reducing sugars (glucose and fructose) is decreased, partially by respiration and partially by conversion to starch.

Cassava, or Manioc, Manihot esculenta, Euphorbiaceae

Cassava, a member of the Euphorbiaceae family, is a species of the tropical lowlands, probably domesticated in eastern Brazil or northern South America. It adapts to poor soil and casual cultivation, and has consequently become a staple food in many of the poor and less-developed parts of the world. Mandioca, yuca, tapioca, sagu are other common names by which the cassava is known, and in Africa more than a half-dozen vernacular names prevail. The Latin name, *Manihot esculenta*, represents a huge complex of cultivars that taxonomists have from time to time tried to separate into distinctive species with but little success. Thus *M. utilissima*, *M. dulcis*, *M. aipi*, *M. palmata*, and other proposed species are generally regarded as synonyms of *M. esculenta*. A number of wild *Manihot* species grow in eastern South America, but it is not known whether any of these are ancestral to modern cassava cultivars; indeed, with widespread abandonment under primitive slash-

and-burn agriculture, it is difficult to distinguish between occasional escapes and wild plants.

Although one of the world's most important food staples, cassava is known in North America and Europe almost solely as tapioca, an occasional dessert. But in remote tropical areas, such as the Amazon valley, it may be almost the sole cultivated plant. World production is reported to be about 90 million tons annually, but the figure may not reflect scattered local production in remote locations. Brazil is the leading producer of cassava, with Indonesia probably second (cassava is the third most important food staple in Indonesia, after rice and maize). Cassava is not as esteemed in Indonesia as are other foods, but it is often the only crop that can be economically grown on poor, exhausted land. Cassava is also an important staple in several tropical African countries, especially in Nigeria and the Congo.

The cassava plant is a shrubby perennial that produces several swollen roots that resemble sweetpotatoes and develop from a central trunk-like stem. The palmate leaves are quite characteristic of the genus, and have been investigated as a tropical source of protein for livestock feed (the foliage is about 30 percent protein). The roots contain almost no protein or oil, but are about 30 percent starch, found chiefly in the voluminous pith. Yielding up to 20 tons of fresh roots per acre with little attention, cassava is one of the more remunerative crops. Its cultivars are vaguely grouped as "bitter" or "sweet," having respectively relatively high or low concentrations of a hydrocyanic glycoside. The poison is found mainly in the periderm and shallow cortex. Sweet cassavas require no special treatment before consumption, but bitter varieties are shredded, squeezed, cooked, and treated in other ways to rid them of the poisonous juice. There is no clear-cut demarcation between these two groups, as the hydrocyanic acid content varies with individual plants and cultivars.

Under the typical slash-and-burn agriculture of the tropics, cassava stem sections are hand planted in holes grubbed into the soil just before the rainy season. Extensive feeder roots grow quickly, making full use of the evanescent soil fertility. If all goes well, within as little as seven months a number of starchy roots can be harvested from each plant, although best yields are not obtained until about 16 months have passed. If allowed to grow for too long a time the roots become rather woody and less edible. Cassava is a heliophile, performing best in full sun. Otherwise it is broadly tolerant, with local selections adapted to a wide variety of soils and moisture conditions. In agriculturally advanced parts of the tropics, such as in Jamaica and parts of Central America, cassava fits well into rotation with maize and other cultivars.

Cassava is used in a number of ways. The whole root may be boiled; it has a sticky, "heavy" consistency, and of itself is rather tasteless. In Brazil it is usually shredded, then heated and dried, to make a meal known as farinha, eaten alone if need be, or mixed with other food and sauces. In Indonesia the roots are sectioned, dried in the sun, and later ground into meal. In the making of tapioca, an important export from Indonesia, the peeled roots are grated, the mass soaked in water for several days, kneaded, strained, dried, and then heated to partly hydrolize the starch to sugar and gel particles into "pearls" while being agitated on a grill. In Jamaica roots are gound into a mush called "bami," or formed into cakes called "casabe." In Africa boiled roots are often pounded into a thick paste called *fufu*. In many parts of the world cassava mush is fermented into a beer that is much like chicha—usually made from maize. The heat and pressure of grinding cassava serves to change or remove poisonous hydrocyanic components. Juices expressed from bitter cassava are usually saved for fermentation to beer, or may be variously treated to yield meat sauces such as "West Indian pepperpot." Starch from cassava, a minor item of commerce, is imported from the Far East into North America and Europe for sizing and adhesives, and is competitive with corn and potato starch.

Sweet potato, Ipomoea batatas, Convolvulaceae

In many respects the sweet potato (Fig. 21-3) reminds one of the cassava root. It contains about 30 percent solids (mostly starch, but some sugar, and not inconsiderable minerals), concentrated especially in the xylem parenchyma. Yields as much as 20 tons to the acre are obtained. The

FIGURE 21-3. *The edible portion of the sweet potato is the tuberous root.* [*Photograph by J. C. Allen & Son.*]

sweet potato is adapted to tropical lowlands, and cultivars are propagated vegetatively. The plant is actually a trailing vine of the morning glory family (Convolvulaceae), quite unrelated to cassava. The cultivated sweet potato, a hexaploid ($2n = 90$), may have been domesticated from *I. tiliacea*, a wild species of tropical America, with which it can be crossed. *I. batatas* is a short-day plant, flowering with a photoperiod of 11 hours or less. It grows best where average temperature is not less than 24°C (75°F).

The spread of the sweet potato from its presumed ancestral home in Central America has been gradual and unspectacular. It was apparently introduced into southern Europe very early in the Colonial period. It was probably carried by navigators of the sixteenth and seventeenth centuries to the Far East, where it has since found greater favor than in its homeland or Europe (but there is evidence of the sweet potato in Polynesia in pre-Columbian times). Especially in China and Japan, the sweet potato has become an important food plant; it is very productive on the intensively managed uplands in rotation with other crops, such as grains. In many parts of the world the sweet potato is used more as a nourishing livestock feed than for human consumption. But in spite of its limited popularity, the sweet potato, whether baked or boiled, is more flavorful than most root crops, especially the moist, reddish cultivars favored in the southern United States, where they are popularly called "yams" (most northerners favor the drier, light-yellow cultivars). The sweet potato is not distinguished from the true yam (*Dioscorea* spp.) in world-wide production statistics; both together exceed 135 million tons annually. China is probably the leading sweet potato center, although the root is extenisvely grown in many other Asiatic countries, and in many Latin American and African countries. Some cultivars rarely flower, and the sweet potato has proven temperamental in breeding programs. It is often self-incompatible; genetic sterility further limits seed production. Both incompatibility and sterility are influenced by environment. Commercially it is vegetatively propagated from numerous sprouts that originate at the head of a root when it is kept moist and warm. A certain percentage of the crop is held from usage to yield new planting "slips" (see p. 178). Although the plant is a perennial, slips grow fast enough to yield roots on an annual basis. Slips are generally planted into hills or rows by hand. In the Far East whatever cultivation and weeding is needed is done by hand, as is digging of the roots about time of killing frost. In the southern United States the same type of mechanization as was developed for harvesting the potato is increasingly being used. Throughout much of the world the sweet potato is a home garden plant that never goes to market. Roots are generally dried several days (to form a protective periderm) and then stored in a cool place. Because sweet potatoes do not keep as well as potatoes, they are sometimes cut into slices and dried, especially when they are to be used for industrial purposes. In Japan and Korea the dried sweet potatoes may be ground into a meal, which in turn can be cooked for human consumption or fermented for alcohol.

Beets, Beta vulgaris, *Chenopodiaceae*

There are two major groups of beet cultivars, the gross sugar beets and the delicate table beets, best adapted to temperate climates. The former are used somewhat as livestock feed, but have been selected principally for the extraction of their high sugar content. Both are biennial, although planted and harvested as an annual. Table beets are generally consumed when young, a few weeks after planting, when the swollen roots are sweet and tender. More than a half million tons are grown annually in the United States. They are about 8 percent carbohydrate and 2 percent protein. Cultivated beets are presumed to be derived from the wild beet of northern Europe, *Beta maritima*. Close relatives are chard and the white-fleshed mangel-wurzels. The leaf of chard is eaten as a "green," which seems to have been the case also with the original cultivated beet in pre-Christian times. The modern table beet was apparently not developed as a garden plant until about the sixteenth century.

Sugar beets have much more commercial value than table beets. Through selection and breeding, sugar content has been increased from the 2–4 percent that is normal in table beets to as much as 20 percent. The annual harvest of sugar beets exceeds 220 million tons; most of the world crop is grown in Europe and the Soviet Union, but sugar beet production was increased in North America soon after the United States boycotted Cuban cane sugar (see Chapter 26). Russia, the leading producer, is followed by the United States, France, Germany, Poland, and Italy.

The sugar beet received its greatest impetus after the German chemist Marggraf proved that sugar from beets was identical with that from sugar cane. The King of Prussia became interested, and in 1802 financed the first sugar beet factory. A few years later Napoleon also became interested, for naval blockades had cut off the supply of imported sugar, and he hoped to supply his armies with sugar produced domestically. The French people never became enthusiastic about growing sugar beets, however, and after Waterloo the industry collapsed in France as the importation of cane sugar was resumed. In the United States sugar beets were first grown in the Philadelphia area in 1830, and shortly thereafter in Massachusetts, Michigan, and finally, with most success, in California. Colorado, Idaho and Michigan are prominent sugar beet producing states today.

Sugar beets are well adapted to a diversity of

soil and climatic conditions, and today are grown in many states, drill planted in early spring. New better-yielding cultivars, and new monogerm seed (which produces a single plant rather than clusters and thus saves thinning) have given substantial boosts to the industry. Machines similar to potato lifters have mechanized harvesting, and the use of by-products (tops, pulp, and molasses for cattle feed) has given the crop further impetus.

Harvested sugar beets, weighing up to 5 pounds are washed at the factory, cut with revolving knives into thin strips (cossettes), and soaked in hot water for diffusion of the sugar. Soluble contaminants are precipitated from the sugary solution with carbon dioxide and lime (clarification) and filtered out. The solution is then evaporated to a syrup, from which the sugar is crystallized by boiling under vacuum. The crystallized sugar is separated by centrifugation and further refined to make table sugar. The pulp may either be dried for stock feed or treated with dilute acid to yield commercial pectin.

Edible aroids, Araceae.

A number of genera of the *Araceae* yield edible corms or tubers—food basic to many tropical areas. Cocoyam, dasheen, eddoe, tannia, taro, yautia are a few of the more familiar common names. The two most prominent genera are *Colocasia*, the taros and dasheens; and *Xanthosoma*, the yautias and tannias. Species of these genera also serve as house plants ("elephant ears") in summer garden pools and greenhouses.

Colocasia esculenta was apparently domesticated in southeastern Asia, and had been spread throughout the Pacific and as far east as the Mediterranean in prehistoric times. It is the source for the famed poi of Hawaii. *Xanthosoma sagittifolium* probably embraces most of the edible yautias or tannias. The species is native to tropical America, but was distributed worldwide in post-Columbian times. *Alocasia*, *Amorphophallus*, and *Cyrtosperma* are other genera prominently used for food, as are a number of others under emergency conditions.

The food value of aroid corms is similar to that of the potato, being about 20 percent starch with little protein and oil. Most of the corms are boiled like potatoes, and then crushed to make "cakes." Corms can also be roasted, steamed, or fried in oil to make "chips." In the making of poi, the crushed corn is allowed to undergo a natural fermentation before final table preparation.

Most edible aroids are simply cultivated in small patches by peasant farmers. They grow best in moist locations, under light shade. Propagation is accomplished typically by means of satellite cormels or by sectioning of a corm. Yields as high as 20 tons per acre have been obtained experimentally, and under favorable conditions a crop can be harvested in as little as six months. Harvested corms can be stored for a few months if kept dry and ventilated. Cropping is mostly by hand in those tropical lands where aroids are an important crop. Accurate statistics on world production are not available, but several millions of tons of edible aroids are probably produced annually, one of the world's basic food resources.

Other important root crops

CARROT, *Daucus carota*, UMBELLIFERAE The species is a wild weed of the Umbelliferae. The cultivated variety (*sativa*) is believed to have been selected in the Near East long ago, but the root was generally accepted as human food only in recent centuries.

HORSERADISH, *Rorippa armoracia*, CRUCIFERAE A crucifer native to Europe, thi member of the mustard family has a white, carrot-shaped root that contains a pungent glucoside used mostly for flavoring.

JERUSALEM ARTICHOKE, *Helianthus tuberosus*, COMPOSITAE This relative of the sunflower has a potato-like rhizome that can be eaten boiled, pickled, or raw.

ONION, *Allium cepa*, LILIACEAE Along with its relatives—garlic, leeks, shallots and chives, all members of the genus *Allium* (Liliaceae)—the onion is an important flavoring and food plant throughout the world. The onion is a true bulb, consisting of condensed food-storage leaves that are rich in sugar and in the pungent compound allyl sulfide. It is a major crop in central and southern Europe and in the mucklands of the Great Lakes area of the United States (Fig. 21-4).

PARSNIP, *Pastinaca sativa,* UMBELLIFERAE
Another relative of the carrot, the parsnip has edible roots that turn quite sweetish after exposure to cold.

RADISH, *Raphanus sativus,* CRUCIFERAE The radish is used chiefly in the Orient, where gigantic roots are grown and often pickled in brine.

RUTABAGA, *Brassica napobrassica,* CRUCIFERAE
Another of the cabbages, the rutabaga is also known as swede, or swede turnip—a cross between the turnip and cabbage.

SALSIFY, *Tragopogon porrifolius,* COMPOSITAE
Also known as oyster plant, salsify has fleshy, parsnip-like roots that are usually consumed boiled.

TURNIP, *Brassica rapa,* CRUCIFERAE Another biennial of the cabbage group, the turnip has been long in cultivation and is probably native to Europe and central Asia. It is often used as stock feed.

FIGURE 21-4. *Hybrid onions grown on muck soil.*

YAM, *Dioscorea,* DIOSCOREACEAE The true yam is not to be confused with the sweet potato, sometimes popularly termed "yam." Mostly tropical vines with worldwide distribution, yams can produce tremendous roots weighing as much as 100 lb, and yields as much as 10 tons per acre. Many resemble potatoes and are consumed in the same fashion; in western Africa the housewife regularly pounds yam afresh to make the prized *fufu* dish. Wild yams, of which there are 600 to 800 species, were much sought as a source of steroidal sapogenins in the development of cortisone, and yams have been the source of diosgenin for birth control pills. Several species are cultivated in China and Africa for food (they contain about the same amount of starch as does the potato). Propagation is chiefly by replanting the top of a harvested root, and trellises are usually provided for the trailing vine. An 8 or 10 month growing season is required to mature a crop, making yams unsuitable for any but tropical climates.

STEM AND LEAF CROPS

Except for a few tubers (underground stems), such as the potato, stems are more important as a source of wood and fiber than as a direct source of food. A few "vegetables," such as asparagus and kohlrabi, and such specialty items as bamboo sprouts are stems that are consumed directly, but none of these is very important as a world commodity. Sugar cane, however, is internationally important. Sugar, of course, is the carbohydrate extract from the cane stem, the stem itself being chewed only incidentally in the tropics. Sugar from cane is chemically identical with that from the sugar beet. But in international importance—and, one might add, in world romance too—cane considerably outdistances the beet. World-wide cane sugar production comes to more than a half billion tons annually.

The tremendous use of plant leaves as forage for animals is, of course, an important indirect source of food for man. But some foliage, often modified (as in the onion), is much utilized directly from human food, typically as "fresh green vegetables" or in salads. Fresh foliage is quite perishable, and it is understandable that little international commerce exists in leafy vegetable and salad plants. In the United States a sizable and very specialized head lettuce industry has developed, in which the national market, at least early in the season, is serviced almost solely from intensively grown irrigated crops in southern Arizona and California. Celery, too, may be shipped widely. But on the whole, bulky and perishable leaf crops are best handled as local truck crops or preserved by canning (and sometimes freezing).

Sugar cane, Saccharum officinarum, Gramineae

Sugar cane (Fig. 21-5) is a crop of the humid tropical lowlands. A little is grown in southern Europe (Spain), a bit more in the southern continental United States (principally Florida and Louisiana), a fair amount in Africa (from a number of countries), and a great deal in South America (principally Brazil), Central America (principally Cuba and other West Indian islands, and Mexico), certain Pacific islands (especially Hawaii and the Philippines), and southern Asia (principally India and Pakistan). World production for 1970 was reported to be 581 million tons (Fig. 21-6).

Although sugar cane is identified botanically as *Saccharum officinarum*, most of the modern cultivars are derived from crosses of an original species (the *S. officinarum* "noble canes"; probably native to New Guinea) with other species. Primitive chewing types were improved first by casual selection, then by intraspecific crop breeding, and finally by interspecific crossing that has yielded the prime clones so widely planted today. "Pure" *S. officinarum* has a diploid chromosome number of 80.

Breeding work with various sugar canes began about 1885, particularly in Java. Crosses and backcrosses involving *S. spontaneum* and *S. sinense* yielded, in 1921, a markedly superior clone (P.O.J.

FIGURE 21-5. *Sugar cane, a member of the grass family, as it appears in flower in Hawaii.* [*Photograph by J. C. Allen & Son.*]

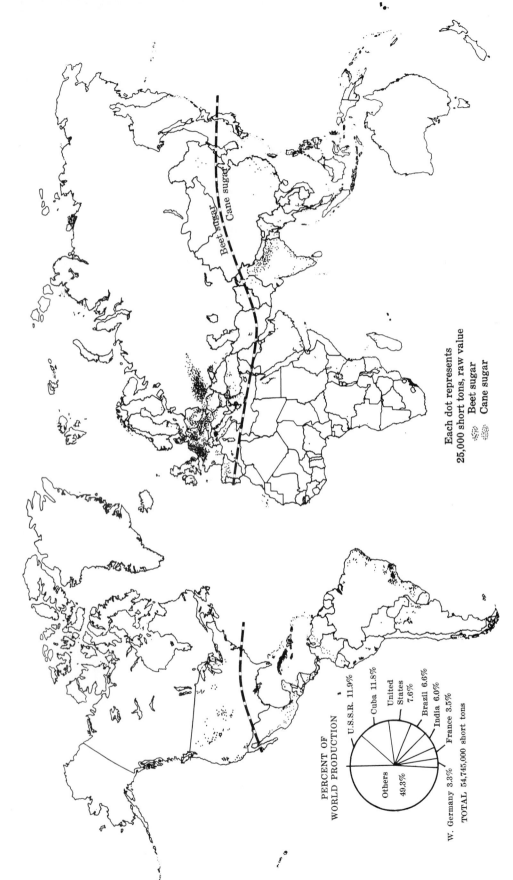

PERCENT OF
WORLD PRODUCTION

U.S.S.R. 11.9%
Cuba 11.8%
United
States
7.6%
Brazil 6.6%
India 6.0%
France 3.5%

Others
49.3%

W. Germany 3.3%
TOTAL 54,745,000 short tons

Each dot represents
25,000 short tons, raw value

Beet sugar
Cane sugar

Beet sugar
Cane sugar

FIGURE 21-6. *World sugar (centrifugal sugar) production, average 1957–1961. [From A Graphic Summary of World Agriculture,* USDA Misc. Publ. 705.]

2878) that changed sugar cane production all over the world. Since then other improved cultivars have been bred, especially in Hawaii, with S. *robustum* "blood" entering the pedigree. The impact on sugar yields has been astounding. The average annual yield of crystal sugar in Java was 2 tons per hectare in 1840, 10 tons by 1910, and 20 tons by 1940. Of course, this phenomenal rate of increase is partly due to improved growing techniques, as well as breeding for higher yields and disease resistance.

Sugar cane was apparently introduced into the Western World from India via ancient trade routes to the Mediterranean, where primitive types are still grown to some extent for chewing canes. Cane was grown in the Canary Islands before 1493, when, it is reputed, Columbus brought it from there to the New World. Soon cane was widely planted in the New World. In 1509 sugar cane was harvested in Santo Domingo, Hispaniola, and had spread to Mexico by 1520, to Brazil by 1532, and to Peru by 1533. Cultivars from Java did not supplant this low-yielding cane until late in the nineteenth century. The first sugar cane raised in what is now the continental United States was in the former French colony of Louisiana, in 1753.

In less developed lands sugar cane has long been crudely planted and tended by hand, with sections of stem merely being stuck in a hole in the ground and weeds kept reasonably in check. But for the important international trade in sugar, such inefficient practices certainly will not do. Few tropical crops have been so thoroughly acclimated to large-scale plantation agriculture with such advanced technology as has sugar cane. Soil testing and customized field preparation, machine planting, efficient cultivation or herbicidal weed control, generous fertilization, preharvest flaming, and mechanized gathering are all commonplace. Twelve to 18 months are generally required to complete a planting cycle. On rich lands subsequent crops may be obtained from rhizomes left in the soil, but, especially in Java, the crop is more often replanted like an annual to better systematize mass production.

Whether harvested by gigantic cutting machines on a modern plantation or hand harvested by machete in a distant jungle, sugar extraction fol-lows much the same procedure. Basically, the cane is brought to a centralized location where the juice is squeezed from it and concentrated. This is accomplished by crushing the stem between fluted rollers, hand or animal powered in small local operations, or part of a well-engineered system in plantation factories. The cane may be subjected to several successive expressions to extract the last bit of juice, the final one being undertaken after the nearly spent cane has been remoistened. The sugar content of cane is about 12 percent, and cane yields as high as 100 tons per acre have been recorded (average yields would be only one-tenth of this figure). The fibrous residue of the cane stem (bagasse) may be burned in the factory boilers, used for making fiberboard or paper, or composted to form an organic fertilizer.

The extracted cane juice is handled in much the same way as is juice from the sugar beet. Successive boilings concentrate the sucrose, which under primitive circumstances may merely be dried to a sticky residue (the "rapadura" of Brazil) containing small percentages of other carbohydrates, proteins, and minerals. As such it is a nourishing staple in some countries where the refinements of technology are lacking. In technological advanced societies the cane juice is treated in modern mills with steam-heated chambers and vacuum pans, and is clarified by chemical precipitation of unwanted materials; the treated extract passes through filter presses to yield a solution that consists almost entirely of sucrose. When this is further concentrated, sugar crystallizes and is usually removed from the remaining syrup by centrifugation. The liquid portion, still containing a few brownish "impurities," is molasses. Molasses is used in foods (both human and animal) or is fermented to make rum, alcohol, or vinegar. The raw crystallized sugar, about 96 percent sucrose, is refined by repeated washings and recrystallizations to satisfy the demand for a pure white sugar, completely free from brown but nourishing "impurities."

Stem and leaf "vegetables"

In this category of convenience are pigeonholed the unrelated plants of relatively minor or specialized importance whose stems or foliage are eaten as human food. For example, asparagus is juvenile stem, broccoli an inflorescence of a cole plant, cabbage a leafy head of a cole plant, celery a petiole, spinach an individual tender leaf. Most species listed are grown in temperate climates as summer annuals. Many others could be included: the bamboo shoots of the Orient; the various potherbs of the tropics, such as "Ceylon spinach," *Basella rubra;* and "New Zealand spinach," *Tetragonia expansa,* which is grown in southern temperate regions.

In technologically advanced countries, especially in North America, many commercial vegetables that don't hold their form well when preserved by canning—for example, broccoli, Brussels sprouts, cauliflower, and spinach—are now being marketed frozen. With a food freezer now in every supermarket, and a freezing compartment commonplace in the home refrigerator, preservation of perishable vegetables by freezing has become an important business. This requires varieties that freeze well. Special handling is also needed to preserve quality and flavor. Most frozen produce, after cleaning and sorting, is blanched (in commercial operations usually by steam) to inactivate enzymes that can cause deterioration or off-flavors. The blanched product may be completely packaged before quick freezing, after which it must be kept frozen until it is sold and used by the consumer.

Most "vegetables" have been known in substantially their present form for centuries. Broccoli, eggplant, kale, lettuce, and mustards (*Brassica*) are typical of species long utilized in the Old World. Spinach seems to have been introduced into China from Nepal in the seventh century. Specialized forms of cabbage (such as Brussels sprouts, kohlrabi, head cabbage) were apparently selected out of more primitive forms (like kale and collards) during the Middle Ages. The stalk of celery was not eaten until the seventeenth century, although the seed was used as a flavoring prior to that time.

The modern vegetable breeder has had to consider a number of factors not previously of great importance. For the commercial market he has had to develop good keeping qualities (that allow long-distance shipping and rough handling), standardization (size and form convenient for canning or packaging), attractive appearance (to lure the urban housewife in some distant market), and acceptable processing qualities (to permit mechanical harvesting, handling, and canning or freezing). And, as with any crop, demand for better yields, improved disease resistance, and so on, is unceasing. Cultivars of the following species are of fair commercial importance today as food plants:

ARTICHOKE, *Cynara scolymus,* COMPOSITAE The artichoke was early domesticated in the Mediterranean region, originally for foliage (kept in the dark before harvest to make it tender). In recent centuries the young flowerhead (Fig. 21-7) has been considered a delicacy; the boiled fleshy bracts are peeled off and scraped clean of soft tissue by the teeth. The fleshy base, the artichoke "heart," is also pickled and preserved by canning.

ASPARAGUS, *Asparagus officinalis,* LILIACEAE Once regarded as a medicinal plant, and not greatly changed from the wild form, asparagus spears (young stems, cut just after they emerge from underground rhizomes in spring) are today considered a delicacy. Asparagus are dioecious perennials, bearing for many years (if planted to fertile ground and if the spears are harvested in moderation). The plant has been cultivated from at least the time of Cato the Elder (234–149 B.C.), who gave directions for its growing in his *De agricultura.*

BROCCOLI, *Brassica oleracea* (Italica group), CRUCIFERAE The first selections of sprouting broccoli were probably made in Greece and Italy before the birth of Christ; the modern heading types were developed mostly in northern Europe within the last three centuries. The fleshy terminal stems and young flowers and bracts are the parts consumed. Cultural requirements are similar to those for cabbage.

BRUSSELS SPROUTS, *Brassica oleracea* (Gemnifera group), CRUCIFERAE Brussels sprouts, another of the "cabbages," are of relatively recent origin.

Leafy buds in the axils of the main leaves become the "sprouts," which are consumed much as though they were miniature cabbages. They are grown in the same way as cabbage.

CABBAGE, *Brassica oleracea* (Capitata group), CRUCIFERAE Ancestral cabbage was evidently brought into cultivation as much as 8,000 years ago, in the low coastal areas of northern Europe, from which it was introduced into the Mediterranean area and the Near East before historical times. From the primitive cabbage have arisen not only the modern heading cabbage (Fig. 21-8), but cauliflower, Brussels sprouts, kale, kohlrabi, turnip, rutabaga, and foliage mustards. In the second century B.C. Cato cited three forms of cabbage, and Pliny enumerated nine others in the first century A.D. Today heading forms of cabbage are among the most popular leaf vegetables in northern Europe and Japan; the loose-leaf "savoy" cabbages are more popular in southern Europe. Although 90 percent water, cabbage is a good source of minerals and vitamins. Sauerkraut is a Teutonic dish, made by fermenting cut cabbage in brine until abundant lactic acid is produced.

Although biennial, cabbage is easily grown from seed, heads being harvested the first year (except when grown for seed). So many cultivars have been selected that head at different times that an almost continuous cabbage harvest is possible by proper selection of varieties (a common practice in Japan). Typically, cabbage seed is started in the greenhouse or cold frame; the young plants are then transplanted to the fields, where they grow rapidly if protected from weeds. Insecticidal spraying is generally needed to prevent damage from such pests as the cabbage looper (voracious larva of the cabbage moth). Biological control through the medium of a virus disease is being investigated for the looper.

CAULIFLOWER, *Brassica oleracea* (Botrytis group), CRUCIFERAE As with broccoli, the fleshy inflorescence of cauliflower is the part consumed; it is sometimes blanched to a lighter shade by tying the outer leaves over the head. Cultural requirements are similar to those for cabbage.

CELERY, *Apium graveolens*, UMBELLIFERAE
Celery is believed to have been domesticated in the Mediterranean region. As a salad plant it is a comparative newcomer, with modern cultivars selected for a milder flavor and whiter stalk color (thus eliminating the need for mounding the stalks to blanch the petioles). In the United States celery

FIGURE 21-7. *The fleshy flower head of the artichoke.* [*Photograph by J. C. Allen & Son.*]

470

FIGURE 21-8. *The cabbage head consists of large fleshy leaves attached to a compressed stem.* [*Photograph by J. C. Allen & Son.*]

FIGURE 21-9. *Celery before and after trimming for shipment.* [*Courtesy USDA.*]

is grown mostly on moist soils in cool climates, as a summer crop in the North or a winter crop in the South (Fig. 21-9).

CHARD, *Beta vulgaris* (Cicla group), CHENOPODIACEAE   Chard is a "beet" without the enlarged root; its fleshy foliage is consumed instead. Like the table beet it is easily grown in the home garden as a source of greens.

CHINESE CABBAGE, OR PETSAI, *Brassica campestris* (Pekinensis group), CRUCIFERAE   Petsai is primarily a crop of the Far East, where there is a bewildering array of cultivars. The species is presumed to be native to eastern Asia, where it has been in cultivation since the dawn of history. It is a cousin to mustards and cabbages, and is grown like them as an annual. The central foliage, which forms into a loose head, is the part consumed.

COLLARDS AND KALE, *Brassica oleracea* (Acephala group), CRUCIFERAE   Collards and kale are primitive, leafy "cabbages," often grown as "greens" in the southern United States. They are very nutritious. In the tropics they are grown as perennials, the leaves harvested periodically or continually. Some types are extremely large leaved, and are grown as a forage crop.

ENDIVE, *Cichorium endivia*, COMPOSITAE   Endive, a relative of chicory, is indigenous to the Mediterranean area. The lower frilled leaves are consumed in the same way as lettuce.

KOHLRABI, *Brassica oleracea* (Gongylodes group), CRUCIFERAE   Kohlrabi, another member of the cabbage clan, is grown for its fleshy lower stem. Growing requirements are as for cabbage.

LETTUCE, *Lactuca sativa*, COMPOSITAE   Lettuce is known to have been grown by the Persians five centuries B.C., and was generally utilized throughout the Old World by Christian times. It has long been a favorite home garden plant, for use in salads and for "prettying up" food at the table. In recent decades it has become a valuable commercial crop, particularly in the southwestern United States, where it is now grown outdoors and shipped across the country (rather than grown in greenhouse near population centers). More than 2 million tons are produced annually in the United States (Fig. 21-10).

Lettuce cultivars have been developed in many varying forms. There are cold-tolerant varieties for mid-winter production and slow-bolting or heat-tolerant ones for summer and autumn growing. Some form heads, and some are loose-leaf. To satisfy the present commercial market, disease-resistant varieties that form a firm head with wrapper leaves for protection are mandatory.

The southwestern crop is intensively managed on irrigated land, with fertilizer practices governed exactly to produce crisp solid heads and delayed bolting (flowering). Side dressings of nitrogen are usually applied. Special machines have been developed for planting, cultivating and harvesting. Modern equipment simultaneously shapes up the planting bed for irrigation, plants the seed, and distributes starter fertilizer. Although much field harvest is still done by hand, gigantic loading

FIGURE 21-10. *Greenhouse-grown Bibb lettuce.*

machines now enter the field for packaging the cut heads. Lettuce is trimmed (dead and discolored leaves removed) in automated sheds on a production-line basis and is then carefully packed and iced for shipment (or vacuum-cooled before shipping in refrigerated cars, a more recent practice).

PARSLEY, *Petroselinum sativum,* UMBELLIFERAE
Parsely has been used as a garnish and flavoring since ancient times. Similar to its relative the carrot, it is a biennial, believed to have been domesticated in the Mediterranean area.

RHUBARB, *Rheum rhaponticum,* POLYGONACEAE
Edible rhubarb is believed to be native to Asia Minor, although a medicinal rhubarb has been used in China since several millenia B.C. Garden rhubarb is a perennial with immense leaves, the succulent petioles of which are cooked with sweetening. It grows best in cool, moist climates, and is more of a home garden plant of rural areas than an item of horticultural commerce. It is typically propagated by dividing the clumps.

SPINACH, *Spinacia oleracea,* CHENOPODIACEAE
The crisp but fleshy foliage of spinach is much esteemed as a cooked green or as a salad leaf. It is a dioecious plant that grows quickly from seed, does well in cool spring weather, but bolts under long day length. Spinach was domesticated long ago in southwestern Asia, but was not introduced into Europe until historical times.

FRUIT AND SEED "VEGETABLES"

In the grocery store you will find this group of foods among the "vegetables," although botanically they are truly fruits, or seeds. Sweet corn, fresh peas and beans properly belong in this category but have already been discussed among cereals and legumes in the previous chapter. "Vegetable" fruits are almost all grown as herbaceous annuals, although in the tropics several may be perennial. This is their chief distinction from typical grocery store fruits—the seed-bearing structures of perennial trees or shrubs, such as apples, peaches, and brambles (bananas and strawberries are also regarded as "fruit", although the plants are herbaceous perennials). Obviously, this is a less than ideal categorization.

As with the leafy vegetables, most vegetable fruits are quite watery, and are not as concentrated a source of nutrients as are the grains, legume seeds, or even root crops. They do provide some vitamins and minerals, however, and above all add variety and flavor to the diet. In the tropics a number of these vegetable fruits are grown only locally, either for home consumption or for sale in the village market. In the technically advanced societies home gardens are less common today, and these "vegetables" are now important summer truck gardening crops near metropolitan areas. In the United States most are grown in the South for shipment to the North in winter, but there is some greenhouse production in the North as a high-priced winter delicacy.

Two plant families stand out as especially important in providing the world with fruit "vegetables." These are the potato family, **Solanaceae,** which includes the tomato, eggplant, and pepper; and the gourd family, **Cucurbitaceae,** which includes various melons, squashes, cucumbers, and other **pepo** fruits (Fig. 21-11). In both families the species grow rapidly under tropical conditions, but are treated as summer annuals in temperate climates. The Solanaceae are erect or sprawling herbs, the Cucurbitaceae trailing vines usually with tendrils and rough, scratchy foliage.

Tomato, Lycopersicon esculentum, *Solanaceae*

The tomato was an obscure and little used fruit when first introduced to Europe from the northern Andes in the early days of New World exploration. At times it was even considered poisonous, as it was in France during the 1600's, when the "pomme d'amour," or love apple, as it was called, was grown strictly as an ornamental. Today the tomato is a major world food plant, the production of which comes to about 18 million tons yearly, mostly in Europe and North America. The leading producing country is the United States, followed by Italy, Spain, and the Arabic countries. Brazil, Japan, and Mexico are not far behind. Little com-

FIGURE 21-11. *The muskmelon is a berry-type fruit called a pepo, typical of the Cucurbitaceae. The rind is exocarp, the edible flesh is mesocarp (see Chapter 7). [Photograph by J. C. Allen & Son.]*

mercial production occurs in the tomato's homeland, the Andes of Ecuador and Peru, although the species is locally grown. In pre-Columbian times the tomato had spread as far north as Mexico. But today's important North American tomato crop stems from the reintroduction into the New World of selected cultivars from Italy, where the tomato had gained much favor as a flavoring for sauces. In the Andean highlands, *L. esculentum* had apparently crossed naturally with other species, such as *L. pimpinellifolium,* and possibly *L. hirsutum,* a wild species used in some modern breeding programs.

The tomato is available in a wide variety of sizes, shapes, and colors. The range is from small cocktail, or cherry, tomatoes familiar in many a home garden, to large, beefy table tomatoes weighing a pound or more. Hybrid varieties are becoming increasingly popular. Most tropical cultivars are relatively unselected, and are neither as large fruited nor as uniform as varieties grown commercially in temperate areas. In the United States, where most of the tomato crop is processed into juice, catsup, tomato paste, or canned tomatoes, the demand is for standardized types suitable for automated handling the same as with other vege-tables.

In the home garden tomato seeds are usually started indoors, or young greenhouse plants are bought and planted in late spring. They will neither grow well nor set fruit until the weather becomes quite warm. Tomatoes are often tied to stakes to save space and to keep fruits from contact with the ground, where they may become "dirty" or rot more readily. Sometimes suckers, sprouts from the leaf axils, are pinched out to regulate growth and fruiting. Greenhouse production is a large industry in northern Europe (Fig. 21-12). In large-scale commercial growing this sort of indi-vidual care is impossible. Typically, tomato plants for the field are started in special nurseries and shipped in for field planting, usually with the aid of planting machines. Much of the processed crop is seeded directly into the field. The fever pitch of mechanization in the growing and harvesting of tomatoes has already been noted in Chapter 15. In California, where nearly 180,000 acres are har-vested annually, several firms are offering auto-matic harvesters. Most of these machines cut the sprawling tomato vines at or below soil level and lift the entire plant onto a conveyor, where debris is shaken loose and the tomatoes are picked and sorted by workers riding on the platform of the machine. Such machines harvest up to 30 tons of field tomatoes per hour, and are eliminating hand picking of processed tomatoes in California.

The Cucurbits, Cucurbitaceae

The several genera and species of the Cucurbita-ceae that constitute the squashes, melons, pump-kins, and so on, can well be discussed together. They are not of great importance to world com-merce, but are mainstay foods in many primitive, tropical societies. Species of *Cucurbita* were major constitutents in the diet of aboriginal American peoples. Squashes and pumpkins were not only eaten as fresh vegetables, but the matured fruits (which store reasonably well) could, like grain, be

FIGURE 21-12. *Greenhouse-grown tomato trained to string, showing fruit borne on trusses along the stem.* [*Courtesy L. Hafen.*]

boiled or roasted after storage. The seeds of some were prized more than the fleshy parts, and were usually eaten roasted. It well may be that early domestications in the New World were grown more for the seed than for the fleshy pepo. Moreover, various gourds provided primitive household utensils.

Not all of the cucurbits of importance are of New World origin. Among the Old World contributions are the muskmelon, *Cucumis melo;* the gherkin, *C. anguria;* the teasel gourd, *C. dipsaceus;* the balsam pear, *Momordica charantia;* the watermelon, *Citrullus vulgaris;* and the sponge gourd, *Luffa cyclindrica* (used as a scouring pad). All of these have been introduced into the New World, and some have naturalized widely throughout the tropics. The bottle gourd, *Lagenaria ciceraria,* is probably native to Africa, but was spread throughout the world tropics many millenia B.C., possibly by ocean currents.

Several cultivated cucurbits of the New World have been and continue to be important. *Curcurbita pepo* includes summer squashes and pumpkins; *C. mixta,* certain winter squashes, pumpkins, and cymlings; *C. moschata,* other winter squashes and pumpkins; *C. maxima,* winter squashes and pumpkins; and *C. ficifolia,* the figleaf gourds. Except for the South American *C. maxima,* these were widely utilized in the primitive agriculture of the southern Mexican-Central American area. They may have arisen from or hybridized with several wild species also found in this region.

Cucurbits of particular importance today include the ubiquitous balsam pear, *Momordica;* the tropical American chayote, *Sechium edule,* with its soft-spined, one-seeded fruit, which is cooked like a squash; the cucumber, *Cucumis sativus,* unknown in the wild but possibly originating in northern India; the gherkin, the small, prickly fruits of which are used mostly for pickles; the muskmelon (cantaloupe) and honeydews, *Cucumis melo,* available in various forms, such as "casaba," "snake," "banana," and other types, possibly indigenous to Africa; the pumpkin, *Cucurbita pepo,* generally consumed boiled or fried after mashing (in America mostly in pies); and the squashes. The more richly flavored squashes, such as *Cucurbita maxima,* generally substitute for the "pumpkin"

in pumpkin pie; summer squashes, similar to pumpkins, are usually eaten at once; winter squashes, allowed to develop a hard rind that aids in their keeping, are generally later baked or boiled (Hubbard, Cushaw, Crookneck, etc.); the watermelon, *Citrullus lanatus,* presumably originated in Africa, where it grows wild and is often an important source of water during drought.

All of these are rank, trailing vines that are rough to the touch, although "bush" forms of squash and cucumber are being bred to facilitate harvest. Gynoecious strains of cucumber are often used in commercial plantings, and the fields kept clean of weeds with herbicides. Over most of the world, however, cucurbits are planted to well-cleared and cultivated fields and then allowed to care for themselves. At one time pumpkins were often interplanted with maize in the United States but today most cucurbits are grown as specialized truck crops, and the melons and pumpkins are shipped widely to distant markets.

Eggplant, Solanum melongena, *Solanaceae*

The eggplant is believed to have been domesticated in northeastern India, and perhaps secondarily in China. It was brought by early traders to Europe, and by the Portuguese colonists to Brazil. The eggplant is an important food plant in Japan and other parts of the Orient, where it is often pickled. It is not as popular in western societies, but is featured in French and Italian cookery. Oftentimes it is grown more as an ornamental; the oblong, round, or pear-shaped fruits are dark purple, white, or green. The species is well adapted to "tropical" growing weather, requiring a long hot season to mature fruit. In temperate climates plants are often started in hotbeds and transplanted outdoors in late spring.

Okra, Hibiscus esculentus, *Malvaceae*

Okra or gumbo is the young pod that quickly follows flowering of this *Hibiscus* plant. It is believed to have originated in northeastern Africa, and is reportedly found wild in the upper Nile area. It has been used in northern Africa for centuries, and was introduced into the New World in the slave-trading days, where it found exceptional popularity in the French and Creole cookery of Louisiana. Most okra is still grown as a home garden plant. The pod, which turns quite slimy on cooking, is most often used in stews, soups, and gravies. Handled much like okra is the pod of the unicorn plant, *Martynia proboscidea*, of the Martyniaceae, often used for pickling in the southwestern United States and Mexico.

Peppers, Capsicum, *Solanaceae*

The capsicum peppers are not to be confused with pepper spice (*Piper nigrum*), the familiar condiment of the pepper shaker. Columbus found the New World capsicum fruits quite as pungent as the pepper he sought in the Orient. Unfortunately, the misnomer "pepper" stuck for these capsicums, although better termed *chilis* to distinguish them from true pepper.

After the discovery of America, capsicum peppers were quickly introduced into other parts of the world. Some were grown as ornamentals, but the genus became best known for the pungent "spicy" types used in making "hot sauces." The less pungent "sweet" (green or red) peppers were grown but apparently not much esteemed, and are listed among the "medicinals" in many herbals.

As a food, green peppers are important for their generous vitamin C content. They are eaten in salads, as hors d'oeuvres, or stuffed. When allowed to ripen until the fruit is red, one cultivar is known as "pimento" (used chiefly for stuffing olives, mixing into cheeses, or in cooking and processing). Paprika, cayenne pepper, and chili powder all come from grinding the dried ripe fruit of certain capsicums. Paprika is used largely as a colorful garnish, and is produced mostly in Spain. Most chili powders include several other flavorings, such as oregano, garlic, onion, and capsicum. Capsicum peppers still find their greatest use in Latin America. The highly pungent types are much favored for making the "hot" sauces so familiar south of the border. The pungent ingredient of capsicum is capaicin, a volatile phenol similar to vanillin; capaicin is so potent that it is said to be detectible to the taste in a dilution of one part in a million.

Some of the strongly pungent capsicums, es-

pecially those grown in Africa, are still used medicinally. They are said to be useful for rheumatism, neuritis inflammations, and diarrhea. And the modern florist, at Christmas season, grows a dwarf form of *Capsicum annuum* as an ornamental.

Exact classification of the capsicum pepper is quite confused. Three species, *C. annuum*, *C. frutescens*, and *C. chinense* probably account for most of the cultivated forms. But many "species" have been described on the basis of such varying characteristics as fruit shape and size. Remains of chili 9,000 years old have been recovered from Mexican caves. The diploid number of chromosomes found in all cultivated species is 24, and sterility barriers between species appear well developed. Capsicums require warm, moist weather and a long growing season to be productive, and considerable hand labor is required for their tending.

FRUIT CROPS

Fruits of perennial plants have been used as food since before the dawn of civilization, and are perhaps more popular than their nutritional quality justifies. No doubt ancestral primates gathered wild fruits as a not inconsiderable part of their food supply (Table 21-1). Symbolic of man's early admiration of fruit is the Garden of Eden scene and its tempting fruit. Fruit crops today, considered together, compare favorably with the world's staple agricultural crops on a tonnage basis.

In spite of their popularity, fruits in general are relatively unimportant to man's sustenance (banana and coconut the exceptions). Most fruits are quite juicy (watery) and have low concentrations of nutrients, but some are excellent sources of vitamin C. Like many "vegetables," however, they do add much appreciated variety and flavor to the diet. Indeed, most fruits have been selected to titillate the palate, and commonly serve as a dessert or special treat.

A few fruits are more or less local staples. This is true of the banana and the coconut in parts of the tropics; avocados have been a prime source of nutrition in sections of the New World, and dates have been gathered in the Old World since time immemorial.

Table 21-1 WILD PLANTS OF THE UNITED STATES, FRUIT OF WHICH CAN BE USED FOR FOOD.

Amelanchier spp., Juneberry; Shadbush—raw, cooked or dried.

Annona glabra, Pond Apple—raw or made into jelly.

Arctostaphylos spp., Manzanita; Bearberry—raw, cooked or dried and ground; also jelly, sirup or cider.

Ardisia escallonioides, Marlberry

Asimina triloba, Papaw—raw or cooked.

Chiogenes hispidula, Creeping Snowberry; Birchberry.

Chrysobalanus icaco, Cocoplum—raw or cooked.

Chrysobalanus pallidus, Gopher Apple.

Chrysophyllum oliviforme, Satinleaf.

Coccoloba uvifera, Seagrape—raw or as jelly.

Cornus spp., Dogwood, Bunchberry—raw or cooked.

Crataegus spp., Haw—raw, cooked or dried.

Diospyros virginiana, Persimmon—raw, cooked or dried and ground as meal.

Elaeagnus commutata, Silverberry—raw or cooked.

Empetrum nigrum, Crowberry—raw, cooked or dried.

Ficus aurea and *F. laevigata*, Wild Fig.

Fragaria spp., Strawberry—raw or cooked.

Gaylussacia spp., Huckleberry—raw or cooked.

Gleditsia triacanthos, Honey Locust—fruit pulp raw; pods fermented for beer.

Grossularia spp., Gooseberry—unripe fruit cooked; ripe fruit raw or cooked.

Lonicera spp., Honeysuckle—raw, cooked or dried.

Lycium spp., Wolfberry—raw or boiled.

Malus spp., Crabapple—raw, cooked or preserved.

Mitchella repens, Partridgeberry

Morus rubra, Red Mulberry—raw, cooked, preserved.

Nyssa spp., Ogeechee Lime: Tupelo—preserved.

Parthenocissus quinquefolia, Virginia Creeper.

Passiflora incarnata, Maypop—raw or as jelly.

Photinia salicifolia, Christmasberry—raw, roasted or dried and ground for mush.

Physalis spp., Groundcherry—raw or cooked.

Podophyllum peltatum, May Apple—raw or cooked.

Prosopis spp., Mesquite; raw, or ripe pods boiled for sirup, or dried and ground for meal or used for beverage, fresh or fermented.

Prunus spp., Cherry, Plum—raw, cooked or dried.

Reynosia septentrionalis, Darling Plum—raw or cooked.

Rhamnus spp., Buckthorn—some species cathartic.

Rhus spp.; red-fruited are not poisonous; avoid *R. vernix* and other species with white fruit.

Ribes spp., Currant—raw, cooked or dried.

Rosa spp., Rose—raw, stewed or as jelly.

Roystonea elata, Royal Palm—raw, boiled for juice, sweetened.

Rubus spp., Blackberry, Dewberry and Raspberry—raw, cooked or preserved.

Sambucus spp., Elderberry—raw, cooked or dried, or made into jelly or wine.

Shepherdia spp., Buffaloberry—raw, cooked or dried.

Smilacina spp., False Solomonseal

Vaccinium spp., Blueberry, Cranberry, Deerberry, Whortleberry—raw, cooked or dried.

Viburnum spp., Black Haw—raw, cooked or dried.

Vitis spp., Grape—raw, cooked or dried.

Yucca spp., Yucca; Spanish Bayonet—unripe cooked; ripe raw, cooked or made into paste.

Source: From Julia F. Morton, *Econ. Botany* **17**(4): 319–330, October–December 1963.

Most fruits are preferred eaten raw. They turn quite perishable once the climacteric (Chapter 8) is reached. Thus fruits have traditionally been preserved through a variety of treatments, such as drying in the sun, cooking in sugary syrups (candied fruits and jellies), "fumigating" with sulfur dioxide (prunes, maraschino cherries), and of course by canning and freezing. Technically advanced parts of the world now enjoy rapid transportation, and increasingly fruits can be marketed fresh even though grown far from the market. A few, such as the coconut and citrus, with something of a protective rind, handle and ship reasonably well. Unfortunately, many of the most delicious fruits, such as the tropical cherimoya (*Anonna*) and mangosteen (*Garcinia*), are very delicate and can hardly be transported efficiently even by air.

In the constant search for better shipping cultivars, flavor is often sacrificed, as with the banana and the avocado, of which the best-tasting tree-ripened cultivars never reach temperate zone markets. Even within a limited market area, poor shipping cultivars, such as the Belle of Georgia peach (thought by many to be the best tasting peach ever discovered), have had to give way to more marketable ones. Unfortunately, most fruits do not adapt well to freezing; they lose their form and substance upon thawing, and consequently this modern technique for providing distant markets with tasty perishables is not of great service so far as fruits are concerned.

Since a fruit is the normal reproductive structure of flowering plants, it is no cause for wonder that an almost unlimited selection of fruits occurs in the wild, from which man has chosen domesticates. Indeed, particularly in the tropics, a wide assortment of wild species is still important to local populations. For example, the wayfarer in Brazil escapes from the heat by stopping under the umbu tree (*Spondias*), not only for its shade but to refresh himself on the tart fruit. In temperate climates the collecting of wild fruits and berries, once commonplace in rural societies, has now largely given way to mass marketing of specially grown cultivars. Even so, it is surprising how relatively little changed most fruits are from the ancestral types, compared to the tremendous change that domestication has wrought on such food staples as the cereals. Over the ages new forms have arisen more from chance crossings or mutations than from genetic design. Breeding programs have dealt mostly with improving characteristics already pretty well established in earlier selections; the reasons are obvious if one considers the time necessary to undertake large-scale selective crossing with long-lived perennials.

Tropical fruits especially offer an exceedingly wide and diverse assortment. The numerically fewer temperate fruits are perhaps better known and more thoroughly investigated as horticultural crops. However, very specialized industries have developed in the temperate zone to grow and market there such tropical fruits as the banana, pineapple, and citrus. The volume of citrus and of bananas entering world commerce exceeds apple production: that of citrus being some 38 million tons annually; that of bananas, about 26 million. But a temperate fruit, the grape, leads all others in total production (more than 50 million tons annually), largely because of its use in making wine. The coconut, at some 30 million tons, ranks third after the grape and citrus, but because it is used mainly as a source of copra it will be discussed later in a chapter dealing with extractives. Olives, another major subtropical fruit, are grown

mainly for processing to oil, and will also be left for later discussion. Pears, peaches, apricots, plums, pineapples, dates, cherries, and figs rank next, essentially in this order of importance, among fruits of world commerce. A tremendous number of fruits are found locally in tropical markets (Tables 21-2, 21-3).

Grape, Vitis vinifera, *Vitaceae*

The grape genus is known in the wild in many species, in both the Old and New World. Of chief importance today is the "bunch" or *"vinifera"* grape from the Old World, with 38 chromosomes. In many areas of the world, however, cultivars of this species must be grafted onto rootstocks of New World species to prevent devastation by the phyl-

Table 21-2 FOOD FRUITS FAMILIAR IN VILLAGE MARKETS OF CENTRAL MEXICO.

SPECIES	COMMON NAME	FAMILY
Acrocomia mexicana	Palm fruits	Palmae
Ananas comosus	Pineapple	Bromeliaceae
Annona cherimola	Cherimoya	Annonaceae
Annona reticulata	Custard apple	Annonaceae
Bumelia laetevirens		Sapotaceae
Calocarpum mammosum	Mamey sapote	Sapotaceae
Carica papaya	Papaya	Caricaceae
Casimiroa edulis	Sapote, white	Rutaceae
Citrus aurantifolia	Lime	Rutaceae
Citrus paradisi	Grapefruit	Rutaceae
Citrus reticulata	Tangerine	Rutaceae
Citrus sinensis	Orange	Rutaceae
Diospyros ebenaster	Sapote, black	Ebenaceae
Hylocereus undatus	Cactus fruit	Cactaceae
Mangifera indica	Mango	Anacardiaceae
Manilkara zapotilla	Sapote	Sapotaceae
Musa paradisiaca sapientum	Banana	Musaceae
Passiflora ligularis	Passion fruit	Passifloraceae
Persea americana	Avocado	Lauraceae
Physalis ixocarpa	Physalis	Solanaceae
Pouteria campechiano	Sapote, yellow	Sapotaceae
Psidium guajava	Guava	Myrtaceae
Punica granatum	Pomegranate	Punicaceae
Pyrus communis	Pear	Rosaceae
Malus pumila	Apple	Rosaceae
Tamarindus indica	Tamarind	Leguminosae

Source: From Whitaker and Cutler, *Econ. Botany* **20**(1): 6–16, January–February 1966.

Table 21-3 FRUITS COMMONLY GROWN IN INDIA.

SPECIES	COMMON NAME	FAMILY
Achras zapota	Sapota	Sapotaceae
Ananas comosus	Pineapple	Bromeliaceae
Annona squamosa	Sweetsop	Annonaceae
Carica papaya	Papaya	Caricaceae
Citrullus lanatus	Watermelon	Cucurbitaceae
Citrus paradisi	Grapefruit	Rutaceae
Citrus reticulata	Mandarin	Rutaceae
Citrus sinensis	Sweet orange	Rutaceae
Cucumis melo	Muskmelon	Cucurbitaceae
Eriobotrya japonica	Loquat	Rosaceae
Ficus carica	Fig	Moraceae
Grewia asiatica	Phalsa	Tiliaceae
Litchi chinensis	Litchi	Sapindaceae
Malus pumila	Apple	Rosaceae
Mangifera indica	Mango	Anacardiaceae
Musa paradisiaca	Banana	Musaceae
Phoenix dactylifera	Date	Palmae
Prunus armeniaca	Apricot	Rosaceae
Prunus domestica	Plum	Rosaceae
Prunus persica	Peach	Rosaceae
Psidium guajava	Guava	Myrtaceae
Punica granatum	Pomegranate	Punicaceae
Syzygium cumini	Jambolan	Myrtaceae
Vitis vinifera	Grape	Vitaceae

Source: From Maheshwari and Tandon, *Econ. Botany* 13(3): 205–242, July–September 1959.

loxera root louse. Europe produces about two-thirds of the world grape production, which amounts to some 52 million tons annually (principally from Italy, France, and Spain). As much as 90 percent of the European crop is used for wine. About a sixth of the crop is dried for raisins.

The wine grape (*V. vinifera*) has long been domesticated. Wild vines are found in Europe, the Near East, and northern India, and fossil imprints of grape leaves have been found in France and Italy that date from before the Quaternary. Grape seeds have been found with the earliest human remains in south-central Europe, and ancient Egyptian inscriptions indicate that the grape was widely grown by 2375 B.C. The Bible mentions the grape as having been planted by Noah; apparently it was introduced into Palestine from the North in about 5000 B.C. Wine was a popular beverage in the Holy Land during Biblical times, and is cited in practically every book of the Bible. Concoctions that include wine have also served as medicinals,

pickling media (as for olives and vegetables), and even for dyeing of clothes. Grapes in summer and raisins in winter were familiar foods. In general, the vineyard was a symbol of wealth and plenty.

The grape was equally as important in ancient Greece. Herodotus (484–425 B.C.) mentions the tremendous commerce in wine sent to Egypt. During the hegemony of Rome terraced vineyards were widely established. Preserved records include instructions for the care of vineyards—clearing away stones, hoeing, manuring, pruning, primitive disease protection, and provisions for irrigation. Today *V. vinifera* remains important throughout Greece (and nearby Turkey), where grafted cuttings of select cultivars are generally planted about six feet apart (1200 vines per acre) in holes punched in the soil. Most of southern Greece is phylloxera-free, but New World rootstocks are generally used as a precaution in newly planted vineyards. Vines are generally pruned to remain low, and are staked rather than trellised, or sometimes simply left to trail on the ground. Since hand labor is rather cheap, girdling (removal of a ring of bark below a grape cluster into which food from the leaves above is to be directed) or its equivalent is not uncommon. Most cultivation is still done by hand.

In North America, centuries before Columbus, Indians were gathering wild grapes of such species as the fox grape (*V. labrusca*), the riverbank grape (*V. riparia*), and the scuppernong (*V. rotundifolia*). The Norse name Vinland lends credence to the claim that North America was discovered by Norsemen. Early settlers were impressed by the profusion of wild grapes, but for 200 years attempts to introduce and grow the more select *V. vinifera* from Europe failed. During these attempts there may have been some natural crossing between *vinifera* stock and the native American species. In 1852, E. W. Bull of Concord, Massachusetts exhibited a seedling grape, which he named "Concord," that was to remake the grape growing industry in the New World. "Concord" came from a seed borne on a fox grape vine, but growing near a catawba in the garden; it may represent a natural cross, or a mutation. In any event, it was far superior to cultivars then available, and even today Concord is still the leading "juice" grape in the northeastern growing areas.

The greatest flowering of viticulture in America, however, awaited the settlement of California. The earliest grapes planted in California were apparently brought from Spain through Mexico by the Jesuit Fathers during the 1600's. The several California missions all had thriving vineyards before 1800. The Mission grape, of obscure ancestry, was widely introduced and is still much grown in California. From 1850 on, with viticulture flourishing, many varieties of *V. vinifera* were introduced from France and other parts of Europe. During the 1960's "French hybrid" grapes (crosses of *V. vinifera* with such American species as *V. rupestris* and *V. lincecumii*, made by French hybridizers) were introduced into the midwestern United States with some success, reviving interest in grape growing there.

In the dry Central Valley of California, where phylloxera and mildew diseases do not have to be contended with, very efficient grape growing is possible on large irrigated ranches. Select varieties are planted at the rate of 400 to 600 vines to the acre, trained to standardized trellises, with weeding and irrigation carefully controlled. Recently there has even been chemical management of fruiting (gibberellin sprays, properly timed, can substitute in large measure for the tedious hand girdling to yield especially full clusters of large grapes for the fresh fruit market). Although grapes must be hand pruned and harvested, irrigation and other growing procedures are increasingly being automated. Mechanical harvesting is past the experimental stage.

Grapes, perhaps more than many fruits, do their best only under intensive care. Although there are many varieties nowadays, including crosses of *V. vinifera* with American species that provide suitable cultivars for most regions, the grape does require a long summer and winters mild enough to avoid winterkill (temperatures generally above $-18°C$ ($0°F$)). Grapes reach best quality when the weather is bright and dry; the sugar content and the various subtle flavorings are then at their peak. Grapes are propagated chiefly by cuttings, and the vines perform best on well-drained soils. As has been noted, the tendency is to graft scions of *V. vinifera* onto phylloxera-resistant rootstocks of American origin. Most modern varieties are self-fertile, but some yield finer fruit when cross-pollinated with other varieties. Grapes flower and fruit on new wood (Fig. 21-13). In the United States canes are generally pruned back to two or four paired branches trained to a wire trellis, on which about 30 buds are allowed to mature into new shoots, each of which may bear two or three bunches of grapes. Depending upon cultivar, use, and market conditions there may be thinning to provide especially select fruit. Wine making is discussed in Chapter 22.

Banana, Musa *spp., Musaceae*

The taxonomy of the banana is confused. Most cultivated bananas are regarded as *Musa paradisiaca* var. *sapientum*, *M. sapientum*, or *M. cavendishii*. A related species, *M. textilis*, abacá, is grown for the petiole fiber, not the fruit. Most of the fruit marketed until recent years has been of the 'Gros Michel' cultivar, the "Poya banana" (so called for its discoverer, Jean Francois Pouyat, a Jamaican planter who came upon this find in 1836 while

FIGURE 21-13. *Abundant fruiting in the grape.* [Courtesy Garth Cahoon.]

strolling through a banana farm in Martinique). Over a century later the world still does not know how the triploid 'Gros Michel' happened to originate. But the 'Gros Michel,' being reasonably fla-

FIGURE 21-14. *The fruiting stalk of the banana consists of many clusters called hands. Each female flower of the basal end of the stem develops into a fruit—the "finger." Male flowers are produced toward the apex of the stem.* [Photograph by J. C. Allen & Son.]

vorful, of good size, and especially durable for shipping, long supplied temperate zone people the world over with the only bananas reaching their stores. Since 1963 new cultivars, including 'Valery' and 'Lacatan,' have gradually replaced "Big Mike" (as the 'Gros Michel' is known) in importance, because of their resistance to the dread Panama disease. The small, thin-skinned "lady finger" bananas, generally thought to be more flavorful, will not stand shipping to distant markets. Another cultivar, the plantain, bears a rather tasteless, large green-skinned or red-skinned fruit that is usually baked, boiled, or fried rather than eaten raw.

The banana is believed to have been first domesticated in Southeast Asia: it has been known in India for more than two millennia. Although commercial banana production has now gravitated mainly to the Western Hemisphere, additional breeding stock is sought in the Malayan area that may yield valuable genetic strains. The banana had been introduced from the Far East into Africa by Arabian traders before historical times; it was encountered there when that continent was first explored in the fifteenth century. The name "banana" apparently became first applied to the fruit in Sierra Leone.

No doubt the banana was introduced into tropical America soon after Columbus' first voyage. Today Latin America accounts for about two-thirds of the recorded world production of bananas, totaling approximately 26 million tons. In the Far East and Africa some banana growing is for export, but much of the crop is consumed locally (in some places the banana being a staple in the diet). Brazil leads the world in banana production, followed by India, Ecuador, Honduras, and Thailand.

Although the banana plant grows as much as 30 feet tall, botanically it is an herb, without woody tissue. It is propagated from a fleshy rhizome that produces a cluster of gigantic leaves, each about 10 feet long; in less than a year a central, pendulous inflorescence forms at the top and, upon fruiting, terminates the plant's life (Fig. 21-14). Commercial bananas have sterile flowers, which develop unfertilized and yield fruit lacking seeds. Thus perpetuation depends entirely upon "suckers" arising from the rhizome.

Most dwellings in the humid tropics have a few banana "trees" constantly in the dooryard, an aid to local subsistence. But the banana is also a remunerative cash crop in the steamy tropical lowlands. On good land it can yield more than 10 tons to the acre annually, with reasonably little attention. Large-scale growing is generally of the shifting agriculture pattern; forest is felled, the vegetation burned or allowed to rot, and rhizome sections planted by hand. Efficient plantations schedule plantings to permit cutting the year round. Bananas are harvested green, since they are quite perishable when ripe (yellow). Experience has indicated just the proper stage for cutting, so that the green bananas, loaded onto refrigerated steamers for transport to temperate zone markets, ripen in just a few days when brought to room temperature in the food store. Unfortunately, the flavor of such bananas is no match for that of tree-ripened fruit in the tropics.

Early in the century banana fruit companies in the American tropics developed into very large, thoroughly organized corporate empires, with extensive plantation lands in diverse countries (often including whole villages) and privately owned railroad and steamship lines for bringing the fruit to market. Of recent years these "banana empires" have retrenched because of the growth of nationalism, and there is increasing tendency to buy fruit from native growers. Even the Indians living in isolated villages, such as in southern Panama and western Ecuador, may bring their green bananas by canoe to larger trading boats that ply the coast on regular schedules to carry produce to the larger ports.

Large land clearing and leveling machinery may be used on well-capitalized plantations to prepare planting sites. Cultivation to restrain weeds until the bananas form a tight canopy is done mostly by hand. Harvesting is also a hand operation: a worker with a knife on a long pole lops the banana plant, settling the fruit cluster onto the shoulder of a second man, who is to carry it, and then severs the inflorescence from the rest of the plant. The carrier may take the bunch to a mule, small rail car, canoe, or other means of transportation for delivery to a major port. There the bananas are carefully loaded into air-conditioned ship holds by means of canvas conveyors; more recently there has been some individual packaging in cardboard cartons before shipping. During the long haul to market the temperature in the ship's hold is maintained at 14°C (57°F), the ideal temperature for preventing premature ripening. A few days at 21°C (70°F) causes the bananas to turn from green to yellow and the starch to hydrolize to fruit sugars. The ripened banana is a wholesome and fairly well balanced source of nutrition, containing various minerals and vitamins, as well as a high complement of carbohydrate with some oil and a little protein.

Citrus, *Rutaceae*

In the genus *Citrus* are the many familiar citrus fruits, of which the orange, tangerine, lemon, lime, and grapefruit are perhaps the most familiar. Oranges are classed as *C. sinensis*; lemons, *C. limon*; limes, *C. aurantifolia*; mandarin orange, or tangerine, *C. reticulata*; grapefruits, *C. paradisi*; citrons, *C. medica*; pomelos, pummelos, or "shaddocks" (after a Captain Shaddock), *C. grandis*; and sour oranges, *C. aurantium*. There are several other species that are seldom used for food. All except the grapefruit (which apparently arose from hybridization between the orange and the pomelo during the 1700's in the West Indies) seem native to Southeast Asia, where the more useful fruit types were selected by primitive man before historical times. Undoubtedly, a number of hybridizations are represented. All species have a diploid chromosome complement of 18.

Most citrus is grown today in tropical and subtropical America, although there is considerable production in southern Europe, Japan, North Africa, and the Near East. Oranges and tangerines are the most popular of citrus fruits, totaling about 30 million tons annually, about four times the production of all other citrus fruits combined. The leading producing areas are the subtropical parts of the United States, Brazil, Japan, Spain, Italy, Argentina, Mexico, and Morocco. Grapefruits find their market almost solely in the United States. Lemons, limes, and other citrus fruits (such as sour orange) are grown mostly in Italy and the United

States. World-wide, recorded citrus production amounts to almost 40 million tons annually.

Citrus fruits are borne on relatively small wiry trees, many of which apparently depend upon root mycorrhizae. Most are temperamental about soils, frequently suffering from trace-element deficiency, undue alkalinity (Texas–California), nematode attack, and similar afflictions. In the United States they do well on the sandy soils of south-central Florida, and in sunny southern California and southern Texas. In these areas highly mechanized orcharding is practiced, often with irrigation, pest control, protection against winter frost, and selection of improved cultivars. Intensive growing is also practiced in Japan, where the mandarin orange, or tangerine, is much esteemed. In less advanced parts of the tropics, however, citrus growing is rather casual, with occasional trees of doubtful origin planted in the home garden. In western Africa, where citrus should do well, fruit may be harvested only from chance trees that get little if any attention.

The citrus fruit, technically a **hesperidium,** is an excellent source of vitamin C and various fruit acids. It is approximately 40 to 45 percent juice, 20 to 40 percent rind, 20 to 35 percent pulp and seeds. Chemically it is 86 to 92 percent water, 5 to 8 percent sugars, 1 to 2 percent pectin, with smaller quantities of acids, protein, essential oils, and minerals. The rind is full of oil glands, and it is the rind of the sour orange that is the part esteemed for production of an essential oil flavoring.

Although the citron was known in the Mediterranean area before Christian times, the western world became acquainted with the more esteemed citrus fruits only during the Middle Ages, when Arabic traders brought them from the East. The Crusades acquainted the northern Europeans with citrus fruits, and the Portuguese and Spanish explorers quickly introduced citrus into the New World. About this same time the sweet orange was introduced to Europe, apparently from India. Citrus was brought into Florida before 1565, and a number of "wild" orange groves (of relatively unpalatable types) were growing in Florida when it became a state in 1821. These wild groves were topworked with buds of superior strains, and Florida was thus started on the road toward leadership in the citrus industry. As with the grape, citrus was introduced into California through the Spanish missions.

The 'Washington Navel' orange sparked the growth of the orange industry in California. In 1869 a missionary, one Reverend Schneider, discovered the navel orange in Baia, Brazil, where it probably had originated as a mutation decades earlier. Budded trees were sent to the United States Department of Agriculture in 1870, and propagations from them were sent to Riverside, California. Two trees planted in the yard of a Mrs. Luther Tibbetts started the 'Baia Navel' (later the 'Washington Navel') orange on its way.

In 1862, near Lakeland, Florida, a William Hancock purchased a farm from one Mrs. Rushing, who had set out three seedling grapefruit trees. One of these happened to bear seedless fruit, and thus originated the 'Marsh Seedless' strain of grapefruit, named after C. M. Marsh, a nurseryman who eventually purchased budwood from this cultivar. Today this variety and selections from it are widely grown in Florida, Texas, California, and in various other parts of the world. Also arising by chance in Florida were the pink-meat grapefruits. A bud sport was found on a Marsh seedless tree near Oneco, Florida in 1913; it became the Thompson Seedless grapefruit, named after the Thompson groves where it was found. A new bud mutation from this stock was discovered at McAllen, Texas, in 1929, and became "Ruby," the first patented grapefruit cultivar, and the one most grown in Texas today.

Although the Temple orange does not keep well, it is one of the more flavorful varieties. The Temple originated in Winter Park, Florida, from trees grafted with budwood obtained in Jamaica. The cultivar may be the result of a cross between a tangerine and a sweet orange (a tangor). A "new" citrus was discovered in Puerto Rico in 1956, and was named the chironja. This is said to combine the better characteristics of the orange and the grapefruit, and peel like a tangerine; it may have originated as a natural cross between the grapefruit and the orange, or between the orange and the tangelo (itself a cross between the tangerine and the grapefruit).

Select cultivars of citrus are usually budded on sour orange stock that is started from seed. Trees are planted about 25 feet apart, often cover cropped or green-manured, sprayed for insects and disease, and irrigated in appropriate fashion. Precautions taken against soil problems and frost have already been mentioned. Occasional serious winter freezings have been disastrous in the southern growing areas of the United States.

Citrus is much used for fresh consumption, since it ships rather well to distant markets. The fruit is generally allowed to ripen almost fully on the tree, except for lemons and limes, which are picked green. Many oranges are not deeply orange, and are artificially dyed or treated with gas before being marketed. The demand for preserved citrus juice is increasing. In the past most juice was preserved by canning, but frozen concentrate (prepared by boiling the juice in vacuum pans and then freezing the concentrate) has captured and revolutionized the industry in the United States. More than half the Florida orange and grapefruit crop is now processed. This has led to intensive study of possible uses for citrus by-products.

The need for disposing of peel, pulp, and seed after juice extraction, has resulted in the use of dried citrus pulp for cattle feed and for production of a molasses that can also be used in stock feed. The processing residues (about 50 percent of the fruit) are shredded in hammermills, limed (which precipitates and coagulates the pectin, causing cells to break down further), cured, and expressed. The expressed liquor is evaporated to yield oils and a molasses liquor (nearly 50 percent sugar), and the press cake is dried separately to an 8 percent moisture content for cattle feed. The essential oils from the expressed liquor, primarily terpenes, are used for flavoring, perfumery, pharmaceuticals, and soap. It is possible also to produce alcohol by fermenting the molasses or to grow yeast (for feed) on these residues. An oil that resembles olive oil can be extracted from the seeds. Sometimes pectin is extracted from the rinds, used chiefly in jellies and confectionary products. Of course, the fruit juices themselves can be fermented to vinegars and liqueurs or be utilized as a source of acetic acid.

Apple, Malus pumila, Rosaceae

Malus pumila (*M. sylvestris*, *M. communis*, *Pyrus malus*) embraces most of the cultivated fruit ap-

ples. Additionally there are many wild species in the genus. The apple was among the first plants sought in the wild; it was gradually adapted to orchards in the northern Mesopotamian center of domestication. Apparently the ancestral apple was indigenous to the Caucasus mountains of western Asia, where wild forms still grow. In the New World the genus is represented by a confusing group of indigenous crabapples, which, along with introduced forms, serve as attractive ornamentals.

World-wide apple production comes to about 22 million tons annually. Europe is the leading apple growing continent (much of the crop is consumed as a mild alcoholic cider), and North America is a rather distant second. France, Italy, and Germany are the leading European apple producers; the United States and Japan are about equally as important in their respective parts of the world. Other than the grape, there is no temperate zone fruit more important than the apple.

The apple fruit lends itself quite well to handling and keeping, in spite of having only a thin skin. A ripe apple is approximately 84 percent water, 11 percent sugar, 1 percent fiber, with fractional percentages of protein, fat, mineral, and fruit acids. It is generally harvested fully ripe, because immature apples ripen poorly after removal from the tree. On the other hand, apples left too long on the tree turn somewhat mealy because of the incipient decomposition of the pectic portions of the cell wall. Ripe fruit kept in cool, properly ventilated storage space will keep for months. A recent development, controlled atmosphere storage (see Chapter 3) permits year-round marketing of apples, which fruit only in summer. A goodly portion of the apple crop is converted into preserves, such as apple butter and jelly, canned apple sauce, cider, and cider vinegar.

Authorities estimate that there are nearly 2,000 named cultivars of the apple today. Not many have become important commercially; it is estimated that in the United States only about two dozen cultivars are much grown. Interestingly, of the two dozen only one, the Cortland, is a product of directed plant breeding. Of the remaining 23, which include many of the best-liked cultivars, most originated as chance seedlings, and some from bud mutations; a few are of unknown origin. The 'Rhode Island Greening' apple, for example, is said to stem from a seedling tree raised by a Mr. Green at his tavern near Newport, Rhode Island, from which admiring travelers obtained scions after tasting the delicious fruit that this tree bore. Other well-known cultivars that seem to have been chance seedlings include 'Delicious,' 'Golden Delicious,' 'Jonathan,' and 'York Imperial.'

The apple is a long-lived, spreading tree, which may remain productive for a century, though becoming gnarled and broken. Apple trees for commercial orchards are propagated by grafting select cultivars onto seedling rootstock or clonally propagated "dwarfing" rootstocks. Different varieties have distinctive fruiting habits, some flowering and bearing abundantly only in alternate years, and a number requiring cross-fertilization from a different cultivar to set fruit adequately. Pollination is accomplished chiefly by bees. The apple fruit is botanically a **pome** (a fruit having an inner cartilaginous core surrounded by fleshy tissue).

Unfortunately, the apple is attacked by many insect and fungus pests. This may not have been of any concern in rural areas, where most apples were probably consumed at home, but for modern marketing blemished fruits become unsalable. Thus the orchardist must undertake a complicated and unceasing series of pesticide sprayings, beginning before buds first appear and ending only just before the apple is harvested. With a half dozen or more such treatments required, apple growers have had to automate lavishly to keep production costs down. Elaborate mist blowers and efficient modern pesticides are commonplace in the orchards of the United States. Not only are such familiar pests as the codling moth controlled by spraying, but chemicals are utilized to thin fruit where too abundant (Fig. 21-15) or to hold ripening fruit on the tree longer (preventing bruising due to preharvest drop).

Pruning of the orchard, to encourage vigorous bearing wood, and to shape trees for ease in harvesting, does not lend itself well to mechanization. European growers are especially adept at training fruit trees to shapes that fit the limited area that can be accorded them. Fortunately, the North American orchardist can handle pruning during winter, when the trees are dormant. If care is taken in pruning to avoid stubs and tearing of the bark, and if hygienic techniques are followed (including the painting of large wounds with a dressing), canker diseases are generally prevented.

One of the most laborious jobs in fruit production is harvesting, still done mostly by hand. Harvesting of apples for processing may eventually become mechanized. One experimental portable harvesting machine has a special catching frame covered with foam plastic onto which apples fall as the tree is shaken by another mechanical device. Other labor-saving approaches are now used, including dwarfed trees for easier harvesting, direct field harvest into 30-bushel pallet boxes, and automatic grading and sorting devices.

The apple is adapted to cold climates, and is quite hardy even where winter temperatures are considerably below freezing. By the same token the apple requires winter-conditioning in order to fruit; it will not last in southern locations where winters are mild. The favorite apple growing locations are to the leeward of bodies of water, or in rolling country where frost does not accumulate in low-lying pockets; thus early flowering is inhibited, and freezing that would ruin the year's crop is avoided. In the United States, Washington, Oregon, and California are leading apple states of the West; New York, Michigan, and Virginia lead in the East. Nearly half the production is consumed as fresh fruit, and much of the remainder is processed for apple juice (sweet cider), which in turn may be fermented to an alcoholic product (hard cider) or by controlled bacterial inoculation yield acetic acid (vinegar). Pectin is an important by-product from the cider mills.

Other fruits

The four fruits that have been reviewed serve as examples of the temperate and tropical types. Quite obviously scores more could be discussed that are of equal interest, and many of them almost as important in world commerce. Brief treatments of a few of those that are fairly important follow.

ADDITIONAL TEMPERATE FRUITS

Apricot, Prunus armeniaca, *Rosaceae.* The apricot was probably native to China, but was

FIGURE 21-15. *Chemical thinning is a standard practice in apple growing. The left side of the tree was thinned; the right side was not. Overbearing may cause broken limbs, small, poorly colored fruit, and may lead to alternate bearing in some varieties. Apples are thinned by materials that burn off the blossoms or cause abortion of the embryo. [Courtesy Purdue Univ.]*

early brought to the Near East, and into California during the eighteenth century. It is a close relative of the peach, but more tender, and is cultivated in much the same way. Southern and central Europe and California account for most of the world production, which exceeds 1 million tons annually. Much of the crop is preserved by sun drying.

Blueberry, Vaccinium *spp., Ericaceae.* The blueberry (Fig. 21-16) occurs in "low-bush" and "high-bush" species, both wild and cultivated, in cultivars ranging up to the hexaploid. Considerable wild fruit is still harvested, especially in New England and eastern Canada. Commercial plantings are mostly on acid soils from Michigan into New England, and in Washington and Oregon. Blueberries have been domesticated from wild forms mostly within the last half century. The larger, high-bush types are used in orchards, now supporting a thriving small fruit industry. The greatest cost is the hand picking of the berries, but machine harvesting is now gaining a toehold in Michigan. Wild fruit is harvested with a specially designed rake that combs the upright stems, gathering many leaves and twigs as well as fruit. Most of the debris is removed by winnowing. Wild blueberries are generally used for processing. Total production, chiefly from the northern United States and northeastern Canada, probably does not exceed 50,000 tons annually.

FIGURE 21-16. *The blueberry has become increasingly important, transforming many formerly worthless acid soils into valuable cropland.* [*Photograph by J. C. Allen & Son.*]

Brambles, Rubus *spp.*, *Rosaceae.* Grouped as "brambles" are the red and black raspberries, blackberries, dewberries, loganberries, and other popular types, each of these common names embracing perhaps a dozen species and botanical varieties. The plants are prickly perennial shrubs (hence the name), but individual canes are often biennial (bearing fruit the second year, then dying). The "berries" are an aggregation of separate pistils on a fleshy receptacle (*aggregate* fruit); in the raspberry the fruit separates readily from its core when picked (giving a "hollow" berry). As with blueberries, brambles are relatively recent domesticates, mostly since the Middle Ages. Appreciable amounts are still harvested from the wild, although cultivated forms are now widely grown, often as a home garden plant, but mainly in commercial orchards in Oregon and the northeastern states. In season, the fruit is consumed fresh locally, but most of the crop is preserved (especially for jellies and sauces), canned, or frozen. Perhaps 100,000 tons are processed annually in the United States. World production of raspberries is about 150,000 tons (mostly from central Europe).

Cherry, Prunus *spp.*, *Rosaceae.* Most grown are the diploid sweet cherries, *P. avium*, and the tetraploid sour or tart cherries, *P. cerasus*. Both are believed native to Asia Minor, where they were well known to the ancients. Wild sweet cherry often serves as a rootstock for grafted cultivars, as does *P. mahaleb*. Cherry orchards are handled in much the same way as apples. In the United States sweet cherries are grown in several western states, sour cherries mostly in Michigan and New York. The most popular sour cherry in the United States is the Montmorency, which originated in the Montmorency valley of France several centuries ago. World production of cherries is a sizable $1\frac{1}{2}$ million tons annually, chiefly from Europe. Italy, Germany, France, the United States, and Turkey are the leading producing countries. The cherry, plum, peach, and apricot, all species of *Prunus*, constitute the so-called **stone fruits.**

Cranberry, Vaccinium macrocarpon, *Ericaceae.* The cranberry is a vine-like relative of the blueberry, grown in acid bogs of northern states. Another recent domesticate, it is still harvested from the wild in some locations, just as it was in colonial times. It is propagated from cuttings set by hand, an expensive procedure considering the difficult bog locations where these tart-fruited vines must grow. The fruit is gathered much like the wild blueberry, by combing with special rakes. Where it is possible, the bogs are flooded to float the berries to the surface where they can be more readily gathered. Cranberries are almost entirely a North American crop, utilized chiefly to make a tart red jelly or paste consumed with fowl at the Thanksgiving and Christmas season.

Currant, Ribes sativum *and other species*, *Saxifragaceae.* Currants and gooseberries are very hardy "bush fruits" grown especially in northern Europe, but used to some extent for jellies and preserves in North America. Their planting has been discouraged in the United States because the plants are alternate host for the troublesome white pine blister rust.

Peach, Prunus persica, *Rosaceae.* The peach, in numerous cultivars, seems to have been first domesticated in China. It was brought to Persia before Christian times (hence the specific name), and then further spread during the era of Roman dominance throughout Europe. The peach is more tolerant of warmth and less tolerant of cold than is the apple; although it is hardly "subtropical," southern climates such as prevail in Georgia (the "peach state") are responsible for considerable production. The peach fruit, botanically a **drupe,** is much prized for its rich, "aromatic" flavor. Select cultivars are usually budded or grafted onto unselected peach or almond rootstock (the almond is a close relative of peach in which the seed rather than the fleshy outer part of the fruit is consumed). Only in the nineteenth century were useful cultivars, such as the Elberta, a freestone seedling from a Chinese cling peach, introduced into the United States to make peach orcharding successful. Active breeding programs have increased adaptation by incorporating winter hardiness, low chill requirements, and an extended harvest season. The world's peach crop exceeds 5 million tons annually, with Europe and North America about equally represented. The United States and Italy are the leading peach growing countries, with no little production in Japan. The **nectarine** is simply a peach without fuzz, grown especially on the Pacific coast. There are a number of cultivars.

Pear, Pyrus communis, *Rosaceae.* The pear is most important in Europe, where there are many cultivars. Total world production exceeds 7 million tons annually, mainly from Italy. The pear is believed indigenous to western Asia where it has long been cultivated. Selection of modern cultivars has been accomplished only fairly recently, mostly in northern Europe. Varieties important to North America, such as the Bartlett, were introduced from Europe (the Bartlett is known as Williams in England; it was discovered by a schoolmaster as a wilding, and brought to America about 1797). The Seckel pear "originated in America," being discovered on a tract of land just south of Philadelphia that became the property of Mr. Seckel, but the original stock had no doubt been brought from Europe. Growing requirements for the pear are much the same as for the apple, although it is less tolerant of extremes. (The pear and apple constitute the so-called **pome** fruits.) When mature, the pear fruit is quite perishable, the stage of edible ripeness coinciding with the peak of the climacteric. In the United States there is a sizable pear canning industry based on the 'Bartlett' cultivar, and in Europe a good portion of the crop is crushed for cider (perry) and wine making. A limiting factor in the United States pear industry is the bacterial disease fireblight used by *Erwinia amylovora.*

Plum, Prunus domestica, *Rosaceae.* The plum is the most extensively grown stone fruit. There are a number of species, some of them wild, and many cross freely. It is estimated that about 2,000 cultivars are grown, selected from 15 different species. Centers of origin are the Orient, Europe, and North America. With such diversity, it is no wonder that plum cultivars can be found suited to almost any climate or soil. Growing requirements are the same as for the peach. Some cultivars are self-sterile, and must be interplanted with other varieties. But other than that, plums are perhaps the least temperamental of the temperate zone orchard fruits, seeming to give at least some yield with but scant attention. The drupe fruits are quite perishable unless refrigerated. Plums are not so much eaten fresh as preserved (jellies and jams), canned, and dried to make prunes. The prune industry is especially important on the Pacific coast, where *"domestica"* plums are dried, "sweated," and "glossed" with a steamy glycerine bath of sweetened juice that produces the sterile glossy skin we associate with prunes. Europe accounts for most of the total world production of plums, about 5 million tons annually; Yugoslavia, Romania, and Germany are the leading producers. In the United States plums are grown for fresh fruit in many scattered locations, but prunes are produced chiefly in California and Oregon.

Strawberry, Fragaria *spp., Rosaceae.* One of the temperate zone's favorite fruits, and probably the most popular "berry," is the strawberry (Fig. 21-17). World production amounts to about one million tons, mostly from the United States, Lebanon, Japan, Mexico, and a host of European countries. Botanically the strawberry is not a true berry, but an aggregate fruit of separate pistils on a fleshy receptacle. *F. chiloensis,* probably native to the southern Andes and possibly the North American coastal ranges, and the native American species, *F. virginiana,* (and possibly other European species) are thought to have been hybridized in Europe after introduction to produce the modern strawberry, often designated *F.* X *ananassa.* Of all fruits discussed, the cultivars of the strawberry stand out as the result of genetic breeding

488

FIGURE 21-17. (*Left*) *The strawberry is one of the most widely adapted fruit crops.* (*Right*) *In California, plantings are made on fumigated soil, and production is stimulated with polyethylene mulch.* [*Courtesy Victor Voth.*]

programs rather than chance selections. Breeders are still making crosses of wild *F. chiloensis, F. virginiana,* and *F. ovalis,* which along with the cultivated *F.* X *ananassa* constitute the four octoploid species of strawberry. Selections have been bred that are adaptable to subtropical climates such as prevail in southern Florida, and for northern states as well. Some varieties now yield tremendous fruits as big as a plum. Most commercial varieties bear but once in the spring, but especially for the home garden there are everbearing sorts that yield almost continuously if the weather remains reasonably cool and moist through summer.

Strawberries are herbaceous perennials that produce clusters of leaves from the condensed stem or crown. Under long days, trailing stems or stolons (runners) spread from the crowns, forming new plantlets. Plantings generally become too crowded after a few years, and the strawberries diminish in size and quality. When this happens new daughter plants must be set in newly prepared beds, or the existing beds must be thinned and allowed to replenish by stolons from adjacent plants. Plants produced during summer will flower and bear fruit the following spring; in this sense the crop is a biennial, although the plants themselves are perennial. Many commercial plantings are replanted annually. In California, responsible for more than half the United States production, improved cultivars are started annually under polyethylene mulch on fumigated soil. The practices are regulated to provide the plants with enough chilling for good fruit yields but not enough to encourage

vegetative (runner) exuberance. Trickle irrigation is increasingly practiced, allowing as much as 50 percent increase in plant population. Yields can be as high as 18 tons per acre.

Several modern herbicides have been found to help control weeds in strawberry plantings, but no mechanized substitute for hand picking of the berries has yet been devised, though this problem is receiving increasing attention and experimental harvesters are under development. Because of the high cost of labor, berries reaching market tend to be relatively expensive. Much of the commercial crop is now processed into jellies, preserves, and sauces. Near urban areas growers may advertise strawberries for a nominal price on a "pick-your-own" basis. Strawberries are soft and quite perishable, but so prized and delectable is this fruit that international air shipment has become a sizable business in recent years.

ADDITIONAL TROPICAL FRUITS

Avocado, Persea americana, *Lauraceae.* The avocado, alligator pear, or aguacate is native to the American tropics, but has now been introduced world-wide. It was among the wild plants first utilized by the early Central American civilizations; in some it was a major source of nutrition. The fruit was strikingly increased in size by prehistorical selection, as archeological remains show. The avocado is still widely grown, esteemed as a nutritious food, although more a delicacy than a staple in the modern diet. The large greenish or purplish fruits contain a single large seed sur-

489

rounded by rich flesh containing as much as 30 percent digestible fat. Compared to most fruits it thus packs quite a nutritional wallop. Many local selections (some of them appreciably more tasty than the commercial avocado) are thin-skinned and ship poorly; they are consumed locally, and do not reach the large temperate zone markets. Cultivars grown in the southern United States (California and Florida) are chosen especially for their keeping and shipping qualities. Even so, once ripe the fruit is quickly perishable, and therefore expensive to market. Avocado oil extraction is suggested for some less developed countries, such as Ghana, which lacks facilities for handling fresh fruit.

Breadfruit, Artocarpus altilis, *Moraceae.* The breadfruit is a handsome East Indian tree, and its large fruit is recognized as an excellent source of food. The species was widely dispersed by Britain during her colonial era. Captain Bligh, against whom the famed mutiny took place on the *Bounty,* was on a mission to gather young breadfruit trees in Tahiti and transport them to the West Indies. The meaty fruit is about one-third starch, and is usually eaten cooked in much the same fashion as potatoes.

Cashew, Anacardium occidentale, *Anacardiaceae.* This small Brazilian tree is better known for the seed (nut) than for the fleshy receptacle (pear) that bears it (Fig. 21-18), and is used in its homeland as a fresh or preserved fruit. The cashew is discussed further under Nuts, later in this chapter.

Coconut, Cocos nucifera, *Palmae.* With world production of coconuts approaching 30 billion nuts annually, this is hardly a fruit that can be overlooked. Chief sources of the coconut are the Philippines, Indonesia, Ceylon, Malaysia, and Mexico. Most of the crop is used to produce copra, a source of oil that will be discussed in the next chapter. Both volunteer and planted coconuts provide food and other essentials for simple maritime societies throughout the tropics.

Custard Apple, Annona *spp.,* *Annonaceae.* Several Annonas, including the cherimoyas or pinhas as well as the custard apple, are flavorful tropical fruits seldom known in temperate countries. They are relatives of the wild papaw (*Asi-*

FIGURE 21-18. *Fresh cashews as harvested in Guatemala. The true fruit at the tip contains the cashew nut kernel; the fleshy pear-shaped pedicel and receptacle can be eaten fresh or preserved.* [*Courtesy USDA.*]

mina). Delicate and quickly perishable, they ship poorly even by air. The delightfully "fruity" sweet types are consumed raw, the pulp (with seeds) being spooned from the inedible rind. Other, more acid Annonas are used mostly to make drinks or flavor frozen desserts. There are relatively few selected cultivars, and often volunteer or wild trees are the source of this delicacy, which is consumed locally.

Date, Phoenix dactylifera, *Palmae.* Although perhaps relatively more important in the past than they are today, the annual production of dates is nearly two million tons, most being grown in Asia Minor and northern Africa. In the Arabic world the date has long been a prime source of subsistence for nomadic tribes, being a rich source of nutrients when dried (about 70 percent carbohydrate, 2 percent protein and 2.5 percent fat). Dates may be eaten fresh, pounded into a paste, mixed with milk products (bolstering the protein content), or fermented to arrak (once described as "the strongest and most dreadful drink ever invented"). The date palm is usually propagated by "suckers" (side shoots from the base of the trunk). Flowering begins about the fourth year, and the flowers of the pistillate (female) tree are dusted by hand with inflorescences obtained from a relatively few staminate (male) trees grown apart from the female trees. In modernized date orchards

FIGURE 21-19. *The fruit of the date palm.* [*Photograph by J. C. Allen & Son.*]

the developing fruit may be protected from birds or other damage by being "bagged" in netting. Date palms are ordinarily planted in oases, or where irrigation is possible. A single tree when mature may yield more than 100 pounds of fruit yearly, the fruit usually ripening in autumn (Fig. 21-19). The fruit must be hand picked, after which it is usually heated or treated chemically to aid in its preservation, and then allowed to ripen further to develop sugar content and precipitate astringent components.

Fig, Ficus carica, *Moraceae.* The fig's relationship with man has been long and involved. *F. carica* is native to the eastern Mediterranean region, and either wildings or escapes are found scattered throughout the Near East. Certainly the fig has been cultivated in the Holy Land for more than 5,000 years, and escapes are found in rocky crevices all along the Jordan River and around the Dead Sea. The fig is cited in Egyptian documents of the fourth dynasty (about 2700 B.C.), and it is mentioned repeatedly in both the Old and New Testaments of the Bible. Figs are self-pollinated and parthenogenetic, producing a nearly continuous sequence of fruits. But the larger, more luscious Smyrna figs (var. *smyrniaca*) bear only pistillate flowers, and so are dependent for fertilization upon the pollen of other figs. For this the wild or semicultivated caprifigs (var. *sylvestris*), possibly ancestral to many of the select cultivars, are used as a source of pollen. The caprifigs are the host for a small fig wasp that carries the pollen from the caprifig to the Smyrna; the process is known as caprification. Caprification seems to have been discovered in Greece, several centuries B.C. In ancient Greece the biological facts of caprification were not clear, but it was realized that to assure good yields caprifig stems had to be hung in the elite fig trees, often with suitable ritual and tribute to the gods. In California the Smyrna figs, termed Calimyrna there, were not a commercial success until caprifigs and fig wasps were introduced in 1899.

The fig fruit is a curious structure called a

syconium, in which unisexual flowers are borne inside a pear-shaped nearly closed receptacle. In figs that bear only pistillate flowers and require pollination, the fig wasp enters the syconium, pollinates the fruit, lays its eggs, and dies. Wasp and eggs are absorbed by the fig. The early fruits are usually larger and consumed fresh, subsequent "second crop" figs often being dried or preserved. Fig trees vary widely in size and shape, ranging from frail climbers to sturdy orchard trees about the size and shape of an apple tree. The latter, propagated by cuttings, are planted commercially. The fruit is generally gathered by hand when fully ripe and is often sun-dried (figs are typically grown in semiarid climates, where orchards must be irrigated). World production of fresh figs totals nearly a million and a half tons, with about 200,000 tons dried. Portugal, Italy, and Turkey are the major fig-producing countries. In the United States production is chiefly in California, with some importation from the Near East. In the Near East figs are utilized for stock feed as well as human food.

Guava, Psidium guajava, *Myrtaceae.* Now widely distributed and naturalized, the guava is believed native to tropical Brazil. The fruits are somewhat gritty, and are more often used to make a dessert paste or jelly than eaten fresh. The fruit is a rich source of minerals and vitamins. Most of the harvesting is done casually and by hand from relatively unselected dooryard trees, although some modern orcharding is practiced, especially in Florida.

Litchi, Litchi chinensis, *Sapindaceae.* Well known in the Far East, the Litchi has been in cultivation since time immemorial. It is a favorite fruit in China, where the species is believed to be indigenous.

Maracujá, Passiflora *spp.*, *Passifloraceae.* Known also as the passion flower, or granadilla, species of *Passiflora* yield tart, aromatic fruits about the shape of a small gourd; the fruit is usually used to make drinks or sherbet-like desserts.

Mango, Mangifera indica, *Anacardiaceae.* The mango is a large tropical tree that bears an oblique, one-seeded drupe with a distinctively aromatic outer pulp. It is apparently native to Southeast Asia, where it has been cultivated for thousands of years. Wild forms are found from the Malay Archipelago to northeast India (it is mentioned in ancient Hindu mythology). There are about forty species in the genus *Mangifera,* and over a thousand cultivars. Many are delectable, tasting like something between a peach and a pineapple; others are quite rich, and have a turpentine taste that must be cultivated to be enjoyed. The diploid chromosome number is 40, and some cultivars are apparently polyploid. Today the mango grows throughout southern Asia, the Pacific islands, and the lowlands of Central and South America. The Portuguese brought the mango from India to Brazil, and it is said that Captain Cook found the fruit abundant in Rio de Janeiro in 1768. Seeds reportedly brought from Rio de Janeiro introduced the mango into the West Indies in 1742. Introductions into Florida, both from Mexico and from India, occurred early in the nineteenth century.

The mango thrives best in climates having a humid rainy season alternating with a dry season (during which the plant flowers and fruits). The species adapts well to a wide variety of soils but should have adequate drainage. Most mango trees are seedlings, and although apomixis occurs variability is high. Selections having superior fruit, made mostly in India, are vegetatively propagated by grafting, budding, marcottage (stem layering), or inarch grafting (bringing into contact partially debarked stock and scion until they grow together—a method widely practiced in India). Plantations are fertilized, often planted to a cover crop or an intercrop until the mango trees are of bearing age, and provided with irrigation when necessary (especially for young trees being established). Harvesting generally involves shaking the fruit from the trees; the fruit is then put into baskets and taken to the market. The fruits are quite astringent until fully ripe; eating mangoes gives many people a rash on the lips, much like that caused by poison ivy (to which the mango is related). Where modern facilities are available, mangos keep well for many days in cold storage. Unfortunately, up to this time the mango has not been sufficiently important in technologically advanced markets to have been accorded the necessary experimentation on how best to keep and ripen the fruit (such as has been given the banana); consequently, much of the market fruit is inferior in taste to tree-ripened mangos locally consumed. The fruit contains as much as 20 percent sugar, plus fruit acids, minerals, and a slight bit of fat and protein. As with other fruits, surplus mangoes can be fermented to vinegar or wines and brandy.

Mangosteen, Garcinia mangostana, *Guttiferae.* This small tree from Southeast Asia yields a fruit of exquisite flavor, often termed the "queen of tropical fruits." Production in the Far East is mostly from dooryard plants, and transplantings to the New World have seldom succeeded for unknown reasons. The mangosteen is quite susceptible to cold.

Olive, Olea europea, *Oleaceae.* Like the fig, the olive was intimately involved in early civilizations in the Mediterranean area, and is frequently mentioned in the Bible. World production reaches about 7 million tons annually, mainly from the Mediterranean area (Spain, Italy, Greece, Turkey, and Morocco). Upwards of 90 percent of these olives are crushed to yield oil, which is discussed in the following chapter on extractives. Olives for table use were first introduced into California through the San Diego Mission in 1769, and the "Mission" cultivar is still a leading one. The olives are carefully hand-picked when straw-colored for green olives, and when black for ripe olives. The fruit is kept for several months in a weak salt solution, where a lactic acid fermentation takes place. The pickled fruit is then treated with sodium hydroxide, washed, and stored in dilute brine. The lactic acid fermentation is omitted for green olives.

There are hundreds of cultivars of the olive tree, a small evergreen native to western Asia. The species has rather exacting climatic requirements, since it is killed outright at temperatures below $-9°C$ (15°F), though a certain amount of winter chilling is required for floral induction. Olives, typically grown in semiarid regions, are apparently tolerant of alkaline soils. For steady and certain production irrigation is required. Select cultivars are budded onto seedling olive trees or propagated from cuttings (often layered). In California softwood cuttings are made under mist, and hardwood cuttings may be set vertically in the nursery row rather than as layers. Several decades are required for olive trees to reach full productivity. Well-

established trees are estimated in Spain to yield about one-third ton of olives per acre. In California olives are more intensively fertilized, and yields are better. Olives, like mangos and many other fruits, are prone to **alternate bearing**—a heavy fruit set one year so exhausts the tree that only a light crop is borne the next. In California this tendency is overcome to some extent by hormonal thinning. Harvesting is increasingly being done with mechanical shakers, especially for olives from which oil is to be extracted.

Papaya, Carica papaya, *Caricaceae.* The papaya or mamão is a fast-growing, short-lived dioecious tree, native to tropical America (Fig. 21-20). It was introduced into India before the seventeenth century and now grows throughout the Pacific islands, where it is more highly esteemed than in its homeland (many Americans think of it as a Hawaiian fruit). The tree propagates readily from seed, and may bear ripe fruit resembling the honeydew melon in as little as a year. The fruit is not very tasty unless eaten at just the proper stage of ripeness, one of the factors contributing to its limited usefulness outside of the area of its immediate production. It is usually consumed fresh in the same fashion as a cantaloupe, but can be squeezed for juices or be pickled, candied, made into jellies, or even cooked like squash. The fruit is 90 percent water, but is rich in vitamins A and C, and contains some carbohydrates and minerals.

A constituent called papain is a digestive agent sometimes used in meat tenderizers; consumption of the fruit is believed to be an aid to digestion and perhaps a vermicide. Papain is extracted commercially by scratching the green, immature fruits, from which a latex exudes and drips to a collecting platform. The papaya tree volunteers readily in warm humid climates, performing best in full sunlight. It is weak, subject to damage where not protected from strong winds, but tolerates a wide variety of soils. Plantations are established by clearing the land of all vegetation, planting seeds in a nursery, transplanting the young plants to the field, and keeping weeds in check (usually by hand labor in the tropics). As with dioecious cultivars, all but a few staminate (male) trees are destroyed; some cultivars have perfect (hermaphroditic) flowers. Since the life of the papaya is only a few

FIGURE 21-20. *Papaya is a popular tropical fruit and the source of the enzyme papain, which is used as a meat tenderizer.* [*Photograph by J. C. Allen & Son.*]

years, rotation planting is often practiced to ensure a continuous supply of bearing trees. Fertilizing and mulching are well rewarded by increased yields.

Pineapple, Ananas comosus (A. sativus), *Bromeliaceae.* The pineapple, also known as ananas, piña, and abacaxí, is one of the outstanding examples of domestication and plantation production of a tropical fruit. So popular has the pineapple become that approximately 3.5 million tons is produced annually, mainly in Hawaii, but also in Brazil, Thailand, Taiwan, Malaysia, Mexico, and South Africa. Yields as high as 30 tons to the acre have been obtained in Hawaii.

The pineapple is a native American species, first discovered by Europeans when Columbus landed on Guadaloupe Island in 1493. At that time the pineapple had spread throughout most of tropical America, where it was cultivated here and there by the Indians. Several essentially seedless types were noted, although the wild forms were quite seedy. This suggests that primitive Americans had already made considerable progress in selection.

Early introductions to Europe were grown in the greenhouse, mostly as a curiosity and the plaything of monarchs. But some selection and improvement was achieved, although it is not clear how much of this was transferred back to the pineapple growing regions. The pineapple was taken by the Portuguese to India in 1548 and was grown in the East Indies soon after. Breeding work was undertaken in southern Florida about 1900, and shortly after in Hawaii. Little came of this, however; most commercially produced cultivars, bearing such varietal names as Cayenne, Queen, Red Spanish, and Pernambuco, have been obtained in various other parts of the world. Red Spanish holds up well for shipping. In Hawaii the Cayenne variety is grown almost exclusively, mostly for canning. Pernambuco is preferred in South America, where it is used as fresh fruit.

The pineapple fruit grows on a central stalk from a rosette of prickly leaves. (The Cayenne variety is a nonprickly, or smooth-leafed type.) The fruit is multiple, consisting of fused individual flowers. It contains up to 15 percent sugar, various acids, vitamins and minerals, and a supposed digestive enzyme (bromelin). Plants are propagated by basal suckers or "slips" that arise from beneath the fruit (Fig. 21-21). Under the careful, automated culture accorded plantation growing in Hawaii, a sucker will bear a new pineapple head in about one year. After the first fruit is cut the plant branches to yield a second crop of two fruits, and when these are cut a further dichotomy to yield four small pineapples, which are used for cut pineapple or juice. After the third crop the field is replanted to avoid any further reduction in fruit size. In Hawaii extensive research has determined just what fertilization, trace elements, and pesticide schedules are needed to give the highest yields of quality fruit. The whole operation, including field harvesting, is almost completely automated. Although the individual pineapple heads are generally still cut by hand, portable conveyor belts carry the fruits to central loading facilities.

Pomegranate, Punica granatum, *Punicaceae.* The pomegranate is a small shrubby tree found wild in eastern Asia and naturalized in the Mediterranean area. The fruit contains many pulpy seeds, each with a red, juicy aril—an additional

494

FIGURE 21-21. *Pineapple may be propagated from slips (leafy shoots originating from axillary buds borne on the base of the fruit stalk), from the crown of the fruit, or from suckers lower down on the stem.* [*Courtesy Dole Corp.*]

covering surrounding the seed. The pomegranate is eaten more in the Old World than in the New, but is grown somewhat in California and through Latin America. There are few select cultivars, and the plants can be propagated from seeds and cuttings or by layering.

NUTS

A number of woody plants (trees and shrubs) of which the seed is the important food constituent are grouped here as "nuts." Nuts are not of great commercial importance, although nutritionally they are quite a concentrated source of food. Unlike most fruits of the previous section, they are generally rich in oil and fat, and contain moderate amounts of carbohydrate and protein. They ship and keep rather well, and can be used in much the fashion as cereal grains and seed legumes. Nuts provide relatively little of the food needed for the world's burgeoning populations. Perhaps they have been too much overlooked and will increase in importance, especially on lateritic soils of the tropics, which cannot stand continuous tillage and annual cropping. Tropical forestlands, yielding little in the way of direct food, would seem to be a logical habitat for more extensive nut cropping. Certainly one would suppose that the seeds of trees could be as important in their adapted habitat as a source of food as are the seeds of grasses and annual legumes.

Most nuts are something of a luxury; they're expensive, and mostly consumed in small amounts as sweetmeats or delicacies. But in certain times and places nuts have been used as staple foods, as the piñon nut was by the American Indian in sections of the Southwest. Indeed, the piñon nut crop is still collected from wild stands of the piñon pine rather than from cultivated orchards. In fact, nuts are more often gathered from the wild than any of the foods discussed in this chapter. Wild trees are the source of most Brazil nuts, cashews, black walnuts, certain hickory nuts and pecans, and of course coconuts, if you prefer to consider it here rather than with fruits or oil sources. Fortunately, others, such as the almond, filbert (hazelnut), and domesticated pecans have been brought into cultivation, and cultivars have been selected that yield vastly improved nuts. Usually these are grown in well-tended orchards with the aid of many of the techniques employed in fruit orchards. There would still seem to be tremendous opportunities ahead for domestication and improvement of nut crops; a handicap, of course, is the long generation time, a problem encountered with all perennial tree crops.

Almond, Prunus amygdalus (P. communis), Rosaceae

In the almond, a species closely related to *P. persica*, the peach, the seed of the drupe is utilized and the flesh discarded. The outer flesh is astringent and tough, unlike that of its near relative the peach. The two species may have had a common ancestor, and some believe the peach was developed from an almond-like progenitor through selection of types having edible flesh. There are two varieties of almond, the sweet (var. *dulcis*), and the bitter (var. *amara*). The former is the source of edible almond nuts used in confectionery; the latter (poisonous to eat because it contains prussic acid) is grown chiefly for the extraction of bitter almond oil. Almonds contain 18–20 percent protein; the "skin" surrounding the nutmeat is rich in minerals and should be consumed with the nut for its food value.

The cultivation of almond trees is similar to that of peaches. They require a subtropical climate that is nearly frost-free, similar to that of their native Mediterranean region. In the United States the center of almond production is in semiarid California; the orchards are irrigated, and harvesting is mechanized. The trees begin bearing within seven years. Planted 25 to 30 feet apart they yield about a half ton of nuts per acre at maturity. The pit or seed separates readily from the split flesh as the fruit ripens and cracks. The pit is mechanically cracked, freeing the seed, which may be bleached, or more commonly is roasted for consumption in candy, mixed nuts, and confectionery.

Almond production in the United States (mainly California) is nearly 100,000 tons (in the shell) annually; in addition, as much as 20,000 tons are imported in some years. Almond production in Italy and Spain is about the same as in the United States; Iran, Morocco, and Portugal produce minor amounts.

Brazil nut, Bertholletia excelsa, *Lecythidaceae*

Known also as the cream nut, or castanha do Pará, the Brazil nut is an important indigenous food plant of northern South America. Closely related is the Brazilian sapucaia, or paradise nut, *Lecythis* spp. Both are magnificent forest trees that grow in the Amazon Valley and bear clusters of seeds (nuts) in horny, urn-like fruits. The tip end of the fruit dehisces, and within it, clustered much like segments of an orange, are the individual brazil nuts that have become so familiar a component

of mixed nuts at temperate zone cocktail parties. The nuts contain as much as 70 percent oil. Most nuts are still collected from fallen pods in the forest. Production amounts to approximately 30,000 tons annually, most of which is exported to the United States and to Europe.

Cashew, Anacardium occidentale, *Anacardiaceae*

The cashew, known also as the marañón, or cajuil (Spanish), cajou (French), and cajú (Portuguese), is a remarkable plant of many uses (see Fig. 21-18). In a previous section it was noted that the fleshy, pear-shaped stalk, or receptacle, which bears the single-seeded fruit, is often utilized fresh or preserved, especially in South America (in India, the chief commercial source of the cashew, the receptacle is usually discarded). In addition to the receptacle and the nut, the species yields a phenolic nut-shell "oil" (90 percent anacardic acid) that is used as a medicinal, an irritant, wood protectant and a source of resin; the young leaves are sometimes used as a flavoring and as a medicinal; a yellow gum obtained as an exudate from incisions made in the bark is used mainly as an adhesive; the wood is utilized in several ways, and the bark to some extent for tanning and a yellow dye.

The cashew is native to coastal parts of northeastern Brazil, but for 400 years has been grown in many parts of the world. It is a sprawling broad-leafed evergreen, well adapted to poor soils and dry sandy locations. Various parts of the tree are quite astringent and toxic, especially the shell that surrounds the nut, which may cause painful blistering, especially before the nut is fully ripe. The tree was first introduced by the Portuguese into India as a shore tree to bind the soil; it flourished, and now grows as an escape throughout India, Ceylon, and Southeast Asia. The Portuguese also introduced the cashew into Africa, where there is considerable nut production (the unshelled nuts are generally shipped to India for processing). Nut production in India may reach 80,000 tons annually, with an additional 60,000 tons from Africa and lesser amounts from South America and various oceanic islands.

The cashew has been propagated chiefly by seed. Typically, three nuts are planted into holes spaced about 20 feet apart; only one of the three seedlings (the most vigorous) is allowed to persist. In recent years cultivars have been selected in India that are propagated by inarching, budding, and other vegetative techniques. The trees may begin bearing in as little as three years, but seldom yield profitably until seven years old. Some mature trees have been reported to yield as much as 200 pounds of nuts. Adapted as the cashew is to sterile, sandy "taboleiros" of Brazil, the trees seldom receive the benefits of any fertilizer.

Nuts are generally harvested from the ground, where they fall with the receptacle when ripe. They are dried in the sun briefly until moisture content is reduced to about 7 percent (marked by a rattling sound when the nut is shaken). At this stage they may be stored for prolonged intervals. Occasionally in India the kernels are eaten in this state, but more frequently the nut is roasted. In India this is customarily done in a shallow, open pan. The shell oil that is exuded by the roasting nuts must be absorbed by shaking the nuts in sand to avoid blistering the hands of handlers and shellers. A few factories in India have mechanized the roasting of cashew nuts and specialized in the recovery of by-product cashew shell oil. The actual shelling is done by hand, typically by women utilizing a mallet. In Africa a machine was developed for cutting into the shell to allow its easy separation, giving a more perfect, undamaged nut. The nuts are chiefly exported to North America and Europe, where they are consumed as table nuts and in confectionery. The kernel is nearly half oil, of which about 74 percent is the oleic type, and about 18 percent protein.

Hazelnut, Filbert, or Cobnut, Corylus *spp.*, Betulaceae

Species of *Corylus* are sometimes also called Turkish nuts or Barcelona nuts, with good reason; Turkey produces as much as 200,000 tons (in the shell) annually, and Spain nearly 20,000 tons. Italy, too, is a major producer, often producing more than 50,000 tons per year compared to 10,000 tons per year in the United States, mostly from Oregon.

There are native species of Corylus in North America and in Eurasia. They make an excellent understory in open woodlands in temperate climates. *C. americana* furnished wild hazelnuts to many a school boy when America was still rural, and several cultivars have been domesticated. But of greater economic importance is the larger European tree, *C. avellana.* Select varieties are propagated vegetatively, often by budding or grafting onto seedlings, which are then transplanted to the orchards. Trees first bear when about four years old, and eventually provide nut yields of nearly a half ton per acre. Nuts contain about 60 percent oil, and 13 percent protein.

Macadamia, Macadamia ternifolia, *Proteaceae*

The macadamia nut, also known as the Queensland, Australian, gympie, bush, or bopple nut, is native to Australia, from which it was introduced notably into Hawaii less than a century ago. In Hawaii it has gained favor so rapidly that it has become one of the most important orchard crops of the islands. Cultivars have been selected both in Australia and Hawaii. Of little consequence in the world food picture, macadamia nonetheless serves as an interesting example of a tropical nut industry in the making.

Macadamia trees are large broad-leafed evergreens that reach heights of 60 feet or more. The fruit is a follicle with a fleshy husk that splits to reveal a spherical seed about an inch in diameter. This seed, or nut, consists of a meaty kernel in a relatively thin enclosing shell that must be cracked to get at the kernel. Machines have been devised to remove the outer husk, after which the unshelled nuts can be kept satisfactorily for several months in dry storage. For final processing the unshelled nuts are dried to approximately 2 percent moisture in a warm draft of air. The shell is difficult to crack by hand, but in commercial production powered nut-cracking machines are used quite effectively. The shelled nuts are "frenchfried" in oil for several minutes, and after being drained are salted and vacuum-sealed in glass containers for export.

Best yields of macadamia are on well-drained lands where rainfall is generous. Harvesting consists of gathering nuts as they fall, a somewhat continuous chore that lasts for several months.

Yields of 100 pounds or more per tree per year are obtained after the trees are 15 years in age.

Pecan, Carya illinoensis, *Juglandaceae*

The pecan was highly regarded as a wild nut tree by settlers in what are now the border states and upper South of the United States. Several species of hickory (the genus *Carya*) yield edible nuts, but *C. illinoensis* is responsible for those pecan cultivars that today support a sizable orchard industry. These thin-shelled ("paper shell") varieties account for about half the domestic production (which in most years well exceeds 100,000 tons), and are grown chiefly in Georgia and Alabama. The harvest from wild and seedling trees exceeds that from select cultivars in Louisiana, Texas, and Oklahoma. There is little growing of the pecan outside of the United States, although successful introductions have been made into Australia and China.

Typical of hickory fruits, the pecan has an outer leathery husk that splits open when mature to reveal an inner shell that surrounds the two kernels. Harvest is by picking the nuts from fallen fruits. Sometimes the fruits are knocked from the trees with long poles, but since a mature pecan tree is of large forest stature, mechanical harvesting aids are not frequently feasible. The nuts are dried, or "cured," for a few weeks, and are then ready for eating without roasting or other treatment. Cracking and shelling machines handle most of today's crop, which is usually marketed shelled.

Pistachio, Pistacia vera, *Anacardiaceae*

The pistachio (or pistache), native to Western Asia, has been widely cultivated since ancient times in the Middle East. It is in the same family with the mango, the cashew, and poison ivy. There are a dozen or more species of *Pistacia*, some serving as rootstock for grafted *P. vera*, but only *P. vera* having the dehiscent fruit that yields the commercially acceptable edible nut. In India and Afghanistan nuts are still obtained from wild trees. In general, the orchard cultivars of Iran, Turkey, and Italy differ little from the wild pistachios, having been selected mainly for larger nut size. A few pistachio orchards have been established in the interior valleys of California, where the locally developed 'Kerman' cultivar yields larger, better-flavored nuts than can be imported. Chief commercial production comes from Turkey, Iran, Italy, Syria, and Afghanistan. World production is difficult to determine, but may approach 50,000 tons annually.

Since *P. vera* does not root readily from cuttings, most cultivars are propagated by budding onto seedling rootstocks. Choice of a suitable species as rootstock can result in a dwarf tree more easily reached for harvesting. Budded seedlings should be planted to the field very early, since they tend to develop a taproot that suffers in transplanting. Transplanting is usually accomplished during the dormant season (December-January). Mature trees are usually spaced about 30 feet apart; trees planted closer together for earlier yield can be thinned. The trees seem to prefer well-drained soils having adequate calcium; the tree thrives on sandy loams, and is quite tolerant of drought. The pistachio is relatively slow-growing and needs little pruning, at least with budded stock. It is one of the few dicotyledonous trees known to attain an age of more than 1,000 years.

The species is dioecious. As low a ratio as one staminate (male) tree to twenty pistillate (female) trees is considered adequate in some orchards. Diploid chromosome number for *P. vera* has been reported as both 30 and 32. Fruit buds are initiated by winter conditioning. Like the apple, the pistachio tends to bear heavily in alternate years. Fruiting begins about the fifth year, but is not really remunerative until the tree is at least 10 years old. Trees 20 years old are reported to yield 150 pounds or more of marketable nuts per tree. The fruit is a small drupe that matures in autumn, at which time the outer "pulp" slips easily from the inner shell. The latter splits longitudinally, making it relatively easy to free the edible seed within. In the Near East picking is done largely by hand. The husks are laid out on a cloth spread under the tree, and the nuts are removed from the husks by hand. These are dried in the sun. In Turkey the dried nuts are often soaked in water, then hulled on stone rollers, accompanied by winnowing.

The nuts are typically consumed roasted and

498

FIGURE 21-22. *Fruit of the English walnut, showing the outer husk and the shell-covered nut.* [*Courtesy USDA.*]

salted, or are absorbed by the confectionery trade. When harvested the nuts contain about 40 percent moisture. Dried kernels are approximately 20 percent protein, nearly 60 percent oil. The oil is approximately 70 percent oleic, 20 percent linoleic, and 8 percent palmitic.

Walnuts, Juglans *spp.*, *Juglandaceae*

There are a number of kinds of walnuts, ranging from the wild black walnut (*J. nigra*) and the butternut or white walnut (*J. cineria*) of the United States, of which there has been only limited domestication, to the English or Persian walnut (*J. regia*), which is extensively orcharded throughout southern Europe, in China, and more recently on the Pacific coast of the United States.

The black walnut is one of the outstanding timber trees of the eastern deciduous hardwood belt in the United States. It has been much decimated for timber and veneer, but some nuts are still collected from wild trees for home consumption, and in certain rural areas for commercial sale. Walnut fruits are drupes. The husk of the black walnut does not readily separate from the shell; it is usually allowed to rot away. Moreover, the husk strongly stains hands and clothing. Add to this the hard-to-crack inner shell, and it is no wonder that the black walnut nut industry has not made great commercial strides. Production of the meat is usually a sideline activity in rural areas for "pin money." The nut fragments are mostly used as confectionery flavoring, for it is difficult to remove the kernels entire.

The English walnut is produced chiefly in California and Oregon (almost 100,000 tons in the shell), France (more than 30,000 tons), and Italy (more than 20,000 tons). Significant additional production occurs in the Near East and China, for which no figures are available. *J. regia* is native to Persia. It is a handsome ornamental tree in warmer climates. In California orchards it is customarily grafted onto American walnut rootstocks. The husks fall away much more readily than those of the black walnut, leaving an attractive shell in which the English walnut is customarily marketed (Fig. 21-22). Harvesting is by collection of the fallen nuts under the trees in the orchard. Walnut meats contain about 15 percent protein as well as abundant oil.

Other nuts

The eight kinds of nuts reviewed so far by no means exhaust the possibilities. These are the chief nuts in commerce today, and serve to exemplify this group of food plants. However, many more are of local interest, especially those harvested from the wild in agriculturally less advanced parts of the world. As noted in the opening paragraphs, a number would seem to have great potential as food plants for the future, especially in tropical climates. Some of the nuts that are well known and were used at least in earlier days are: acorns (fruit of the oak tree, *Quercus*); African locust beans (*Parkia*); beechnuts (*Fagus*); chestnuts (*Castanea*); sal-nuts (*Ginkgo*); hackberry (*Celtis*); several hickories (*Carya*); oysternut (*Telfairia*); pignolia and piñon (seeds of certain pines. *Pinus;* the former European, the latter American); pili (*Canarium*, of the Far East); souari (*Caryocar*, of South America; also known as butter, paradise, or guiana-nut); *Terminalia;* wingnut (*Pterocarya*); and many others.

Selected References

Barnes, A. C. *The Sugar Cane.* Interscience (Wiley), New York, 1964. (A broad picture of the world sugar cane industry.)

Chandler, W. H. *Evergreen Orchards.* Lea & Febiger, Philadelphia, 1950. (See next reference.)

Chandler, W. H. *Deciduous Orchards.* Lea & Febiger, Philadelphia, 1959. (The two books by Chandler are companion volumes on temperate and subtropical fruit growing.)

Collins, J. L. *The Pineapple: Botany, Cultivation, and Utilization.* Interscience (Wiley), New York, 1960. (An excellent treatment.)

Condit, I. J. *The Fig.* Chronica Botanica, Waltham, Mass., 1947. (A thoroughgoing technical review of an age-old crop.)

Darrow, G. M., and others. *The Strawberry: History, Breeding and Physiology.* Holt, Rinehart and Winston, New York, 1966. (A readable review of this small fruit, involving history, breeding, and culture.)

Dawson, V. H. W., and A. Aten. *Dates—Handling, Processing and Packing.* Food and Agriculture Organization, Rome, 1962. (The United Nations takes an overall look at the date industry.)

Eck, P., and N. F. Childers (editors). *Blueberry Culture.* Rutgers University Press, 1967. (A broad compilation of information on this special crop, including its botany, breeding, and growing.)

Edmond, J. B., and G. R. Ammerman. *Sweet Potatoes: Production, Processing, Marketing.* AVI Publishing Co., Westport, Conn., 1971. (A remarkably comprehensive, factual review distilling, in understandable language, the sum total of knowledge about the sweet potato to time of publication).

Herklots, G. A. C. *Vegetables in South-East Asia.* George Allen and Unwin, Ltd., London, 1972. (Encyclopedic coverage about a subject on which there is little reference literature in the western world; the opening chapters on "principles" are followed by successive chapters on greens, cabbages, beans and peas, cucurbits, other fruits, root crops, and market vegetables).

Hulme, A. C. (editor). *The Biochemistry of Fruits and Their Products.* Academic Press, New York, 1970 (Vol. 1), 1971 (Vol. 2). (Volume 1 deals primarily with the chemistry of fruits responsible for flavor, nutrition, climacteric, and so forth; Volume 2 discusses individual fruits specifically in Part I, and processing and preservation in Part II. Of interest primarily to food scientists).

Jones, H. A., and L. K. Mann. *Onions and their Allies: Botany, Cultivation, and Utilization.* Interscience (Wiley), New York, 1963. (A monograph on the cultivated *Alliums.*)

Kefford, J. F., and B. V. Chandler. *The Chemical Constituents of Citrus Fruits.* Academic Press, New York, 1970. (This book by Austrailian authorities is supplement 2 to *Advances in Food Research,* elaborating upon the composition of citrus fruits in detail; primarily of interest to the citrus specialist).

Mortensen, E., and E. T. Bullard. *Handbook of Tropical and Subtropical Horticulture.* United States Department of State, Agency for International Development, Washington, D. C., 1964. (A manual written for the nonspecialist.)

Reed, C. A., and J. Davidson. *The Improved Nut Trees Of North America and How to Grow Them.* Devin-Adair, New York, 1958. (Semipopular presentation of a subject seldom covered comprehensively, with references.)

Reuther, W., H. J. Webber, and L. D. Batchelor (editors). *The Citrus Industry.* University of California Div. of Agricultural Sciences, Berkeley, 1967 (Vol. 1), 1968 (Vol. 2). (A monumental work on citrus. Volumes 3 and 4 of the new edition are in preparation.)

Salaman, R. N. *The History and Social Influence of the Potato.* Cambridge University Press, Cambridge, 1949. (The potato throughout its history; a delightful treatment.)

Shoemaker, J. S. *Small-fruit Culture.* (2nd ed.). Blakiston, Philadelphia, 1948. (A standard reference through the years.)

Shoemaker, J. S., and B. J. E. Teskey, *Tree Fruit Production.* Wiley, New York, 1959. (Discusses orcharding with apples, pears, peaches, cherries, plums, apricots, nectarines, quinces, and citrus.)

Simmonds, N. W. *Bananas* (Tropical Agriculture Series). Longmans, London, 1959. (An authoritative monograph.)

Singh, L. B. *The Mango: Botany, Cultivation, and Utilization.* Interscience (Wiley), New York, 1960. (A treatise on one of the most popular and luscious of tropical fruits.)

Smith, J. R. *Tree Crops.* Devin-Adair, New York, 1950. (A plea for greater use of tree fruits and nuts, popularly presented but bringing diverse information into a single volume.)

Teskey, B. J. E., and J. S. Shoemaker. *Tree Fruit Produc-*

tion (2nd ed.). AVI Publishing Co., Westport, Conn., 1972. (Comprehensive treatment brought up-to-date on the production of commercially important deciduous tree fruits—apple, pear, peach, cherry, plum, apricot and nectarine. A good reference and guide for orchard practices).

Thompson, H. C., and W. C. Kelly. *Vegetable Crops.* McGraw-Hill, New York, 1957. (An excellent text and reference for vegetable crops of temperate horticulture.)

Tressler, D. K., and M. A. Joslyn, editors. *Fruit and Vegetable Juice Processing Technology* (2nd ed.). AVI Publishing Co., Westport, Conn., 1971. (De-tailed discussion of the processing of commercially important fruit and vegetable juices; emphasis is more on the "engineering" of processing than on the plant sources).

Woodroof, J. G. *Coconuts: Production, Processing, Products.* AVI Publishing Co., Westport, Conn., 1970. (Woodroof and collaborators describe the multiple uses for coconuts and coconut products, and detail processing practices in many parts of the world. Chapters are devoted to the coconut palm, its varieties and propagation, and its growing, as well as to production of copra, oil, and so forth).

CHAPTER 22

Plant Extractives and Derivatives

This chapter is intended as an overview of the ways many plants support civilization in addition to the sustenance provided by the food plants discussed in the preceding two chapters. Direct consumption as food is, of course, the most obvious and important use for plants. But after food wants are satisfied, man turns to somewhat more sophisticated usage of plant products. Extractives and derivatives from plants supply him with the raw materials to make other products, many of them of a complicated industrial nature requiring much technical skill to fabricate—paints, automobile tires and plastics, for example. In other cases the derived products are little different from some of the foods reviewed in the previous chapter, and are rather arbitrarily discussed here rather than there—beverages, seed oils, and spices in particular.

Yet even these food-like products are processed to such an extent that they bear no resemblance to the natural plant structures from which they are derived. Or in other cases specialized components are isolated and often concentrated—perhaps useful for flavoring, but with little real nutrient value. Thus the grape was included in Chapter 21, but wine, derived by fermenting grape juice, receives mention here. There is no compelling reason for having lumped soybeans and sugars among the food plants rather than in this chapter, except that they are both commonly thought of as foods (even though in technically advanced countries the soybean is mostly processed for oil and meal, and only secondarily made into foods; while sugar is extracted from the cane stem or the beet root, which are consumed directly only in minor quantity).

Organic chemists have learned to synthesize many of the products once derived from plants, and have provided excellent substitutes for others, particularly those from limited or obscure wild plant sources. This has been very much the case with medicinals, for example. Thus a change in emphasis has been occurring in recent decades, away from wide exploratory efforts with a diversity of limited derivatives and extractives, to mass production of a few crops yielding steady, economical supply of substances the organic chemist can use to mold what modern civilization demands. Consequently, there have arisen gigantic enterprises based upon huge volumes of such things as soybean and palm oil, processed starch, and resins or essential oils obtained as by-products of pulping. The chemical industry depends even more upon the residues from ancient plants—petroleum and natural gas—than upon living plant derivatives. But as these fossil raw materials become exhausted, civilization must turn ever more to raw materials from renewable crops.

Because of obvious similarities to the food plants just discussed, beverage plants will be the first group discussed. Then will come oils, fats and waxes from vegetational sources; next spices and other products important for essential oils; then medicinals and fermentation products. Next we examine some of the blandishments man has bestowed upon himself, fumatories and masticatories, such as tobacco, betel, and psychotic drugs. It is only a step further to those extractives employed almost entirely industrially (i.e., not taken internally), such as latex products, resins, tannins, and dyes.

BEVERAGE CROPS

Even stronger than the need to satisfy hunger in man and animal is the need to satisfy thirst. Water has always been and will continue to be the prime quench for thirst. But since the beginning of civilization inquiring man has sought additives for water that might make it more tasty and zestful, or has turned to various juices and fermentations for his enjoyment as well as to satisfy his thirst.

Man's earliest "beverage" was probably juice sucked or squeezed from fruit. What with spores of wild yeast blowing everywhere about, it was only a small step further to the discovery that natural fermentation could convert these sweet juices into wines. Nor would it take a great deal of insight to discover that hard, dry grains, which must have been difficult and unpleasant to consume under primitive cooking conditions, could be softened by soaking, and that the resulting liquor would ferment to a pleasant and nutritious beer. (Sumerian tablets more than 5 millenia old describe in detail the making of several varieties of beer.) These are still the basic techniques by which

man makes his alcoholic beverages, the products of fermentation; wines and beers are among the oldest and most cherished of man's beverages.

Unfermented fruit juices, of course, contain no alcohol. Modern "soft drinks" are, essentially, synthetic fruit juices, compounded of sugar, fruit acids, and other flavorings. But more universally utilized than these are coffee and tea, in which tasteful substances (including stimulating caffeine) are diffused from plant parts steeped in water. Cocoa, or chocolate, is perhaps consumed less as a beverage these days than it is as chocolate candy. But in its ancestral American homeland it was highly esteemed as a ceremonial beverage, and is still much used in hot chocolate or chocolate sodas.

Almost any sugary or starchy substance can be fermented. Primitive tribes in various parts of the world have long used locally available substances for fermentation to alcoholic beverages—the potato in the Andes (for chicha), cassava in the tropical lowlands (manioc beers), palm toddy in the Eastern tropics, and various obscure plant materials "secretly" chosen by the shaman for ceremonial purposes. Some of these products are hallucinatory and even semipoisonous. A few secondary beverage plants can be listed, but coffee, tea, and chocolate; beers and wines must serve here as examples of beverages derived from plant sources.

Coffee, Coffea arabica, *Rubiaceae*

Unlike most of the crop species discussed in this chapter, coffee is a relatively recent domesticate, having become popular as a beverage only since the eighteenth century (although the berry may have served as a stimulant and "medicinal" much earlier in tribal cultures of the Near East). So quickly and widely has coffee become popular that today it ranks among the leading international agricultural commodities. Total world production of dry "beans" (Fig. 22-1) reaches about 4 million tons annually, more than one-fourth of which comes from Brazil. Colombia is the next most important coffee-raising country, followed by several tropical African nations (Ivory Coast, Uganda, Angola, Ethiopia, and others), by a number of

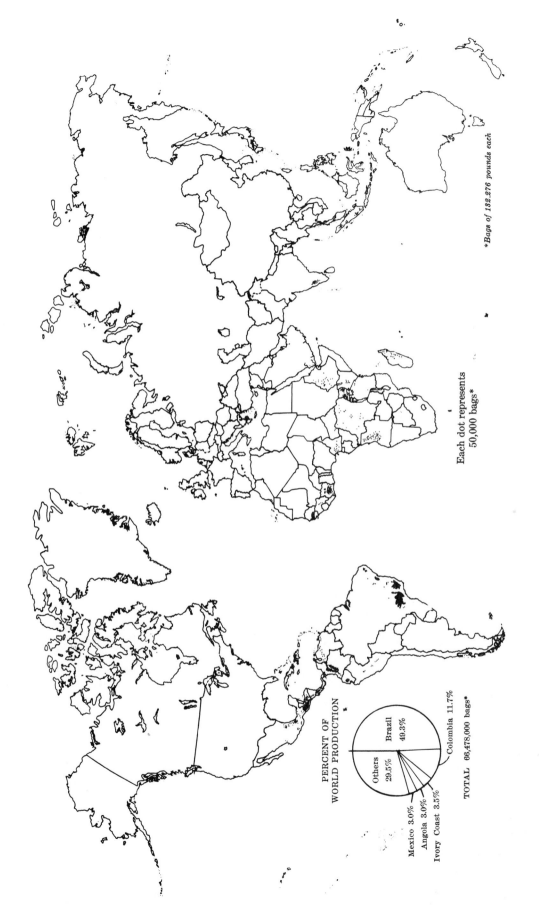

PERCENT OF
WORLD PRODUCTION

Brazil
49.3%

Colombia 11.7%

Ivory Coast 3.5%

Angola 3.0%

Mexico 3.0%

Others
29.5%

TOTAL 66,478,000 bags*

Each dot represents
50,000 bags*

*Bags of 132.276 pounds each

FIGURE 22-1. *World coffee production, average 1957–1961.* [*From* A Graphic Summary of World Agriculture, *USDA Misc. Publ. 705.*]

Central American nations (Salvador, Guatamala, and Costa Rica), and by Mexico and Indonesia.

The genus *Coffea* is a large one, numbering between 40 and 70 species. Only two of these species have attained commercial importance for the fruit or berry: *C. arabica* (accounting for about 75 percent of the coffee crop), and *C. canephora* ("robusta" coffee, mainly produced in Africa, recently found suitable for making valuable "instant" coffees, and accounting for about 25 percent of the crop). The species *C. arabica* is tetraploid with 44 chromosomes and is self-compatible; all other species are diploid with 22 chromosomes and are apparently self-incompatible.

Recent investigations leave little doubt that *C. arabica* is native to and still grows wild in the mountainous sections of Ethiopia. It is difficult to tell the escapes from indigenous plants, so little has domestication altered the coffee species. Perhaps during the Middle Ages the first coffee plants were brought from Ethiopia to Yemen, where cultivation may have begun, and these yielded the planting stock taken to Java by the Dutch in the seventeenth century and to the New World early in the eighteenth century.

A serious leaf rust disease, *Hemileia*, has accompanied coffee wherever it is grown, and native breeding stock capable of resisting the disease is today being sought in the Ethiopian homeland. The rust does not seem too troublesome on the robusta coffees in Africa, provided the trees grow in the shade of an over-story of such species as *Albizzia*. Because coffee has attained such great commercial importance within the last century, a number of elite clones have been selected in various parts of the world for vigor, yield, and degrees of resistance against disease. Selection work to provide *Hemileia* resistance has been especially investigated in Portugal, since coffee is of basic importance to the commerce of the Portuguese colony Angola.

Coffee thrives best in a rainforest habitat at moderate elevations. The southwestern Ethiopian homeland has an average annual rainfall of nearly 63 inches, with the months April to September having the heaviest rainfall. Coffee cannot stand freezing, and performs best where seasonal temperature variation is small and the average temperature is about 20°C (68°F). It grows best on deep soils, which in its native African homeland are generally volcanic. When first introduced into Brazil it was grown on low valley land, and was only later found to prosper far better at the higher elevations in Saō Paulo, on deep soils rich in iron and potassium.

In Africa the coffee trees are often grown as dooryard plants for local usage. But a number of large coffee plantations have been established, in which crop production is more efficient. Yields have gradually been increased to an average of nearly a half ton per acre (experimentally, a ton or more). Research has shown that coffee yields could be nearly doubled by mulching (usually alternate rows are mulched each year). In Africa coffee plants are usually started from seed in nurseries and transplanted to the field. In Brazil direct field planting is common, with 3 or 4 seeds sown to a hole to provide several "stems." Recommended spacing for mature multiple-stemmed trees is about 10 feet. In Africa the tendency is to shade the plantation (thus reducing bearing, which would otherwise have to be controlled by pruning; also helping inhibit rust); in Brazil plantings are generally in the open. In most tropical lands weeds are controlled by hand, and fertilization limited. The desirable flavor in coffee seems to result more from the location in which the coffee is grown than the kind or amount of care. Colombian, Blue Mountain Jamaican, and highland Central American coffees are usually considered superior to African coffees, even though the same cultivars may be planted.

Coffee berries are harvested by hand. In much of Africa and the Near East "dry" harvest is practiced: individual berries are dried in the open long enough to prevent deterioration of the pericarp. At a later time the hard seeds, commonly called beans, are separated mechanically from the pulp, often by special mobile machines circulated from village to village. "Wet" processing is more common in the Americas, the berries first being floated to separate the defective berries from the good ones. The good berries settle to the bottom of the flotation tanks, and pulping machines remove most of the fleshy outer parts. The remaining fragments of pulp are fermented for approximately a day, and the beans are then washed again. The beans are then dried in open paved areas, either on mats or in trays, and polished in special mills. After the beans have been polished they are ready for roasting and grinding.

The roasting process develops the coffee aroma; the intensity of roast varies from location to location according to the preference of the market. Roasted and ground coffee gradually loses its flavor, and eventually turns rancid, so that "fresh-roasted" coffees have particular appeal to the connoisseur. When the ground coffee bean is steeped in hot water it loses about 25 percent of its weight as soluble constituents, including caffeine, the stimulant for which coffee is noted. To varying degrees, caffeine relieves fatigue and promotes a feeling of well-being, hence the modern "coffee break."

The coffee industry has been beset by cyclic periods of glut. This has had serious consequences for countries like Brazil, which depend upon coffee for a large share of their foreign exchange. Encouraged by the opportunities that exist when coffee prices are high, growers make extensive plantings that in a few years oversupply the market and depress the price. Brazil has attempted to regulate production and demand by destroying huge surpluses of coffee from time to time or by subsidizing removal of coffee trees. With most of the good coffee land in Brazil now already planted, overproduction may not be so voluminous as in the past; but there are still ample opportunities for increasing coffee production in Africa and the Far East.

Tea, Thea (Camellia) sinensis, *Theaceae*

Tea, like coffee, is a beverage prized for the stimulative effects of the caffeine it contains. Tea is made from young leaves of the tea plant, a tropical broadleafed evergreen that may reach 40 feet in height if not pruned. It grows best in equable, moist environments where the rainfall is at least 60 inches annually and the temperature varies between 21° and 32°C (70° to 90°F). The best tea-growing locations are Assam (northeastern India) and Ceylon, which together account for half of the world production, which exceeds a million tons annually. Other important tea-

producing countries are China, Japan, Indonesia, and Kenya (Fig. 22-2).

The exact origin of the tea plant is unknown, but the art of growing tea has been practiced in China for 4,000 years, and presumably the ancestral tea plants originated there. Tea drinking was not prevalent in Europe and England (where perhaps it is now most used) until the Dutch, and eventually the English East India Company, initiated the tea trade during the 1600's. How important it became as an item of international trade is known to every American school child who has studied about the "Boston Tea Party" and the American Revolution.

Tea plants are usually started in nurseries as seedlings and then transplanted to the plantation fields. Clonal selections are increasingly being propagated vegetatively. In Ceylon average yields of about 800 pounds to the acre are said to increase to 1,500 pounds where improved cultivars are planted. Tea seeds don't preserve well, and after ripening in October and November they are usually sown immediately. Fields are generally planted on about a 4-foot spacing and in locations with ample drainage. Taller trees are sometimes maintained over the plantings for shade, and the tea plants themselves are top-pruned from an early age to give the plants a low, spreading structure that facilitates leaf picking. The tea plant is a "heavy feeder," and in the normal course of harvesting 50 or more pounds of nitrogen are lost per acre. For good production this and other nutrients must be restored through manures and fertilizers. Weeds are generally controlled by hand cultivation, and insecticide application is carefully regulated to avoid affecting salability of the tea leaves.

Women circulating through the plantation pluck the leaves by hand, removing the terminal bud and the two or three leaves immediately below it. The tea bush generally yields about 2 pounds of plucked green shoots per year. The basketfuls of green leaves are brought to a receiving station and subjected to a "withering" (fermentation) that develops taste and aroma. Leaves rich in tannin are considered superior. The withered leaves are rolled to rupture the cells, freeing juices and enzymes that play a part in further fermentation, for which the leaves are spread in cool, moist rooms

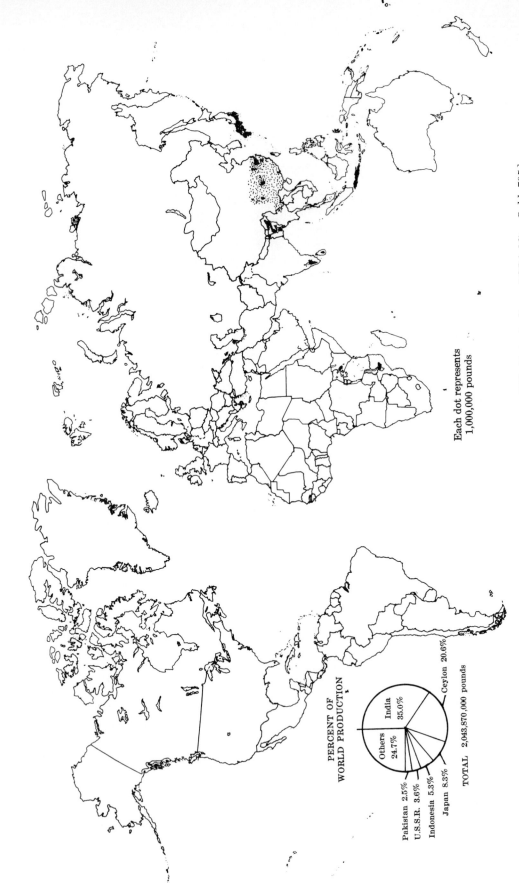

PERCENT OF
WORLD PRODUCTION

Others
24.7%

India
35.0%

Pakistan 2.5%

U.S.S.R. 3.6%

Indonesia 5.3%

Japan 8.3%

Ceylon 20.6%

TOTAL 2,043,870,000 pounds

Each dot represents
1,000,000 pounds

FIGURE 22-2. *World tea production, average 1957–1961.* [*From A Graphic Summary of World Agriculture, USDA Misc. Publ.* 705.]

for several hours. Here the brown color and pleasing aroma develop, after which the leaves are dried by warm air until their moisture content is reduced to about 3 percent and the leaf turns black. "Orange Pekoe" is a tea of this type, highly esteemed in the western world. Green tea may be preferred locally, but is a minor export item. Green tea is not withered.

As with the coffee berry, the quality of the tea leaves varies with the area of production (and also with the care taken in its cultivation and processing). Expert tea tasters can almost pinpoint the location from which a sample comes. Obviously much hand labor is involved in tea growing, and the industry remains largely confined to countries with an abundant source of cheap labor. In China and throughout Southeast Asia isolated dooryard trees yield appreciable tea for home consumption. On the large tea plantations waste leaf and dust fragments serve as a source of commercial caffeine. Tea leaf ready for marketing contains up to 45 percent of soluble materials, including dextrins, pectins, and essential oils as well as caffeine (theine). Tea accumulates several parts per million of fluorine, an element with the reputation of preventing dental caries.

To avoid a market glut, tea production has been regulated by international agreement, with export quotas assigned. On the other hand domestic shortage of tea is anticipated in India, as preference for tea drinking increases, but most suitable land is already being cropped.

Cacao, Theobroma cacao, Sterculiaceae

Cacao (Fig. 22-3), the source of cocoa and chocolate, had already been domesticated when the first Europeans reached the New World. There is speculation that the species may be native to the South American lowlands just east of the Andes, but certainly it was well known (and most famed) as the prized "chocalatl" of Montezuma and his Aztec empire of Mexico. Cacao was soon introduced by the Portuguese into Africa, chief seat of modern day production, which totals well over a million tons annually (of cocoa beans). About a third of the world production comes from tiny Ghana, one-sixth from Nigeria, and one-ninth from

Brazil. There is also appreciable cacao growing in the Ivory Coast, the Cameroons, and other equatorial areas of Africa, and in the Dominican Republic, Mexico, Equador, and other Latin American countries (Fig. 22-4).

The cultivated cacaos have been divided into three groups: the *Criollo* (with the finest flavor), the *Forastero* (the South American cacaos that were introduced into Africa and now account for most production), and the *Trinitario* (grown in Trinidad, possibly a hybrid of Criollo and Forastero). Although there is some vegetative propagation of cultivars, most of the cacaos are grown from seeds and exhibit typical seedling variability. Selection for yield and disease resistance has, however, provided superior populations to those found in the wild. The trees are believed to be partially self-pollinated and partially cross-fertilized, with at least the widely planted Amelonado populations reputed to be uniformly self-compatible.

Cacao is planted as an understory, beneath the high shade of larger trees. Plantings in full sun usually suffer more from weed competition and pests, and grow poorly. Fertilization improves yields. Virus and "black pod" diseases are the main hazards, and resistant cultivars are being sought. The cacao trees bear small flowers along the trunk, which when fertilized develop into pendulous, football-shaped fruits that hang from the larger branches. The fruits have a tough outer husk and a soft pulp that contains several seeds. A single prime tree may bear 70 or more fruits in the course of a year, providing annual yields of cacao beans that average 300 pounds per acre (and in select locations as much as a ton). Cacao fruits are carefully removed by hand to avoid injuring the flowering "cushions" along the trunk. Usually the pods are split by hand, and the seeds hand-scraped from the pulp.

The seeds are cured by subjecting them to a period of fermentation (during which unpleasant odors develop as temperature mounts in the mounds of seeds); it is this step that helps develop the characteristic chocolate aroma. No longer viable after being cured, the seeds are then cleaned or polished further, and sent on to market. Final processing includes roasting, mechanical cracking of the seed shell, separation of the kernels, and

508

FIGURE 22-3. *Cacao fruits.*

expression of the meats to yield the oils and fats that constitute the "cacao butter" of the trade. This is one of the most valuable edible fats known, and a constituent of the finest chocolate candy. The "cake" which remains after expression is a source of the caffeine-like alkaloid theobromine, much used in so-called cola drinks. The cake may also be employed as flavoring for various products of the grocery and confectionery trade.

Other nonalcoholic beverage crops

As noted earlier, a great many plant species yield beverage flavorings. These range from the "tonics" of folklore to stimulants and even hallucinatory substances that often play a part in primitive (and present-day) rituals. Some are as important in limited regions as are coffee, tea, and chocolate

the world over—for example, maté (Paraguayan tea), in southern South America. Others are of only occasional use, and little or no commerce.

COLA, *Cola nitida*, STERCULIACEAE   Although most of the flavorings and stimulants added to modern cola drinks are now derived from such sources as caffeine and theobromine, the cola tree, native to western Africa and now widely introduced throughout the world tropics, started it all and gave its name to several of today's popular soft drinks. Most commercial production is from Africa and Jamaica. Cola seeds are relatively rich in caffeine, essential oils, and alkaloids. They may be chewed by tribesmen or be pulverized and boiled to make a beverage that is said to inhibit fatigue and stall hunger. The seeds occur in star-shaped fruits, and are removed by hand. They are sun-

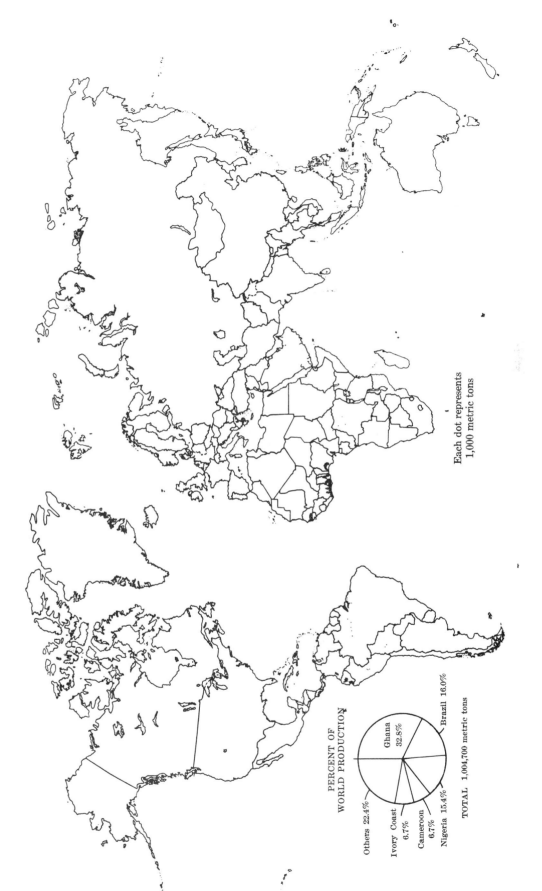

PERCENT OF
WORLD PRODUCTION

Others 22.4%

Ghana
32.8%

Ivory Coast
6.7%

Cameroon
6.7%

Nigeria 15.4%

Brazil 16.0%

TOTAL 1,004,700 metric tons

Each dot represents
1,000 metric tons

FIGURE 22-4. *World cacao production, average 1957–1961.* [*From A Graphic Summary of World Agriculture, USDA Misc. Publ. 705.*]

dried before shipment to commercial markets. Propagation is from seed.

GUARANÁ, *Paullinia cupana*, SAPINDACEAE Guaraná is the "cola" of Brazil. The seeds yield a paste considerably richer in caffeine than either coffee or tea, and small quantities of guaraná are reputedly sufficient to counteract fatigue. *Paullinia* is a trailing shrub found in the wild and casually cultivated in Brazil. The fruits are hand picked, pulverized and dried for sale to make the guaraná beverage especially popular in northern Brazil.

MATÉ, *Ilex paraguariensis*, AQUIFOLIACEAE
Known also as yerba maté, or Paraguayan tea, maté was apparently long used by the Guarani Indians of Paraguay before South America became settled. The maté tree, a relative of holly, is frequent in the broken forests of southern Brazil, Paraguay, and northern Argentina. Harvest is from both wild and planted trees. The Jesuit missionaries of the seventeenth century encouraged maté planting, and are probably responsible for having spread the species widely in this part of South America. At harvest time the smaller branches are cut by hand and usually dried over an open fire. When dry the leaves are stripped from the branches and crushed, after which they are ready for use in the same fashion as tea leaves. Travelers in this part of South America generally carry a pouch of maté tea leaves and a straw-like "bombilla" with a perforated sieve-like base. From time to time, when crossing a stream, someone makes an infusion in a hollowed gourd, and the beverage is passed around to all members of the party, as each in turn sip from the bombilla. A national institute to supervise the production and trade in maté was created in Brazil in 1938.

Beer

Beer, next to water the earth's most popular beverage (in 1970 1.6 billion gallons were drunk), is not the product of a single plant species, but rather the end result of fermentation by yeast (*Saccharomyces*) of any of a number of carbohydrate sources; modern beers are usually flavored by hops (*Humulus lupulus*). There are a number of local "beers," such as **chicha,** fermented from potatoes in the South American Andes; **sake,** fermented from rice in Japan; **pulque,** fermented from Agave leaves in Mexico; and **mead,** fermented from honey in Africa and Scandinavia. Even such natural fermentations as hard cider and palm toddy might loosely be considered beers. Beer brewed from malt, as is typical in North America and Europe, will serve as our example here. The various steps and the several plant products involved in making beer are shown in the flowsheet on page 511. The integration of the various products and procedures to make the flavorful beverage bought as bottled and draft beer is as much an art as a science. Brewers still zealously guard their particular methods and their prized yeast strains. Regionally there are wide differences in public taste for beers: the Irish like it dark, Berliners light; the British drink it warm, the Americans cold; the Swedes view it as intoxicating, the Russians as an antidote for drunkenness. Bavaria can be considered the center for brewing, both on the basis of per capita consumption (said to be about 1.3 liters per day, on average) and number of breweries (reported as 1400, one-fourth of those in existence). Munich, Germany, is responsible for one of the two great types of beer, the **Münchener**—a golden lager, heavy-bodied and mild. The other great type is **Pilsner** (after Pilsen, Czechoslovakia), lighter in color and body, more tart and alcoholic, with an especially creamy head.

Conventional beer is brewed from malt. Malt is derived by germinating certain cereal seeds, especially barley. During sprouting, enzymes are produced in the grain that are capable not only of hydrolyzing the remaining carbohydrates of the barley seed, but other starchy products added to the malt before fermentation. In the United States barley used for malting is primarily of the two-row type, 90 percent of which is produced in the Midwest. After being thoroughly cleaned it is washed and then steeped in huge vats of water for 2 days. It is then germinated in rotating drums, where temperature and humidity are carefully controlled. At about the stage when the primary root has emerged, but before the coleoptile has ruptured, germination is halted by drying in kilns at about 82°C (180°F). For bock beers the drying temperature is a bit above this, caramelizing the malt to give a darker, sweeter beer.

In the typical brewing procedure an adjunct—

rice, corn, wheat, potatoes, or even cassava—is added to the malt before brewing, more economical than using malt alone. Munich, however, prohibits by law the use of adjunct in beer brewed there, it is said. Many brewers use about 65 percent malt, 35 percent adjunct. This combination becomes the mash, which is soaked in hot water to provide the wort. During mashing many soluble substances are released from the malt and adjunct under the influence of the malt enzymes. After the wort is drawn off for fermentation, the spent grains are retrieved, dried, and usually utilized for stock feed. The wort is boiled with hops for two or three hours to provide the characteristic bitter flavor of beer, and help clarify the liquor.

Hops, *Humulus lupulus*, Moraceae, have long been used by man as a medicinal and salad green. There is no evidence that hops were used to flavor

Brewing flowsheet

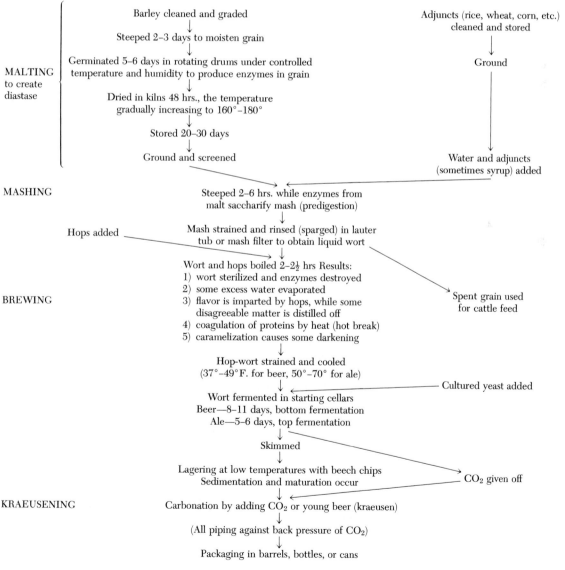

MALTING
to create
diastase

Barley cleaned and graded
↓
Steeped 2–3 days to moisten grain
↓
Germinated 5–6 days in rotating drums under controlled temperature and humidity to produce enzymes in grain
↓
Dried in kilns 48 hrs., the temperature gradually increasing to 160°–180°
↓
Stored 20–30 days
↓
Ground and screened

Adjuncts (rice, wheat, corn, etc.) cleaned and stored
↓
Ground
↓
Water and adjuncts (sometimes syrup) added

MASHING

Steeped 2–6 hrs. while enzymes from malt saccharify mash (predigestion)
↓

Hops added

Mash strained and rinsed (sparged) in lauter tub or mash filter to obtain liquid wort
↓
Wort and hops boiled 2–2½ hrs Results:
1) wort sterilized and enzymes destroyed
2) some excess water evaporated
3) flavor is imparted by hops, while some disagreeable matter is distilled off
4) coagulation of proteins by heat (hot break)
5) caramelization causes some darkening

BREWING

Spent grain used for cattle feed

Hop-wort strained and cooled (37°–49°F. for beer, 50°–70° for ale)
↓
Wort fermented in starting cellars
Beer—8–11 days, bottom fermentation
Ale—5–6 days, top fermentation
↓
Skimmed
↓
Lagering at low temperatures with beech chips
Sedimentation and maturation occur
↓

Cultured yeast added

CO_2 given off

KRAEUSENING

Carbonation by adding CO_2 or young beer (kraeusen)
↓
(All piping against back pressure of CO_2)
↓
Packaging in barrels, bottles, or cans

Adapted from R. W. Schery, Plants for Man *Prentice-Hall, New York 1952.*

beer until the eighth century, though the species was in cultivation before that time. During the Middle Ages many monasteries became famous for the hopped beers they produced. Today world production of hops is about 100,000 tons, chiefly from Europe (Germany, United Kingdom) and the United States (Pacific coast states).

The hop plant is a dioecious vine, with female plants bearing spike-like clusters of bracts called "cones," which become the commercial hop. The vines are usually propagated from rhizome cuttings and trained onto specially constructed trellises. In Europe the cones are hand-harvested; in the United States they are frequently machine picked (the entire cone-bearing branches being fed into the picking machine by rotary drums). Freshly picked hops contain as much as 80 percent moisture, and must be dried until their moisture content is around 12 percent. Usually this is done in kilns or by blowing warm air through the hops. After being dried the hops are cured for about two weeks, during which time the bracts become tough and pliable and develop a characteristic aroma. In the United States each 31-gallon barrel of beer utilizes less than a pound of hops, but European beers may be hopped as much as 4 pounds per barrel, which gives a more bitter brew. Hops without seeds (that is, female plants grown in the absence of male plants) bring a premium, for it is thought that the seeds sometimes contribute objectionable odors, and in any event add to the mass of spent material that must be discarded after final brewing of the wort.

The cooled wort is piped under antiseptic conditions to the starting cellars, where in huge tanks similar to swimming pools it is inoculated with a select strain of yeast (*Saccharomyces cerevisiae*, Ascomycetae), sometimes obtained from the sediment of a previous brew. The yeast ferments the wort to produce carbon dioxide, alcohol, and minor amounts of organic compounds that impart the characteristic flavors to the beer. After a day or two of fermentation in the starting cellar, the beer is transferred to lagering vats, where a secondary fermentation takes place for a number of days and various substances settle out. In traditional brewing practice the beer is then aged in closed vats for weeks or even months, where it becomes naturally carbonated, since the carbon dioxide is unable to escape. In modern, faster brewing the beer may be artificially carbonated; in any event, before it is bottled or kegged a bit of new green beer is added to give a final fillip of fermentation. The beer is then filtered and usually pasteurized (draft beer is unpasteurized, and is best kept refrigerated if stored for long). Finished beer contains about 90 percent water, 5 percent ethyl alcohol, and minor quantities of maltose, gum, dextrins, and nitrogenous substances.

Other malt beers include **ale,** an amber beverage fermented by floating colonies of yeast and usually with higher hop and alcohol contents than beer; **kvass,** or **quass,** a Russian beer made from barley and rye with peppermint flavoring; **pombe,** or **bousa,** an African beer made from millet; **porter,** a dark, sweetish ale like bock beer; **stout,** a porter of higher alcoholic content and strong hop flavor; and **weiss,** a light malty ale made mostly from wheat.

Wine

Grape growing was reviewed in Chapter 21, under discussion of Fruits. It was mentioned there that most of the grape crop is not consumed fresh but is fermented to wine. Wine making is nearly as ancient as is civilization itself in the Near East. Wine consumption has long been a cherished tradition in western culture. Wine making continues as an impressive European art, consuming about 40 million tons of grapes annually, largely in France and Italy. World wine production amounts to about 30 million tons annually.

The principle of wine making is relatively simple; the juice of crushed grapes (called "must"), rich in fermentable sugars, is inoculated with the proper strain of yeast (or under primitive circumstances is allowed to ferment naturally), until the alcohol content reaches 10 to 14 percent (when the yeast is automatically killed by the alcohol). Such wines are called beverage wines, or table wines, and are consumed with meals. Fortified wines are those to which additional alcohol is added to raise their alcohol content to about 20 percent. They are generally termed apertif, or desert, wines. Sherry and port are examples of fortified wines.

Wines are further classified as **red** (the skins

included when the grape is expressed) or **white** (generally the skins not included); **sweet** (in which not all of the sugar is fermented) or **dry** (in which almost all of the sugar is fermented); **sparkling** (if final fermentation is in an enclosed container so that carbonation results) or **still** (fermented with escape of the carbon dioxide); and so on. Subtle differences in flavor can be attributed to variety of grape, strain of fermenting yeast, soil and growing conditions, and suitability of weather (summers with dry bright weather yield vintage wine grape crops).

The yeast responsible for wine fermentation is generally considered to be *Saccharomyces ellipsoideus*. As in beer fermentation, a poor strain of yeast or the invasion of unwanted bacteria and molds will lower the quality of the wine, producing musty and undesirable flavors. Grape juice for wine is generally mechanically expressed. The must may be pasteurized or perhaps treated with sulfur dioxide to control unwanted organisms, and is then inoculated with a pure culture of yeast. Fermentation continues for many days, accompanied by settling out of solids, tannins, and pigments. Final fermentation usually takes place in closed vats, slowly (perhaps for a period of years), at reduced temperatures, to produce suitably mature wines. Final fermentation of champagne and sparkling burgundy takes place in sealed bottles from which the carbon dioxide cannot escape, making a carbonated wine. Some wines, such as vermouth, have bitter and aromatic flavors blended in to provide a particular taste.

Broadly considered, wine can be fermented from fruits other than the grape, to make, for example, cherry, blackberry, or elderberry wines. All of these are 70 percent or more water, about 10 to 14 percent alcohol, with highly variable amounts of sugar and trace organic substances. (Most wines have up to 1 percent acids, around 0.1 percent tannin, and a pH of 3.0–3.5.)

Distilled spirits

As noted with wine making, natural fermentation can continue only until the alcohol content is about 10 to 14 percent. To make beverages with a higher alcohol content these beers and wines are distilled to yield highly concentrated potions or even pure ethyl alcohol (industrial alcohol). Distillation involves heating the beer or wine, usually in a copper still, with fractionation of the alcohol and other volatile substances in suitable condensers. Depending upon the type of beer or wine distilled, characteristic components accompany the alcohol, to yield whiskeys, rums, brandies, vodka, and so on. The process sounds simple enough, as indeed it is in backwoods distilleries; but for choice spirits the process requires great skill and experience. Proper blending of minute quantities of ingredients imparts the characteristic flavor esteemed in a distilled beverage.

Brandy is made by distilling wine or fermented fruit mash. The distillate is generally aged in wooden casks to develop its own characteristic flavor, and usually contains 40 to 50 percent alcohol. Legend has it that brandy was first invented by a Dutch shipmaster attempting to devise a wine concentrate to lower shipment costs. French brandies are named for the region in which they are produced (Cognac, Armagnac), and are often said to be unduplicatable because of distinctive characteristics of the fruit of the region and characteristic yeast strains in the casks that have developed through the centuries, to say nothing of "secret" formulations.

Whiskey is to cereal beers what brandy is to wine. Fermented grain is distilled, and the distillate customarily aged in charred white oak casks. Most whiskeys are about 50 percent alcohol, the "blended" types being brought up to approximately this alcoholic strength by the addition of ethyl alcohol (grain neutral spirits). Scotch whisky is made from a barley mash in which the malt is cured in the smoke of peat fires to impart the characteristic flavor. Irish whisky is similar, but the malt is kiln-dried. Bourbon is a corn whiskey (not less than 51 percent corn grain in the mash), and rye is a rye whiskey (not less than 51 percent rye in the mash).

Gin and **vodka** are made by distilling the mash to obtain a nearly pure alcohol. In making gin, aromatics, such as juniper "berries" or sloe (*Prunus spinosa*) "berries," are added to impart the flavor. The type of grain fermented is relatively immaterial, for it will have little affect on the flavor (ethyl alcohol is ethyl alcohol no matter what the source of the mash).

Rum is the distillate from fermented sugar cane juices or molasses, suitably aged in wood. Rum distillation began in the Orient, but was quickly picked up in the Caribbean area during the days of the Yankee Clipper trade, and is now produced mainly in the West Indies and the mainlands around the Caribbean sea. Characteristic flavorings (even a bit of brandy) may be added. Raw rums are often locally fermented and distilled in primitive areas, and broadly considered may be made from other carbohydrates besides sugar or molasses.

Liqueurs, or **cordials,** were originally devised as elixirs, to which mystical properties were attributed. They are usually produced by combining brandy with a variety of flavorings. Often dried fruit of a particular type is soaked in brandy to produce such things as "cherry liqueur" or "apricot cordial." For example, creme de cacao is flavored with chocolate; creme de menthe is flavored with mint. Absinthe is made from a strong brandy flavored with wormwood (*Artemisia absynthium*) and other aromatics. Tequila is the distillate of the Mexican pulque (from juices of *Agave* spp.).

VEGETABLE OILS, FATS, AND WAXES

Vegetable oils, fats, and waxes, are derivable from a great many sources; economics—the price of the oil in its market—usually governs which crops will dominate the vegetable oil industry. Because of this there is relatively little glamour and excitement connected with vegetable oils—little exploration in search of new sources, and no wars fought to command the trade. Alternative supplies are too readily available. Indeed, today's vegetable oil chemist is so adept at changing oils (as by hydrogenation) and substituting one for another that price differential often overshadows inherent qualities. And not only can one vegetable oil be made to substitute for another, but petroleum or mineral derivatives often take over markets from botanicals, and vice versa—for example, in the paint industry synthetic resins and latices are appreciably substituting for once dominant linseed oil and expensive tung oil.

About a dozen crop plants make up the main vegetable oil seed market. Most are traditional sources, for which an established industry has been built, which gives these plants considerable commercial advantage. It takes a while to work out details with new crop introductions and create grower and market acceptance; but government agencies such as the USDA are on the lookout, and several newer oil sources such as *Crambe* (Cruciferae) and safflower (*Carthamus,* Compositae) show promise. Since oils are relatively abundant in almost any seed, it is no wonder that researchers have looked at such an assortment as ironweed (*Vernonia*), *Cuphea,* wild carrot (*Daucus*), *Limnanthes, Comandra;* and even spruce (*Picea*), *Tribulus,* grape seed (*Vitis*).

Vegetable oils, fats, and waxes serve many uses. Several, such as corn, cottonseed, peanut, olive, and soybean oil, are used for cooking oils, margarine, and salad dressings; some are incorporated into many food products and animal feeds. But equally important are the industrial uses of these and other oils, including coconut and palm oils, linseed oil, castor oil, tung oil, grape seed oil, and many others. The major industrial uses are for paints, varnishes, and lacquers, soap, detergents, plastics, and as components in linoleums and oilcloths. Other industrial uses are for lubricants, hot-dip tinning of metal products, hydraulic fluids, manufacture of synthetic fibers, oxidation to resin, food coatings, cosmetics, medicinals, printing inks, plastic foams, and fatty acid raw materials.

Of course, many plant parts contain oils, fats, or waxes, which are chemically quite similar. Oils occur in small droplets within the plant cells, and are typically most abundant in seeds. The globules are removed from the oil-bearing tissue by crushing (expression) or a combination of crushing and solvent extraction. Vegetable fats differ only slightly from oils, having acidic constituents that are more or less solid at ordinary temperatures rather than liquid. They, too, are usually most remuneratively garnered from seeds. Waxes are the fatty acid esters of monohydroxy alcohols rather than of trihydroxy glycerol, as are oils. They are found mostly as protective coatings on the leaves and stems of plants, where they function to retard water loss. Waxes are a less voluminous (and typically more expensive) commodity than the vegetable oils and fats.

The oil seeds utilized commercially generally have an oil content of 40 percent or more, some as high as 70 percent. With 35 million tons of soybeans, 22 million tons of cottonseed, and 18 million tons of peanuts produced annually, most of which is subjected to oil extraction, it is apparent that we are talking about sizable quantities of vegetable oils. In addition, about 4 million tons of linseed, more than 3 million tons of copra (dried coconut meat, used for oil extraction), about 2 million tons of extracted palm oil, and about 1.5 million tons of olive oil are produced annually.

The production of vegetable oil is not concentrated in any single part of the world; rather, production of vegetable oil is important to less advanced agricultural economies as well as to countries that are agriculturally mechanized and industrialized.

Vegetable oils are fundamental to the functioning of the modern industrialized world. Thus they will be in continuing, steady demand, and their importance can be expected to increase as nonrenewable mineral resources become exhausted. In this respect the crop plants we are reviewing here are equally as basic as are food plants.

There would seem to be little cause to dwell on the uses of vegetable oils as foods. Their main use, as has been noted, is for making cooking fats and margarines. There has been a great deal of commotion in recent years about substituting unsaturated oils and fats for saturated fats in the diet, because of the concern about cholesterol in the blood. Thus a moderately unsaturated oil, such as corn oil (rich in oleic and linoleic components), carries something of a premium for the manufacture of margarine. But almost any edible oil can

be converted to a cooking fat or margarine by chemical manipulation, such as changing fluid oils into fats by elevating them to a higher degree of saturation through **hydrogenation.** This is accomplished by introducing hydrogen under pressure into the heated oil in the presence of catalysts, such as pulverized nickel compounds. Margarine is formulated through such techniques, with the addition of small percentages of milk products, vitamins, and other ingredients, to make a spread that is nutritionally the equal of butter and scarcely different in taste. As a matter of fact, margarines are generally advertised as being less fattening than butter, particularly when formulated from unsaturated vegetable oils.

Industrially, many things can be made from vegetable oils (Fig. 22-5). The oils, being fatty acid esters of glycerol, can readily be broken down to fatty acids and glycerol. In fact most fats and oils are naturally unstable, and when kept for considerable lengths of time become rancid due to breakdown of the glycerides into various lesser products. When oils are largely glycerides of saturated acids, such as the oleics characteristic of palm, peanut, olive, and grape oils, they are said to be **nondrying** oils. If the oil is high in glycerides of the unsaturated type, particularly linoleic and linolenic with little oleic, the oil is said to be a **drying** oil (capable of forming an elastic film upon absorption of oxygen from the air). Linseed, safflower, soybean, and particularly tung oil are of this type. Drying oils are more important commercially, and are used in such products as paint and varnish. The chief industrial uses of nondrying oils (and an intermediate group termed "semidrying," such as cottonseed, corn, and sunflower oils) are for soap, detergent, and other industrial raw material.

Saponification is the process whereby soap is made. Vegetable oils treated with alkalis, such as potassium or sodium hydroxide, split into three molecules of the metal salt of the particular fatty acid and one molecule of glycerin. Early civilizations mostly utilized animal fats boiled with wood ashes from the fire to make soap. (Wood ashes are rich in potash, potassium carbonate.) Gradually vegetable oils, such as olive oil, were substituted for animal fat, and sodium hydroxide (which makes a harder soap than does its potassium equivalent) for wood ashes. The by-product, glycerin, has

FIGURE 22-5. *One of the major sources of vegetable oil is the soybean. Represented here are the proportions of drying oil and protein meal resulting from milling. A bushel of soybeans produces about 11 pounds of oil and 48 pounds of meal. The price of beans depends largely on the value of oil and meal in the market place.*

found many industrial uses, including the making of explosives (trinitroglycerin, or TNT).

Linoleum was invented about a century ago, when it was discovered that boiled drying oil made an excellent binder for pulverized cork (and other particles). Today many plastic materials substitute for the original linoleum (the name derived from "linum" for flax, and "oleum" for oil). Likewise, drying oils made an excellent vehicle for pigments of various sorts, which hardened into a resinous protective coating, or paint. Varnishes and enamels are basically similar to paint.

Compared to oils, vegetable waxes are of relatively minor importance. The waxes are used chiefly for polishes and carbon paper, and to a minor extent for such diverse products as candles and chewing gum.

Nondrying oils

OIL FROM PALM FRUITS, PALMAE As sources of commercial oil, two palms stand out, the coconut, *Cocos nucifera*, and the African oilpalm (often known just as "palm," or dendé in Brazil), *Elaeis guineensis*. Throughout the tropics the palm family is as ubiquitous as grass is in temperate climates, and it is no wonder many other palm species contribute oil-bearing fruits of greater or lesser local importance. For example, in the lower Amazon Valley, there is considerable traffic in murumuru palm nuts, *Astrocaryum murumuru*, and in northern Brazil the babassú, *Orbignya speciosa*, used much in the Brazilian soap industry. The fruits are harvested from extensive wild stands, neither species being cultivated. The ouricuri, *Cocos coronata*, of eastern Brazil yields not only an oil from the fruit, but a wax scraped from the leaves, much like the wax of the carnauba palm. Extensive stands of the caranday, *Copernicia australis*, and the mbocayá, *Acrocomia*, grow in Paraguay, and supply many materials for rural living: they could become a sizable source of vegetable oil if economic conditions warranted their harvest.

Usually transportation, marketing, and processing facilities are primitive or lacking in such tropical locales, which prevents much greater production from wild sources. One palm abundant in Venezuela, the bataua, *Jessenia bataua*, contains as much as 24 percent oil in the pulp (rather than the nut); the oil is chemically very similar to olive oil (about 7 percent palmitic, 9 percent stearic, 73 percent oleic, and 5 percent linoleic acids). It has been under investigation since 1910, but there has been insufficient economic stimulus to warrant plantings that might lead to commercial production.

In contrast are the oilpalm and coconut (copra). Nearly 2 million tons of palm oil reach commerce annually, chiefly from west-central Africa but significantly also from Indonesia and Malaysia, and from scattered parts of the New World, where *Elaeis* has been introduced or naturalized. World production of copra is about $3\frac{1}{3}$ million tons, chiefly from the South Pacific region. Nearly 30 billion coconut fruits are harvested annually, again mostly from the tropical Far East but also from Mexico and Brazil.

The African oilpalm is one of the world's most useful oil-yielding species. The "palm oil" of commerce comes from the outer husk or pericarp of the fruit; a less prized kernel oil is obtained from the nut or seed. The "pulp oil" is used in metal plating (to protect the metal surface from oxidation before plating) and in foods and dentifrices. The kernel oil is used mostly in the making of soap and as a source of glycerin. Under favorable circumstances, fantastic annual yields that exceed 2 tons of oil per acre can be achieved.

The palm grows naturally in a coastal belt about 150 miles deep, from Angola to Senegal. The palm fruit is harvested largely from wild stands, but some extensive plantations have been developed, particularly in the Congo, where attempts at mechanization have been made. In 1910 the African oilpalm was introduced into Sumatra and Malaya, where it has been widely interplanted with rubber trees. A great many cultivars were selected, and Dutch agriculturists sent expeditions to Africa in search of potentially more valuable breeding stock. As a result, in the Far East, and to a limited extent in the New World, there has been considerable plantation growing of the African oilpalm.

Under primitive circumstances the palm fruit is boiled, the free oil skimmed off the cauldron, and residual amounts squeezed from the pulp. The nuts are marketed or processed separately. On managed plantations the fruit is subjected to a preliminary digestion, after which the pulp oil is centrifuged or pressed free and purified; the nuts are cleaned, dried, and cracked, and the freed kernels further dried.

The exact home of the coconut is uncertain, for it has been widely scattered throughout the tropical coastal environments of the world. These picturesque, graceful trees of the sea coast are often sources of food, thatch, and income for island people. As many as 400 coconut fruits may be produced annually from a single tree under favorable circumstances. The husk of the coconut is usually removed and discarded (although there is a limited market for it as a fiber, coir). Copra oil, from the coconut, is of course a kernel oil, being derived from the endosperm of the coconut seed. In making copra, the inner shell is cracked and the "meat" removed to be smoked or sun-dried before sending it to the extraction plants. Most of the harvesting and splitting is done by hand. The copra is gathered to a central location (or even imported into industrialized countries without processing), where it is usually expressed while heated to yield coconut oil. The press cake is sold as stock feed.

OTHER IMPORTANT NONDRYING OILS The castor-bean, *Ricinus communis*, Euphorbiaceae, is potentially an important industrial oil, but is of uncertain supply and poses difficulties in adapting it to mechanization. Presumably native to Africa, it is today grown throughout the world tropics. It is adaptable as a summer annual in temperate climates. Efforts have been made to breed cultivars that will not shatter seed so readily, making them more suitable for mechanical harvesting. The seeds, borne in spiny capsules, have an oil content as high as 50 percent or more. The oil is removed mainly by expression; whatever remains in the press cake is extracted with solvents. Because the castor bean contains a toxic alkaloid, ricin, the press cake is unsuitable for stock feed. The oil is used mostly as a medicinal (purgative) and lubricant, and to some extent in soaps, linoleum, plastics, inks, and various finishes. Through chemical alteration castor bean oil is often made to serve

other purposes, such as substituting for drying oils and in the plastics industry. India and Brazil are the largest producers of castor bean.

The olive was discussed at some length as a fruit in the previous chapter. About 7 million tons of olives are produced annually, from which come about 1.5 million tons of olive oil. Oil production is chiefly from Spain, Italy, and Greece. Many of the cultivars have been selected with oil production in mind rather than edibility of the fruit. Olives grown for oil may contain as much as 40 percent oil. The fruit, usually gathered by hand on small, private holdings, is crushed to yield the oil. Several expressions are made, the first giving the more valuable "virgin" oil suitable for use without further refining. The press cake is generally treated with solvents to get the final measure of oil, and the press cake may serve as cattle feed or as a fertilizer for the orchards. Olive oil finds its greatest use as a salad and cooking oil.

Also discussed at some length in Chapter 20, was the peanut, grown for its edible seed as well as its oil. Much of the African and Indian production is imported into Europe, where the oil is extracted. The nuts contain as much as 45 percent oil, which may be recovered by expression or extraction with solvents. The oil is used mostly in edible products, and the press cake, rich in protein, is a valuable livestock feed.

Semidrying oils

The major semidrying oils of commerce originate as by-products from crops grown primarily for other purposes. Especially important are maize (mainly a feed cereal, yielding corn oil) and cotton (mainly grown for the fiber, but yielding cottonseed oil). Cottonseed production generally exceeds 20 million tons annually, chiefly from the United States, India, China, and Brazil. Most cottonseed contains 30 to 40 percent oil, and the remaining press cake contains approximately 10 percent nitrogen.

Corn oil production is the "other half" of wet-milling the grain for starch (see discussion under Corn, Chapter 20). The oil-rich "germ," or embryo, constitutes 6 to 13 percent of the grain. In wet milling this is freed by "cracking," which follows the initial steeping (which yields "corn steep liquor," used in industrial fermentations). The germ is removed by flotation, and is then subjected to expression or solvent extraction. Refined corn oil is used principally as a cooking oil and in margarine.

Huge quantities of cottonseed are left after the ginning of cotton fiber. In oil production the hulls are first removed from the seed, and may be incorporated into cattle feeds as roughage. The kernels are then crushed between rollers, heated (to expel moisture, make the oil fluid, coagulate albumenoids), and subjected to hydraulic expression or solvent extraction. After refining, the oil may be hydrogenated for margarine or used to some extent as a liquid cooking oil and in the manufacture of soap.

OTHER SEMIDRYING OILS   Rape, seeds from various species of the cabbage group, *Brassica*, Cruciferae, enter commerce to the tune of approximately 6.5 million tons annually. Canada, China and India are the chief producing nations. The seeds contain as much as 45 percent oil, largely of erucic, oleic, and linoleic types.

Sesame, *Sesame indicum*, Pedaliaceae, yields nearly 2 million tons annually of commercial seed, mostly from India, China, and the Sudan. The seed is often used as a garnish on bakery products. But it does contain as much as 55 percent oil, largely of the oleic and linoleic types. Oil is extracted by expression, and used to a great extent as a salad or cooking oil. Most production comes from areas that are not agriculturally mechanized, and harvesting is often done by hand, by flailing the pods with a stick. Under mechanized agriculture it is said that this oil seed can yield more per acre than any other annual oil crop; efforts have been made to develop nonshattering cultivars suitable for mechanized growing in the United States. Sesame oil is said to keep for several years without turning rancid, and the press cake is an excellent livestock feed.

The sunflower, *Helianthus annuus*, Compositae, also produces an oily seed that may be toasted for direct consumption. The oil is expressed for margarine and other food uses, or for industrial purposes. Nearly 10 million tons of sunflower seeds are produced annually, chiefly in Russia and southeast-ern Europe, but in appreciable quantities in Argentina too. The oil is resistant to rancidity, and has attracted attention as an extender for more expensive oils. The press cake contains over 50 percent protein.

Drying oils

Linseed or flax, *Linum usitatissimum*, Linaceae, is both a source of fiber and of a valuable oil (Fig. 22-6). However, cultivars selected for the stem fiber generally differ from those used for production of oil seed (where flax fiber is the "by-product"). Total world production of linseed approximates 4 million tons annually, chiefly from Canada and the United States, but also from Argentina and India. Linseed oil has long been the chief ingredient in oil paints, but its importance there has been declining in recent years as various latex paints and plastic-like substitutes have been developed. Linseed contains as much as 43 percent oil. It is extracted by expression, and when refined is said to keep almost indefinitely in sealed containers. Typical of drying oils, however, when exposed to air (oxygen) it turns into the tough, impervious coating that makes it so valuable for paints, oilcloths, linoleum, and suchlike.

Tung, *Aleurites fordii*, Euphorbiaceae, has

FIGURE 22-6. *Flax plants with maturing seed pods ready for harvest and extraction of linseed oil.* [*Photograph by J. C. Allen & Son.*]

gained recognition as a significant source of one of the world's best drying oils. A. montana, the mu oil tree, produces tung oil of equally high quality. A. fordii is native to central and western China, A. montana to southern China and the Malaya peninsula. The former has been introduced into the New World, and to some extent is plantation-grown along the Gulf coast of the United States. It is a small monoecious tree, with fruits that contain two to five heavy-shelled seeds having a white kernel that is about 65 percent oil. The oil is used especially for paint and varnish. World production comes to more than 130,000 tons annually, chiefly from China, but also from Argentina and Paraguay. A. montana has been introduced into tropical Malawi, where 20,000 acres have been planted to this species.

A. fordii is highly susceptible to frost kill once budded out. But it can stand mild winter frosts, and indeed some cold is needed to initiate full flowering and fruit production. The species does best on well-drained, fertile soils in climates having ample rainfall. The tree starts bearing at about three years of age, and has a productive life of about 30 years. It tends toward biennial bearing. In the United States and on African estates high-yielding cultivars are propagated, usually by bud grafting onto open-pollinated seedlings. The species responds well to fertilization, especially nitrogen fertilizer. Since the tung cake resulting from expression is toxic as a feed, it is usually returned to the fields for fertilizer, often accompanied by application of ammonium sulfate. Tung is especially sensitive to zinc deficiency, and soils must frequently be treated with zinc sulfate for survival of the orchards. Yields generally run between 1 and 2 tons of nuts per acre. The plantations are usually clean cultivated.

The tung fruit is collected after it has fallen to the ground. The seed should be collected immediately, and dried to prevent any hydrolysis of the oils. Often the fruits are hulled in the field; transporting the husks is eliminated by the use of portable hulling machines. Nut shells are removed mechanically at centralized locations, after which the kernels are heated, ground, and expressed, usually in a screw expeller. Solvent extraction is also feasible. Except for filtering, no refining is necessary. Tung oil dries rapidly into a hard, water-proof coating, resistant to both acids and alkalis. In this respect it is said to be superior to linseed oil.

OTHER DRYING OILS   Scores of other species are either minor items of commerce or are potentially useful as sources of drying oils. By and large, only safflower, among the new entries, and soybean, among established agricultural crops, have been able to compete or promise to compete effectively. The safflower, Carthamus tinctoria, Compositae, has been a favorite candidate of researchers seeking new crops for North American agriculture. Only recently have there been signs that it might live up to its promise; as much as 60,000 acres are planted to it in California. Originally the species was grown for the yellow dye that the flowers yield, a substitute for saffron. It has long been cultivated in India, and more recently in Russia as an oil crop.

In the United States cultural requirements for safflower are just being worked out. Seedlings grow slowly and tend to be overrun with seeds where proper cultivation is not possible. And if there is insufficient moisture at flowering stage, seed set may be poor. Unfortunately, oil content of the seed is not great, averaging about 37 percent in present cultivars. This tends to make its production more costly in competition with other drying oils, and there is need for breeding higher-yielding strains. Safflower is planted and grown in about the same way as some of the small grains, such as barley. Combines like those used for barley have also been used to harvest safflower seed. Yields run from 350 to 1,500 pounds to the acre, and yields as large as 3,400 pounds per acre have been obtained experimentally in Germany. More than 2 tons per acre should be possible in California under irrigation. The oil is either extracted by solvent or expressed in an expeller (a screw press).

Safflower oil has long been used for cooking and illumination in India, and as an edible oil in Europe. But in the United States the chief interest in safflower oil has been as a raw material for making certain synthetic resins, and as an all-purpose drying oil. So far supply has not been sufficient to really prove its commercial usefulness. It has a higher linoleic acid content (77 percent) than either linseed or soybean oil, but has a lower linolenic content. The press cake is quite suitable as a livestock feed, but is not very palatable when used alone.

The soybean, *Glycine max*, Leguminosae, has been discussed in several sections of this book. We will not discuss it further here other than to note that one of its chief uses is for the oil extracted from the seed. It has had a dramatic rise in the United States from the status of a virtual unknown to one of the leading present-day agricultural crops. Its oil has come to outrank cottonseed as the chief vegetable oil produced in the United States; world production probably exceeds 10 million tons annually. The oil is expressed either by expeller or hydraulic press, or is extracted by solvents. In the United States the most frequent method is continuous expelling, in which the beans are crushed or ground, steam heated to about 150°F, then fed into the expeller. The oil is chiefly used for shortening, margarine, and salad oil, but to some extent for making soap, paint, ink, linoleum, and oilcloth. We have already noted how useful the protein cake remaining after oil expulsion is likely to be in a world where food proteins are in short supply.

Waxes

Carnauba, *Copernicia cerifera*, Palmae, is the world's premier wax plant. Both wild and planted carnauba palms grow in arid northeastern Brazil, the chief area for the production of carnauba wax. Mention has already been made that several other palms of the tropics yield waxy substances as well as oil seeds. For example, the ouricuri, *Cocos (Syagrus) coronata*, of eastern Brazil, and the caranday, *Copernicia australis*, of Paraguay, yield leaf wax that approaches carnauba in quality. Wax from various desert shrubs or even from sugar cane stems may substitute for carnauba when prices are high or supply short.

Carnauba is a slow-growing fan-palm that develops thick secretions of wax on the leaves in response to the seasonal drought and scorching winds of its native habitat. But the tree prefers to root in moist soil, and is commonly found in low areas that, in certain seasons, have a high water table or even standing water. Today harvest

of the leaves is regulated, to prevent overcutting and exhaustion of the plants. The leaves of young trees, which have been planted, are cut by machete; those from wild trees, which are older and taller, are cut with the aid of long pole-saws and ladders. The leaves are taken to a drying shed where they are fed into shredding machines. The shredded leaves are left to dry for several days, during which they wither and the waxy coat becomes loose, falling to the floor as a whitish powder. Laborers further flail the shredded leaf parts to remove all adhering wax. The loose wax is swept up and put into containers for further processing. Spent leaf is returned to the field as a mulch.

The dust-like wax is melted, strained, and poured into molds, where it hardens for shipping to market. Most crude wax goes to the United States, where it is the most prized natural wax for polishes and floor coatings; but it is also much used for phonograph records, plastics, cosmetics, films, and other products that require a very hard, high-melting wax. It is sometimes added to inferior waxes to increase toughness and luster, or to reduce stickiness and plasticity.

Candelilla, *Euphorbia antisyphilitica*, Euphorbiaceae, is a native desert plant of the southwestern United States and Mexico. The leafless candelilla secretes an epidermal wax in adaptation to the arid environment. For years candelilla has been gathered in rural Mexico, in some places to the point of exhaustion. Plants are pulled by hand, transported on burros to an extraction camp, where they are boiled in water in sunken tanks. About 8 pounds of sulfuric acid is added for each 100 pounds of candelilla in the brew. The wax floats to the surface as a light-colored foam, and is skimmed into other containers for further refining. It eventually solidifies as a cake, which can be cut to convenient size for shipping to market. This crude wax, containing much debris, must be further refined before being put to commercial use.

In Mexico candelilla wax is often mixed with paraffin to make candles. Although of limited and erratic industrial supply in the United States, it often substitutes for more expensive beeswax and carnauba. Candelilla finds occasional use in coatings and adhesives, automobile polishes, printing inks, chewing gum, cosmetics, explosives, and various dressings or polishes. About 13,000 tons is consumed annually in the United States, mostly imported from Mexico (although there are restrictions against its exhaustion). Since only wild plants are harvested, candelilla production is destined to be a declining industry until such time as the plant can be brought into cultivation and be grown in an economically efficient fashion.

Jojoba, *Simmondsia chinensis*, Buxaceae, is native to the southwestern United States and Mexico, in spite of the name chinensis. It bears a seed that contains about 50 percent liquid wax. Jojoba has been suggested as a substitute for sperm whale oil, which is declining in availability. By hydrogenation, liquid wax of jojoba can be transformed into a substitute for carnauba and other hard waxes of commerce. It seems the equal of hard waxes for furniture polishes, and as an extender for other waxes in a variety of end uses.

Wild jojoba is found from southern California and Arizona south into northwestern Mexico, usually at elevations between 2,000 and 4,000 feet. Since it is adapted to barren desert localities, where it has little competition, and because it grows slowly through the seedling years, possibilities for introducing the species into cultivation are not encouraging. The plants are dioecious. As much as $2\frac{1}{2}$ pounds of seeds, which contain about 50 percent "oil," can be harvested from a single plant. Jojoba has been experimentally planted directly to the field, but is usually transplanted from nurseries. In cultivation it responds to proper spacing, pruning, and irrigation, but so far seems poorly adapted to mechanical harvest of the seed, which is essential if it is to become a cultivated oil crop in modern agricultural systems. Jojoba is mentioned here more to show the many problems encountered in trying to develop a wild plant into a cultivated crop than as an example of an established source of vegetable wax.

SPICES, FLAVORINGS, PERFUMES, AND OTHER ESSENTIAL OILS

Extractives important for their essential oil content obviously interrelate with other extractives discussed in this chapter (namely, medicinals, resins, and masticatories), and of course with foods, the subtle flavors of which are frequently due to essential oils.

Essential oils are highly aromatic substances, chemically simpler than the oils of the preceding section (see Chapter 8). Mostly they are benzene or terpene derivatives or straight-chain hydrocarbon compounds of intermediate molecular length (seldom more than 20 carbon atoms long). The highly aromatic nature of this diverse group of substances provides the "zing" to spices, the taste to flavorings, the fragrance to perfumes, and the "clean" smell to antiseptics and medicinals. Usually only very small quantities of an essential oil are needed to produce a noticeable effect.

The political importance of essential oils, especially as represented in the spices, has been out of all proportion to their limited usefulness and present minor economic value. Historically, spices have been responsible for the making and breaking of empires, the exploration of far corners of the globe, and, of course, the discovery of America by Columbus. But even before this heyday of Western interest in spices and aromas, which reached its zenith between the fifteenth and nineteenth centuries, essential oils have had a compelling appeal to almost all civilizations. Aromas and flavorings have been part of the magical rites of the shaman; they were used in the earliest temples, and have been in demand for various "purification" ceremonies throughout the centuries. Essential oils were used by the early Egyptians for mummification, and were thus important even in afterlife. A number of primitive cultures believed that spices of one kind or another were aphrodisiacs, and many of the ancient wines were flavored with essential oils. The early healers used spices and fragrances in the first medicinals, and the Romans demanded aromatics for the bath. One could recite at length the ways in which essential oils have had a bearing upon the rituals and developments of various cultures. But they have seldom been necessities, in the same sense as are food and drink, though they may have motivated man to a greater extent (being the cause for the first extensive trade between the East and the West, and their procurement instigating numerous wars both great and small).

Sesame, already mentioned as a seed oil, seems

to have also been the source of one of the earliest essential oils. Hieroglyphics deciphered from ancient Egyptian tombs mention sesame. But certainly almost as early in use were such prized spices of the pre-Christian world as ginger, cardamon, cassia, and cinnamon. These were the mainstay of early Arabian commerce with the Malabar Coast of India, and were vaguely thought by the European and Mediterranean world to be produced somewhere in Arabia ("the spices of Araby"). Also highly prized by the Mediterranean and Near Eastern civilizations, and treated with much respect in the Old Testament of the Bible, are such exotic essences as anise, basil, balm, coriander, cumin, fennel, frankincense, mace, marjoram, mint, myrrh, mustard, nutmeg, thyme, and many others.

In Assyria and the Fertile Crescent, about 3000 B.C., many plants were grown for their essential oils, as in the gardens of the kings of Babylon. In an era having little concern for sanitary procedures, their fragrance would no doubt be thought a counteraction to the evil and foul smells of the developing cities. The streets were often fumigated with spices before royal visits, and essences were used to "counteract" the plague for centuries. Herodotus mentions their use in embalming, the bodies of the deceased being filled with "the purest myrrh, cassia, and every other sort of spicery except frankincense."

The spice trade has few parallels for wild romance. The Biblical story of the sale of Joseph to the spice merchants indicates that the trade was well established very early. So far as the Western World was concerned, it was chiefly in the hands of Arabian peoples as late as the fourteenth century. The sources of these spices lay in the Far East, with centers in southern China, Malaysia, and India. Apparently there was commerce by sea from the Red Sea and the Persian Gulf to the Malabar Coast of India, to which intermediate point Chinese and other East Indian merchants delivered their spices, often procured in the mysterious East Indian islands. The Arabs, who controlled the traffic in these spices, intentionally spread false rumors of their sources and of the "insurmountable" difficulties in procuring them (requiring death-defying ventures against war-like tribes and hostile environments). During the centuries of Arabian hegemony there was, of course, no little overland caravan trade from India and the Persian Gulf, but early records indicate that much of the Egyptian trade, at least, was consummated along the Red Sea. In later centuries the spice traffic extended throughout southern Europe, with the Arabian peoples of the Near East serving as intermediaries.

Most of the spices used today were already items of commerce at the time of the Greek and Roman empires. Indeed, taxes and tribute on their transshipment sustained the royal coffers of many a sheikdom and kingdom. Around the time of Christ, the Egyptians apparently tried to break the Arabian monopoly of the spice trade, and sent expeditions from Alexandria to India. But is was not until the fifteenth century that the Arabians really lost their dominance. By then information about the Far East had filtered back to the West, especially after Marco Polo's travels to China. Polo had mentioned cloves in the Nicobar Islands, sesame, pepper, ginger, and cinnamon in Ceylon and the Malabar Coast. He spoke of ships arriving at the port of Aden from India, "loaded with spices which they sell to great advantage." In the East Indies this spice trade was of considerable moment too, with Javanese and Chinese merchants doing a thriving business in cloves, nutmeg, pepper, cassia, and other spices destined for shipment to southern India. The Indian way-station for Far Eastern spices was to become gradually less important, as mariners from Europe reached the eastern spice lands. Even in the fifteenth century Venice still lay astride the spice routes through the Mediterranean, and her merchants and bankers gained their profits from the transshipment of spices from the Near East. This was indeed a thriving business, for spices were becoming more and more in demand in the West for such uses as the preservation and flavoring of food on shipboard during increasingly long voyages and for seasoning the favorite dishes of the European nobility.

The control of Venice and Constantinople over the spice trade was broken when the dauntless Portuguese went to the Far East themselves. At first they crossed the Mediterranean and re-embarked on Moslem ships in the Arabian Sea. But between 1497 and 1499 Vasco Da Gama skirted Africa, reaching India to find "Calicut the centre of a region producing cinnamon, ginger and pepper,—and learned that many other aromatic and pungent spices came to this port from more distant spice lands." The handwriting was on the wall for the princes of Venice and the sultans of Egypt.

The voyage of Columbus, which failed to find the sources of Eastern spices but resulted in the discovery of America, is well known. With Spain looking West, Portugal soon dominated the sea routes from Borneo to Europe. Tiny Portugal now had a stranglehold on the spice trade, to be exercised even more zealously than by the Arabians. The monopoly was maintained by subjugating coastal lands in the East, and by curtailing trade by native islanders in that area. No spices were to leave the East Indies except in Portuguese ships!

The reaction of other European powers to the Portuguese monopoly could be predicted. By the 1500's English mariners, and then Dutch, were challenging the Portuguese. In 1574 there was an uprising against the Portuguese in the Moluccas, and in 1579 Sir Francis Drake reached the East Indies by sailing westward around South America. Soon England was in control of India and the nearby spice islands, establishing the famed East India Company. Meanwhile, the Dutch had broken Portuguese power in the East Indies, and were in control of the spice trade there. It is reported that a shipload of cloves and pepper loaded in the spice islands at that time, would bring twelve times its purchase price a few months later in northern Europe.

The Dutch East India Company, patterned after the English, eventually gained control of the producing regions for pepper, most cinnamon, cloves, ginger, mace, and nutmeg. In an effort to maintain the monopoly, spice trees were extirpated in all except certain proscribed areas. For a short time the Dutch were the undisputed masters of the East Indian spice trade. But such avarice was bound to backfire, for economic pressure to break the Dutch monopoly steadily mounted. Before long both the French and the English succeeded in smuggling spice species from the Dutch islands into those that they controlled, including many in

the New World. After nearly 200 years of operation, the Dutch East India Company was dissolved. British ascendancy over the Far East was well established by the early 1800's, and the spice trade was no more to be a monopoly.

Thus did essential oils influence the ebb and flow of empire, and the search for spices create such giants as Diaz, Da Gama, Columbus, Magellan, Drake, Cavendish, and Van Houtman. While all this history-making was taking place, it is strange how little information accumulated about the spice plants themselves. The intentional confusion spread by the Arabian traders was understandable, but even until the recent century much uncertainty existed about the exact identity and growing requirements of the major spices from the East. Table 22-1 indicates the families, species, and common names of many oriental spices as determined by Newcomb.

Some noteworthy spices

BLACK PEPPER, *Piper nigrum*, PIPERACEAE Commercial pepper is obtained from the fruits of *Piper nigrum*, a perennial climbing vine that is indigenous to Ceylon and south India but now extensively cultivated throughout much of Southeast Asia. Clusters of the hard berries (peppercorns) are hand picked, piled in heaps, and left in the sun for several days to dry. Natural fermentation causes the fruits to turn black. Occasionally the clusters are dipped in boiling water to accelerate the blackening and drying. If the weather turns completely unsuitable, artificial drying may be used. The dry berries are detached from the stalks by beating with sticks and by rubbing with the hands. They are then screened to remove impurities and readied for shipment. The black pepper for temperate zone "pepper shakers" is derived by grinding these peppercorns. White pepper is produced by immersing ripe pepper fruits in running water for about two weeks. After the outer skins are softened they are removed easily by hand or by treading. The smooth white kernels are washed and dried on mats in the sun (or sometimes artificially in small smokehouses).

In the producing countries the pepper plant is usually propagated by making cuttings that are

Table 22-1   BOTANICAL SOURCE OF MAJOR ORIENTAL SPICES.

SPECIES	COMMON NAME	FAMILY
INDIAN CENTER		
A) Himalayan foothills		
Cinnamomum tamala	Indian Cassia	Lauraceae
Piper longum	Long pepper	Piperaceae
Murraya koenigii	Curry leaf tree	Rutaceae
Amomum aromaticum	Bengal cardamom	Zingiberaceae
Amomum subulatum	Greater or Nepal cardamon	"
Curcuma zedoaria	Zedoary	"
Zingiber officinale	Ginger	"
Zingiber zerumbet		"
B) South India (Malabar Coast and Ceylon)		
Cinnamomum cassia	Cassia	Lauraceae
Cinnamomum zeylanicum	True Cinnamom (Ceylon)	"
Myristica fragrans	Nutmeg, Mace	Myristicaceae
Myristica malabarica		"
Areca catechu	Betel-nut Palm	Palmae
Piper betle	Betel pepper	Piperaceae
Piper cubeba	Cubeb pepper	"
Piper longum	Long pepper	"
Piper nigrum	Black pepper	"
Curcuma longa	Turmeric	Zingiberaceae
Curcuma zedoaria	Zedoary	"
Elettaria cardamomum	True cardamom	"
Zingiber officinale	Ginger	"
Zingiber zerumbet		"
MALAYSIAN CENTER		
Cinnamomum cassia	Cassia	Lauraceae
Cinnamomum loureirii	Saigon cinnamon	"
Cinnamomum zeylanicum	True cinnamon	"
Myristica fragrans	Nutmeg and Mace (Moluccas)	Myristicaceae
Syzygium aromaticum (*Eugenia caryophyllata*)	Clove (Moluccas)	"
Areca catechu	Betel-nut palm	Palmae
Piper betle	Betel pepper	Piperaceae
Piper cubeba	Cubeb pepper	"
Piper longum	Long pepper	"
Piper nigrum	Black pepper	"
Piper retrofractum (*Piper officinarum*)	Javanese long pepper	"
Amomum kepulaga	False cardamom	Zingiberaceae
Amomum xanthioides	False cardamom	"
Curcuma longa	Turmeric	"
Curcuma zedoaria	Zedoary	"
Elettaria cardamomum	True cardamom	"
Zingiber officinale	Ginger	"
Zingiber zerumbet	"	"
SOUTH CHINA CENTER		
Cinnamomum cassia	Cassia	Lauraceae
Illicium anisatum	Chinese anise or Bastard star anise	Magnoliaceae
Illicium verum	Star anise	"
Amomum xanthioides	False cardamom	Zingiberaceae
Curcuma longa	Turmeric	"
Curcuma zedoaria	Zedoary	"
Curcuma zerumbet	Zedoary	"
Zingiber officinale	Ginger	"

Source: From Newcomb, *Econ. Botany* **17**:127–132, April–June 1963.

stuck in the ground around the base of shade trees, which later support the vines. Layering, marcottage, and separation of rhizomes are occasionally practiced as more certain means of propagation where nurseries are available. Apparently seeds germinate poorly. Production is chiefly a home operation by small land owners; the total crop seldom exceeds 100,000 tons annually, almost entirely from Malaya, Indonesia, and India.

The long pepper is generally considered to be *Piper longum*, although in a loose sense the name refers to a number of species. Burr-like clusters of fruits constitute the item of commerce. The long pepper is considered more a medicinal than a condiment, since it contains an alkaloid that induces salivation and numbness. In India it is often cultivated along with *P. nigrum*.

CINNAMON, *Cinnamomum zeylanicum*, LAURACEAE   As we have seen, cinnamon is one of the classic spices in the Eastern spice trade. It comes from the bark of cinnamon trees native to Ceylon and India. "Cassia," also a famed name in the early spice trade, is from another species, *C. cassia*. In Biblical stories the two may have been confused under the name of cinnamon.

Bark of the small cinnamon tree contains as much as 1 percent of essential oil, which is in turn nearly three-fourths cinnamic aldehyde. The bark is stripped from the trees just after the monsoon season (from May to August), preferably from second-year wood. An active cambium in the twigs makes peeling relatively easy then. This is a hand operation, accomplished with a knife. The best grade of bark forms hollow tubes, which after thorough drying are scraped free of epidermal particles to become the cinnamon "quills" of commerce. Fragments and less perfect quills are ground to yield powdered cinnamon.

Most cinnamon is produced on small native-owned plantations. The trees may be propagated by seed or by cutting; seedlings are started in the nursery and later planted at spacings suitable to land conditions.

CLOVE, *Eugenia caryophyllata*, MYRTACEAE
Cloves are another of the exotic spices once under complete control of the Dutch monopoly. Their

production was confined then to a single island of the Moluccas (the Spice Islands), where the tree was grown under close supervision. Clove trees had been eliminated from all other East Indian territory.

The highly aromatic clove condiment is the dried unopened flower bud and twig tip. Such cloves were used early in China, where it was customary for court officials to perfume their breath with cloves before addressing the sovereign. Clove trees are small, and the buds are generally harvested by women who hand-pick them as they begin to color. The plucked twig tips are sun-dried and are then ready for marketing as the familiar spice. Most production comes from Zanzibar and Madagascar, but there is some growing in Indonesia, where "oil of cloves" is extracted by distillation. Clove oil consists largely of eugenol, which in turn can be used as raw material for the synthesis of vanillin. Clove oil finds considerable use in disinfectants, as a flavor in dentifrices, and in perfumery blends.

VANILLA, *Vanilla planifolia*, ORCHIDACEAE   This climbing orchid (Fig. 22-7) and other related species yield the world-famous "vanilla bean," the fully-grown but unripened and fermented seed pod of the species. Fermentation develops the characteristic vanilla odor, and the essential oil is extracted with alcohol to yield the natural vanilla extract of commerce. Since the active fragrance, vanillin ($C_8H_8O_3$), can now be synthesized from other materials, the vanilla plant is perhaps more of historical than present-day importance.

Bernal Diaz, an officer under Cortés, was perhaps the first European to note the vanilla spice when he observed its use by Montezuma in compounding the chocolate drink "chocalatl." In the Aztec empire vanilla beans were valued as tribute. The Spanish were quick to import vanilla beans, and even today cultivation and production of vanilla beans is a reasonably important minor industry in Mexico. During the 1800's vanilla was introduced into the Far East and into Madagascar, where plantation growing of vanilla is still part of the agricultural economy. Most of the present-day production comes from islands of the Western Indian Ocean, and from Uganda, where perhaps

FIGURE 22-7. *Commercial vanilla extract is made from the pods of* Vanilla planifolia. [*Courtesy USDA.*]

upwards of 50,000 acres are under cultivation and perhaps 20,000 workers engaged in vanilla production.

The horticulture of vanilla growing is reasonably complicated. Land must be cleared, except possibly for trees that are to serve as supports for the climbing vanilla plants, and mechanical supports provided. Vanilla prefers shade, so suitable shade trees and windbreaks are necessary. After cultivating the soil, cuttings of the vanilla vine are planted, and the vines trained to climb over the supports. Since vanilla roots grow mostly near the surface, it is sound practice to maintain a layer of humus about the plant. A certain amount of pruning is necessary, and a watch must be maintained for parasites and diseases. The flowers are often hand-pollinated and the fruits reduced to the proper number to allow full development. With this much attention required, vanilla tends to be grown in

areas of inexpensive labor. Vanilla requires a hot, moist, tropical climate. It has done very well in Madagascar, in various parts of the West Indies, and in Tahiti, Fiji, and other Far Eastern islands.

THE MINTS, *Mentha* SPP., LABIATAE Distillation of oil from peppermint (Fig. 22-8), *Mentha piperita*, was known to the Egyptian pharaohs. The species may be a natural hybrid, and it seldom sets seed. Other important mints are spearmint (*M. spicata*) and *M. arvensis*, a source of menthol. Commercial production of mint oil is not great, totaling only somewhat more than 1,000 tons annually in the United States. But mint plants are often utilized as garden plants and herbs around the home. Commercial mint production is primarily on muck soils, today principally in the Columbia River basin of Oregon and Washington.

Both spearmint and peppermint are propagated vegetatively from rhizomes or young plants. Mint is usually planted in rows, which are cultivated to control weeds the first year. Typically the mint is "plowed down" before frost, and spreads from the buried parts into a complete field of "meadow mint" the second year. This mint is cut with a mowing machine, and left to dry in the field until moisture content is reduced to approximately 35 percent. It is then taken to a still for extraction of the oil by steam distillation. Distillation usually takes about 45 minutes, and the spent hay is spread back on the fields.

Peppermint oil, consisting largely of menthol, is used for antiseptics, lotions, dentifrices, medicinals, and culinary flavors. Yields are reportedly not much over 30 pounds of distilled oil to the acre.

OREGANO, LABIATAE Oregano can serve as our example of savory herbs—those used in cookery. Several plants turn up as commercial oregano, among them *Origanum vulgare* and various species of *Lippia*. *Origanum*, now called wild marjoram, was brought to America by the early colonists from the Mediterranean area. American oregano seems to lack the rich flavor typical of Old World oregano. Many diverse species of both the *Labiatae* and the *Verbenaceae* have been identified as sources of oregano since the early days of interest in the spice. Origanum oil has been field-distilled

FIGURE 22-8. *Peppermint* (Mentha piperita) *contains an essential oil in the foliage. The oil, produced under long days, is extracted for flavors and medicinals.*

from *Coridothymus capitatus*, which grows wild in Andalusia, Spain. Syria has also been a source of commercial oil.

OTHER SPICES   Allspice is derived from *Pimenta officinalis*, Myrtaceae, produced mainly in Jamaica; angelica oil is distilled from the roots of *Angelica*, Umbelliferae; anise is derived from the seed of *Pimpinella anisum*, Umbelliferae; balm comes from distillation of the leaves of *Melissa officinalis*, Labiatae, native to the Mediterranean

area; cardamon is the aromatic dried fruit of *Elettaria cardamomum* of the Zingiberaceae (ginger family); caraway, celery, coriander, cumin, and fennel are all seed spices from genera of the Umbelliferae; ginger is obtained from the rhizome of a large perennial herb, *Zingiber officinale*, Zingiberaceae, native to tropical Asia; mace is derived from the aril surrounding the nutmeg kernel, the fruit of *Myristica fragrans*, Myristicaceae; mustard is a spice obtained from the seeds of various species of *Brassica*, Cruciferae, an important genus among food plants; nutmeg is the seed of *Myristica fragrans*, Myristicaceae, native to the Molucca islands; parsley is from the foliage of *Petroselinum hortense*, Umbelliferae; sage is obtained from the foliage of *Salvia officinalis*, Labiatae; turmeric serves as a yellow dye and a kitchen flavoring, obtained from rhizomes of *Curcuma longa*, Zingiberaceae.

Essential oils for perfumes and cosmetics

Essences used in perfumery are most frequently obtained from flowers; often those used in soaps and cosmetics are derived from vegetative parts. Most flowers have nectar glands that secrete essences attractive to insects—an aid in pollination. Foliage oils generally occur in epidermal glands or hairs of uncertain usefulness to the plant.

Oils cherished for their scent are often extracted by absorption into cold fat, a method termed **enfleurage.** Enfleurage is typically used for delicate essential oils when the harsher distillation process may have deleterious effects. Enfleurage is the chief means of extracting floral odors in the famous perfume centers of Provence, France. There huge volumes of flowers are grown for the perfume industry, which, when harvested, are applied to plates covered with cold fat that absorbs the floral essences. Later the saturated fat is subjected to alcoholic extraction of the essential oil.

More commonly, and especially with fragrances derived from vegetative portions of the plant, distillation, solvent extraction, or expression are practiced, much as they are with edible oils. In distillation the plant is generally boiled in water, the distillate separating into an immiscible oil layer that floats on top of the condensed water. Sometimes steam treatment substitutes for the boiling. Direct extraction with solvents is used occasionally with flowers and commonly with other aromatic plant parts. It is usually the simplest and least expensive extraction method; the solvent dissolves the fragrance and is then evaporated itself, leaving the essential oil behind. Expression involves the squeezing out of the essential oils, usually practiced with gross plant parts, such as rinds of citrus or whole seeds and fruits of various kinds.

Most essential oils deteriorate or polymerize to resins rather readily. Delicate essential oils like those used for perfumery are consequently stored in hermetically sealed bottles kept in dark cool cellars. These essences are blended according to secret recipes to yield the famed perfumes of commerce. As we have seen, fragrances used to mask foul odors have been in demand since the earliest days of civilization, and have often been involved in the religious rituals of the times. The art of perfumery has discovered through the ages various additives that extend and make more permanent the expensive floral essential oils. Sometimes these are animal oils, or the fatty oils and resins from plants. (Ambergris, a fatty substance from the intestine of the sperm whale, is a highly valued ingredient.) Synthetic materials are coming more and more into use.

ROSE OIL, *Rosa* SPP., ROSACEAE Rose oil, or attar of roses, is chosen here as an example of an expensive floral perfume oil. It is obtained from flowers of various species of *Rosa*. Shrub roses such as *R. damascena* are much grown in southern Europe and Asia Minor. They bloom in late spring, at which time the flowers are collected for distillation of the essential oil (Bulgaria) or for enfleurage and solvent extraction (France). Less than one-half gram of oil is obtained from each 1,000 grams of flowers. The principle constituent of rose oil is citronellol; lesser amounts of such essential oils as geraniol, nerol, linalol, and other organic compounds are also present. Because pure rose oil is so expensive, it is generally extended with less expensive essential oils from other sources. Flowers are generally harvested by women and children when the rose is in the late bud stage.

PATCHOULI, *Pogostemon cablin*, LABIATAE Pat-chouli is a mint native to the Philippines, widely grown in the Orient since time immemorial. In India it has served as an insect repellent and a constitutent of wardrobe cachets. Patchouli was once prized as a fine perfume, but later came to be associated with women of low repute and fell into disfavor. It is still regarded as one of the finest fixatives for heavy perfumes, and is used for scenting soaps and cosmetics. Patchouli is much grown in the Seychelle islands, and in various parts of Southeast Asia. It seldom flowers, and is propagated vegetatively by cuttings. It is grown under light shade and is generally mulched to preserve moisture and provide some fertilizer. Harvesting takes place about six months after planting, and is continued every three or four months; a planting is said to last for several years before replanting is necessary. The plucked leaves are kept overnight in small heaps to induce a fermentation that is reputed to increase the yield of oil. They are then spread out to dry in sheds or barns. Steam distillation is utilized to retrieve the oil, which makes up 3 to 5 percent of the dry weight of the leaf. The oil consists mainly of sesquiterpenes, which upon aging develop a cedar-like odor. Probably no more than 100 tons of oil is produced annually, world-wide.

OTHER ESSENTIAL OILS OF IMPORTANCE TO PERFUMERY Bergamot is expressed from the peel of *Citrus aurantium bergamia*, cultivated in southern Italy where about 150 tons of oil are produced annually. Camphor oil is distilled from the heartwood of *Cinnamomum camphora*, the camphor tree, which grows wild and under cultivation from Japan to southern China. Cananga or ylang-ylang is the famed essence of *Canangium odoratum*, Annonaceae, which is prized for fine perfumes and is much grown in Java, where the essence is derived from the flowers. Cassia is the essence derived from *Cinnamomum cassia*, Lauraceae, a tree that grows in China and was especially important in the early spice trade; the oil, mostly cinnamic aldehyde, is distilled from terminal foliage. Citronellol is obtained mainly by distillation of the citronella grass, *Cymbopogon nardus*, Graminae, much grown in Ceylon and the Far East; the oil has been substituted for as an insect repellent by modern synthetics, and can also be fractionated from other sources of essential oil, such as menthol; there is considerable production from Formosa and from Guatemala. The Eucalyptus tree provides several medicinal and perfumery oils from distillation of the foliage, chiefly from *E. citriodora*, Myrtaceae, native to southern Australia; the oil is produced in Spain, Portugal, and Brazil as well as Australia, and similar Eucalyptus oils are extracted from other species in Southeast Asia.

Geranium oil is distilled from the flowers and foliage of *Pelargonium*, Geraniaceae, mostly on the island of Reunion and in Algeria. Lavender oil comes from *Lavandula officinalis*, Labiatae, a fragrant herb much cultivated in southern France, from which more than 100 tons of production is obtained in some years. Lemon grass is another of the *Cymbopogon* species, *C. flexuosus*, grown and used much like citronella in both the East and West Indies; a component of the oil is used for commercial synthesis of vitamin A. Palmarosa oil is distilled from another *Cymbopogon*, *C. martini motia*, grown in central India. Petitgrain is the oil distilled from the foliage of the bittersweet orange, *Citrus aurantium amara*. Rosemary oil is produced chiefly in Spain and Tunisia, from *Rosmarinus officinalis*, Labiatae. Thyme oil is distilled from the flowering tops of several species of *Thymus*, Labiatae, chiefly in Spain. Tuberose yields an essence obtained by enfleurage or solvent extraction from the flowers of *Polianthes tuberosa*, Amaryllidaceae.

Industrial essential oils

A number of aromatic oils that have at least essential oil components serve industrial rather than food and esthetic purposes. Turpentine and camphor stand out as two of the most consistently useful, and will serve here to exemplify essential oils used industrially.

CAMPHOR, *Cinnamomum camphora*, LAURACEAE The camphor tree is a close relative of the cinnamon spice tree (*C. zeylanicum*), but yields camphor from the wood rather than from the bark. The camphor molecule is not uncommon, and is also obtained from various species in the Labiatae (mint), Compositae, and Dipterocarpaceae families. It has long been known that camphor can be synthesized by oxidation of borneol, a fairly common ingredient of essential oils.

The true camphor tree is native to Japan and eastern China. Originally the chipped wood was crudely distilled locally. Such destruction of the trees of course threatened the industry, and it is perhaps well that other sources of camphor have been found. It is said that as much as three tons of camphor could be derived from a single tree, and that the wood chips contained as much as 5 percent of the essence. Camphor finds use in medicinals, liniments, and insecticides, but was especially important for making celluloid from nitrocellulose (an industrial use that once accounted for most of the camphor production). Celluloid has now been supplanted by a wide variety of plastic materials; consequently, the demand for camphor has diminished.

TURPENTINE, *Pinus*, PINACEAE Turpentine is a mixture of resins and essential oils that was originally derived by injuring the sapwood of coniferous trees, especially the longleaf pine, *Pinus palustris*, and the slash pine, *P. elliotti*. It is the basis of the **naval stores** industry, so named because in the days of wooden sailing ships the rosin residue left after turpentine removal was vital for the caulking and treatment of hulls and rigging. Naval stores were first collected by digging a pit and filling it with pine logs. They were then set afire and immediately covered with soil. The heat from the smoldering fire caused oleoresins to drip from the wood and accumulate in the bottoms of the pit. When cool, dirt and ashes were removed and the dirty resin dug out of the waste to provide the gum resin. Men working in these pits got it on their feet, hence the name "tar heels," which is used for North Carolinians even today. The old pits used in this manner are still to be seen throughout the slash and longleaf pine region of the southeastern United States.

In the early years of American colonization, when forests seemed limitless, turpentine was collected in the southeastern pine belt by "boxing" the tree—that is, cutting a deep hollow in the base, into which the exudate drained from chipping the bark above. Even today some boxing continues, but within recent decades there has been gradual change to more progressive and less wasteful means of collection. Today only trees 9 inches or

FIGURE 22-9. *Gutter* (apron) *and cup for collecting turpentine in pine.* [*U.S. Forest Service.*]

greater in diameter are tapped. The bark near the base of the tree is shaved off, and into a shallow cut made a few inches above the ground an apron or gutter is fitted. Below this is placed a collecting vessel, into which the gutter drips. A shallow wound is made in the bark above the gutter, from which drips the oleoresin containing the turpentine (Fig. 22-9). During the turpentining season (spring and summer), the "chipper" makes approximately weekly rounds to each tree and renews the wound by slicing away a thin section of bark or "streak" above the last one. The streak is now routinely treated with sulfuric acid, found to increase the duration of turpentine flow. If sulfuric acid spray is used, chipping need be done no oftener than every second or third week.

Formerly, the oleoresin was locally distilled in wood-fired iron kettles. The turpentine was condensed in iron coils, and the rosin that remained behind was independently barreled. In later years water came to be added to the oleoresin, and distillation was accomplished in copper stills, in

effect becoming steam distillation. Turpentine is immiscible with water, and can be separated quite readily. More recently collected oleoresin has been taken to processing centers, where modern stainless steel batch stills and flash stills effect the separation. The rosin remaining after extraction of the turpentine is used in preparing paints and varnishes, as a sizing for paper, in polishes and waxes, and for many other purposes. It keeps smooth surfaces from being slippery, hence its use by players of string instruments and by baseball pitchers.

Turpentine production in the United States runs somewhat more than 600,000 barrels (50 pounds each) annually; minor quantities are imported. Only an eighth of this is "gum turpentine," obtained by streaking pine bark as described above. Nearly another fourth is obtained from distillation of wood, primarily old pine stumps and roots. More than half of the production is now recovered from sulfate paper processing, utilizing southeastern pines for pulp. Spain, Greece, Russia, India, Yugoslavia, Austria, Indonesia, and Mexico also produce turpentine oleoresin from various species of pine, but exact production figures are not available.

Wood turpentine contains about 80 percent alpha pinene. The pine oil fraction of the wood distillate (a secondary product of the distillation process) is a complex mixture of terpene hydrocarbons, aldehydes, alcohols, ethers, ketones, and phenols. The pinene fraction is used chiefly as a solvent; the pine oil fraction is used in the textile industry as a wetting agent, and as an inhibitor of bacterial growth. It is also used in paper making as a wetting agent during the coating stages. Sulfate pulp mill turpentine is condensed from the vapor that arises during the cooking of the pine chips. Depending upon species and condition of the pine wood, as much as 4.3 gallons of crude turpentine may be obtained for each ton of air-dried pulp produced. Sulfate turpentine must usually be further washed or treated to remove the sulfate odors. During the sulfate processing the resinous constitutents are saponified, and the solution is known as **tall oil;** this oil is composed essentially of oleic and lineoleic acids, and of course serves such purposes as were described in the section on vegetable oils.

OTHER VOLATILE OILS USED INDUSTRIALLY Volatile oils of relatively minor importance, usually recovered by steam distillation from tree parts, include: cedarwood, from *Juniperus virginiana,* used in soaps, polishes, and perfumes; conifer leaf oil, from the foliage of various spruces, hemlocks, and junipers, used mainly for ointments, soaps, and shoe polishes; sweet birch oil, from the bark of *Betula lenta,* used in drugs, candies, and for flavorings; sassafras oil, from various parts of *Sassafras albidum,* used in dentifrices, carbonated beverages, and medicinals; eucalyptus oils, previously mentioned; and sandalwood, from the heartwood of *Santalum album,* a tree of southern India, used chiefly in cosmetics and soaps.

MEDICINALS AND RELATED DERIVATIVES

Among medicinals, too, there is much overlap with other sections of this chapter. For example, castor oil has been a purgative since the earliest times; gum arabic, primarily an adhesive, serves as a demulcent too; olive oil has traditionally been used in ointments and as a lotion; clove spice and clove oil serve as stomachics and are used in a wide variety of remedies; camphor has long been used as an antiseptic and as a stimulant; betel (*Areca*) is an anthelmintic; nicotine (from tobacco) makes an effective natural insecticide.

Perhaps the earliest extractive used medicinally was gum arabic, derived from *Acacia* spp. (Leguminosae), which grow wild in northern Africa; it is recorded by Herodotus as being of medicinal importance before the fifth century B.C. Interestingly, gum arabic is still listed in the Pharmacopoeia, although today it is used only as a vehicle (emulsificant) in such remedies as cough syrups and tinctures. Gum tragacanth (*Astragalus,* Leguminosae), a spiny shrub of the Mediterranean area, yields a similar resin that has been used medicinally since time immemorial.

Mankind has been especially concerned with and awed by curatives, which until recent years were almost exclusively derived from plant sources, often from species found only in the wild. The fear, mysteriousness, and highly personal nature of disease has led man to try almost any remedy offering hope. It is no wonder that much superstition surrounds the choice of medicinals, especially under primitive conditions where symptoms rather than causes must be attacked. And who is to say that if the patient is convinced of medicinal efficacy that this of itself is not therapeutic! In modern experiments a placebo has proved beneficial in many cases. Civilized man, of course, no longer regards disease as "divine punishment" or "visitation of the devil"; but he is still as intensely interested as ever in finding more effective controls for disease. His search has involved systematic screening of many plant species. Among other things, the remedies discovered for various diseases have helped increase life expectancy from an average of about 20 years at the time of the Roman Empire, to something over 70 years in medically advanced parts of the world today.

The fall of Troy, Hannibal's failure against Rome, the defeats suffered by Lee during the Civil War, and the French abandonment of their Panama Canal project have all been attributed to malaria; had quinine been known earlier it might have changed the course of history! Morphine, derived from opium, has long eased man's pain—and also led to drug addiction. Hippocrates was well acquainted with oil of wintergreen (mainly methyl salicylate), recommended for many maladies; its investigation led eventually to the synthesis of acetyl salicylate, aspirin, in 1853, of which today about 40 tons are consumed daily in the United States alone! Lysergic acid diethylamide, LSD, was accidentally discovered years ago in the dust from an obstetrical drug made from ergot (*Claviceps purpurea,* Ascomycetes); today its use as a hallucinating agent creates social problems, even though it may prove promising in the treatment of mental illness. Domesticated animals frequently die from toxic principles consumed in graze.

It is more and more recognized that the folklore about medicinal plants and the home remedies of an earlier era are not all based on sheer superstition. The biological influence of extracts ranges from the extreme toxicity of such substances as prussic acid (as in manioc juice) or potassium fluoroacetate (in *Dichapetalum cymosum*) to the relative inertness of oils and sugars. The search continues for unusual substances that have profound physiological effect when administered in small doses. Important discoveries in recent decades have produced the antibiotics, such as penicillin, and the plant growth regulators, such as gibberellin, both from molds. Useful plant sources of the rather commonplace emollients, cathartics, and even more specialized medicinals such as cardiac glycosides (digitalis), have been known for many years. The alkaloids, a class of nitrogenous bases, have been outstanding for the profound physiological responses they produce in animal systems. Alkaloids occur commonly in such plant families as the Solanaceae and Apocynaceae, especially in tropical genera. Many of the medicinals we list are important for the alkaloids they yield; for example, quinine (from *Cinchona*, Rubiaceae) and reserpine (from *Rauwolfia serpentina*, Apocynaceae) (see Table 8-3).

A recent list of medicinal plants used by the Maori of New Zealand runs to well over 100 species and indicates well how comprehensive is the search for healing botanicals even in primitive and circumscribed cultures. Table 22-2 compares plants included in the first edition (1820) of the United States Parmacopoeia (U.S.P.), and the sixteenth revision of 1960. Note how many of the old folk remedies have been dropped over the years. It cannot be assumed that all these older remedies were without merit; possibly the modern chemist should analyze more carefully species once thought effective, isolating components that might be medicinally useful. Compared with the many scores of species dropped there have been only 20 new additions to the most recent revision.

For want of a better place to pigeonhole them, insecticides, growth regulators, and similar biologically active agents obtained from plant sources are mentioned in this section too. Like medicines, they have profound influence on biological systems in small doses. They are found in a wide variety of plants and plant parts, from simple molds to the flowers of advanced Spermatophytes. Some discoveries have resulted directly from biochemical research, and in turn the isolation of natural regulators has helped elucidate the functioning of biological systems. Fundamental work on auxins and other growth hormones led directly to the immense modern pesticide industry, of which today most herbicides, insecticides, and fungicides are synthesized rather than obtained from plant materials.

In general, many medicinals and related products are today chemically synthesized rather than obtained from higher plants (almost the only source even a few generations ago). A sophisticated chemical industry has made possible economical substitution of even more potent products patterned after the natural ones; settlement of the world's land and the development of an increasingly efficient agriculture have built the demand. There is now a revival of interest in botanicals, especially those that might yield antibiotics, and in garden pesticides that would not be as lethal to higher organisms as are some of the potent chemicals used today.

Aloe, Aloe *spp.*, *Liliaceae*

Various *Aloe* species have been used medicinally since antiquity. Juice from the fleshy leaves is used as a laxative, well known to the pre-Christian Greeks. In the West Indies, juice of *A. barbadensis* is used as an antiseptic applied to minor cuts and skin injuries. (Sometimes, mixed with rum and honey, it is employed to treat colds.) The pulp has at times been popular for treatment of x-ray burns. There has been considerable confusion about the identity of medicinal Aloes. *A. perryi*, native to islands in the Arabian Sea is one of the most valued. *A. barbadensis* is probably of Old World origin, but is widely naturalized in the West Indies (and often cited as *A. vera* or *A. vulgaris*). *A. ferox*, a distinctive arborescent species of southern Africa, has been considered as a source of cortisone.

A. ferox was brought into cultivation in England as early as the middle 1700's, and "gum aloe" was offered by London pharmaceutical houses before the nineteenth century. Leaves are cut to release the juice, which drips from the butt end; the juice is collected and the bitter principle concentrated by boiling. In the West Indies cut leaves are stacked in V-shaped troughs while the juice drains. Production comes almost entirely from wild plants in Africa; in the West Indies the naturalized *E. barbadensis* is frequently propagated by offsets, especially in Aruba, where the industry has been an important source of livelihood. A small industry has developed in Florida for aloe pulp, used in ointments, lotions, and similar products.

Table 22-2 A BOTANICAL COMPARISON SAMPLING OF THE UNITED STATES PHARMACOPOEIA OF 1820 AND 1960.

OFFICIAL TITLE	SCIENTIFIC NAME 1820	SCIENTIFIC NAME 1960	FAMILY
Acacia	*Acacia vera*	*A. senegal*	Leguminosae
Aconite	*Aconitum neomontanum*	NLO	Ranunculaceae
Allium	*Allium sativum*	NLO	Liliaceae
Aloe	*Aloe spicata*	*A. perryi*	Liliaceae
	A. barbadensis	*A. barbadensis*	"
		A. ferox	"
		A. africana	"
		A. spicata	"
Ammoniacum	*Heracleum gummiferum*	NLO	Umbelliferae
Almond	*Amygdalus communis*	NLO	Rosaceae
Angustura	*Bonplandia trifoliata*	NLO	Polemoniaceae
Anise	*Pimpinella anisum*	*P. anisum*	Umbelliferae
		Illicium verum	Magnoliaceae
Anthemis	*Anthemis nobilis*	NLO	Compositae
Armoracia	*Cochlearia armoracia*	NLO	Cruciferae
Assafetida	*Ferula assafoetida*	NLO	Umbelliferae
Aurantii Cortex	*Citrus aurantium*	*C. sinensis*	Rutaceae
Avenae Farina	*Avena sativa*	NLO	Gramineae
Belladonna	*Atropa belladonna*	*A. belladonna*	Solanaceae
Benzoin	*Styrax benzoin*	*S. benzoin*	Styracaceae
		S. parralleloneurus	"
Cajeput	*Melaleuca cajuputi*	NLO	Myrtaceae
Camphor	*Laurus camphora*	*Cinnamomum camphora*	Lauraceae
	Dryobalanops camphora		Dipterocarpaceae
Canella	*Canella alba*	NLO	Canellaceae
Capsicum	*Capsicum annuum*	NLO	Solanaceae
Cardamom	*Elettaria cardamomum*	NLO	Zingiberaceae
Caraway	*Carum carvi*	NLO	Umbelliferae
Caryophylli	*Eugenia caryophyllata*	*Eugenia caryophyllus*	Myrtaceae
Cascarilla	*Croton eleutheria*	NLO	Euphorbiaceae
Cassia Fistula	*Cassia fistula*	NLO	Leguminosae
Cassia Marilandica	*Cassia marilandica*	NLO	Leguminosae
Catechu	*Acacia catechu*	NLO	Leguminosae
Chenopodium	*Chenopodium anthelminticum*	NLO	Chenopodiaceae
Cinchona	*Cinchona lancifolia*	*Cinchona* spp.	Rubiaceae
	C. oblongifolia		"
	C. cordifolia		"
Cinnamomum	*Laurus cinnamomum*	*L. cinnamomum*	Lauraceae
Colchicum	*Colchicum autumnale*	*C. autumnale*	Liliaceae
Colocynthis	*Cucumis colocynthis*	NLO	Cucurbitaceae
Conium	*Conium maculatum*	NLO	Umbelliferae
Copaiba	*Copiafera officinalis*	NLO	Leguminosae
Coriandrum	*Coriandrum sativum*	NLO	Umbelliferae
Cornus Florida	*Cornus florida*	NLO	Cornaceae
Crocus	*Crocus sativus*	NLO	Liliaceae
Cubeba	*Piper cubeba*	NLO	Piperaceae
Digitalis	*Digitalis purpurea*	*D. purpurea*	Scrophulariaceae
Dolichos	*Dolichos pruriens*	NLO	Leguminosae
	Pubes leguminis	NLO	"
Dracontium	*Dracontium foetidum*	NLO	Araceae
	Ictodes foetidus	NLO	"
	Symplocarpus foetidus	NLO	"
Dulcamara	*Solanum dulcamara*	NLO	Solanaceae
Elaterium	*Momordica elaterium*	NLO	Cucurbitaceae

Source: From Hershenson, *Econ. Botany* **18**:342–356, 1964.
NLO = no longer official.

Box 22-1 THE MODERN "DRUG CULTURE"

On a per capita basis the addictive "drug problem" was probably as serious some generations ago as it is today, although it caused relatively little stir until recently. Beginning in the middle 1960's considerable interest in psychedelics or "mind-expanding" drugs developed, particularly among younger people. Curiosity centered not only on the age-old natural hallucinogens such as marijuana and peyote, but on synthetics such as LSD (lysergic acid diethylamide). Even the sniffing of glue or other aromatics having a physiological influence was undertaken by the unsophisticated.

Emotions, both pro and con, have been intense, forcing political action that was often based on rather little sound information. Highly restrictive laws have been passed in some localities, leading to such incongruities as an elderly judge (at liberty to enjoy martinis at the end of his day) perhaps sentencing some youth to decades of imprisonment merely for possession of a marijuana cigarette. All evidence indicates that marijuana is less habituating and no more addictive than are tobacco and alcohol.

However, serious human degradation has resulted from use of depressants such as the narcotic heroin, even though used only by a small minority (perhaps one in a thousand). Whether the harsh measures taken to limit availability and use of addictive drugs are wise is a matter for debate. When availability is limited, prices rise, and more serious crime is required for the addict to gain funds enough to satisfy his craving. Great Britain has tried regulated availability, treating the matter as a medical rather than a criminal problem, thus eliminating the potential for large profit that draws organized crime.

Whatever the answers may be, drugs have had an extraordinary influence on contemporary "life styles." In America it is said that nine out of ten adults drink alcohol, and that at least one of them becomes a "problem drinker." Some ten percent of the people are thought to be smoking maijuana. Equally as many take tranquilizers; a like percentage, barbiturates almost to the point of addiction (ostensibly to relieve insomnia). Many have tried amphetamines (to prevent drowsiness or to lose weight). Almost everyone drinks the stimulant caffeine, in coffee, tea and colas. More than half smoke tobacco in spite of medical hazards. "Pill taking" without medical supervision is widely accepted; some sort of pill seems to be offered for every problem!

Psychoactive drugs have been part of human culture since the beginning of time; it is unrealistic to suppose that people will suddenly forsake their euphoric and relaxing qualities. Society's problem may well be to learn how best to live with a modicum of "drug culture," avoiding police-state repressive tactics that could be even more injurious to individual well-being than are the drugs.

The main types of drugs in use are these:

DEPRESSANTS: habituation very strong; addiction, moderate to strong (except essentially nonaddictive inhalants).

    alcohol—natural fermentation products, such as beer, wine, whiskey.

    barbiturates—synthetic chloral hydrate, phenobarbital, etc.

    inhalants—synthetic aerosols and solvents.

    narcotics—mostly natural opiates (heroin, morphine).

    tranquilizers—synthetics such as Miltown, Librium, etc.

PSYCHEDELICS: habituation, moderate to low; not addictive.

    marijuana—in various forms (hashish, etc.).

    hallucinogens—various natural products such as scopolamine or mescaline, and synthetics such as DMT, LSD, and STP.

STIMULANTS: habituation, strong (except essentially nonhabituating antidepressants); not addictive.

    amphetamines—synthetics, such as Benzedrine and Methedrine, for depression and fatigue.

    antidepressants—synthetics such as Elavil, for relief of anxiety.

    caffeine—naturally occurring in coffee, tea, cola.

    cocaine—alkaloids in foliage, for feeling of exhilaration.

    nicotine—alkaloids from tobacco (nicotine is very toxic).

Antibiotics

Natural antibiotics have generally held their own in competition with synthetics. True, it is possible to synthesize such well known "active ingredients" as tetracycline. But so efficient are modern methods of growing various microorganisms on prepared media under controlled conditions that economics has usually favored this means of production even where alternatives exist. This is especially so when high-yielding strains of organisms are developed.

The Chinese realized 4,000 years ago that green mold applied to skin ulcers aided in their cure and stopped the festering. Pasteur, too, realized that his disease-causing bacterial cultures might be destroyed by other microorganisms. But practical application of this ecological fact of life among microorganisms had to await Sir Alexander Fleming's observation that a green mold (*Penicillium*) contaminating his bacterial cultures destroyed the bacteria. From this chance observation, made in a British hospital in 1928, arose the modern concept of pitting one microorganism against another to prevent disease. Penicillin is the name given the active ingredient produced by *Penicillium notatum* (and other species), Ascomycetae, which started the rash of antibiotic discoveries. Since then useful molds and microorganisms of many genera have been unearthed, to yield such well known antibiotics as aureomycin, bacitracin, chloromycetin, neomycin, polymixin, streptomycin, terramycin, and many others.

Many technological problems had to be worked out in order to make the production of antibiotics practical. Growing media (generally utilizing corn steep liquor) had to be devised and kept sterile yet properly aerated to support the mass growth of what are essentially surface-living molds. Suitable methods of solvent extraction, purification, and crystallization into stable form had to be found. To increase yields of antibiotics new species or strains of microorganisms had to be sought continuously, or new mutants had to be created by irradiating microorganisms with ultraviolet light. As disease organisms became resistant, through mutation, to antibiotics to which they were originally susceptible, new antibiotics had to be found. The original penicillin was simply the filtrate of the nutrient broth in which *Penicillium notatum* was grown. Eventually it was proven that such solvents as ether were suitable for extracting the active penicillin, and that through subsequent precipitations a purified product could be produced that was far more potent than Fleming's original mold.

Modern production of antibiotics is not unlike the fermentation of beer, discussed in an earlier section. Highly pedigreed molds (analogous to the yeast in brewing) are kept in a sterile environment for starting cultures in small flasks. These are used to inoculate larger "seed tanks," which in turn start the mold in huge vats containing thousands of gallons of nutrient broth. At a determined point of maximum growth the solution is filtered, subjected to solvent extraction, and the purified antibiotic precipitated, all under aseptic conditions. Sources of the most useful antibiotics have been fungus molds and Actinomycetes, yielding substances mainly effective for controlling bacteria. Since not all disease-producing bacteria are equally affected, a great deal of empirical screening goes into selecting and proving-out an antibiotic. Under modern conditions millions of dollars may be invested before any income derives from the medicinal. And for every useful antibiotic discovered, thousands of others have been dropped in the screening because they are insufficiently effective, competitively impractical, or because they produce undesirable side effects.

Belladona, Atropa belladona, *Solanaceae*

Called also deadly nightshade and death's herb, belladona was well known to the herbalists, who wrote much of medicinal plants (both real and fanciful) during the Middle Ages. The juice of belladona was first used in salve preparations, and apparently not until the nineteenth century were extracts and tinctures much used. Belladona alkaloids are most abundant in physiologically active tissues; atropine, scopolamine, and hyoscyamine are medicinally the most important. In preparation of alkaloidal extracts the plant parts are typically dried for several weeks, after which solvent extraction (with ether or ethyl acetate) and crystalization of the alkaloids is undertaken. These serve as stimulants to the sympathetic nervous system (and hence as antidotes for opium alkaloids), for dilation of the eye pupil in optical treatment, in ointments, for asthmatic paroxysms, as a circulation stimulant, and in many other ways.

Cascara, Rhamnus catharticus, *Rhamnaceae*

Both *R. catharticus* and *R. purshiana* are sources of the cathartic usually referred to as cascara. The former was used in the Middle Ages, but the latter is the official drug of the United States Pharmacopoeia; it is called "cascara sagrada." The bark of these and perhaps other buckthorns is pulverized to yield the medicinal. *R. purshiana* is native to the Pacific Northwest of the United States, where it was discovered by the Lewis and Clark expedition.

Cocaine, Erythroxylon coca, *Erythroxylaceae*

Cocaine is almost as renowned a sedative as opium. It appears capable of suspending emotion and perception, and has been used as a local anesthetic. But perhaps its greatest fame rests upon its use as a stimulant for porters traversing the rigorous Andean mountain trails. It is said that workers who chew coca can carry on with superhuman endurance for days, even with little food. The coca leaf is typically chewed with powdered lime, which is said to be required to achieve the full effect. It is perhaps significant that lime is also used in commercial extraction of the alkaloid.

In the Andean countries coca is one of the secondarily cultivated crops. The foliage is commonly sold in the local market for local consumption. The world's commercial supply comes chiefly from Java, where cocaine was introduced. Most is utilized for "cola" flavorings, after removal of much of the alkaloid. Since *E. coca* is a small tree, the foliage can be readily collected by hand, after which it is fermented briefly and dried. The cocaine alkaloid is usually extracted with solvents. Chemists have learned to synthesize novocaine, which is very similar to cocaine and is used to a large extent by dentists and doctors as a substitute for the natural product. In its Andean homeland, however, cocaine continues to be used as a stimulating narcotic, which, like opium, can become habit-forming.

Digitalis, Digitalis purpurea *Scrophulariaceae*

Digitalis, or foxglove, was recorded by the herbalists as useful for treating epilepsy, wounds, and other ailments. One William Withering, an Englishman, first introduced it to modern medicine in the eighteenth century, when he obtained samples from the garden of "the Witch Woman of Shropshire." Withering used digitalis successfully in treating dropsy, and published a treatise on his results, which was soon recognized and followed on the continent. Used for many purposes during the nineteenth century, it is used today almost exclusively as a heart stimulant and to regulate the pulse, for which digitalis ranks as one of the world's most important drugs. Several glycosides, all quite similar, are the active ingredients.

Although a common garden plant, digitalis grown for drug extraction is usually started under glass and transplanted to the fields. It is a biennial, but the leaves (which yield the glycosides) can be gathered at any time. These are dried in the shade, sometimes pulverized for direct usage, or more commonly subjected to solvent extraction.

Ipecac, Cephaelis ipecacuana, *Rubiaceae*

The rhizomes and roots of this and *C. acuminata*, herbaceous perennials of Latin America, yield alkaloidal extracts that have been used to treat intestinal amoebas, and as emetics. Wild plants, gathered from the rainforest, are a minor item of commerce in South America.

Marijuana, Cannabis sativa, *Moraceae*

Marijuana (the word is said to be a corruption of the Portuguese *mariguango,* signifying "intoxicant") has long been in mankind's service. Chinese writings nearly 5,000 years old indicate medicinal and psychic uses of marijuana, and it figured importantly in Indian religion as early as 1000 B.C. It had spread through the Near East and into

Europe before 500 A.D. A hashish cult was fashionable in Paris during the middle nineteenth century, at about which time a British study in India determined that Cannabis was not harmful (Christian missionaries had objected to it). Marijuana became a focus of contention in America, after its widespread adoption as a social amenity during the 1960's. It is estimated that about 10 percent of the population smokes marijuana in spite of rigorous attempts to stamp out the custom (the current situation thus resembling that of the "prohibition era" of the late 1920's and early 30's, when alcoholic beverages were illegal).

Of course *Cannabis sativa* is the source of hemp fiber, for which it can legally be cultivated; hemp oil and oil cake are legal commodities. The species grows readily in temperate regions, and is often found naturalized as an escape along roadsides and in old fields. All soft tissues contain active principles, although these principles are most concentrated in the young foliage and especially in the bracts surrounding the female inflorescence. Subtropical strains of marijuana are said to have higher potency than northern-grown plants. Hashish refers to the concentrated resin scraped from the pistillate inflorescences; it is several times as potent as typical marijuana "leaf". From the "Hashishins"—a Mohammedan sect of the Middle Ages ready to eliminate enemies by assassination, presumably while under the influence of marijuana—come the English words "hashish" and "assassin". Other designations for Cannabis are bhang (the cut-off tops), Ganja (a purified Indian bhang), and kif (the term for all forms of marijuana in northwest Africa).

Among other ingredients, marijuana contains tetrahydrocannabinol (THC), which is hallucinatory under certain circumstances, producing illusions, dreaminess, and euphoria. Marijuana smokers report enhanced perceptivity and sensitivity as the chief pleasures from social use of the drug. Marijuana smoking is estimated to be 50 percent efficient in the picking up of the approximately 1 percent of THC that marijuana leaf contains. Equally effective is a beverage made from the powdered upper leaves and seedheads. The oil may also be inhaled after vaporization.

Marijuana is not considered addictive, although it may be mildly habituating. Efforts have been made to curb marijuana under the supposition that its use may lead to addictive drugs such as heroin; studies, however, show that only a small proportion of marijuana users ever progress to opiates. The legal penalties for possessing marijuana seem unduly severe, and the assessment of criminal penalties for its use and possession seems hardly justifiable for a euphoric that authorities judge to be socially less harmful than alcohol.

Natural plant growth regulators

Many extractives have profound hormonal effects on plant growth. It has been realized for some time that leachates from plants may inhibit seed germination and the growth of nearby plants; this is especially true of desert plants. But the widespread development and use of herbicides had to await the results of basic research on the hormones (auxins) present in growing shoot tips. Chemical analysis showed that such substances as indole butyric acid had profound effects upon cell growth, a line of investigation which led eventually to the remarkable growth-influencing compound 2,4, dichlorophenoxyacetic acid (2,4-D). Here was a growth regulator that could be synthesized economically and was highly effective in minute amounts; very small dosages stimulated plant growth, but slightly larger onces caused plants, especially dicots, to twist, curl, and finally die, their physiology disrupted. Here was the basis for selectivity; certain plants, such as grasses, are not injured at doses that kill broad-leafed weeds. Intensive research on the way in which such herbicides affect plant physiology, and extensive empirical screening, have since yielded scores of specific, selective, and highly effective general herbicides, to the lasting benefit of modern agriculture. Unhappily, this also ushered in a new type of chemical warfare—a means of killing the crops that support the population of an enemy, by aerial application of a potent herbicide.

Equally of interest are some of the other growth regulators, a number of them derived from botanical sources. One that has received much scientific attention in recent years, though its practical applications are limited, is gibberellin. It is obtained by methods similar to those used for producing antibiotics from the fungus *Gibberella fujikoroi*, a parasite that attacks rice plants in the Orient. Mild attack by *Gibberella* causes rice to grow abnormally tall. Japanese scientists isolated the biologically active ingredients responsible and named them, collectively gibberellin. Very small quantities of gibberellin applied to growing plants (usually in solution) markedly increase growth rate and, in some plants, size, or differentially alter the growth pattern. The response differs with species; sometimes there is no response. Perhaps the most significant results from commercial application have been obtained in grape production in California, where spraying with gibberellin has produced larger, fuller clusters of grapes on shorter stems.

Other growth-controlling chemicals have been discovered, and botanical science may be on the threshhold of controlling facets of plant growth by means other than breeding programs. The prospects for the development of medicinals for plants as well as for animals seem promising for agriculture's future.

Opium, Papaver somniferum, *Papaveraceae*

Opium in its crude form contains about 11 percent morphine and 1 percent codeine (as well as many other alkaloids, resins, and oils); it is perhaps the most renowned of all botanicals used medicinally. Morphine is world famous for relieving pain. Smoking or eating opium leads to addiction, which in turn leads not only to mental and physical deterioration, but often forces the addict to turn to crime to support his habit. The opium poppy is believed to have originated in Asia Minor, where if often grows as an escape; what its wild ancestral form might be is not known, so long has the species been a ward of man. Opium taking appears as ancient as is written history, and the opium poppy seems to have had a fearful fascination for all civilizations. The drug is prepared in the Balkans, the Near East, and Southeast Asia, as well as in Japan and China. Commercial supplies for medicinal use come mostly from the Near East, under careful regulation.

The opium poppy is a spiny-leafed annual. In commercial plantings several successive sowings are generally made in spring to ensure that a crop will result, since the small seeds and seedlings are temperamental about unfavorable weather. By early summer, about two weeks after the petals have fallen, the large, urn-shaped capsules of the poppy are ready for opium production. The capsules are incised by hand (dictating that opium production be confined to cheap labor areas), usually in late evening. The next morning the coagulated drops of milky exudate are scraped from the capsule. The exudate is dried and eventually kneaded into balls of crude opium. Yields as high as 40 pounds per acre are possible under good growing conditions. Yields this size may require 20,000 poppy plants and untold hours of human labor. In Western countries opium can be sold legally only for medical purposes. Some 200 tons are imported into the United States annually, even though newer sedatives and synthetic hypnotics are slowly replacing opium.

Pyrethrum, Chrysanthemum *ssp.*, *Compositae*

The daisy-like flowers of several species of *Chrysanthemum* yield a natural insecticide, pyrethrum, especially useful because it is nontoxic to warm-blooded animals. The plants are perennial herbs; the active ingredient, pyrethrum, is most abundant in the mature ovaries. The most important species seems to be *C. cinerariaefolium*, native to southeastern Europe, now widely cultivated as a source of insecticide in Japan, Africa (the Rift Valley of Kenya and Tanzania), and New Guinea (Fig. 22-10). Pyrethrum plantings may be started from seed in nursery beds, but the chief means of propagation is through crown separation. Planted at a 2-foot spacing, as much as one-quarter ton of flower heads can be harvested annually per acre (usually by hand in the inexpensive labor areas where most pyrethrum is grown). The dried flowers yield about 3 percent crude pyrethrum, the active ingredients of which are several pyrethrins and volatile oils. Pyrethrum is much used in aerosol sprays for use around the house, in dusts for garden produce, and in protection of food stuffs, such as grain stored in commercial elevators.

FIGURE 22-10. *The dried flowers of* Chyrsanthemum cinerariaefolium *are the source of the natural insecticide pyrethrum.* [*Courtesy E. R. Honeywell.*]

Quinine, Cinchona *spp., Rubiaceae*

Quinine, principally derived from the bark of such species as *C. lancifolia, C. oblongifolia,* and *C. cordifolia,* was probably in use by the South American Indians before colonization. Jesuit missionaries are believed to have first taken the bark to Europe. The name "chinchona" derives from the successful treatment of the wife of the Spanish Viceroy of Peru, the Countess Ana of Chinchón, who suffered a severe fever there in 1638. By the middle 1600's "peruvian bark," as quinine was called, was utilized for making a variety of extracts, infusions, and tinctures. In addition to quinine's usefulness in treating malaria, the component alkaloid quinidine is a cardiac depressant. Synthesis of quinine-like substances has been successful, particularly in Germany, where quinachrine hydrochloride (atabrine) was developed in 1932. Still later, primaquine, chloroquine, and plasmoquine were synthesized. Malaria seemed pretty well under control, until the war in Vietnam exhibited forms of the disease more or less resistant to all existing malarial drugs.

Quinine species are native to the Andean highlands of South America. But after *Chinchona* was introduced into Java for plantation growing, Java accounted for about 95 percent of the world's commercial supply until captured by the Japanese during World War II. The several associated alkaloids constituting crude quinine occur in the bark of native trees in amounts up to about 7 percent (Fig. 22-11). Selection of superior stock on the plantations increased yields to nearly 20 percent with high-yielding clones budded onto seedling root stock (Fig. 22-12). The trees are planted under shade, rather closely spaced; harvesting gradually thins the stand until the entire planting is consumed after a number of years. The trees are uprooted to provide bark from roots, trunks, and branches. The bark is first beaten with a mallet to loosen it, and is then peeled by hand and quickly dried (in drying ovens on plantations). It is usually sent in crude form to the consuming country for solvent extraction. Powdered bark is typically mixed with lime, and the alkaloids dissolved in amyl alcohol or ether. The alkaloids are removed from the solvent by acidified water, and then precipitated by increasing the alkalinity.

FIGURE 22-11. *Bark from wild* Cinchona calisaya, *harvested in Bolivia.* [*Courtesy USDA.*]

540

FIGURE 22-12. *Seven-year-old plantation of* Cinchona *in Guatemala.* [*Courtesy USDA.*]

Reserpine, Rauwolfia serpentina, *Apocynaceae*

Derived from *R. serpentina,* reserpine has long been used in India as a medicinal for treatment of the mentally ill. Today this is widely employed against hypertension and for the lowering of blood pressure. It is effective as a tranquilizer in cases of insanity, and has even been used to counteract insect bites, fevers, and dysentery. Reserpine is chemically similar to a substance in the brain called serotonin, as well as LSD. One hypothesis suggests that an imbalance in the metabolism of serotonin is responsible for schizophrenia.

Demand for *Rauwolfia* roots increased so much during the 1950's that most of the wild stands were exterminated in the Indian and East Indian forests. Most production comes today from field plantings. *Rauwolfia* is a perennial shrub about a meter tall,

capable of growing under a wide range of climatic conditions. It is propagated by seeds, cuttings, or root separations. Plants started from seed, usually in nurseries, are transplanted to the field. Satisfactory growth requires ample rainfall, or in lieu of this, irrigation. Roots are ready for harvest after 2 or 3 years of growth. As much as 4 tons of fresh root may be harvested per acre. The alkaloid content of fresh roots is somewhat less than 2 percent. Alkaloid content is higher in the bark than in the wood. India, the main producer, exports the root to Europe and America.

Rotenone, Derris *and* Lonchocarpus, *Leguminosae*

Rotenone, along with pyrethrum and nicotine, is one of the most widely used natural insecticides.

It is primarily employed where there is fear of chemical contamination of edibles or of human habitations. Long known as a fish poison in the jungles of South America, it was recognized that this "barbasco" in no way poisoned the fish for human consumption, or in any way caused ill effects to warm-blooded animals. *Lonchocarpus nicou*, in tetraploid form ($2n = 44$), was primitively cultivated and vegetatively propagated by South American Indians. In the Far East, *Derris*, very similar to *Lonchocarpus*, served as a source of rotenone. In recent decades there has been interchange of planting stock, usually propagated by cuttings, often in small plantations. Planting methods may amount to little more than sticking a live stem into a hole. About half the stems will grow into sprawling shrubs that are uprooted 2 or 3 years later. Hand-weeding must be carried on in the interim; sometimes an intercrop is planted in the first year. For harvest the tops are cut away, the roots dug, sectioned, and dried. Rotenone extract is a mixture of aromatic compounds that are soluble in oils but not water. Concentration in the dry roots ranges from almost nothing to as high as 20 percent. The root may be pulverized for direct use as an insecticide (usually mixed with talc or clay), or an extraction may be made with solvents.

Strychnine, Strychnos nux-vomica, *Loganiaceae*

Strychnine, like opium, is another of the awesome names among plant extractives. Originally it was a component of arrow poisons (curare) made by primitive tribes in the South American jungles. Until even more potent synthetics were developed it was best known as a "poison," even in modern societies. Yet properly used it is a tonic, stimulant to the central nervous system, and a relief for paralysis. Actually, many species of *Strychnos* are used; the one cited here, however, is the source of most commercial production, now centering in the Orient. The alkaloid strychnine and a less poisonous one, brucine, are the active components; they are extracted mainly from the seed, where they are most concentrated. Production comes chiefly from uncultivated plants. The large orange-like fruit is collected in off hours, the seeds removed, washed, and dried. These are bought by local merchants and eventually end up in industrialized centers, where they are pulverized and the alkaloids extracted. The exact compounding of curare to tip arrows and blowgun darts is uncertain; plants other than *Strychnos* are included in the brew, and probably there is no exact formula.

Other botanicals

The foregoing are random examples from thousands of species that have been or are used in analogous ways.° Many included in the latest Pharmacopeia have not been listed, such as benzoin (*Styrax*, Styracaceae), colchicine (*Colchicum*, Liliaceae), licorice (*Glycyrrhiza*, Leguminosae), mandrake (*Podophyllum*, Berberidaceae), and tolu balsam (*Myroxylon*, Leguminosae). One could search the herbals for hordes of remedies that have been part of early folklore. These are fairly well known for Europe and the New World, but information on the Far East is less accessible.

INDUSTRIAL FERMENTATIONS

Fermentation,† broadly considered, is the degrading of energy-rich molecules by microorganisms. As such it is a pervading biological fact of life. It is responsible for the decomposition of organic wastes and residues (detritus) and the return to the soil and the atmosphere of elements that will be used again in plant growth. A myriad of saprophytic microorganisms live by fermenting; they are

---

° More than one hundred are discussed briefly as being of some drug value in Schery, *Plants for Man*, Prentice-Hall, New York, 1972, including some as well recognized as *Arnica*, Compositae; Chaulmoogra, genera of the Flacourtiaceae used to treat leprosy; ephedrine, from *Ephedra*, Gnetaceae; ergot, from *Claviceps*, Ascomycetes; ginseng, from *Panax*, Araliaceae, mystical Chinese remedy; peyote, or mescal buttons, *Lophophora*, Cactaceae, containing another hallucinatory alkaloid similar to LSD; quassia, from genera of the Simarubaceae, both an anthelmintic and insecticide.

† Fermentation to the physiologist refers more specifically to the anaerobic degradation of carbohydrates, often to CO_2 and an alcohol or an organic acid; the degradation of protein is called putrifaction.

capable of utilizing almost any energy source, yielding end products of tremendous variability. By means of a specific enzyme the microorganism catalyzes the decomposition of a complex source molecule, splitting it into simpler end products with the release of energy for growth of the fungus, becterium or other microorganism. One of the most familiar examples is the fermentation of a sugar solution to alcohol and carbon dioxide:

$$C_6H_{12}O_6 \xrightarrow[\text{enzymes}]{\text{yeast}} 2C_2H_5OH + 2CO_2.$$

In a nontechnical sense "fermentation" refers to the bubbling off of gas during such microorganismal decompositions.

Several of agricultural's most important fermentations have already been mentioned in this chapter. The production of alcoholic beverages by the action of yeast is especially important. The production of antibiotics is another example. Mention has also been made of the several "curing" processes in the development of flavors and aromas in such things as cacao, tea, and several essential oils. In the next chapter we will see that microorganisms are useful in **retting**—freeing fibers from surrounding tissues. They are important in producing acids during the fermentation of fodder to ensilage, which keeps the silage preserved. Fermentation makes bread dough rise, converts milk curd to cheese, cabbage to sauerkraut, and cider to vinegar. Many plant diseases (such as rots) involve fermentation, as does food spoilage and decay of structural timber. Fermentation is used in sewage disposal; gardeners prepare compost by fermenting plant residues.

All of this, of course, is not "industrial." But these examples do serve to illustrate the ubiquitous activity of microorganisms. Because these applications relate so closely to the discussions in other sections and chapters, industrial fermentations need only brief coverage here. As a direct source of materials usable in a modern economy, fermentation industries are relatively unimportant (compared to agricultural crop growing or chemical synthesis from such raw materials as natural gas and petroleum). Indeed, some of the major industrial fermentations of a former era, such as the production of ethanol, acetone, and butyl alcohol, have lost much of their once dominant position in competition with chemical synthesis from petrochemicals. This, of course, is a matter of economics; with cost of molasses, grain, or other raw materials suitable for fermentation rising, and with the increased demand for these products as human and livestock food, petrochemical production has generally proven less expensive. But since petroleum and natural gas are not renewable resources, industrial fermentations of renewable crop materials would seem to have time on their side.

Production of industrial ethanol, or ethyl alcohol, is one of the more important industrial fermentations. Up until World War II most alcohol was obtained by fermentation, molasses being the chief raw material. Production exceeded a half billion gallons in 1945, a year of unusual demand because of World War II. Since then total demand has exceeded 200 million gallons annually, but only about 20 percent of this production has been by fermentation, the rest being synthesized from ethylene. The story is much the same with acetone and butyl alcohol. During World War II about two-thirds of the butyl alcohol production was by fermentation, and around 1940 nearly half the acetone. As with ethanol the source material was cheap blackstrap molasses. Since World War II synthetic production of these materials has supplied most of the market. The reason is that the cost (calculated on a sugar basis) of molasses has increased from about one cent per pound before World War II to several times this price since.

A number of organic acids are also produced by fermentation, and have fared better economically. Citric and lactic acids are both produced almost entirely by fermentation. The demand for citric acid approaches 100 million pounds annually in the United States. Acetic acid (vinegar) is another fermentation product that has held its own. Among the less well known but important acids manufactured by fermentation are gluconic, fumaric, itaconic, and tartaric, whose esters find some use in the plastics industry.

Vitamins are also produced largely by fermentation. Production of ascorbic acid alone amounts to several million pounds in the United States, and riboflavin and vitamin B_{12} to lesser amounts. Similar growth factors are components of yeast cultivation, which is often undertaken to make use of by-products in other industries. For example, *Torula* yeast may be grown on pulp liquors from the paper industry, and constitutes a nourishing source of proteins and vitamins, which are utilized to some extent in animal food, and even baby food.

Of course, enzyme preparations other than malt are frequently the by-products of fermentation. And production of antibiotics, discussed in the previous section, is a billion dollar industry. Dextran has been prepared by fermentation; it is used in food products, as a blood plasma extender, and as an additive to well-drilling muds. It can be produced by fermenting almost any sugar source with *Leuconostoc mesenteroides*. This is an example of a potentially important new field—the production of polymers by microbial fermentation.

Not commercially of much consequence, although also potentially useful, is the production of resin substitutes from microorganisms: the same form of polyisoprene that occurs in natural rubber has been recovered in small amounts from such fungi as *Lactarius* and *Peziza*. Other fungi are capable of converting carbohydrate into fats that may possibly be useful in special diets (although it is unlikely that they will ever compete with such inexpensive products as soybean oil). Microorganisms can also aid in the production of amino acids, or even proteins: lysine has been commercially produced by fermentation. Since about half the dry weight of bacteria, molds, and yeast is protein, they are a potential source of animal feeds.

These examples merely indicate the industrial possibilities of fermentation: new strains of microorganisms capable of executing specific fermentations are discovered almost daily. As in the improvement of antibiotics, the selection of new strains of microorganisms should increase the efficiency of even the most well-known fermentation processes. All that seems necessary is to find fermentable raw materials that are less expensive than those now being used. For the present, however, industrial fermentations are confined chiefly to nonfood specialties and to the utilization of waste and by-products.

FUMATORIES AND MASTICATORIES

Plant products smoked or chewed for pleasure are not uncommon. Coca leaves are chewed in the Andean highlands for the stimulative effect that they have; marijuana and opium are smoked for the feeling of euphoria they induce. But one plant in particular is smoked more than any other—tobacco. From time to time tobacco has been considered to have medicinal value, and it would be difficult to make a hard and fast distinction between it and the many plants whose medicinal value is established.

Tobacco, in the form of chewing tobacco, also serves as a masticatory, although this unappealing custom is less and less practiced in sophisticated societies. In the Orient, perhaps more people chew the betel nut, a locally produced stimulant, than smoke tobacco. Betel will serve as our example of a much used masticatory, and tobacco as an example of a fumatory that is smoked throughout the world. Chewing gum, a popular masticatory in the United States, is reserved for discussion along with latex products.

Other than tobacco, fumatories and masticatories are not important items of commerce. Many, such as coca and betel, although intensively used regionally, enter market almost solely as local items of trade. But tobacco has for centuries been an important international commodity, produced significantly on all continents. Nearly 5 million tons of tobacco leaf are grown annually, almost all of which is used for smoking.

Tobacco, Nicotiana, *Solanaceae*

Tobacco, grown and used moderately by the American Indians, was discovered by the explorers of the New World. It was first introduced into Europe much as a curiosity, and was claimed to possess medicinal values. Throughout the seventeenth and eighteenth centuries tobacco was recommended as a sort of "cure-all," for treating many common afflictions ranging from the toothache and rheumatism to indigestion and blood poisoning. For a while the use of tobacco in the form of snuff was socially accepted. Gradually this use gave way to smoking. Tobacco was used almost exclusively by men until the advent of the readymade cigarette in the twentieth century, when it came to be used by women as well. The use of tobacco continues to increase in spite of health warnings. The chief center of production is North America, where nearly a million tons of leaf are grown annually. China, India-Pakistan, Brazil, the U.S.S.R., Japan, Turkey, Greece, South Africa, and Bulgaria are other prominent tobacco producing countries (Fig. 22-13).

Tobacco is one of the most pampered of crops. It must be started in sterilized nurseries and tended intensively from planting until harvest. A glance at the technical literature about tobacco suggests that it is prone to more ailments than even the human hypochondriac; the tobacco breeder endlessly searches for genetic resistance to anthracnose, black rootrot, blackshank, bluemold, brownspot, cyst nematodes, frogeye leaf-spot, fusarium wilt, Granville wilt, powdery mildew, root knot nematodes, rattle virus, tobacco etch, tobacco mosaic virus, tobacco streak virus, wildfire, and several others. Since the tobacco genus is, fortunately very diversified and genetically variable, potential sources of disease resistance are always at hand, even though the situation is complicated by the general sterility of F_1 interspecific hybrids.

Nicotiana contains in the neighborhood of 65 species. *N. tabacum* is the specific name applied to most cultivated tobaccos. In its many cultivars, it is a widely variable and polymorphic species. A review of the genus by Goodspeed indicates that *N. tabacum* may have originated as a chance cross between ancestral forms of *N. sylvestris* and some distant relative of *N. otophora*. There is no definite record of *N. tabacum* in the wild state, and it appears as though the species may be of relatively recent origin, having been brought quickly into domestication after formation as an amphidiploid, possibly in central South America. *N. tabacum* is a tetraploid, with 48 chromosomes, as is *N. rustica*, a second species utilized by the North American Indians (and perhaps the first to be introduced into

Europe; it is still grown in Africa and the Near East).

The tobacco plant is grown as a perennial in tropical environments, but it is customarily grown as an annual in temperate localities (Fig. 22-14). Although some cultivars of tobacco are grown ornamentally for flowers, the inflorescence is removed as it develops in commercial leaf production. The cut leaf is fermented ("cured") to develop the desired aroma and taste, which is raw and harsh in fresh tobacco. The alkaloid nicotine is perhaps the most distinctive constituent of tobacco leaf. It volatilizes readily when the leaf burns and is responsible in part for the irritating vapor in tobacco smoke or nicotinic insecticides. There are, of course many other organic components in tobacco smoke; in recent years these have been analyzed repeatedly in the search for carcinogenic factors. Tobacco "tar," which includes resins, essential oils, organic acids, and phenols, is usually implicated as the health hazard.

Nicotiana is named after the French Ambassador to Portugal during the late 1500's, one Jean Nicot, who was responsible for introducing tobacco to the court of France. Although tobacco had been sent to Europe from the West Indian colonies of Spain for half a century, it remained for John Rolfe to initiate the first significant commercial planting, in the Virginia colonies of England, in 1612. Ever since, tobacco growing and export have remained an important agricultural enterprise along the Atlantic seaboard. In the early years production was sent to Europe, for cigars (the Spanish custom), pipe tobacco (the English custom), snuff (the French custom), and eventually cigarettes (the American custom, adopted from the Turkish). It is perhaps historically significant that cigarette manufacturers pride themselves upon tobacco blends that include small-leaved, untopped, aromatic Turkish cultivars.

The growing of tobacco varies in the intensity of care given in various parts of the world, but follows a similar pattern wherever it is cultivated. Tobacco growing in Rhodesia is typical of the intricate procedures practiced in major growing regions, including the United States. Seedbeds (usually about 100 square feet, sufficient to supply seedlings for several acres) are meticulously culti-

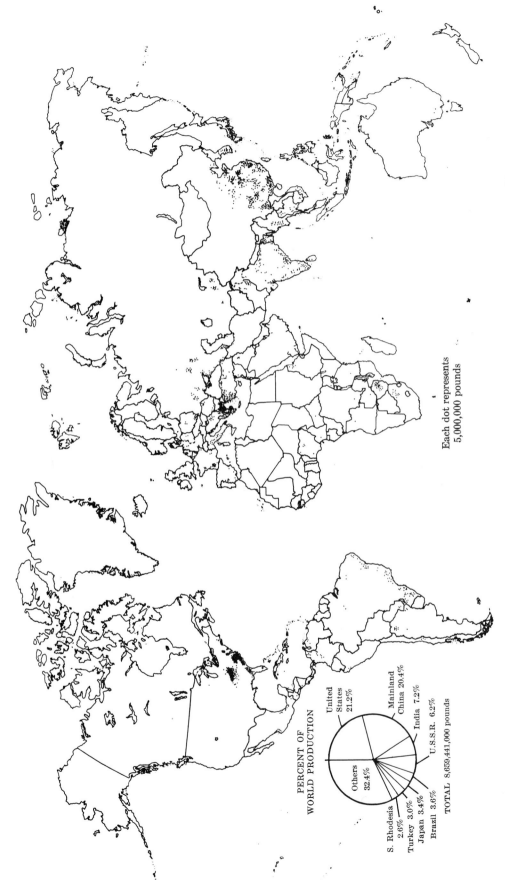

PERCENT OF
WORLD PRODUCTION

United
States
21.2%

Mainland
China 20.4%

India 7.2%

U.S.S.R. 6.2%

Brazil 3.6%

Japan 3.4%

Turkey 3.0%

S. Rhodesia
2.6%

Others
32.4%

TOTAL 8,659,441,000 pounds

Each dot represents
5,000,000 pounds

FIGURE 22-13. *World tobacco production, average 1957–1961.* [*From A Graphic Summary of World Agriculture, USDA Misc. Publ. 705.*]

FIGURE 22-14. *Tobacco culture. (Above) Fields of Broad Leaf Burley growing in terraced and contoured fields in Kentucky. (Below) Shade houses for cigar wrapper production in Connecticut. [Upper photograph courtesy USDA; lower photograph courtesy Conn. Ag. Expt. Sta.]*

vated and fumigated with methyl bromide (to kill nematodes, weed seeds, and soil-borne diseases). Fertilizer rich in phosphate is liberally mixed into the top 2 or 3 inches. Tobacco seed is extremely small and is generally extended with an inert material to help achieve uniform distribution or is mixed in water and applied in a sprinkling can; an ounce seeds more than 1,200 square feet with a density of about 60 seedlings per square foot. Hybrid seed, although very expensive, is mostly used in the burley belt of the United States, and is produced through hand pollination of a male-sterile line. After sowing, a grass mulch is applied and the bed sprinkled several times daily until seed is sprouted and the seedlings are well established. In the United States some growers cover the seedbed with polyethyelene plastic to give the seedlings more adequate protection and maintain better control of environmental changes than does the more general practice of shading seedlings with cloth. The nursery is sprayed regularly with fungicides to control disease, and no smoking is allowed nearby for fear of introducing diseases. As the seedlings become large enough for transplanting to the fields, watering is withheld to harden the transplants.

In the meanwhile the growing fields have been thoroughly worked, fertilized, and perhaps fumigated. The seedlings are typically planted in ridges, the rows about 3 feet apart, the plants spaced about 2 feet apart in the row. The plant population runs approximately 7,000 to the acre. Transplanting is usually a hand operation in Africa (there is some mechanization in the United States), and the transplant is given a measured amount of water as it is set. This is sufficient to carry it until rains take over. Such weeds as develop are normally controlled by cultivation. As the plants begin to flower the inflorescence is topped by hand. This stimulates side branching or "suckering," which must be checked repeatedly (either by hand or by applying maleic hydrazide sprays) if leaves are to achieve their full potential.

The leaves are harvested progressively from the bottom upward on each stalk, two or three at a time, as they reach the proper degree of ripeness. They are gently collected, covered for protection from the sun, and taken to a "tying shed" for fastening to sticks with twine. They will be suspended by these sticks in the curing barn, in a framework that accommodates many thousands of leaves at a time. The curing barns are fired with wood (or occasionally coal), and the heat is radiated from a series of flues. Barns are ventilated at roof level. Curing procedures vary with the kind of tobacco and with the region of production, but may start with a temperature of 29°C (85°F) for one day, after which the temperature is raised to about 32°C (90°F) until the leaf is well colored, and then increased to 38°C (100°F) until no more green shows in the leaf. Subsequently the temperature may be raised even higher, with a final drying at about 71°C (160°F). Curing takes several days, after which the dried tobacco is allowed to absorb sufficient moisture to make it pliable for convenience in handling. Grading, baling, and eventually auctioning follow. Oriental and Turkish tobaccos may not be so elaborately handled. Moreover, they are generally cured in the sun. The curing process reduces leaf weight about 80 percent. After auctioning, tobacco is generally further aged, typically in huge casks, for a period that may extend into years.

The tobacco industry has developed many individualized customs, some of which have even hindered acceptance of new disease-resistant cultivars. The classification of tobacco as "burley," "fire-cured," "piedmont," "filler," "cigar wrapper," and so on, is confusing to those not conversant with the tobacco industry. These designations may be based as much upon region where produced, techniques of culture, and method of curing as on genetic characteristics of the tobacco. The blending of different tobacco types constitutes the professional secret of the manufacturing company, and provides such distinctiveness as there is among the different brands. Cigarette manufacture, the greatest use of tobacco, is especially competitive. Manufacturers vie for public acceptance with tremendous advertising programs and varying types of cigarettes (with or without filter, menthol, and so on). The basic operations involved in cigarette manufacturing are: conditioning the fully aged tobacco to a uniform moisture content, removing coarse parts (petioles and large veins), and chopping the leaf into "strip." The strip may be further aged, then blended with other tobaccos (such as Turkish). Humectants are applied to keep the tobacco moist (often glycerine, glycols, or apple juice), and, depending upon the type of cigarette, flavorings may be added (sugars, coumarin, rum, and so on). The different tobaccos are thoroughly blended in rotating drums, and the mix sent to shredding machines. After remixing it is routed to the "making" machine, which mechanically rolls the tobacco in cigarette paper. The slender strands of rolled tobacco are automatically cut into proper lengths, affixed with filters if required, and packaged by elaborate machines capable of handling thousands of cigarettes per minute.

Betel, Areca catechu, *Palmae*

Among the dense populations of Southeast Asia, betel chewing is as commonplace as is the chewing of chewing gum and the smoking of tobacco in the Western World. It is said that one-tenth of the world population chews betel. The chief component of betel is the seed, or "nut," of the betel palm, *Areca catechu*. The betel nut is typically chewed along with a fresh or processed leaf of a native pepper, *Piper betle*, Piperaceae. The combination of betel nut, betel leaf, and the small quid of lime gives the mouth and saliva a wine-red coloration. The mass is worked in the mouth without swallowing, which of course stimulates a copious flow of saliva that is continuously expectorated. After a time the teeth of a betel addict turn blackish. Betel contains a narcotic that provides mild stimulation. Although its use seems disagreeable to westerners, it is comparable to smoking and chewing tobacco in other parts of the world. In India, Malaysia, Indonesia, and eastern Africa the practice is socially quite acceptable.

The betel palm has been domesticated for so long that betel is mentioned even in the earliest Hindu mythology. Early writings vaunted the medicinal virtues of betel, claiming that it could be used to "expel wind, remove phlegm, kill germs, subdue bad odors, beautify the mouth, remove impurities, and induce love," among other things. Arabian visitors to India in the Middle Ages reported, "the betel brightens the mind and drives away the cares. . . ." Marco Polo commented upon the Oriental habit of betel chewing in his writings of 1298. Exactly where the betel palm may have originated is uncertain—possibly the Malay peninsula or the Philippines. The trees are now widely distributed from India to the Philippines and Japan, all seeming to be the cultivated form. *Areca* has also been introduced into the Mediterranean area and into Florida as an ornamental. A number of botanical varieties of *A. catechu* have been described in the Far East, and three other species are recognized in the genus (only one of which is occasionally chewed). The various forms of betel nut found in the Eastern markets represent seedling races or local cultivars.

Areca palms reach heights of about 100 feet and bear large pinnate leaves. They are generally planted in gardens or small plantations, flourishing best where rainfall is abundant and temperature rather unvarying. Since *Areca* stands shade well, the palms are often planted mixed with fruit trees such as the mango, citrus, coconut, etc. Propagation is by seed, for which the seed nuts are allowed to ripen completely on the tree before drying and sowing. Sowing is generally to nurseries, with transplanting to permanent site. The palm flowers in about its seventh year, and reaches full bearing potential in somewhat more than a decade. The fruits form in large bunches, and are hand harvested (by climbing the tree) when ripe (bright red). A tree that yields well will supply 500 to 1,000 nuts a year. Like other crops, betel palms suffer from various diseases and insect pests, although steps are seldom taken to control them.

Betel nuts may be consumed raw or cured, following hand removal of the outer husk. Surplus nuts may be stored in pits in the soil, or under water in earthenware jars. Sometimes nuts are dried in the sun and then marketed. Sliced nuts are often boiled for several hours before drying. The endosperm contains several alkaloids, as well as tannins, fats, carbohydrates, and proteins. At least three different alkaloids seem involved, constituting as much as 0.5 percent of the endosperm. Tannin content may be as much as 18 percent, and contributes to the astringency during mastication. As is typical of palms, the betel palm has much oil in its endosperm, and the husks constitute a potential by-product.

LATEX PRODUCTS

Latex is a milky, colloidal secretion found especially in species of the Apocynaceae, Asclepiadaceae, Compositae, Euphorbiaceae, Moraceae, and Sapotaceae families. Latex occurs in specialized cells, or cavities, that drain when ruptured, thus permitting its flow when a plant is tapped (usually by gouging or slicing the bark). Lactiferous plants are found in all parts of the world, in many sizes and shapes, from the humble dandelion (*Taraxacum*) and milkweed (*Asclepias*) of temperate lands to the impressive lactiferous trees of the tropical forests. You may recall that several of the species reviewed earlier yield a latex, particularly the opium poppy capsule (drug) and the papaya fruit (meat tenderizer).

The colloidal particles in latex consist mainly of hydrocarbons and resins, but also contain oils, essential oils, various carbohydrates, proteins, enzymes, organic acids, minerals, salts, alkaloids, sterols, tannins, and other substances. It has been theorized that latex functions in the plant to seal wounds and that it serves as a disinfectant and a food reserve. It may, however, consist of nothing but waste materials. The coagulated latex (coagulum that results when the colloidal particles are precipitated) is of two general types: **rubber,** which contains a high percentage of cis-polyisoprene, a highly elastic polymer, or **balata,** which contains high proportions of trans-polyisoprene and resin, yielding an inelastic coagulum. Rubber is, of course, the best known. Until synthetics became developed it was derived chiefly from the Pará rubber tree (*Hevea brasiliensis*). To a lesser extent *Castilla elastica*, species of *Parthenium*, various figs (*Ficus*), *Manihot, Hancornia, Cryptostegia, Funtumia, Landolphia,* and many others have served as sources of rubber. The principle balata gums are chiclé, or chewing gum (*Achras zapota*), chilte (*Cnidoscolus*), gutta-gum (*Couma* and *Manilkara*), gutta-percha (*Palaquium,*) and jelutong (*Dyera*).

Latex extraction procedures vary. Most commonly these days the bark of growing trees is tapped, as with *Hevea*. Sometimes the tree is felled and the latex freed by girdling, as with wild *Castilla* and *Manilkara*. Solvents can be used to extract balata from foliage, as with *Palaquium*. Or maceration can be used as a physical means of freeing the polymers, as with *Parthenium*. Only with *Hevea* has tapping been well investigated and an efficient technique worked out. This has been possible because of concentrated plantation cultivation. Most other latices are gathered from wild sources in remote areas by uneducated tappers lacking supervision.

Before a latex can be used the solids must be separated from the liquid supernatant "serum". **Coagulation** can be induced in various ways. If the latex is merely left standing for a day or two, coagulation will sometimes occur naturally, due to enzymatic changes in the proteins or to acidification through fermentation. Usually, however, the latex is heated or boiled, often with the addition of salts or acids (in primitive locations fruit juices or sea water serve). Such treatments break the repelling forces between colloidal particles, causing them to coagulate. The fresh coagulum is whitish, soft, and pliable. As it ages or is further treated rubber becomes tough and elastic; balata is usually dark, hard, and brittle. The many uses of rubber are well known, principally for pneumatic automobile tires; balatas are used mainly for insulating materials, for making molded products, and for chewing gum.

Rubber

RUBBER, *Hevea brasiliensis,* EUPHORBIACEAE In intrigue and romance the rise (and partial fall) of the natural rubber industry challenges that of the spice trade. Rubber suddenly became more than a curiosity when Charles MacIntosh learned to waterproof fabric with it, and especially when Charles Goodyear learned to vulcanize it early in the nineteenth century. Interest first centered on *Landolphia* and other Apocynaceous vines of equatorial Africa. The latex was obtained by cutting the vines at ground level, a practice that soon depleted supplies and, of course, wrote a finish to the African rubber boom. While the boom lasted, however, the exploitation of native peoples became a worldwide scandal. Peaceful tribes were taxed ever increasing quotas of rubber by their colonial overlords. If the quotas were not met women were seized and men put in chains—even mutilated and left to die a lingering death as new sources of the rapidly disappearing rubber trees in the Congo jungles were sought. Soon the boom spread to the New World, where the species destined to become the world's main source of rubber, *Hevea brasiliensis,* grew wild in the Amazon Valley. *Castilla,* which grew in Central and northern South America, had its moment of glory, and later *Funtumia* was introduced into the West Indies from Africa, but only to become another of the West Indian casualties as *Hevea* gained the day. With only wild rubber to be had, and demand soaring because of the newly burgeoning automobile industry, rubber prices increased astronomically. In 1910 the price of rubber reached the all time high of $3.06 per pound, an especially heady price considering the value of the dollar in those days. No wonder those were tumultuous times.

As early as 1875 the competent British Colonial Service had sent Henry A. Wickham to Brazil to procure seeds of *Hevea brasiliensis,* and Robert Cross to Central America to bring back seeds and plants of *Castilla* (Cross later went to Brazil for young trees of *Hevea*). These seeds were germinated in England, and the seedlings eventually shipped carefully to Ceylon, where they were received initially with little enthusiasm. These few seedlings were to spawn the tremendous rubber plantation industry of the Far East! But in the early 1900's the world still relied upon wild rubber from the Amazon. At that time the rubber capital of the world was Manaos, situated in the sweltering jungles thousands of miles up the Amazon River. At the height of the rubber boom, so great was the wealth flowing into Manaos that a tremendous opera house was erected there from construction materials imported from Italy. The boom was to be short-lived, however, for by the end of World War I the once-scorned rubber trees of the British colonies in the Far East were widely planted and of tappable age. These plantations yielded more rubber than the market could then consume. The price broke to 14 cents per pound, rose again as the Stevenson plan restricted rubber production, then fell again to an all-time low of 3 cents in the depression years. Today, with synthetic rubber available, a ceiling exists on rubber prices; never again is price instability in the rubber industry likely to result in the chaos, the political maneuvering, and the amassing of fortunes that characterized the early days of the industry.

Some wild rubber is still harvested in the Amazon Valley, mostly by lonely "seringueros," living in thatched huts and crudely tapping distant trees that may still bear the tapping scars of the boom years. The unused opera house at Manaos is but a reminder of the former glory. There was a revival of interest in wild rubber during World War II, when Japan controlled the Far Eastern plantations, and a wild search took place for alternate plant sources for latex. Chemists, however, synthesized rubber-like polymers from isoprene and butadiene in 1945, and things have never been the same with rubber since. Today both Brazil and the Far East must compete with petroleum sources for the rubber market. In this battle natural rubber has generally come out second best; in competition with synthetics it seldom accounts for more than 20 percent of annual world production. Of the 3 million tons of natural rubber that are produced annually, more than 90 percent comes from plantations in the Orient (principally Malaysia and Indonesia). Brazil, rubber's ancestral homeland, produces only 25,000 tons, from both wild and cultivated sources. And all of Africa yields only 190,000 tons (mainly from Nigeria and Liberia).

Tapping of wild rubber trees in the Amazon Valley has always been a wasteful and expensive procedure. The trees are not easily accessible, and much time and effort are spent in collecting the latex. Gouging of the bark with crude tools mutilates the trees and interferes with efficient continuing production. There are few processing centers, a necessity for economical mass handling of latex; instead, individual tappers have traditionally smoked the latex over open fires on paddles, forming it into huge balls, which in turn must be transported tremendous distances to market. Such a system can hardly compete with the highly efficient plantation growing of rubber in the Far East, where abundant, skilled, and inexpensive labor can tap and tend concentrated plantings. Moreover, trained botanists from Europe early initiated breeding and selection programs to develop high-yielding clones of *Hevea* that are tractable to tapping; these clones are much more productive than the wild trees of the New World. So much has tree quality been improved that during and after World War II, in an effort to re-establish rubber growing in Central and South America, planting stock was zealously sought in the Far East. In less than a century that little-appreciated introduction from the New World had been thoroughly domesticated, one of the few documented cases of domestication of an important crop species.

The **jebong system,** the most efficient system of

tapping, was developed on the Far Eastern planta- tions. A "panel" is opened half way round the tree, from the upper left to the lower right (since the latex canals in the bark spiral to the right, more canals are opened by a cut made in this way). Thin slices of bark are removed with each new tapping from the lower edge of the panel, to renew the flow of latex. Laborers are trained not to cut too deeply into the tree, so that the cambium will not be injured and will be able to regenerate new bark (Fig. 22-15). By the time tapping reaches ground level on one side, and the opposite side then worked, the bark will have healed over the original panel and be ready for tapping again. Tapping is done early in the morning, when the flow of latex is greatest and showers are unlikely. A small spout is inserted at the lower end of the panel, which guides latex into an affixed cup. With high-yielding trees as much as a ton of latex per year can be obtained per tree, although the intensity and sea- son of tapping varies somewhat with local condi- tions. Good trees on well-tended plantations can endure tapping as frequently as every second day, except for a rest period during change of foliage.

The latex is carefully kept free of bark and dirt. At each tapping latex coagulated on the cut is removed by hand and kept separate as "scrap." After tapping his quota of trees, the tapper retraces his route and carefully empties each cup of latex into a clean, covered pail, which is then taken to a central processing plant. The treatment of the latex varies with locality. In some places latex is treated with an anticoagulant (usually ammonium hydroxide) and shipped in bulk tankers to the country where it will be processed; more com- monly, however, it is coagulated locally by addi- tion of acid. The soft coagulum is squeezed free of "serum," and the resulting thin sheet is typically "smoked" in a special smoke house to dry and preserve it. Properly smoked rubber will not dete- riorate, and can be shipped to consuming markets as needs and prices dictate. This is a far cry from the grimy smoking over an open fire, or coagu- lation by fermentation in ill-provided jungle camps. In the mid-1960's "heveacrumb," com- pressed granular rubber, began to replace smoked sheet in the trade.

High-yielding *Hevea* clones are perpetuated vegetatively, especially by budding. So technically

FIGURE 22-15. *Rubber tapper using special jeboug knife that scores the bark deep enough to cause latex to flow without damaging the tree.* [*Courtesy USDA.*]

specialized has propagation become that "com- posite" trees are often created from separate, exceptionally good root stocks, high-yielding trunks, and vigorous tops. A trunk from a high- yielding clone is budded to a proven root stock and grafted with a vigorous top that is especially efficient photosynthetically. A plantation tree can be tapped when 5 years old but peak yields don't come until it is at least 12 years of age. Usually the tree is abandoned when it is about 25 years old, mainly because better types will have been bred by then. Research has indicated just which are the best fertilizer regimens and cover crops for the particular locality. Usually, in addition to the major fertility elements, iron and manganese are required for vigorous growth, and occasionally other minor elements. Thus, in contrast to what is generally true of tropical tree crops, as much is known about the requirements of plantation- grown rubber trees as is known about the require- ments of almost any agricultural annual. It has even been found that very light applications of 2,4-D or growth regulators just below the tapping cut sometimes increase the yield as much as 30 percent without damaging the tree!

Fortunately for the Far Eastern plantations,

Hevea diseases inherent in the Brazilian homeland were not transported with Wickham's seeds to England, and thence on to Ceylon in 1876. As improved stock is reintroduced into the New World, it often suffers from leaf blight and other diseases. Thus for the New World rubber industry, additional research is necessary to develop clones adapted to New World conditions. Because it is expensive and difficult to control diseases with fungicidal sprays, the most practical approach has been to develop disease-resistant clones through selection and breeding.

PANAMA RUBBER, *Castilla*, MORACEAE Several species of *Castilla* have been utilized for rubber. Indeed, the first reports of rubber from the New World, by La Condamine, who explored the Amazon in 1734, probably referred to *Castilla* rather than *Hevea*. La Condamine called the extract "caoutchouc," but reported the Guiana Indian name of "heve," which later led Aublet to name a different genus *Hevea*, mistakenly believing it to be the species reported by La Condamine. The word "rubber" gained acceptance over caoutchouc when Priestly used samples for rubbing out pencil marks.

In the heyday of the natural rubber boom, high hopes were held for *Castilla*. Royal Botanic Garden botanists, testing rubber species in Trinidad, felt that the rapid growth and wide tolerance of *Castilla* indicated that it might be of greater promise than *Hevea*. But they were proved wrong. Following enthusiastic (and often black market) distribution of planting stock in the West Indies, the combination of inexperience in growing and tapping, the lack of knowledge about handling latex, the vagaries in the price of rubber, and insufficient research, led to complete demoralization of not only the *Castilla* rubber industry, but of all plantation rubber growing. *Castilla elastica* still grows as an escape in the West Indies and wild throughout Central and northern South America. In booms or times of emergency there is some tapping. Usually, however, the wild trees are felled and the latex drained. The latex canals in *Castilla*, unlike those in *Hevea*, are continuous, and not amenable to frequent tappings as with the jebong system. A large tree can yield several gallons of latex, which

can be processed into as much as 70 pounds of rubber.

MANICOBA, *Manihot*, EUPHORBIACEAE Several species of this genus, especially *M. dichotoma* and *M. piauhyensis*, grow wild in the arid sections of northeastern Brazil and yield a good rubber latex, but only in small quantities. The trees have been exploited only in times of emergency and when rubber prices are high. There have been only desultory attempts to cultivate the tree, so that tapping has largely been of an exploitative unscientific nature, involving a series of gashes from top to bottom of the tree in a fashion that causes the latex to drain progressively downward into a receptacle at the base. The bark is hard, not suitable to jebong tapping; and in the desert-like "caatinga" growth is slow, replenishment of tree stands poor.

MANGABERIA, *Hancornia*, APOCYNACEAE Like manicoba, mangaberia (especially *H. speciosa*, native to the sandy coastal belts from eastern Brazil south into Paraguay) has served as a source of a rubber latex during times of emergency. The quality is not as good as that of other rubbers, and the small trees of the scrub forest provide relatively small yields for the effort expended in gathering the latex.

PALAY RUBBER, *Cryptostegia grandiflora*, APOCYNACEAE This and other species from Madagascar have been introduced into Central America as a promising source of rubber. The plants are sprawling vines, and the latex is typically gathered by cutting off the tips of the stems and hand-gathering the small drops of exudate. Attempts made during World War II to grow the plant extensively and to develop mechanized harvesting proved a failure; consequently, interest in this widely naturalized "weed" as a source of rubber has died.

GUAYULE, *Parthenium argentatum*, COMPOSITAE Guayule, native to Mexico and the southwestern United States, was extensively tested during World War II as a possible source of rubber. It was hoped that extensive stands could be cultivated as field crops, and the rubber retrieved by maceration of the harvested foliage. Although progress was made in developing strains of acceptable yield, many practical difficulties have prevented guayule production from becoming competitive with other sources.

OTHER RUBBER SOURCES A great many other genera have been tried from time to time as possible sources of rubber. In the Apocynaceae, *Alstonia, Forsteronia, Funtumia, Landolphia, Mascarenhasia, Odontadenia,* and *Tabernaemontana* have all been tried. So have various species of *Ficus,* of the Moraceae. In the Compositae, Russian dandelion, *Taraxacum koksaghyz,* and various goldenrods, *Solidago,* have been investigated. *Euphorbia, Sapium,* and other genera of the Euphorbiaceae have at times shown promise, too.

Balata gums

CHICLÉ *Achras zapota,* SAPOTACEAE Chiclé, the latex from the sapodilla tree of Central America, provided the original chewing gum. It is still of some importance, but other synthetic and natural balatas now substitute for or extend chiclé. Chiclé has always been harvested from wild trees of *A. zapota,* found scattered throughout the dense forests of Latin America, especially in southern Mexico and Guatemala. Because of this remoteness, the harvest of chiclé follows much the same pattern as the harvest of wild rubber in the Amazon valley. Because the tree cannot be repeatedly tapped, but must recover for many months before yielding appreciable quantities of latex again, the balata is collected by itinerant individuals or small groups who scour the jungles for untapped trees. Tapping is done by cutting a series of gashes from treetop to base. It is practiced principally during the rainy season.

CHILTE, *Cnidoscolus,* EUPHORBIACEAE Species of *Cnidoscolus* are characteristic of semiarid environments like those of northern Mexico and northeastern Brazil. Only two species, the highland or red chilte, *C. elasticus,* and the lowland chilte, *C. tepiquensis,* have served in a minor way as commercial sources of latex, the latter since remote times in southern Mexico, where it has been used

for forming ornamental articles. The latex from both species is high in resin and unsuitable for rubber. All production is from wild trees, red chilte being found in the mountains of west-central Mexico, and white chilte a little farther south along the coast into the state of Jalisco. The small trees are typically tapped by gouging the soft bark in a series of channels that form a sort of herring-bone pattern; several tappings are made over a period of three months, after which the trees are given a nine-month resting period. Latex coagulates naturally if left standing for a day or so, and is then pressed into rectangular blocks for export. Dried coagulum is nearly half resin, half hydrocarbon. Early peoples in southern Mexico used white chilte to mold various ornamental articles, but today chiltes are used chiefly as extenders for more important elastomers, such as gutta-percha.

GUTTA-GUM, *Couma,* APOCYNACEAE *Couma* is found throughout northern South America and Central America. *C. rigida,* known as mucugé in Brazil, and *C. macrocarpa,* leche-caspi, or sorva, have been tapped for latex from time to time; during rubber booms it has been used to a certain extent as an adulterant for rubber latices. The smooth bark of *Couma* makes it possible to tap the tree with the same tools used for *Hevea* in the Amazon area. *Couma* is not, however, believed to be amenable to repeated tapping, as with the jebong system. Often the trees are felled and completely drained by girdling, as with *Castilla.* The latex is usually coagulated by boiling, being only partially coagulated by most acids and alum. Coagulated latex is often stored immersed in water until ready for sale; exposure to air oxidizes it and makes it friable. The dry coagulum has a very high resin content. Much of the gutta-gum coming to the American market is utilized in chewing gum. Locally it may be used as a caulking compound. The fresh latex is potable, and may be used as "cream" in coffee or as a digestive aid.

GUTTA-PERCHA, *Palaquium,* SAPOTACEAE Gutta-percha is a well-known, traditional balata, supplied principally from *P. gutta. Palaquium* grows naturally in India and throughout the South and

Central Pacific. The balata becomes pliable when heated, turns hard, and darkens when exposed to air. It is an excellent nonconductor of heat and electricity, and is impervious to water; thus it has found widespread use as an insulating material, particularly for oceanic cables, but also for such miscellaneous purposes as golf ball centers, acid-resistant receptacles, denture plates, and certain types of transmission belts. In many of these uses it has been superceded by modern synthetic and plastic materials.

Long before it was known to Europeans gutta-percha was used as a molding material by the native peoples of Malacca, Borneo, and Sumatra. British colonials in the Far East found it useful in the humid tropics for handles, splints, and other articles, since it is inert and not subject to deterioration. Not until the middle 1800's was its botanical identity recognized, about which time a German officer discovered its excellent insulating qualities for submarine telegraph cables. This sparked quite a market demand—a demand that resulted in extending the tapping of wild *Palaquium* trees into the remoter sections of Borneo, the Philippines, and other parts of the southeastern Pacific. The original gutta-percha came from felling *P. gutta* trees on the island of Singapore. As supplies became scarce other species were utilized, and no doubt there was some dilution with the latices of other Sapotaceous genera.

In recent decades the felling of trees has been abandoned in favor of tapping, and there has been some planting on estates. Tapping consists in cutting a series of channels in the bark to conduct the latex to some receptacle at the base of the tree. Several tappings a year are possible, but the tree is usually so injured in the process that it may die. Much of the latex coagulates directly in the gashes in the bark, and is gathered as scrap. At a central camp the latex is heated to complete coagulation.

Recently, in order to preserve the trees, a method of extracting the balata from leaves has been developed. Twigs with foliage are ground, washed, and treated with cold water, causing small threads of balata to adhere to each other and float to the surface. This mass is skimmed off, purified, softened in hot water, and formed into compact blocks of nearly pure balata. This method of extraction is now preferred commercially, especially since the felling of gutta-percha trees has been prohibited over much of the Far East. Nevertheless, the gradual extermination of trees has prevented any sizable capital investment in facilities for extraction from leaves. Gutta-percha is generally preserved under water, to avoid oxidation of the surface. The balata softens on heating, and then may be molded into any shape, or even drawn into threads by hand. Production of gutta-percha is no longer a sizable industry, having declined over the years from a level as high as $\frac{3}{4}$ million pounds annually during World War I. This has been due partly to substitution by gutta-gum from South America.

JULETONG, *Dyera*, APOCYNACEAE Juletong is obtained from several species of *Dyera*, tropical forest trees of Malaysia and the East Indies, especially *D. costulata*. Indiscriminate tapping causes high mortality, although the trees are now protected in many areas. Improved tapping schemes have now been developed, one of which is similar to the jebong panel system utilized for *Hevea*. As much as a pound of latex may be obtained per tapping from top-yielding trees. Coagulation occurs naturally, but is more frequently stimulated with acid or other chemical agents. As with gutta-percha, juletong is plastic when warm, but hardens upon cooling and exposure to air. It is generally stored under water to prevent brittleness and crumbling. Only a few million pounds of juletong are produced annually, most of which is mixed with other balatas in chewing gum. Juletong can also serve to extend gutta-percha or inexpensive rubber.

OTHER BALATA SOURCES Many other latex-bearing species, particularly tropical trees, yield balata. Their tapping can be locally important if the market warrants. Almost any can substitute for those named. Genera of the Sapotaceae are especially important, including *Manilkara* and *Mimusops*, as well as *Calocarpum, Sideroxylon, Dipholis, Bumelia, Lucuma, Chrysophyllum*, and others. In the Apocynaceae are *Stemmadenia*,

Tabernaemontana, Thevetia, Plumeria, Cameraria, and others. *Brosimum* and *Pseudolmedia* of the Moraceae are sometimes used.

RESINS, GUMS, AND RELATED EXUDATES

This is something of a catch-all section. Some of the products discussed here are related to those discussed in various other sections. For example, resins are an important component of latex, especially balata types. Many medicinals owe their effects to their resin content. Rosin, the important industrial resin obtained as a by-product of turpentining, was mentioned among essential oils.

When civilization was young, natural resins and gums were much more important than they are now. They were often the only source of materials for waterproofing, caulking, and incense, and were used in emollients, paints, medicinals, and embalming compounds. As with spices, they were often procured in distant lands at considerable effort. Almost all were extracted from wild plants, and some still are today. But natural resins and gums have largely given way to synthetics in the modern economy. As with rubber, less expensive chemical substitutes have been found. And, as with oils, modern technology creates a variety of specific end-products from simple materials.

Resins constitute a more diversified and complicated group than do the gums. Basically they are polymerized from carbohydrates that combine with various organic substances. They are amorphous, brittle, generally clear or translucent, and insoluble in water (but often soluble in alcohols or other organic reagents). Most burn with a smoky flame (one of their primitive uses was for all-weather torches). As with latex, their usefulness to the plant is obscure. They occur as secretions in special ducts in a wide variety of plants and plant parts, often mixed with other materials, such as latex or essential oil. Many resins seem to be oxidative products of an essential oil. They may be antiseptic and prevent desiccation, thus aiding in healing plant wounds. Commercial supplies are typically derived by bruising or cutting the bark of stems and roots. Because there is no truly satisfactory way to classify resins, they are usually identified by source. The more important natural resins come principally from the Pinaceae, the Leguminosae, and the Dipterocarpaceae families. Those from the Pinaceae are often "oleoresins," like crude turpentine; resins from legumes include not only the "copals," but those that contain benzoic or cinnamic acid, loosely termed "balsams"; the "damars" are Far Eastern resins from the Dipterocarpaceae.

One of the more interesting and valuable resins is amber, an exudate from an extinct pine (*Pinus succinifer*) or related species. Miocene amber from tropical America probably comes from *Hymenaea* of the Leguminosae. Amber occurs in subterranean deposits, particularly off the northern coast of Europe, where it frequently washes up on the beaches. In ancient times it was quite a mysterious substance, with many wild surmises as to its origin. It has always been prized for ornaments and charms. Within recent centuries it has been too valuable to use in varnishes and in other utilitarian ways, as are other resins, even though it has excellent workability and insulating properties. That amber was once fluid is attested to by the fact that fossil insects found trapped in fragments of amber are preserved just as if embedded in a transparent plastic.

Today's dominant synthetic resins are derived from inexpensive base chemicals such as phenol, formaldehyde, and glycerin. They substitute for natural resins in adhesives, plastic coatings, paints, varnishes, safety glass, waterproofing compounds, and so on. But natural resins still find some market in many of these products, as well as in paper and textile sizings, in some pharmaceuticals, coatings, inks, and incenses. The compounding of resins with an oil or solvent yields a **varnish**. Varnish leaves a hard and essentially waterproof coating as the oil oxidizes or the solvent volatilizes.

Gum-like hydrocolloids (including pectin) are chemically simpler than the resins. By the same token they are much more widespread in plants, and thus more easily substitutable. Although widely used, they have generally commanded neither the price nor the human interest that resins have. Gums and pectins occur naturally as cementing substances in plant tissues, showing up especially upon injury, disease, or tissue disintegration. Most gums are amorphous colloids without any definite melting, freezing, or boiling point. They consist largely of polymerized carbohydrate constituents, and in contrast to resins are hydrophillic, being soluble or dispersable in water (but not in familiar organic solvents and oils). Their greatest use stems from their capacity to form viscous colloidal gels used industrially as sizings, thickeners, and stabilizers. Gelation is due to chemical cross-linkage of adjacent polymers to give a three-dimensional network; many food gels are firm when they are no more concentrated than 0.1 percent. Mucilaginous types are mainly polysaccharides, with little resin content.

One of the best-known gums is agar, an alginate obtained from seaweed, mostly in Japan. Seaweeds such as *Gracilaria* (Rhodophyceae) are boiled, and the agar gum flaked out by alternate freezing and thawing. Attempts have been made to discover economical substitutes for this among crops adaptable to North American farming. Over 300 species were reviewed by Tookey and Jones, who concluded that four plant families in particular contain promising genera—the Leguminosae, Plantaginaceae, Cruciferae, and Convolvulaceae. In all of these families there are species in which the seed endosperm contains more than 18 percent gum (36 species in the Leguminosae alone). Leguminous gums are mostly galactomannans. Extraction usually involves no more than separation of the endosperm and its grinding to the fineness of flour. Galactomannans are important as paper additives, in the waterproofing of explosives (their swelling prevents further penetration of water), as a textile size, as flocculants in sewage treatment, or as thickeners and stabilizers in cosmetics and drugs.

Pectins bind adjacent cell walls in the plant, and occur with greater or lesser frequency in all parts. Commercially they are extracted chiefly from fruit by-products, such as citrus rind and apple pomace from cider milling (vinegar fermentation). The most familiar use of the methoxyl pectins is to insure solidification of jellies. But they are also widely used in pharmaceuticals, cosmetics, medicinals, candies, ice creams, and various food prod-

ucts. Industrially, pectins are incorporated into adhesives, sizes, used for the "creaming" of rubber latex, in the quench-hardening of steel, and in other ways. Apple pomace and citrus residues contain as much as 6 percent pectin. This is extracted by mild acid hydrolysis, followed by precipitation with alcohol or other hydroxides.

It is evident that resins and gums are derivable from hordes of botanical sources; price and competition with synthetics determine which will enjoy economic success. By-products are especially low-cost raw materials, not all of which have been exploited. A useful resin can be derived from cashew nut shells, for example, and a gum that is used in water paints and mucilages from solvent-extracted flax seed meal. Western larch (*Larix occidentalis*) waste can be made to yield an arabogalactan gum, substitutable for gum arabic. Resinous substances are derived from tall oil, mainly a by-product in kraft paper-pulp production. Of course, in less affluent regions, many little-used wild plants may possibly prove to be economical sources of gum, such as tropical trees of West Africa (*Anogeissus* and *Kahya*). Gum tragacanth, a classical gum-resin, is obtained from wild sources, and guar is a potential new crop source for mechanized agriculture.

Gum tragacanth, Astragalus, *Leguminosae*

Tragacanth is a traditional gum, produced chiefly in Iran. On occasion production has reached nearly 2,000 tons annually, although production has declined in recent years because of the exhaustion of wild stands, the only gum source. The several species of *Astragalus* that yield gum tragacanth grow in the arid highlands, where they are often used for badly needed firewood in spite of a national conservation law designed to protect them. Even the larger species grow only to moderate shrub size in the inhospitable environment from the Zagros mountains northward, where the gum is collected seasonally by nomadic peasants.

Perhaps as an adaptation to the arid environment, species such as *A. gossypinus* and *A. echidenaeformis* develop an amorphous cylinder of gum in the center of the stem and root, where pith would normally be expected. This gum is under pressure in summer, and will exude if the stem or root is cut. Twenty-three different species are known to produce such gum. Originally the bark alone was tapped; eventually it was discovered that incisions to the central gum cylinder provided better yield. Tapping is accomplished by digging the earth away from the taproot and by incising the upper part of the root deeply enough to reach the gum cylinder. "Ribbons" of gum exude, progressively more slowly as the days go by, and are collected each several days for a week or two, after which a new tapping may be initiated. Tapping starts in June, and tapers off before autumn rains come. A large, healthy bush may yield as much as 10 grams of gum on first tapping, although the average is nearer 3.

As was mentioned under the discussion of medicinals, gum tragacanth has been collected since before historical times. Initially it may have served as a food during times of stress, and later as a medicinal because of its bitterness. Greek physicians recommended it as early as the seventh century B.C. Today the better quality "ribbon gum" is used in pharmaceuticals, liquors, cosmetics, dental creams, and confections. Bitter gum of lesser quality is often used in dyes, sizes, and water proofing materials. With synthetic substances available, tragacanth will decline in importance.

Guar, Cyamopsis tetragonoloba, Leguminosae

Guar, an ancient crop of India, is also grown in Africa. In general appearance it is very similar to its relative the soybean. In India the seed, source of the colloidal guar gum, is consumed as human food. The gum, a polymer of galactose and mannose, is extracted from the endosperm. Guar gum exhibits high viscosity at low concentration, swells well in cold water, is useful over a wide pH range, and has excellent film-forming and stabilizing properties. It is used as a flocculent and filter-aid in mining; it is also used in cosmetics, drugs, paper manufacture, and foods.

Guar seems to be the kind of plant that would adapt to mechanized agriculture. Moreover, it is drought resistant, and should grow well in the arid southwestern United States, where new crops are needed and where, indeed, it was grown to a limited extent decades ago for soil improvement and forage. During World War II, when imported gum supplies were cut off, improved selections were rogued from old fields in Arizona and New Mexico. These provided an economic toehold, when after the war competitive pressures again made growing of uncertain profitability. Several "bugs" still remain to be worked out, and wider acceptance must be achieved to establish guar as a cash crop. Federal aid has helped, through sponsored research and acceptance of guar on restricted acreage.

By and large, guar can be handled by existing farm equipment. There is some planting in Texas, where facilities have been developed to handle and process the crop. This gives hope that interest in guar will spread, as with the soybean before it. Guar fits well into rotation with cotton, benefitting the soil and generally increasing cotton yields. It has been estimated that market potential for guar gum in the United States should exceed 50,000 tons annually, principally for food processing, but also in the paper, mining, textile, and explosive industries.

Other resins and gums

A list of plants would be almost unending were it to include most past and potential sources of gum and resins. Pectins, for example, can come from almost any fruit pulp, in almost any part of the world. The costs of gathering, processing, and shipping determine which ones are economical.

Gum arabic is another of the traditional gums from wild species of *Acacia*, Leguminosae, native principally to north Africa and the Near East; gum arabic was mentioned among the medicinals. **Karaya gum** is the dried exudate from *Sterculia urens*, Sterculiaceae, of India. **Locust bean gum,** or **carob gum,** is derived from the fruit of *Ceratonia siliqua,* Leguminosae, of the Mediterranean area.

Other resins are **Balsam of Peru,** *Myroxylon balsamum,* Leguminosae; **Canada balsam,** *Abies balsamea,* Pinaceae, North America; **copaiba balsam,** from *Copaifera,* Leguminosae, South and Central America; various **copals,** from Leguminous genera, such as *Trachylobium* in eastern Africa and

Hymenaea in South America; various **damars** from Dipterocarpaceous genera, such as *Balanocarpus, Hopea, Shorea,* and *Vateria;* **elemi,** from genera of the Burseraceae, in various parts of the world; **frankincense,** from *Boswellia,* also of the Burseraceae, of much interest to the ancients in the Near East; **kauri resin,** from *Agathis australis,* Araucariaceae, New Zealand; **lacquer resins** from genera of the Anacardiaceae, such as *Rhus verniciflua,* of China; **mastic,** from a shrubby pistachio, *Pistacia lentiscus,* also of the Anacardiaceae; **myrrh,** from *Commiphora,* Burseraceae, mainly from Africa, companion to frankincense in the olden days; **sandarac,** a varnish resin from *Tetraclinus,* Pinaceae, of North Africa; and **storax,** from Oriental and North American species of *Liquidambar,* Hamamelidaceae.

TANNINS AND DYES

As with several of the natural products discussed in previous sections, the heyday of trade in tannins and (especially) dyes lies in the past. Products from chemical synthesis have almost entirely substituted for natural dyes, and more and more leather is being tanned with chemical agents or being substituted for by synthetic leather or rubber and composition shoe soles. Nevertheless, vegetable tannins are still sufficiently important to merit some discussion, and dyes should be mentioned if for nothing more than their historical importance.

Tannins are astringent substances basically of carbohydrate origin. They characteristically produce a dark coloration when combined with iron salts; precipitate gelatins, proteins, and alkaloids; and combine with skin proteins to form leather. Natural tannins have been divided into two groups, catechols and pyrogallols, largely on the basis of their origin rather than on the basis of their rather obscure chemical structure. Some tannins are hydrolized by acids and enzymes, and others are not; this difference forms the basis for still another grouping.

Tannins occur in small amounts in most plant tissue. You will recall that tannin is relatively quite abundant in tea leaves, and that tea of high quality generally has a high content of tannin. Tannins are also relatively abundant in tree bark, certain types

of wood, most foliage, and many young fruits and often are in relatively high concentration in insect galls. As with rubber and resin, the usefulness to the plant is obscure; tannin may be merely a by-product of plant metabolism, or it may serve some antiseptic purpose.

Bark and wood have been the chief commercial sources of tannin. In North America the supply long came from chestnut wood, now a thing of the past, since chestnut blight largely eliminated the species. Hemlock, oak, and Douglas-fir bark have all been of some importance. But since domestic production has seldom met demand, there has been much importation of tannin—quebracho especially, derived from the wood of a South American tree; mangrove tannin, from the Far East; and wattle tannin, from *Acacia* in South Africa.

It may be that tanning was discovered when pelts were soaked for softening in bogs charged with leaves and twigs. In any event, since time immemorial hides have been treated with tannin to yield an especially durable product, leather. The Romans used various barks, berries, galls, and leaves as sources of tanning agents. Since tannins leach readily, they can be extracted and concentrated, to give more certain and uniform tanning results. In tanning, hides are first soaked in lime solution to loosen the hair, which is removed by special machines. Then it is immersed in tanning solution, and afterwards washed, bleached, oiled (and in some cases rolled or polished and coated with various substances).

In gathering bark for tannin, trees are typically girdled and the lowermost bark removed while the tree is still standing. The trees are then felled, and additional sections of bark are removed, each about 4-feet long. Dry bark is shipped to processing centers, where it is shredded (if wood is used, it is chipped), and the tannin extracted by various methods of water diffusion or percolation (often under pressure in autoclaves, especially with wood chips). Usually the extraction liquor is heated by steam, and circulated progressively from exhausted bark to the newest charge. Chemicals are added to some extraction liquors; sodium sulfite is added in quebracho tannin extraction.

Tanning liquor from the initial extraction is concentrated by vacuum evaporation. Liquid concentrates contain up to 35 percent tannin, solid ones up to 70 percent. Tannins are shipped in coated tanks, to prevent the tannin from reacting with the metal. Natural tannin is primarily used for heavy leathers, but an appreciable quantity is used to control viscosity in oil-well drilling mud. It was formerly used to produce ink. The mining industry uses small quantities for flotation of ores, and a little is used as an astringent in medicinals.

The importance of natural dyes in the ancient world is evidenced by their use in decorating mummy cerements in Egypt; untold toil went into the collection and extraction of yellow from saffron flowers, blues from indigo, and reds from madder. The aboriginal peoples of Britain even colored their skin with the blue dye woad. Everywhere, until the middle 1800's, dyes from vegetable sources were coveted nearly as much as spices. Almost every region had its favorite dyes, for plants capable of yielding dyes are legion. But in 1856, when an Englishman named William Perkin developed a coloring material much like indigo, from formerly "useless" coal-tar substances, the death knell sounded for natural dyes. Particularly in Germany was synthesis of coal-tar dyes pursued, until the civilized world eventually came to rely almost entirely on aniline derivatives for its coloring materials. Not only are the synthetics less expensive, but they are in more certain supply, truer, and more lasting.

In early times, most dyes were not very permanent; they washed out of the fabrics rather readily and easily faded or discolored. It is understood now that dyes applied in soluble form must be made "fast" to prevent their running. This is accomplished with chemical **mordants,** which combine with the dye and fiber molecule and fix the color on the fabric. Most mordants are salts of metals, and often they are effective in helping develop the characteristic color of the dye. In fact different mordants can cause a particular dye to yield different colors.

Quebracho, Schinopsis lorentzii, Anacardiaceae

Quebracho is probably the most important tannin derived from wood, now that the chestnut is nearly extinct in eastern North America. Development of

the quebracho industry has occurred largely within the last half century. The tree is native to southern South America, where *Schinopsis* grows in scattered stands on the savannahs and lowlands of Argentina and Paraguay. The hard wood has long been used for posts and cabinet wood, preserving well since it contains about 22 percent tannin. For tannin production logs are felled in the interior and transported to local extraction plants along the rivers after trimming them of bark and sapwood (which contains relatively little tannin). The heartwood, about 30 percent tannin, is chipped on special machines, the tannin extracted by steam under pressure to yield a heavy dark liquor that is evaporated to produce tannin cake. Quebracho tannin is well thought of, acting quickly to give a strong, tough leather.

Mangrove, Rhizophora mangle, Rhizophoraceae

This abundant small tree grows world-wide along tropical seacoasts. In the Far East, where abundant and inexpensive manpower is available, cutting the trees has often proved profitable. The tannin is extracted from the bark, which is stripped from the small stems and transported to local extraction factories. After drying it is pulverized and boiled to yield the commercial extract.

Wattle, Acacia, Leguminosae

Several *Acacias*, especially species native to Australia (but planted in various parts of the world), yield wattle bark, from which tannin is extracted. The bark contains as much as 50 percent tannin. Pulverized bark may be used directly, or concentrates can be made as previously described. Wattle production is chiefly from South Africa and Brazil.

Canaigre, Rumex hymenosepalus, Polygonaceae

Canaigre, looking much like the familiar dock weeds of lawn and garden, has a fleshy taproot abundant in tannin. It is mentioned here because it should be adaptable to mechanized agriculture were ecomonics to favor its growing. The root resembles a parsnip and contains up to 35 percent tannin. The tannin is extracted by pulverizing the root and eliminating soluble sugars by fermentation.

Sumac, Rhus, *Anacardiaceae*

Various sumacs, in the same family with quebracho, yield tannin rather abundantly from the foliage. *R. coriaria* is said to contain 35 percent tannin in the dry leaves, which have been a minor source of tannin in world trade for a number of years. No doubt many of these shrubs and small trees could be adapted to cultivation, were demand for the tannin sufficiently strong.

Hemlock, Tsuga canadensis, *Pinaceae*

Hemlock bark was once wastefully stripped from the trees by the North American colonists as an important source of tannin. The bark contains up to 50 percent tannin, easily extracted, and useful for tanning heavy leathers such as sheepskins. With supplies depleted, colonists soon turned to chestnut wood, and with that now gone too, some bark is harvested from the western hemlock, *T. heterophylla*, mostly as a sideline to logging operations in the Pacific Northwest.

Oaks, genera and species of the *Fagaceae*

Several oaks provide both tannins and dyes. Examples are the chestnut oak, *Quercus prinus*, which grows in the eastern United States, and the tan oak, *Lithocarpus densiflorus*, which grows in the Pacific forests. Acorn cups of the Turkish oak, *Quercus aegilops*, abundant in Asia Minor, yield as much as 45 percent tannin, known in the trade as valonia.

Indigo, Indigofera tinctoria, *Leguminosae*

Indigo was once one of the world's most important dyes, possessing unusual fastness. It does not achieve its characteristic color until undergoing aerobic fermentation. Preparation involves holding chopped indigo plants in water for some hours, followed by aeration of the solution (by agitation) until the blue color develops. After a precipitate settles out, further fermentation is stopped by heating. The filtered sludge, indigo, is formed into cakes for shipment. Indigo was especially important to the commerce of India in the days of the spice trade. It is reported that in 1631 three Dutch ships brought to Holland a cargo of indigo worth more than half a million dollars. Some indigo was grown in the English colonies in North America during the 1700's, but the industry was rather short-lived in competition with less expensive production in the Orient.

Madder, Rubia tinctorum, *Rubiaceae*

Another of the classical dyes was madder, a brilliant red ("turkey red") obtained from the root of this species. The plant is a weak herbaceous shrub, the dye found principally in the cortex of the slender roots. Typically 2-year-old plants are dug, the root washed, dried, pulverized. An infusion was made from the powder as needed. Different mordants give decidedly different colors with madder. Madder was once grown in India, but introduction into Europe in the late middle ages made Holland a center both of dye production and textile fabrication.

Other tannins and dyes

Gambier, *Uncaria gambier*, Rubiaceae, native to the Far East, yields a resinous extraction from the leaves that contains as much as 40 percent tannin; myrobalan, the unripe fruit of *Terminalia*, Combretaceae, is a tropical source of tannin somewhat used in India; divi-divi is the fruit of *Caesalpinia coriaria*, Leguminosae, a small tree of lowland tropical America; the pods of this plant yield up to 50 percent tannin and are collected from wild trees for export, principally to Europe.

Many are the ancient dye plants. Woad, *Isatis tinctoria*, Cruciferae, yields from its leaves the same colorless glucoside as indigo; the species was generally cultivated in Europe and Russia before the advent of aniline dyes, and for a while was protected in England by prohibition of the importation of indigo. Extraction involved fermentation, and only after oxidation of the dyed fabric did the blue color develop. Logwood dye is obtained from

Hematoxylon campechianum, Leguminosae, a small thorny tree of tropical America. The red heartwood is a source of the dye, still somewhat used in the staining of laboratory slides. Annatto, the seed pulp of *Bixa orellana*, Bixaceae, native to Brazil but widely naturalized throughout the tropics, yields a yellow dye that was once much used in food coloring (margarine). Saffron was the principle yellow dye of the ancients, derived from the stigmas of the saffron crocus, *Crocus sativus*, Iridaceae. It is said that about 4,000 stigmas are needed to yield an ounce of the dye; no wonder saffron was supplanted by synthetic substitutes! Safflower, *Carthamus tinctorium*, Compositae, is known today as an increasingly important source of edible oil. Originally, however, a yellow dye was obtained from the flowers, a substitute for saffron.

Selected References

Adrosko, R. J. *Natural Dyes in the United States.* National Museum Bulletin 281, Smithsonian Institute, Washington, D.C., 1968. (A review of the many sources of dyes which could be utilized in this country in place of synthetics.)

Akehurst, B. C. *Tobacco.* Humanities Press, New York, and Longmans, London, 1969. (An in-depth review of "the weed" and its uses.)

Amerine, M. A., H. W. Berg, and W. V. Cruess. *Technology of Wine Making.* (3rd ed.). AVI Publishing Co., Westport, Conn., 1972. (Concerned chiefly with manufacture in the wine regions of the world.)

Arctander, S. *Perfumes and Flavor Materials of Natural Origin.* AVI Publishing Co., Westport, Conn., 1960. (Comprehensive. Part one discusses processing and evaluation; part two, 537 raw materials sources.)

Cook, A. H. (editor). *Barley and Malt: Biology, Biochemistry and Technology.* Academic Press, New York, 1962. (Comprehensive technical review.)

Eckey, E. W. *Vegetable Fats and Oils.* (American Chemical Society Monograph 123). Reinhold, New York, 1954 (Chemical composition and sources related to botanical group.)

Eden, T. *Tea* (2nd ed.). Longmans, London, 1958. (Discussion of history, botany, planting, care, harvest, and processing of this important beverage, with references, by the Director of the Tea Research Institute in Ceylon and Africa.)

Goldberg, H. S. (editor). *Antibiotics, Their Chemistry and Non-Medical Uses.* Van Nostrand, Princeton, N.J., 1959. (Antibiotics in nutrition, plant disease control, food preservation, and so on.)

Guenther, E. *The Essential Oils.* Van Nostrand, Princeton, N.J., 1948. (A classical reference.)

Haarer, A. E. *Modern Coffee Production.* Leonard Hill Ltd., London, 1962. (On the history, genetics, physiology, and practical growing of coffee by an English expert with wide practical experience in E. Africa.)

Harler, C. R. *The Culture and Marketing of Tea* (3rd. ed.). Oxford University Press, London, 1964. (A widely recognized review of tea, from the viewpoint of a society in which tea has long been a very important accouterment.)

Haslam, E. *Chemistry of Vegetable Tannins.* Academic Press, New York. 1966. (Of possible reference value, although of limited interest to the general reader.)

Hill, A. F. *Economic Botany* (2nd ed.). McGraw-Hill, New York, 1952. (A textbook on cultivated plants.)

Hoffer, A., and H. Osmund. *The Hallucinogens.* Academic Press, New York, 1967. (A detailed compendium about drugs of especial importance to the "drug culture").

Howes, F. N. *Vegetable Gums and Resins.* Chronica Botanica. Waltham, Mass., 1949. (A definitive discussion.)

Howes, F. N. *Vegetable Tanning Materials.* Butterworth, London, and Chronica Botanica, Waltham, Mass., 1953. (A thorough discussion of tannin sources, which have changed little since publication of this book.)

Jacobson, M. *Insecticides from Plants.* USDA Agricultural Handbook 154, 1958. (A review of the literature 1941–1953 reveals use of thousands of species, here catalogued according to botanical classification with brief discussion.)

Joyce, C. R. B., and S. H. Curry (editors). *The Botany and Chemistry of Cannabis.* Proceedings of conference Inst. Study Drug Dependence, Ciba Foundation, J. & A. Churchill, London, 1969. (An unbiased effort by numerous authorities to provide information on the emotional marijuana subject;

the work is somewhat technical, and reveals that much still remains to be learned about marijuana.)

Kaplan, J. *Marijuana—The New Prohibition.* World, New York, 1970. (Marijuana viewed more from the social than the technical standpoint; of broad, general scope.)

Kingsbury, J. M. *Poisonous Plants of the United States and Canada.* Prentice-Hall, Englewood Cliffs, N.J., 1964. (700 species are organized systematically, with descriptions and many illustrations for identification, mention of the poisonous principle, and toxicity.)

Leggett, W. F. *Ancient and Medieval Dyes.* Chemical Publishing Co., Brooklyn, 1944. (Written especially for their historical interest.)

Mantell, C. L. *The Water-Soluble Gums.* Reinhold, New York, 1947. (Source material for a limited group.)

Merory, J. *Food Flavoring.* AVI Publishing Co., Westport, Conn., 1960. (Section 1 discusses natural food and beverage flavorings; later sections review synthetics and technology.)

Parry, J. W. *Spices* (2 vols.). Chemical Publishing Co., New York, 1969. (Chiefly directed to food chemists, this work emphasizes morphology, histology, and chemistry of the familiar spice plants.)

Polhamus, L. G. *Rubber: Botany, Production, and Utilization.* Interscience (Wiley), New York, 1962. (An excellent monograph on rubber and rubber-bearing plants.)

Schery, R. W. *Plants for Man* (2nd ed.). Prentice Hall, Englewood Cliffs, N.J., 1972. (A work on economic botany by one of the authors of the present book.)

Schetky, E. J. (guest editor). *Dye Plants and Dyeing.* Plants and Gradens Handbook, Brooklyn Botanical Gardens, 1964. (A popularized collection of articles with a crafts flavor, bringing an age-old industry to date.)

Singer, R. *Mushrooms and Truffles: Botany, Cultivation, and Utilization.* Interscience (Wiley), New York, 1961. (A fascinating review of the industry throughout the world.)

Snyder, S. H. *Uses of Marijuana.* Oxford University Press, New York, 1971. (A concise, readable book, which, after reviewing marijuana's history and pharmacology, comes to the conclusion that "moderate social use of marijuana . . . is unlikely to have grave harmful consequences.")

Society for Economic Botany. *Economic Botany.* New York Botanical Garden, New York. (A journal, published since 1947, devoted to applied botany and plant utilization with much information on minor as well as major crops.)

Taylor, N. *Plant Drugs That Changed the World.* Dodd, Mead, New York, 1965 (Catalogue of plants and the drugs they contain, related to use.)

Tso, T. C. *Physiology and Biochemistry of Tobacco Plants.* Dowden, Hutchinson & Ross, Stroudsburg, Pa., 1972. (An authoritative survey of worldwide scope, by a leading tobacco scientist from the USDA; discusses various types of tobacco, their growth and handling, metabolic factors, and regulation of the tobacco industry.)

U.S. Department of Agriculture. *Tobacco in the United States.* Misc. Pub. 867, USDA, 1966. (A brief historical sketch is followed by a thorough technical discussion on growing, curing and marketing; bibliography.)

Urquhart, D. H. *Cocoa* (2nd ed.) Wiley, New York, 1961. (A thorough review of cacao worldwide, with especial attention to its growing in different regions.)

Weiss, E. A. *Castor, Sesame, and Safflower.* Leonard Hill Books, London, 1971. (A comprehensive work on these three oilseed crops, with numerous illustrations and citations. The most thorough resume of these crops available to time of publication.)

Wellman, F. L. *Coffee: Botany, Cultivation and Utilization.* Leonard Hill, London, 1961. (Another of the thorough "world crops" series, dealing with all aspects of coffee and knowledge concerning the crop to the time of publication.)

Williams, L. O. *Drug and Condiment Plants.* USDA Agricultural Handbook 172, 1960. (A brief resume in alphabetical sequence of major species grown in the United States; bibliography and index.)

Winkler, A. J. *General Viticulture.* University of California Press, Berkeley, 1962. (Textbook treatment of grape culture, especially as practiced in California.)

CHAPTER 23

Fiber, Forest, and Ornamental Crops

Chapter 22 is primarily concerned with the plant extracts that are important for their chemical and combining properties. In this chapter, emphasis is on direct employment of relatively permanent plant substances for their mechanical, structural, or esthetic qualities. Strength and rigidity are implied; cell wall and the cell-to-cell cementing substances rather than the cell content or living protoplast are mainly involved. The cell wall is durable and chemically rather inert. It consists mainly of a polymerized carbohydrate, cellulose. It is deposited by the protoplast, which is typically dead and long gone by the time the plant tissue is utilized by man.

This is not to say that our subject area involves no chemical transformations. In the making of paper, for example, the fibers may be separated by dissolving the cementing lamellae between cells with chemical solutions. The substances dissolved from the wood in such operations may be hydrolized to simpler carbohydrates and concentrated for certain uses. In the preparation of fiberboard, for example, solutions used to soften and separate

the fiber can yield a "wood molasses" that is equally as useful for feeding cattle (or for industrial purposes) as is the traditional molasses obtained from sugar cane.

Especially noteworthy is the changing forestry picture. In the modern slaughterhouse, it is said, "everything is used but the squeal"; forestry, once wasteful of a goodly portion of the tree, now takes a similar tack. Where formerly only wood of lumber grade was used, the industry now integrates operations to utilize more fully all of a tree. Chapter 15 pointed out how forestry is fast changing into a planned cropping system, albeit on a long-term cycle as compared to annual cropping on the farm. With the end of virgin timber in sight, forest management must simulate farming by making trees a renewable crop. The economy already views wood substance as a convertible raw material, to be made into the necessities of modern life by whatever processing is most economical. If trees cannot supply enough good lumber, perhaps even scrawny stock will make good plastic. If the fiber needed for paper pulp cannot

Box 23-1 THE ONCE "LIMITLESS" FORESTS: ANOTHER NON-RENEWABLE RESOURCE?

Man is exhausting the world's forest bulk (biomass) as relentlessly as he is using up accumulated fossil fuels (petroleum, coal). The end of virgin forest tracts of any great size is in sight, what with a new highway existing now in Amazonia. On a crowded earth good land able to produce forest fairly quickly and abundantly is being delegated to cash crops with an annual or short cycle, not left to bulky, slow-maturing trees (which, for the most part, now are only allowed to regenerate on or are planted to rocky slopes, these often degraded by erosion after earlier forest felling).

The situation seems especially critical for the tropical rainforest, the ecology and succession of which are poorly known. Since the rainforest has been traditionally subjected to no greater pressure than the limited clearing inherent in shifting agriculture, regeneration seems to have occurred without mass extinctions. Now, as part of the tropical Green Revolution, much land is being brought into permanent use for crops.

What regeneration is allowed to occur is mostly of the secondary heliophiles ("weedy" species), and succession to the lordly primary species of the forest is never attained.

As the primary species continue to be felled on neighboring virgin land, even seed sources are eliminated. And becuase of repeated land disturbance, these species face increasing competition with more weedy growth and, consequently, extinction. Indeed, in much of the virgin rainforest, the soil and surroundings become permanently unsuitable for tall forest, at least that containing very many of those species that have been most prized for wood and fiber. Again, man's proclivity for exploitation seems to be killing the goose that lays the golden egg! This seems more applicable to the tropical rainforest (which has a fragile ecology poorly adapted to serious disturbance) than to temperate forests (which are less rich in species).

be had economically from trees, perhaps perennial bamboos or fibrous stems of annual crops can be substituted. Already technology has shown ways to utilize what were formerly "useless" species and waste tree parts, for the making of pulp and other products, such as charcoal briquets and soil conditioners.

The beauty and "warmth" of wood can hardly be duplicated by synthetics. Thus, while it is hard to imagine the abandonment of wood for esthetic uses, one can visualize the day when houses, for example, might be constructed mostly with plastic materials, perhaps molded to exact size and shape— "plastic lumber" as it were. Whether the raw material for such plastics would come from trees, from annual crops (such as soybean oil), or from mineral sources (such as petroleum) will be determined by economics. No plant substance is these days entirely secure in its market. We can be fairly confident that certain plants now managed as crops will yield the raw materials for future civilizations,

since exhaustable mineral resources must diminish and thus become more expensive. But we can't be sure which species will be used, nor to what extent. The products discussed in this chapter are particularly amenable to competition from synthetics and other structural materials, such as metals; pronounced change seems in store for many of the traditional uses.

Plants used as ornamentals are also reviewed in this chapter. Although man has tried to beautify his surroundings in one way or another since time immemorial, only in recent decades has the affluence provided by industrialization created a mass market for ornamental plants. We don't often think of ornamentals as a "crop," ranking in dollar value in the United States on a par with a number of important food and industrial species such as rice and citrus. But market research predicts that the "outdoor living" market will soon amount to seven or eight billion dollars annually. Of course, this includes swimming pools, patio furniture, and maintenance equipment, as well as lawns, shrubs, trees, and products for their care. Nonetheless, the market is based on esthetic employment of plants. Where once agronomy and horticulture dealt largely with farm crops, today many specialists give full attention to lawn and home garden problems. With only some 5 percent of today's population on the farm in the United States, it is to be expected that increasing interest will center on urban and suburban horticulture.

FIBERS

The word "fiber" has many meanings. To the anatomist a fiber is a slender, tapered cell with an unusually thick wall and simple pits. String-like masses or entrainments of cells are also called fibers, at least commercially. Single-cell fibers are most familiar as constituents of hardwood xylem. Fibrous bundles are more often found in the pericycle, phloem, or cortex of the stems of such plants as flax, hemp, and ramie. Cotton, the world's most important natural fiber, consists of individual, elongated cells borne on the seed. A wide assortment of plant substance is sometimes used as stuffing material along with wool, silk, feathers, and similar products that are loosely termed "fiber" in the trade. Examples are dried Spanish moss (*Tillandsia*), coconut husks (**coir**), broomcorn inflorescences, frayed bark (especially of palms), and leaf thatch.

Chemically, the fiber cell wall is chiefly cellulose. Most cell walls contain an appreciable percentage of lignin, the dark residuum evident in humus as plant tissues decompose. Other substances, too, are contained in minor quantities, including tannins, dyes, and various soluble or readily converted components. But the better fibers are nearly pure cellulose, a complex polymer of linked glucose residues $(C_6H_{10}O_5)_n$ of great molecular length. Although man has had no reason to synthesize cellulose, so amply and inexpensively do plants provide it, he has nevertheless created similar polymers from a variety of raw materials (mostly from petrochemical sources) to yield such synthetic fiber materials as rayon, nylon, acrylics, polyesters, and polypropylenes.

So successful have these tailor-made synthetics been that they have greatly supplanted natural fibers in the modern market. The use of several major synthetic fibers has more than doubled in the past decade. All told, synthetic fibers account for some 5 billion pounds, more than half the fiber market in the United States.

Moreover, because of certain characteristics, such as the wrinkle resistance inherent in the polyester fiber, synthetic fibers have generally maintained a premium over cotton, and represent the more profitable trade item. Excessive production of synthetic fibers occasionally drives the price down sufficiently to present strong competition to natural fibers on price alone. Without government subsidy and protection, it is doubtful that the cotton industry could have held as much of the market as it does in the United States today. Fortunately for the cotton industry the tendency is to combine fiber types in fabrics, such as half-and-half cotton-polyester blends. That such trade names as Acrilan, Dacron, Fortrel, Kodel, Lycra, Orlon and Vycron are today household words is ample evidence of the inroads that synthetic fibers have made into the fiber market.

Rayon deserves special mention, both because it was the first important synthetic fiber, and because it is a chemurgic derivative of plant cellulose. Antedating nylon, granddaddy of the "new" synthetics, rayon at one time accounted for most of the synthetic fiber market. Nowadays rayon, like many natural fibers, is losing position to nylon, acrylics, polyesters, and polypropylenes.

Rayon traditionally has been made either from wood cellulose or from lint removed from cottonseed after ginning of the cotton fiber. The cellulose is dissolved in any of several solvents to make a viscous solution, after which it is drawn out into long strands by forcing it through apertures or dies, and is then hardened again into a cellulosic ester by plunging it into a chemical hardening bath or by evaporating the solvent. A number of filaments are twisted together to form thread. A process was patented as early as 1855 for making thread from dissolved nitrocellulose, but commercial production of synthetic fiber did not begin until 1891 in France, and not until 1911 in the United States. Rayon production exceeded that of silk by 1923.

Today more than 4 million tons of wood pulp are used annually for the production of rayon, mostly in Europe, North America, and Japan.

The viscose process of making rayon is the most commonly employed. Clean cellulose is steeped in sodium hydroxide, shredded, and aged under controlled conditions. It is then treated with carbon disulfide to make the xanthate. This is dissolved in diluted hydroxide to provide the spinning solution, to which pigments may be added. The solution is forced through a spinneret, made from precious metal, each orifice yielding a thin fiber filament. The filaments emerge into a hardening bath of sulfuric acid, various sulfates, and other materials. A rotating wheel twists the several filaments into a yarn. The yarn is washed, bleached and then spun and woven in the same way as any other fiber.

Natural fibers are often grouped according to their botanical origin. **Surface fibers** are obtained from the surface of leaves, fruits, seeds; cotton and kapok are the two chief examples. **Soft,** or **bast, fibers** develop in the bark tissues (i.e., external to the xylem), and are typically found in clusters that make strong, durable strands, as do those from flax, ramie, hemp, and jute. **Hard fibers** are strands of vein-like supportive cells (typically fibrovascular bundles consisting of both xylem and phloem plus ensheathing cells) found primarily in the leaves of monocotyledonous plants; they are somewhat more lignified, rougher and weaker than soft fibers. **Wood fiber** will be left for discussion in the section on forest products.

Fibers are also classified according to use. Those used for stuffing (upholstering, mattresses, etc.). caulking, or for reinforcement and insulation (e.g., wall board) are designated as **filling fibers.** Those used to weave mats and baskets or to make brooms are termed **brush, broom,** or **braiding fibers.** Fibers used to make rope or twine are called **cordage fibers,** and those used to weave cloth are called **textile fibers.**

Surface fibers are usually separated mechanically, by the process called ginning. The principal behind the cotton gin, invented by Whitney in 1793, is to catch the fibers onto toothed discs or combs and pull them through slots too narrow for the seed to follow, thus tearing the fiber free for subsequent spinning into yarn. Hard fibers

frequently, and soft fibers rarely, are separated mechanically, too, by machines that scrape away soft surrounding tissues; this work is usually done by hand where labor is cheap. Hand scraping is often the practice with abacá, sisal, and New Zealand hemp, although mechanical cleaning may be combined with biological retting. **Retting** is a fermentation process that rots away weaker cells and leaves the strong fiber strands; it is employed mainly with soft fibers. In some parts of the world retting is accomplished merely by immersing sheaves of stems in ponds or rivers; this method is frequently used in India in the retting of jute. Flax fiber is often separated by dew retting, which involves nothing more than leaving the stems in the field, where dew and other natural moisture allow microbes to do the job. In some parts of the world elaborate retting ponds are built, in which the stems undergo a more precisely controlled retting process that yields a more uniform fiber.

Fibers are today very much taken for granted, but early man did not have things quite so easy. Thongs of hide may first have served in bows, snares, nets, and skin garments. Long before recorded history, however, primitive man had turned to plant fibers, not only for tying and thatching, but eventually for weaving more suitable and lighter fabrics than animal skin could provide. The invention of the knot was almost as monumental as the invention of the wheel! Fragments of woven flax fiber have been found in Swiss lake dwellings 10,000 years old. Egyptian cerements almost as old document the use of both flax and papyrus. Hemp was grown in southeastern Asia before 2000 B.C., and the pre-Christian armies of the Near East were clothed in cotton cloth. Palm leaf fibers are found among 12,000-year-old artifacts in the Tehuacan diggings of Mexico. *Typha, Agave,* and *Yucca* fibers almost as old are found in archeological sites in western North America. A beautifully crafted net of extracted *Apocynum* bast fiber, found in Danger Cave, Utah, dates back to about 5000 B.C.

Before fabric can be woven, fiber must be spun into yarn. Until just a few centuries ago this was done entirely by hand, by straightening and thinning the fibers between the fingers and then twisting them into yarn. The yarn becomes a continuous thread of twisted, interlocking fiber strands. The distaff and the spindle, among the first hand tools used to spin yarn, gradually gave way to the familiar spinning wheel, so much a part of the cottage textile industry of the Middle Ages. The spinning wheel consisted of a hand- or foot-powered wheel that rotated a spindle and twisted the fiber into yarn—a foretaste of eventual automation. The principle behind the immense powered spinning machinery utilized by today's textile industry remains the same.

Since fiber cells are found in most plants, and in all parts of the world, it is evident that there is no lack of candidate species for exploitation as sources of fiber. One could gather bast fibers from milkweed stems as well as from flax, surface fiber from dandelion seeds as well as from cotton, hard fibers from grass as well as from abacá. Plants from 44 different families have yielded fiber of greater or lesser commercial value. Which fiber is to prove most fruitful for commerce depends not only upon its natural qualities, but upon cropping and mechanization possibilities (including suitable machinery for cleaning and spinning, labor, market familiarity, and economical mass production (which may call for genetic improvement leading to higher yield and better quality). Only a favored few natural fibers have attained world stature by satisfying these requirements, and even these are hard pressed by synthetics.

We will concentrate here chiefly upon cotton, still the world's single most important natural fiber. Flax, hemp, jute, ramie, and kenaf among the soft fibers will be mentioned; abacá and sisal among the hard fibers. Nor can the brush fibers be entirely overlooked. This is not to ignore the fact that dozens of other plants have been and still are grown for fiber: some have consistently served specialty purposes. The quality of most of these fibers is satisfactory; they are not generally marketable, however, because they cannot compete economically with other major fibers. Everything considered, fiber is an abundant and low-cost commodity, but the expense involved in harvesting, cleaning, shipping, spinning, and weaving some fibers often overshadows their basic virtues. Add a touch of politics, such as government subsidies, tariffs, and the export duties entailed in world-wide commerce, and it is apparent that distantly

related events may often determine success or failure for any natural fiber. The possibilities for substitution and interchangeability among fibers are sufficient to preclude runaway prices for any single one.

Surface and soft (bast) fibers

COTTON, *Gossypium*, MALVACEAE   The cotton genus, *Gossypium*, is represented by approximately 30 tropical and subtropical species of both the Old and the New World. There has been considerable crossing, both directed and natural; five diploid genome groups ($2n = 26$) and one tetraploid combination (of the A and D groups, $2n = 52$) can be distinguished. Of the more familiar species, *G. arboreum* and *G. herbaceum* are Old World diploids; *G. hirsutum* and *G. barbadense* New World tetraploids. Collectively these species account for by far the greatest world production of any natural fiber, nearly 12 million tons annually. The crop is grown world-wide, but the United States and the U.S.S.R. are the leading producers, followed by China, India, Brazil, Mexico, Pakistan, and the United Arab Republic (Fig. 23-1). Although *Gossypium* is tropical by nature, it can be cultivated as an annual as far north as 47° in the Ukraine, and, in the United States, into southern Illinois in the Mississippi Valley. Cotton is a heliophile and relishes warm weather, growing best at temperatures around 90°F. Wild cottons are semixerophytic. Cultivated cottons have become so pampered that they are today beset by an increasing array of ills and pests that require both skilled and persistent attending. Cotton is one of the more temperamental crops with respect both to weather and to local ecological conditions.

The world importance of cotton is relatively recent. Although the fiber was known in the Near East before Christian times, it was singularly little mentioned by the Greeks or the Romans; both of these cultures depended mostly upon flax among botanical fibers, and wool and silk. The tropical nature of cotton no doubt prevented its cultivation over much of the area where Greek and Roman hegemony prevailed. Cotton had been grown in Africa, however, for many centuries, and when the Moors invaded Spain at the beginning of the tenth

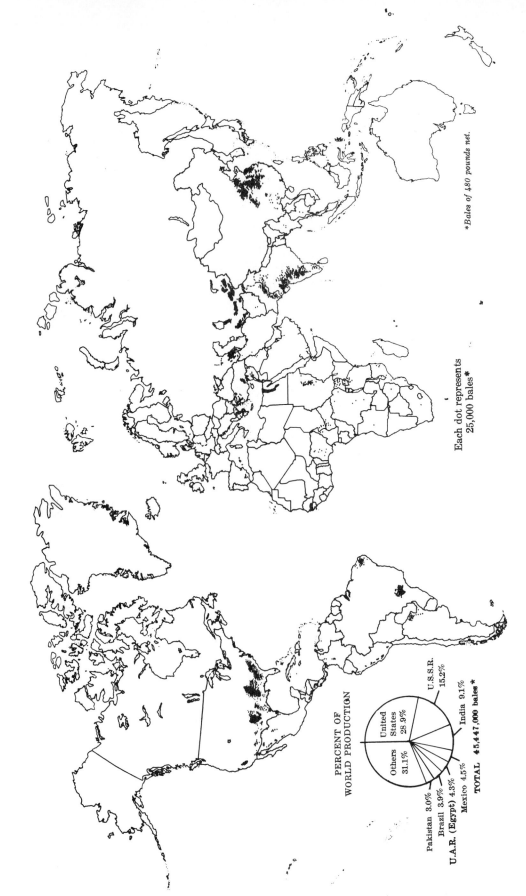

PERCENT OF
WORLD PRODUCTION

United
States
28.9%

U.S.S.R.
15.2%

India 9.1%

Mexico 4.5%

U.A.R. (Egypt) 4.3%

Brazil 3.9%

Pakistan 3.0%

Others
31.1%

TOTAL 45,447,000 bales *

Each dot represents
25,000 bales*

*Bales of 480 pounds net.

FIGURE 23-1. *World cotton production average 1957–1961. [From A Graphic Summary of World Agriculture,*
USDA Misc. Publ. 705.]

century, they introduced an already highly developed cotton industry. Barcelona became a center of textile manufacturing, especially of sailcloth for the Spanish fleet.

By the fourteenth century some cotton was being shipped from Asia Minor into northern Europe for looming. The establishment of trade with the Far East brought cotton cloth manufactured in India to Europe. During the 1600's a textile industry sprang up in England, first based upon a mixed fabric using linen as the warp and cotton as the woof. Weaving of cloth was still largely a cottage industry, until the invention of several successful spinning and looming devices in the mid-nineteenth century brought about mechanization. With machine weaving, demand for cotton fiber rose astronomically. Cotton growing was pursued assiduously in the United States, eventually causing the virtual ruination of much land in the South. The longer, finer fiber from the American species *G. barbadense* and *G. hirsutum* became the main source upon which the dominant British textile industry depended. The political and social ramifications of the cotton industry in the United States, involving slavery and the Civil War, attest to the influence such a plant product can have on human affairs. During the American Civil War, when cotton supplies were largely cut off, the British introduced the famed Sea Island and better upland strains of cotton into their colonies throughout the world.

New World cottons seem to have contained Old World genomes long before the crop achieved world importance. How this may have occurred is quite obscure. Archeological findings suggest that *Gossypium* is of considerable antiquity in the New World. Bolls have been identified in the Tehuacan excavations that date back to before 500 B.C.; they are scarcely distinguishable from modern cottons. The plant was probably cultivated by the Indians of ancient Tehuacan, but there is no certainty that the fiber was used for more than twisting into simple strings, more flexible than cord made of such hard fibers as the even more anciently cultivated maguey (*Agave*). Twine fabrics may have been produced shortly thereafter, and certainly cotton textiles are known from Peru before 2500 B.C. Cotton fibers from *G. arboreum*

were woven into cloth in Pakistan equally as long ago.

In some parts of the world cotton is still grown and handled in a relatively unmechanized fashion. In Ethiopia, for example, cotton still supports a cottage industry. American varieties have been introduced and are gaining favor. These have crossed with Old World cottons, perhaps first introduced from India during Roman times. All are grown as short-lived perennials, the planting lasting for about five years. The fibers are harvested mostly by hand, and the spinning and weaving are largely done at home. Such hand crafting permits the use of variable cotton fiber, which can be separated for staple length and selected for different spinning qualities. It is interesting that machine-ginned cotton is rejected by spinners as inferior to cotton separated by hand. The preferred hand "ginning" is accomplished simply by pulling the fiber from the seed with the fingers (often with aid of a smooth stone or wooden block). Blankets and garments are woven from the hand-spun yarn. The cotton plant is also used for other than its fiber. Leaves, flowers, seeds, and roots are all reputed to have medicinal value, used both internally and externally. A fermented cake that contains cottonseed is eaten in certain parts of Ethiopia.

Quite in contrast is the highly automated and business-like production of cotton in parts of the southern United States, where the plant is grown as an annual and the land clean-cultivated (Fig. 23-2). The most remunerative growing has shifted from the traditional southeastern producing region to the irrigated lands of the Southwest. Remunerative cotton growing demands a long frost-free season, moderate moisture, and either ample labor or well-capitalized mechanization. Without extensive mechanization it is estimated that cotton requires about 150 man-hours per acre, a greater labor input than for any other major crop.

Top growers in such locations as the Imperial Valley of California average between 5 and 6 bales of cotton to the acre, more than twice the yields on good cotton land in the Southeast. Even higher yields are possible under skip-row planting patterns, such as have been fostered through government restriction in the United States to control production. Skip-row planting takes various forms, such as planting two rows, skipping one (common in the Southeast), or planting four and skipping four (common in the Southwest). Yields may thus be increased as much as 50 percent, provided only the rows planted are counted as acreage. Solid planting is generally more favorable economically, however, considering land amortization, better opportunities in cultivation, spray coverage, and concentrated harvest. Throughout the Southwest narrow-row planting is supplanting conventional 40-inch row seeding, with plant populations reaching 60 thousand to 80 thousand per acre in 10- to 14-inch rows. Environmental factors such as proper timing of planting and pest control are extremely important in determining cotton yields.

In the Imperial Valley of California, and in southern Arizona, cotton planting begins about the first week of March, and is often staggered somewhat to avoid overtaxing of facilities at harvest time. About 50 pounds of delinted seed of a modern cultivar (such as Deltapine), treated with fungicide, is mechanically planted to each acre, many rows at a pass. After the sprouts appear the rows are mechanically thinned to approximately one plant to each three inches. About two hundred pounds of phosphate and a hundred of nitrogen (as anhydrous ammonia) per acre are worked into the seedbed ahead of planting. About six weeks after planting, supplementary fertilizer side dressings are made, and later fertilizer is applied in the irrigation water about every two weeks. Up to six

FIGURE 23-2. *Cotton. (Left) Commercial cotton as it is grown in the southern United States. (Right) Open cotton boll ready for picking. [Photograph at left courtesy USDA; photograph at right by J. C. Allen & Son.]*

hundred pounds of nitrogen per acre is used to bring home a crop. Herbicides are increasingly employed to save labor; Dacthal, for example, is incorporated into the soil ahead of seeding as a pre-emergent. Diuron or something similar may be used as a directed post-emergence spray. Arsonates are often used to control johnsongrass. Insecticides are sprayed as needed, to the tune of nearly 50 dollars per acre annually (Fig. 23-3). They are usually applied by airplane at night, when application seems to be most effective. Efficient sled cultivators weed within an inch of the row, a half-dozen rows at a time, for weed control and soil loosening in the early stages of growth. Harvesting, which is completely mechanized, usually consists of going over a field twice with a cotton picker (Fig. 1-9), a machine with rotating spindles onto which the cotton fiber sticks and is drawn from the boll. This is generally done after spraying with a defoliant to cause leaf drop. "Scrapper" machines usually follow the pickers once or twice over, to salvage an additional quarter or half bale per acre of fiber that escaped the picker. Integration of all these complex procedures requires constant programming and business-like management of what was a century ago a hand operation tended by slaves.

Research shows that heterosis in cotton can increase fiber yields greatly. Unfortunately, cotton does not lend itself well to regulated crossing; in the field it exhibits quite a mixture of self and cross-pollination (the degree of cross-pollination often depending upon bee population). There are indications that suitably irradiated cottonseed grows into plants that show some pollen sterility; it has been suggested that a fair degree of hybridization could be obtained either by this technique or by chemical manipulation, providing hybrid seed economically, and more uniform F_1 populations.

The seed hair that constitutes a cotton fiber is but a single cell, from one thousand to six thousand times as long as it is wide in those species grown for fiber. In 1 pound of cotton there are some 90 million seed hairs! As the hair grows, wall thickness increases daily; growth is partly influenced by the weather and other momentary environmental influences. By the time the fiber is mature,

many secondary wall layers have formed, and very little lumen (cavity) remains. As the seed ripens the hair collapses into a twisted filament of nearly pure (90 percent) cellulose. Its unadulterated cellulosic nature makes for a fiber of surprising durability and strength; and its twisted form equips it well for spinning.

After picking, ginning, and grading, cotton goes to a mill, where the fiber is further fluffed and cleaned. It is then carded, combed, and drawn—a sequence of operations accomplished by feeding the fiber onto a series of barbed cylinders that progressively draw it out and thin it. One calculation estimates that during this process an inch of the product, after carding and combing, is drawn to a length of many miles. The final combing ensures parallel orientation of the fibers, which in pedigreed cotton are nearly of uniform staple length. These are twisted (spun) into yarn or thread, which in turn is used for everything from cordage to the weaving of fabrics. Depending upon its use, the fiber may be chemically treated, dyed, or coated. Most fiber, of course, goes for the weaving of cotton cloth. A considerable amount of cotton is used for cordage; cotton was formerly used to make automobile tire cord and carpeting.

FIGURE 23-3. *The boll weevil, shown here on a forming cotton fruit, was a pest even on prehistoric cottons. Boll weevils are being fought to a standstill with modern techniques and sprays.* [*Courtesy USDA.*]

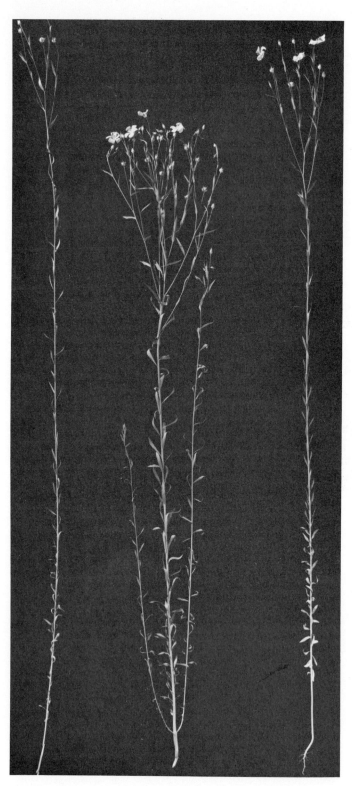

FLAX, *Linum usitatissimum*, LINACEAE Flax (Fig. 23-4) is perhaps the world's finest bast fiber. For centuries it was Europe's chief fiber, and in some places the only textile fiber. Egyptian mummies dating back four millennia are wrapped in cerements made of linen, attesting to flax's antiquity in the Near East. Its discovery among artifacts from the ancient Swiss lake dwellings was noted earlier in this chapter. Today flax growing is confined mostly to temperate parts of Europe. The leading flax country is Russia, accounting for more than half of the world production of considerably less than a million tons annually. France, Poland, Czechoslovakia, and other northern European nations account for most of the rest. Growing flax for fiber has practically ceased in North America, and is but a minor industry in Japan and other parts of Asia and in the Near East.

The once pre-eminent position of linen, the weaving of which centered in Belgium, has gradually given way under competition, first with cotton and wool, and more recently with synthetic fibers. Even so flax fiber still has few equals for quality. It is often used for fabrics that must be especially tough yet flexible, such as the fiber for making fire hose. Different flax cultivars are used for fiber than for linseed oil (Chapter 22). Growing techniques differ too, including field spacing. Some of the early investigations on cytoplasmic male sterility were done with flax, and there seems to be little doubt that acceptable hybrid seed could be produced. Normally flax is mostly self-pollinated. With cytoplasmic male sterility and proper vectors to facilitate crossing, hybrid seed should be more readily procurable than with many other crops.

Flax does best in cool climates with rather moist summers. The crop is handled as an annual, its planting and care being much the same as with small grains. Like grain, it lends itself well to mechanization up to the point of retting, for which considerable labor is needed. In much of Europe plants are still harvested by hand-pulling rather than cutting, to take advantage of full stem length.

FIGURE 23-4. *Branched and unbranched flax. The phloem fibers of the stem are used to make linen, most easily extracted from stems that have few branches. (See also Fig. 22-6.)*

The harvested stems are generally left in the field to dry until foliage shrivels and drops, after which retting takes place. In dew retting the stems are simply left in the field for several weeks until the soft tissues decay, freeing the bast. Where ponds (or tanks, often with controlled water temperature) are available cold water retting is practiced; tied bundles of flax stems are completely immersed. It is said that controlled retting in tanks produces a superior fiber in just a few days, as compared to the several weeks required for dew retting. After retting, the fiber is subjected to scutching. This consists of feeding the newly dried stems between fluted rollers that crush and break the central tissues and allow the peripheral bast fibers to be worked free by various forms of shaking (as in revolving drums).

The flax fiber as it comes to market is actually a bundle or group of fiber cells cemented together. The bundles may be up to three feet long in well-grown flax. They are remarkably flexible, tough, lustrous, strong, and durable. Moreover, they are stronger wet than dry, which makes linen an excellent material for towels and fabrics subjected to weather. Just as with other fibers, linen goes through an elaborate series of cardings and combings to clean and draw out the fiber before spinning it into a yarn. Naturally, the quality of the fibers ("slivers") determines the quality of the thread; over-retted fiber tends to be of an inferior blackish appearance. Spinning can be done either by a wet or a dry process. In the former the fiber is first drawn through warm water to soften it and is then further attenuated into particularly delicate strands. Spun yarn may be bleached or dyed before weaving. Today, most weaving is done on elaborate power looms, but there is still some traditional craft weaving in Ireland and other countries. Because of its strength linen thread is still in demand for shoe sewing and for making fine embroidery or lace. The stems of oil flax find some use in paper making, but its fiber is not of sufficient quality for weaving. Cigarette papers are often made from flax because the nearly pure cellulose burns evenly with no objectionable taste or odor.

HEMP, *Cannabis sativa*, MORACEAE   At least in China, hemp was domesticated for its fiber as long ago as flax was in other parts of the world. For many purposes these two fibers are interchangeable, although linen is of higher quality for fabrics. Cloth is made from hemp almost nowhere but in the Far East, where a good bit of flax fiber is imported from Europe. The world hemp crop is only about half that of flax, amounting to no more than $\frac{1}{3}$ million tons annually. Most hemp is grown in regions slightly to the south of those in which flax is grown; the leading producing countries are Russia and India. Very little hemp is grown in either the Americas or Africa.

Hemp is of course the plant from which marijuana is obtained (see Chapter 22); it also yields a seed oil of minor importance. Apparently a native of Asia, it had been introduced into the Near East and Europe by the Middle Ages. It is grown and harvested in much the same way as flax. Occasionally, as in Korea, the fiber is separated by steam retting, a process taking only a few hours. The quality of steam-retted fiber, however, is not as good as that of dew- or pond-retted fiber. Hemp fibers occur in bast bundles, which may be as much as 6 feet long. The thick wall of the cells consists largely of cellulose, but not as pure a cellulose as that in flax. By and large, hemp has been displaced by other fibers in most of its traditional uses, although hemp twines are still much used, as are hemp fabrics for summer clothing and shrouds in parts of Asia. Gibberellin treatment of growing hemp not only produces a greater number of fibers per plant, but produces individual cells that are larger in diameter and up to ten times as long as normal.

There has been some selection of higher-yielding hemp cultivars, and these have been crossed to give simple hybrids showing some heterosis. But in less advanced areas where much of the hemp is grown, there has been little genetic improvement; a few plants are simply left to mature in a corner of the field for seed. Hemp yields from primitive growing in the East are about 620 pounds per acre, some 20 to 50 percent less than where more advanced methods are used, as in parts of southern Europe. About 110 days are required to mature hemp, which may cause some of it to be harvested prematurely in the Far East to permit planting of food crops. In the Orient hemp is often alternated with rice, both in paddies and in upland locations. Seed is generally broadcast at a rate of about 45 pounds per acre, even though tests indicate that row planting and proper spacing provide higher yields and better fiber. In recent years some fertilization has been practiced, nitrogen markedly increasing yields. Weeding is generally done by hand with a hoe. Insecticidal sprays may be used to control borers.

Hemp is usually harvested by hand with sickles. Steam, dew, or pond retting follows. After retting, the hemp stems are dried and then "broken" (partly crushed), either by hand or in primitive machines. The woody core and bark fragments are typically shaken away by hand. The freed fibers are twisted into yarn on hand spindles and woven into a rather coarse fabric at home. The fabric may be "degummed" by soaking in hydroxide. Hemp-cloth garments are traditional for mourners in the East, and for wrapping corpses.

JUTE, *Corchorus*, TILIACEAE   Jute is the world's foremost bast fiber—foremost in quantity, not quality. From the standpoint of quality it is among the poorest; it is rough, weak, and its cellulose is of relatively low purity. Jute is derived from two domesticated species of *Corchorus*, *C. capsularis* and *C. olitorius*, both of which have a diploid chromosome number of 14. The genus embraces about forty species. Considering the tremendous comsumption of a fiber that has so little to recommend it, jute must claim its markets because of low cost. It is produced in regions where labor is cheap; Pakistan and India account for two-thirds of the world's supply, which amounts to about three million tons annually.

C. olitorius is believed to be native to Africa; *C. capsularis* is probably native to eastern Asia. Fiber from *C. olitorius* is frequently finer and softer but more yellowish than that from *C. capsularis*. But there is much variation among the many cultivars of both species. Both are chiefly self-pollinating. Jute is grown as an annual; the plants mature within 3 to 5 months, by which time they have developed stems as tall as 15 feet. The fibers occur in long wedge-shaped bundles in the bark, the individual fiber cells seldom being longer than 2 or 3 millimeters. The cell walls are considerably more lignified than those of flax or hemp.

Jute is grown during the rainy season, and performs best on deep alluvial soil in tropical climates with about sixty inches of annual rainfall. Seed is planted to well-cultivated, preferably fertilized (manured) soil, and is usually broadcast at a rate of 6 to 10 pounds per acre. Whether broadcast or row planted, stands are generally thinned down (by hand) to populations of less than 200,000 plants per acre as growth progresses. At the same time weeds are hoed. Land used for raising jute is generally planted once or even twice during the year to some other crop, such as rice.

As seed pods form, the stems are hand-cut close to the ground with a sickle. In flooded areas this may entail working under water. The stems are dried to rid them of foliage and then tied into bundles for retting. The bundles are submerged in pools or ditches until the tissues binding the fibers disintegrate. Depending upon temperature this may take from 1 to 4 weeks. A number of bacteria and some of the fungi most active in retting have been isolated. Experiments have shown that these microorganisms are capable of hydrolyzing the pectins (thus loosening the fiber) in as little as five days. Workers separate the fiber strands from the retted stalk by hand while standing in the pools or ditches, so that the freed fiber can be floated on the surface while fragments of the adhering bark are swished away and picked off. Fiber yield averages about 6 percent of the green weight of the stem; chemically, the fiber is about 75 percent cellulose, 11 percent lignin, and 12 percent xylan. It is stronger dry than when wet; prolonged bleaching reduces its rather poor strength even further. After retting, the fiber is dried and usually sold to small buyers circulating within the community. Most of the crop is eventually transported to Calcutta, the chief weaving and export center for jute. Mills purchasing jute characteristically test for fiber strength as an indication of quality. Careless retting is the chief cause of poor fiber.

Jute fiber, used in India since ancient times, became important as sackcloth during the late sixteenth century. Yarns were spun by peasants and generally woven into cloth at home on hand looms. During the 1800's the commerce of British merchants in India created a great demand for gunny sacks made of jute. Jute mills sprang up near Calcutta, and hand looming declined. Much raw jute fiber was exported to Scotland for spinning and looming there. The earliest American mill (at Ludlow, Massachusetts) is still making jute fabric for such specialty purposes as roadside ditch matting. Burlap sacks still accommodate much grain, seed, and produce (even protecting bales of a rival fiber, cotton); burlap is favored because it is not torn even when perforated by sharp hooks or other implements in handling. Jute fiber wastes are sometimes used as stuffing material, and as an extender or raw material for plastics.

KENAF AND ROSELLE, *Hibiscus*, MALVACEAE

Kenaf, *Hibiscus cannabinus,* and the related roselle, *H. sabdariffa,* are bast fibers of minor commercial importance. Kenaf has shown promise as a jute substitute in the western hemisphere, and roselle has substituted somewhat for jute in Southeast Asia. Crash research was undertaken on kenaf during World War II, when jute and other fiber imports were cut off from the United States; the species proved well adapted to a variety of growing conditions, yielding a crop quickly (90 to 120 days) with reasonably little care. Roselle requires 180 days. Kenaf is a diploid with 36 chromosomes, roselle a tetraploid. Kenaf appears to have been domesticated in western Africa, where, after cotton, it is perhaps the most widely cultivated fiber plant. Roselle in particular serves other uses in addition to yielding fiber. The seeds may be roasted or ground into meal; the leaves and shoots serve as a vegetable; the flowers are used in jellies and confections; and a fermented drink is made from the juice. Both kenaf and roselle are also thought to have medicinal virtues.

RAMIE, *Boehmeria nivea,* URTICACEAE Ramie, or "Chinagrass" has always shown much promise but little fulfillment as a fiber crop. The bast fiber is strong, very durable, and has many of the attributes of flax. Growing and fiber extraction have unfortunately proved difficult and espensive. Consequently the crop has remained a casual one; it is grown in scattered parts of the Orient, where it sometimes substitutes for hemp, jute, and other fibers.

The biggest difficulty with ramie is that the cementing gums binding the bast bundles are not easily destroyed by retting. Separation of the fiber has generally had to be done at least in part mechanically, by pounding, scraping, and other means. Even then the freed fiber is "gummy" and must be chemically treated to meet flexibility standards. The fiber cells themselves are among the longest and strongest of natural fibers, often being 30 centimeters long. But they are quite smooth, which is a disadvantage, because spinning machines for other fibers are ill-fitted to handle a smooth yarn.

Ramie, like jute, which it much resembles, is adapted to moist tropical climates and deep soils. Unlike jute, however, it is a perennial, and thus must occupy the land the year around, preventing the growing of a second crop. On an acreage basis it is thus generally less remunerative than either hemp or jute. Where grown in the Orient it is typically harvested by hand as canes mature. In areas of inexpensive labor the bast fiber is still removed by hand and generally cleaned by repeated pulling across a knife edge. Most fiber is locally spun and woven into coarse cloth, often without degumming. Degumming is accomplished by boiling in lye or acid, which frequently weakens the fiber, but the processing is necessary to provide yarn adaptable to machine spinning. In spite of the inherent excellence of ramie fiber it finds only a limited market. Attempts to introduce a ramie industry in southern Florida have been commercially unsuccessful, in spite of the considerable research that has been directed toward mechanization.

Hard fibers, Monocotyledonous genera

A preliminary word is in order about hard fibers as a group. They are really no "harder" than the bast fibers of the several genera just discussed, but they are thicker and stiffer because they constitute the entire fibrovascular bundle derived from leaves of tropical and subtropical Monocotyledons. Of course they are not as pure in cellulose as the bast fibers, and they are generally employed for rough use, such as making rope. *Agave* (principally sisal and henequen) accounts for some four-fifths of the approximately one million tons of hard fiber produced world-wide annually, with abacá accounting for most of the remainder.

ABACÁ, OR MANILA HEMP, *Musa textilis*, MUSACEAE *Musa textilis* is closely related to the banana, which it resembles. As the alternative name would suggest, it is primarily a crop of the Philippine Islands, where it is native. Ninety-five percent of world production of somewhat less than 100,000 tons comes from the Philippines, where the fiber has even been used for weaving textiles.

Abacá was not known to the western world until 1686, and did not significantly enter commerce until the nineteenth century. Since then it has gained an excellent reputation, especially for cordage. The fiber strands range in length from 6 to 12 feet, and have a high tensile strength. Moreover, they are light in weight and resistant to water, especially salt water; hence their demand for rope and ship caulking. Originally, the fiber was produced from wild plants as a home industry. In recent centuries, however, plantation growing, similar to banana growing in the American tropics, has accounted for most of the harvest (Fig. 23-5). Abacá is truly tropical, performing best where annual rainfall exceeds 70 inches annually and mean annual temperatures are around 27°C (80°F).

Although abacá, unlike banana, produces viable seed, it can be propagated more quickly from rhizomes or from "suckers" (side shoots of a clump). Abacá is usually planted in rows, individual holes often being dug among fallen brush and trees after land clearing. Later the fields may be cultivated, even interplanted with food plants. As is typical of slash-and-burn agriculture, the soil is generally exhausted within a decade (unless fertilized). Potassium appears especially useful in maintaining the crop. Harvesting consists in cutting the outer stems as soon as the inflorescence begins to emerge, care being taken not to damage young shoots and developing suckers. The internal leaf sheaths yield a higher quality fiber. The sheaths are cut to approximately five-foot lengths for transporting to decorticating centers. Under primitive circumstances the leaf sheaths are simply sectioned longitudinally into strips and the gross pulp discarded. The strips are run between a wood block and a knife to scrape the fibrovascular bundles clean. Powered decorticators are used on plantations. A mature abacá rootstock may pro-

FIGURE 23-5. *Abacá harvest in Panama. The stalk is cut at the base after the plant has flowered, and the leaves are cut from the top.*

duce as many as twenty separate stems rising to nearly thirty feet. About one-third reach flowering stage each year, and are then harvestable.

In addition to its main use for making rope, abacá is much used for fishing nets in the Philippines. There is also some use of the fiber there for weaving table "linens," hats, purses, and other handicraft. Waste abacá, such as from old rope, is suitable for making paper. In fact, the term "manilla," as used in "manilla envelope" or "manilla folder," originated from abacá being used to make a heavy paper when fibers were in short supply during the American Civil War.

Agave SPP., AMARYLLIDACEAE Several species of *Agave* are important sources of hard fiber, their production amounting to nearly a million tons annually. Sisal, from *A. sisalana*, is the most important, accounting for two-thirds of this. Sisal comes mainly from East Africa (Tanzania, Kenya, Angola, Mozambique, Madagascar) and Brazil. Henequen, from *A. fourcroydes*, is produced primarily in Mexico and Cuba, and accounts for less than one-fourth of the world's total *Agave* fiber production. Other fiber *Agaves* of lesser importance are cantala, *A. cantala;* lecheguilla, *A. lophantha poselgaeri;* Mexican henequen, *A. lurida;* and letona, *A. letona.* Most of the commercially important *Agave* species are polyploids and are sterile. The basic chromosome number appears to be 30, but various investigations of sisal and henequen show chromosome numbers ranging from 137 to 149. Probably a good deal of ancient crossing is represented in modern cultivars.

Most *Agaves* are native to warm arid environments, but perform best as fiber sources where rain is reasonably abundant. Soil type is not too important, but it should be well drained. Sisal is generally grown in equatorial plantations. It is propagated by hand-planting bulbils that form on the inflorescence or basal "suckers" separated from mature plants. Sisal, native to Central America, was first introduced into Florida in 1836. It did not thrive, and success awaited introduction into the West Indies and Africa. African production has dominated the sisal industry in the twentieth century.

During the life of an *Agave* plant about three hundred leaves are produced. Upon flowering the plant dies; in favorable environments plants may flower in as little as five years, but in less favorable environments plants may need as much as fifty years. When a leaf is old enough to have attained an almost horizontal position it is ready for cutting. First cutting of sisal in Africa is generally about $2\frac{1}{2}$ to 3 years after planting; with henequen in Yucatan about five years after planting. Cutting is done by hand with a machete. With spiny-leaved *Agave*, the spines are removed before bringing the leaves to a central decorticating center. As with abacá the pulpy tissues are scraped away mechanically. On plantations this is done with special machines after the leaf has been crushed. After initial separation the mass is washed free of loose tissue, dried, and further cleaned by shaking, rubbing, and brushing, often on special machines. The fibers consist chiefly of sheathing cells around the vascular strands; the internal xylem and phloem are generally destroyed in the processing. There will be differences in size of the fiber strand and in its degree of lignification, depending upon the part of the leaf from which it is taken, as well as upon species and upon growing conditions. Waste material from decortication is a valuable source of fertilizer nutrients, and is generally returned to the fields. Short and off-grade fibers unsuitable for twine may be made into paper or "fiberboard."

OTHER HARD FIBERS   There are numerous additional sources of hard fiber similar to that produced from *Agave*, world production amounting to about 50 thousand tons annually. Perhaps most encountered as minor items of world commerce are palma istlé, from *Samuela carnerosana*, Liliaceae, a relative of yucca, and often popularly so-called, produced chiefly in Mexico; formio, or New Zealand flax, from *Phormium tenax*, Liliaceae, produced in Argentina, New Zealand, Chile, and some of the oceanic islands near Africa; fiqué, from *Furcraea macrophylla*, Amaryllidaceae, produced in Colombia; Mauritius fiber, *Furcraea gigantea*, Amaryllidaceae, mainly from the islands of Mauritius and Reunion; and caroá, *Neoglaziovia variegata*, Bromeliaceae, from Brazil.

Brush and filling fibers

As noted earlier, a miscellaneous assortment of materials, ranging from the delicate seed hairs of kapok, to coarse twigs or fragments such as split bamboo, are also marketed as fibers. Most of these are used as stuffing materials or for making brooms, and as with textile and cordage fibers, they are feeling the competition of synthetics such as foam rubber, urethane plastics, and nylon bristles. Even excelsior—shredded wood—serves the same purpose as many filling fibers, and effectively establishes a low price level for this group of unspecialized materials.

One filling fiber that has found a successful specialty market is kapok, or "silk cotton," the seed hair from fruits of a giant tropical tree native to Central America, *Ceiba pentandra*, Bombacaceae. The fiber is of high quality; like cotton, it is nearly pure cellulose, but because of its smooth, untwisted form it does not spin well. The fiber is very resistant to wetting, however, and has long filled a need for a resilient stuffing material in life preservers, outdoor cushioning, sleeping bags, and so on. Many years ago kapok was introduced into the Far East, where it is rather extensively cultivated, today supplying most of the kapok fiber entering commerce. Fruits are gathered by hand, often being cut with knives on long poles, the fiber typically separated with the fingers. In the United States kapok is now often replaced by such plastics as expanded polystyrene and polyurethane.

Another tropical fiber long used as filling material is coir, the fibrous outer husk of the coconut, *Cocos nucifera*, Palmae. Coir is generally loosened by retting in salt water for prolonged periods, after which it is beaten and manipulated to separate the fibers. The stiff fibers are used mostly as an inexpensive stuffing or mat-making material.

Strips of the palm-like leaf of *Carludovica palmata*, Cyclanthaceae, are woven into the famed Panama hats (incidently, produced mostly in Ecuador).

Broomcorn, *Sorghum vulgare*, of the Graminae, mentioned among grains and forages, is grown for a limited fiber market. Cultivars used for making brooms have a dense, bristly inflorescence which, when cut, dried and assembled, makes the typical household and whisk brooms.

Several tropical palms have fibrous accumulations at the base of the petiole, which is collected by hand and marketed as piassava fiber. This is something of an export from eastern Brazil, gathered chiefly from wild *Attalea* palms.

FOREST CROPS

Wood

As a twig elongates the apical meristem lays down a train of primary tissues with relatively little bulk. One of these is a thin cylinder of secondary meristem, the cambium, located between the phloem and other outer tissues (the bark), and the xylem (the wood). Throughout the life of the tree the cambium continues to produce more xylem to the inside, more phloem to the outside. The xylem produced by the cambium is secondary xylem, or wood, and makes up the great mass of the tree. Xylem cells nearest the cambium remain physiologically active for some time as **sapwood;** when the sapwood grows old it is infiltrated with additional lignin, dyes, tannins, and other substances to become **heartwood.**

Activity of the cambium varies primarily with the season. Even in tropical forests there is apt to be seasonality. In temperate climates there is a very pronounced seasonal activity, responsible for xylem growth rings, which give the characteristic banded appearance to wood. New xylem is usually produced very abundantly as growth resumes in spring after winter dormancy. The cells produced then are relatively large in diameter and have comparatively thin walls. As spring turns into summer, xylem formation slows, and cells have comparatively thicker cell walls. The abrupt change from the thick-walled cells to the new xylem of the next year marks an annual growth ring. The relative thickness of a tree's annual ring (that is, the abundance of xylem produced) is a reliable indication of yearly growth, and in turn a reflection of whether environmental conditions are favorable. For example, trees shaded by an overstory grow very slowly and have extremely narrow annual rings; but when overstory trees are removed or die, **release** occurs, and the wider annual rings are a reflection of a greater growth rate.

Visualized as a whole, the cambium of a tree forms a narrow cone that ensheathes the wood of each stem. The base of the trunk usually contains all growth rings since the tree was a seedling. When the tree was a sapling it bore branches low to the ground; as it matured these were shaded and died. The branches leave scars, which become knots. Large trees have deposited many additional layers of xylem outside of the old sapling branch scars, and hence lumber from large trees is relatively knot-free (**clear timber**) toward the outer portion of the trunk. Typically the rings deposited in youth and middle age are wider than when the tree has matured. The best quality lumber comes from steady, moderate growth, in which all annual rings are the same width. Wide rings from temporarily lush growth may be weak, tending to shrink more and split.

Drawings of blocks of woods are given in Figure 7-6, representing both a softwood and hardwood. Figure 23-6 indicates how this wood might be utilized in a log. The disposition of the cells and their appearance not only delineate the annual rings, but are responsible for cell pattern, or **figure.** The way a log is sawed to expose transverse, tangential, and radial faces will determine its figure; the type, size, and arrangement of the cells, however, will determine the inherent characteristics of the wood. Growing conditions are reflected in variations of annual rings, knots, scars or blemishes, and so on. The diagrams of softwood and hardwood illustrate some of the distinguishing differences between the two types. Large vessel cells characterize some hardwoods, softwoods consisting entirely of tracheid cells with bordered pits. The horizontally aligned cells are rays, running from the outside of the xylem to the center. The scattering of the large vessels through the wood varies according to species. For example, in sugar maple the vessels are small and rather uniformly distributed between spring and summer xylem, so that this type of wood is said to be **diffuse-porous.**

FIGURE 23-6. *Disposition of wood in a sawlog.* [*Courtesy St. Regis Paper Co.*]

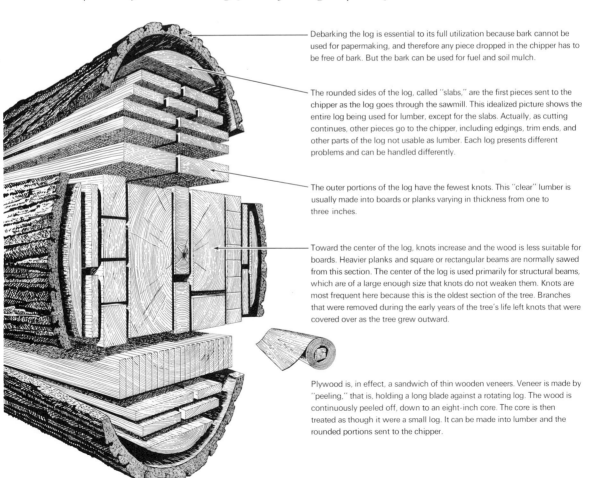

Debarking the log is essential to its full utilization because bark cannot be used for papermaking, and therefore any piece dropped in the chipper has to be free of bark. But the bark can be used for fuel and soil mulch.

The rounded sides of the log, called "slabs," are the first pieces sent to the chipper as the log goes through the sawmill. This idealized picture shows the entire log being used for lumber, except for the slabs. Actually, as cutting continues, other pieces go to the chipper, including edgings, trim ends, and other parts of the log not usable as lumber. Each log presents different problems and can be handled differently.

The outer portions of the log have the fewest knots. This "clear" lumber is usually made into boards or planks varying in thickness from one to three inches.

Toward the center of the log, knots increase and the wood is less suitable for boards. Heavier planks and square or rectangular beams are normally sawed from this section. The center of the log is used primarily for structural beams, which are of a large enough size that knots do not weaken them. Knots are most frequent here because this is the oldest section of the tree. Branches that were removed during the early years of the tree's life left knots that were covered over as the tree grew outward.

Plywood is, in effect, a sandwich of thin wooden veneers. Veneer is made by "peeling," that is, holding a long blade against a rotating log. The wood is continuously peeled off, down to an eight-inch core. The core is then treated as though it were a small log. It can be made into lumber and the rounded portions sent to the chipper.

In contrast, red oak wood has its large vessels confined to the spring xylem, and is thus said to be **ring-porous**.

CHARACTERISTICS AND QUALITY OF WOOD   The character of wood is determined by the nature of its cell walls and the structural configurations of its cells. It burns well, so has been mankind's most important fuel since time immemorial, only in recent centuries giving way to fossil fuel and today to atomic energy. It is strong, yet easily cut, nailed, and formed—qualities that make wood highly suitable as a construction material. Esthetically, wood is highly pleasing; it polishes well to exhibit beautiful design, hence its great popularity for furniture, interior trim, and other decorative purposes. Consisting as it does of individual cellular air pockets, wood is a good insulator, comfortable to the touch and efficient in resisting temperature changes. It is a poor conductor of electricity, and is more flexible and resilient than metals. Along with these virtues it does have drawbacks; it decays when certain combinations of moisture, warmth, and air prevail (although it can be treated to prevent this). It shrinks and swells appreciably with changes in temperature and particularly humidity, and occasionally twists or cracks.

The physical qualities of wood are reflected in certain measurable ways. Some woods are so dense that they will not float (chiefly tropical hardwoods such as ironwood, lignum vitae, and quebracho). But most woods, when dry, are less dense than water. They float well, having a specific gravity generally between 0.3 and 0.7. But if soaked in water long enough so that all air space in the cells is replaced by water (**waterlogging**), then wood will sink. Cell wall substance itself is about half again as dense as water. A piece of wood compacted to eliminate all intracellular space, would sink in water readily.

Durability is a feature of wood that varies greatly. Natural durability depends mainly upon resistance to decay, which in turn depends upon the kind of deposits in the cell walls of the heartwood. Where tannins, dyes, and other organic substances are prevalent, the fungi and bacteria that cause decay grow poorly. Woods that absorb little water may remain too dry for most decay

organisms to flourish. Most woods preserve almost indefinitely if kept dry; witness the wooden artifacts taken from Egyptian tombs. When it is waterlogged, it is not likely to decay, since most decay fungi must have air to grow. Wood decays most rapidly when it is moist and exposed, as in contact with soil or in locations where humidity is consistently high. The peculiar impregnations of heartwood of members of the Taxodiaceae, such as redwood (*Sequoia*) and baldcypress (*Taxodium*) make these species especially resistant to decay; to a greater or lesser extent so are various cedars and junipers (*Juniperus*), black locust (*Robinia*), Osage-orange (*Maclura*), black walnut (*Juglans*), chestnut (*Castanea*), certain mulberries (*Morus*), and catalpa (*Catalpa*). At the other extreme, weak, soft woods such as poplar, willow, and basswood decay readily. Of course almost any wood can be made more durable by impregnating it with preservatives such as creosote, pentacholorophenol, or sodium arsenate.

Hardness of wood is a reflection of the sturdiness of the individual cell walls, the amount of lignin they contain, and other factors. The harder the wood, the better it will resist wear. Some species, such as black locust (*Robinia*) and oak, are extremely hard. Other woods are so soft they can be dented even with a thumbnail. Balsa (*Ochroma*), perhaps the most familiar of the unusually soft woods, has been imported into the United States for fashioning lightweight model airplanes. Tensile strength and resistance to shear are other fairly important characteristics of wood. In some species the cells separate much more readily than they do in others; easy cleavage may be an advantage in splitting firewood, but a disadvantage where toughness is needed, which is what was needed for wagon wheel spokes, and still is for handles, athletic equipment (skis, bats, hockey sticks), and so on. Hickory, elm, maple, ash, and persimmon are strong woods, with considerable bending strength and shock resistance.

A characteristic of wood that gives certain types particular appeal is the grain (orientation of cells), which determines the figure (design). As already mentioned this is determined not only by the inherent arrangement of the cells in the wood, but by the direction of cutting. If a log is sawed along its radius (quarter sawing), the rays are revealed as broad patches, but if it is cut tangentially (plane sawing) the rays, like the annual rings, will be cut across and be seen end view as narrow bands. Figure is chiefly of importance for cabinet woods; other characteristics of esthetic importance are color, luster, polishability, and capacity to hold glue.

Porosity of wood may be of importance in its ability to take paint, or even to resist decay. White oak (*Quercus alba*), esteemed for barrels in which to age whiskey, has low porosity in spite of being a hardwood with large vessel cells; it develops balloon-like **tyloses** that fill the cell lumen, making the wood essentially impermeable. Tyloses help make black locust resistant to decay.

Shrinkage occurs as cell walls dry. This is especially manifest in sapwood, which may shrink as much as 20 percent in certain dimensions. Greatest shrinkage is across the grain (at right angles to the long dimension of the cells). Such shrinkage causes twisting or warping due to differences in orientation of cells or in types of cells in different parts of a board. Quality lumber is seasoned before being used, to dry it out slowly and uniformly, holding shrinkage and strains to a minimum. Since seasoning in open air requires a considerable length of time, much lumber is seasoned in kilns (**kiln-dried**) under controlled temperature and humidity. In most instances moisture content is reduced to around 6 to 10 percent; there is no point in drying wood more than this, since it will absorb at least this much moisture from the air in most localities.

LIGNIN The importance of cellulose (40 to 50 percent of wood) has been emphasized, but **lignin,** the second most abundant component of wood (constituting as much as 25 to 30 percent of the wood) is often dismissed as a humus-like residue left after wood decomposes. Considering the volume of wood consumed, the amount of lignin wasted annually in pulping operations is tremendous. There have been many efforts to decipher the chemical nature of lignin, and to find worthwhile uses for it. To a limited extent it has been broken down to yield vanillin-like intermediates of minor value, and itself has served as a plastic-

like binder for road construction soil and in fertilizers. But the cost of handling and transportation is such that it is unprofitable to move lignin very far to market; moreover, as a residue from pulping operations it is dirty, sticky, and foul-smelling. Hence, lignin continues to be more a problem than an asset; means have to be found for disposing of it without polluting streams. Great reward will await the man who discovers a large-scale profitable use for lignin!

Lignin has been investigated chemically for nearly a century, but its exact structure is still not certain. As a matter of fact, there are at least three, possibly more, lignins of slightly differing nature, with chemical structures almost as complicated as that of protein. It is generally agreed that they are polymers of phenylpropanoid units. Studies with radioactively labeled carbon dioxide show that they are slow to form in developing wood, and are thus considered to be an end-product in cell wall formation. Apparently cellulose forms the basic structural framework around which the lignin infiltrates and cements the lattice. Very little lignin is found in young cell walls, but perhaps 20 to 35 percent is found in old heartwood. Lignin is chemically bonded to cellulose, and serves in strengthening and stiffening the cell wall.

Forests

Because different tree species are adapted to different climates and soil types, many kinds of forest occupy the earth today. This has not always been the case. Primitive forests of several hundred million years ago consisted of fewer kinds of trees. In fact, the earliest "trees" (Psilophytales), which grew nearly a half billion years ago, were like giant clubmosses. They lacked true roots and consisted of a tangle of specialized branches that clambered over rocky ground. Fifty million years later came the dense forests of tree ferns that prevailed in tropical climates of that era. The forerunners of modern conifers were on the scene 300 million years ago, when plant life abundantly colonized marshy land, building the tremendous coal and oil reserves so important today. Two hundred million years ago seed plants had evolved, and many trees assumed an appearance quite similar to those of today. By the time the dinosaurs roamed the earth some 180 million years ago during the Cretaceous, seed-bearing trees had evolved that shed their leaves in winter; from these have sprung the Angiosperms and our present deciduous forests.

The ability to produce seeds that can survive unfavorable seasonal cycles was a significant evolutionary advance that occurred at about the end of the Paleozoic. These primitive plants (related to the seed ferns) bore seeds not entirely unlike those of the modern needle evergreens, members of the Gymnospermae. Consider the familiar pine tree as an example of the Gymnospermae. Two ovules are born on the upper side of each pine cone scale, fertilized by pollen from separate, smaller male cones, carried by the wind directly to the exposed ovule. The ovule matures into a free seed, which drops to the ground when ripe. The fundamental distinction between such gymnosperms and the later-evolving Angiospermae is that the angiosperm ovule is surrounded by a fleshy protective tissue through which a pollen tube must grow, and which becomes a fruit that encloses the seed. Typical of the Angiospermae are such deciduous trees as the oaks (*Quercus*), maples (*Acer*), ashes (*Fraxinus*), and palms (*Cocos,* and others).

Gymnosperms and angiosperms are distinguished by the lumbermen as well as the botanist; the gymnosperms are the "softwoods" of the lumber industry, the dicotyledonous angiosperms the "hardwoods." In temperate climes, hardwoods are readily distinguishable in the forest from softwoods; they usually have flat, broad leaves rather than needles or scales, and they shed their leaves annually. There are discernable differences in wood anatomy, too (see p. 120). The wood of gymnosperms has a less complicated structure, lacking certain distinctive cells (vessels) characteristic of angiosperms and often having resin canals that hardwoods lack. (Some hardwoods develop resin canals when injured, however.) Gymnosperms comprise the bulk of commercial temperate zone forests; angiosperms generally predominate in tropical forests. There are fewer gymnosperm species than angiosperms, and the commercially important trees are confined almost entirely to the order Coniferales (the needle-bearing gymno-

sperms, or conifers). In the United States about two-thirds of the commercial forest land is in softwoods, accounting for nearly 80 percent of the timber cut. Softwoods also predominate at high elevations in the tropics and at high latitudes in northern Europe, Siberia, and Japan.

In temperate regions, hardwoods such as birch (*Betula*) and aspen (*Populus*) mingle with softwoods in forests that are predominantly gymnospermous. Hardwoods predominate only at middle and subtropical latitudes. The great diversity of angiosperms, especially in tropical forests, leads to some practical difficulties in managing and marketing. Total yields per acre are frequently great, but production consists of small lots of a myriad of varying species, and for some there is not a ready market. Temperate hardwood forests predominate in a central belt of the United States, roughly between Tennessee and the Great Lakes, and in central and southern Europe eastward to southern Siberia and Korea; tropical hardwood forests are circumtropical wherever there is sufficient rainfall. Tropical hardwood forests are perhaps the greatest remaining timber reserve in the world, and no doubt will become increasingly more important as a source of forest products.

In the humid tropics, luxuriant forests represent the ecological climax; it is difficult and usually uneconomic (particularly in view of the limited human and material resources) to reject the wealth of the forest in favor of annual cropping systems that are frequently inappropriate for the soil and the climate. Yet that is the all too familiar approach. Since primeval times the tropical forest has been a vast storehouse of natural wealth, not only for wood, but for extractives. A genus such as *Manilkara* yields not only excellent timber, but a latex (balata) of export value, and fruit useful at least to wildlife. Rubber, resins, waxes, oils, nuts, and medicinal or insecticidal alkaloids long have been important forest products. Among timber trees West Indies mahogany, *Swietenia mahagoni,* has for centuries been a prized cabinet wood; rosewood (*Dalbergia*), Spanish cedar (*Cedrela*), balsa (*Ochroma*), and several others likewise have been selectively much harvested but seldom planted. Only the vastness of the forest has generally prevented their complete exhaustion.

Nonetheless, it is encouraging that tropical authorities are beginning to adopt a more farsighted approach. In Trinidad, for example, after centuries of cutting and depletion, reforestation with select species is being undertaken. Teak (*Tectona grandis*), introduced from India, is especially used. In salubrious tropical climates the cycle from planting to harvest is probably much less than the 50 to 150 years deemed necessary for maturing a forest in temperate localities. For the teak plantations, scrub forest is first cleared (the wood often made into charcoal) and the land cropped with food plants for a year before teak trees from a nursery are planted. Food cropping may be continued between the trees until they are tall enough to shade the crop. Thinnings useful for poles and fencing are had in as little as four years, and small lumber stock not long after. Perhaps someday, as transportation and development progress, tropical forests will be more meaningfully managed in such a way that their tremendous potential will be realized.

As nations developed from infancy to a high level of technological advancement, the regard in which they held their forests seemed to pass through three distinct stages. First, the forests seemed to be obstacles to colonization—hurdles to be surmounted. Land was cleared, the trees piled and burned in order to provide necessary living space and agricultural land. During the next stage, they became storehouses to be exploited. They were sources of energy and construction materials.

Box 23-2 FOREST ECOLOGY

The influences that shape forest regeneration are only beginning to be understood. Forest change is often a continuing process, and foresters attempt preservation of what is a successional stage. This, of course, is not possible without some form of management. In many instances there may be no stable climax, since the overall environment is being altered by man's activity and perhaps by climatic change.

Fire as an influence on forest succession has been recognized as having both beneficial and negative effects. It has always been an intregal part of the forest environment, and has contributed to a mosaic of successional sub-climaxes within an overall species climax. For example, fire may induce grassland in a redwood forest, which is only slowly reconverted to redwood depending upon frequency of firefall.

The succession that occurs in the northern Rocky Mountain forests is influenced by fire. Lodgepole pine is the offspring of fire, occurring in even-aged stands after burning. If fire does not reoccur, species of *Picea* and *Abies* come to dominate, and, if fire control is exercised, become the climax. Interestingly, the fact that lodgepole pine will not reproduce under its own canopy helps to explain the even-aged stand.

On the other hand wildfire can alter soil conditions so much that it prevents succession. This has been demonstrated in the ponderosa pine stands in the Southwest, where no new ponderosa catch occurs after burning even though reproduction takes place on land not burned. The soil constituency following fire—or for that matter after clear-cutting and the exposure this procedure brings—may materially depauperate the biological composition and influence forest quality far into the future

In northcentral Colorado an outbreak of spruce beetles caused death of large-diameter spruce trees in patches, and altered species composition in favor of fir and lodgepole pine. The outbreak developed in logging slash (and was thus an upset caused by man), but it remained epidemic for only two years, until natural controls (for example, parasites such as nematodes, and woodpeckers, which lent great assistance) responded.

As man assumes greater influence in the forest through harvesting and attempted management, the balance and buffering characteristic of virgin forest can be expected to diminish. The rapidity with which natural forests are being cut allows little time for the research necessary to avoid habitat degradation.

In the last stage, forests were finally recognized as a renewable resource and were husbanded and protected so that they could be maintained. But the developing nations of today must take heed of past mistakes and handle their forests with care. They cannot permit the luxury of waste or mismanagement.

A summarization of several forest regions and types will help make clearer the major kinds of forest and the state of forestry in various parts of the world today.

FORESTS OF THE SOVIET UNION The Soviet Union (including Siberia) serves well as an example of a region with tremendous forest reserves only partly exploited. The Soviet Union has more than 2.7 billion acres in forest, of which about one-quarter is managed. Seventy-eight percent of the forests are coniferous, accounting for nearly half of the world's total. Approximately two thirds of the forested land is in mature timber, although the rigorous environments near the tundra in the north and in the deserts in central and southern areas cause trees to be small and sparse. About 60 percent of Soviet territory is considered densely forested, most of the remainder either having no trees or a scattered cover of less than 5 percent.

Trees grow best in European Russia, as far south as the northern periphery of the Black Sea, where rainfall is ample and the climate not overly severe. In this thickly populated area, however, there is a deficit of forest products and there are many demands for other land use. Hence timber is imported from abundant but less productive forests to the east. Where the land remains in forest in the western part of the Soviet Union, reserves of wood are as great as 3,000 cubic feet per acre, the prevailing condition in the Caucasus and Carpathian highlands and just north and east of the Black Sea. North and east of the Caspian Sea dry climate limits the forest, and wood reserves there range from less than 350 cubic feet per acre to around 700. In the Ural Mountains, north-central Siberia, and southeastern Siberia (just north of Korea), forests are reasonably luxuriant, with wood reserves averaging 1,400 to 3,000 cubic feet per acre. In northeastern and northwestern Siberia, and in central sections north of Mongolia, wood reserves are only 700 to 1,400 cubic feet per acre, and growth is slow.

As would be expected, forest growth is slower and less luxuriant in the cold northern belts, where winters are long and the soil frozen much of the year. Much of the forest here is scrubby taiga. But the productive forests of the southwestern belts show an annual increment of 140 cubic feet or more per acre. The annual increment in Siberia (where half the forest reserves lie) totals about 35 billion cubic feet. The northern taiga consists mainly of pine, spruce, fir, larch, and "cedar" among the conifers; birch is the only significant hardwood. The forest is often interspersed with peat bogs and open areas of permafrost. The soil is a heavy podzol. The chief species are *Picea excelsa* and *P. obovota*, *Pinus sylvestris* and *P. siberica*, *Larix sukaczevi*, and a few *Abies*.

South of this northern belt there is a zone of mixed softwood-hardwood forest. The main conifers are again *Picea excelsa* and *Pinus sylvestris*, but mixed among them are oaks (*Quercus*), maples (*Acer*), *Tilia*, and *Corylus avellana*. In the Far East, near the Pacific a much richer assortment of hardwoods occurs, including *Phellodendron*, *Juglans*, *Fraxinus*, *Carpinus*, and *Fagus*. The climates in the extreme southern zones of the Soviet Union are often quite mild, and support such broadleaf evergreens as *Buxus*, *Laurus*, *Magnolia*, *Ilex*, and *Laurocerasus*. Prominent forest trees are *Juglans regia*, *Castanea sativa*, and several species of *Fagus*.

Pines constitute the most prized timber group in the Soviet Union, of which *P. sylvestris* is the most important. Spruces probably rank second. In terms of timber reserves, however, the larches (*Larix*) lead, constituting about 40 percent of the forest cover. The pines rank second at about half this figure, followed closely by spruce, with birch the ranking hardwood and third most abundant genus in terms of cover (about 13.5 percent). Following these come cedars, firs, aspens, oaks, beeches, lindens, alders, and poplars, about in that order.

FORESTS OF SOUTHEAST ASIA From China south through the Malay Peninsula and on into the East Indies, luxuriant tropical hardwood forests once prevailed. On much of the Asian mainland, China especially, and on the heavily populated East Indian Islands, most of the most productive forest has long been exhausted. Even though timber needs are very low in this section of the globe, under the low standard of living that prevails, timber must be imported. Much of Borneo's forest land is still unexplored, and not all the timber trees have even been identified. The most important family is the Dipterocarpaceae, which supplies about 90 percent of the export timber. Much of this reaches the United States as "Philippine mahogany." There are at least 150 species in the family.

By modern forestry standards the immensely diverse and little-explored forests of Borneo are underutilized. Transportation and manpower are both limited in the jungle, and careless procedures are employed in harvesting the trees. Undoubtedly many fine timber trees are wasted , simply because the wood is not known in outside markets and because familiar species (such as of the Dipterocarpaceae) continue to be called for in specifications. Facilities for local processing are crude, and in spite of an abundance of some of the world's finest woods, quality lumber (as is needed for making furniture) must be imported from traditional trade sources.

Under these primitive circumstances it is no wonder that very little is known about regeneration of cutover land or the quality that can be expected from second growth forest. The prevailing custom is to girdle and kill all large trees (above six foot in girth) left after selective logging in the belief that they will be overmature by the next cycle of rotational harvesting. It is not known which species will regenerate under prevailing ecology, what the shade requirements of desirable species might be, nor what rotational cycle is best fitted to these particular forests (the forest authorities presume an 80-year cycle). In Malaya, where forestry is more advanced than in Borneo, and the virgin forest similar, it has been found that smaller trees of the Lauraceae, Myristicaceae, Rosaceae, Leguminosae, and oaks, once considered useless, are of considerable value in providing cover that allows a new stand of dipterocarps to emerge. When the forest composition has been changed on cocoa lands, there have been serious consequences,

such as lowering of the water table and slow debilitation of the plantation crop. Much still remains to be learned about soils and ecological relationships in the tropics!

Aside from teak (*Tectona grandis*, Verbenaceae) timber trees from the tropical hardwood forests of Southeast Asia are largely unknown in the western world. Useful and commercially quite important to Borneo are *Shorea, Parashorea, Dryobalanops, Dipterocarpus, Koompassia, Cratoxylon, Anthocephalus*, and several other genera, including species of the Leguminosae and Sapotaceae as well as the Dipterocarpaceae. But all in all, much remains to be done in acquainting the commercial world with available supplies and in standardizing and improving harvesting and handling. Much research is needed to discover the best techniques for managing these very diverse and luxuriant forests.

FORESTS OF NORTH AMERICA   Forests in North America exhibit general latitudinal banding as do those of Asia. The northernmost belt is chiefly coniferous, extending from Alaska and western Canada through northern and eastern Canada into the Great Lakes and New England sections of the United States. It also extends southward at the high elevations of the Appalachian Mountains as far as North Carolina. South of this belt, between the Great Plains and the Atlantic Coast, is hardwood deciduous forest; it stretches roughly from east-central Texas through northern Georgia and northward along the Piedmont into Maryland. South of this is a belt of southern coniferous forest that stretches into middle Florida and is dominated by several so-called southern yellow pines (except in the bottomlands, where hardwoods predominate). Subtropical and tropical forests are of limited importance in the continental United States, existing only in southern Florida and in a narrow band along the Gulf Coast from Florida to Texas; southward in the lower-lying lands of Mexico they become fairly extensive, but even there they are of limited economic usefulness. In the extreme western part of North America are found the most majestic forests, especially as a coastal band on the moist Pacific slopes from middle California north to Alaska—magnificent stands of redwoods, coastal

Douglas-fir and other species. Inland is a drier Rocky Mountain forest belt, extending from western Canada through Colorado into central Mexico. Important and extensive forests occur, although confined chiefly to mountainous outcroppings that punctuate otherwise desert-like country. The forest belts in the United States are depicted in Figure 23-7.

In contrast to most tropical forests, and to some extent Siberian forests, the forests of North America are well known, the tree species well identified. Outstanding in the eastern quarter of the northern coniferous forests, at least as the early colonists found it, was the white pine, *Pinus strobus*. White pine alone supported almost the entire lumbering industry in the northeastern United States for nearly two centuries; regrowth of white pine makes it still of importance. However, lumbering has now shifted to the virgin forests of the West and the managed forests of the Southeast. When the United States was still a colony of Great Britain, so important was the eastern white pine that an act was passed for its preservation on crown lands, withholding the finest trees for masts of the sailing ships of the Royal Navy. This was another of the irritations that eventually led to the American Revolution. In addition to white pine there are several other species of *Pinus* in the eastern portion of the northern forests, of which the red pine, *P. resinosa*, is the only large one of great commercial interest. There are several important spruces (*Picea rubens, P. glauca, P. mariana*), hemlock (*Tsuga canadensis*), fir (*Abies balsamea*), arborvitae, or northern white cedar (*Thuja occidentalis*), and a scattered interspersion of hardwoods, including principally the aspen (*Populus tremuloides*) and birch (*Betula lutea = B. alleghaniensis*) and other species.

The arctic portion of the northern forest belt is composed largely of *Abies balsamea* and *Picea glauca* with the *A. balsamea* disappearing as one goes west. It merges westward with those of the Pacific slopes, which are similar in general composition but marked by different species. The white pine there is western white pine (*P. monticola*)

FIGURE 23-7. *Forest types as they occur in the United States. [Courtesy St. Regis Paper Co.]*

Northern Forest covers a region of abundant lakes, river valleys, and hills. Some areas are predominantly coniferous (mainly pine, spruce, balsam, fir, hemlock), some hardwood (mainly maple, birch, beech), and some mixed, as shown here.

Central Forest is characterized by hardwood species. Much of this region was long ago converted to farmland. Important species include oak, hickory, yellow poplar, as well as such conifers as yellow pine and red cedar.

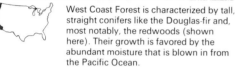 West Coast Forest is characterized by tall, straight conifers like the Douglas-fir and, most notably, the redwoods (shown here). Their growth is favored by the abundant moisture that is blown in from the Pacific Ocean.

Mountain Forest of the West has less rainfall because moisture from the Pacific is barred by the Cascades and Coast Ranges. Prominent species are ponderosa pine (front), western white pine, sugar pine, Englemann spruce, and Douglas fir.

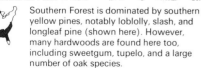 Southern Forest is dominated by southern yellow pines, notably loblolly, slash, and longleaf pine (shown here). However, many hardwoods are found here too, including sweetgum, tupelo, and a large number of oak species.

Tropical Forests of the United States are found only in the southern tip of Florida and a small area in southeastern Texas. Mangrove (shown here in foreground) and mahogany are the principle trees of our tropical forests.

FIGURE 23-8. *The California Big Tree, one of the giant redwoods in Sequoia National Park, California.* [*Courtesy U.S. Forest Service.*]

and sugar pine (*P. lambertiana*); western hemlock is *T. heterophylla;* spruce is mainly Sitka spruce (*P. sitchensis*); the arbovitae is *T. plicata;* and the firs are represented by an important group of different species (*A. grandis, A. nobilis, A. magnifica,* and *A. amabilis*). But the Pacific coast forests are primarily noted for genera having no close counterpart in the East. In recent decades the most important timber tree has been the Douglas-

fir (*Pseudotsuga menziesii*), and in northern California the most dramatic forests in the world consist of almost pure stands of redwood (*Sequoia sempervirens*) containing the tallest trees in the world, some more than 370 feet tall! Some stands of redwood contain nearly a half million board feet to the acre, but unfortunately in the felling of such gigantic trees much breakage and waste occurs. Although most of the finest stands of coastal Douglas-fir and redwood have already been cut, significant old growth still exists in remote sections of California, Oregon, Washington and British Columbia. This forest belt is the richest timber reserve in North America today.

A number of other distinctive trees of no little economic importance are found in the Pacific forest belt, too. Incense-cedar (*Libocedrus decurrens*) is found in Oregon and California, where it is often a companion to the giant sequoia or big tree (*Sequoia gigantea*), the largest tree in the world, some 4,000 years old and weighing 6,000 tons (Fig. 23-8). The few giant sequoias remaining in the California Sierra are today protected, and in fact have never been of any great commercial importance because of their limited number, weak wood, and tremendous size, which makes cutting and handling difficult. There are several "cedars," too, in the Pacific forest belt, of the genus *Chamaecyparis*, although of secondary importance compared to Douglas-fir.

The Pacific forest belt merges with that of the Rocky Mountains in western Canada; in the United States these two zones are separated by the arid Great Basin. An inland subspecies of Douglas-fir grows in the Rocky Mountain forests as far south as middle Mexico, although seldom so luxuriantly as in the coastal forest belt. The western white pine grows in the northern Rockies, but the pine species of chief importance as a Rocky Mountain timber tree is the ponderosa, or western yellow, pine (*P. ponderosa*), also well represented in the Pacific belt. On the more arid southerly plateaus piñon pine (*P. edulis*) assumes minor importance; it is too small for general timber use, but the seeds (piñon nuts) have long been important to the Indian tribes as a staple, and are marketed as a modern-day confection. Lodgepole pine (*P. contorta*), another small pine, is as widespread as ponderosa but not as useful for lumber. The western larch (*Larix occidentalis*) is found abundantly in the northern Rocky Mountains, and two spruces (*Picea engelmannii, P. pungens*) at higher elevations throughout this forest region. Several firs also grow in the Rocky Mountain forests, of which the most important species is probably the white, or Colorado, fir (*Abies concolor*).

As is typical of hardwood forests (Fig. 23-9), great diversity is found in the forest belt just south of the coniferous zone, from southern New England and the middle Great Lakes states (where hardwoods mingle with the conifers) southward to central Texas and the northern parts of the states to the south and east of Tennessee. The most humid eastern portion of this forest belt is characterized especially by yellow birch, beech, sugar maple, and hemlock in the more northerly portions; by several oaks, yellow-poplar (*Liriodendron*), and sweetgum (*Liquidambar*) in the more southerly portions. Oaks, hickories, and maples dominate the drier westerly and southeasterly portions. Although these genera characterize the forest, diverse other hardwoods that grow along with the dominants are often of even greater economic value (black walnut, for example, or old growth black cherry). There is almost no virgin timber left.

Chestnut (*Castanea dentata*) was once an important component of the eastern hardwood forest, but, except for short-lived stump sprouts, is nearly extinct due to introduced chestnut blight. Today oaks in the northern and western portion of this belt, and yellow poplar in the southeastern portion, are probably the two most important trees. There are many species of oak, but the white oak (*Quercus alba*) is the most prized, not only for lumber but for such specialty purposes as barrel staves. Black walnut (*Juglans nigra*) and black cherry (*Prunus serotina*) are perhaps the most valuable individual species; they serve primarily for making cabinet and furniture veneers. Large trees are now very scarce. Among the other hardwood trees of commercial importance are ashes (*Fraxinus* spp.), basswood (*Tilia americana*), beech (*Fagus grandfolia*), the hickories (especially the shagbark hickory, *Carya ovata*, and the pecan,

FIGURE 23-9. *Second-growth deciduous forest in Indiana, composed mainly of oak and hickory.* [*Courtesy Purdue Univ.*]

590

FIGURE 23-10. *Old-field loblolly pine 50 years old, Arkansas.* [*Courtesy U.S. Forest Service.*]

C. illinoensis), maples (especially the hard or sugar maple, *Acer saccharum*), black locust (*Robinia pseudoacacia*), honeylocust (*Gleditsia triacanthos*), sycamore (*Platanus occidentalis*), the tupelos (species of *Nyssa*), sweetgum (*Liquidambar styraciflua*), yellow-poplar, or tulip tree (*Liriodendron tulipifera*), many species of oak (*Quercus*), walnuts (the butternut, *Juglans cinerea*, as well as the black walnut), and such commercially unimportant (but abundant) trees as willows (*Salix*), poplars (*Populus*), and elms (*Ulmus*).

The final forest belt is that of the Southeast, characterized especially by longleaf, slash, and loblolly pines (Fig. 23-10). This coniferous forest has assumed major economic significance in the eastern United States. Major holdings by large integrated companies are being built up in this zone rather than in northern regions because the trees grow fast and the population density is relatively low. Much cutover land is being reforested solely for production of pulp stock and small timber. In some areas second growth of sufficient size for lumber is available for harvest, but only in very inaccessible locations are there any virgin forests.

The longleaf pine (*P. palustris*) is probably the most famous of the southern yellow pines. It is a major source of lumber and pulp, and was once a major source of turpentine. Regeneration of longleaf pine is made difficult because seedlings are slow growing and do not survive well when planted. Slash pine (*P. elliottii*) was originally more restricted in its southeastern range than the longleaf pine, but has been successfully established in forest stands over the entire deep South; it is probably the chief source of natural naval stores today. Loblolly pine (*P. taeda*) is the most important yellow pine species in the mid-South. It colonizes abandoned agricultural land quickly, and is the principal commercial pine of the Piedmont region today. Shortleaf pine (*P. echinata*) is the widest-ranging of the southern yellow pines, occurring as far north as Missouri and Connecticut. Another important timber species of the Southeast is the baldcypress (*Taxodium distichum*), typically found in swamps and bayous along the coast and river bottoms from southern Virginia to Texas.

In the eastern United States, most forest land is under private ownership, whereas most western forests are in federal hands (mostly as national forests or as holdings of the Bureau of Land Management). In volume of growing stock the New England forests contain somewhat more than 30 billion cubic feet, of which slightly more than half is softwoods. The central hardwood belt comprises over 100 billion cubic feet of growing stock, almost entirely hardwoods. The southeastern forests contain about 135 billion cubic feet, about evenly divided between softwoods and hardwoods. The Pacific forests contain nearly 260 billion cubic feet, almost entirely softwoods; and the Rocky Mountain forests nearly 100 billion cubic feet, also almost entirely softwoods. The West accounts for more than half of the country's total of about 628 billion cubic feet, the Southeast for about a fifth, and the central and northern forests together for about a fifth. In terms of growth potential, the South is far ahead of the rest of the nation.

In total reserves Douglas-fir is by far the leading species, with over 600 billion board feet of sawtimber. Next comes western hemlock with about 270 billion board feet and ponderosa pine with about 242 billion board feet. All firs together total about 235 billion board feet, all spruces about 155 billion board feet. In the east shortleaf and loblolly pine together constitute the biggest reserve, amounting to about 161 billion board feet. The once abundant white pine runs less than 20 billion board feet. Among the hardwoods the red oaks as a group provide the greatest reserve, about 90 billion board feet, followed by the white oaks as a group at about 73 billion board feet. Hickory, maple, sweetgum, black tupelo, and yellow-poplar follow in about that order.

Forest products

Forest trees have traditionally been the most abundant and economical source of cellulose. Cellulose is today a staple of the economy, used for many varied purposes. Even throw-away clothing has been marketed, made of especially attractive but nonetheless disposable paper. World production of wood products exceeds 2.1 billion cubic meters annually, almost half consumed for fuel, about one-third for lumber, approximately one-eighth for pulp. Cellulose consumption is an indicator of a nation's economic conditions and material standard of living. So many goods come to market packaged in paper or paperboard that pulpwood consumption usually foretells shipments to market. Similarly, consumption of newsprint is a rough indication of adequacy of communication and degree of literacy. Paper and paperboard are used at the rate of about 600 pounds per person per year in the United States, no more than 200 pounds in Britain, less than 100 pounds in France, about 50 pounds in Japan, 25 pounds in the Soviet Union, and only 2 pounds in China.

Forest products account for more than 11 billion dollars worth of the world's exports, and represent nearly 5 percent of world trade. Forest products are especially important to the Scandinavian countries (accounting for nearly half the trade of Finland, for example); Canada exports nearly as much in the way of forest products as all of Scandinavia (about half the world's newspapers are printed on paper from Canadian pulp). Although much of South America is forested, it exports only modest amounts of forest products, and Asia exports relatively little. The absolute value of forest products from tropical Africa is not

great, but proportional to the total trade in this undeveloped area, it is quite appreciable. The greatest importers of forest products are Japan, the United States, Great Britain, and Germany.

Forest products of greatest importance in our modern economy are treated in the following discussion.

LUMBER, PLYWOOD, AND VENEER World-wide the greatest use for wood (next to fuel) is for lumber. Nearly a billion cubic meters of sawlogs for lumber and veneer are cut annually, most of it from the United States and the U.S.S.R., about 80 percent softwood! The United States harvests nearly 38 billion board feet, almost a quarter billion cubic meters. Home construction provides the greatest outlet for lumber. Even houses with brick or other exteriors utilize appreciable lumber for studding, floors, doors, windows, and trim. Properly constructed to resist rotting and termite damage, a frame house can last indefinitely. In New England frame houses centuries old are still in use. Techniques are at hand for bonding small dimension stock that is light in weight and conservative of wood, yet just as effective for many facets of house construction as traditional lumber. The mid-twentieth century witnessed appreciable change to plywood from use of solid planks, especially for subflooring and sheeting. Wooden crating, too, once consumed sizable quantities of rough lumber, but this is being replaced by fiberboard and fiber (cardboard) containers. Never have there been widely accepted substitutes for wood in making furniture, although veneer is now accepted as being superior to solid wood for most cabinetry because it will not check and split so readily and because it is usually cheaper.

Veneers provide an economical use of fine logs, since a thin layer of wood is peeled and cemented to the surface of a less-expensive, more-abundant core wood. Cut fairly thick, and bonded several odd-number plies together, veneers become plywood, an increasingly important construction material. Consumption of timber for veneer is nearly 20 percent that for lumber, amounting to over a billion cubic feet annually in the United States alone. Fine veneers for furniture are made from expensive and often scarce woods, frequently imported. Examples are walnut, bird's-eye maple and cherry, mahogany, teak, and rosewood. Plywood, however, utilizes abundant supplies of inexpensive wood, chiefly Douglas-fir, although southern pines are beginning to compete for the plywood market. Japan has, in recent years, become an important producer of plywood.

Both veneers and plywood achieved public acceptance only when suitable bonding agents (phenolic resin glues) had been developed that would solidly and lastingly cement the thin veneer. Today's excellent bonding agents set stronger than the wood itself, and are suitable for exterior as well as interior use. Successive plys are positioned with the grain at right angles, thus reinforcing one another and making the finished panel far stronger and much more dimensionally stable than an equivalent panel of solid wood. It also comes in far larger sizes (normally $4' \times 8'$) than is obtainable in lumber (and in various thicknesses and numbers of plies). Construction is speeded up greatly because fewer individual pieces are handled. For balance the center ply is made twice as thick as the adjacent ones. Inner plies are generally of less select grades of wood; those containing no visible defects are reserved for facing. However, special machines cut out such blemishes as knots from a plywood veneer and replace them with a duplicate insert that is scarcely noticeable. Plywood is assembled by elaborate machines that automatically apply the adhesive and affix adjacent plies. With the familiar thermosetting adhesives, the plywood can be heated to a specified temperature for curing in a few seconds.

Veneers may be produced in many ways, such as by sawing, slicing, or peeling. The most frequent method is the economical rotary-peeling. The log (**bolt**) is softened by boiling or steaming for a number of hours. It is then fastened securely to a rotating shaft, which works against a large stationary knife much like a lathe or the reel of a lawn mower works against the bedknife. The knife is set to adjust continuously inward and maintain a uniform width of veneer, which peels off the bolt much as a roll of paper unwinds. Except for the very center of the bolt there is almost no waste of wood (and this is used for pulp). Rotary cutting is employed for most inexpensive veneers and all

plywood, although sawing and slicing to obtain a special figure may be practiced with certain expensive cabinetry veneers.

There has been a tremendous rush into plywood production, causing cyclic overcapacity when construction demands decline. Douglas-fir plywood plants tripled in the Pacific Northwest during the 1950's, where the bounteous supplies of old growth Douglas-fir assured economical raw materials for some years to come. Vertical integration has been inaugurated, in most plants, including chip handling facilities to dispose of by-product wood profitably. Technological developments have · also made it possible to use peeler logs of smaller dimension ("shallow-clear" logs), thus reducing dependence on butt sections of old growth timber ("deep clear" logs), which can be reserved for top quality lumber and timbers. With modern steaming methods, even the sapwood makes acceptable veneer. This in turn has made greater care necessary in the handling of logs, to prevent, for example, jabbing them with spiked poles, the traditional way of moving them about the millpond. Plywood is being increasingly made from southern pines, too; as of 1966 there were more than a score of plywood plants in Arkansas, Mississippi, and Louisiana, with announced capacity of over a billion square feet ($\frac{3}{8}$ inch equivalent).

With technological improvements manifest, it is no wonder that plywood capacity has risen about fivefold from the approximately 2-billion-square-feet capacity of 1950. It is estimated that within the next quarter century the present annual production will rise tenfold again, providing by the turn of the century enough plywood yearly "to roof the state of Connecticut." Newer techniques permit plywood to be formed into graceful curves, used much like molded plastics. Modern glues permit use of plywood outdoors and are said to make it stronger than steel, pound for pound. The average new house today utilizes approximately 3,000 square feet of plywood, estimated to rise to nearly 9,000 square feet eventually. With plastic overlays on the surface, plywood well may be employed increasingly for attractive exteriors. The chief uses today are for roof and wall sheeting, subflooring, interior cabinet work, boxes and crates, tanks and vats, boxcar lining, pallets, farm structures (includ-

FIGURE 23-11. *Poles ready to go into pressure chambers for treatment with preservatives.* [*Courtesy U.S. Forest Service.*]

ing fences), boats, concrete forms, barricades and outdoor signs.

POSTS, POLES, PILINGS, AND TIES   In spite of the diminishing availability of large timber in many regions, wood poles are still used for stringing utility lines, for pilings and ramps, and in many parts of the world for fence posts and mine timbers. Worldwide this usage is about one-twentieth that of all lumber and veneer. In technically advanced parts of the world, both fenceposts and utility poles are treated with preservatives to improve their lasting qualities. At one time poles were merely soaked in a preservative, which provides a peripheral barrier against fungi but does not protect the center wood of the pole. More thorough impregnation of preservative is achieved by pressure and vacuum treatments in sealed chambers, in which the preservative is forced into all cells (Fig. 23-11). Creosote is still the most frequently utilized preservative, but chemicals such as pentachlorophenol are equally useful.

Although railroad trackage is scarcely expanding in settled parts of the world, the continuous replacement of old railroad ties causes a minor drain on timber. So far no metal or synthetic tie has been developed that is as useful and inexpensive as wood for undergirding the rails. Wood is resilient, easily fashioned, and amenable to the driving of spikes to hold the rail. Ties were once hand hewn, but today are mostly sawn. The modern railroad tie, treated with preservatives, has a potential life of nearly a half century. The kind of wood used for ties depends partly upon local availability, and in the United States red oaks, yellow pines, gums, and Douglas fir have all found favor; however, the various oaks account for about 55 percent of the million ties used in the United States.

COOPERAGE   Cooperage accounts for another minor use of wood. Although such commodities as cement and flour are no longer transported in wooden barrels or tubs, as they once were, and though most industrial liquids are shipped in steel drums, wooden barrels still find some specialty use, especially for the storage and aging of liquors. **Tight cooperage** for holding beer, whiskey, wine, syrups, and corrosive chemicals must be carefully constructed of woods chosen for their impermeability. Hardwoods have been favorites, especially white oak, the heartwood of which is made impermeable by tyloses. White oak imparts no disagreeable odor or taste, and is required by statute for the aging of bourbon whiskey. The manufacture of white oak barrels is of local importance in the midwestern hardwood forest belt. Staves cut to the proper curvature by special saws are made flexible by steam heating for the "raising" of the barrel. After the barrel is assembled it is flame-charred within; the charred wood gives whiskey its amber color. Redwood and baldcypress have also been much used for tight cooperage, both species being noted for their natural resistance to decay.

MISCELLANEOUS WOOD PRODUCTS Limited quantities of excelsior are still made from weak, soft woods that shred well, such as poplar. Excelsior has declined in importance as a packing material, but there is new interest in it as a mulch; it has proved quite effective for mulching newly seeded roadside berms. Modest interest continues to be shown in wooden shingles and shakes for roofing, mainly for esthetic reasons, even though synthetic roofing materials are less expensive and more practical. Wood is used for toys, novelties, picture frames, gunstocks, and a multitude of other purposes. Wood chips and wood flour are compressed into fiberboard for insulation and construction materials. They are also compressed into fireplace "logs" and made into charcoal. Wood flour is used as an abrasive for cleaning oily floors, as a bulking material in the manufacture of linoleum, in explosives, and as a diluent or raw material for plastics. The use of wood cellulose to make acetate and viscose rayon was discussed in the previous section of fibers. Charcoal has many uses, from the purification of sugar to fuel for barbecue fires. In Brazil tremendous amounts are used as a source of carbon in the manufacture of steel.

PULP AND PAPER Wood is the chief raw material for conversion to pulp, principally used for making paper (Fig. 23-12). World pulp production exceeds 0.3 billion cubic meters annually. In technologically advanced parts of the world pulpwood consumption is nearly half that of lumber. In the United States about 8 billion cubic feet of pulpwood are consumed annually, about three-fourths of which is domestically produced. It is estimated that in the United States nearly 600 pounds of paper and fiberboard are consumed annually per capita. Some two-thirds of this is derived from softwoods. At one time only select species such as spruce were thought fit for making paper but these species now account for less than one-fifth of production. Technological improvements have made many short-fibered hardwoods acceptable and they now account for about 50 percent of production. The resinous yellow pines are used for heavy kraft and linerboard papers. Synthetic sponges, honeycomb insulation, and disposable linens, draperies, and clothing are all being made from pulp these days. In mild climates, we may one day see paper houses!

Papermaking is an ancient art. The modern concept of paper—a felt or mat of fibers—appears to have originated among the Chinese during the first century A.D. Stems, bark, and silk were pulverized, the fiber dispersed in water, than sieved out onto a porous mold of reeds. Bamboo and the inner bark of the paper mulberry (*Broussonetia papyrifera*) were early sources of fiber. The principle behind modern papermaking is still the same: fibers are separated and dispersed and then matted into a thin layer that is filled, coated, and compressed on elaborate machines.

The name "paper" is derived from the word "papyrus," a sheet made by pasting together thin sections of an Egyptian reed (*Cyperus papyrus*). But almost any sort of fiber can be used for making paper. In Europe textile fibers were used during the Middle Ages, principally flax derived from old linen cloth, but cotton and hemp rags were also used. *Agave* or *Ficus* bark were used in the early Spanish colonies of the New World. With Gutenberg's invention of the printing press in the middle fifteenth century, the demand for paper increased greatly, and the supply of old rags could no longer satisfy the demand. Straw and wood were tried in England and France as early as 1800, but it was

FIGURE 23-12. *Wood ready for conversion to pulp.* [*Courtesy U.S. Forest Service.*]

not until about the middle of the nineteenth century that practical techniques were developed for converting wood into pulp. In Germany a wood grinding mill was built at about this time, and in 1854 a patent was issued in the United States for chemically reducing wood to pulp (the soda process). Sulfite chemical pulping was introduced right after the American Civil War, and some years later the sulfate process now so much employed for making kraft paper was brought out. These processes have made wood the most economical source for paper pulp, which accounts today for more than 90 percent of all paper made. A few special papers, such as those used for cigarettes and banknotes, are still made from flax fiber, but no other fiber is as abundant, consistently available, and economical as wood fiber. Moreover, wood supplies have generally been near water, which is needed in abundance for pulp processing.

In the **mechanical,** or **groundwood, method** of preparing pulp, short debarked logs are pressed sideways against a revolving grindstone sprayed with water to produce a slurry of short fibers and fragments. The quality of groundwood pulp varies with the kind of wood used and with the care taken in grinding. But inasmuch as the lignin and other noncellulosic components are not removed, groundwood paper "ages" and discolors easily. Groundwood pulp is chiefly employed for making newsprint or inexpensive sheeting of secondary quality. It is the most economical of the pulping methods, and produces the highest yields. A ton of air-dry pulp can be derived from little more than one cord of wood, nearly double the yield from chemical methods. Although power costs are high, the number of man-hours of labor required is generally only two-thirds that for chemical methods, and less-skilled supervision is needed.

Better quality paper is made from pulp produced by chemical methods. Logs with the bark removed are mechanically chopped into small chips that are screened for size. The chips are chemically treated to dissolve most of the lignin, leaving nearly pure cellulose fiber. This in turn may be bleached for white papers. In the **sulfite process** the chips are digested in a solution of bisulfites at temperatures from 130° to 150°C (266° to 302°F). The amount of cooking determines the degree of lignin removal; prolonged cooking may cause some degradation of the cellulose. Highly resinous woods cannot be converted in this process. In the **sulfate process** the chips are digested in a solution of sodium hydroxide and sodium sulfide at temperatures of 170° to 180°C (338° to 356°F). The process is particularly suitable for pulping resinous pinewood, and is used to make kraft wrapping papers and paperboard. Tall oil, sulfate turpentine, and alkaline lignin are by-products of sulfate pulping. The **soda process** uses sodium hydroxide alone as the cooking agent; it finds limited use for making book and magazine papers from softer hardwoods. A so-called **semi-chemical process** removes only a portion of the lignin from the wood by chemical cooking, after which the softened chips are mechanically pulped.

Tree species used for pulping vary widely around the world. In the Southeastern coniferous belt of the United States the yellow pines are much used; in the northern coniferous zones various spruces, firs, and hemlocks are used; in the West hemlocks, firs, Douglas-fir, larches, and several pines and spruces are employed. Canada, the world's largest producer of groundwood pulp for newsprint, processes principally softwoods, such as spruces, pines, hemlocks, and firs. In Latin America sugar cane bagasse is often used instead of wood, but the use of tropical hardwoods and conifers for pulp is increasing. Paraná pine (*Araucaria*) from southern South America supplies the Brazilian mills. In northern Europe pulp is made chiefly from softwoods, much the same as in Canada. In Australia, Chile, and Spain *Eucalyptus* is used, as well as herbaceous fibers. Japan utilizes spruce, fir, and pine pulp, as well as several textile fibers. China depends more upon bamboo, rice straw, cotton stalks, and bagasse. In South Africa introduced *Eucalyptus* and pines, as well as native *Acacia*, are utilized. It is apparent that almost any locally abundant and inexpensive fiber serves as a source of pulp. In India bamboo is grown as a crop and is an important pulping species. Should wood sources become scarce or too costly, annual crops or high yielding "shrubs" could substitute as a source of pulp. In the United States many species have been studied as possible future sources of pulp. Sunn, or sann hemp (*Crotalaria juncea*, Leguminosae), has specifically been suggested as a potential annual pulp crop.

In the making of printing papers, fillers (such as clay, chalk, calcium sulfate, and talc) are added to the bleached and processed pulp to give weight, opacity, and "body" to the finished product. The forerunner of the modern paper-making machine was invented in France by Nicolas-Louis Robert in 1799. In the early 1800's improvements were made by Henry Fourdrinier, whose name is still applied to the endless belt of wire cloth onto which the pulp is discharged—the **Fourdrinier screen.** This screen vibrates in two directions, dispersing the pulp fibers and intermeshing them as the pulp is moved continuously from the discharge box to the drying and calendering rolls at the other end. During the interval on the Fourdrinier screen most of the water drains from the pulp; a series of suction boxes near the terminal position removes all remaining free water. The incipient paper so formed is picked up by another endless belt, on which it is further dried and consolidated by heated rollers. **Sizing,** the addition of rosin or some other substance, is done to give paper resistance to the penetration of water. **Coating,** the application of materials consisting of some combination of mineral pigments and adhesives, is done to increase opacity, surface smoothness, and whiteness. Coatings give paper a smooth surface that is impervious to ink. Clay is the most common coating agent; others are chalk, titanium dioxide, and zinc oxide. The more common adhesives are animal glue, casein, starch, soybean protein, various synthetic resins, and cellulose derivatives. The final step is an intense squeezing between huge rollers of the calender, which **finishes** the paper. Finished paper emerges from the calender as a continuous sheet, usually rolled, although it may be cut to a desired size, stacked, and warehoused.

ASSOCIATED FOREST INDUSTRIES There are many forest products and industries of specialized interest. The production of tannins and naval stores (turpentine, rosin, and pine oils) have been discussed in the preceding chapter. A number of food crops, particulary nuts, but also such things as maple syrup and sugar are significant local industries in various parts of the world. Typical small industries linked to the forest are cork and Christmas trees.

Cork. Commercial cork comes from the bark of the cork oak, *Quercus suber,* Fagaceae, a subtropical evergreen. The cork oak is native to the western Mediterranean region, and is important principally to Portugal, Spain, and northwestern Africa, although it is utilized to some extent in France, in Italy and in Corsica and other islands in the Mediterranean. The bark is stripped from trees that are sufficiently old in a way that does not damage the tree; new bark grows from the uninjured cambium, and renews sufficiently within a decade to be stripped again. Cork has been harvested in this region for nearly two millennia. Forests containing cork oak cover about 5 million acres; most of these forests are natural rather than planted. Only recently have management steps been taken to control fire in these forests and to reforest select areas. Government regulation now extends to stripping, to ensure that the operation be done carefully, in a fashion to protect and perpetuate this important resource.

Commercial cork continues to find a steady world market for both century-old and new-found uses. Whether it can continue to be used for seals, gaskets, flooring, and so on will depend upon its ability to compete economically with synthetic materials, such as plastics, which are not of great importance in the rural areas where cork is produced. So far, no synthetic has been substituted completely for cork, whose isodiametric cells are like so many resilient, air-filled balloons, each coated with a wax that makes them impervious to liquids. Cork is resilient, light of weight, easy to hold when wet, and imparts no odor or flavor to substances with which it comes in contact. It is slow to catch fire, lasts almost indefinitely, is an excellent insulator, and is a nonconductor of electricity. An outstanding attribute is its chemical inertness; cork not only makes an excellent closure, but does not deteriorate in contact with most liquids, although strong alkalis or halogens do affect it.

Cork was known to the early Greek and Roman civilizations; there are references, for example, to its use as a float when soldiers traversed rivers. It has also been used as a seal for jars and casks since early times, although it remained for the invention of the glass bottle in the fifteenth century to extend this use throughout the world. There were some beginnings of cork oak cultivation in the eighteenth century, as its commercial importance became increasingly apparent. The thick bark of the cork oak adapts it well to the arid climate and desiccating winds of the western Mediterranean. The species grows about 60 feet tall and has a bole circumference of 10 feet or even greater. Generally the trees have not been pruned for a straight bole, and consequently exhibit a spreading, branching habit. Careful examination of cork bark shows that the cells are laid down in annual layers like those of wood, although cork does not have a layered appearance. The outer part of the bark, which is quite rough and fissured, must be scraped away to provide the smooth cork used for closures.

Cork bark is stripped in summer, when the cork is most easily separated from the trunk. The dead cork cells contain no sap. Circular cuts are made at the base of the tree and just below the first branches, and then a vertical slit is made the length of this section. Using special hatchets, skilled workmen pry away the cylinder of bark. They climb into the tree to remove sleeves of bark from the larger branches in the same fashion. Bark must be at least an inch thick to be worth stripping. A skilled stripper generally collects a few hundred pounds of bark per day; exceptionally large trees may yield as much as a ton of cork. Stripped bark is stacked and left to season for a few weeks before being boiled in huge tanks to soften it (incidentally removing much tannin and extraneous matter), after which the rough outer surface is scraped away. It is then dried, trimmed, graded, and baled, coming to market in flat sheets several square feet in size.

In modern economy cork is used for gaskets, seals, floats (including life preservers), nonslip walkways, handles, corkboard, floor tile, and linoleum. The manufacture of composition cork is today one of the more important phases of the industry. Pulverized cork is mixed with a suitable adhesive, pressed into large molds, and heated until cured. Corkboard made in this fashion is lightweight and nonconducting, making it an excellent insulation for coldrooms and refrigerator compartments. The original linoleum was made by bonding a mixture of ground cork, resins, and linseed oil to a burlap backing. Today many less expensive filling materials, such as wood flour, are used, with a minimum of cork.

Portugal accounts for about half of the world's cork commerce of about one-third million tons annually. Spain and Spanish Morocco account for most of the remainder. The United States frequently imports more than 100,000 tons of cork in a year, and attempts have been made to establish the cork industry in California, where the tree grows well. It seems unlikely, however, that cork can flourish as a crop in the United States, requiring as it does so lengthy a production cycle and so much hand labor in its harvest. If inexpensive imported cork is not available, it appears likely that such materials as plastic foams and improved synthetic rubbers will substitute for it.

Christmas trees. The custom of buying a small evergreen tree for ornamentation and display in the home at Christmas season has gained widespread acceptance in Christian societies. The idea was initiated during the Middle Ages in Germany, where individual families would select and harvest trees from the forest. From that simple beginning the production of Christmas trees has grown into an important crop, for which trees are planted and pruned ("sheared") to make dense growth. In the United States the Christmas-tree industry yields more than $200 million. Many species are used, preferences varying with region. Spruces are generally not favored, since they shed needles rather readily, although one Minnesota firm harvests bog-grown black spruces only a few feet tall though perhaps 30 years old. Harvested in summer, each tree is dipped in a paint-pentachlorophenol solution and the butt placed in a hollow stand containing a glycerin solution. The trees can be kept for months without desiccation or needle loss. Scotch pine, an increasingly favorite plantation-grown species usually started from seed imported from Europe, is much used in the eastern United States. Large numbers of balsam firs are imported from Canada, three or four million harvested mostly from the wild (although shearing and shaping may have been practiced). Eastern red-cedar (*Juniperus virginiana*) is used somewhat in the Southeast. Various pines and firs are used in the West, the noble fir being one of the most esteemed species.

Land that is to yield a Christmas tree crop is generally earmarked for that purpose, and is not meant to be used for future timber. One- or two-year-old purchased seedlings are generally planted with a mechanical tree planter, and weed control is assiduously practiced for the next few years. As stature is attained, trees are sheared each spring–summer to achieve pyramidal shape and good density. Measures for the control of pests, ice and snow, and fire are undertaken as necessary, and some species (such as Douglas-fir) are typically fertilized. Harvest is usually a hand operation since it is necessary to practice some selectivity as well as avoiding damage to foliage. Pines are usually harvested the seventh or eighth year; spruces and firs, the eleventh or twelfth year. After a tree is cut close to the ground, side branches will develop, of which one of the more promising can be trained for a future harvest four or five years later. Many conservation departments offer conifer stock for reclaiming waste land, and permit the stock to be harvested for Christmas trees. In some areas, where reforestation is desired, thick initial plantings will actually benefit from thinning for Christmas trees. Harvest of the Christmas trees in the early years can subsidize the cost of reforestation.

In addition to Christmas trees, greenery, often obtained from the forest, is also marketed at Christmas. Holly (*Ilex*), a traditional decorative greenery, is grown in favorable climates, such as western Oregon, where orchards are established primarily to supply decorative material for sale during the Christmas season.

RECREATIONAL PUBLIC BENEFITS Public uses of forest lands must be considered along with forestry. Recreation (Fig. 23-13), hunting, wildlife, habitat, and such are increasingly important to burgeoning civilizations, even if they cannot easily be assigned a dollar value. Nor can indirect benefits

FIGURE 23-13. *Winter sports at Bitterroot National Forest, Montana. Forests have become important recreation centers.*

from the forest, such as protection of watersheds, control of flooding, and the contribution to the atmosphere be accorded a monetary value. These may well be the most important uses of the future!

The business of forestry

The tremendous reserves of wood in tree trunks throughout the world are a capital asset that has long seemed inexhaustible. Yet in the twentieth century we are witnessing the final stages of exploitation of the forests of the temperate zone. This has been brought about by extensive colonization and ever-increasing demands for timber and pulp. Forest land has become cropland, the tree crops cut and planted or replenished every 15 to 100 years. Consequently, there are few virgin forests left except in isolated reserves.

In Chapter 15 the impact of recent changes in the forest industries was discussed. Whereas forests were once thought of as an "enemy" of civilization, to be felled and disposed of as rapidly as possible in order to permit field cropping, they are today regarded as an important raw material source to be managed and husbanded. Today even secondary species and residues formerly wasted are now utilized. Scrub land and cut-over holdings are potential sources of cellulose, now that the wood chemist has learned to transform cellulose into such products as plastics, fibers, and even food. In short, chemurgy has come to the forest and we can expect to witness economic battles over whether trees or some other crop can serve most economically as the source of cellulosic raw materials. Land suitable for agriculture will likely be pre-empted for raising food crops as populations increase, with less productive lands managed to yield forest crops.

Forest cropping is still an unsophisticated business in many parts of the world. Half of the trees felled in the world are used for fuel rather than timber, and many prime logs are wasted in slash-and-burn preparation of agricultural lands, especially in the tropics. The selective cutting and high-grading that are practiced in tropical forests leaves only cull trees and species of lesser value to perpetuate the forest. Remoteness, lack of ready market, and a lack of appreciation for conservation perpetuate this wastefulness. Even in technically advanced parts of the world, forest resources are not always efficiently employed. Witness what is happening in the midwestern United States, where the few remaining stands of mature white oak are often cut for firewood and where sizable trees are girdled in local woodlots to increase graze, while lumber is transported from western states for repair and construction on the farm.

But the trend is unmistakable: the forest products industry is becoming more businesslike and conscious of overall values, especially in technically advanced parts of the world. Research, lavish by former standards, is sponsored by the large and well-capitalized firms in the forest products industry, as well as by public agencies. Projects range from practical investigations on tree breeding, forest fertilization, pest control, fire prevention, and selective weeding to fundamental studies on the influence of the forest on the watershed, weather and climate, snow melt and water yield, energy requirements of various species, and so on. Forest management increasingly leans upon research, and its practices are tempered by overall economic considerations. Unfortunately, enlightened forest management is apt to be slow in coming to the Amazon Valley, central Africa and Southeast Asia, where immense tropical forests still stand.

THE INTEGRATED FOREST PRODUCTS INDUSTRY Nowhere is the change from a "cut and move on" forest philosophy to a firmly established business base more manifest than in the integrated, business-managed paper and timber corporations that today dominate the forest products industry in North America. Massive investment demands long-range planning for sustained yield. The chart on pages 602 and 603 depicts the diversity of operations in a modern integrated forest products corporation. Net sales of such a corporation exceed a billion dollars annually, and its properties are capitalized at more than a half billion dollars.

Much company forest land is acquired outright, inventoried, managed, and improved by modern techniques. This particular company has several plants in the Pacific Northwest, where it cuts both private and government timber, stressing complete utilization of the harvest (peeler logs for veneer, saw logs for lumber, small logs and veneer cores for studs, mill residues such as chips, sawdust, and planer shavings for pulping, and bark for garden mulch), all processed at a single location. It maintains tremendous automated log-handling equipment and storage yards. In the East it has private holdings from Maine to Florida, where professional foresters manage its forests on a sustained-yield basis. Its timberland holdings in North America alone run into the millions of acres, primarily in the three major coniferous forest belts (Northwest, Northeast-Great Lakes, and Southeast). As part of its public relations, it permits public use of some of its timberland for recreation. Overseas it has investments in Latin America, Europe, Africa, and Australia. It sponsors its own research and development centers, which look into lower-cost higher-yielding pulping processes, higher grade use of chemicals derived from the noncellulosic components of the tree, better manufacturing and finishing processes. Computers are used to monitor and analyze data and provide accurate operating information. By-products carry over into the food field (yeast production), and into chemicals for plastic coatings and bondings (for waterproof wrappers and all-weather plywood). Large land holdings must be managed on a sustained-yield basis. This is a far cry from the small operations of family farms and private timber holders of the past.

THE CHANGING FORESTRY PICTURE In many areas forestry has begun to change from being concerned mostly with husbanding and protecting (for example, fire prevention) to being more concerned about a future when all forests, like today's farm plantings, will be an intensively managed crop. In Europe and in the United States this change is already manifest. Elsewhere, especially in the vast, largely unexploited tropical forests, enlightened

Box 23-3 IMPEDIMENTS TO HIGH-YIELD FORESTRY

As the demand for prime timber continues to mount, the forest products industry dips more and more into its remaining capital of virgin or old-growth timber. Foresters assume that increment can be brought into balance with depletion by utilizing highly productive management techniques and genetically superior trees. Draw-down of existing forest is predicated upon the assumption that an era of high-yield forestry lies ahead. Studies are made of potential productivity of new inbred and clonal lines, requirements for quick regeneration in all sorts of habitats, and the stimulation that can be expected from fertilization, thinning and efficient harvesting under foreseeable economic and mechanization conditions. All these variables are fed into the computer, which suggests an acceptable harvesting rate.

Immense regional "programs" such as "third forest" expectataions for the Southeast have been launched: (the first forest was the virgin wilderness that yielded a trillion board feet of lumber; the second forest was that which regenerated naturally after the virgin forest was felled, and which has supplied most of the region's needs for the past century; the "third forest" is to consist of tailor-made plantings for enhanced crop-like yields in this productive climate). In these projections, their proponents visualize millions of superior trees, planted and grown by revolutionary means, eventually logged by gigantic machines that mop up the slash after harvest and set out thousands of new seedlings per hour (these perhaps even protected by "mini-hothouses" which soon disintegrate). Fertilization, stand density, pest control, and so on are all optimized.

An intensive program, yes, but what of the problems resulting from such a completely regulated ecological situation? A switch to intensive tree monoculture should have consequences similar to those encountered in the cultivation of annual crops (Chapter 15).

One conflict that has already surfaced is that initiated by the conservation-minded, who question the wisdom of converting most natural forest to a cropping system likely to diminish esthetic enjoyment, wildlife habitat, and recreational values. No matter how excellently the trees are growing, a "walk in the woods" loses its savor beneath unvarying trees all in a row, so obviously the ward of man. Gordon Robinson, staff forester for the Sierra Club states, "Biologically the forest that is most secure against fire, insect and disease is one with a variety of ages and species." He prefers selective cutting of individual trees as they reach their prime, and contends that the cutting of old trees in anticipation of a new harvest that may never result is an exercise in brinksmanship.

In the "third forest" area, seed from the select bloodlines intended to repopulate the forest has for several years diminished in yield and germination. The exact cause of this appreciable reduction of quality seed as the seed-bearing trees grow older is not entirely clear. It seems not to be disease, and is probably not insect damage. Some experts believe that, as the seed trees grow large enough to produce pollen abundantly, greater and greater inbreeding occurs (with non-clonal pollen less and less represented in ovule fertilization). This may reduce seed vigor. Will it become necessary to control pollination of those "super" trees somehow, and if it can be done, how costly will it be?

forest management is still many years away. But more and more, attention is being given to fitting the forest to overall land use patterns. There is much interest in practical management techniques, of course, to keep forest production efficient and wood competitive. Progress and exciting change are evident in many fields, including tree improvement, fertilization, and pest control.

Tree improvement. Because of the prolonged cycle for tree generation, 25 years or more for some species, it is no wonder that tree breeding has generally lagged behind breeding of annual crops. There have been some notable exceptions, as with rubber and some fruit crops (peaches, for example). Here the emphasis has been upon selection of naturally occurring superior genotypes that are then perpetuated vegetatively. In many countries (the United States, for example) the finest forest trees have been harvested, leaving the culls and unwanted types behind to reproduce and

ST. REGIS PAPER COMPANY'S PRODUCTS AND PLANTS

PRINTING PAPERS
Publication and commercial printing papers.

PULP AND PAPER MILLS
Bucksport, Maine
Deferiet, New York
Harrisville, New York
Sartell, Minnesota

Technical assistance is rendered under contracts with paper manufacturers in France and Germany in connection with on-machine coated printing papers.

KRAFT PAPERS
Bleached and unbleached kraft pulp, paper and board. Coated and uncoated kraft food board. Specialty papers.

PULP AND PAPER MILLS
Jacksonville, Florida
Pensacola, Florida
Tacoma, Washington

PAPER MILLS
East Pepperell, Massachusetts
Carthage, New York
Herrings, New York

St. Regis owns 50% interest in North Western Pulp & Power Ltd. which produces bleached sulphate pulp in its mill at Hinton, Alberta, Canada.

St. Regis owns 20% interest in Companhia Industrias Brasileiras Portela which has a paper mill at Jaboatao, Brazil.

St. Regis owns 22% interest in Tennessee River Pulp & Paper Company which is building a kraft pulp and board mill at Counce, Tennessee.

SPECIALTY PAPERS
Glassine and greaseproof papers. Waxing papers. Foil mounting papers. Reproduction papers.

PULP AND PAPER MILL
Rhinelander, Wisconsin

PAPER MILL
Columbus, Ohio

St. Regis owns 50% interest in R-W Paper Company which produces glassine and greaseproof papers in its mill at Longview, Washington.

CONTAINERBOARD
(Other than Kraft)
Folding and set-up boxboard. Corrugating medium. Chipboard. Specialty paperboards.

BOARD MILLS
Coshocton, Ohio
Milwaukee, Wisconsin
Cornell, Wisconsin

St. Regis owns 49% interest in Fome-cor Corporation which produces Fome-Cor board in plants at Addyston, Ohio and Mt. Wolf, Pennsylvania.

PLASTICS
Industrial and decorative laminates. Extruded, vacuum-formed, injection and compression-molded parts and assemblies. Plastic pipe.

LAMINATING PLANTS
Trenton, New Jersey
Kalamazoo, Michigan

MOLDING PLANTS
Cambridge, Ohio
Dexter, Michigan

EXTRUDING AND VACUUM-FORMING PLANT
Richmond, Indiana

Panelyte plastics are also produced by licensees in Great Britain, Germany, France, Spain, Italy, Sweden, Brazil, Colombia, Australia, Japan and Union of South Africa.

LUMBER
Dried and planed lumber. Plywood. Doors. Siding. Moulding. Flooring. Timbers. Treated poles and posts. Railroad material.

SAWMILLS
Libby, Montana
Troy, Montana
Klickitat, Washington
Tacoma, Washington

PLYWOOD PLANTS
Tacoma, Washington
Olympia, Washington

DOOR PLANT
Tacoma, Washington

BY-PRODUCTS
Crude tall oil. Sulphate turpentine. Torula yeast. Smoked yeast. Calcium lignosulphonate. Pres-to-Logs. Waste-wood briquets.

TALL OIL PLANTS
Jacksonville, Florida
Pensacola, Florida

YEAST AND CHEMICAL PLANT
Rhinelander, Wisconsin

WOOD BY-PRODUCTS PLANT
Libby, Montana

BAGS
Multiwall bags.
Small bags for sugar, etc.
Slit cases (BAX).
Semi-rigid containers.
Burlap and cotton bags.
Open-mesh and specialty bags.

BOXES
Corrugated boxes.
Folding boxes. Set-up boxes.
Solid fiber boxes. Wooden boxes.
Wire-bound veneer boxes.
Display containers.
Metal-edge and metal-stay boxes.

WRAPPERS
Bread wrappers and end labels.
Frozen food wrapers.
Printed waxed paper,
cellophane and
polyethylene wrappers.

PACKAGING MATERIALS
Polyethylene-coated papers.
Corrugated sheets.
Single-face corrugated.
Gummed papers and tapes.
Film and foil laminations.
Wire ties and tying tools.

MULTIWALL BAG PLANTS
Nazareth, Pennsylvania
Franklin, Virginia
Savannah, Georgia
Louisville, Kentucky
Pensacola, Florida
Toledo, Ohio
Kansas City, Missouri
Houston, Texas
Salt Lake City, Utah
Tacoma, Washington
San Leandro, California
Los Angeles, California
Ponce, Puerto Rico (60% Ownership)

INTERNATIONAL
Tubize, Belgium
Buenos Aires, Argentina
Soa Paulo, Brazil
Recife, Brazil
Belo Horizonte, Brazil
(under construction)

TEXTILE BAG PLANTS
Houston, Texas
Lubbock, Texas
Salt Lake City, Utah
Kansas City, Missouri

Multiwall bags are also produced by
licensees in the United States, Great
Britain, France, Spain, Holland,
Germany, Eire, Austria, Norway,
Sweden, Italy, Greece, Union of
South Africa, Indonesia, Australia,
New Zealand, Mexico, Colombia
and Peru.

CORRUGATED BOX PLANTS
Cohoes, New York
Buffalo, New York
Mt. Wolf, Pennsylvania
Pittsburgh, Pennsylvania (2 plants)
Hagerstown, Maryland
Atlanta, Georgia
Jacksonville, Florida
Birmingham, Alabama
Grafton, West Virginia
Canton, Ohio
Cleveland, Ohio
Coshocton, Ohio
Newark, Ohio
Chicago, Illinoise (2 plants)
Milwaukee, Wisconsin
Dubuque, Iowa
Houston, Texas
Garland, Texas
Dallas, Texas
Salinas, California
Fullerton, California
Tacoma, Washington

FOLDING AND SET-UP
BOX PLANTS
Cleveland, Ohio
Marshall, Michigan
Birmingham, Alabama
Pensacola, Florida
Milwaukee, Wisconsin
Chicago, Illinois
Dubuque, Iowa
Dallas, Texas (3 plants)
Forth Worth, Texas
Mineral Wells, Texas

WOODEN BOX PLANTS
Savannah, Georgia
Jackson, Mississippi
Maringouin, Louisiana
Libby, Montana
Klickitat, Washington

Corrugated, folding and set-up
boxes are produced in South
Africa by companies, partially
owned by St. Regis, that operate
seven plants at Johannesburg,
Capetown, Port Elizabeth,
Maitland and Durban.

CONVERTING PLANTS
Atlanta, Georgia
Birmingham, Alabama
Columbus, Ohio
Middletown, Ohio
Dallas, Texas
Houston, Texas
Seattle, Washington
San Jose, California

CONVERTING PLANTS
Pensacola, Florida
Troy, Ohio
Chicago, Illinois

WIRE TIE PLANT
Cleveland, Ohio

PACKAGING MACHINERY
Bag-making machines.
Bag-filling and closing machines.
Automatic scales.
Packaging machines.
Box-making and closing machines.

OTHER PAPER PRODUCTS
Envelopes.
School supplies.
Shipping labels and tags.
Paper plates.

MACHINE SHOPS AND
ASSEMBLY PLANTS
East Providence, Rhode Island
Emeryville, California

Packaging machinery is also
produced by licensees in Great
Britain, France, Holland, Germany,
Austria, Sweden, Italy, Australia.
Denmark and Japan.

CONVERTING PLANTS
Detroit, Michigan
Marion, Indiana
Birmingham, Alabama
Des Moines, Iowa
St. Louis, Missouri
Houston, Texas
Pico Rivera, California

dominate. Much genetic degradation occurs as this process is repeated over time.

Modern tree improvement is the empirical application of genetic principles to the improvement of forest trees. At present, most practical tree improvement is based on phenotypic rather than genetic information. But forest geneticists are rapidly accumulating an abundance of genetic information about trees.

Unlike fruit trees, which are usually propagated vegetatively, most forest trees are propagated by seed. Where efficient means of vegetative propagation can be achieved for forest trees, improvement programs will be geared to isolating unique genotypes, as is the practice for most fruit tree breeding. This is now true for improvement programs in *Populus* species (aspen, cottonwood) in the United States and Europe and for *Cryptomeria* in Japan. Where seed propagation should prove to be the only practical alternative the improvement program must be geared to the identification and production of superior seed trees. This is slow work because progeny testing should proceed until maturity, or at least until evidence can be obtained that the performance of young trees can be used to predict the productivity at normal harvestable age. Thus the extent to which hybridizing, backcrossing, inbreeding, ploidy manipulation, and so on, can be practiced usefully with forest trees remains to be determined for most species. There is no doubt that in dealing with anything so bulky and slow to reproduce as forest trees, there are formidable obstacles to merely making available a suficient amount of the proper habitat to permit a comprehensive breeding program. Nevertheless, some of the sophisticated techniques developed for annual crops (see Chapter 19) may well prove adaptable for forest tree crops. Improvement similar in magnitude to those achieved by hybrid corn can be expected (Fig. 23-14), but speedy progress seems unlikely until new seedling evaluation techniques are developed.

Mapping the world's soils and climates has progressed far enough so that we have a pretty good idea of where to look in the world for potential stock adaptable to a given location. The Chinese chestnut, *Castanea mollisima*, has already been introduced into the eastern United States as a blight-resistant species, and its resistance can be transmitted to offspring of crosses with the native chestnut, *C. dentata*. It may be possible to introduce a disease-resistant clone in time to permit natural crossing, with the hope that some progeny may exhibit disease-resistance. More needs to be known about interspecific heterosis in trees. This should be a continuously rewarding field of investigation, in spite of the awkwardness of handling tree populations.

Already some seed is produced in orchards developed by forest geneticists. When seed of a high, uniform quality is produced in quantity, much of it will go into the national and international market. A guarantee of quality is therefore necessary to the buyer. For field crops, certification

FIGURE 23-14. *Hybrid vigor in pine. Large tree on left is a Jeffrey-Coulter pine backcross,* Pinus jeffreyi × (P. jeffreyi × Coulteri); *tree on right is* P. jeffreyi. *Both trees are 13 years old.* [*Courtesy U.S. Forest Service.*]

programs have proved to be of great value in keeping the sale of undesirable seed to a minimum (see Chapter 9). The same device, certification of seed by a public agency, is being used in some states in the United States to insure quality. It seems likely that all forest tree seed in international commerce may eventually be certified as to source (progeny-tested parents, seed orchards, isolated open-pollinated trees, or merely selected trees).

Seed collected from superior parents for seeding nurseries and for restocking is open-pollinated, and the progeny is therefore variable. However, off-type seedlings can be eliminated in the nursery. Such factors as rapidity of seedling growth, suitability to climate and cultivation, even resistance to disease, can often be noted in a relatively few years, making it possible to avoid the long wait until maturity to judge these characteristics. Species that do mature sufficiently fast to produce seed in a few years offer the opportunity for controlled crossing analogous to that undertaken for herbaceous crops. Lending urgency to programs for improving tree stock is the need to incorporate disease resistance in once-important species now sadly decimated by introduced diseases, such as the American chestnut (chestnut blight, caused by *Endothia parasitica*), white pine (blister rust, caused by *Cronartium ribicola*), and the American elm (Dutch elm disease, caused by *Ceratostomella ulmi;* and phloem necrosis, caused by a virus). Poplars, oaks, maples, locusts, sycamores, and many other genera, too, are today threatened by serious disease or insect afflictions, both native and introduced. It is frequently impossible to control epidemics in the forest, and even if possible the cost may be exorbitant. The advantage of breeding resistant stock for reforestation is obvious.

Although the tree breeder experiences certain disadvantages compared to the breeder of annual farm crops, there may be compensations. An exactly uniform plant population may not be so important in the forest as it is with crops grown in more restricted areas and which must be all harvested simultaneously. Thus the advantages of heterogeneity and variability can be utilized, lessening the chance of a forest population being suddenly wiped out by a new affliction.

Systematic screening of tree introductions has been nowhere so intensive as the screening of herbaceous plants. There would seem to be considerable possibility of using exotic trees in many parts of the world, in locations where they are well adapted and perhaps resistant to prevailing insects or pathogens.

Forest fertilization.   Until recently there has been little reason to encourage more rapid growth through fertilization. Moreover, timber has been traditionally a bulky material, and it has been doubted whether the returns from fertilization would justify the cost. Other than on orchards and plantations, relatively little tree fertilization is undertaken. But as economical, concentrated fertilizers and mass means for their dispersal are developed, forest fertilization may in many instances become remunerative.

In the United States, forest tree fertilization has generally been limited to nurseries and occasionally new field plantings, although some fertilization with urea has been practiced from the air. In Scandinavia the practice of fertilizing the forest from the air seems to be growing rapidly. With forest products so important to the economies of Sweden, Norway, and Finland, all are feeling pressure to increase the output of wood products in order to remain fully competitive in world markets. If Sweden is to realize its goal of raising pulp production from about 90 million tons to 250 million tons, forest fertilization will certainly be required. About 200,000 acres of forest was fertilized from the air in 1966, mostly with pelleted urea applied at a rate of about 200 pounds per acre. Fifty thousand acres were similarly fertilized in Finland, with plans projected to cooperatively fertilize more than a million acres in the next few years. Phosphate is utilized on the damp peaty soils in Finland, as well as urea on the drier lands. It is recommended that the forest be fertilized every 5 years. Swedish foresters estimate a profitable annual return on the outlay in fertilizing the forest, a practice apt to become more commonplace in other parts of the world where mineral nutrition is a limiting factor.

Forest pest control.   Control of pests in the forest has traditionally been slow and expensive, because of the immense area to be covered and the difficulty of reaching inaccessible parts. Some-

606

FIGURE 23-15. *Spraying a young western white pine stand with an antibiotic fungicide to arrest damage caused by white pine blister rust.* [*Courtesy U.S. Forest Service.*]

thing on the order of 58 million dollars has been spent over the years on government programs to eradicate wild gooseberry by hand; the plant is the alternate host for white pine blister rust in the northern coniferous forests of the United States. Now with slow, stable, low-flying aircraft available, spraying of the forest is becoming commonplace. In the United States large-scale spraying programs carried out by airplane or helicopter have been limited to extensive acreages away from croplands (Fig. 23-15); herbicides have been used for such purposes as killing the hardwood overstory in the southeastern coniferous belt to encourage growth of young pines, and pesticides have been used to stop insect epidemics in remote northern or western forests.

A large business has been developed around "brush killing" along roadsides and powerline rights-of-way with 2,4-D, 2,4,5-T, and other herbicides. In general, the intention has been to eliminate all vegetation rather than to selectively remove certain species. Proximity of farmland often makes air application unfeasible, and drift, contamination of runoff, and possible air pollution are becoming matters for concern. Ground-spraying

with low-volatility materials permits better pinpointing of the treatment. Soil applications of such herbicides as picloram are often used, lethal to many weed species in dosages as low as a few ounces per acre. Other sterilants and nonselective herbicides are often used along fence rows, around buildings, under guardrails, and so on. Sometimes sufficient labor is available for frilling (partially girdling the tree and applying the herbicide to the girdle), basal treatment (hand spraying with a portable sprayer, the herbicide usually in an oil carrier, made around the tree base or stump, typically to prevent resprouting), or even injection of herbicide into the tree for systemic effect.

ORNAMENTAL AND HORTICULTURAL SPECIALTY CROPS

Many crops are employed chiefly for esthetic purposes—for beautification rather than for food, fiber, or timber. It is difficult to decide, sometimes, whether a nut- or fruit-bearing tree, for example, is more important for its yield or to the landscaping! But, in general, any noncommercial plantings, especially around the home, are regarded as ornamental (Fig. 23-16).

Economically, ornamental plant industries "sit at the second table" compared to the food crop industry; ornamental plants are one of many luxuries enjoyed by affluent societies, where basic needs are easily filled and where there is leisure time as well as surplus money for the enjoyment of plants. Today North America and Europe constitute the main commercial market for horticultural specialties. Individual properties throughout the world enjoy ornamental plants; in many parts of the world home plantings also have a practical bent, yielding fruit, for example, along with beauty. Japan is an outstanding example of a country in which the use of ornamental plants has become highly specialized and where landscaping is meticulously kept and often designed for miniaturized gardens. Throughout the tropics ornamentals are more casually handled, often being locally propagated and passed from party to party rather than supporting a significant commercial operation.

In the United States the sale of ornamentals has achieved a high order of importance, linked with extensive population moves from city and farm to the suburbs. The value of all horticultural specialties has been estimated at 1 percent or more of all farm crops; about 1 percent of the gross national product is spent in the "outdoor living" market. In some regions the retail value of horticultural specialties averages 0.7 percent of total retail trade. The lion's share of the "outdoor living" market goes to tools, power equipment, outdoor furniture, and recreational devices associated with ornamental landscaping. Every year, more than 50 million home gardeners, tending nearly 10 million acres, directly consume perhaps 150 million pounds of lawnseed, and more than a billion dollars worth of "green goods" (bedding plants, bulbs, deciduous shrubs, evergreens, perennials, shade trees, and so on). Related sales of fertilizers, soil amendments, and pesticides amount to nearly as much.

Ornamentals

Horticultural specialties used for home landscaping and plant materials used for outdoor display (such as along the burgeoning highway system) are

FIGURE 23-16. (Above) In the garden of this suburban home native trees, introduced ornamentals, and well tended lawn are combined for pleasing landscaping effect. (Below) Ornamental plantings are also considered vital for cities. [Above photograph courtesy The Lawn Institute; lower photograph courtesy American Association of Nurserymen.]

here grouped with ornamentals. This is primarily a "discretionary spending" market, apt to vary appreciably depending upon whether times are "good" or "bad." Since the end of World War II use of ornamentals has expanded rapidly. The industries dealing in ornamentals are consumer-oriented, with high markups, promotional advertising, appeal to fashion and status, and impulse buying. Competition is keen. Most industries have specialized either in production or in merchandising, but competitive pressures have brought both specialization and integration into large operations to better enable them to capitalize the extensive and complicated activities required.

THE SEED INDUSTRY   During the twentieth century the farm seed industry has evolved into two segments, one primarily supplying the home gardener, the other supplying the farmer. The home garden seed industry has divided into two specialties, one supplying seeds for flowers and vegetables, the other supplying seed for lawns and fine turf. The market for home garden seeds has not grown greatly, since the move to surburbia brought less rather than more need for home food production; even annual flowers often are purchased from the local nurseryman as seedlings rather than started at home. The lawnseed market, however, has grown extravagantly. A lawn is fundamental to the landscaping of many a suburban home, whether the seed is planted in situ or purchased indirectly as sod. A sizable proportion of the lawnseed consumed is used for rejuvenating deteriorated turfs. New demands have been created for seed; for example, plans for the extensive interstate highway system call for nearly 30 acres of landscaped right-of-way per linear mile.

The lawnseed industry.   Grass seed was once supplied by the field seed industry, the main concern of which is forage grasses. Right after World War II demand arose for new quality standards and special turf cultivars solely for lawns. Today highly specialized farmers, utilizing up-to-the-minute production techniques (for example, weed control with new herbicides, and maintenance of genetic identity in select cultivars) produce lawnseed essentially free of weed and contaminating "crop" seed.

The most important lawn grasses grown from seed in the United States are the Kentucky bluegrasses, *Poa pratensis;* the fine or red fescues, *Festuca rubra;* bentgrasses, *Agrostis,* in several species; and Bermudagrass, *Cynodon dactylon.* The first three are so-called cool-season grasses, well adapted to the northern two-thirds of the United States. Bermudagrass is adapted from Tennessee southward, and in the deep South, near the Gulf of Mexico, bahiagrass, *Paspalum notatum,* carpet grass, *Axonopus,* and centipedegrass, *Eremochloa ophiuroides,* are also planted from seed. A limited amount of *Zoysia* seed is available, genetically mixed because of sexuality. (Many grasses are apomictic.) Special cultivars and hybrids of both zoysia and bermuda are propagated vegetatively, with plugs, sprigs, and sod sold through nursery-type outlets. Another lawngrass that is widely used in the South is st. augustine grass, *Stenotaphrum secundatum;* because it produces few seeds it is marketed entirely as vegetative material.

Not very many years ago most lawnseed was secured from pastures and meadows left ungrazed and unmowed through the seeding season. To a great extent bahiagrass is still so produced in Florida and Texas; limited amounts of natural Kentucky bluegrass may be produced in this way in the Midwest, and fine fescues in the Peace River country of Canada. Mostly, however, select cultivars are agronomically grown in fields free of genetically unlike strains. Once sufficiently large and steady markets for lawnseed became available, such intensive growing proved more profitable than gathering seed from pasture and meadow. The sizable plantings in the Pacific Northwest, heavily fertilized for high yields and carefully tended to avoid weeds and competing vegetation, justified their high capitalization once the demand for select cultivars of lawnseed was large enough. The transition from the harvesting of "natural" Kentucky bluegrass in Kentucky and other sections of the Midwest to agricultural production in the Northwest occurred in less than a decade beginning in the late 1950's. A resumé may be of interest, for the economic pressures that have affected bluegrass mark a characteristic trend.

Kentucky bluegrass seems to have been introduced from Europe by the first North American colonists, perhaps as chance seed in hay and cattle bedding aboard ship. Wherever the soil was turned, newly arrived Kentucky bluegrass volunteered. Evidently it was spread as far west as Illinois by the early French missionaries, and was well established in the Ohio Valley by the 1800's. Early in the nineteenth century it gained the name "Kentucky" bluegrass because it flourished so well on the rich phosphatic soils of north-central Kentucky. By the 1900's it had spread westward into Missouri and northward into Canada, being utilized chiefly for pasture and hay.

It was from such fields that Kentucky bluegrass

lawnseed was harvested. There is much to recommend this grass; natural selection should have proven out the diverse ecotypes found in every field, which when blended for lawnseed provide a widely adaptable genetic base. In a sense such a seed crop is subsidized by the livestock industry, since after seed harvest in June or early July the bluegrass is grazed or cut for hay until the next May. But savings in grass maintenance are now more than offset by increased harvesting and handling costs. Instead of harvesting bluegrass from scattered pastures and meadows, it is more economical to grow it strictly for seed under intensive management that gives yields of as much as a thousand pounds of seed per acre, compared to an "excellent" yield of 200 pounds per acre from pasture or meadow. Moreover, the soils of Oregon and Washington, virgin to bluegrass, permit easy maintenance of genetic purity now that proprietary cultivars have become important.

Special stripping machines harvest natural Kentucky bluegrass in the Midwest (Fig. 23-17). These have to be capitalized for but a few weeks of activity during the harvest season. In the "western district," from Missouri northward into Canada, portable stripping machines have been used, these are moved (at considerable expense) from location to location as the seed ripens and contracts for its gathering are negotiated. Little wonder that sedentary production in the Pacific Northwest is favored, where conventional farm machinery can be used, and be capitalized for diversified and year-round activity rather than sitting idle most of the year. Today most lawnseed is forced to maximum yields by heavy fertilization, and even if lodging occurs, it is lifted, cut, windrowed, and field combined at appropriate seasons (Fig. 23-18).

Where fine lawnseed is grown agronomically it is sometimes feasible to establish local cleaning plants to remove chaff and contaminating seed and thus avoid the expense of shipping rough seed. The larger farms even build their own seed-processing equipment. The end result is an integrated operation for the particular area, from planting to finished seed. In a few favored regions, established channels of trade and experienced farmers and handlers make lawnseed growing exceedingly efficient. This small specialty crop has become established in these regions not only because growing

FIGURE 23-17. *Natural Kentucky bluegrass seed harvest in Kentucky. These seed stripping machines are pulled by a tractor.* [*Courtesy The Lawn Institute.*]

conditions are favorable (weather, soil, rainfall), but because of human and economic factors. It has become, on the whole, less expensive to produce quality seed in Oregon and Washington and ship it all the way across the country to eastern markets than to harvest seed from eastern fields.

The garden seed industry. Although many large seed firms have both garden seed and lawnseed divisions, the respective operations are not alike, and the products usually marketed separately. Garden seed is produced mainly in the western states—most flower seeds in California, much vegetable seed in Idaho. The main market, as with lawnseed, is in the more populated eastern states. More so than with lawnseed, cultivars of garden plants must be grown in isolation to maintain their distinctive characteristics. The production of hybrid garden seed has increased rapidly in recent years, partly for the same reason that hybrid corn proved more profitable to the seedsman than did open-pollinated seed (that is, exclusive control through inbred lines). Generally, F_1 hybrids are superior because of heterosis, and the control of parent lines assures premium prices for several years. The industry has initiated a number of merchandising flairs, such as the annual "All-American" selections of new varieties. It is stand-

FIGURE 23-18. *Combining Highland bentgrass in Oregon for lawn seed.* [*Highland Bentgrass Commission.*]

ard practice to allocate a percentage of the profits from new All-American originations for promotion; competing firms must pay a royalty for the privilege of also offering these new cultivars.

Seed production requires a great deal of technical competence, as well as meticulous attention to planting time, controlled pollination, and customized harvesting techniques for each crop. The flower seed industry centers in the interior valleys of California, where a mild climate with little rain during the growing and harvesting seasons permits fairly certain production and harvest under irrigation. It was not until World War I, however, that garden seed production in the western United States gained much momentum; prior to that time much seed was imported from Europe, some of it poorly adapted to the United States. Today domestic garden seed is produced by a relatively small number of firms, most of which are rather large; each has its own research department for breeding new cultivars, widespread trial grounds, and a corps of experts. Once a new cultivar is considered to be ready for market it may be grown under contract by independent farmers of the area. By the time a new cultivar is ready to be marketed, the investment in research, in building up supplies of foundation seed, in storage, handling, and so on, is appreciable. Moreover, the investment is staked on presumed but uncertain market acceptance. The risk is apparent, should public taste suddenly change, or some unforeseen circumstance endanger a lengthy commercial life for the new cultivar.

Most of the home garden seed traditionally has been sold as packet seed, from catalogs. (This, however, is a small portion of the vegetable seed business.) Several large firms specialize in this activity, one vying with another to see who can produce the most attractive, superbly illustrated catalog each year. The highlight of the winter season for many gardeners is to receive the attractive catalogs, which are increasingly expensive for the seed houses to produce and account for much of the overhead. Seed houses operate under two economic disadvantages. One is that seed unsold at the end of the season is generally destroyed, since by law there is a time limit to the germination guarantee (it is obviously impractical to re-work small lots of seed in sealed packets). The other is that the sales season lasts but a few months at the most, usually oriented to spring planting.

Since World War II the garden seed industry has increased the use of glass and screened structures, particularly for production of high-value items such as F_1 hybrid petunias. Controlled crossing can be better regulated under technically precise conditions. It is obviously necessary to screen out insects that may contaminate a cross with outside pollen, and precise emasculation of perfect flowers is more easily handled on benches than in open fields. Because of the high cost of labor much hybrid seed production is subcontracted to Japanese firms.

The processing, storage, and testing of both lawn and garden seeds is an exacting and highly technical procedure. Seed-testing laboratories manned by registered seed technologists operate under precise rules governing each species in order to provide the guarantees for germination and purity that are the basis for trade. In general seed remains viable longest if kept cool and dry; this is often best accomplished by keeping the seed in sealed containers in a cold room. Government control officials constantly sample seed supplies on dealers' shelves to determine whether the seed meets the purity and germination claims on the label.

Growers typically overplant somewhat, to assure ample supply regardless of weather. This is not entirely unselfish, since markets once sacrificed are regained with difficulty. Large firms generally hedge against the weather by growing seed in several locations. The entire genetic stock of prized cultivars is seldom risked in a single planting, in a single year. In fact so valuable is the germ plasm, the real wealth of the industry, that seed stocks are usually stored in fireproof vaults, to be withdrawn from time to time only as a field planting demands. Touchy items such as asters, onions, and parsnips, are generally packaged in special foils, plastics, or cans to control humidity.

VEGETATIVE MATERIALS AND THE NURSERY INDUSTRY Living ornamental stock is fundamental to landscaping. Seasonal demand for corms, bulbs, and rhizomes, and for herbaceous perennials, has been steady if modest. Even more important to home landscaping is shrubbery; evergreens (principally yews, junipers, and arborvitae among needle types; azaleas, hollies, *Mahonia*, boxwood, and so on among broad-leafed types) constitute the chief segment of the nursery industry. Production of select forms of deciduous shrubs and shade trees has enjoyed rapid expansion in recent years, what

with the native forest almost gone and needing replacement in the mushrooming subdivisions, and with the roadside beautification program in full force. Because of their tremendous popularity, the production and sale of roses is almost equally as important, and has been settled upon as a specialty by several national firms. The offering of vines, ground covers, bramble and tree fruits, perennial "vegetables" (such as asparagus and rhubarb) rounds out the ornamental nursery trade.

Bulbs. In the horticultural trade the word "bulb" is used for corms, tubers, rhizomes, and swollen roots as well as true bulbs. Thus tulips, hyacinths, narcissus, and even iris, daylily, and dahlia fall in this category. There is limited trade in nonhardy bulbs for planting indoors or outdoors in summer only. Included are amaryllis (*Hippeastrum*), *Anemone*, various tuberous begonias, *Caladium*, *Canna*, *Clivia*, *Dahlia*, *Freesia*, *Gladiolus*, *Hymenocallis*, *Montbretia*, *Ranunculus*, *Tigrida*, tuberose (*Polianthes*), *Zephyranthes*, and a few others. Hardy bulbs generally left in the soil outdoors include various crocuses (*Colchicum* and *Crocus*), *Fritillaria*, snowdrops (*Galanthus*), hyacinths, lilies, daffodils (*Narcissus*), *Lycoris*, *Scilla*, tulips, iris, and several others.

Many species of these genera are of Old World origin, introduced into horticulture long ago, subjected to unrecorded selection and crossing through the years to yield the many modern cultivars. One of the most popular is the tulip, *Tulipa*. Tulips are widely grown in gardens as botanical species, but are especially prized in select forms of the garden tulip (which arose from crosses between thousands of cultivars representing several species). Garden tulips are roughly grouped as early tulips, breeder's tulips, cottage tulips, Darwin tulips, lily-flowered tulips, triumph tulips, Mendel tulips, parrot tulips, and so on. The garden tulips seem to have been developed first in Turkey, but were spread throughout Europe and were adopted enthusiastically by the Dutch. Holland has been the center of tulip breeding ever since the eighteenth century, when interest in the tulip was so intense that single bulbs of a select type were sometimes valued at thousands of dollars. Holland remains today the chief source of tulip bulbs planted in the United States; nearly 200,000 are imported annually. Holland has also specialized in the production of related bulbs in the lily family, and provides tens of thousands of hyacinths, narcissus, crocus, and others to the United States market annually. The Dutch finance extensive promotion of their bulbs in the United States to support their market.

Years of meticulous growing are required to yield a commercial tulip bulb from seed. Thorough soil preparation, high fertility, constant weeding, and careful record-keeping are part of the intensive growing in Holland, employing much hand labor. Bulbs sent to market meet specifications as to size and quality, which assure at least one year's bloom even if the bulb (having been subjected to cold induction) is supplied nothing more than warmth and moisture. The inflorescence is already initiated, and the necessary food stored in the bulb. Under less favorable maintenance than prevails in Holland, a subsequent year's bloom may be smaller and less reliable; it is not surprising that tulip bulb merchants suggest discarding bulbs after one year's performance and replanting with new bulbs. For featured display, such as at botanical gardens, this is the general practice, although there is no reason why the homeowner cannot derive several years of reasonable satisfaction from a single bulb planting if he allows maturation of the tulip foliage after flowering.

Bulbs generally grow best when planted fairly deeply (several times the vertical dimension of the bulb itself) in friable, loamy soil supplied with phosphatic fertilizer. Autumn, a few weeks before freezing weather, is the recommended planting time for tulips, hyacinths, daffodils, and crocus. The bulbs have a chance to root and become established, and will flower in spring as soon as the soil warms. After the foliage has died back it is often recommended that the bulbs be dug up and reset, a practice that is of some value for rearranging the planting pattern and determining which bulbs are in good condition if nothing else. Some bulbs, such as daffodils and hyacinths, proliferate; digging them up provides additional stock.

Herbaceous perennials. Garden perennials include a number of herbaceous species grown for their flowers, or occasionally used as vegetative groundcovers. Under adequate growing conditions the plants of course persist and increase year after year. The biggest drawback to perennials as compared to annuals is that they must be maintained throughout the growing season, but have only a limited flowering period. Typical perennials are hollyhock (*Althea rosea*), columbine (*Aquilegia*), bellflower (*Campanula*), *Chrysanthemum, Delphinium*, pinks (*Dianthus*), coralbells (*Heuchera sanguinea*), *Phlox*, poppies (*Papaver*), primroses (*Primula*), many speedwells (*Veronica*), and scores of others.

Perennials are often produced and sold as a sideline to other nursery activities; some are sold through seedhouses. Perennial production could be undertaken on a massive scale, with attendant economies, but the market is neither large enough nor predictable enough (except for the greenhouse growing of such cut flowers as chrysanthemums and carnations) to interest most growers. The homeowner himself can provide his long-term needs by dividing a few plants and perhaps exchanging cultivars with friends and neighbors. Perennials can of course be started from seed, but flowering is more immediate through vegetative propagation. Most perennials are planted in spring, kept properly weeded, fertilized, and watered. In northern climates, many varieties must be mulched in autumn to protect the plant from winter kill.

Shrubs. Production of ornamental shrubs is the backbone of the nursery trade, and amounts to approximately half the dollar value of florist crops. The nursery industry is especially important in the Great Lakes and Middle Atlantic states and in Florida and California. In some of these areas nursery sales (including sales of roses, shade trees, and small fruits for the home) account for nearly 40 percent of the retail trade in horticultural specialties. The nursery business is about equally divided among the production of (1) coniferous evergreens such as yew, juniper, spruce, and pine; (2) broad-leafed evergreens such as rhododendron, camellia, holly, and boxwood; (3) deciduous plants such as *Forsythia, Viburnum*, barberries, privets, lilacs, and flowering vines; and (4) roses, which will be discussed separately.

Fields of specialization have evolved within the ornamental shrub industry. Some firms confine activity mostly to the production of "lining out" stock—seedlings and rooted cuttings for sale to growers for field plantings that must be tended several years before reaching salable size. Such firms will generally offer not only the bare root liners themselves, but small plants started in plant bands or slightly larger ones in containers. They may also offer small grafted plants of expensive or rare cultivars. Lining out stock of a typical evergreen, such as the Pfitzer juniper (*Juniperus media pfitzeriana*), is relatively inexpensive, generally several for a dollar; it is more economical for a grower to purchase these ready to plant in the field than to undertake the specialized propagation himself.

The field grower may, in turn, specialize in mass growing for the wholesale trade only. The field plantings are tended until they attain marketable size, having been shaped and sheared as necessary, provided with pest control, irrigated, culled, and so on. Because of the time required to produce a marketable crop, and because of appreciating labor costs, this phase of the nursery industry can easily be caught in a profits squeeze, especially if landscaping preferences should change and there were no quick market for a field of mature shrubs. But wholesale growing escapes the high overhead of retail marketing in urban areas, and although many growers do sell retail at the nursery they generally avoid the expensive merchandising required of the typical urban-area garden center. Growers are especially interested in new technology that will lessen labor requirements, and are turning to herbicidal control of weeds (as with dichlobenil) and shortcut methods for transplanting (in substitute for the traditional hand digging and burlapping).

Most shrubs are usually marketed **balled and burlapped** ("B and B"). The plants are dug with soil and wrapped with burlap or other suitable material. Recently a trend has developed toward **container-grown stock**—nursery stock grown in the container in which it is sold. This allows year-round sales of plant material. Deciduous shrubs may be marketed bare-root. There are obvious economies in not having to dig and wrap the plant carefully and handle added weight and soil volume. Fairly satisfactory techniques have been developed for holding bare-root ornamental shrubs in viable condition until marketed. Usually this involves digging after leaf fall in autumn and storing the plant through winter in cold cellars with regulated humidity for spring sale. Unfortunately, the plants are often handled carelessly. The chief cause of loss during storage is desiccation, but respiration also plays a part when temperatures rise. But even if the plants are properly cared for during cold storage, they often suffer from warming and drying at the sales outlet. Bare-root plants ordered by mail often arrive in poor condition because of warming and drying in transit. It is apparent that the great effort spent to control humidity and temperature during winter storage can be defeated if there are not cold storage facilities at retail outlets.

In recent years a breed of "nurseryman" that is more merchandiser than plantsman has taken over much of the retail trade. Exemplifying this change are the numerous diversified garden centers that sell all sorts of horticultural merchandise and maintain display yards of shrubs, trees, and roses purchased wholesale from large growers. Many such firms have associated landscaping services, and offer the homeowner counsel, product, and service.

Roses. The production of roses is even more specialized than that of other shrubs; the rose grower often deals solely in rose plants. Whereas other shrubs are typically started from seed or cuttings on their own roots, most roses are budgrafted onto rootstock (typically multiflora). This is the only way to achieve rapid and economical increase of a new selection to meet market demands. Large-scale production of roses has tended to center in the southwest (from Texas to southern California), where the nearly year-long growing season makes rapid production possible. Rose production is also carried on near the important consuming areas, such as northern Ohio.

Partly because of widespread promotion, and partly because of the great popular interest in the rose, the marketing of roses is nearly as important a facet of the nursery industry as is the production of other classes of shrubs. Mail order sales from attractive catalogs that picture the roses in color account for an appreciable part of the market. Enthusiasm for new selections is heightened by an elaborate scheme for testing and selecting "All-

American" roses (only a few roses receive the coveted award each year), and by a strong national rose society with headquarters in Shreveport, Louisiana, that encourages rose clubs and sponsors rose shows throughout the nation. Each year accredited judges compile and issue a rating sheet that ranks all available named cultivars on a numerical performance scale. Thus even the neophyte can determine readily which have been the most successful named roses through the years, and purchase his 'Peace' or 'Tropicana' on the basis of authoritative testing rather than just catalog hearsay.

Because the budding operation calls for skilled hand labor, and because field maintenance is expensive, few economies can be practiced in the production of roses. But techniques have been developed for the distribution of roses that do offer certain economies. In the mail order handling of bare-root roses coated paper or plastic bags are now used to retain humidity, in contrast to the former practice of wrapping the roots in damp moss. For mass marketing through chain outlets the stems of dormant stock are coated with wax to prevent desiccation during prolonged storage and handling, and the root system is "potted" in a moss-filled carton on which the name and a picture of the rose appear. Although much stock held dormant for long periods is not of first quality, this method of penetrating the mass market is well established in the nursery trade today.

Trees. Ornamental shade trees are usually grown and marketed in conjunction with shrubs. Deciduous trees and shrubs together have a somewhat greater value than the rose crop. While the value of fruit trees around the home has been recognized since colonial times, only in recent decades has there been much of a market for nursery-grown shade trees. Again, as part of the mid-twentieth century migration to the suburbs, and the construction of houses on cleared land, shade trees have become an increasingly important part of the nursery picture. The growing of shade trees was especially stimulated by the interstate highway program, which, for the first time, undertook the landscaping of highway rights-of-way rather than merely protecting them from erosion. As interest in shade and ornamental trees in-

creased, creation of improved cultivars followed. There is still some activity in transplanting native trees from the woodlot, and some trees are still grown from genetically unselected seed or cuttings. But more and more trees, like roses and shrubs before them, are being offered as named cultivars and as patented horticultural varieties when grafted, budded, or propagated vegetatively.

There are some obvious drawbacks to the tree-growing facet of the nursery industry. The crop takes a fairly long time to reach marketable size, and trees not saleable within a year or two become too large for convenient handling. The investment in a tree is high, and consequently so is the price. Risks are entailed, too, in trying to anticipate what the market will be for a crop planted today but not usable for five or six years. Yet so indispensible is an ornamental tree to attractive landscaping that sales in suburban communities have been quite rewarding. Some mortgage lenders require landscaping, including shade trees, for new homes (Fig. 23-19)! There has been a notable trend toward the use of small trees in small yards and under utility lines, especially trees that bear colorful flowers or fruits, such as crabapples, redbuds, dogwoods, hawthorns, and magnolias. Larger shade trees have been selected for shape and form, color of foliage, and other characteristics, so that there are available both spreading and upright cultivars of hard maple, ash, linden, and other favorites. Thornless honeylocust, of a type whose new foliage is yellow rather than green; dense-growing Scotch and Austrian pines; and once rare introductions, such as the Ginkgo and dawn redwood (*Metasequoia glyptostroboides*), can readily be had today, although only a generation ago selection was limited and trees came as often as not from some rural woodlot.

The digging, handling, and transporting of trees has always been difficult. Except in the sapling stage seldom are trees sold bare-root. A good deal of experienced manpower is needed to dig a tree and wrap its root system in burlap, the traditional way of moving larger trees. Through the years, however, special machines with slings and hoists have been developed for handling trees. A powered tree spade is now widely used, first to dig a hole where the tree is to be planted, then to dig the tree so that the root ball is the exact size of the previously dug hole, and finally to transport the tree to the hole. Thus there is no need for wrapping or separate handling. Such a machine makes it possible to move relatively large trees easily, but of course the tree must be contracted for in the field rather than selected from a nursery display.

Whether indigenous or transplanted from the nursery, trees about the modern home usually encounter an unnatural environment. Indigenous trees usually suffer damage to the roots from grading and compaction during house construction. The root systems of indigenous trees are often buried by soil from basement excavation. In order to save such trees dry wells are built around the trunks, but this is by no means an ideal solution. The leaves of deciduous trees are raked rather than allowed to decompose in situ to recycle nutrients, although in most yards there is compensatory fertilization from the lawn. Because of size, lack of familiarity, and less understanding of the physiology of trees than of herbaceous plants, trees are often poorly tended. For example, when trees are pruned, instead of merely removing dead, moribund, or interfering branches, they are often indiscriminately pollarded or topped, regardless of natural shape or growth habit. When branches are removed the base is often left as a stub, preventing healing of the scar. In urban and industrial areas air pollution may be serious enough to damage trees, especially many of the needle evergreens, which cannot be maintained in a city. Shade trees seem ever more susceptible to "decline" from uncertain causes, such as the wilt of oaks and the debilitation of hard maples, and to the complete loss by disease of such once proud species as the American elm and the American chestnut. Because of their size, trees are especially difficult to treat for disease; spraying is difficult with the usual home equipment, and it is expensive to hire tree specialists to do the job.

FIGURE 23-19. *Landscaping has become an integral part of the home construction business.* [*Courtesy Ford Motor Co.*]

tural specialties market. It is a competitive and sophisticated business, requiring both technical skill and appreciable capitalization. Maintenance of a greenhouse is itself a demanding preoccupa-

Floriculture

The growing of plants in the greenhouse for cut flowers, potted plants, and greenery has long been an important industry (especially in Europe), and even today accounts for about half of the horticul-

FIGURE 23-20. *Substances that reduce the height of ornamental flowering plants without reducing flowering time or flower size are finding promise in florist crop production. The chrysanthemums on the left were dwarfed with foliar applications of succinic acid-2,2-dimethylhydrazide.* [*Uniroyal.*]

tion. Even with the best of greenhouses, the grower must choose his specialty crop in light of local demand, provide efficient operating conditions, and be constantly alert against infestations of pests, heat failure, or other inadvertancies that could doom his crop overnight. Exact scheduling must be arranged, because many floral crops are only seasonally in demand (such as poinsettias at Christmas, lilies at Easter, and roses and carnations on Mother's Day and other occasions). Because the keeping quality of the crop is limited, established trade channels or auctions must function smoothly for quick disposal.

The chief cut flowers are roses, carnations, chrysanthemums, gladiolas, orchids, and gardenias. Some of these, such as gladiolas and orchids, are produced outdoors in the south and in Hawaii, and are shipped to market by air freight. Potted flowering plants include *Hydrangeas,* poinsettias, lilies, African violets, *Cyclamens,* and others. Foliage plants include ferns, palms, ivy, *Philodendron,* and succulents. Bedding plants such as begonias, geraniums, *Vinca,* and so on may be started in pots or plant bands and sold to homeowners for planting out. Seasonal annuals such as petunias, marigolds, salvia, and home garden vegetables may be started indoors for sale to the homeowner as seedlings. In most instances the grower will concentrate on but a few such items, arranging his seasonal schedule in such a way that valuable greenhouse space will not stand idle.

As in the nursery business, the floriculture trade involves both the grower, who mass-produces flowers for the wholesale market, and the retail florist, who markets to the public and includes such services as delivery, floral arrangements, and custom orders for funerals and various occasions. As is characteristic of other industries in the horticultural specialties field, the retail florist is perhaps more specialized as a business man than as an expert in floriculture. Purchased cut flowers are his most important product, since they have about three times the value of potted plants. Among potted plants foliage types rank first. There is also some greenhouse growing of vegetables (especially tomatoes) and mushrooms in specialized habitats.

While it would seem impossible to automate greenhouse operations completely, the production of plants under glass is constantly becoming more complex technically. Soil sterilization and proper hygiene in the greenhouse (insect and disease control with fumigants and pesticides) are now taken for granted. Optimum fertilization and pH, related to temperature, water regimen, and illumination, have been determined for most crops, and must be understood by the grower. Where appropriate, hydroponic culture can be practiced, or at least subirrigation of the growing bed. Automatically timed misting is commonplace today in beds for rooting cuttings. The modern floriculturist must have training and practical experience in chemistry, plant physiology, and related subjects; the occupation today demands a trained craftsman, not an amateur. In the era that lies ahead, the grower will probably be capable of controlling plant growth even more precisely than he is now. Already chemical growth regulators determine shape and size (Fig. 23-20), rate of production, and even affect disbudding (in chrysanthemums). Some growers already enrich the greenhouse atmosphere with carbon dioxide to increase luxuriance of growth. Greenhouse floriculture is in many ways the most sophisticated part of crop agriculture.

Selected References

Baule, H., and C. Fricker. *The Fertilizer Treatment of Forest Trees.* Stechert-Hafner, New York, 1970. (A subject concerning which there is little literature, translated from the German. Forest fertilization is becoming routine in Europe, often accomplished through the planting of legumes as well as by the application of fertilizer.)

Braun, E. L. *Deciduous Forests of Eastern North America.* Hafner Publishing Company, New York, 1967. (A reprinting of Lucy Braun's definitive ecological examination of the eastern North American forests, first published in 1950.)

Brown, H. B. *Cotton* (3rd ed.). McGraw-Hill, New York, 1958. (Textbook on the various phases of cotton production.)

Calkin, J. B. (editor). *Modern Pulp and Papermaking.*

Reinhold, New York, 1957. (Fundamental principles of papermaking, including processes and equipment, in a single volume.)

Cooke, G. B. *Cork and the Cork Tree.* Pergamon Press, London and New York, 1961. (Perhaps more than you will want to know about all aspects of cork and its production in the Mediterranean area.)

Davis, K. P. *Forest Management* (2nd ed.). McGraw-Hill, New York, 1966. (A standard textbook concerning accepted forestry practices.)

Duvigneaud, P. (editor). *Productivity of Forest Ecosystems.* UNESCO, Paris, 1971. (Proceedings of the 1969 Brussels Symposium providing international opinion in the subject area.)

FAO. *Yearbook of Forest Products 1969–70.* United Nations, Rome, 1971. The Food and Agriculture Organization's statistical compilation on production and trade in forest products worldwide.)

Forbes, R. D. *Woodlands for Profit and Pleasure.* American Forestry Association, Washington, D.C., 1971. (A rather popularly presented review of forest basics designed for the small property owner.)

Haden-Guest, S., J. K. Wright, and E. M. Teclaff (editors). *A World Geography of Forest Resources.* Ronald Press, New York, 1956. (A partial replacement for the classical Zon and Sparhawk *Forest Resources of the World,* now outdated; 35 authorities discuss species, ecology, and products, providing a general picture of world forests.)

Harlow, W. M., and E. S. Harrar. *Textbook of Dendrology* (5th ed.). McGraw-Hill, New York, 1967. (An authoritative work on American tree species.)

Hawthorn, L. R., and L. H. Pollard. *Vegetable and Flower Seed Production.* Blakiston, New York, 1954. (The technology of seed production.)

Kirby, R. H. *Vegetable Fibres.* Leonard Hill, London, and Interscience Publishing Co., New York, 1963. (This technical monograph in the World Crops series deals in detail with a large number of important fiber crops excepting cotton.)

Kozlowski, T. T. *Growth and Development of Trees;* Vol. 1: *Seed Germination, etc.;* Vol. 2: *Cambial Growth, etc.* Academic Press, New York, 1971. (A technical reference attempting to convey broad biological principles in the subject areas from the forester's viewpoint.)

Laurie, A., D. C. Kiplinger, and K. S. Nelson. *Commercial Flower Forcing* (7th ed.). McGraw-Hill, New York, 1968. (A textbook of greenhouse florist crops with the emphasis on practice.)

Lock, G. W. *Sisal.* Wiley, New York, 1962. (An important, comprehensive treatment.)

Mirov, N. T. *The Genus Pinus.* Ronald Press, New York, 1967. (An "in depth" treatment of one of the most important forest genera: paleo-record, geography, genetics, morphology, physiology, ecology, chemistry, and taxonomy.)

Osmaston, F. C. *The Management of Forests.* Stechert-Hafner, New York, 1968. (Another broad view of forest management.)

Panshin, A. J., Carl De Zeeuw, and H. P. Brown. *Textbook of Wood Technology,* Vol. 1 (2nd ed.). McGraw-Hill, New York, 1964. (The first volume of this two-volume work deals with structure identification, uses, and properties of the commercial woods of the United States; the second volume (H. P. Brown, A. J. Panshin, and C. C. Forsaith, 1952) deals with the physical, mechanical, and chemical properties.)

Penfold, A. R., and J. L. Willis. *The Eucalypts: Botany, Cultivation, Chemistry, and Utilization.* Interscience (Wiley), New York, 1961. (A survey of the genus *Eucalyptus.*)

Post, K. *Florist Crop Production and Marketing.* Orange Judd, New York, 1949. (A standard text on the florist industry.)

Randall, C. E. (editor). *Enjoying our Trees.* American Forestry Association, Washington, D.C., 1969. (A popular review designed to improve enjoyment and understanding of American trees.)

Reichle, D. E. (editor). *Analysis of Temperate Forest Ecosystems.* Springer-Verlag, Berlin and New York, 1970. (An attempt to understand the fundamentals of forest regeneration.)

Roberts, E. H. (editor). *Viability of Seeds.* Syracuse University Press, 1972. (Of interest for the emerging field of tree and shrub cultivar selection.)

Sarkanen, K. V., and C. H. Ludwig (editors). *Lignins. Occurence, Formation, Structure and Reactions.* Interscience (Wiley), New York, 1971. (A definitive text for the first time sifting widely disbursed literature on the subject to provide a good overview.)

Stamm, A. J. *Wood and Cellulose Science.* Ronald Press, New York, 1964. (Discusses microscopic and molecular qualities of wood, relating them to its mechanical behavior; wood physical chemistry related to processing.)

Stoddard, C. H. *Essentials of Forestry Practice* (2nd ed.). Ronald Press, New York, 1968. (A broad consideration of forestry and its applications.)

Tsoumis, G. *Wood as a Raw Material: Source, Structure, Chemical Composition, Growth, Degradation and Identification.* Pergamon Press, Elmsford, New

York, 1968. (Useful for technical detail regarding wood.)

U.S. Department of Agriculture. *Landscape for Living.* Yearbook of Agriculture, Washington, D.C., 1972. (A typical U.S. Department of Agriculture yearbook, with many contributing authors, emphasizing the department's increasing concern for urbanites. Some of the book is sociological in nature, but appropriate sections give useful instructions on home landscaping and gardening.)

Wackerman, A. E., et al. *Harvesting Timber Crops* (2nd ed.). McGraw-Hill, New York, 1966. (Details on forest cropping.)

Watkins, J. V. *Florida Landscape Plants.* University of Florida Press, 1971. (A useful review of subtropical conditions, increasingly of concern what with the current influx of people into the southland.)

Zimmerman, M. H., and C. L. Brown. *Trees Structure and Function.* Springer-Verlag, Berlin and New York, 1971. (Considers the controversial and relatively little known nature of large woody plant physiology; the relation between growth and habit is discussed, followed by a highly technical consideration of translocation and circulation of assimilates.)

Zivnuska, J. A. *U.S. Timber Resources in a World Economy.* Published for Resources for the Future, Johns Hopkins Press, Baltimore, 1967. (An evaluation of where the United States stands vis-a-vis timber demands.)

PART VI

THE MARKETPLACE

CHAPTER 24

The Economics of Crop Production and Distribution

The food and fiber industry encompasses the whole range of economic activity involved in manufacturing and distributing the industrial inputs used in farming: the farm production of crops, animals, and animal products; the processing of these materials into finished products; and the provision of products at a time and place demanded by consumers.

In the low-income developing countries a large share of the total value added to the national product by the food and fiber industry is contributed by farm production. Farmers in poor countries use quite limited quantities of industrial inputs, such as fertilizer, pesticides, and machinery. The value added in marketing, by processing and distribution, is low. A high percentage of farm output is consumed at the farm or in villages, near the source of production. Only a small share of final output reaches urban consumers in a packaged or processed form.

In the high-income countries, farm level production makes up a much smaller share of the value added to national output by the food and fiber industry. The purchased industrial inputs represent a larger share of the expenses incurred in farm output. Farm produce itself constitues a relatively small share of the total value of food and fiber production. Consumers want a wide variety of fresh and processed foods throughout the year. And they demand that these products be packaged, frozen or dehydrated, and processed for easy preparation in the home. Both the farm and the household supply a smaller proportion of the value of the total output of the food and fiber industry than in the past, or than in today's less developed countries.

In 1971, sales of final product by the food and fiber industry amounted to $263 billion, or about one-fourth of the United States gross national product. Farm level production, the gross value of farm output, amounted to $56 billion, or just slightly more than 20 percent of the value of output by the entire food and fiber industry (Fig. 24-1).

Economic decisions are involved at every phase of production, marketing, and utilization in the food and fiber industry. Decisions by producers concerning what to produce, how much to pro-

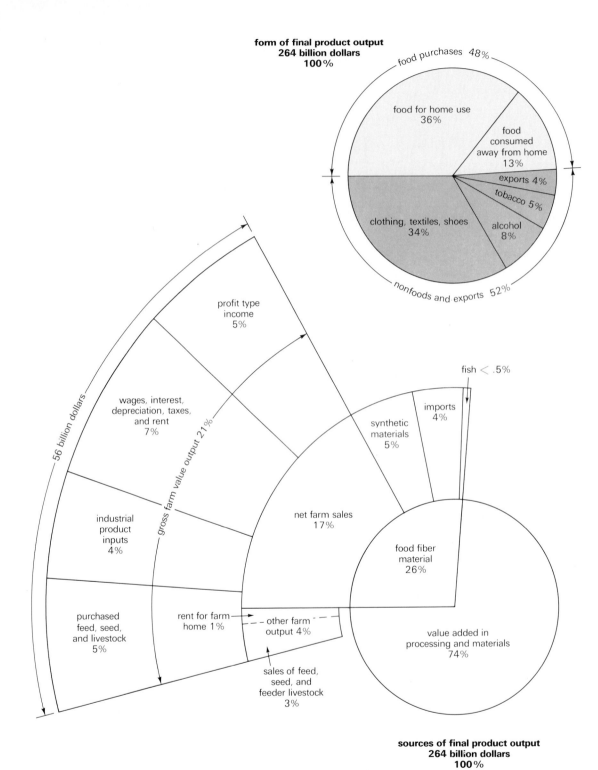

FIGURE 24-1. *The United States food and fiber industry in 1971. After Daly,* National Agricultural Outlook for 1972. *USDA, ERS, 50th National Agricultural Conference, February 22, 1972, pp. 4, 9.*

duce, and what inputs to use depend both on technical input-output ratios, or relationships, and on prices of inputs and products. When the consumer decides what to consume and how much to consume, he is influenced not only by his tastes and preferences but also by his income and the relative prices of the products available in the market. Decisions by the marketing, processing, and distribution firms are influenced by the decisions of the crop producer and the consumer and by the prices associated with these decisions.

ECONOMICS OF CROP PRODUCTION

Crop-production decisions fall into three general categories:[*]

1. What combination of inputs should be used in producing a particular crop? For example, should straight nitrogen or some nitrogen-phosphate-potash formulation be used on corn? What combination of intensity of land preparation and cultivation, herbicide use, and hand labor should be used to control weeds?
2. How much of each input should be used? For example, should 60 or 120 pounds of nitrogen per acre be applied on corn? How many applications of irrigation water should be used on tomatoes?
3. What combination of crops, or of crops and livestock, should be produced? How many acres should be planted to corn? How many acres to soybeans? Should feed grain be sold in the market or fed to livestock?

To answer these questions two types of information or data are needed. Data are needed, either from records or experiments, on the effects of decisions to change input combinations and/or the product mix. It is also necessary to know as accurately as possible the prices of the inputs and of the products that are affected by such decisions. Failure to evaluate the technical or economic data accurately leads to inefficient production decisions.

[*] The authors are indebted to Randolph Barker of the International Rice Research Institute (Philippines) and to Earl Fuller of the University of Minnesota for assistance in the preparation of this section.

Partial budgeting of costs and returns

When these data are available or can be obtained it is possible to calculate the costs, the gross returns, and the net gain from production decisions. The simplest procedure for making such calculations is a technique called **partial budgeting.** Partial budgeting focuses attention only on those inputs, products, and prices that are expected to change as a result of a particular decision. This is in contrast to the more detailed procedure of **whole farm budgeting,** which is frequently used to summarize costs and returns to the total farm operation over a particular time period, such as a year. In the generalized partial budget format presented below, any change in use of inputs or in the amount or mix of products can be classified under one of the four headings.

Partial budget of estimated change in annual net cash farm income from a change in resource use		
1. Added costs _____		3. Added returns _____
_____		_____
2. Reduced returns		4. Reduced costs
_____		_____
Subtotal A _____		Subtotal B _____
Estimated change (B − A) =====		

The simplicity of partial budgeting has made it a valuable planning tool for agricultural research, marketing, and extension organizations, as well as for producers. The applied research organization needs to select, from among the preliminary results obtained by its research staff, those lines of research that should be developed and tested more completely. To conserve time and effort certain alternatives must be eliminated because they lack scientific, technical, or economic interest. After completing more intensive development, and testing, private research organizations must decide which results should be incorporated into new products and placed on the market. Public research agencies need to decide which results should be recommended to farmers or to agencies working with farmers. Even after rather thorough screening by researchers, extension workers or distributors may frequently find it desirable to conduct trials and to budget the costs and returns in their own country or market area.

Determining how much fertilizer to use

Use of partial budgeting to determine how much fertilizer to use is illustrated in Figures 24-2 to 24-4 and Tables 24-1 and 24-2. Figure 24-2 presents data on the relationships between inputs of nitrogen and yield of rough rice from experiments conducted at the Maligaya Experiment Station in the Philippines by the Philippines Bureau of Plant Industry and The International Rice Research Institute. Table 24-1 presents partial budgets showing the change in costs, gross returns, and net returns from increasing the level of nitrogen from 30 to 60 kg per hectare (27 to 54 lb per acre).* Table 24-2 shows similar changes when the level of nitrogen is increased from 60 to 90 kg per hectare.

* A useful rule of thumb is that one metric ton (1,000 kg) per hectare is roughly equivalent to 1000 pounds per acre. One kilogram is equivalent to 2.2046 pounds; one hectare is equivalent to 2.471 acres. One kilogram per hectare is, therefore, equivalent to approximately 0.89 pounds/acre. See appendix for more detailed conversion factors.

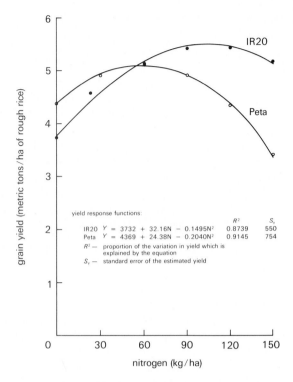

FIGURE 24-2. *Yield response of two varieties (cultivars) of rough rice to nitrogen at the Maligaya Experiment Station (Philippines) during the 1970 dry season. [Data from IRRI, March 1972.]*

The results of a larger series of partial budgets are presented in Figure 24-3. Note that the net return curve (B − A) reaches its maximum while both the gross return curve (B) and the cost curve (A) are still rising. (Geometrically, the tangent to the net return curve is horizontal at the point where the tangents to the gross return curve and the cost curve are parallel.)

Net returns are, in this example, quite sensitive to the price of rough rice. Figure 24-4 presents net return curves obtained by budgeting the yield response data of Figure 24-2 at three alternative price levels for rough rice. Note also that the most profitable amount of fertilizer to apply per hectare rises as the price of rough rice rises relative to the price of nitrogen fertilizer.

Net returns are, in this example, also highly responsive to the tenure arrangements under which the crop is produced. Figure 24-5 compares the

Table 24-1 APPLICATION OF 60 KG OF NITROGEN VERSUS 30 KG OF NITROGEN TO RICE, VARIETY IR20, 1970 DRY SEASON, MALIGAYA EXPERIMENTAL STATION, PHILIPPINES.

1. Added costs		3. Added returns	
Materials[1]	$ 6.00	Change in rough rice produced[5(a)]	$40.00
Interest[2]	0.36		
Equipment[3]			
Application[4]	0.38		
Harvesting[5(b)]	6.65		
2. Reduced returns		4. Reduced costs	
Change in rough rice produced			
Subtotal A	$13.39	Subtotal B	$40.00
		Estimated change (B — A)	$26.61

[1] 30 kg of nitrogen at $0.20/kg ($4.53/50 kg bag of urea containing 45% N).
[2] 6% for 6 months.
[3] Broadcast, no equipment cost.
[4] Broadcasting: 4 additional man-hours/ha at $0.094/hour = $0.38.
[5] (a) Increased production of 571 kg valued at $0.07 per kg ($3.12/cavan of 44 kilos) = $40.00.

(b) Harvest cost (at $\frac{1}{6}$ of 571 kg), 95 kg valued at $0.07 per kg = $6.65 (one U.S. dollar = 6.40 pesos).

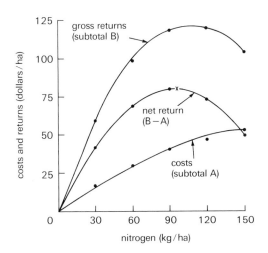

gross returns (subtotal B)

total	$0.00	$58.44	$98.44	$118.88	$120.88	$103.88
change		$58.44	$40.00	$20.44	$2.00	$-17.00

costs (subtotal A)

total	$0.00	$16.48	$29.89	$40.04	$47.11	$53.85
change		$16.48	$13.41	$10.15	$7.07	$6.74

net return (B−A)

total	$0.00	$41.96	$68.55	$78.84	$73.77	$50.03
change		$41.96	$26.59	$10.29	$-5.07	$-23.74

FIGURE 24-3. *Cumulative costs and returns from fertilizing rice, variety IR20, dry season (based on fertilizer response reported in Fig. 24-2). (X = optimum fertilization level.) [Data from IRRI, March 1972.]*

Table 24-2 APPLICATION OF 90 KG OF NITROGEN VERSUS 60 KG OF NITROGEN TO RICE, VARIETY IR20, 1970 DRY SEASON, MALIGAYA EXPERIMENTAL STATION, PHILIPPINES.

1. Added costs			3. Added returns	
Materials[1]	$ 6.00		Change in rough rice	
Interest[2]	0.36		produced[5(a)]	$20.44
Equipment[3]				
Application[4]	0.38			
Harvesting[5(b)]	3.43			
2. Reduced returns			4. Reduced returns	
Change in rough rice				
produced				
Subtotal A	$10.17		Subtotal B	$20.44
			Estimated change (B — A)	$10.27

[1] 30 kg of nitrogen at $0.20/kg ($4.53/50 kg bag of urea containing 45% N).
[2] 6% for 6 months.
[3] Broadcast, no equipment cost.
[4] Broadcasting: 4 additional man-hours/ha at $0.094/hour = $0.38.
[5] (a) Increased production of 292 kg valued at $0.07 per kg ($3.12/cavan of 44 kilos) = $20.44.
(b) Harvest cost (at $\frac{1}{6}$ of 292 kg), 49 kg valued at $0.07 per kg = $3.43 (one U.S. dollar = 6.40 pesos).

farmer's net return from the use of fertilizer under different tenure arrangements. When the landlord shares the cost of the fertilizer and the increased production with the tenant, the rate of return on the tenant's fertilizer investment is the same as that of the owner-operator, but the absolute size of his return is reduced. When the tenant pays the full cost of the fertilizer and shares the increase in output with the landlord, the tenants rate of return is reduced, and the absolute size of his return is lowered even further. The effect is to further reduce the tenant's incentive to use fertilizer.

Other elements also enter into even a simple decision, such as how much fertilizer to use. Some uncertainty regarding the yield response to fertilizer always exists. Even under carefully controlled experimental conditions some of the observations lie well off the calculated response curve (see Fig. 24-2). Under these circumstances a farmer having

FIGURE 24-4. *Cumulative net return from fertilization at alternative prices for rough rice (based on IR20 fertilizer response reported in Fig. 24-2). (X = optimum fertilization level.) [Data from IRRI, March 1972.]*

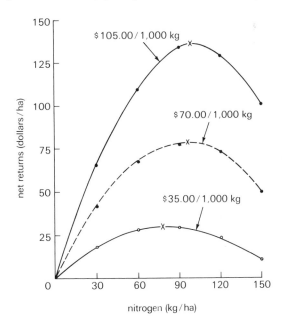

626

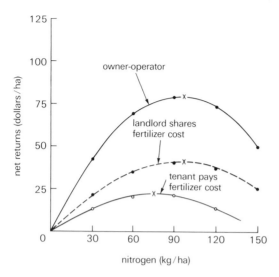

FIGURE 24-5. *Implications of tenure on net returns and optimum level of fertilization of rough rice (based on fertilization response curve for IR20 reported in Fig. 24-2 and a price of $70/1000 kg).* (X = *optimum fertilization level.*) [*Data from IRRI, March 1972.*]

a strong aversion to risk, or whose access to credit is inadequate, may use less fertilizer than the calculated "optimum." Decisions that affect the use of farm resources in the course of a long time, such as the acquisition of additional land or investment in equipment and buildings, are even more difficult because of the greater risk and uncertainty elements. In recent years economists have developed and implemented rather elaborate computer programming models for use by farmers in these more complex decision processes. The basic principles employed and information requirements are, however, essentially similar to those that have been illustrated in the partial budgeting analysis of decisions on fertilizer use (Figs. 24-2, 24-3, and 24-4; Tables 24-1 and 24-2).

MARKETS AND PRICES

Prices play an important role in guiding farmers' decisions regarding how much of each input to use, what combination of inputs to use, and how much of each product to produce.

If production is to be organized so that farm commodities are produced in the proportions desired by consumers, some mechanism must be available for transmitting information about consumer demand to producers. Similarly, some mechanism is needed to make consumers aware of differences in the costs incurred by farmers in producing different commodities, so that consumption choices can be made in a manner that enables individual consumers, and society as a whole, to obtain the greatest satisfaction from the use of individual income and from the use of national resources.

The role of prices in an exchange economy

In an exchange economy this function is performed by market prices. When prices are not regulated, either by government intervention or private economic power, they function as a thermostatic or feedback device to achieve a balance between production and consumption.

If production of a product such as soybeans fails to expand as rapidly as the demand for soybean meal and oil, the price of soybeans will rise. As a result farmers find it more profitable to use their land and other inputs to produce soybeans rather than other crops, such as corn. At the same time the rise in the price of soybeans causes users of soybeans to substitute other sources of protein feed and vegetable oil for soybean meal and oil. The effect of these two simultaneous responses to the initial rise in the price of soybeans—the stimulation of production and the curtailment of consumption—acts to bring about a new balance between production and consumption.

Demand and supply curves

The functioning of the supply-demand thermostat, with its system of information feedback through prices, is illustrated graphically in Figure 24-6. In this figure, curve SS is referred to as a supply curve, or function. It describes the quantities of a particular commodity, such as soybeans, that farmers as a group are willing to produce and offer for sale at different prices during a particular time period, such as a crop year. It may shift to the left

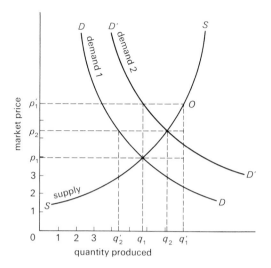

FIGURE 24-6. *Supply and demand functions illustrating the effect of a shift in demand on price and output.*

or to the right in response to changes in the availability of resources or the level of technology in the region or country to which the supply curve applies. It rises to the right because in order to produce and market larger quantities of a particular commodity it is necessary to bring inferior land into production, or compete with other crops such as corn for labor, capital equipment, credit, and other inputs thus forcing the costs of producing upward.

The curve DD is referred to as a demand curve, or function. It describes the quantities of a commodity, such as soybeans, that buyers as a whole are willing to purchase at different prices during a particular time period. Its position on the chart is determined by such factors as the level of income and the size of the population of a particular country. For many commodities international demands are also important. The steepness with which it slopes downward to the right depends on the availability and willingness of users to substitute other commodities for the particular commodity in question as prices change. If there are no close substitutes, or if users are unwilling to use substitutes because of rigid preferences, the curve will have a steep slope: small changes in the amount supplied will be accompanied by large changes in price. If close substitutes are readily

available, and are acceptable to users, the curve will be relatively flat: relatively large changes in the amount supplied, will be accompanied by small changes in price. With supply and demand represented by DD and SS the quantity supplied and the price that would just clear the market—that is, the price that would balance farmers willingness to produce and the consumers willingness to use—can be represented by q_1 and p_1.

The market can be forced out of adjustment if, as a result of some combination of population growth and increase in income, the demand curve shifts to $D'D'$. It could be said to be out of adjustment, or in a disequilibrium state, because while producers would be willing to produce the quantity q_1 for a price of p_1, consumers would be willing to pay a price of p_1' if only the amount q_1 were supplied. On the other hand, if consumers would be willing to pay a price of p_1, producers would be willing to produce and market the amount q_1'.

This is the point at which prices begin to function as "feedback" devices. As prices begin to rise toward p_1' producers begin to expand production. If the price system functions with great precision, the system will rapidly shift to a new "equilibrium" that just balances producer and consumer interests with an amount q_2 produced and marketed at a price p_2. With supply and demand represented by $D'D'$ and SS the quantity supplied and the price that would just clear the market—the price that would balance farmers' willingness to produce and consumers' willingness to use—can be represented by q_2 and p_2. If, however, the demand-supply system is not highly sensitive or efficient—if it works like an unreliable thermostat—prices and quantities may fluctuate around the equilibrium values of q_2 and p_2 to produce a "cobweb" pattern of price and output as is shown by "hog" cycles (Fig. 24-7).

Spillover effects, externalities, and market failure

The importance of an effective supply–demand feedback mechanism for efficient market behavior can not be overemphasized. When the value that society places on the inputs used in production or

628

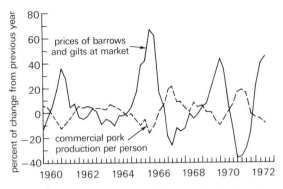

FIGURE 24-7. *Changes in hog prices and pork production. The prices are those of barrows and gilts at seven markets. The percent change is from the previous year.* [*After U.S. Dept. of Agriculture, 1972 Handbook of Agricultural Charts, Economic Research Service, Agricultural Handbook 439, October 1972.*]

on final products are not accurately reflected in the market, the production decision that is "efficient" for the producer—that maximizes his return—may become socially "inefficient," and thereby suboptimal for the entire economy.

The failure of society to reform the market system so that it accurately reflects social values is one of the causes of the excessive environmental impact of many agricultural production practices. Farmers have typically made their decisions about levels of pesticide use in the same manner as their decisions on seed, fertilizer, and labor use. But most pesticide prices have not accurately reflected the "spillover" effects of their use: they have not included the cost of potential damage to neighbors' crops, the health hazards to farm workers and consumers, and the damage to wildlife. The effect has been to encourage the farmer to use larger quantities of pesticide per unit of production, and to prefer chemical methods of pest control in general rather than other methods.

The sensitivity of pesticide use to price is illustrated in Figure 24-8. The yield response curve for pesticides on most crops typically rises sharply and then flattens out. This is reflected in the shape of curve R in the figure. The lines p_1 and p_2 reflect the price ratio of the pesticide and the crop per unit of output. Line p_1 reflects a relatively high pesticide price per unit and line p_2 reflects a lower

price. The optimum level of pesticide use is the point at which the price line is tangent to the gross returns line. There, the increment to cost is the same as the increment to returns. Because the upper range of the response curve is quite flat, a small reduction in the pesticide price induces a large increase in the optimum level of application. And the farmer may be tempted to apply amounts much larger than the optimum to insure against uncertainty, since typical pesticide costs per unit of output are low, relative to the price of the commodity.

Market reforms that could lead to greater consistency between farm production decisions and efficient use of resources for the entire economy would include a combination of changes in property rights, tax incentives and subsidies, regulation of pesticide use, and public management of inte-

FIGURE 24-8. *Sensitivity of pesticide use to price. The slopes of p_1 and p_2 represent alternative ratios of $\frac{\text{price of pesticide per unit}}{\text{price of crop per unit}}$. [After Headley and Lewis, The Pesticide Problem: An Economic Approach to Public Policy. Baltimore: The Johns Hopkins Univ. Press, 1969, p. 19.]*

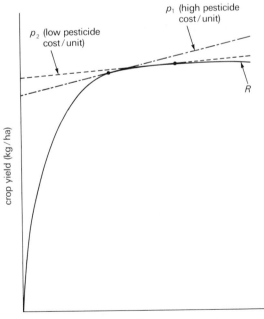

grated pest control programs. In Michigan, for example, the intorudction of an experimental pest management program for apples has permitted a reduction in the level of pesticide application by as much as three-quarters. Under this program, complex insect detection, monitoring, and scouting methods have been developed, and the cooperation of the Michigan ate University Cooperative Extension Service, the Michigan Department of Agriculture, and the U.S. Department of Agriculture has been enlisted. The Michigan program is publicly supported. An alternative would be to establish a pest management district and pay for the service by an assessment or a direct charge to the producers who benefit from the program.

MARKETS

Markets are the institutions through which the price-making forces, which have been summarized in terms of demand and supply schedules, are channeled and interpreted. Markets perform this role by converting (a) information on changes in yield and changes in the area planted, and (b) information affecting the quantities of a commodity desired by consumers or other users into price signals that trigger the production-consumption feedback system described above. Markets are cultural artifacts. They were invented by man. Their effectiveness depends on human inventiveness and ingenuity in market organization and development.

Marketing institutions

Markets in underdeveloped countries tend to be localized. Buying, selling, and market-information activities are typically concentrated at a single place, such as a village market or a transportation terminal (Fig. 24-9). At these locations buyers and sellers meet each other face to face, and sales are consumated by the physical transfer of commodities and money. In a modern economy with a sophisticated information system, the concept of a specific market "place," where producers and consumers meet to bargain over prices and quality, is largely obsolete. Instead, producers and consum-

FIGURE 24-9. *The village market concentrates the buying, selling, and market information functions at a single location.* [Courtesy USDA.]

ers are linked by a large number of intermediaries, each of whom may perform one or several market services or functions.

In addition to the buying, selling, and market-information services, modern market institutions perform a number of other services, including storage, transportation, and whatever services are necessary to facilitate buying and selling, such as establishing standards, arranging financing, and assuming risks. These services are provided by retailers, wholesalers, brokers, and commission agents, credit agencies, transportation firms, and others. The organizations that provide these services are also frequently engaged in other activities, such as the processing or manufacturing of basic commodities into consumer goods. The processing or manufacturing phase is sometimes done by separate firms. For many commodities more resources are used in the marketing and processing activities than in crop production itself. In the conversion of wood to paper the value added to the original forest product represents perhaps 90 percent or more of the final price to the consumer.

Marketing efficiency

There are two bases on which to evaluate the efficiency of the marketing system. The first is in terms of pricing efficiency. This is measured by the speed and precision with which the equilibrium or marketing-clearing process is discovered and made known to consumers, producers, and processors. The second is in terms of technical efficiency. This is measured by the resources utilized in marketing and processing agricultural commodities into finished consumer goods.

Technical efficiency should not, however, be confused with the percentage share of the total consumer dollar that goes to pay for marketing services and processing. The fact that farmers in the United States receive 57 percent of each dollar the consumer spends for eggs and only 16 percent of each dollar the consumer spends for bread says nothing about the efficiency of the marketing system. Bread is a much more highly processed product than eggs. Technical efficiency can only be determined by a careful examination of whether the marketing and processing functions that are preferred by the consumer could be performed at a lower cost.

Market structure

Markets are, of course, not the only method of transmitting information about consumer preferences and production alternatives between producers and consumers. Markets are in some ways very inefficient devices for transmitting information. In the process of translating information about the forces that affect demand and supply into market prices, much relevant information about producers' plans, market stocks, and quality preferences, for example, may be lost.

When large numbers of buyers and sellers are involved, when the commodity is not perishable, and when quality differentials can be accurately specified and measured, organized commodity markets, such as the grain exchanges in Chicago, Minneapolis, and Kansas City, appear to be relatively efficient.

When the commodity is perishable or if technical efficiency is increased by agreements between producers and processors about quality and timing of delivery, as in the canning industry, open markets have frequently been replaced by contractual arrangements. Farmers agree to delivery schedules and quality specifications in return for price and quantity guarantees by processors. In some arrangements the terms of sale are established by bargaining between producer organizations and processor organizations.

One of the most important factors leading to new patterns of marketing and distribution has been the rapid development of transportation facilities and communication media. The development of national railway systems led to the establishment of great concentrations of marketing organizations at such rail centers as Kansas City, Minneapolis-St. Paul, and Chicago. Central wholesale markets, futures markets, and processing facilities were established at these centers because the produce from large regions was forced to pass through these centers on the way to population centers and port cities.

With the development of more efficient long-distance communication systems, the construction of a national highway system, and the development of modern motor freight equipment the necessity to move goods through a few central locations has been reduced. Many of the old established central markets for fruits and vegetables, livestock, and dairy, and poultry products that developed during the railway era have declined in importance. Many products are now being processed closer to the point of production. The major food-distribution organizations now send their own buyers directly to the producing areas. Many go directly to growers for such fresh products as fruits and vegetables.

GOVERNMENT AND MARKETING

In addition to private marketing institutions government plays an important role in the marketing of crops in most countries. Governments attempt to improve pricing efficiency—the sensitivity and precision with which the marketing system reflects producer and consumer interests and behavior. And governments attempt to overcome inequities in the distribution of income that result from the effects of prices on the allocation of resources.

Programs to improve pricing efficiency

A number of conditions must be met if marketing procedures are to efficiently perform their function in establishing prices and if prices are to act as efficient guides for determining the future quantities to be marketed. Effective standards for weights and measures and for quality must be established and adhered to. Information on prices and quantities must be equally available, as nearly as possible, to all market participants—producers, marketing agencies, processors, and consumers. In all advanced economies government plays an important role in establishing effective market standards and in providing market information (Fig. 24-10).

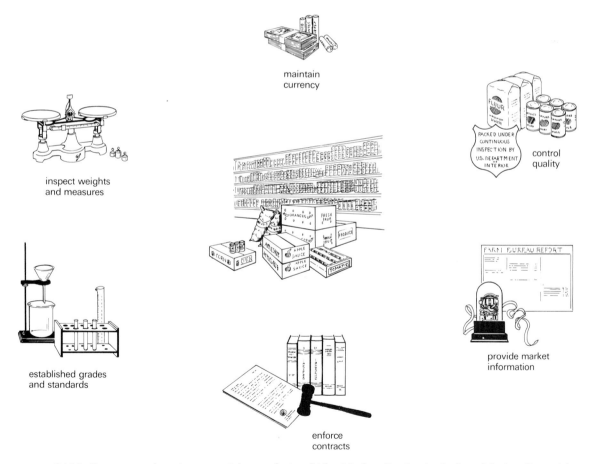

maintain currency

inspect weights and measures

control quality

established grades and standards

enforce contracts

provide market information

FIGURE 24-10. *Governmental services essential to marketing.* [*After Mosher,* Getting Agriculture Moving: Essentials for Development and Modernization. *New York: Praeger, 1966.*]

Weights and measures

The standardization of weights and measures is one of the oldest recognized functions of government. Without such standardization internal trade within countries could hardly be conducted on any extensive scale. As international trade has expanded, the need for greater standardization of weights and measures has been increasingly recognized, although discrepancies still exist, even between countries employing more or less the same systems. For example, liquid and dry volumetric measures are not the same in the United States and Great Britain. The Imperial gallon used in Canada and Great Britain to measure liquid volume is about one-fifth larger than the United States gallon. And in both systems the dry volume measures are different than the liquid—the dry pint is larger than the liquid.

The English system of weights and measures has gained wide use throughout many parts of the world because of British dominance of trade during the last half of the nineteenth century and the early part of the twentieth century. The metric system, developed by a commission of French scientists and adopted as the legal system of weights and measures in France in 1799, has gained increasing acceptance because of its great simplicity and logical structure. It is widely used for scientific work even in countries that continue to use the English system for commercial and industrial use, and is continuously achieving even wider commercial and industrial acceptance.

Grades and standards

Establishment of effective quality standards for agricultural commodities represents a much newer function of governments. Standardization of volume measurement has little meaning if the content of the container varies in quality. With increasing commercialization of agriculture, commercial standards tended to evolve. At a later stage government efforts were often exerted to achieve uniformity or increased precision of such standards. In the early days, for example, each grain market had its own grades and grading methods. Different early standards for No. 2 corn required that the

corn "be dry," "reasonably dry," or "have not more than 16 percent moisture." Lack of accepted fruit and vegetable standards resulted in especially chaotic trade conditions. In the United States the passage of the Cotton Futures Act of 1914 and the Grain Standards Act of 1916 initiated a series of laws that have gradually resulted in broadening federal and state responsibilities in the establishment of uniform grades and standards for agricultural commodities.

It is clear, however, that much of the grading system that has evolved from early commercial practices and been given broader legitimization by government are no longer as relevant as in the past. Improvements are being made in many areas. Mechanical and chemical tests are being used to replace sensory tests, which often vary with the sensory ability of the individual grader. Less progress has been made in developing criteria that consumers can effectively use to distinguish between different qualities for which they are willing to pay different prices.

Market information

The establishment of prices will clearly be inefficient or distorted if information on the quantity produced and the prices at which commodities are sold is unavailable or available to only some of the participants in the marketing process. In recognition of the importance of such information most governments have sharply increased their efforts to provide producers, or consumers, and marketing agencies with increasingly adequate data on commodity production, storage, movement, and prices. The following list illustrates the type and extent of the information available from the Market News Service of the United States Department of Agriculture.

Cotton
1. Prices of lint cotton, cotton linters, and cottonseed.
2. Daily price quotations from principle markets.
3. Quality data during the ginning season.
Fruits and vegetables
1. Prices and supply movement for the large city markets.

2. Shipping area data on prices, volume of loadings, quality, and market trends.
3. Cold storage holdings in principle centers.

Grain

1. Daily grain prices and supply movements (cash and futures) for the principal markets.
2. Prices, market movements, and production data of feedstuffs.
3. Prices at country shipping points.

Similar information is provided by governments in an increasing number of countries. In the less developed countries such data are more frequently provided for export crops, on which the countries' foreign trade earnings depend, than for the crops used for domestic food consumption. Provision of economic information is also an important function of the Food and Agriculture Organization of the United Nations.

Government agencies are increasingly undertaking the responsibility of analyzing and interpreting the information they collect. Major efforts are made by such agencies as the Council of Economic Advisers and the United States Department of Agriculture to analyze the economic outlook and its implications for the several sectors of the economy and for individual commodities over the years ahead. At the state level the Agricultural Extension Service holds area and county "outlook" meetings to acquaint farmers with the economic forces that will affect the outlook for prices during the forthcoming year.

At one time it was hoped that such economic information activities would, in themselves, help to stabilize commodity output and prices. By anticipating the forces that would affect prices, farmers were expected to restrict production if low prices were anticipated and to expand production in anticipation of rising prices.

Many students of commodity price behavior have been convinced, however, that even with adequate information the price system could never become a completely reliable guide upon which to base decisions governing the allocation of resources. Most analysts agree, however, that the system does do an effective job of determining the prices that will effectively clear the market of whatever is produced. The major criticism that has been made is that the price fluctuations that result from unregulated market processes are too great and that farmers need more definite guides in making their production decisions. An implication that has frequently been drawn from this criticism is that the government should actively intervene in the market to stabilize prices and protect commodity producers from the price fluctuations resulting from the free play of market forces.

Programs to overcome inequities in income distribution

When the pricing system is functioning efficiently, it is completely impersonal in its operation. When consumer tastes shift away from a product for which demand was formerly strong, or when technological change creates industrial substitutes for products formerly produced by agriculture, or when transportation or natural resource development results in the opening up of new regions, supply and demand functions shift and prices of products or inputs are affected. The production adjustments induced by the price change also result in income gains and losses for individual families, for whole producing regions, and for the workers and owners of resources in the industrial and commercial sectors that serve agriculture.

The income adjustments forced upon families and regions as a result of such changes represent a second reason why farmers and their representatives frequently demand that governments intervene in the market process to prevent production, consumption, or prices from freely responding to changes in supply or demand.

Dissatisfaction with the supply-demand feedback system, which relies on price as a signal for changes in production or consumption activities, therefore rests on two grounds. The first is that the system is inefficient—that it does not operate with sufficient speed and precision. The second rests on equity grounds—that when it works efficiently some families and regions are unfairly penalized by income losses and other families and regions receive unearned "windfall" gains. Both sources of dissatisfaction were present in the demands for "fair" or parity prices for agriculture and the associated acreage control and price support programs that were instituted in the United States following passage of the Agricultural Adjustment Act of 1933. Parity price in the Agricultural Adjustment Act of 1933 was the price that would give agricultural commodities (say one bushel of wheat) the equivalent purchasing power over articles that farmers bought, as in the base period (for most commodities) from August 1909 to July 1914. The 1909–1914 base was chosen because of the relatively favorable prices for agricultural commodities that prevailed at that time. Technological and other changes that have modified the relationship between production costs and the parity price have made this earlier base obsolete. Today, there is no clear relationship between "parity price" and "parity income" for farm and nonfarm families.

The problems posed by the efficiency and equity criticisms of the supply-demand feedback system is that efforts to improve the pricing efficiency of the system are frequently inconsistent with efforts to use the price system to achieve income distribution objectives. Figure 24-6 can be used to illustrate this problem. Suppose that the demand for a crop, such as potatoes, is initially represented by $D'D'$ and the supply by SS. Price is equal to p_2 and production is q_2. Suppose, as incomes rise, that consumers' tastes shift away from potatoes and can, after a time, be represented by DD. In order to protect the incomes of potato producers, the government could attempt to hold the price at p^2. At p_2 prices consumers will now be willing to consume only an amount q'_2, whereas producers will be willing to produce an amount q_2. A surplus equal to $q_2 - q'_2$ will appear unless the government also takes action to restrict the amount of potatoes that can be produced through acreage restrictions or marketing quotas.

The difficulty of using the supply-demand feedback system to simultaneously achieve pricing efficiency and equity in income distribution has been increasingly recognized by policy makers in recent years. Extension of Social Security to farmers in 1956 represents one attempt to provide for greater equity in income distribution through direct income payments rather than through commodity prices. Agricultural programs of the 1960's

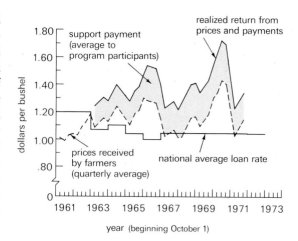

FIGURE 24-11. *Corn prices and support rates. The area in gray indicates the share of the realized price of corn which the producer receives in the form of a direct payment.* [After USDA, 1972 Handbook of Agricultural Charts, *October 1972.*]

continued the trend of recent years to shift toward more direct payments to commodity producers and to place less reliance on price supports to transfer income to farmers (Fig. 24-11).

In the Food and Agriculture Act of 1965 there were two types of direct payments. Farmers who participated in the program by holding 20 to 50 percent of their base acreage out of feed grain production were eligible for a payment of 30 cents per bushel on the "projected" yield on 50 percent of their base acreage. Additional "diversion payments" could be obtained by diverting more than 20 percent of the base acreage (but not more than 50 percent). These "diversion payments" were equal to half the "projected" yield multiplied by the support level ($1.00 + .30) in any acres diverted from production beyond the 20 percent (but not over 50 percent) necessary to participate in the program.

The basic program remained essentially unchanged until 1971. Diversion requirements and options, and payment levels, were adjusted annually to induce a quantity of output in line with anticipated utilization at some target price (see Fig. 24-11). The Set-Aside Program covering crops for 1971, 1972, and 1973 differs in one critical respect from the earlier program. Under the Set-Aside Program receipt of price support payments for feed grains was not tied directly to planting restrictions on feed grains. Farmers were eligible for price supports on feed grains if they diverted a specified percentage of their total cropland from production. The diverted acreage could come out of any crop, not just feed grains. Therefore, for a given level of acreage diversion, the Set-Aside Program was much less restrictive on feed grain producers than were previous programs. To some extent precision of output control was traded for greater freedom to plant.

In 1971 the only requirement for government diversion payments and price support loans was to idle cropland equal to 20 percent of the participant's base acreage. In the 1972 program the minimum set-aside was increased to 25 percent. The loan rate was set at $1.05 and the diversion payment at 80 cents. Additional payments were also made available as an inducement to divert additional acreage beyond the mandatory minimum. In 1973 further modifications were made in the set-aside program in order to encourage expansion of soybean and wheat production.

It seems apparent that the dominant trend in the development of policy governing agricultural commodities in the United States is toward utilization the supply-demand feedback system to achieve greater precision and efficiency in allocating resources. Programs to achieve equity in income distribution between farmers and nonfarm families, and among farm families, are more likely to rely on use of direct payments and extension of such government services as Social Security,

Medicare, family assistance, and other programs that do not directly affect commodity prices.

PERSPECTIVE

This chapter has presented a framework for understanding the role of economic forces in crop production and marketing decisions. Particular emphasis has been placed on (a) the role of supply-demand feedback systems in influencing production and marketing decisions, (b) the role of markets and marketing institutions in the functioning of this system, and (c) the contribution that government makes in increasing marketing efficiency and in balancing efficiency with equity to the many individuals and organizations that contribute to the production and marketing of agricultural commodities.

The general principles developed in this chapter are treated in greater depth in the next five chapters. Chapter 25 explores in greater detail the forces that affect the position and slope of the demand curve for individual crops and how demand changes with economic development. Chapter 26 discusses some of the domestic market and international trade policies that nations have developed to regulate prices and market activity. Chapter 27 discusses the organization and management of agricultural research systems. Chapter 28 examines the economics of technical advance in crop science and in crop production. In Chapter 29 specific attention is given to the dynamic role that technological change in agriculture plays in national economic development.

Selected References

Barker, Randolph. *"The Economics of Rice Production."* A/D/C *Teaching Forum,* Farm Operations and Management, No. 5. The Agricultural Development Council, Inc., New York, June 1971. (An excellent discussion of farm management decision processes and information needs and simpler analytical methods used in making farm management decisions.)

Cochrane, Willard W. *Farm Prices: Myth and Reality.* University of Minnesota Press, Minneapolis 1958.

(Price-income problems of commercial agriculture in the United States. Section one deals with farm price-income behavior, in particular the myth of an automatically adjusting agriculture; section two presents an analysis of farm-price behavior; and section three discusses certain aspects of economics and policy such as the economist as a policy adviser, blind alleys, and hard policy choices.)

Greig, W. Smith. *The Economics of Food Processing.* Westport, Connecticut, The AVI Publishing Co.,

Westport, Connecticut, 1971. (Discusses the characteristics of the food processing industry in the United States, including its structure, growth regulation, technological base, locational characteristics, vertical integration, and future trends.)

Headley, J. C., and J. N. Lewis. *The Pesticide Problem: An Economic Approach to Public Policy.* The Johns Hopkins University Press, Baltimore, 1967. (A useful discussion of the economic aspects of pesticides. Pesticide use is discussed from the perspectives of both farm production and public policy.)

Heady, Earl O. *A Primer on Food, Agriculture and Public Policy.* Random House, New York, 1967. (A useful elementary introduction to the problems of agriculture in developed economies. Considers public policies for those who must leave agriculture, for those who remain in agriculture, and in relation to world food needs.)

Kohls, Richard L., and W. David Downey. *Marketing of Agricultural Products* (4th ed.). Macmillan, New York, 1972. (A marketing text written for the beginner in economics. Its approach is mixed—partly functional, partly institutional, and partly by commodities. Part one develops the framework of the marketing problem; part two discusses some functional problems—that is, agricultural prices, market organization, and market functions; and part three presents commodity marketing problems on a commodity by commodity basis.)

Ruttan, Vernon W., Arley D. Waldo, and James P. Houck. *Agricultural Policy in an Affluent Society.* W. W. Norton, New York, 1969. (A collection of articles on the economics of U.S. agricultural policies by leading agricultural economists and policy makers. The book includes articles on political power and agricultural policy, price and income policy for commercial agriculture, food and fiber marketing policies, rural poverty and income distribution, and agricultural trade, aid, and development policies.)

Tomek, William G., and Kenneth L. Robinson. *Agricultural Product Prices.* Cornell University Press, Ithaca, New York, 1963. (This book provides a firm understanding of how commodity prices are determined and how changes in commodity prices are forecast.)

Tweeten, Luther. *Foundations of Farm Policy.* University of Nebraska Press, Lincoln, 1970. (A historical and analytical treatment of agricultural policy issues. Provides an in-depth treatment of the policies for commercial agriculture and the policies and programs designed to overcome rural poverty.)

CHAPTER 25

The World Market for Plant Products

During the years since World War II, a fundamental restructuring of the world market for agricultural products has occurred. The general direction of agricultural trade has shifted from a North–South direction to an East–West direction. The less developed countries' share of agricultural trade has declined. These have, as a group, shifted from being net exporters to net importers of basic food crops. Their exports of agricultural raw materials to the developed countries have also lagged. Among the developed countries feed crops have replaced food crops as the dynamic factor in agricultural trade. New opportunities for trade between the U.S.S.R. and Eastern Europe, particularly in food grains, feed grains, and oilseeds, have emerged.

In discussing the factors that have brought about these shifts in market patterns, attention will be focused on the food and feed crops, tropical beverages, agricultural raw materials, and forestry products (Table 25-1). The extent to which these crops are traded among countries or are consumed domestically varies widely. Only about 5 percent of world rice production enters international

trade, whereas 93 percent of rubber production is traded internationally.

WORLD MARKET TRENDS

In the international economic system that emerged during the nineteenth century, and persisted until World War II, raw materials, particularly food and fiber, were exported from the most recently settled countries of the temperate region and from the tropical-colonial areas to the developed countries. Industrial products were exported by the industrial countries to the less developed countries. It was considered that it would be to the economic advantage of both the developed and the less developed countries if each country pursued its comparative "natural advantages." Furthermore, by capitalizing on their natural advantages in the production and export of raw materials, the less developed countries could earn the foreign exchange, or attract an inflow of foreign capital, that would permit them to evolve first from an underdeveloped to a developed agricultural economy and then from

638

Table 25-1 IMPORTANCE OF WORLD TRADE
IN TERMS OF PERCENTAGE OF WORLD
PRODUCTION OF SELECTED CROP CLASSES,
1964–1966.[1]

CATEGORY	PERCENT
Food and feed	
Wheat	21.0
Rice	5.4
Coarse grains	8.5
Sugar	28.0
Fats and oils[2]	21.7
Citrus fruit	21.1
Tropical beverages	
Coffee	67.0°
Cocoa	87.2°
Tea	50.9°
Wine	7.6
Tobacco	18.7
Agricultural raw materials	
Cotton	34.4
Rubber	93.1°
Jute and allied fibers	33.0°
Hard Fiber	68.0°
Forestry products[3]	5.4°

Source: From *Commodity Projections 1970–80.* UN FAO, Rome, 1971.

[1] Asian centrally planned economies excluded. Trade data for 1964–66 are exports.

[2] Including oil content of oilseeds traded as such.

[3] Production excludes fuelwood.

° Calculated from production and export data (1964–66 average), *FAO Commodity Review and Outlook 1970–1971.* UN FAO, Rome, 1971.

a highly developed agricultural economy to a diversified agricultural and industrial economy.

This was the pattern that was actually followed by the United States, Canada, Australia, and a few other countries during the last half of the nineteenth century and the early half of the twentieth century. Before 1890 the United States, Russia, and the Danube countries of Eastern Europe were the major grain exporters. After 1890 grain exports from Australia, Argentina, and Canada expanded rapidly. India was a major exporter of wheat between 1880 and 1900. Burma and Thailand became major exporters of rice following the opening of the Suez Canal. Exports of such tropical products as coffee, tea, cocoa, and rubber expanded rapidly. At the beginning of the twentieth century the system of comparative "natural" advantages appeared to be working well.

This system gradually broke down after World War I and was essentially reversed after World War II, with many raw materials and foods being exported from the developed to the less developed countries. Three major factors brought about these changes. The first was the erection of trade barriers by most of the industrial nations on the imports from less developed countries that might compete with the products of their own farmers. The United States for example, found it necessary to erect barriers to the importation of wheat, cotton, corn, tobacco, and other products as farm price programs were developed that held prices above the world market level. The temperate region countries of Europe and North America erected trade barriers against the imports of raw sugar from tropical areas in order to encourage domestic production of sugar beets.

The result was a shift away from a system of free trade based on agricultural commodities to one that placed severe limitations on imports of tropical agricultural products other than beverages, such as coffee, tea, and cocoa, raw materials, such as rubber, and food crops, such as bananas, that did not compete directly with the agricultural products grown by farmers in the developed countries. This in turn depressed the prices and the growth of export earning by the countries that traditionally depended on agricultural exports for foreign exchange earnings (Fig. 25-1). The lag in prices and export earnings forced the exporting countries to reduce their own development efforts.

The second factor, the rapid growth of population in the less developed countries (see Chapter 4), has resulted in a sharp reduction in the exports of basic food crops, particularly food grains, from the less developed to the more developed countries. The rise in the rate of population growth has reduced the internal supply of food available for export at the same time that trade barriers were reducing the demand for food exports from the less developed countries.

The third factor, technological change in the developed industrial countries, has contributed to a reversal in the direction of trade (see Chapter 28). A developed industrial economy is capable of making fertilizer, farm machinery, and various mechanical, biological, and chemical inputs available to agriculture in large quantities and on favorable

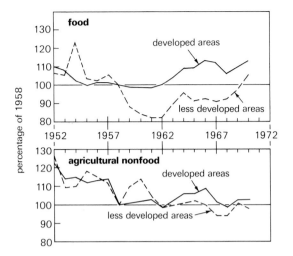

FIGURE 25-1. *Prices of food and nonfood agricultural commodities in developed and underdeveloped areas.* [*UN, Statistical Yearbook, various issues.*]

terms. The relatively high prices enjoyed by farmers in the industrial countries as a result of trade restrictions added further incentives toward adoption of the new technological inputs. The strong technological and economic incentives to expand production, plus the slow growth in demand, particularly in the United States, combined to produce agricultural surpluses in the developed countries. Policies were then adopted that subsidized the export of these surpluses to the major commercial markets and to the countries that had rising food deficits. Some subsidized exports from the developed countries have been directly competitive with the products produced for commercial export by the less developed countries. Subsidized grain exports from the United States to India and Indonesia have competed with commercial exports from Thailand and Burma.

The net effect has been a fundamental restructuring of the world trade pattern for food crops. This is particularly true of food grains (Fig. 25-2). Before World War II, from 1934 to 1938, the less developed countries exported 11 million metric tons of grain annually. Between 1948 and 1952 the less developed countries became a net importer of grains. Grain imports by the less developed countries continued to rise and exceeded 25 million metric tons in the mid-1960's. By the early 1970's there was some hope that the long term

trend might be reversed. The introduction of new fertilizer-responsive varieties of wheat and rice was resulting in rapid growth in yield per hectare in some less developed countries. By 1970 net imports of food grains by these nations had declined to approximately 15 million metric tons. This decline in developed country exports to the less developed countries was at least partially offset by rising exports to the socialist block countries.

During the time that the less developed countries shifted from being net exporters to net importers of food grains, they continued to expand their production and exports of tropical beverages and agricultural raw materials. The less developed countries have continued to rely heavily on agricultural and raw materials exports in order to earn the income they need to purchase imports from the developed countries. Yet exports of their agricultural commodities have tended to be smaller than their imports of the commodities produced by the developed countries (Fig. 25-3). In part this lag has been due to relatively slow growth in the demand for tropical beverages (tea, coffee, cocoa) and raw materials (rubber, cotton, jute, sisal). The slow growth in exports has also reflected, in some nations, limits in supplies, since the production of agricultural raw material, beverage, and food crops for export has become increasingly competitive with the use of the same resources, particularly land, for domestic food production.

FIGURE 25-2. *Net flow of world grain (in millions of metric tons).* [*After* Food for Freedom, *U.S. Departments of State and Agriculture, 10 February 1966; World Grain Trade Statistics 1969/70, FAO, Rome, 1971.*]

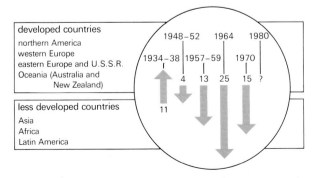

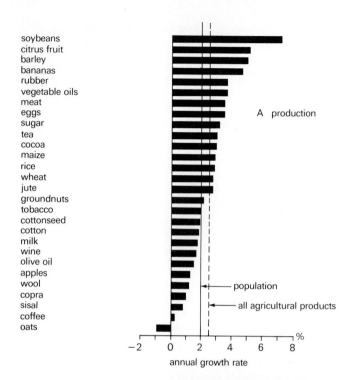

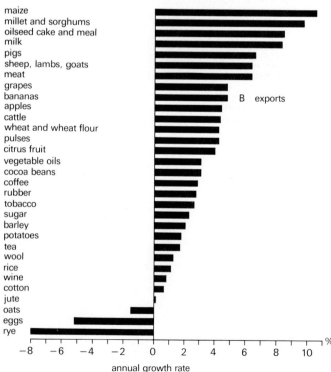

FIGURE 25-3. *Annual growth rates in (A) world production and (B) exports of the major agricultural commodities (for production, excluding mainland China; for exports, including exports to the USSR, eastern Europe, and mainland China, but excluding exports from these countries), 1958/60 to 1968/70.* [*After* The State of Food and Agriculture, 1971. *FAO, Rome, 1971.*]

World production and trade in forest products have also undergone substantial changes in recent years. The output of wood for industrial purposes (plywood, fiberboard, paper and paperboard, and pulp) has risen rapidly (Fig. 25-4). Exports have expanded even more rapidly, and those of forest products from the less developed countries—tropical hardwood logs, sawn hardwood, and plywood and veneers from Asia and Africa—have expanded much more rapidly than forest product exports from the developed countries.

Therefore, although exports of their agricultural products may have declined, the less developed countries have continued to increase their share of total world trade in forest products. However, a disproportionate share of total wood exports is still roundwood and sawnwood rather than the more highly processed wood products.

INCOME AND THE DEMAND FOR FOOD

In the early phases of economic development, growth in the demand for food and fiber is highly responsive to changes in both population and income. In countries with relatively low incomes the demand for agricultural products is highly responsive to increases in per capita income. As incomes continue to rise the proportion of the increase that is spent on agricultural products rises less rapidly. And at very high levels of income, additional increases in income may be associated with no real increase in the quantity of food consumed.

Total food demand

At any given stage in a nation's growth the rate of growth in demand for food—that is, the rate at which the demand curves are shifting to the

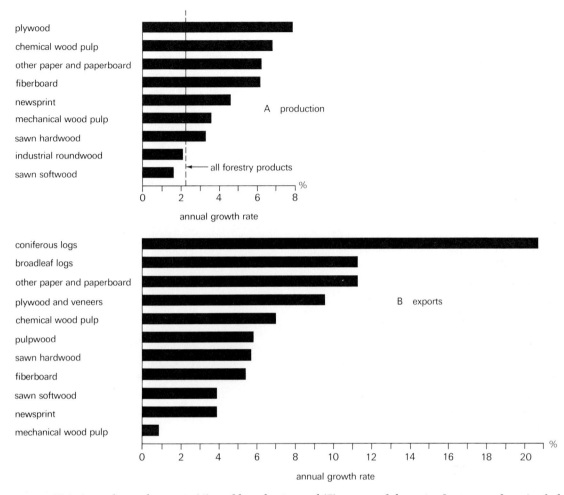

FIGURE 25-4. *Annual growth rates in (A) world production and (B) exports of the major forestry products (excluding mainland China), 1958/60 to 1968/70.* [*After* The State of Food and Agriculture, 1971. *FAO, Rome, 1971.*]

right (Figs. 24-6 and 28-8)—can be estimated by summing the population effect and the income effect. The population effect depends only on the rate of growth of population. An annual population growth rate of 1 percent adds 1 percent to the growth of demand. The income effect is more complicated. It depends on both the rate of growth in per capita income and the **income response,** or **elasticity,** of **demand.** If, for example, the annual per capita income growth rate is 1 percent per year and if a 1 percent increase in income is accompanied by a 0.5 percent increase in expenditures on food (that is, if the income elasticity of

demand is 0.5) the income effect adds 0.5 percent per year to the growth of demand for food. In this example, the total growth in demand obtained by summing the population and income effects would be 1.5 percent per year.

The importance of the income and population effects is quite different, depending on the population growth rate, the per capita income growth rate, and the income elasticity of demand. The rate of growth in demand for agricultural products is greatest in a country like Mexico, which is experiencing both a rapid rate of population growth and a rapid growth in per capita income combined with a high income elasticity of demand. In low-income countries, where a rapid growth in per capita income is combined with a high income elasticity of demand, the income effect may be of essentially the same magnitude as the population effect. The income effect, however, may be small despite a high-income elasticity of demand if per capita income is growing slowly. If the income elasticity of demand is low, as in the United States, the income effect may also be small, even if per capita income is rising rapidly.

Probably the most important factor in determining the level of the income elasticity of demand for total food is income itself. It appears that the income elasticity of demand for all food is about 0.8 when annual per capita incomes are around $100; about 0.5 when annual per capita incomes are around $500; and close to zero when annual per capita incomes are in the neighborhood of $2,000.

Table 25-2 illustrates the changes in total demand, and in the population and income components of demand, during a nation's economic and demographic transition from low to high per capita income. Note that the growth in food demand is highest relatively early in the development process.

Variations in demand among foods

Increases in demand that are due to population growth are typically proportional in the sense that the demand for each agricultural commodity expands by the same percentage. Growth of demand arising out of the income effect are not proportional. As per capita income rises the demand for some components of the diet expand rapidly, whereas others expand slowly or even decline. Responses in both quantity and quality of food occur as per capita income rises. The response of quality can be expected to continue even after the response of quantity has decreased to zero.

Data on the income elasticity of demand for food, in the form of calories, animal protein, and farm value, are presented for major economic regions in Table 25-3. The calorie elasticities are, in a sense, a measure of the pure quantity elasticities of demand. The farm value elasticities represents not only growth in demand for calories but the qualitative changes in diet that are reflected in a shift to foods of higher value. The elasticity of demand for animal protein reflects the effect of income growth on the demand for the highest value items of the diet.

Note that even in the relatively low-income countries the income elasticity of demand for energy foods, as measured in calories, is relatively low. Even at relatively low levels of income a substantial share of increased food expenditures is used to improve the quality of the diet. This is reflected by the relatively high farm value and animal protein elasticities. Note also that as incomes rise to around $500 and above the farm value elasticity of demand falls to the calorie elasticity level for the lowest income countries, and at relatively high income levels even the animal protein elasticities fall to relatively low levels.

Table 25-2   COMPARISON OF GROWTH OF DEMAND FOR AGRICULTURAL COMMODITIES, AT DIFFERENT STAGES OF DEVELOPMENT, HYPOTHETICAL CASES.

LEVELS OF DEVELOPMENT	PERCENT OF POPULATION IN AGRICULTURE	RATE OF POPULATION GROWTH	RATE OF INCOME GROWTH	INCOME ELASTICITY OF DEMAND	TOTAL GROWTH IN DEMAND	TOTAL GROWTH IN DEMAND, WITH 3% POPULATION GROWTH
Very low income	70%	2.0	.5	1.0	2.5	3.5
Low income	60%	3.0	1.0	.9	3.9	3.9
Medium income	50%	3.0	3.0	.6	4.8	4.8
High income	35%	2.0	4.0	.5	4.0	5.0
Very high income	20%	1.5	3.0	.1	1.8	3.3

Source: From Mellor, *The Economics of Agricultural Development.* Ithaca: Cornell Univ. Press. Copyright © 1966. Used by permission of Cornell Univ. Press.

Table 25-3 INCOME ELASTICITIES OF DEMAND FOR FOOD IN THE FORM OF CALORIES, ANIMAL PROTEIN, AND FARM VALUE OF ALL FOOD, FOR VARIOUS GEOGRAPHIC AREAS.

	ASIA AND FAR EAST (EXCLUD-ING JAPAN)	NEAR EAST AND AFRICA (EXCLUD-ING S. AFRICA)	LATIN AMERICA (EXCLUDING ARGENTINA AND URUGUAY)	MEDITER-RANEAN EUROPE	JAPAN	EUROPEAN ECONOMIC COMMUNITY	OTHER WESTERN EUROPE	UNITED STATES AND CANADA
GNP per capita (U.S. dollars,[1] 1957–1959)	165	260	491	575	910	1,285	1,440	2,190
Income[2] elasticities of the demand for all food expressed in terms of:								
Calories	0.6	0.4	0.3	0.2	0.2	0.1	0.01	−0.03
Animal proteins	1.5	1.2	0.8	0.9	0.9	0.6	0.3	0.23
Farm value	0.9	0.7	0.6	0.55	0.6	0.5	0.2	0.16

Source: From *Agricultural Commodities—Projections for 1970. Special Supplement FAO Commodity Review 1962.* UN FAO, Rome, 1963.
[1]Converted into U.S. dollars at 1955 prices according to purchasing power parity.
[2]Value of the income elasticity in the base period 1957–59. The values were derived from the projected demand for each major food group in each country or group of countries.

Changes in the composition of the diet as incomes rise have sharply different effects on the growth of demand for individual farm products. Direct consumption of calories in cereals and starchy food declines as incomes rise. Starchy foods are replaced in the diet by fats and oils, proteins, and fruits and vegetables. Nevertheless, the per capita utilization of cereals and starches continues to rise with income. They are used as feed for the larger numbers of livestock that are required to produce the animal proteins and fats that enter into the diet of higher-income families and nations (Fig. 25-5).

This shift in the demand for cereals and starches from a direct demand for food consumption to a derived demand for livestock feed also implies a change in the composition of grain production. The demand for feed grains, such as corn and sorghum, rises in relation to the demand for food grains, such as wheat and rice. The demand for oil crops such as soybeans, which are particularly valuable as a livestock feed because they have a high content of protein, rises in relation to the demand for such oil crops as oil palm and copra (coconut), whose by-products do not have a high protein content.

Shifts from direct food consumption to indirect consumption in the form of feed that enters as an input in animal production does not represent a feasible alternative for many crops. As incomes rise, per capita consumption of such crops as dry beans and peas, potatoes, and sweet-potatoes tend to decline both relatively and absolutely. In many countries the income elasticity of demand for these foods is negative (see Table 25-3).

FIGURE 25-5. *Per-capita consumption of cereals and starches in relation to per-capita income. Cross-sectional comparison among countries.* [*After* The State of Food and Agriculture, FAO, Rome, 1957.]

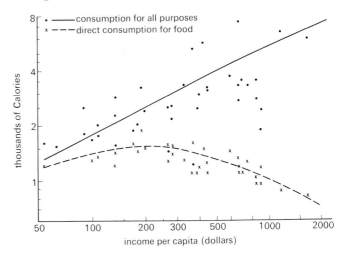

For the United States, it is possible to group the foods based on direct crop consumption into the following three categories with respect to their income elasticity of demand.

Negative Elasticity Foods
Dry beans and peas
Potatoes and sweet-potatoes
Flour and cereal products
Sugar and syrups

Low Elasticity Foods
Beverages (coffee, tea, and cocoa)
Most vegetables and fruits

High Elasticity Foods
Tomatoes and citrus fruits
Some tree nuts
Some fresh, canned, and frozen vegetables

These same rankings would apply with slight modification for much of Western Europe, Canada, Australia, and New Zealand. In the less developed countries, however, the demand for many of the foods ranked in the low or negative elasticity classes in the above list may be highly elastic.

The tropical beverages, primarily coffee, cocoa, and tea, are unique, in that production is concentrated heavily in the less developed countries of the tropics, and subtropics. Nevertheless, a substantial share of total consumption is concentrated in the higher income countries of the temperate regions. In the consuming countries demand is typically relatively inelastic with respect to income. Consumption expands with population growth, but very little in response to income growth. Consumption also tends to be relatively inelastic (that is, unresponsive) with respect to changes in price. As a result failure to achieve a close balance between the rate of growth in supply and the rate of growth in demand has a tendency to result in violent price fluctuations. Among the three products the demand for cocoa appears to be more elastic with respect to both prices and income because of its nonbeverage uses.

DEMAND FOR AGRICULTURAL RAW MATERIALS AND FOREST PRODUCTS

The factors affecting the demand for agricultural raw materials are quite different from those that affect the demand for basic food crop and tropical beverages. The agricultural raw materials are valued, not for their direct utility in consumption, but as raw materials for the manufacture of other products or consumer goods.

The demand for the agricultural raw materials depends, therefore, on (1) the growth in total demand for the end product in which they are used, (2) the availability of close substitutes, and (3) the price of substitutes. The demand for agricultural raw materials may, therefore, be sharply modified by technological changes that bring new substitutes into the market or change the prices of agricultural raw materials relative to substitutes. In recent years two major factors have combined to limit growth in demand for agricultural raw materials.

First, a series of technological changes has resulted in a decline in the raw material inputs of finished products. Bulk handling has reduced the need for cotton and jute bags, for example. The cumulative effect of such changes has been substantial. In 1900 the United States used approximately $22\frac{1}{2}$ cents worth of raw materials for each dollar of gross national product (GNP). By 1960 we were using only about $11\frac{1}{2}$ cents worth of raw materials per dollar of GNP. Second, the "chemical revolution" of the last half century has resulted in a massive substitution of synthetic materials for natural ones. Plastics, synthetic fibers, and synthetic rubber have been substituted for cotton,

Table 25-4 CHANGE IN RELATIVE IMPORTANCE OF WORLD'S PRIMARY FOREST INDUSTRIES, 1960 TO 1969.

INDUSTRY	PRODUCTION (MILLIONS OF CUBIC METERS)		PERCENT OF TOTAL	
	1960	1969	1960	1969
Sawnwood	337	400	68.5	59.3
Wood-based panels	22	54	4.5	8.0
Wood pulp	59	98	12.0	14.5
Paper and paperboard	74	123	15.0	18.2
Total	492	675	100.0	100.0

Source: From *Yearbook of Forest Products 1969–70.* UN FAO, Rome, 1971.

jute, hard fibers (abacá and sisal), natural rubber, and wood.

The effects of technological change have not all been in one direction. Biological research has increased the efficiency with which certain agricultural raw materials are produced. Breeding work carried on at the Rubber Research Institute of Malaya has produced new budded (grafted) varieties of rubber that yield two to three times as much as previously available varieties. As a result the rate of substitution of synthetics for natural rubber appears to be decreasing.

The demand for forest products (Table 25-4) has been subject to much the same forces as for industrial crops. In the industrial countries there has been substantial substitution for sawtimber of such industrial products as cement, aluminum, plastic, and others. Synthetic fibers have reduced the demand for wood fibers formerly used in rayon production. There have also been technological changes that have resulted in an increased demand for roundwood, particularly for sheet materials, such as plywood, fiberboard, and particle board.

There is also apparently a high-income elasticity of demand for pulp products. At per capita income levels of around $100, income elasticity is apparently in the 2.5 to 3.0 range; at per capita levels of $200 to $400 it ranges from 1.5 to 2.5. At European income levels, roughly $500 to $1,000, it is well above 1.0. Only at per capita income levels above $2,000, as in the United States, does demand for paper and board drop below unity.

The net effect of these several forces is expected to result in a world growth of demand for industrial roundwood of about 3 percent per year—slightly greater than the rate of population growth. In the developed countries, however, this rate is expected to be substantially lower, probably falling below the rate of population growth.

Even though consumption of wood and wood products in the developed countries is rising more slowly, relative to income growth, than in the less developed countries, the demand for wood in developed countries will remain important in the 1970's. The absolute size of the market in the developed countries is so large that, even if the rate of growth is somewhat slower, these nations can provide a dynamic market for wood and wood products from the underdeveloped world.

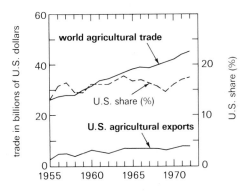

FIGURE 25-6. *U.S. and world trade in agricultural products, 1955–72, and trend projections to 1984.* [*After West, "World Trade Prospects for U.S. Agriculture,"* American Journal Agricultural Economics, *Vol. 54, December 1972, p. 830.*]

FIGURE 25-7. *The composition of agricultural exports from the United States.* [*After West, "World Trade Prospects for U.S. Agriculture,"* American Journal Agricultural Economics, *Vol. 54, December 1972, p. 828.*]

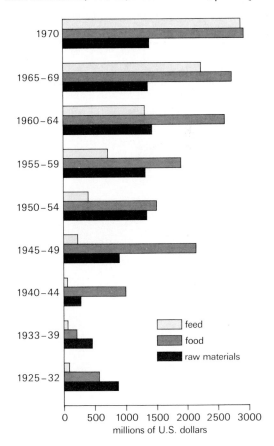

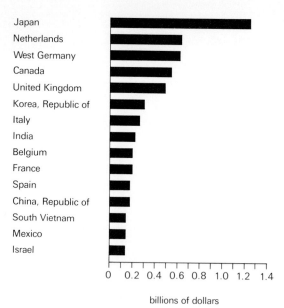

FIGURE 25-8. *U.S. agricultural exports by country, 1971.* [*After U.S. Department of Agriculture, 1971 Handbook of Agricultural Charts,* Economic Research Service, *November 1971.*]

FIGURE 25-9. *U.S. agricultural imports by origin, 1970–71.* [*After U.S. Department of Agriculture, 1972 Handbook of Agricultural Charts,* Economic Research Service, *October 1972.*]

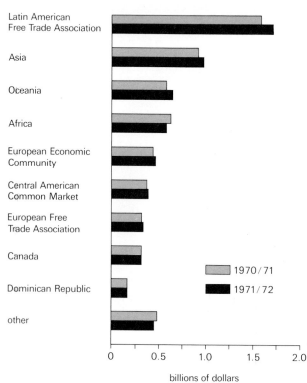

U.S. AGRICULTURE IN THE WORLD MARKET

U.S. agriculture is highly dependent on the world market. Exports account for more than half of the annual sales of wheat, soybeans, and rice. Approximately one of almost every $7.00 of U.S. farm sales comes from foreign markets. Agricultural trade is an important factor in the U.S. balance of trade. In 1971 the United States sold the rest of the world $1.9 billion more agricultural commodities than it purchased from the rest of the world, and supplied approximately one-sixth of the world's agricultural exports (Fig. 25-6). The United States has participated in, and is partially responsible for, the massive shift in the composition of agricultural exports from one primarily of food crops and industrial raw materials (such as rubber and cotton) to one of crops used mostly for animal feed, such as maize and soybeans (Fig. 25-7).

The continued importance of agricultural exports in an economy in which agriculture used only about 4 percent of the labor force and contributed 3 percent of national output is due to a very significant extent to the rapid gains in productivity that have resulted from technical change in American agriculture. The shift from a resource-based to a science-based agriculture has enabled agriculture in the United States to compete more effectively with other sectors of the economy for labor and capital. The growth of U.S. agricultural exports has reflected also the rapid growth of income and the changing composition of demand in Japan, Western and Eastern Europe, and the U.S.S.R. As their incomes have risen, consumers in these nations have increased their consumption of meat and poultry. This increase has created a rising demand for imports of feed grains, and the developed countries of Western Europe and Japan now account for a very large share of U.S. exports (Fig. 25-8). The agricultural imports into the United States, however, come primarily from the less developed countries (Fig. 25-9), but the growth in demand for these commodities—cocoa, coffee, tea, bananas, sugar, and natural rubber—is slow. Consequently, U.S. trading patterns for agricultural commodities tend to reinforce the long-term disequilibrium in agricultural trade patterns between the developed and less developed countries.

Selected References

Food and Agriculture Organization. *The State of Food and Agriculture, 1971.* United Nations, Food and Agriculture Organization, Rome, 1971. (An annual report published by the FAO on the world food and agricultural situation. Each annual issue presents an in-depth review in a specific area of food and agricultural development or policy, in addition to a review of current developments.)

Food and Agriculture Organization. *Wood: World Trends and Prospects.* United Nations, Food and Agriculture Organization, Rome, 1967. (An in-depth survey of the situation, problems, and prospects for wood and the wood-using industries and their products for the period 1950–75.)

Houck, James P., Mary E. Ryan, and Abraham Subotnik. *Soybeans and Their Products: Markets, Models and Policy.* University of Minnesota Press, Minneapolis, 1972. (Examines the behavior of the U.S. and world soybean markets since World War II. Examines foreign and domestic markets for soybeans and their products in relation to competing oilseeds, oils, and feeds.)

26

Commodity Market and Trade Policy

Trade in agricultural products constitutes a very large proportion of the total world trade volume. Because of the inelastic supply and demand relationships discussed earlier, international prices for agricultural commodities tend to be highly variable. When Cuba withdrew from the world sugar market in 1963, the free market price of sugar rose from the 3 to 4 cents per pound range to more than 8 cents in a few months. By 1965 the price had fallen below the 1963 level. After several years of relative stability, sugar prices again rose to a new peak of more than 9 cents per pound in early 1972 (Fig. 26-1). Price movements of this magnitude have relatively little effect on the foreign exchange position of the developed countries. They can have extreme implications for the economic life of exporting countries that depend on sugar for a major share of their foreign exchange earnings.

Achievement of a satisfactory answer to the problem of price instability of internationally traded products has come to represent a major policy objective of many of the less developed countries. The United States has also committed itself at a number of international conferences and conventions to work toward policies that "should contribute toward stabilization of agricultural prices at levels fair to producers and consumers alike."

It is clear that the United States, in the past, has not always adopted market and trade policies designed to serve our national interests in the broadest sense. We have frequently adopted policies that sacrifice broad national and international objectives in the interests of particularly vocal or politically strategic groups of agricultural producers or processors. We have, since World War II, engaged in massive development efforts, directed toward the recovery of Western Europe and Japan from the effects of the war and toward the development of the less developed nations of the world. These development efforts have frequently been frustrated by our lack of experience and lack of knowledge of development processes. Our efforts have also frequently been hindered by the difficulty of reconciling our own short-term national interests with the development objectives of the countries we have been attempting to assist.

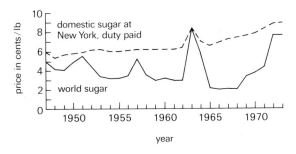

FIGURE 26-1. *Spot prices of raw cane sugar. Note that "world sugar" refers to Cuba, Raw 96°, export price; destinations other than the United States (No. 4 contract), f.o.b. From 1961, prices are No. 8 contract, bagged, f.o.b., and stowed at Greater Caribbean ports, including Brazil. [Sources for world sugar: FAO Production Yearbook, 1947–1966; USDA, Sugar Reports, 1967–1972. Sources for domestic sugar: USDA, Sugar Situation, 1965–1972; Sugar Reports, 1965–1972.]*

More recently, however, United States development efforts appear to reflect increasing sophistication in the technical and economic aspects of development.

PRINCIPLES OF COMMODITY PRICE STABILIZATION

There are only three general policies that can be adopted to stabilize market prices, or prices paid to producers, whether in the domestic or the international market. These are policies that would (1) control or influence supply, (2) control or influence demand, or (3) require the payment of subsidies (incentives) and/or the collection of taxes (penalties). Within each of these general categories there are limited alternatives.

Control of supply may be achieved through control of the amount produced or marketed, destruction of output, restrictions on imports or exports, and storage.

Control of demand may be undertaken by instituting income supplements to consumers, such as the Food Stamp Plan; multiple price systems, which permit some individuals or countries to purchase at lower prices than others; and forced utilization of products, such as mixing require-

ments that specify a certain percentage of flour from domestic wheat to be mixed with flour milled from imported wheat.

Subsidies and taxes can be used as incentives or penalties to induce desired producer or consumption behavior. Subsidies can be paid to induce desired changes in land use—for example, to induce Malay farmers to plant rubber trees or American farmers to remove cropland from production. In the United States, taxes can be imposed as a penalty, to prevent planting of wheat and feed grains beyond acreage allotments.

In addition to the techniques of price stabilization institutional arrangements must also be considered. The two main methods employed are agreements between two or more governments and arrangements organized through an international organization. The arrangements by which the United States regulates sugar imports are an example of the first method. The International Wheat Agreement is an example of the second. Since these two commodity arrangements also illustrate the use of several of the major price stabilization techniques, they will be discussed in greater detail.

SUGAR TRADE AND PRICE POLICY

World trade in sugar is characterized by special marketing arrangements between exporting and importing countries. Only about half of total exports enter international trade channels without any preferential arrangements. The governments of nearly every sugar-producing country exercise some degree of control over the production, refining, and marketing of sugar. A number of importing countries have preferential arrangements with dependent overseas territories or independent countries with whom they have close political ties. In addition most importing countries impose tariffs or import quotas to protect their domestic producers.

For almost 150 years—from 1789 to 1934—the United States sugar industry was protected and regulated solely by tariff duties. During 1934 the United States established more restrictive import policies under the Jones-Costigen Act in order to protect domestic producers and foreign suppliers from the economic disorganization in world trade, that accompanied the Great Depression.

The basic structure of United States sugar policy since 1934 has involved (1) establishment of an annual domestic marketing quota for sugar based on an estimate of consumption requirements, (2) a division of this quota between domestic and foreign producers, and (3) an allocation of the domestic quota among mainland and offshore (Hawaii and Puerto Rico) sugar beet and cane producers and of the foreign quota among supplying nations. The Jones-Costigen Act has been modified nine times since 1934, but the basic structure of policy has remained unchanged.

The program is managed by the United States Secretary of Agriculture in accord with the detailed rules and regulations set forth in the legislation. The most important decision that faces the Secretary of Agriculture is the estimation of the total sugar requirements. The Jones-Costigen Act contains a formula for determining the target price that the Secretary of Agriculture considers, along with other factors, in establishing the total sugar requirements. In 1972 the formula or target price was approximately 9.0 cents per pound.

The pressure exerted by producers to maintain a price at least as high as the target price are typically stronger than pressures that consumers exert to prevent the price from rising above the target price. The Secretary of Agriculture frequently adopts a strategy of establishing a low consumption estimate early in the year and raising it later if it becomes necessary to allow increases in domestic marketing or imports to prevent prices from rising too far above the target prices.

Allocations of the marketing quota to domestic producing areas and foreign countries are made with reference to guidelines established by the Congress (Table 26-1 and Box 26-1). In domestic areas the Secretary of Agriculture establishes both processor allotments and individual farm proportionate shares, except in years of shortage, when he may remove restrictions.

This places the Congress in a position to exert a major impact on an important component of the foreign exchange earnings of many of the tropical exporting countries by making relatively small changes in their quota. The result is that domestic producers, foreign suppliers, and sugar processing and refining organizations engage in extensive public relations and lobbying activities designed to influence the size of the quota.

The effect of the revisions of the sugar legislation during the 1950's was to increase the share of the total quota going to domestic producers and to

Table 26-1 1972 SUGAR QUOTAS AND PRORATIONS.

AREA	SHORT TONS, RAW VALUE	PERCENT
Domestic areas		
Domestic beet sugar	3,500,000	29.66
Mainland cane sugar	1,643,000	13.92
Hawaii	1,218,238	10.33
Puerto Rico	175,000	1.48
Total domestic	6,536,238	55.39
Foreign countries with specified tonnages		
Philippines	1,401,761	11.88
Ireland	5,351	0.05
Total foreign with specified tonnages	1,407,112	11.93
Total domestic and foreign with specified tonnages	7,943,350	67.32
Foreign countries with temporary quotas and prorations[1]	3,856,650	32.68
Grand total	11,800,000	100.00

Source: From USDA, *Sugar Reports*, June 1972, p. 14.

[1] Quotas of 30 foreign countries receiving prorations on percentage basis of quotas withheld from Cuba and Southern Rhodesia.

Box 26-1 ADMINISTRATIVE ACTIONS RELATING TO 1972 SUPPLIES

Sugar Requirements Decreased and Deficits Redetermined

A decrease of 200,000 short tons, raw value, in the determination of sugar requirements (quotas) for the continental United States was announced on June 5 by the U.S. Department of Agriculture. Quotas for 1972 now total 11.8 million tons.

At the same time, deficits of 99,864 tons were declared and reallocated to foreign countries able to supply the sugar. Of this total, 30,000 tons represented an increase in the Puerto Rican deficit. Other deficits were the Bahamas, 29,192 tons; the West Indies, 15,940 tons; Uganda, 15,252 tons; Bolivia, 7,028 tons; and Panama, 2,452 tons.

Among the domestic areas, the quota for the mainland cane area was reduced by 34,667 tons as a result of the decrease in total requirements, and the quota for Puerto Rico was reduced 30,000 tons to reflect the further deficit declaration. The quotas for the domestic beet sugar area and Hawaii were undisturbed. The basic quota for the beet sugar area was reduced by 95,333 tons and the previously declared deficit decreased by the same amount. Hawaii's quota is not affected by a change in requirements.

Total foreign quotas are decreased by 135,333 short tons, raw value, 65,333 tons because of reduced domestic deficits, and 70,000 tons as the foreign share of the decrease in requirements. However, because of the reallotment of additional foreign deficits, the quotas of the 25 countries able to supply additional sugar were reduced by only a total of 59,650 tons.

The Sugar Act requires USDA to revise to the extent necessary the annual sugar requirements determination whenever the average price of raw sugar for seven consecutive marketing days is four percent or more above or below the price objective of the Act. For the seven consecutive market days ended June 2, the average price of raw sugar was 8.64 cents per pound and thus four percent below the average price objective of 9.00 cents per pound.

Source: Quoted from USDA, *Sugar Reports*, June 1972, p.13.

---

reduce the share available to foreign producers. When Cuba, previously the largest United States supplier, began diverting her production to the Soviet Union and China in 1960, the share going to United States producers was again raised. Quota shares of the traditional foreign quota holders were also raised, and a number of additional countries were added to the quota list.

The quotas that are made available to foreign suppliers are particularly valuable to them, since they receive the United States quota price (recently 9.0 cents) less transportation costs and an import duty of 0.625 cents per pound. This is more than double the price at which sugar was available on the international market during most of the last decade and well above the 3 to 4 cents per pound range within which sugar prices have typically fluctuated. Under the old tariff system that prevailed prior to 1934 world prices and United States raw sugar prices, excluding duty, were generally quite similar. Since the advent of the quota system, however, United States prices have been at a premium above the "free market" price. Furthermore the difference between domestic and international prices has tended to widen (see Fig. 26-1). During the years 1934–1938, the premium averaged about 1.12 cents, and during the decade 1952–1962, about 1.84 cents. Following the break in world prices in the mid-1960's, the premium averaged well above 4.00 cents until world prices began rising again in 1968. In 1972, after a substantial rise in world prices, the premium had declined to a little more than 1.00 cent per pound.

The sugar policy of the United States has clearly been successful in achieving high and relatively stable prices for domestic sugar producers (Fig. 26-1). It has also resulted in relatively stable prices to United States consumers. Prices to foreign quota holders have also been above the level that would have prevailed in the absence of a quota system.

It is also clear that this has been an expensive program for United States consumers and taxpayers and that it has worked to the long-term disadvantage of the tropical exporting countries. Sugar prices paid by domestic comsumers have been far higher than they would have been if target prices had been set at or near world market levels. The high target prices have acted as an incentive for domestic producers to use their political power to increase their share of the total quota, thus reducing the size of the market available to the tropical exporting countries.

There have also been other undesirable consequences of the program. A requirement that imports of sugar be in a raw or semiprocessed form has led to a legitimate concern on the part of many less developed countries that United States policy is directed to limiting their industrial development. Unilateral decisions regarding the size of quotas for individual countries by the Congress has sometimes led to the assertion that we were using these quotas to force compliance with United States political objectives. The existence of such a restrictive trading system has also acted to weaken the U.S. bargaining position in efforts to achieve freer access to the European Common Market for agricultural products in which the United States has a comparative advantage in production.

INTERNATIONAL STABILIZATION EFFORTS: WHEAT

The history of international action in the primary commodity markets is a long one. Attempts to regularize production and sales of some commodities were made in the late nineteenth century. The great proliferation of international stabilization efforts came after World War I, first as a reaction to the violent price fluctuations of the early postwar years, and then to the depression of the 1930's.

The action during the interwar period was taken by producers in defense of prices or markets. At first they formed producer cartels, imposed restrictions on production, and established division of the market in an effort to sustain or restore prices. The procedure typically led to consumer or importer resistance in the form of purchases from nonparticipants in the control schemes. This had a tendency to undermine the price structure and led to the breakdown of the arrangements.

The experience of stabilization efforts in the interwar years made it clear that successful stabilization of international commodity markets would require participation by both exporting and importing countries. This meant that any agreements would not only have to protect producers from excessive declines in market prices and guarantee market access, but that importing countries would have to be protected from undue price increases and from shortages in supplies. The first International Wheat Agreement (IWA) to be instituted after World War II, in 1949, embodied this general principle. In essence, the 1949 International Wheat Agreement was a long-term, multilateral commodity contract that specified the basic maximum and minimum prices at which "guaranteed quantities" of wheat were to be offered for sale by participating export countries or were to be purchased by participating importing countries. The selling obligations of the exporters were effective only if prices rose to the maximum. Similarly, the buying obligations of the importers came into effect only at the minimum price.

The agreement did not impose any direct restrictions on the total trade or production of any country or on the level or method of pricing wheat sold to countries that did not sign the agreement. As long as a participating exporting country stood ready to sell its guaranteed quantity at the maximum price, it was free to export to any destination, at any price, as much additional wheat as it desired. Similarly, a participating importing country was free to buy any amount of wheat at any price as long as it stood ready to fulfill its agreement obligations to purchase at the minimum price. No restrictions were placed on domestic wheat pricing policy, except that such policies were not to impede the free movement of prices between the maximum and minimum prices in the international market.

The world wheat market has been dominated by the five leading producers: Argentina, Australia, Canada, France, and the United States. These nations have typically supplied in excess of 80 percent of the world's wheat. Canada and the

United States alone have generally provided about 60 percent of the world trade. Six other countries, the United Kingdom, West Germany, Poland, Brazil, the Netherlands, and Japan have accounted for about half of the "commercial" world wheat imports, with the remainder accounted for by more than 50 small importing nations. Of the major producing countries only Argentina did not participate in the initial agreement. Every sizable "free world" importer subscribed to the agreement. The agreement has been reused periodically since it became effective in 1949. The revisions are summarized in Table 26-2.

The most significant revision took effect in 1968. Trade in agricultural products formed an important part of the negotiations designed to achieve a general reduction in tariffs among trading nations that were organized under the General Agreement on Tariffs and Trade (GATT) in Geneva between 1963 and 1967. A new International Grains Arrangement was negotiated in 1967. Its Wheat Trade Convention (WTC) entered into force on July 1, 1968.

This arrangement differed in several important ways from the agreement that had been in force since 1962. Previously the Canadian Lakehead (Fort William) had served as the basic point of origin for the calculation of equivalent prices for other shipping points. The disadvantage of this location was the fact that the Greak Lakes are closed to shipping during the winter months. In the new WTC, U.S. Gulf ports were chosen as the basic point of origin on which to base the price range. An attempt was made to give greater precision to the price range negotiated under the agreement by the establishment of minimum and maximum prices for 14 different wheats rather than only one (Manitoba No. 1) as in the former agreements (Table 26-2). A Prices Review Committee was established to consider adjustments in maximum and minimum prices when world market conditions warranted them.

In retrospect it appears that the new price levels established under the 1968 agreement were too high. Wheat was being exported from U.S. Gulf ports at prices lower than the grade minimum prices during much of the 1968–1971 period. The negotiations leading to the 1971 International Wheat Agreement were conducted in an environment in whch U.S. producer and trade groups were questioning whether the United States should enter into a new agreement. It proved impossible to reach agreement on a price range. No concensus could be reached on which wheats should be used as "reference wheats" to establish quality price differentials. There was also disagreement on which port or ports should be designated as the basic point of origin. Consequently the International Wheat Agreement has, in effect, been inoperative since 1971.

Actually, the International Wheat Agreement was in trouble almost from the beginning. During the initial agreement (1949–1953) world market prices were consistently above the maximum. The agreement did stabilize the price to importers, who were perfectly happy to fulfill their import obligations. When the agreement was revised in 1953 the exporting countries insisted on raising the price range. As a result the United Kingdom, the world's largest importer, refused to participate, and was only brought back in the fourth agreement, when the exporters agreed to lower the price range.

Between the mid-1950's and the mid-1960's the major threat to the agreement was excessive wheat supplies in the exporting countries. The fact that the world price generally remained within the limits of the agreement was not the result of the buying and selling commitments of the member countries but of the surplus management operations of the major exporters, primarily the United States. The wheat stocks held by Argentina, Australia, Canada, and the United States quadrupled between 1952 and 1959. By the early 1960's the United States was holding over 35 million tons— one year's production—off the market. By holding such a large buffer stock and limiting both internal production and the flow of wheat into the commercial market the United States was in effect acting as the world wheat trade stabilization agency (Fig. 26-2).

By the late 1960's a combination of more effective supply management and surplus disposal activities had brought the stock of U.S. wheat in government hands to the lowest levels since the early 1950's (Fig. 26-3).

Table 26-2 INTERNATIONAL WHEAT COMMODITY ARRANGEMENTS: PRINCIPAL CHARACTERISTICS.

ENTRY INTO FORCE	DURATION YEARS	PARTICIPATION	INSTRUMENTS OF CONTROL	REMARKS
1933	2	Major exporters and importers	Export quotas; acreage restrictions	Poor compliance; outdated by bad harvests. Wheat Advisory Committee established
1942	0[1]	Argentina, Australia, Canada, United States (exporters) and United Kingdom	Export quotas; production control; buffer pool, maximum and minimum exporter stocks; maximum and minimum prices	Draft Convention set up an International Wheat Council
1949	4[2]	Major exporters (except Argentina) and importers	Guaranteed export and import quotas at the limits of a specified price range	Multilateral contract, Price range per bushel: $1.50–$1.80, declining over the four years to $1.20–$1.80
1953	3	Major exporters (except Argentina) and importers (except United Kingdom)	As in 1949–1953	Price range per bushel: $1.55–$2.05
1956	3	Major exporters and importers (except United Kingdom)	As in 1953–1956	Price range per bushel: $1.50–$2.00
1959	3	Major exporters, major importers	Reciprocal buying and selling obligations within the price range; exporters undertake to sell minimum quantities at upper price limit	Obligations at the price limits no longer symmetrical. Price range per bushel: $1.50–$1.90
1962	3	As in 1959, plus the Soviet Union	As in 1959–1962	Price range per bushel: $1.625–$2.025
1965	1	Same	Same	Same
1966	1	Same	Same	Same
1967	1	Same	Scope of agreement limited to administrative arrangements	Trading provisions inoperative
1968	3[3]	Major exporters and major importers, except Soviet Union	Importers agree to import not less than a stated percentage of commercial purchases from member exporters; exporters agree to make available, collectively, quantities to satisfy commercial requirements of importing countries	Price level raised, quality differences recognized. United States Gulf ports replace Canadian Lakehead prices as base point
1971	3[4]	Membership expanded to include U.S.S.R. (exporter), Brazil (importer), and several smaller exporters and importers. Includes 51 nations, 90 percent of world trade	Essentially the same	No agreement on pricing provisions

Source: From *United Nations Conference on Trade and Development* (Vol. 3, Commodity Trade). New York: United Nations, 1964, p. 88; *Wheat Situation*, ERS-USDA, Washington, D.C., August 1968, pp. 15–20, and subsequent (quarterly) issues.
[1] Draft Convention of the Washington Wheat Conference did not enter into force.
[2] First International Wheat Agreement actually to enter into force.
[3] International Wheat Agreement replaced by International Grains Arrangement.
[4] International Wheat Agreement (IWA) replaces International Grain Agreement.

The price support level was reduced in order to permit wheat, for example, to move freely in international trade without substantial subsidies. Farmers who participated in the program received marketing certificates (valued at $1.57 per bushel in 1970) on their "domestic allotment"—the share of wheat programmed for domestic food use—designed to bring their returns on the domestic allotment to 100 percent of parity. The average return to program participants for wheat sold for domestic food use, for export, and/or for other uses (Fig. 26-4) reflects both the price received for wheat in the market and the value of the certificates.

As wheat stocks built up again in the early 1970's, world wheat prices again declined. In 1972, however, U.S. grain exports rose to more than a billion bushels, compared to 600 million in 1971, primarily as a result of very large purchases by the U.S.S.R. World market prices rose sharply, and stocks held by the government declined. Export demand continued to be exceptionally strong in 1973. Prices rose to more than $4.00 per bushel.

In the early 1970's it appeared that the conditions that had fostered considerable stability in international wheat markets in the 1950's and 1960's had essentially broken down. The instability resulting from this deterioration of the institutional relations governing international trade in agricultural commodities was reinforced by poor grain crops in the U.S.S.R. and in South Asia. The precise patterns of the new trade relationships and institutional arrangements that will emerge from the next series of negotiations in the mid-1970's are, at this writing, still unclear. Until new arrangements are worked out it seems reasonable to anticipate substantial instability in both domestic and international wheat markets and in the markets for other food grains, feed grains, and oilseeds.

FOOD AID:
FROM DISPOSAL TO DEVELOPMENT

At the same time that the United States has been participating in the International Wheat Agreement it has also utilized programs designed to facilitate the noncommercial disposal of surpluses

FIGURE 26-2. *United States wheat stocks have had a major stabilizing influence on wheat prices.*

to poor countries in order to reduce the surpluses of wheat and other commodities that accumulated during the period of high domestic prices and ineffective control programs of the 1950's and early 1960's. This program is carried out primarily through the Agricultural Trade and Development

FIGURE 26-3. *U.S. wheat supply. The data for 1972 are preliminary. [After U.S. Department of Agriculture, 1972 Handbook of Agricultural Charts, Economic Research Service, October 1972.]*

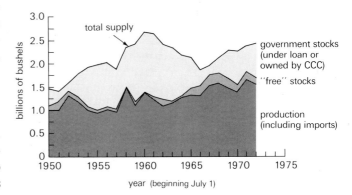

656

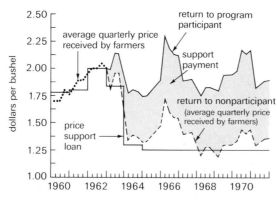

FIGURE 26-4. *U.S. wheat prices and support rates.* [*After U.S. Department of Agriculture,* 1972 Handbook of Agricultural Charts. *Economic Research Service, October 1972.*]

Act of 1954, generally known as Public Law (PL) 480.

The original PL 480 consisted of three titles: Title I, local currency sales; Title II, government-to-government emergency food aid; and Title III, domestic food distribution, foreign food distribution, and barter. It was first developed as a temporary program to eliminate surplus stocks in storage. It was hoped that United States farm prices would increase as sales and disposals come into closer balance with production and that this would permit the government to retire from intervention in agriculture. As it became clearer to the Congress and to the Administration that the problem of excess capacity could not be solved with the control programs then in force, PL 480 was periodically extended but continued to be considered a short-term, or temporary, program.

Over time other goals began to supplement the surplus-disposal objectives of PL 480. Increased attention was given to the possibilities of using food as a means of supporting economic development in low-income countries. In 1960 and 1961 greater efforts were made to implement this concern through long-term sales contracts to such countries as India, and partial wage payments in the form of food to workers on development projects. Furthermore, the emphasis of the pro-

gram, particularly the sales for foreign currencies, was shifted toward those countries with heavy or rapidly growing populations and limited foreign exchange earnings.

In addition, the humanitarian objectives of the program have been expressed in the form of (1) assistance programs after earthquakes, floods, drought, and other natural disasters; (2) support of CARE (Cooperative for American Relief Everywhere) and UNICEF (United Nations Childrens Fund) efforts to provide food and other aid; and (3) efforts by religious organizations to improve economic well-being.

Food aid to less developed countries has been a substantial portion of U.S. agricultural exports since the program was initiated in 1954. During most of the 1960's annual program expenditures were roughly between $1.5 billion and $1.6 billion (Table 26-3). Program expenditures declined to approximately $1.0 billion in 1968–1969 and remained near that level into the 1970's.

Sales for foreign currencies

Since the largest part of the foreign surplus disposal programs has been the sale of United States farm products for foreign currencies, it is important to understand how this works. The first step is for the United States government and a foreign government to reach an agreement that enables that government to pay for some farm products with its own currency rather than with the short supply of dollars it earns from exporting to the United States or other countries. The agreement specifies the quantity of the product involved, the price, and, usually, the use to be made of the foreign currency acquired by the United States. The farm products are usually sold to the foreign government by a private United States exporter who is paid in dollars by our government. The United States government pays the exporters in United States dollars, which it has obtained as part of tax revenue. In return it obtains, from the importing government, payment in the currency of that country.

The foreign government can then sell the imported farm products to its population and recoup the funds paid to the United States for the im-

Table 26-3 UNITED STATES AGRICULTURAL EXPORTS BY MEANS OF FINANCING.

	FISCAL YEAR										
ITEM	1960–1961	1961–1962	1962–1963	1963–1964	1964–1965	1965–1966	1966–1967	1967–1968	1968–1969	1969–1970	1970–1971
	Billions of dollars										
Assisted exports											
Food aid[1]	1.5	1.6	1.5	1.6	1.7	1.6	1.6	1.6	1.0	1.0	1.1
Subsidized commercial sales	1.3	1.5	0.6	1.4	1.1	1.2	1.3	0.8	0.6	1.2	1.6
Total	2.8	3.1	2.1	3.0	2.8	2.8	2.9	2.4	1.6	2.2	2.7
Unassisted exports	2.1	2.0	2.9	3.1	3.3	3.9	3.9	3.9	4.1	4.5	5.1
Total exports	4.9	5.1	5.0	6.1	6.1	6.7	6.8	6.3	5.7	6.7	7.8
	Percent										
Total exports accounted for by:											
Food aid	26	31	30	26	28	24	24	25	18	15	14
Total assisted exports	57	61	42	49	46	42	43	38	28	33	35

Source: From USDA, *Foreign Agricultural Trade of the United States,* various issues.
[1] PL 480 and Mutual Security (or AID).

ported commodities. The United States has several choices of what to do with the currency acquired from the foreign government. For example, it can spend the money in the importing country to run our embassy or it can use it to purchase military supplies if we maintain bases in the country. Such uses for funds are usually quite limited. The United States may also loan the foreign currency back to the country for use in economic development. This has been the major use of foreign currency accumulated under the PL 480 program.

This is, in effect, no more than a roundabout way of lending the country the money to buy the agricultural commodities. If the country had borrowed the money from the United States to buy the food, it could have paid for the development projects with its own money rather than using its money to pay the United States for the food and then borrowing its own money back for development. The primary reason why importing countries are willing to engage in these "fictional" transactions is that they realize that it is unlikely that the United States will ever use a large share of their money for its own purposes, like running the embassy, or call on them to repay the development loans. The primary reason the United States government engages in such "fictional" transactions is that they appear more acceptable on a political basis than outright grants of food or of dollar aid.

Sales in exchange for foreign currency should, therefore, be regarded as substantial price reductions or discounts to the recipient countries. It is this price discount that creates most of the problems of the program.

One problem is to ensure that sales in exchange for foreign currency represent additional demand rather than purchases that might otherwise have been made for dollars.

Another problem is to ensure that such sales do not come at the expense of commercial purchases that the importing country might have made from other friendly nations. For example, when the United States sells rice to India for rupees, does it mean that India reduces her imports of rice from Thailand? If this happens Thailand can legitimately accuse the United States of dumping rice on the world market at bargain prices.

A third problem arises if the imported farm products are used to depress the prices received by farmers in the importing country when sold on the local market. United States farmers are protected from such action on the part of foreign countries by such programs as the sugar program discussed earlier in this chapter. If the country is in danger of serious inflation associated with rising food prices, the PL 480 imports may be extremely useful in keeping prices from rising too fast. But if the food shipments are too large it may dampen the farmers incentives to produce in the importing countries and retard agricultural development.

Food grants for wage payments on development projects: Tunisia

A number of special efforts have been made to ensure that aid would represent a net addition to demand rather than a substitute for domestically produced food. Among the more successful efforts of this kind was a program in Tunisia in which food grants (not sales for foreign currency) were made to employ rural people on development projects.

Tunisia is basically an agricultural country, with a shortage of water in the central and southern parts of the nation. Most of the projects selected were designed to conserve or provide more water through the construction of dams and cisterns, so that water would be trapped for later use in irrigation and livestock production; to remove a weedy shrub called "jujibier," so that idle land could be grazed; and to plant trees and repair roads and firebreaks.

The early program guaranteed a daily wage of four kilograms of American hard red winter wheat supplied under Title II of PL 480 and a small cash payment. This wage amounted to about $0.71 per day, with about one-third paid in cash and the remainder in wheat. The payment was equal to the basic wage in rural areas. Since the Tunisian national dish, "couscous," was made best from durum wheat, which was ground into a coarse meal called semolina, the American hard winter wheat was unsuitable for direct use by the Tunisian peasant. Therefore, approval was secured to permit the exchange of American hard wheat on a local value basis for semolina, which was made from local durum wheat. This semolina was distributed in bags bearing the phrase in Arabic "Tunisian Semolina donated by the people of the United States of America." The ratio of semolina to cash was changed several times, and in January of 1960 the workers were paid a daily wage of $0.68, of which two-thirds was in cash.

The program employed some 50,000 to 70,000 unemployed rural workers on a ten- to fifteen-day rotation basis, equivalent to about 24,000 workers full-time until November 1959. After that workers were employed on a full-time, 48-hour week. The number had been increased to 120,000 by July 1, 1960, and to nearly 200,000 by November, 1961.

The program was evaluated as extremely successful by the Tunisians. According to American experts some of the features accounting for the success of the experience were: local planning of projects, good local administrative ability, and a close relationship between the local projects and the office of the Presidency through a work relief administrator.

The quality of local planning and administration are extremely important variables in a food-for-wages program. A major difficulty of repeating such efforts on a large scale is the difficulty of achieving effective program administration in underdeveloped nations that are short of skilled manpower.

Food aid and development

Two factors were important in shifting the emphasis of United States agricultural trade and food aid programs away from a price stabilization and commodity disposal focus and toward a development focus in the mid-1960's.

One important element was the increasing effectiveness of the commodity programs of the early and mid-1960's in balancing United States food and fiber production with the growth of domestic and foreign demand. Farm surpluses were reduced to near the levels required for long-term price stabilization policy (see Figs. 26-3 and 26-4). This permitted a shift in emphasis from simply disposing of accumulated surpluses to programming the use of U.S. agriculture's capacity for production to meet the demands of the domestic market, the requirements for profitable commercial export, and the need for food aid.

A second factor was the growing realization that in the course of time the food needs of the less developed countries would have to be met, for the most part, by the countries themselves expanding their own capacity to produce. This change in perspective was reflected in the extension of the

PL 480 Act by Congress in 1966. The Act directed the President to place particular emphasis on assistance to countries that were taking appropriate steps to improve their own productivity. Consistent with this objective the Congress authorized the use of local currencies generated by PL 480 sales to promote food production in the less developed countries. This change in perspective by Congress, from disposal to development, was instrumental in enabling the U.S. Agency for International Development (AID) to focus a much more significant part of its total effort on assistance in the improvement of food production capacity in the less developed countries.

At the same time that the food aid efforts of the United States were becoming focused on development, changes were taking place in the less developed countries that would, by the late 1960's, begin to reduce their dependence on food aid. By the late 1960's substantial progress was being made in the development of new high-yielding fertilizer-responsive varieties of wheat, maize, and rice suitable for the tropics (see Chapter 27 and 28).

The views of the policy makers in many of these countries regarding the potential role of agriculture in national economic development also changed. In India, for example, the drought of 1965 and 1966 impressed on the Indian government its vulnerability to external political pressures as a consequence of its heavy reliance on food aid. As a result the Government of India embarked on a "new agricultural strategy" designed to take full advantage of the new cereals-production technology. The achievement of self-sufficiency in food grain production in India was delayed by a drought in 1972. However, it does appear that India, once the world's largest recipient of food aid, has been able to eliminate the need for food aid except during years of great climatic distress.

The net effect of all of these changes, then, both in the United States and in some of the less developed countries, was, by the early 1970's, to reduce substantially the absolute volume of both bilateral and multilateral food aid, and to focus the remaining food aid more directly on the support of developmental and nutritional goals.

COMMODITY TRADE POLICIES IN THE 1970's

Since the 1930's, efforts to release trade from the protectionist barriers have been centered primarily on the negotiation of reductions in tariff barriers. As tariffs have been reduced, agricultural trade has expanded. Although tariffs still exist, they are no longer the major obstacles to agricultural trade expansion. Non-tariff barriers, such as import quotas, are the chief inhibitors to world trade.

In order to ease the adjustment problems of rural people that have resulted from agricultural modernization, the developed countries have erected non-tariff barriers, which have restricted the diffusion of the gains from productivity growth. For example, countries, or blocks of countries such as the European Economic Community (EEC) have instituted export subsidies, variable levies, import quotas, and other non-tariff devices. Some of these barriers were constructed to "protect" domestic farmers and consumers from low-cost imports, as were the U.S. sugar policies and EEC wheat policies. Other devices have been certain actions taken to "dump" surpluses on the world market, such as the U.S. wheat policies and French dairy policies during the 1960's.

In the 1970's the world trading community will have to decide whether it will develop a set of trade policies, including an effective world monetary system, that will permit producers and consumers in the developed and less developed countries to have equal access to the gains in agricultural productivity that are now feasible. There are indications that such steps may be less constrained by ideological barriers than in the 1950's and 1960's. Both the U.S.S.R. and China appear to be ready to participate more actively than in the past. At the same time the emergence of new trading blocks or common market arrangements may place considerable stress on international trading arrangements while intra-block "free" trade arrangements are being developed.

The potential gains resulting from a shift to freer trading patterns in agricultural products are large. Consumers would clearly benefit from lower prices. Production and consumption of animal protein would be stimulated by lower feed grain prices. In contrast, if the more restricted agricultural and trade policies that prevailed in the industrial countries in the early 1970's are maintained for the rest of the decade, it seems likely that the demand for agricultural commodities will grow less rapidly than the capacity to produce. However, if the degree of protectionism can be substantially reduced, it may be possible to return to cultivation much of the land that has been idled by government programs in the United States. In Canada freer trade will permit the relaxation of marketing controls. And the less developed countries that are experiencing the "green revolution" may, too, have an opportunity to acquire some of the gains from a greater productivity, stimulated by higher export earnings.

Selected References

Commission on International Trade and Investment Policy. *United States International Economic Policy in an Interdependent World: Report to the President and Papers* (2 vols). U.S. Government Printing Office, Washington, D.C., 1971. (Explores the entire range of international economic policy. A major section is devoted to an examination of the new approaches needed in agricultural trade policies.)

Horton, Donald C. "Policy Directions for the United States Sugar Program," *American Journal of Agricultural Economics*, Vol. 52, No. 2, May 1970, pp. 185–196. (History and evaluation of U.S. sugar policies.)

Johnson, D. Gale. *World Agriculture in Disarray*. St. Martin's Press, New York, 1973. (An exploration of the consequences of the trend toward greater agricultural protectionism. Outlines the policy alternatives for dealing with world agricultural trade problems.)

Menzie, E. L., *et al. Policy for United States Agricultural Export Surplus Disposal*. University of Arizona Agricultural Experiment Station Technical Bulletin 150, Tucson, 1962. (The focus of this bulletin is on farm export programs of the 1950's, but the chief concern is with the decisions on trade and agricultural policies which will be made in the 1960's. The publication is intended to lay the groundwork for a better understanding of such problems as the impact of PL 480 on receiving nations, effects of export programs on competing nations, and the effects of PL 480 on sending nations.)

CHAPTER **27**

The Organization and Management of Agricultural Research Systems

A belief that the application of science to the solution of practical problems was a sure foundation for human progress has been a persistent theme in American intellectual and economic history. During the two decades after World War II this belief was seemingly confirmed by the dramatic association between the progress of science and technology and rapid economic growth. The technological revolution in American agriculture (see Chapter 28), the growth of industrial productivity, and the contributions of science to military and space technology seemed to reinforce this perspective.

By the late 1960's however, the formula that had permitted the United States to move into a position of scientific, economic, and political leadership in the world community was being seriously challenged. The opposing view that has emerged is that the potential consequences of the power created by modern science and technology—as reflected in the cataclysm of war, the degradation of the environment, and the psychological cost of rapid social change—are obviously dangerous to the modern world and to the future of man. Sci-

ence, including agricultural science, is regarded by increasing numbers as a new vested interest exerting an irrational claim on financial support, rather than as an efficient source of national growth. The result has been to question seriously the significance of scientific progress, technical change, and economic growth for human welfare.

This change in perspective has also posed a new challenge to science administrators in universities, government, and business, and consequently the impetus to redirect funds to assure their use in ways that are scientifically valid, economically efficient, and socially viable is stronger than ever before. In agriculture, then, as well as in other disciplines, this shift is leading to a search for new allocative methodologies and for wider definitions of the objectives of scientific and technical effort.

A HISTORICAL PERSPECTIVE

Public support for education and research as an instrument of economic progress is a relatively recent institutional innovation. In Great Britain, the basis for modern agricultural science was established in the later part of the eighteenth and the first half of the nineteenth centuries, the responsibility for agricultural improvement being carried out by the owners of the great estates. For example, the famous Rothamsted Experimental Station was established in 1843 and financed personally by Sir J. B. Lawes for the rest of the nineteenth century. The Edinburgh Laboratory (founded in 1842) was supported by the Agricultural Chemistry Association of Scotland, a voluntary agricultural society.

In contrast, a publicly supported agricultural experiment station was successfully established in Germany (at Möckern, Saxony) in 1852. A group of Saxon farmers drafted a charter for the station, which the Saxon government legalized by statute, and secured an annual appropriation from the government to finance the experiment station operations. Although the German system of agricultural research evolved later than the British, it provided a more effective environment for the "enlargement" of new scientific and technical knowledge. The development of publicly supported agricultural research institutions in Germany reflected the emergence of a social and political climate that fostered a perception of science and technology as instruments of economic growth, and their advancement as a major responsibility of the state.

In the United States the institutionalization of public sector responsibility for research in the agricultural sciences and technology can be dated from the 1860's. The Act of May 15, 1862 "establishing the United States Department of Agriculture" and the Act of July 2, 1862 "donating public lands to the several states and territories which may provide colleges for the benefit of American agriculture and the mechanic arts" became the first federal legal authority under which a nationwide agricultural research system was to develop.

The institutional pattern that emerged for the organization of agricultural research in the United States drew heavily on the German experience. Samuel W. Johnson of the Yale Analytical Laboratory, who had studied agricultural chemistry in Germany (under Liebig), was familiar with the German experiment station movement.° He became America's first advocate of state agricultural experiment stations. The first one, the Connecticut State Agricultural Experiment Station, was not established, however, until 1877. When the Hatch Act, which provided federal funding for the support of the land-grant college experiment stations, was passed in 1887, only six state experiment stations had been established.

In institutionalizing agricultural research the United States created a dual federal-state system. The federal system, administered by the United States Department of Agriculture, developed more rapidly than the state system. Yet it was not until the later years of the nineteenth century that the Department of Agriculture achieved any significant capacity to provide the scientific knowledge needed to deal with the urgent problems of agricultural development. The ability of the land-grant colleges to produce new scientific and technical knowledge for agricultural development was even more limited than that of the Department of Agriculture. It was well after the turn of the century before the new state experiment stations could be regarded as productive sources of new knowledge or significant contributors to productivity growth in U.S. agriculture. In 1900, after 50 years of "institution building," expenditures for agricultural research by the state experiment stations and by the Department of Agriculture were less than $2.0 million (Table 27-1).

The U.S. federal-state agricultural research system developed rapidly after 1900. In 1914, upon the passage of the Smith-Lever Act, a firm institutional basis, in the form of a cooperative federal-state extension service, was established for the educational functions of the Department of Agriculture and the land-grant colleges. By the early 1920's, a national agricultural research and exten-

°The German experience also played a seminal role in the development of agricultural science in Japan. Japan initially (1870) invited British and U.S. scientists and educators to assist in the development of Japanese agricultural colleges and experiment stations. The failure of these efforts lead to their replacement by German agricultural scientists and an emphasis on German agricultural chemistry and soil science rather than on mechanized agriculture of the Anglo-American type. The result was a direction of scientific effort to achieve increases in output per hectare rather than output per worker.

Table 27-1 EXPENDITURES ON AGRICULTURAL RESEARCH AND EXTENSION IN THE UNITED STATES (IN MILLIONS OF CURRENT DOLLARS) 1880–1970.

YEAR	PUBLIC EXTENSION	STATE EXPERIMENT STATION RESEARCH FEDERAL FUNDS	NON-FEDERAL FUNDS	USDA RESEARCH	RESEARCH BY PRIVATE INDUS-TRIAL FIRMS[1]	AGRICULTURALLY-RELATED RESEARCH[2]
1880	—	—	.1	.5	—	—
1890	—	.7	.2	.8	—	—
1900	—	.7	.3	.8	—	—
1910	—	1.3	1.3	4	—	—
1920	1.5	1.4	6	7	—	—
1930	2.4	4	12	15	—	—
1935	20	4	10	12	—	—
1940	33	7	13	22	—	—
1945	38	7	18	23	—	—
1950	75	13	47	47	—	—
1955	101	19	71	54	—	—
1960	142	31	111	92	325	—
1965	175	47	180	167	460	365
1970	291	61	246	190	—	—

Sources

Robert E. Evenson, *The Contribution of Agricultural Research and Extension to Agricultural Production*, (unpublished Ph.D. dissertation), Univ. of Chicago, 1968, p. 3.

USDA, *Funds for Research at State Agricultural Experiment Stations and Other Institutions*, CSRS 15–16, Cooperative State Research Service, January 1971, p. 20.

USDA Legislative and Financial Division.

[1] The estimated expenditures for research by private firms for 1960 and 1965 are the only estimates available for this effort. The 1960 estimate includes some agriculturally related research.

[2] Research indirectly related to agriculture. This includes research by private foundations and non-land-grant universities.

sion system had been effectively institutionalized at both the federal and the state levels. The institutional innovations in the organization of public sector agricultural research, development, and extension between 1860 and 1920 proved capable, during the 1925–1965 period, of absorbing new resources at a relatively rapid rate and of generating rapid growth in new scientific and technical knowledge. Figure 27-1 shows two USDA installations that are currently in operation.

The rapid growth in public resources devoted to agricultural science and technology after 1900 was clearly a response to economic demand. Throughout the nineteenth century, U.S. agriculture had experienced rapid growth of output and labor productivity as a result of expansion in the area cultivated and the "horse mechanization" of land preparation and harvesting operations. Industry was able to supply the necessary advances in mechanical technology with little help from public sector research. Between 1900 and 1920,

however, the constraints on agricultural growth imposed by an increasingly inelastic supply of land became more severe. One index of this problem was a more than doubling of corn prices, relative to the general price level, between 1895 and 1920. The economic conditions had been established for the development of a new biological and chemical technology to complement the increasingly effective mechanical technology. The flow of resources into the state and federal research system in response to the need for this development was large. And the returns from this research have been higher than from any alternative investments available to the American society (Table 27-2).

By the mid-1960's, however, the agricultural research and education system was beginning to encounter a series of challenges to its scientific and economic viability. It was being criticized by its detractors as a servant of the "vested interests" in agriculture and agribusiness, and even by its defenders as an "academic ghetto" out of touch

664

FIGURE 27-1. *Installations of the United States Department of Agriculture. (Above) Agricultural Research Center, Beltsville, Maryland; (Below) Federal Experiment Station in Puerto Rico, Mayaguez.*

Table 27-2 ESTIMATES OF SOCIAL RETURNS FROM INVESTMENT IN AGRICULTURAL RESEARCH.

	ANNUAL RATE OF RETURNS		
	INTERNAL RATE[1]	EXTERNAL RATE[2]	BENEFIT-COST RATIO[3]
United States			
Aggregates			
Research and extension, 1945 to 1959[4]	53	—	—
Research and extension, 1938 to 1963[5]	48	—	—
Individual commodities			
Hybrid corn research, as of 1955[6]	37	690	69
Hybrid sorghum research, as of 1957[6]	—	360	36
Poultry research (public research only), as of 1960[7]	21	140	14
Mexico			
Aggregates, 1943 to 1963[8]	—	290	29
Individual commodities			
Wheat research, 1943 to 1963[8]	—	750	75
Corn research, 1943 to 1963[8]	—	300	30

[1] Discount rate which equates earnings from the outcome of research to costs on research.

[2] Percentage of annual earnings from the outcome of research relative to total past cost, assuming 10 percent interest rate.

[3] Benefit-cost ratio = $\dfrac{\text{Annual rate of return}}{100 \times \text{interest rate}}$

[4] Zvi Griliches, "Research Expenditures, Education and the Aggregate Agricultural Production Function," *American Economic Review*, Vol. 54, December 1964, pp. 967–968. Adjusted for private research by assuming the magnitude of private research expenditures equal to public research. Griliches estimated the marginal product of research expenditure, but he himself did not estimate the annual rate of returns. The above cited figure is the estimate by Peterson based on the Griliches study. See Willis L. Peterson, "The Returns to Investment in Agricultural Research in the United States," in Walter L. Fishel (ed.), *Resource Allocation in Agricultural Research*, Minneapolis: Univ. of Minnesota Press, 1971.

[5] Robert E. Evenson, *The Contribution of Agricultural Research and Extension to Agricultural Production* (unpublished Ph.D. dissertation), Univ. of Chicago, 1968. Adjusted for private research expenditure.

[6] Zvi Griliches, "Research Costs and Social Returns: Hybrid Corn and Related Innovations," *Journal of Political Economy*, Vol. 66, October 1958, pp. 419–431.

[7] Willis L. Peterson, *Returns to Poultry Research in the United States* (unpublished Ph.D. dissertation), University of Chicago, 1966.

[8] L. Ardito Barletta, *Costs and Social Returns of Agricultural Research in Mexico* (unpublished Ph.D. dissertation), Univ. of Chicago, 1970.

with current scientific and intellectual developments.°

THE AGRICULTURAL SCIENCE ESTABLISHMENT

The successful "socialization" of agricultural research necessitated the establishment of viable organizational patterns, both federal and state, that were capable of generating effective financial support. In the U.S. Department of Agriculture

° Hightower, Jim, *Hard Tomatoes, Hard Times; The Failure of the Land Grant College Complex*, Preliminary Report Task Force on the Land Grant College Complex, Agribusiness Accountability Project, Washington, D.C., 1972.

the result was the organization of scientific bureaus, such as the Bureau of Plant Industry and Bureau of Animal Industry, focusing on a particular set of problems or commodities. These innovations served to generate long-range visibility for the research effort, and support by client groups. In the states, experiment station research efforts were oriented to meet local and regional needs. Experiment station directors were able to generate financial support from state legislatures by directing their research in such a way that it had a direct impact on state economic development. Minnesota farmers were provided with corn and soybean varieties adapted to shorter growing seasons, for example. Consequently, an effective feedback relationship between the effectiveness of the applied research effort and the economic viability of the state experiment station evolved. And continuous competition, as well as cooperation, developed between the state and federal research systems.

This system did not lead to uniform growth rates in the public resources devoted to agricultural research. In fact, a marked disparity in the growth of the funding of research has existed among states. In general such funding lagged in the South, the Appalachian region experiencing the greatest retardation. North Carolina, Louisiana, and Texas are important exceptions to this general pattern, however. In the Pacific states and the Midwest, agricultural research has generally been well sup-

FIGURE 27-2. *Agricultural research in the United States.* (A) *shows the sources, by amount and proportion, of public funds available for agricultural research in 1969.* (B) *shows agriculture's share of federal funds for research for the years 1940 and 1970; figures for 1970 are estimates.* (C) *indicates the amount of funds devoted to agricultural research by the public and private sectors in 1969.* (D) *shows the amounts and sources of funds available to state agricultural experiment stations for research in 1969.* [*After USDA,* A National Program of Research for Agriculture, *Washington, D.C., October 1966.*]

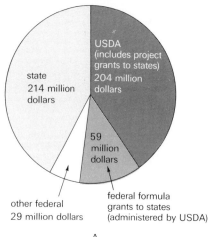

A

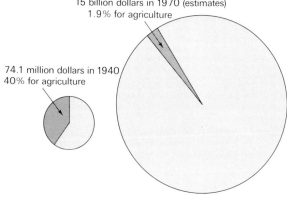

B

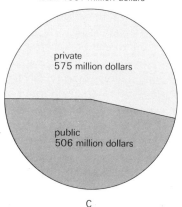

C

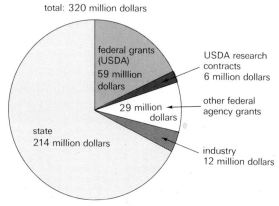

D

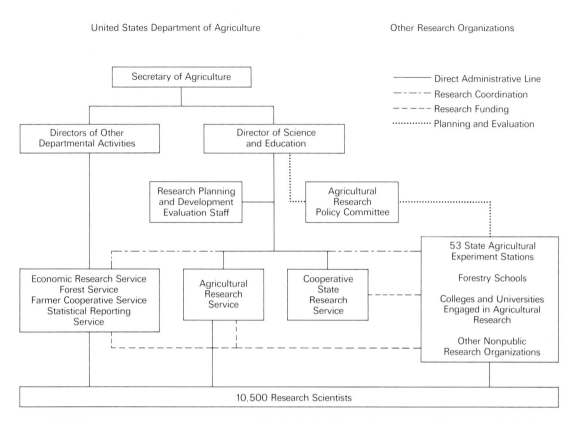

United States Department of Agriculture ⸻⸻ Other Research Organizations

Secretary of Agriculture

⸻⸻ Direct Administrative Line
—·—·— Research Coordination
— — — — Research Funding
············· Planning and Evaluation

Directors of Other Departmental Activities

Director of Science and Education

Research Planning and Development Evaluation Staff

Agricultural Research Policy Committee

53 State Agricultural Experiment Stations

Forestry Schools

Colleges and Universities Engaged in Agricultural Research

Other Nonpublic Research Organizations

Economic Research Service
Forest Service
Farmer Cooperative Service
Statistical Reporting Service

Agricultural Research Service

Cooperative State Research Service

10,500 Research Scientists

FIGURE 27-3. *The organization of research supported by the United States Department of Agriculture.* [*After Bayley, "Research Resource Allocation in the Department of Agriculture," in Fishel, editor,* Resource Allocation in Agricultural Research. *Minneapolis: Univ. of Minnesota Press, 1971, p. 221.*]

ported. In the East, support for agricultural research has generally not been strong, except in New York and Pennsylvania.

As the agricultural research system has grown, the source of funding has become increasingly diverse and the management of the system has become increasingly complex. By the late 1960's, industry was conducting more than half of the agricultural research; the U.S. Department of Agriculture, 20 percent (Fig. 27-2) and the state agricultural experiment stations, 26 percent. Of the research conducted at the state stations, approximately two-thirds came from state appropriations.

The problems of administration, decision making, and resource allocation have become increasingly complex. Within the public sector the major linkages between the federal-state system are illustrated in Figure 27-3. In addition, the agricultural and appropriation committees of the U.S.

House of Representatives and the Senate and the President's Office of Budget and Management have an important influence in determining how much money is devoted to the progress of agricultural science and which of the several areas of research receive priority. At the state level, there are similar problems of coordination and responsibility. As the size of the research effort in the private sector has expanded, the problem of coordination between the public and private sectors has become increasingly important and complex.

THE AGRICULTURAL EXPERIMENT STATION

The basic functional unit in any national agricultural research system is the experiment station, research center, or laboratory.° The productivity of the human and material resources that are devoted to the development and operation of an agricultural research station, whether operated as part of the federal-state system or in the private sector, must be evaluated in terms of the station's contribution to social progress. The experiment station can be viewed as a system for transforming raw inputs (human and physical capital) into outputs (new knowledge). This new knowledge becomes available in the form of research papers, books, bulletins, information releases, and consultation with other scientists and producers. It is frequently embodied in blueprints and formulas. And its social and economic impact is ultimately realized in the form of technical or institutional change.

Research system output

One useful depiction of an institution, such as a research station, is shown in Figure 27-4. Stock

° This section is adapted from Melvin G. Blase and Arnold Paulson, "The Agricultural Experiment Station; An Institutional Development Perspective," *Agricultural Science Review*, Vol. 10, Second Quarter 1972, pp. 11–16.

and flow resources are combined into intermediate products. These products are, in turn transformed into final outputs. The three final outputs can be categorized as *information, capacity* and *influence.*

The most important and visible output of an experiment station or a research laboratory is the *information,* or new knowledge, that is generated and released. In the long run, the use of resources for agricultural research must be justified in terms of the economic value of the output of the new knowledge that it produces.

If a research institution is to remain a valuable social asset, it must also devote resources to the enlargement of its own *capacity.* This means diverting some resources from the production of information that has immediate application. It means expanding the capacity of its scientific staff by allowing more time to be designated for graduate education, study leaves, and "basic research." It also includes modernization of facilities, administrative structure, and the ideology that serves as the rationale for a research program. The frequent arguments on the degree of emphasis that should be placed on "basic" research, and the degree on "applied" research, often reflect a more funda-

FIGURE 27-4. *A systems model of experiment station performance and development.* [*After Blase and Paulson, Agricultural Science Review, Vol. 10, Second Quarter 1972, p. 13.*]

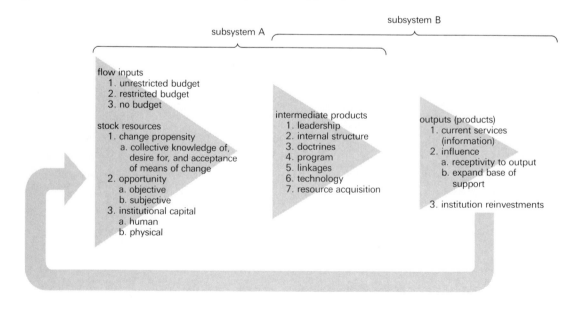

mental disagreement on how much emphasis should be given to the production of current services relative to investment in future capacity.

All successful research institutions devote significant amounts of resources to increasing their *influence*. In order to establish a successful claim on current and future resources, a research organization usually finds it necessary to maintain effective relationships with funding agencies—legislative bodies, operating bureaus and divisions, and private foundations. The organization may also find it useful to devote effort to building a public image as a valuable institutional resource. Although the resources devoted to the production of influence may have little direct value to society, such activities are essential to the maintenance and continuity of both public and private research institutions. Institutions that have outlived their social function as producers of *information* frequently devote excessive effort to organizational maintenance activities.

Intermediate products

The intermediate products or services identified in Figure 27-4 are for internal use. They have little direct value to society. But they are indispensable to the research institution itself. They represent the "engine" that determines the efficiency with which the research system makes use of the resources available to it. By and large, these intermediate services must be produced by the research institution itself rather than be purchased from external sources.

The *linkage* services include the establishment and maintenance of contacts and relationships with individuals and institutions outside the research station or laboratory, with other scientists, with the clients who use the services of the station, and with sources of support. The linkages carry messages in both directions. It is through the linkages with the outside that a research institution influences technical and institutional change and is in turn influenced by the external changes in society and in science and technology. The state agricultural experiment stations have an exceedingly complex set of linkages with the external environment—with state crop improvement asso-

ciations, community and regional development councils, producer and consumer interest groups, the extension service, and the entire hierarchy of local, state, and federal administrative and political institutions. The problem of maintaining effective communication with outside groups without become either a captive or an adversary is exceedingly difficult.

Leadership is an extremely important intermediate product. The myth that all a research director needs to do is hire good men and let them "do their thing" has only minimal support at a time when the solution to significant technical and social problems requires concerted multidisciplinary research effort. Leadership must be sensitive to changing social goals and effectively transmit their implications to the scientific staff. Leadership also includes *resource acquisition*. It must be able to visualize and communicate effectively the potential contribution of the experiment station or research laboratory to the solution of emerging problems. Then, it must be capable of mobilizing and allocating both financial and scientific resources to produce the high returns from scientific and technological effort that society has come to expect. To do so, not only must leadership acquire the necessary manpower and finances, but also it must create an environment in which they can be used productively. The *technology*, or the methodology, of research is in continuous flux. The research program must be organized in such a way that the research staff is aware of the advances in its own and closely related fields and contributes to them.

Doctrine is reflected in the articulation of institutional goals and philosophy and in the operating style of an institution. The traditional production-oriented doctrine of the state and federal agricultural research system has experienced severe stress as a result of the increasing pressure to give more emphasis to the environmental "spillover" effects of technical change in agriculture and to the problems of personal and community development in the formulation of research priorities. The operational manifestation of a shift in doctrine is the reformulation of a research station's *program*. Modifications in doctrine are sterile unless accompanied by the allocation of research resources in such a way that they contribute to the realization of the revised priorities. Many modifications in program imply drastic revisions in the *internal structure* of a research station. Problem-oriented interdisciplinary "centers" erode the decision-making authority of discipline-oriented departments. During periods of tension, the reformulation of doctrine, the redirection of program, and the reorganization of internal structure may absorb substantial resources and seriously compete with the production of information. These efforts must be justified primarily in terms of their impact on the future productivity and viability of the experiment station.

THE RESEARCH AND DEVELOPMENT PROCESS

What are the processes by which an experiment station or a research institute generates new knowledge or information? The generation of information and its embodiment in the form of a new crop variety, a synthetic fiber, or a new communication device is a complicated process that requires the cooperation of large numbers of individuals and organizations. Educated people—scientists, engineers, technicians, literate farmers, and laborers—are critical to the process that produces technological change and economic growth.

A striking feature of modern research and development is the extent to which it has become institutionalized. In applied research and development, the industrial laboratory and the research institute has increasingly replaced the individual inventor or the academic scientist who did research in the free time he could divert from teaching.

The process of innovation

The design of effective institutions for research and development requires a clear understanding of the processes by which new things—**innovations**—emerge in science and technology. This subject has received much attention in applied technology, sociology, and history. In economics it has occupied only a peripheral, though expanding, role.

The distinction between the role of insight and the role of skill in the emergence of a new concept or technique has posed major difficulties in developing a theory of innovation. A. P. Usher proposed the following distinction:

> Acts of skill include all learned activities whether the process of learning is an achievement of an isolated adult individual or a response to instructions by other individuals . . . acts of insight are unlearned activities that result in new organizations of prior knowledge and experience. . . . Such acts of insight frequently emerge in the course of performing acts of skill, though characteristically the act of insight is induced by the conscious perception of an unsatisfactory gap in knowledge or mode of action.

Usher distinguishes among the several concepts of how innovation occurs on the basis of the relative prominence given to insight and skill.

At one extreme is the **transcendentalist approach,** which attributes the emergence of invention to the inspiration of the occasional genius who from time to time achieves new insight through intuition. This theory lies behind the great inventor concept—the assertion that Fulton invented the steamship or that Marconi invented the radio. At the other extreme is the **mechanistic-process approach,** which assumes that "necessity is the mother of invention." In this, the individual inventor is one of a number of individuals with the appropriate skills who makes the final modification that is acclaimed as an invention.

A more comprehensive view is to regard major innovations as emerging from the cumulative synthesis of relatively simple innovations, each of which requires an individual "act of insight" by a person who also possesses a high degree of skill, acquired through training and experience in the particular area of investigation. In this view, the individual invention involves four steps:

1. *Perception of the problem,* in which an incomplete or unsatisfactory pattern or method of satisfying a want is perceived.

2. *Setting the stage,* in which the elements or data necessary for a solution are brought together through some particular configuration of events or thought. Among the elements of the solution is an individual who possesses sufficient skill in manipulating the other elements.

3. *The act of insight,* in which the essential solution of the problem is found. Large elements

of uncertainty surround the act of insight. It is this uncertainty that makes it impossible to predict the timing or the precise configuration of an innovation in advance.

4. *Critical revision*, in which the newly perceived relations become more fully understood and effectively worked into the entire context to which they belong. This may again call for new acts of insight.

A major or strategic invention represents the cumulative synthesis of many individual inventions. It will usually involve all the separate steps that may be found in the case of the individual invention. Many of the individual inventions do no more than set the stage for the major invention, and new acts of insight are again essential when the major invention requires substantial critical revision to adapt it to particular uses. A schematic presentation of the elements of the individual act of insight and the cumulative synthesis as visualized by Usher are presented in Figures 27-5 and 27-6.

A major advantage of the cumulative-synthesis approach is that it clarifies the points at which conscious efforts to speed the rate or alter the direction of innovation can be effective. The possibility of affecting the rate or direction of innovation is obscured by the transcendentalist approach, with its dependence on the emergence of "the great inventor," and is denied by the mechanistic process approach, with its dependence on broad historical trends or forces.

The focus of much conscious effort to affect the speed or direction of innovations centers on the second and fourth steps in the process—in "setting the stage" and in "critical revision." By consciously bringing together the elements of a solution—by creating the appropriate research environment— the stage can be set in such a manner that fewer elements are left to chance. It would be inaccurate to suppose that it is yet possible to set the stage in such a manner as to guarantee a breakthrough in any particular area, although the probability of a breakthrough may be increased as more resources are devoted to research on a given problem. As more is learned about the effectiveness of various research environments, the probability that breakthroughs will be achieved should increase. For the

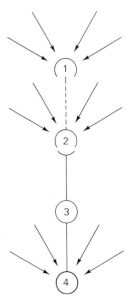

FIGURE 27-5. *The mergence of novelty in the act of insight. Synthesis of familiar items: (1) perception of an incomplete pattern; (2) setting the stage; (3) the act of insight; and (4) critical revision and full mastery of the new pattern.* [After Usher, A History of Mechanical Inventions, *Cambridge: Harvard Univ. Press, 1954.*]

present, it can only be emphasized how little is known about the administration of basic research.

At the level of "critical revision," considerable progress has already been made in bringing economic and administrative resources to bear. Many of the elements of critical revision require "acts of skill" in contrast to "acts of insight." The effectiveness of modern research procedures in shortening the time span from the test tube to the production line testifies to our ability to exert conscious direction at the applied research level.

*The role of skill and insight
in a biological innovation:
The rice blast nursery case*°

° This section was developed with the assistance of Dr. S. H. Ou, plant pathologist, The International Rice Research Institute. For a description of the blast nursery, see S. H. Ou, Some Aspects of Rice Blast and Varietal Resistance in Thailand, *International Rice Commission Newsletter*, Vol. 10, No. 3, Sept. 1961, pp. 11–17. See also, Report of the Committee on the Establishment of Uniform Blast Nurseries, *International Rice Commission Newsletter*, Vol. 11, No. 1, March 1962, pp. 23–32.

672

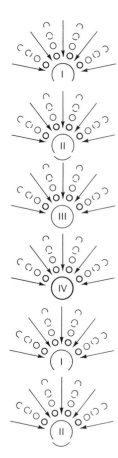

FIGURE 27-6. *The process of cumulative synthesis. A full cycle of strategic invention and part of a second cycle. Roman numerals I–IV represent steps in the development of a strategic invention. Small figures represent individual elements of novelty. Arrows represent familiar elements included in the new synthesis.* [*After Usher,* A History of Mechanical Inventions. *Cambridge: Harvard Univ. Press, 1954.*]

The four stages in the process of innovation—(1) *perception of the problem;* (2) *setting the stage,* bringing together the elements of a solution; (3) *the act of insight,* which permits solution of the problem; and (4) *critical revision,* including refinement and improvement of the original solution—are ideally illustrated in the development of a recent biological innovation, the rice blast nursery. The example also illustrates the complex interactions that frequently accompany technological change.

Rice blast is an airborne fungus disease caused by *Piricularia oryzae,* which attacks the leaves and stem of the rice plant and substantially reduces yield. It has emerged as an important disease in Southeast Asia during the last several decades. In the past a combination of natural selection and farmer selection resulted in the use of rice varieties that were relatively resistant to local strains of the rice blast fungus. Most of these varieties were also adapted to conditions of low fertility, but unfortunately their yields were usually low and they were not responsive to improved fertility. Attempts to increase yields by the use of nitrogen fertilizer and by introducing new varieties have been accompanied by increased virulence of the rice blast fungus in those areas where farmers had adopted new varieties and cultural practices most extensively.

The first step, perception of the problem, was the recognition of the rising incidence of the rice blast disease following the introduction of new rice varieties. Increased recognition that rice blast disease was an important factor in limiting attempts to achieve higher rice yields in tropical areas led to the search for an economically and biologically efficient technique for screening rice varieties to be used in rice breeding programs and the new varieties produced by such programs.

The object of the screening was to select varieties of the host (the rice variety) that were highly resistant to infection by the fungus. Initial screening efforts involved inefficient laboratory techniques or inexact field observations. To complicate matters further, the fungus spores were found to be present in the atmosphere in concentrations that exceeded those that could be effectively produced under laboratory conditions.

The second step, setting the stage, involved a high degree of technical skill but relatively little new knowledge. The problem was to create an optimum environment for the growth of the rice blast fungus spores that provided, at the same time, favorable conditions for the mass production of rice seedlings. The components of such an environment, well known at the time, are:

1. *Adequate moisture in the environment.* The fungus spores multiplied most rapidly and attacked the rice plant most successfully under conditions of adequate moisture and high humidity. The

importance of direct contact between the plant and the water meant that the optimum time of water application occurred in the evening. The influence of humidity and the requisite skills required for controlling humidity were well known by many competent researchers.

2. *High levels of nutrition.* The same variety of rice tends to be more susceptible to rice blast disease when grown under high levels of nutrition, particularly nitrogen, than when grown on poor soils. This was also well known to competent research workers.

3. *Upland nursery conditions.* The rice blast develops more rapidly when grown under upland (that is, nonflooded) conditions because of the slower accumulation of silica. Susceptibility to the rice blast disease is inversely proportional to the amount of silica in the plant. This information was obtained through recent experimental work.

4. *Tropical temperature.* The blast spores multiply and attack the plant more effectively in the range 20° to 30°C (68° to 86°F). As tropical temperatures are typically within this range, special temperature controls are not necessary to create a favorable environment for spread of the disease.

The third step, the act of insight, was facilitated by the increasing concern with the damage caused by the rice blast disease in Thailand during the mid-1950's, and by the presence of a pathologist in Thailand who possessed the professional skill necessary to combine the elements of the solution into the design for a blast disease nursery that would permit effective identification of blast resistant varieties and identification of the races of the blast pathogen.

The fourth step, critical revision of the solution, involved the addition of refinements aimed at improving the effectiveness of the solution—for example, the addition of border rows of susceptible cultivars, and the installation of automatic moisture-control devices to increase the concentration of spores and to maintain the required moisture conditions.

RESOURCE ALLOCATION IN AGRICULTURAL RESEARCH

As the agricultural research system has grown increasingly complex, the problem of the use of research resources—of scientist man years and research laboratories and field plots—has also become increasingly complex. Yet the fundamental issues can be condensed into two simple questions.

What is it technically or scientifically feasible to accomplish?

What value does society place on the results?

The first question can be answered by only the most imaginative and skillful research scientists in each field. There is no alternative source of information on potential outcomes (in terms of new knowledge or new technology) from the allocation of research resources to a particular project or problem area. Furthermore, only the most productive research scientists, who are most intimately involved in an area of research, are able to supply the estimates of the resources and time required to achieve the objectives of a particular research project or program. Finally, even under the best of circumstances, the precision of these estimates is subject to great uncertainty.

Although the research scientist is in a strong position to outline what can be done, he is in a weak position to evaluate the priority that society places on research in a particular field or on a particular scientific or technological innovation. Evaluations of the scientific significance and the social or economic significance of new knowledge may not be the same. A low-income underdeveloped country, such as India, may place a relatively high value on research leading to increases in crop yields or more efficient water use, whereas a high-income country, such as the United States, may place more emphasis on solving the medical problems of the aged (cancer, heart disease, and arthritis). These kinds of choices affect not only the allocation of funds for applied research, but also the selection of those areas of basic research in which research effort may have substantial application in applied research.

Both the efficient selection of research alternatives and the efficient design of a national agricultural research strategy require that both sets of considerations—the resources needed to accomplish feasible alternatives and the value of the research product to society—be taken into account. This decision-making system—in which the individual scientist, the research team, the academic department, the experiment station, the government research bureau, the corporate research division, the planning and budgeting offices of the university, the federal government or the corporation, and the state and federal legislatures all participate—has become so complex that more systematic, more formalized procedures are now being sought. This search is also a response to the new skepticism regarding the role of science in society.

The demand for more efficient decision-making in the allocation of research resources has emerged simultaneously with the evolution of a set of "third generation" planning-oriented budgeting techniques. The *first* generation of techniques consisted of efforts to develop increasingly effective systems of expenditure control (1920–1935). The second generation was management or performance ori-

ented, with its major thrust devoted to (a) casting budget categories in functional terms and (b) providing work cost measurements to facilitate the efficient performance of prescribed activities. The thrust of the *third* generation program budgeting approach is planning oriented. As such it has shifted the planning perspective toward a more careful evaluation of the purposes to be fulfilled by the allocation of resources to economic activity—toward systems efficiency in the place of the more limited objectives of process efficiency.

The elements of any systematic research planning and resource allocation system, regardless of its level of formalization, can be outlined in a systems model such as Figure 27-7. The steps outlined in the figure are implicit even in the more intuitive traditional decision systems.

In the mid and late 1960's there were a number

FIGURE 27-7. *A modular form of research resource allocation.* [*After Fishel, editor,* Resource Allocation in Agricultural Research. *Minneapolis: Univ. of Minnesota Press, 1971.*]

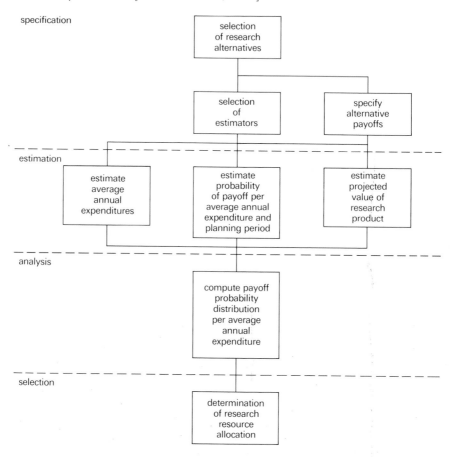

of attempts to design and implement research evaluation and planning procedures that were more sensitive to both the perceptions of research scientists regarding the opportunities for productive research effort and changing social values and priorities.

The national program of research for agriculture projections

One of the most ambitious research planning attempts was the joint effort of the U.S. Department of Agriculture and the state experiment stations to project future needs for publically supported research. The impetus for the study originated within the U.S. Senate Appropriations Committee. It was conducted by a joint U.S. Department of Agriculture–State Experiment Station Agricultural Research Planning Committee. It was carried out by a series of task forces composed of a broadly representative group of research administrators and scientists.

One of the first concerns of the task force was to establish an orderly system for the classification of agricultural research. The classification system was designed in such a way that it would permit the efficient organization and retrieval of the information to be used in describing the total current program of agricultural research, establishing research priorities, and projecting agricultural research needs for the period 1965–1977. The classification system that was finally adopted was a three-way categorization of information on research according to (a) type of research activity, (b) the commodity or resource focus of the research, and (c) the scientific and professional fields contributing to the research. This classification scheme is reproduced in Table 27-3. It has since gone through several modifications.

A second major task was the establishment of the relevant national goals for agricultural research. The problem of translating a set of broad national goals into a numerical rating system for evaluating individual projects was one of the more difficult problems of the entire exercise. Although such a numerical system was in fact developed (Table 27-4), many of the participants in the long-range study have seriously questioned whether the scoring scheme had any appreciable effect on the

final estimates of research needs in 1972 and 1977. Another major deficiency of the study was that it was not able to incorporate effectively scientific judgment on the feasibility of achieving specific goals. The procedure used was to work backward from broad social goals rather than forward from judgements on feasibility under different levels of budgetary support.

The projections that emerged from the study are summarized, in terms of scientific man years, in Figure 27-8. In general the 1972 and 1977 goals suggest a modest shift in priorities from the commodity-production, protection, and product-marketing categories to and toward a community- and resource-development orientation. However, the greatest absolute growth would continue to be in the commodity-protection and production areas.

In retrospect it appears that the projections developed in the National Program study have not been closely followed. Perhaps the most significant long-term value of the study was that it provided the impetus for the establishment of a computer-based Current Research Information System (CRIS) for the storage and retrieval of U.S. Department of Agriculture and state research information. The establishment of this system, and its gradual improvement, are producing an information base for improved agricultural research planning at both the federal and state levels.

PPB in the Agricultural Research Service

At the same time that the long-range National Program studies were in progress, the U.S. Department of Agriculture was also engaged in a series of pilot operations to implement the new Program Planning Budgeting (PPB) methodology in research resource allocation. This effort was implemented by the Program Planning and Evaluation Staff in the Office of the Administrator of the Agricultural Research Service (ARS). The implementation of this system was based much more firmly on the information inputs supplied by the individual scientist.

The objective of the program budgeting system is to evaluate the merits of alternative programs more effectively. The relevant decision units must, therefore, be programs that manage the inputs

Table 27-3 CLASSIFICATION CODESHEET FOR REPORT OF AGRICULTURAL RESEARCH FOR FISCAL YEAR ENDING JUNE 30, 1965.

ACTIVITY

Conservation, development and use of soil, water, forest and related resources

1. Resource description and inventory
2. Resource conservation
3. Resource development and management
4. Evaluation of alternative uses and methods of use

Protection of man, plants, and animals from losses, damage, or discomfort caused by

5. Insects
6. Diseases, parasites, and nematodes
7. Weeds
8. Fire and other hazards

Efficient production and quality improvement

9. Biology of plants and animals
10. Improving biological efficiency of plants & animals
11. Increasing consumer acceptability of farm and forest products
12. Mechanization and improvement of physical efficiency
13. Management of labor, capital, and other inputs to maximize income

Product development and processing

14. Chemical and physical properties of food products
15. Developing new and improved food products & processes
16. Chemical and physical properties of non-food products
17. Developing new and improved non-food products and processes

Efficient marketing, including pricing and quality

18. Identification, measurement & maintenance of quality
19. Improving economic & physical efficiency in marketing, including analysis of market structure and functions
20. Analysis of supply, demand and price, including interregional competition
21. Developing domestic markets, including consumer preference and behavior
22. Foreign trade, market development, and competition

Improvement of human nutrition and consumer satisfaction

23. Nutritional values, consumption patterns, and eating quality of foods
24. Quality of family living, including management and use of time, money, and other resources

Development of human resources and of economies of communities, areas, and nations

25. Description, inventory, and trends
26. Economic development and adjustment
27. Improvement of social well-being, including social services and facilities and adjustment to social and economic changes

98. Evaluation of public programs, policies & services
99. Research which cannot be allocated to one or more of the above activities

COMMODITY OR RESOURCE

Natural resources

1. Soil and land
2. Water
3. Watersheds and river basins
4. Air and climate
5. Recreational resources
6. Timber and forest products
7. Range
8. Wildlife and fish

Crops and crop products

9. Citrus and subtropical fruit
10. Deciduous and small fruits and tree nuts
11. Potatoes
12. Vegetables
13. Ornamentals and turf
14. Corn
15. Grain sorghum
16. Rice
17. Wheat
18. Other small grains
19. Pasture
20. Forage crops
21. Cotton
22. Cottonseed
23. Soybeans
24. Peanuts
25. Other oilseed crops
26. Tobacco
27. Sugar crops
28. Miscellaneous & new crops

Animals and animal products

29. Poultry
30. Beef cattle
31. Dairy cattle
32. Swine
33. Sheep and wool
34. Other animals
35. Bees and honey

Manmade resources used on farms or by people

36. General purpose farm supplies and facilities, including equipment, structures, fertilizers, and pesticides
37. Clothing and textiles
38. Food
39. Housing, household equipment & non-textile furnishings

Human resources, organizations, and institutions

40. People as individual workers, consumers, and members of society
41. The family and its members
42. The farm as a business enterprise
43. Communities, areas, and regions, including counties and States and their institutions and organizations
44. Agricultural economy of United States & sectors thereof, including interrelationships with the total economy
45. Agricultural economy of foreign countries and sectors thereof, including interrelationships with the total economy
46. Farmer cooperatives
47. Other marketing, processing, and farm supply firms
48. Marketing systems and sectors thereof

99. Research which cannot be allocated to one or more of the above commodities or resources

FIELD OF SCIENCE

Biological

1. Biochemistry and Biophysics
2. Biology—Environmental, Systematic, and Applied (Botany, Ecology, Zoology, etc.)
3. Biology—Molecular
4. Entomology
5. Genetics
6. Immunology
6. Immunology
7. Microbiology
8. Nematology
9. Nutrition and Metabolism
10. Parasitology
11. Pathology
12. Pharmacology
13. Physiology
14. Virology

Physical

15. Chemistry—Analytical
16. Chemistry—Inorganic
17. Chemistry—Organic
18. Chemistry—Physical
19. Engineering
20. Geology and Geography
21. Hydrology
22. Mathematics and Statistics
23. Meteorology
24. Physics

Social and behavioral

25. Anthropology
26. Economics
27. Education and Communications
28. History
29. Law
30. Political Science
31. Psychology
32. Sociology

Source: From USDA and Association of State Universities and Land Grant Colleges, *A National Program of Research for Agriculture*, Washington, D.C., 1966, p. 26.

Table 27-4   CRITERIA USED IN
ESTABLISHING RELATIVE PROGRAM
PROJECTIONS FOR THE 91 RESEARCH
PROGRAM ACTIVITIES IN THE NATIONAL
PROGRAM OF RESEARCH FOR AGRICULTURE.

CRITERIA	SCORING WEIGHT[1]
Extent to which the research meets state experiment station, department, and national goals	9
Scope and size considering area, people, and units affected	8
Benefits of research in relation to costs	7
Urgency of research	10
Contribution to knowledge	9
Feasibility of implementation and likelihood of successful completion in a reasonable period of time	5
Likelihood that research results will not be available elsewhere	6
Likelihood of extensive and immediate adoption of results	6
	60

Source: From USDA and Association of State Universities and Land Grant Colleges, *A National Program of Research for Agriculture*, Washington, 1966, p. 29.

[1] Program activities were given a score ranging from 1 to 5, based on the degree to which each criterion was met. The score was then multiplied by the weight assigned to each criterion and summed over all criteria to give a total score.

necessary to reach some applied goal. Early in the effort to adopt the PPB methodology for research decision-making in the ARS, the concept of a three-link chain consisting of science, technology, and benefits was adapted. Science was perceived as advancement by inquiry about cause and effect relationships; technology, as the methods in the various segments of the agricultural industry employed to produce, market, and process commodities, benefits, as the measure of the advantage of proceeding in some visualized new way.

The research program directed toward the achievement of a piece of visualized new technology is referred to as a research activity. If a visualized technology is to yield benefits, it must be possible to substitute it for a current technology in a practical manner. As an example, the development of an improved cultivar of corn that could replace a current cultivar would be considered a logical goal for a research activity.

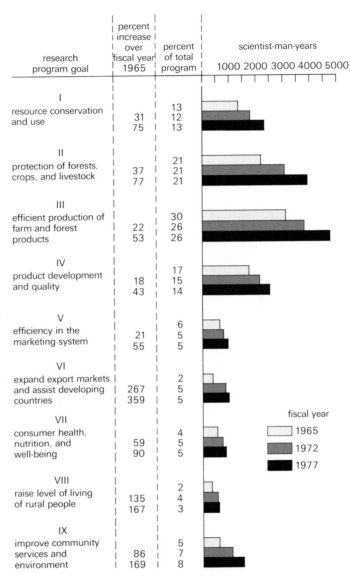

FIGURE 27-8. *The number of scientist-man-years in the SAES-USDA program allocated to each research goal in 1965, and the numbers recommended for 1972 and 1977. [After U.S. Department of Agriculture and Association of State Universities and Land Grant Colleges, A National Program of Research for Agriculture. Washington, D.C.: U.S. Government Printing Office, 1966, p. 12.]*

The implementation of a PPB system consisted of (1) a change from an organizational budget, largely based on scientific disciplines, to a problem-oriented budget; (2) a change from a budget with a given base, subject to annual increments or reductions, to a zero-base budget in which the entire activity, and not just the annual change, would be evaluated; (3) a change from an annual budget to a multiyear budget, the duration determined by the relevant time period required for the research activity. Implementation of the program budget that relates program inputs to objectives requires more explicit planning of research programs and objectives than does the traditional organizational budget, which tends to be structured more in terms of organization maintenance and growth.

A number of difficulties in implementing the PPB approach in the ARS occurred. The procedure itself seems rather straightforward: expected benefits are estimated by a comparison of the potential new technology with the current technology, influenced by an estimate of the probability of success for a given total expenditure. Quantifiable benefits are based on estimates of the extent to which a product can be produced or a process conducted at lower cost. Measures of quality improvement are reflected by the price differential which the higher quality product might be expected to command in the market. Two major difficulties have arisen in the attempts to evaluate research activities. One is the assessment of "incommensurable benefits"—those that may improve the welfare of one group at the expense of another: for example, a research project might produce benefits for consumers but losses for farmers, or benefits for farm owners but losses for hired farm workers. The other problem is the evaluation of "intangible benefits"—benefits that are real but can not be measured: an example might be the reduction of a health hazard by the substitution of a biological for a chemical system of pest management.

Studies on research resource allocation

At the same time the attempts to adopt more efficient research decision and resource allocation systems were being implemented, research was underway to develop the principles that should guide the further development of such systems. These principles concern primarily the implications of the extreme uncertainty that exists with respect to the costs, the feasibility, and the benefits of achieving research objectives. Research itself is a sequential learning process. The new information generated in the process of conducting a research program frequently leads to changes in research procedures and objectives that may sharply alter the estimated costs and benefits of the research program.

Researchers have identified a variety of techniques for reducing the uncertainty facing research decision-makers at the initiation of a research program. Three of these are particularly important.

LOOSE CONNECTIONS AMONG PROJECT COMPONENTS The structuring of relatively "weak" interrelationships among the several components of a research program—that is, the avoidance of too rigid "coordination"—is one important method of dealing with uncertainty. A loose interrelationship among the components of a research program implies that failure of one particular activity will not endanger the total program. Clearly achievement of the objectives of modern applied research in agriculture does require coordination. The lone researcher guided only by his own "interests" is no longer as effective a contributor to the progress of agricultural science or technology as in the past. Yet a research program in which coordination is too tight—in which the methodology and objectives are strictly delineated—is not readily able to take advantage of the new knowledge generated in the research process itself or to make use of those components of the research program which are successful should some components not be successful.

PARALLEL APPROACHES Since the achievement of research objectives is highly uncertain it is frequently wise to compensate for the uncertainty by the use of parallel approaches. By not concentrating every effort on one single aspect, it is possible to take advantage of the new knowledge developed in the research process of the costs, benefits, and feasibility of alternative approaches. In pest management research, for example, it would seem wise to explore alternative chemical and biological control methods until the uncertainty that surrounds the direct costs and returns of alternative strategies, as well as the possible effects on the environment, becomes clearer.

DURATION OF THE RESEARCH PROGRAM Estimates of costs, feasibility, and development-time tend to improve as the research program proceeds. The accumulation of new information, often at relatively low cost, frequently helps to determine the relative priority that should be given to the several alternatives for achieving a certain research program objective—reducing the environmental impact of a specified level of pest damage in apple production, for example. Studies on time span indicate that "crash" research programs tend to result in inflated costs relative to the research objectives, in contrast to programs designed to achieve similar objectives in the course of a longer period of time.

MODERNIZING THE NATIONAL RESEARCH SYSTEM

The new research decision-making and planning methodologies are making important contributions to research decision-making. At present, however, they are more appropriately viewed as *communication* rather than as *decision-making* methodologies, intended largely to relay the relevant information through an increasingly complex decision-making system with as little distortion as possible. Furthermore, some of the major issues confronting the future of agricultural science are only marginally amenable to the algebra of the new PPB type methodologies. A number of currently existing problems with broad philosophical overtones will condition the evolution of the U.S. agricultural research system.

Articulation of the federal-state system

When the federal-state system of agricultural research initially evolved, the relationship between the activities of the two systems was quite loose. In the course of time, the systems have become more closely articulated at both the professional and administrative levels. Given the pressures for efficiency in resource utilization it appears likely that there will be considerable pressure for the "rationalization" of the management of two systems. An important aspect of this rationalization has to do with the size and location of research facilities.

Recent research has indicated wide variations in the returns to research by the several components of the federal-state system. (a) The returns to research tend to be lower in the small states than in the large; (b) a training component is an essential element in the long term viability of a research institution; (c) a decentralized system tends, within limits, to compensate in relevance what it loses in potential efficiency. There is little doubt that, in the federal-state system, research is being conducted at too many small stations with inadequate logistical and professional resources. System efficiency could be enhanced by somewhat greater centralization and coordination. One model for a more efficient system would incorporate both federal and state research under a regional director. In the early 1970's steps were taken, both by the USDA and the state agricultural experiment stations, to move toward such a pattern of research organization.

Articulation of public and private research

The case for public sector agricultural research has traditionally rested primarily on a "spillover" effect rationale—the inability of an individual farm to capture more than a small share of the benefits from private research. Individual farms were too small to conduct research. If comprehensive areas of new knowledge happened to be developed, particularly in biological science and technology, a private firm could capture only a small share of the benefits from research leading to new crop production practices and superior cultivars. Therefore, exclusive reliance on the private sector would have resulted in underinvestment in agricultural research.

This rationale is clearly less valid then in the past. Where the results of research can be embodied in proprietary products, as in hybrid corn and poultry, the private sector has become a dominant supplier of new knowledge and new technology. In 1971, "breeders rights" legislations was passed that will encourage private firms to invest even more heavily in the field of plant breeding. In the 1960's approximately half of agricultural research was done in the private sector (Fig. 27-2, C). In the future the private sector may well conduct as much as 75–80 percent of agricultural research in the traditional areas of plant and animal science.

The role of the private sector in agricultural research can be expected to vary considerably among commodities and among scientific disciplines or problem areas. Expansion of private sector agricultural research can be expected to be greatest in those areas in which the new knowledge can be embodied in proprietary products—in new cultivars, for example. In the more fundamental areas on which private applied research is based (such as plant pathology, physiology, and genetics), it seems likely that the private sector will continue to rely primarily on research at the universities and in the U.S. Department of Agriculture for advances in knowledge. Thus a higher percentage of the total public sector research budget should logically be directed toward research on intermediate knowledge (i.e., basic research) and those areas of applied research which have substantial social value but are unprofitable for the private sector.

Redirection of public sector research priorities

The demand for public sector applied research in certain other areas of the agricultural sciences is rising. The growing public consciousness of the impact of the rapid growth in agricultural productivity on the environment is leading to a heightened demand for more information on these effects of the new technology. It will be difficult to create the incentives necessary to induce the private sector to respond to this demand. If the universities and the federal research agencies remain committed to the principle of directing re-search resources to those areas receiving inadequate private sector research investment, the outcome will be a substantial redirection of research effort away from the area of production science and technology and toward the problems of environmental stress resulting from the more intensive systems of agricultural technology.

INTERNATIONAL SPECIALIZATION AND COOPERATION IN AGRICULTURAL RESEARCH

A major reason for low productivity in agriculture in the less developed countries of the tropics had been the lack of capacity to produce an agricultural technology adapted to the local resource endowments and the economic environment. This lack of research capacity was particularly limiting in traditional food crop production.

Consequently, one of the dramatic developments of the 1960's was the growth of agricultural science capacity in some of these countries. A major factor in this growth has been the establishment of an international research, training, and communications network centered on a new set of international research institutes. These new institutes have been particularly significant in strengthening agricultural science capacity in the smaller nations of the tropics. A small nation with either a specialized or a limited agricultural or industrial base cannot capture as high a proportion of the benefits of its investment in basic research as a large nation with a diversified economic base. The highest returns from research investment in countries with a limited economic base will typically be found in applied problem-solving research, which draws on the knowledge produced by research institutions with a broader base. This is the major reason for the development of a new set of international research institutes by such philanthropic organizations as the Ford and Rockefeller Foundations, and for the support of international research programs by the Food and Agriculture Organization (FAO), the World Bank (IBRD), and by national aid agencies such as the U.S. Agency for International Development (USAID).

The series of steps that led to the establishment of the new international institutes began in 1943 with the establishment of the Office of Special Studies (Oficina de Estudios Especiales) in the Mexican Ministry of Agriculture with the assistance of the Rockefeller Foundation. Field research programs were first initiated with wheat and corn. The program was later expanded to include field beans, potatoes, sorghum, vegetable crops, and animal sciences. Norman Borlaug (Fig. 27-9), who received a Nobel Prize in 1971 for his research on wheat, was stationed in Mexico by the Rockefeller Foundation in 1944 and continues as leader of the wheat program.

By 1973, four international research centers had been established and two others were in the planning or development stage (Box 27-1). For an example of the way the new international research institues are organized and function, it is useful to examine in some detail one of these new institutes, the International Rice Research Institute (IRRI).

Because of its importance throughout Southeast Asia, research on rice was particularly well adapted to an international cooperative effort. Basic work at the IRRI has found broad application by researchers throughout the rice-growing regions of the tropics—in Asia, Africa, and Latin America. The interchange of knowledge fostered by the Institute's fellowship program and international conferences could be used to disseminate basic research results, applied research methods, and research skills rapidly throughout the region.

Because of these considerations, the Ford and Rockefeller foundations, in cooperation with the Government of the Philippines, established in the Philippines a world center for the study and improvement of rice (Fig. 27-10). The Ford Foundation contributed a capital grant of $7,150,000. The foundations together agreed to provide annual operating funds in the neighborhood of $1,250,000.

The Institute, dedicated in February 1962, assembled an international staff of scientists from Australia, Ceylon, Taiwan, India, Japan, the Philippines, and the United States. In addition to its research program, the IRRI conducts a training program for rice technicians and scientists from the rice-producing regions of the world in coopera-

FIGURE 27-9. *Norman E. Borlaug, 1970 recipient of the Nobel Peace Prize, working with hybrid wheats at the CIMMYT Toluca station, Mexico.*

tion with the University of the Philippines. Participants in the training program include students who are interested in pursuing graduate work at the University of the Philippines while conducting their research at the IRRI, plus more mature scholars who are primarily interested in expanding their research experience. Scholars from most of the countries of Southeast and South Asia, plus Latin America and Africa, study at the IRRI.

A major aim of the research program at the International Rice Research Institute has been to breed new high-yielding rice varieties that are adapted to tropical conditions. By 1966 the first new IRRI varieties were available for field tests under actual farm conditions. The yields of the new IR8 were typically almost double the yields of a number of widely used local cultivars grown under essentially the same conditions.

Box 27-1   THE 29: NEW GROUP AIDS AGRISEARCH

The Consultative Group is a unique consortium of international banks, assistance agencies, governments and private foundations. This year alone it has raised over $15 million for the 1972 operations of four international agricultural research and training centers that were originally established by the Ford and Rockefeller Foundations. For 1973 it hopes to marshal some $23 million for expanded activities of these four institutes and for the creation of two new ones. The major objective of the Consultative Group, and of the international centers it finances, is to assist the poorer nations to rapidly increase output of basic food crops both to meet the food needs of growing populations and to speed the economic development that is needed if the living standards of both rural and urban people are to be improved.

Among the centers being supported are:

The International Rice Research Institute (IRRI) founded in 1960 in the Philippines by the Ford and Rockefeller Foundations in cooperation with the government of the Philippines. This institute over the past decade has produced the widely heralded "miracle" rice varieties and their related technology. It has trained hundreds of Asian scientists and technicians and has provided direct technical assistance to national research and development organizations in most of the rice-growing nations of tropical Asia.

The International Maize and Wheat Improvement Center (CIMMYT, from its name in Spanish) established by the Rockefeller Foundation and the Government of Mexico in 1966. From CIMMYT, and from the earlier Rockefeller Foundation-Mexico cooperative agricultural program, have come the high-yielding dwarf wheats now in worldwide usage. CIMMYT also has significant work underway internationally in corn improvement. The center's work in wheat is directed by the Rockefeller Foundation's Dr. Norman E. Borlaug, the recipient of the 1970 Nobel Peace Prize.

The International Institute of Tropical Agriculture (IITA) in Nigeria, established in 1967 by the Ford and Rockefeller Foundations in cooperation with the government of that country. It serves particularly the low, humid areas of Africa, concentrating its work on cowpeas and other legumes, the long-neglected root crops, corn, rice and tropical cropping systems.

The International Center for Tropical Agriculture (CIAT, from its name in Spanish) near Cali, Colombia. Established in 1967 by the Ford and Rockefeller Foundations and the government of Colombia, CIAT is attempting to speed the agricultural development of the humid tropics, especially in the Americas, for human benefit. It concentrates particularly on beef-production systems and on improved production of cassava, field beans and other important crops.

The International Crops Research Institute for the Semi-Arid Tropics (ICRISAT) at Hyderabad, India. This new institute will be concerned with the improvement of four crops especially important to farmers in the low-rainfall areas of Asia—sorghum, millets, chick-peas and pigeon peas. It was established in mid-1972 by the World Bank and UNDP in cooperation with the government of India. It was organized for the Consultative Group by the Ford Foundation and is based on an original design by the Rockefeller Foundation's Dr. Clarence Gray.

The International Potato Center (CIP) in Peru. Organized initially by USAID and the North Carolina State University in cooperation with the government of Peru, this center seeks to intensify production of the white potato, a staple food of people in high elevations in the Andes and in many other regions of the world.

The Consultative Group membership comprises the three sponsors (World Bank, UNDP, FAO) the Inter-American Development Bank, the Asian Development Bank, the African Development Bank, the European Fund for Economic Development (FED), and the governments of 13 nations: Australia, Belgium, Canada, Denmark, Federal Republic of Germany, France, Japan, Netherlands, Norway, Sweden, Switzerland, United Kingdom and the United States. Also members are the International Development Research Center of Canada, the Kellogg Foundation, the Ford Foundation and the Rockefeller Foundation. Representatives of each of five developing regions of the world participate in Group meetings.

Source: RF Illustrated, Vol. 1, No. 1, October, 1972. New York: The Rockefeller Foundation.

FIGURE 27-10. *The International Rice Research Institute, established at the University of the Philippines, Los Baños, Laguna.*

The stage had been set for the development of the new IRRI varieties by the successful development of similar varieties by Japan in Taiwan between 1900 and 1930. The much more rapid development of the new varieties at the International Rice Research Institute was the direct result of an IRRI breeder's insight, which led to the identification of a specific plant type as a breeding objective. The new varieties are fertilizer responsive, lodging resistant, and nonphotoperiod sensitive. The resulting biologically efficient rice plant permitted the IRRI plant breeders to design a whole group of new higher yielding varieties with much greater precision and in much less time than was believed possible by many rice breeders.

By 1970 the new rice cultivars developed at the IRRI, or similar cultivars developed in other countries as a result of the stimulus of the Institute, were planted on approximately 20 million acres in Asia. Similar cultivars were being developed and adopted in Africa and Latin America. At the IRRI itself increasing attention was being given to problems of pest management, and to broadspectrum resistance to fungus, bacterial, and virus diseases, such as rice blast, bacterial blight and

streak, and tungro virus. Increased attention was also being given to nutritional and cooking quality, to the thermodynamics of flooded rice soils, to the physiology of plant nutrition and growth, to the economics of rice production, marketing, and trade, and to the economics of the rural communities in the rice-producing areas of Southeast Asia.

The new international institutes do not, of course, represent completely new concepts of research organization. Commodity-centered research institutes established in the tropics under British, French, Belgian, and Dutch colonial auspices have been responsible for substantial productivity gains in the production of tropical export crops, such as rubber and sugar. The Rubber Research Institute in Malaysia, an industry-supported research institute, is one of the world's outstanding research organizations. And the sugar station established in West Java under the Dutch (Proefstation Oost Java, P.O.J.) was the major source of the advances in sugarcane breeding that led to dramatically higher yields in almost all major sugarproducing countries.

The significance of the new international institutes does, however, extend well beyond their direct impact on the availability of new wheat, corn, and rice technologies. The most important contribution has been the evolution of an institutional pattern for the organization of scientific resources that can be replicated for work with a wide variety of crops, in different localities, with a reasonable probability of success. It is now possible to organize a multidisciplinary team of biological, physical, and social scientists capable of developing or adapting a new biological and chemical technology for crop production to local growing conditions and to make that technology available to farmers in a form that they are capable of accepting within a period of five to ten years.

Another important contribution of the new international centers has been the evolution of a technique for establishing a set of linkages with national and local education and research centers. The technique includes activities such as exchanges of staff, professional conferences, support of graduate and postgraduate training, personal consultations, and exchanges of genetic materials. This communications function of the international institutes is particularly important for the experiment stations located in the smaller countries where the feasibility of the development of a broad-based national agricultural research system is limited.

Selected References

Dupree, A. Hunter. *Science in the Federal Government: A History of Policies and Activities to 1940.* Harvard University Press, Cambridge, 1957. (This book contains an excellent discussion of the emergence of agricultural research in the U.S. Department of Agriculture. Particular attention is given to the search for a method or organizing scientific effort in agriculture to achieve scientific and financial viability.)

Fishel, Walter L. (editor). *Resource Allocation in Agricultural Research.* University of Minnesota Press, Minneapolis, 1971. (This book presents the views and findings of a number of experts concerned with the problems, issues, and procedures involved in the allocation of resources for agricultural research. The twenty-one contributors include agricultural and general economists, public administration officials, scientists, and other specialists. The chapters are based on papers given at a symposium sponsored jointly by the Minnesota Agricultural Experiment Station and the Cooperative States Research Service of the U.S. Department of Agriculture.)

Knoblauch, H. C., E. M. Law, and W. P. Meyer. *State Agricultural Experiment Stations, A History of Research Policy and Procedure.* USDA Misc. Publ. No. 904, 1962. (A history of the development and growth of the U.S. Department of Agriculture land-grant college system of agricultural research in the United States.)

Moseman, Albert H. *Building Agricultural Research Systems in the Developing Nations.* Agricultural Development Council, Inc., New York, 1970. (An excellent introduction to the strategy and practice of developing agricultural research capacity in new nations.)

U.S. Department of Agriculture and Association of State Universities and Land Grant Colleges. *A National Program of Research for Agriculture.* Washington, 1966. (This report presents the results of a study of agricultural and forestry research programs in the United States. It provides an inventory of research programs in 1965. It projects recommended allocations of resources for agricultural research in 1972 and 1977.)

Usher, A. P. *A History of Mechanical Inventions.* Harvard University Press, Cambridge, 1954. (The classical study of the evolution of western technology from the mechanical equipment of antiquity to modern machine tool and power industries. Usher's discussion of the role of skill and insight in the generation of new ideas in science and technology, Chapter IV, is particularly valuable.)

CHAPTER **28**

Technological Change and Agricultural Development

Economic growth is generally thought of in terms of a rising level of consumption or of real income per person. Economic growth can occur as a result of advances in the techniques of production (in technological progress) that result in a greater output with the expenditure of a constant total quantity of resources. It can also occur as a result of an increase in the quantity of other factors of production, such as capital equipment and raw materials per unit of labor, in such a manner that real income per person rises even though the ratio of output to the total quantity of resources used in production remains unchanged or even declines.

A crucial element in the rise in the material well-being of most rapidly developing countries has been the progressive use of a growing store of knowledge. The fundamental significance of **technological change** for the growth of the less developed countries is that it permits the substitution of knowledge and skill for resources. The effect is to permit the substitution of less expensive or more abundant resources for more expensive or relatively scarce resources, and to release the constraints on growth imposed by inelastic resource supplies.

To be sure, technological change is only one of a number of interrelated elements that contribute to economic growth. The stock of capital and the availability of raw materials have already been mentioned. Others are: (a) the quality of the labor force, including its general education, health, occupational skills, incentives, and motivations; and (b) the way that economic activity is organized—the ability of the government to provide facilities and services that individual firms find difficult to provide themselves; the efficiency with which the markets for factors of production and products are organized, and the manner in which economic and social institutions, such as the land tenure system, the tax structure, and the monetary system, reinforce or dampen personal incentives.

It is clear, however, that technical change is an essential element in the growth of agricultural production and productivity from the very beginning of the agricultural development process. It is also evident that the process of technical change in agriculture occurs, to a substantial degree, as a dynamic response to the resource endowments and the economic environment in which a country finds itself at each stage in its development history.

The design of a successful agricultural development strategy is based on a unique pattern of technical change and productivity growth consistent with the resource endowments in each country or major region. It may also require the modernization of social and political systems to bring about an economic and social environment conducive to an effective response by firms and individuals to the new technical opportunities.

The previous chapter describes the organization and management of scientific and technological resources for the generation of new scientific knowledge and the progress of agricultural technology. This chapter discusses the various aspects of technical change: its role in agricultural development; its diffusion from farm to farm, from region to region, and in the course of time; its implications for the development of market institution. Linkages between technical change and the supply-demand feedback system described in chapter 24 are critical to producers' decisions to adopt new technology. Also, the "spillover effects" of technical change on environmental quality must be carefully considered in evaluating the implications of a particular change for resource utilization.

ALTERNATIVE PATHS OF TECHNICAL CHANGE

In any attempt to develop a meaningful perspective on technical change in agricultural development,° the first step is to abandon the naive view of agriculture in premodern or traditional societies as essentially static. Even in premodern times, agriculture underwent a continuous, though relatively slow, development—in agricultural tools and machines, in cultivars and animal breeds, and in crop and animal husbandry practices (see Chapter 1).

The conservation model

The emergence of modern agricultural science itself was in part generated by an attempt of the

---

°This section draws heavily on Yujiro Hayami and Vernon W. Ruttan, *Agricultural Development: An International Perspective*, The Johns Hopkins University Press, Baltimore, 1971.

founders of agricultural science such as Arthur Young (1741–1820) in England and Albrecht Thaer (1752–1825) in Germany, to interpret and extend the English agricultural revolution of the late eighteenth and early nineteenth centuries. The English agricultural revolution consisted of the evolution of an intensive, integrated, crop-livestock husbandry system of agriculture. More intensive crop-rotation systems evolved to replace the older system in which arable land had been allocated between permanent cropland and permanent pasture. Crop rotation, in turn, led to the introduction and heightened use of new forage and green manure crops and an increase in the availability and use of animal manures to maintain soil fertility. This "new husbandry" practice included an expansion in the cultivation of new crop and livestock production by the recycling of plant nutrients, in the form of animal manures, to maintain soil fertility. The advances in technology were accompanied by the enclosure of the "common" pastures, by the consolidations of fields, and by investment in drainage, fencing, and other forms of land development. The net effect was a substantial growth both in total agricultural output and in output per acre. The inputs used in this conservation system of farming were supplied largely by the agricultural sector itself, with little dependence on the industrial sector. Power, in the form of labor and animals, was produced primarily within the community in which it was used. The organic fertilizers, which enhanced output per acre, were produced as green manures or animal manures, primarily on the farm on which they were used. Even the capital improvements were, for the most part, produced by the labor available within the farm or the local community.

The conservation model, or approach to agricultural development—based on the development of increasingly complex land- and labor-intensive cropping systems, the production and use of organic manures, and investment in physical facilities to utilize land and water resources more effectively—has been clearly capable in England, and in many other areas of the world, of sustaining a growth in agricultural production, of about 1.0 percent per year, for relatively long periods of time. Agricultural growth rates that are this low, however, are inadequate for the requirements of less developed countries today, with their prevailing conditions of explosive population growth and rising per capita incomes.

The induced innovation model

Modern agricultural growth is heavily dependent on the use of new technical inputs that are the product of research in agricultural science and technology and that reach the farmer in the form of items that must be purchased from the industrial sector of the economy. Three types of investment are essential if society is to achieve sustained growth in agricultural productivity: (a) in the capacity of agricultural experiment stations to produce new scientific and technical knowledge; (b) in the capacity of the industrial sector to develop, produce, and market new technical inputs; and (c) in the capacity of farmers to use modern agricultural inputs effectively.

This does not imply, however, that there is only one path to the achievement of a modern agricultural economy undergoing rapid technical change and productivity growth. On the contrary, multiple paths of technological development are available to a nation. Technology can be developed to relieve the constraints on growth imposed by a scarcity of certain factors of production. In some societies this constraint may be a limited supply of labor. In others it may be a scarcity of land. In countries having high population densities and a scarcity of land, as in East and South Asia, increases in output depend on the development of an agricultural technology, including fertilizer-responsive crops, that can release the constraints on growth imposed by an inelastic supply of land. Likewise, in an economy with a scarcity of labor, as in the United States and the other countries of recent settlement, substitution of land and capital for labor is made possible primarily by improving agricultural implements and machinery. The constraints imposed on agricultural development by scarcity of land may be offset by advances in biological technology. Those imposed by a limited supply of labor may be offset by advances in mechanical technology. The capacity of a country to achieve rapid growth in agricultural output and

productivity seems to hinge on its ability to make an efficient choice among the alternative paths. Failure to choose a path that effectively relaxes the constraints imposed by a shortage of resource endowments can depress the whole process of economic development.

Two alternative patterns of technical change, influenced by shortages of two diverse resources, are illustrated in the recent histories of the United States and Japan. In the United States it was primarily the progress of mechanization that facilitated the expansion of agricultural production and productivity by increasing the area operated per worker, so compensating for the labor shortage. In Japan it was primarily the progress of biological technology, resulting in seed improvements, that increased the yield response to larger quantities of fertilizer, thus permitting the rapid growth in agricultural output in spite of severe limits in the supply of land. In Japan the impact of biological technology has exerted a dramatic influence on the growth of production and productivity throughout much of the period since 1870. U.S. agriculture has undergone significant biological innovations since the 1930's, and farm mechanization in Japan has been progressing at an accelerating pace since the 1950's.

These contrasting patterns of productivity growth and factor use in U.S. and Japanese agriculture can best be understood if they are thought of as processes of dynamic adjustment to changing relative factor prices—dynamic in the sense that production technology changed in response to the alterations in relative factor prices, which in turn reflected changes in resource endowments and accumulation in the two economies. In the United States the long-term decline in the prices of machinery relative to wages (Fig. 28-1) encouraged the substitution of land and power for labor. This substitution was dependent on continued progress in the development of larger-scale mechanical technology. For example, an optimum factor combination with the reaper (such as the McCormick or Hussey), assuming two weeks for harvesting and two shifts of horses, was approximately five workers, four horses (two horses for original models), and 140 acres of wheat. Only when a new technology, in the form of the binder, was intro-

688

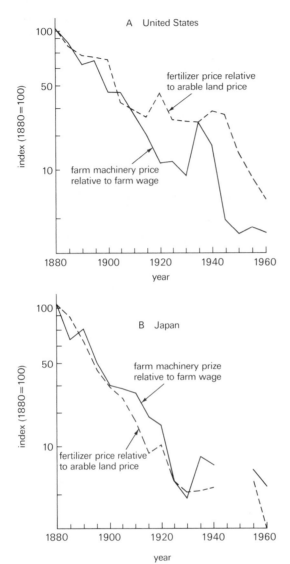

FIGURE 28-1. *Farm machinery price relative to farm wage, and fertilizer price relative to arable land price (1880 = 100), in logarithms, (A) in the United States and (B) in Japan, 1880–1960. Prices are the averages for the five years preceding the year shown. [After Hayami and Ruttan,* Agricultural Development: An International Perspective. *Baltimore: The Johns Hopkins Univ. Press, 1971, p. 123.]*

duced was it possible for the farmer to change this proportion to two workers, one reaper (binder), four horses, and 140 acres. This process has continued. Introduction of the tractor permitted the

replacement of human and animal power by mechanical power. This greatly raised the possibilities of substituting power for labor by making it much easier to command more power per worker. Substitution of more highly powered tractors for low-powered tractors has had a similar effect.

In Japan the supply of land was limited and the price of land rose relative to wages. It was not, therefore, profitable to substitute land and power for labor. Instead, the new opportunities arising from continuous declines in the price of fertilizer relative to the price of land were exploited by advances in biological technology. Seed improvements were directed to the selection of more fertilizer-responsive cultivars. The traditional cultivars had equal or higher yields than the improved varieties at the lower level of fertilization, but they did not respond to higher application of fertilizer. The enormous changes in fertilizer input per hectare, that have taken place in Japan since 1880 and in the United States since the 1930's, reflect not only the effect of a decline in the price of fertilizer but the effect of crop improvement efforts designed to achieve a yield response to larger amounts of fertilizer in order to take advantage of the decline in the real price of fertilizer.

In Japan, then, the declining fertilizer prices relative to land prices were a strong inducement for farmers and experiment station workers to develop the biological innovations such as the high-yielding, fertilizer-responsive crops. It is significant that in the United States the biological innovations, the first important result being hybrid corn, began about ten years after the rate of increase in arable land area per worker decelerated (around 1920), and that biological innovations and fertilizer application were accelerated after acreage restrictions were imposed by the government. The changes in the land supply conditions coupled with a dramatic decline in fertilizer price induced a more rapid rate of biological innovation in the United States after the 1930's.

Adjustments in factor proportions in response to changes in relative factor prices can be thought of as movements along the isoproduct surface of an innovation possibility curve or a metaproduction funtion. This is illustrated in Figure 28-2. *U*

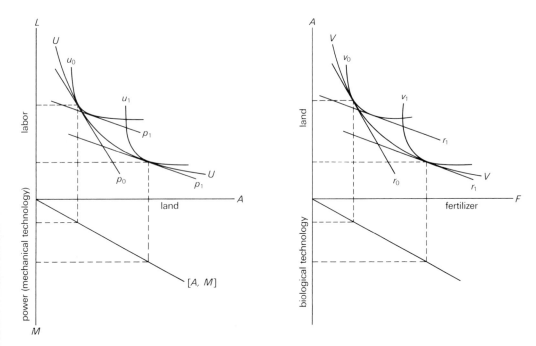

FIGURE 28-2. *Graphs illustrating the ways in which factor prices induce technological change.* [*After Hayami and Ruttan,* Agricultural Development: An International Perspective. *Baltimore: The Johns Hopkins Univ. Press, 1971, p. 126.*]

in Figure 28-2 (left) represents the land-labor iso-quant, or constant output curve, of the innovation possibility curve, which is the envelope of less elastic isoquants such as u_0 and u_1 corresponding to different types of machinery or technology. A certain technology represented by u_0 (e.g., the reaper) is created when a price ratio, p_0, prevails for a certain length of time. When the price ratio changes from p_0 to p_1, another technology represented by u_1 (e.g., combine) is induced.

The new technology represented by u_1, which enables enlargement of the area operated per worker, generally corresponds to a higher intensity of power per worker. This implies a complementary relationship between land and power, which may be drawn as a line representing a certain combination of land and power [A, M]. In this simplified presentation, mechanical innovation is conceived as the substitution of a combination of land and power [A, M] for labor (L) in response to a change in wage relative to an index of land and machinery prices, though, of course, in actual

practice land and power are substitutable to some extent.

In the same context, the relationship between the fertilizer-land price ratio and biological and chemical innovations, such as the development of crops that are more responsive to application of fertilizers, is illustrated in Figure 28-2 (right). V represents the land-fertilizer isoquant of the innovation possibility curve, which is the envelope of less elastic isoquants, such as v_0 and v_1, corresponding to varieties of different fertilizer responsiveness. A decline in the price of fertilizer relative to the price of land from r_0 to r_1 makes it more profitable for farmers to search for cultivars or alternative crops described by isoquants to the right of v_0. The farmers also press public research institutions to develop new cultivars. Through a kind of dialectic process of interaction between farmers and experiment station workers, a new cultivar, such as that represented by v_1, is developed.

Such movements along the innovation possi-

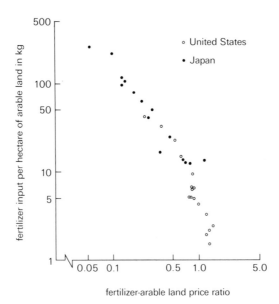

FIGURE 28-3. *The relationship between fertilizer input per hectare of arable land in the fertilizer-arable land price ratio, in logarithms, for the United States and Japan, based on five-year observations taken from 1880 to 1960. This relationship is expressed in terms of hectares of arable land that can be purchased by 1 ton of $N + P_2O_5 + K_2O$ contained in commercial fertilizers.* [After Hayami and Ruttan, Agricultural Development: An International Perspective. *Baltimore: The Johns Hopkins University Press, 1971, p. 127.*]

bility curve is consistent with the U.S. and Japanese data on the relationship between fertilizer use per hectare of arable land and the fertilizer-land price ratio as plotted in Figure 28-3. Despite the enormous differences in climate and other environmental conditions, the relationship between these variables is almost identical in both countries. As the price of fertilizer has declined relative to the price of land, both Japanese and American agricultural scientists have developed more fertilizer-responsive crops. Farmers in both countries have responded similarly to the increased availability of fertilizers and of fertilizer-responsive crops.

It should be recognized, of course, that the identification of mechanical technology as "labor saving" and of biological technology as "land saving," as we have done in this section, can be

oversimplified. All mechanical innovations are not necessarily motivated by incentives to save labor, nor all biological innovations by incentives to save land. For example, in Japan horse plowing was developed as a device to cultivate more deeply, to increase yield per acre rather than primarily to save labor. In the United States in recent years attempts have been made to develop crops more suitable for mechanical harvesting. For example, in order to facilitate mechanical harvesting, tomatoes have been developed that have a sturdier skin and concentrated ripening. Actually, the most sophisticated technological progress may depend on a series of simultaneous advances in both biological and mechanical technology. Thus in the mechanization of tomato harvesting, the plant-breeding research and the engineering research were conducted cooperatively in order to invent new machines capable of harvesting the tomatoes that were specifically bred for mechanical handling.

DIFFUSION OF NEW TECHNOLOGY

Once a technological innovation has been developed and introduced by a research organization or a commercial firm and adopted by one or a few farms, how long does it take another farm to adopt the innovation? What factors determine the speed with which other farms imitate the innovator? By what process is technical change diffused among farms, regions, and nations?

Diffusion among farms

The speed with which innovations are adopted affects both the private and the social return to investment in research and development. In economies with high interest rates, the passage of only a few years before a new practice or a new product is adopted may sharply reduce the return on investment in research and development.

Even under relatively favorable conditions, the diffusion of new technology is a fairly slow process. The introduction of such a relatively simple innovation as hybrid corn in the relatively sophisticated farming community of Iowa took eight years from the time commercial hybrids were offered

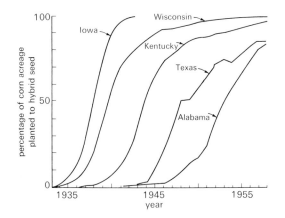

FIGURE 28-4. *Percentage of all corn acreage planted to hybrid seed. [After Z. Griliches, "Hybrid Corn and the Economics of Innovation," Science, Fig. 1, Vol. 132, pp. 275–280, 29 July 1960. Copyright © by the American Association for the Advancement of Science.]*

for sale until it was adopted by 90 percent of the Iowa farmers (Fig. 28-4). The same pattern, with even longer delays, has been characteristic of the more complex agricultural and industrial innovations (Fig. 28-5). There are, of course, substantial differences in the rates of adoption of the several innovations identified in Figures 28-4 and 28-5.

Studies of the diffusion process identify five important generalizations that explain many of the differences in rate of adoption of innovations.

FIGURE 28-5. *Machines in use on farms in the United States, 1940–1959. [After Z. Griliches, "Hybrid Corn and the Economics of Innovation," Science, Fig. 2, Vol. 132, pp. 275–280, 29 July 1960. Copyright © by the American Association for the Advancement of Science.]*

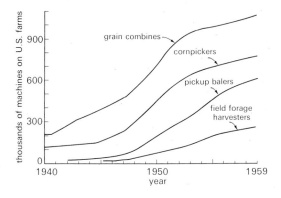

1. *As the number of farms in an area adopting an innovation increases, the probability of its adoption by a nonuser increases.* As experience and information about a particular innovation accumulates, the element of risk associated with its introduction diminishes. Competitive pressures mount, and general adoption of the innovation is encouraged. Where the profitability of an innovation is difficult to assess, the mere fact that a large proportion of neighboring farms have adopted the innovation may prompt those who have not adopted the innovation to consider its adoption all the more seriously.

2. *The expected profitability of an innovation influences the probability of its adoption.* The more profitable the investment in an innovation promises to be, relative to other available investments, the greater will be the probability that its potential profitability will compensate for the risk involved in its adoption.

3. *For equally profitable innovations, the probability of adoption tends to be smaller for innovations requiring relatively large investments.* Firms will be more cautious in committing themselves to large, expensive changes, and they will have more difficulty in financing such changes. The same principle applies to relatively complicated inventions. Adoption of innovations that require a complete reorganization of the farm layout and equipment or impose heavy demands on managerial ability are adopted more slowly than innovations that just replace one piece of equipment or one crop variety with another.

4. *The rate of adoption of an innovation is influenced by the geographic region or type of farm enterprise in which the innovation is introduced.* For equally profitable innovations requiring the same investment, the rate of adoption in one region or in one commodity sector might be higher than in another because (a) farmers in that region or producing a particular commodity are more inclined to experiment and take risks, (b) the area or commodity markets are more competitive, (c) the area or commodity is healthier financially, and (d) the area or commodity is less free to make investment decisions because of government regulations. Observation suggests that such differences between one area or commodity and another affect rates of diffusion.

5. *Differences in adoption rates among industries and firms also depend on a number of social and personal factors.* Sociologists, in particular, have studied the factors that cause some individuals to be classified as (a) innovators, (b) early adopters, (c) early majority, (d) late majority, and (e) laggards in the adoption of new ideas or techniques. By and large, the early adopters are more venturesome. They tend to have a wide range of social relationships; their circle of acquaintances and contacts tends to have a wide geographic radius. They also tend to be younger and to have higher social status than those who are slow to adopt.

International diffusion

The diffusion of technical change among regions and among countries depends not only on the personal and economic factors identified previously but on the speed with which the innovation is adopted and made available to farmers by experiment stations, extension services and agribusiness firms. For example, when hybrid corn was developed, the lag in the initial adoption by farmers in Texas and Alabama (Fig. 28-4) was not due to a resistance to new technology, but rather to the fact that the agricultural experiment stations and the seed producers did not produce hybrids suitable for the South as soon as they did hybrids for the Midwest. Hybrid corn cultivars are quite "location specific" and must be individually "tailored" to a particular environment. The primary factors limiting geographic adaptability are sensitivity to temperature (see Chapter 11) and to "photoperiod" (see Chapter 10). These factors limit adaptability in a north-south direction but not necessarily in an east-west direction. However, other ecological factors, such as rainfall and pathogens, combine with the first two to make most hybrid corn cultivars location specific from east to west as well. The technical effort of both the public experiment stations and the commercial seed companies was initially concentrated in the Midwest where the intensity of corn production indicated the greatest social and private returns from the development and marketing of the higher-yielding hybrids.

The international diffusion of agricultural technology is not new. The diffusion of better husbandry practices and of crops and livestock between countries and continents was a major source of productivity growth even in prehistory. It is well known that transfer of new crops (potatoes, maize, tobacco, and so on) from the new continents to Europe, after the discovery of America, had a dramatic impact on European agriculture. Before agricultural research and extension were institutionalized, this diffusion took place as a by-product of travel and communication undertaken for other purposes. In the course of a long gestation period (several decades or centuries) exotic plants and techniques were gradually adapted to local conditions. These relatively slow processes are clearly not consistent with the requirements for technical change in agriculture in today's developing countries where the rate of growth in demand for agricultural output is typically from 3 to 6 percent per year.

It appears useful to distinguish the three phases of transfer of international technology as: (a) **material transfer;** (b) **design transfer;** and (c) **capacity transfer.** The first phase is the simple transfer or import of new materials, such as seeds, plants, animals, and machines, as well as the techniques needed for these materials. Local adaptation does not take place in an orderly and systematic fashion. The naturalization of plants and animals tends to occur primarily as a result of "trial and error" by farmers.

The second phase is the transfer of certain designs (blue prints, formulas, journals, books, and so on). During this period exotic plant materials and foreign equipment are imported by the less developed country in order to obtain new plant-breeding materials or to copy equipment designs, rather than to use in direct production. New plants and animals are subject to orderly tests and are propagated by means of systematic multiplication. Domestic production of the machines imported in the previous phase is initiated. This phase usually corresponds to an early stage of evolution of publicly supported agricultural research in that experiment stations are established to conduct, for the most part, simple tests and demonstrations.

In the third phase, the scientific knowledge of

technology and the capacity to implement it are transferred, so enabling the production of locally adaptable technology, modeled on a "prototype" of that already existing in another country. Increasingly, plants and animals are bred locally to adapt them to local ecological conditions. The imported machinery designs are modified in order to meet climatic and soil requirements and factor endowments of the economy.

An important element in the process of capacity transfer is the travel and at least temporary migration of agricultural scientists. In spite of advancements in communications procedures, diffusion of the ideas and craft of agricultural science depends heavily on extended personal contact and association. The participation of scientists in international research conferences can be of critical importance in counteracting the limited supply of scientific and technical manpower in the less developed countries, and it is essential if these countries are to enter into the capacity transfer phase (see Chapter 27) as soon as possible.

The development of sugar cane

The development of improved sugar cane cultivars is a useful example of the international diffusion and transfer of biological technology in agriculture.° It is of particular interest because the process has evolved from a simple transfer of biological materials to the capacity transfer phase. There have been four stages in the development of sugar cane.

STAGE I: DIFFUSION OF INDIGENOUS CULTIVARS Sugar cane was cultivated in India as early as 400 B.C. Cane and the art of sugar making were diffused from India to China, to Arabia, and to the Mediterranean region very early. Shortly after 1400 A.D. sugar cane was introduced on Madeira

° The material in this discussion is based on the following sources: J. J. Ochse, M. J. Soule, Jr., M. J. Dijkman, and C. Wehlburg, *Tropical and Sub Tropical Agriculture* (2 vols.) MacMillan, New York, 1961, pp. 1197–1251; R. E. Evenson, J. P. Houck, Jr., and V. W. Ruttan, "Technical Change and Agricultural Trade: Three Examples—Sugar Cane, Bananas and Rice" in Raymond Vernon (editor), *The Technology Factor in International Trade*, Columbia University Press, New York and London, 1970, pp. 415–480.

and in the Azores. Columbus took it with him to Hispaniola on his second voyage to the new world in 1493, and soon thereafter it was carried to Cuba and Puerto Rico, and later to Mexico, Peru, and Brazil. Throughout this period the sugar cane cultivars used in commercial cultivation were two closely related species indigenous to India, *Saccharum sinense* Roxb. and *S. barberi* Jesw. In 1791 Captain Bligh introduced the thick-stemmed *S. officinarum* from Tahiti to Jamaica. The thick-stemmed *S. officinarum* rapidly replaced the thin forms within a few years after it was introduced. The most prominent cultivars (clones) were 'Bourbon,' 'Lahaina,' 'Cana Blananca,' and 'Otaheito.' They diffused widely throughout the world and were the basis for commercial production for nearly a century.

STAGE II: SEXUAL REPRODUCTION In nature the cane plant reproduces only asexually. Until methods of sexual reproduction were discovered this limited the selection of superior clones from indigenous cultivated or wild cultivars. Procedures for the sexual reproduction were discovered independently in Java (1887) and in Barbados (1888). The cane plant can be induced to flower and produce seedlings under appropriate temperature and light control. Each new seedling is then a potential new cultivar since it can be produced asexually. The Java station (Proefstaten Oost Java) was the first to develop a new cultivar of commercial significance (P.O.J. 100). It later introduced many more important cultivars. Important commercial varieties were also developed by 1900 at experiment stations in Hawaii, Barbados, and elsewhere. The Coimbatore station in India released the first of its improved cultivars in 1912. By 1930 P.O.J. and Coimbatore cultivars were planted commercially in every major cane-producing area of the world. Only simple tests and demonstrations, if any, were required for recipient countries to propagate these cultivars. Many of them, however, were susceptible to local diseases, and their initial yield advantages were lost.

STAGE III: INTERSPECIFIC HYBRIDIZATION Breeding for disease resistance became a dominant concern, since many of the new cultivars were found to be susceptible to local diseases and pests. The P.O.J. station played a leading role in the development of disease-resistant varieties. In 1921 the variety P.O.J. 2878 was produced. This was achieved by crossing a 118-chromosome disease-resistant thin-stemmed wild cane, S. *spontaneum*, with one of the 80-chromosome S. *officinarum* canes. By a series of crosses and back crosses, new interspecific hybrids were developed that incorporated the hardiness and disease resistance of the noncommercial species. Later, the Coimbatore station in India developed a series of tri-hybrid canes by introducing a third species S. *barberi*. Use of the S. *barberi* species resulted in new cultivars adapted to local climate, soil, and disease conditions. The high-yielding, disease-resistant stage III cultivars were transferred to every producing country in the world. Although this international transmission was widespread, it was much more dependent on local experiment station capacity than was the transmission of the stage II cultivars.

STAGE IV: LOCATION-SPECIFIC BREEDING The Coimbatore station set the stage for modern breeding activity directed toward the development of cultivars suited to the specific soil, climate and disease conditions, and management practices of smaller regions. For the most part, this breeding is conducted by experiment stations located in the specific producing region. More than 100 sugar cane experiment stations exist in the world. Very little international transfer of varieties is taking place. Almost every important cane-producing country is now using locally adapted stage IV cultivars. Yet the cultivars developed at P.O.J. and Coimbatore are important sources of genetic materials in these locally adapted canes.

It seems possible to allocate sugar cane variety transfers of stage I to the material transfer phase, and those of stage IV to the capacity transfer phase. Stage II appears to be a transition from the material transfer to the design transfer, and stage III a transition from the design transfer to the capacity transfer. This sequence indicates the important function that public research has had, not only in producing and "naturalizing" sugar cane cultivars but also in developing increasingly effective modes of transfer, from the initial international diffusion of superior canes to the international diffusion of the capacity to "invent" location-specific cultivars superior to the "naturalized" cultivars.

The green revolution

The most dramatic example of the international transfer and diffusion of agricultural technology during the last several decades has been the development and rapid diffusion of new high-yielding varieties (HYV's) of rice and wheat in the tropics (Table 28-1). The development and diffusion of the new cereals technology was made possible by a series of institutional innovations in the organization, financing, and management of agricultural research in a number of less developed tropical countries (see Chapter 27).

The development of the HYV's illustrates the process of agricultural technology transfer from the temperate zone to tropical and subtropical zones by means of the transfer of scientific knowledge and capacity. Within the tropics the diffusion of the new cereals technology from Mexico (wheat) and the Philippines (rice) was that of an initial materials transfer phase. The initial impact of this diffusion on grain production in Pakistan, India, Malaysia, Turkey, and certain other countries resulted from the direct transfer of seed from Mexico and the Philippines, and from transfer of fertilizer, insecticides, and fungicides from Japan, the United States, and Western Europe. In other countries, such as Thailand, the impact was delayed until the design and capacity transfer stage could be achieved, in order to maintain the quality characteristics of the Thai rices, which are important in the export market. In the countries that benefited initially from capacity transfer (Mexico and the Philippines), there has been a rapid movement to develop the capacity of the local experiment station so that it will permit them to move to the design-transfer and capacity-transfer phases in the development of ecologically adapted cultivars. There is also, in many countries, a move toward the development of a domestic fertilizer and agricultural chemical industry based primarily on models existing in developed countries.

Table 28-1 ESTIMATED AREA, IN THOUSANDS OF ACRES, PLANTED IN HIGH-YIELDING VARIETIES OF RICE AND WHEAT IN WEST, SOUTH, AND SOUTHEAST ASIA.

COUNTRY	RICE					WHEAT				
	1966–1967	1967–1968	1968–1969	1969–1970	1970–1971	1966–1967	1967–1968	1968–1969	1969–1970	1970–1971
Iran	—	—	—	—	—	—	—	25	222	312
Iraq	—	—	—	—	—	—	16	103	482	309
Turkey	—	—	—	—	—	1	420	1,444	1,343	1,184
Afghanistan	—	—	—	—	—	5	54	302	361	574
India	2,195	4,408	6,625	10,729	13,593	1,270	7,270	11,844	12,133	14,559
Nepal	—	—	105	123	168	16	61	133	187	243
Pakistan (E)	1	166	382	652	1,137	—	—	20	22	24
Pakistan (W)	—	10	761	1,239	1,548	250	2,365	5,900	6,626	7,288
Burma	—	8	412	356	496	—	—	—	—	—
Ceylon	—	—	17	65	73	—	—	—	—	—
Indonesia	—	—	488	1,854	2,303	—	—	—	—	—
Korea	—	—	—	—	7	—	—	—	—	—
Laos	1	3	5	5	133	—	—	—	—	—
Malaysia	104	157	225	238	327	—	—	—	—	—
Philippines	204	1,733	2,500	3,346	3,868	—	—	—	—	—
Thailand	—	—	—	—	400	—	—	—	—	—
Vietnam	—	1	100	498	1,240	—	—	—	—	—
Total	2,505	6,486	11,620	19,105	25,293	1,542	10,186	19,771	21,376	24,493

Source: From Dalrymple, *Imports and Plantings of High-Yielding Varieties of Wheat and Rice in the Less Developed Nations,* USDA in cooperation with Agency for International Development, Washington, D.C., February 1972, pp. 48, 49.

TECHNOLOGICAL CHANGE AND PRODUCTIVITY

A basic characteristic of technological change is that it permits the use of a given set of resources in a way that better satisfies human wants. It includes the application of new organizational and managerial concepts as well as advances in science and technology.

It is useful to distinguish between technological changes that reduce the cost of producing already-existing private and public consumer goods and those that create new or substantially improved products that enlarge our choices in production or consumption. Radio, penicillin, nylon, and the automobile are examples of new products from technological changes. The development of transistor radios and the automatic transmission represent qualitative improvements in existing technology. Improved rices and chemical nitrogen fertilizer are the result of technological advances that reduce the cost of producing goods or services already available in the market. By and large, technological change in agriculture reduces the cost

or improves the quality of existing commodities.

Because of the complex ways in which technological change interacts with the other components of economic growth, its effects have been described or measured in many ways. Individual inventions or innovations may be described in terms of changes in the blueprints or the specifications for individual items of capital equipment. A new cultivar may be described in terms of the genetic stock from which it was derived or in terms of its differential response to alternative levels of fertilizer application (Chapter 24). For economic purposes, however, it is most useful to measure technological change in terms of input-output, or productivity relationships—that is, in terms of such ratios as yield per acre, output per man hour, and pounds of gain in weight per pound of feed.

A program to increase output at the national level involves a hierarchy of potential changes in products, inputs, and techniques of production. Such a system for achieving increased food output for a tropical rice producing economy is illustrated schematically in Figure 28-6. The changes on the

696

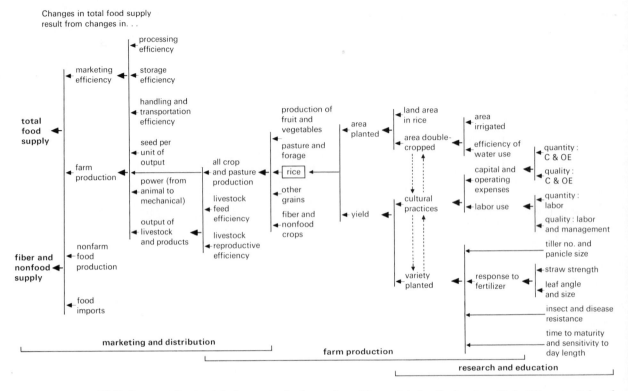

FIGURE 28-6. *Sources of potential change in food supply, with special details for rice. (C & OE = capital and operating expenses.)*

right side of the chart are made by individuals and by organizations engaged primarily in research, development, and education. The changes in the center of the chart fall within the area of farm production and management decisions. The changes on the left are made primarily in the distribution, or marketing, sector.

Each arrow represents a technical input-output linkage. Changes in technology introduced at these points change the factor-product, or output-input, ratios that produce the total and partial productivity changes for the entire food-producing sector of the economy.

Measures of partial and total productivity

The adoption of innovations by individual farms and industrial firms results in more efficient use of resources throughout the entire economy. These industry-wide and economy-wide gains in efficiency are frequently measured by partial productivity ratios or indexes.

Changes in partial productivity ratios, such as output per man hour or yield per acre, are the most frequently employed measures of technological change. Partial productivity measures, such as yield per acre or output per man hour, have several advantages. They are relatively easy to compute and they summarize a great deal of information in a single measure.

Although useful for some purposes, partial productivity measures have some limitations. They are not precise indexes of technological change. A rise in output per man hour may occur as a result of mechanization of operations that were formerly performed manually—for example, the substitution of mechanical cotton pickers for hand picking. If this is the case, the resource savings resulting from the rise in output per unit of labor input are at least partially offset by an increase in the use of capital, which results in a decline in output per unit of capital.

Dissatisfaction with the precision of partial productivity measures as indexes of technological change has led to the development of total productivity measures. Total productivity indexes measure the change in output per unit of total input. They are therefore a more valid measure of technological change than partial productivity measures. In constructing total productivity measures, the index of total input is derived by adding together the inputs of labor, land, capital, and raw material. Each factor is given a weight proportional to its relative importance in the production process. As a result, substitution of capital for labor has less tendency to result in a biased measure of technological change.

The relationship between partial and total productivity measures is illustrated in Figure 28-7 (see also Tables 29-3 and 29-5). The total productivity measure indicates the growth in output relative to the growth in total input. This can be compared to the two partial productivity measures: (1) **labor productivity,** which measures the change in farm output relative to the change in labor inputs, and (2) **land productivity,** which measures the change in farm output relative to the change in land use. In addition to land and labor the total input measure includes inputs of capital equipment (buildings and machinery) and operating expenses (fertilizer, insecticides, and fuel).

Note the wide differences in the rate of change in the two partial productivity measures. During recent years use of either the land or the labor productivity measure, instead of the total productivity measure, would overstate the true rate of technological change. This is because the total productivity measure takes into account the substitution of other resources—in the form of capital equipment and operating expenses—for part of the land and labor that has been withdrawn from production.

MARKET IMPLICATIONS OF TECHNOLOGICAL CHANGE

How does technological change influence, and how is it influenced by, the supply and demand relation-

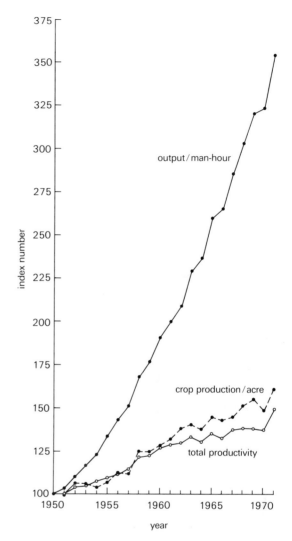

FIGURE 28-7. *Index numbers of output per man-hour, crop production per acre, and output per unit of total input (total productivity) in U.S. agriculture, 1950–1971 (1950 = 100). Data for 1971 are preliminary.*

ships that underlie price determination and behavior, which were studied in Chapter 24 (see Fig. 24-6).

The effect of technological change is to introduce a dynamic element into the supply relationships studied earlier. A rise in output per unit of total input results in higher net returns to farm operations at one of the input-output linkage points of Figure 28-8—for example, increased

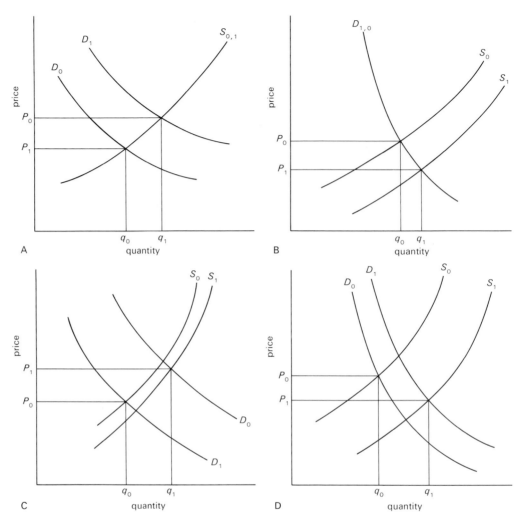

FIGURE 28-8. *Price effects of four alternative demand and supply shifts. (A) Rapid population growth, income growth, or high income elasticity results in rising demand; resource availability and technology remain unchanged. Prices rise. (B) Resource availability or advancing technology shifts product-supply curves rapidly to the right; unchanged population and income holds demand unchanged. Prices fall. (C) Resource availability or advancing technology shifts product curves only moderately to the right, whereas rapid population and income growth (or high income elasticity) result in rapid growth in demand. Prices rise. (D) Resource availability or advancing technology shifts product-supply curves rapidly to the right, whereas slow population growth or income growth (or low income elasticity) results in slow growth in demand. Prices fall.*

returns to fertilization—and shifts the supply curve to the right, as in Figure 28-8, B.

Changes in population and income also introduce dynamic elements into the demand relationship. A rise in population with a constant or rising per capita income shifts the demand curve to the right, as in Figure 28-8, A.

The net impact of these two dynamic forces depends on their relative growth rates. If technological change occurs more slowly than growth of demand for the products of an industry, prices tend to rise, as in Figure 28-8, C. If technological change proceeds more rapidly than the growth of demand for an industry's output, prices tend to fall, as in Figure 28-8, D.

The price system also serves to keep the growth of demand and the growth of supply in balance. If demand grows more rapidly than supply, product prices rise more rapidly than input prices. This tends to stimulate production and curtail demand for the product. If demand grows less rapidly than supply, product prices fall relative to input prices. This tends to curtail production and stimulate demand for the product.

The price system is also an important mechanism in transmitting the impact of technological change from one industry to another. A technological change in the fertilizer industry shifts the fertilizer supply curve to the right and reduces the price of fertilizer relative to the price of crops. This increases the use of fertilizer by producers and, in turn, reduces the cost of producing crops.

The most favorable economy for rapid adoption of new technology in agriculture is one that is growing rapidly enough to create labor shortages or rising wage rates in agriculture and which is capable of generating new technology rapidly enough to make capital and operating inputs available to agriculture at declining relative prices. This can only occur if nonfarm employment is expanding at a sufficiently rapid rate to provide alternative employment opportunities for agricultural workers and produce a labor "shortage" in the agricultural sector. Indeed, a labor shortage in agriculture is one of the most reliable indicators of economic development.

Selected References

Hayami, Yujiro, and Vernon W. Ruttan. *Agricultural Development: An International Perspective.* The Johns Hopkins University Press, Baltimore, 1971. (Identifies the capacity to develop a technology that is consistent with environmental and economic conditions as the single most important variable that explains the growth of agricultural productivity. Analyzes the historical and economic factors in the development of agriculture in the United States and Japan, and draws the implications of the U.S. and Japanese experience for less developed countries.)

President's Science Advisory Committee. *World Food Problems* (3 vols.). U.S. Government Printing Office, Washington, D.C., 1967. (A comprehensive treatment of technical, economic, and institutional aspects of agricultural development in the less developed countries. A valuable reference.)

Rogers, Everett M., and F. Floyd Shoemaker. *Communication of Innovations: A Cross Cultural Approach.* The Free Press (Macmillan), New York, 1971. (The definitive sociological treatment of the diffusion of technical and institutional innovations. Includes a very thorough review of the theoretical and empirical work on diffusion of innovations.)

Schultz, Theodore W. *Transforming Traditional Agriculture.* Yale University Press, New Haven, 1964. (Major emphasis is placed on the importance of research and development in making new and more profitable inputs available to farmers.)

Spicer, Edward H. (editor). *Human Problems in Technological Change.* Russell Sage Foundation, New York, 1952. (Case studies of the introduction of new ideas in agriculture, industry, and medicine to areas deficient in these technologies. Particular emphasis is placed on the social and cultural dynamics involved in the adoption of new technologies.)

CHAPTER 29

Agriculture and Economic Development

Economic growth was previously defined as a "sustained rise in the level of consumption or real income per person." **Economic development,** a broader concept, relates to progress along a wide spectrum of technological, economic, social, and cultural dimensions. Thus it is much more difficult to characterize a country in terms of development than of growth. In many respects India is a highly developed low-income country, whereas the United States is an underdeveloped high-income country.

One index of development is the extent to which an economy can free its resources from the production of basic food and fiber requirements. In a society in which a high proportion of resources must be devoted to meeting basic subsistence requirements, most people are engaged in the production of food and fiber for themselves and for their neighbors. They do not have the time, the ability, or the tools to produce other goods and services. In a highly developed economy only a small proportion of the population is engaged in agriculture, agricultural productivity is relatively high, and most of the working population is engaged in the production of other goods and services.

There are four ways in which the agricultural sector can contribute to national economic development. First, and most essential, it can provide the food and fiber necessary for an expanding population that is growing in income and wealth. Second, it can release workers needed for the production of nonagricultural goods and services. Third, it can serve as a market for nonfarm goods and services, thus providing a stimulus for expansion of employment and output in the nonfarm sector. Finally, it can provide a source of capital that can be invested in improved productive facilities in the rest of the economy.

A progressive urban-industrial economy contributes, in turn, to the rapid development of agriculture by expanding the markets for agricultural products; by supplying the farm machinery, chemical fertilizers, etc., that raise the level of agricultural technology; by expanding productive employment opportunities for workers released from agriculture by technological change; and by making possible improvements in the quality of rural life by raising standards of consumption in both urban and rural areas.

The net effect of this dynamic interaction between the agricultural and nonagricultural sectors in a developing economy is the complete transformation of the agricultural sector. Indeed, the term agriculture itself may have little meaning for the modern food and fiber industries that are in the process of emerging from the economic and technological revolution now occurring in the United States.

AGRICULTURAL DEVELOPMENT UNDER LABOR SCARCITY: THE UNITED STATES

From the time of the Plymouth and Jamestown settlements until the closing years of the nineteenth century, the encounter with the frontier represented a dominant theme in American agricultural development. This encounter created an opportunity for the evolution of an agriculture based on an abundance of land and a relative scarcity of labor. This, in turn, stimulated the development of an agricultural technology that was primarily directed toward achieving gains in labor productivity rather than gains in land productivity.

Since the closing of the frontier, in the last quarter of the nineteenth century, the encounter with an increasingly dominant urban-industrial sector has emerged as a major theme in American agricultural development. By 1880 nonagricultural employment exceeded agricultural employment. By 1929 manufacturing employment alone exceeded agricultural employment. By 1970 agricultural employment was less than total unemployment in the United States (Table 29-1).

We can best visualize the interactions between the farm and the nonfarm sectors that led to this fundamental restructuring of the American economy by looking in turn at each of three sets of market relationships: the **product market,** through which the output of the agricultural sector is transmitted to the nonfarm sector and through which incomes are generated in the farm sector; the **input markets,** through which move the manufactured inputs, equipment, and capital used in agricultural production; and the **labor market,** through which labor is allocated between the agricultural and nonagricultural sectors and among firms in each sector.

The product market

Throughout most of American economic history the product market—the market for things farmers sell—represented the primary link between the farm and the nonfarm sectors of the economy. It was the dominant channel through which international shifts in the terms of trade, national fluctuations in nonfarm income, and local variations in nonfarm demand have been channeled into the agricultural sector.

In most low-income countries, where a substantial share of increases in per capita income are devoted to dietary improvement, the product market is still the main link between the peasant and the urban-industrial sector of the economy. It has already been pointed out that as income per person rises, consumption of agricultural products expands less rapidly (Chapter 25). At very high income levels there may be no additional food consumed as income continues to rise. In the United States, the declining response in consumption of food and fiber to increases in nonfarm income has almost eliminated the commodity market effects of growth in economic activity in the nonfarm sector.

In addition, agricultural trade and commodity policies have been designed which insulate agricultural commodity prices, particularly crop prices, from normal trade and market fluctuations.

As a result of these changes the rate of growth of domestic demand for food and fiber products in the American economy now barely exceeds the rate of population growth. Opportunities for growth in U.S. agricultural output have become increasingly dependent on the growth of demand in Western Europe and Japan and on food aid to the developing countries (Chapter 28).

Technological change in agricultural production has been accompanied by institutional changes in the product markets that link the agricultural to the urban-industrial sector. The production of some products has become essentially indus-

Table 29-1 EMPLOYMENT BY SECTOR IN THE UNITED STATES, 1880–1970 AND
PROJECTIONS TO 1980 (IN THOUSANDS OF WORKERS).[1]

	AGRICULTURE	TOTAL NONAGRICULTURE	MANUFACTURING	UNEMPLOYMENT
1880	8,585	8,807	NA	NA
1929	10,450	37,180	10,534	1,550
1950	7,160	51,760	15,241	3,288
1970	3,462	75,165	19,369	4,088
1980 (est)[2]	3,156	95,444	22,133	3,944

Sources:

1880: United States Bureau of the Census, *Historical Statistics of the United States, Colonial Times to 1957*, Washington, D.C., 1960.

1929 to 1971: United States Department of Labor, Bureau of Labor Statistics, "Employment and Earnings," Vol. 18, No. 9, March 1972, pp. 21, 49.

1980 Projections: United States Department of Labor, Bureau of Labor Statistics, "Patterns of U.S. Economic Growth," Bulletin 1672, 1970, p. 6.

NA = Not available.

[1] Before 1950 data includes persons of age 14 and over; 1950 and beyond includes persons of age 16 and over.

[2] Assumes 4 percent unemployment rate.

trialized. Broiler "factories" have almost entirely replaced farm production of poultry meat. Commercial production of fruits and vegetables is becoming highly concentrated. As a result of technological and organizational changes in processing, transportation, and distribution, regional specialization in the production of fruit, vegetables, and animal products has reduced the impact of local urban-industrial development on the demand for locally produced farm products, Milk, protected by a series of local market trade barriers, remains a major exception to this generalization.

These developments in the farm sector have been accompanied by the replacement of small-scale retail stores served by local wholesalers with large chain store distribution systems served by integrated wholesale and retail operations. There is an increasing tendency for these large distributing organizations to establish direct linkage with the large specialized fruit and vegetable producers, poultry producers, and livestock feeders, and to bypass traditional marketing channels.

This type of development has not yet had any substantial impact on the major agricultural commodities, such as wheat, cotton, corn, soybeans, and most hog and beef production. But the emergence of highway truck transportation as a major competitor with the railways in the hauling of bulk agricultural commodities has eliminated many of the older terminal markets.

The markets for purchased inputs

The markets for manufactured capital equipment and current inputs have become increasingly important in transmitting the effects of changes in the nonfarm sector to the agricultural sector. Much of the new agricultural technology is embodied in the form of new capital equipment or more efficient fertilizers, insecticides, and other manufactured inputs. In 1870 the typical American farm was still a subsistence unit, with inputs purchased from the nonfarm sector amounting to less than 3 percent of the value of farm production. By 1900 nonfarm inputs still amounted to only 7 percent. But by the early 1970's they were more than 30 percent of farm expenditures.

The use of purchased inputs has been closely related to developments in the labor market. The demand for labor, resulting from rapid urban-industrial development, reinforced the economic pressure for substitution of capital equipment for labor in American agriculture at precisely that period when the frontier was disappearing as a major factor in agricultural development. It was during this period that the tractor, first functioning as the source of motive power for formerly horse drawn equipment, and later as the power unit around which new equipment was designed, became the symbol of technological change, both in American agriculture and around the world.

Table 29-2 ACREAGES, YIELDS, AND PRODUCTION OF GRAINS FOR THE UNITED STATES.

PERIOD	ACREAGE (MILLIONS)	PRODUCTION (MILLIONS OF) (METRIC TONS)	YIELD (METRIC TONS) (PER ACRE)	INDEXES (1880–1889 = 100)		
				ACREAGE	PRODUCTION	YIELD
1880–1889	135.0	72.2	0.535	100	100	100
1890–1899	162.8	86.8	0.533	121	120	100
1900–1909	185.2	103.5	0.559	137	143	104
1910–1919	208.1	111.2	0.534	154	154	100
1920–1929	218.2	120.2	0.551	160	169	103
1930–1939	219.6	102.8	0.468	163	142	87
1940–1949	212.5	134.7	0.634	157	187	119
1950–1959	187.9	143.3	0.762	139	198	142
1960–1964	160.7	176.7	1.099	119	244	205
1965–1969	152.9	215.3	1.408	113	298	263

Sources: From USDA, *Agricultural Statistics*, 1938, 1952, 1957, and 1964; and *Crop Production, 1964 Annual Summary* and *1971 Annual Summary*. (The crops included are wheat, rye, rice (from 1900), buckwheat, corn, oats, barley, and grain sorghums (from 1920). In 1900 the acreage of rice was 0.36 million, while in 1919 the grain sorghum acreage was 3.63 million. Acreage are in terms of harvested area.)

The rapid growth of labor productivity in American agriculture was not, at first, accompanied by parallel changes in land productivity. Grain yield per acre in American agriculture remained essentially unchanged from the end of the Civil War in the 1860's until well into the twentieth century (Table 29-2 and Fig. 29-1). By the mid-1920's, however, the production of fertilizer and other agricultural chemicals was beginning to be reflected in higher yields. Higher yield potentials were also emerging as a result of the application of advances in genetics to plant breeding by scientists in the Land Grant College system and the United States Department of Agriculture.

This emergence of a new chemical and biological technology in American agriculture in the 1920's was itself the product of a long sequence of institutional development. As early as 1862 the United States Congress passed legislation granting public lands to the states, to be used for the "support and maintenance of at least one college where the leading object shall be . . . to teach such branches of learning as are related to agriculture and the mechanic arts." It was not until 1887, however, that an act was passed that established agricultural experiment stations in each state, usually at the "land grant" colleges.

By the mid-1920's substantial progress in agricultural research, embodied in crops, production practices, and other forms of new knowledge, was beginning to flow from the agricultural education and research system to the farmer via the newly organized Federal-State extension service. Hybrid corn was a particularly dramatic product of this research.

The combination of rapid advance in biological research plus the high volume of relatively inexpensive agricultural chemicals created a new dimension in agricultural productivity in United States agriculture. Land productivity, which had experienced no real growth between 1900 and 1925, rose by 1.4 percent per year for the period 1925–1950, by 2.5 percent per year between 1950 and 1965,

FIGURE 29-1. *Current corn yields in selected countries related to United States' historical trend.* [After Brown, Increasing World Food Output, *USDA, Foreign Agricultural Economic Report No. 25, April 1965.*]

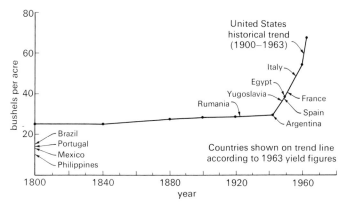

704

Table 29-3 ANNUAL AVERAGE RATES OF CHANGE
(PERCENT PER YEAR) IN TOTAL OUTPUTS, INPUTS, AND
PRODUCTIVITY IN UNITED STATES AGRICULTURE,
1870–1971.

ITEM	1870–1900	1900–1925	1925–1950	1950–1965	1965–1971
Farm output	2.9	0.9	1.5	1.9	2.3
Total inputs	1.9	1.1	0.3	−0.2	0.7
Total productivity	1.0	−0.2	1.2	2.1	1.6
Labor inputs[1]	1.6	0.5	−1.8	−4.3	−3.0
Labor productivity	1.3	0.4	3.3	6.6	5.4
Land inputs[2]	3.1	0.8	0.1	−0.8	0.5
Land productivity	−0.2	0.0	1.4	2.5	1.8

Sources: From USDA, *Changes in Farm Production and Efficiency,* Statistical Bulletin 233 (revised), Washington, D.C., June 1972; and D. D. Durost and G. T. Barton, *Changing Sources of Farm Output,* USDA Production Research Report No. 36, Washington, D.C., February 1960.

[1] Number of workers, 1870–1910; man-hour basis, 1910–1971.
[2] Cropland used for crops, including crop failure and cultivated summer fallow.

and by 1.8 percent per year between 1965 and 1971. This higher output per acre combined with continued mechanization to produce a rate of growth of labor productivity of 6.6 percent per year between 1950 and 1965 and by 5.4 percent per year between 1965 and 1971. (Table 29-3).

The labor market

The labor market has become an increasingly important channel of interaction between the farm and the nonfarm sectors. Technical and economic developments have made it increasingly profitable to substitute inputs purchased from the industrial sector for farm labor. The slow growth in domestic demand for farm products, the insulation of domestic markets from changes in demand in other countries through import restrictions and export subsidies, and the rapid growth in labor productivity combined to place most of the burden of balancing the rate of growth of agricultural output with the rate of growth in demand for agricultural products on the labor market.

With the demand for agricultural output expanding at about 2 percent per year and labor productivity rising by about 6 percent per year, the burden of adjustment in the labor market has been extremely heavy. It has been particularly difficult in the low-income agricultural regions where local nonfarm employment has not expanded fast enough to absorb both the excess agricultural labor force and the new entrants to the labor force from rural areas.

So large has the labor surplus been, and so great the obstacles to migration for the older and less well-educated members of rural communities, that migration has generally not been sufficient to narrow income differentials between farm and nonfarm workers, except where there has been substantial growth in the local nonfarm labor market. It has not been possible to realize the full potentials of rapid growth in agricultural productivity and income per farm worker or per farm family throughout much of the southeastern United States, and in scattered areas throughout the rest of the nation, where local urban-industrial development has lagged (Fig. 29-2).

Progress and poverty

American agricultural development policies have been uniquely successful in meeting national farm output and productivity objectives. These policies have been less successful in meeting the income objectives of all of the families who are engaged in the production of agricultural commodities. One observer has pointed out that "behavior of rural people, their representatives and their institutions implies a materialistic bias in favor of plants, land, and animals and against people." Although this is perhaps overdrawn, it is true that the policies of the past were designed primarily to solve technological and commodity problems rather than to solve the income problems of rural people.

It can be argued that this was a valid choice at the time these policies were established. It was important for United States economic development that the agricultural sector achieve sufficiently high rates of output and productivity growth to meet national food and fiber requirements and at the same time release substantial numbers of farm workers for nonfarm employment.

An unanticipated by-product of these changes has been the emergence of a dual structure in the well-being of rural families. There is no sector of American agriculture that can be properly classed as a peasant sector. There is, however, substantial poverty in rural areas. The poverty problem has several dimensions. There is a regional dimension, an occupational dimension, and a racial dimension (Table 29-4).

There is no economic reason, however, for this dual structure in American agriculture to continue further. In fact, agricultural policies that are less commodity oriented are now emerging. There is a shift away from the relatively unsuccessful attempts to improve the income distribution within agriculture through policies designed to maintain or improve commodity prices. And there is growing recognition that rapid technological change, although desirable on grounds of efficiency, does impose severe financial and personal adjustment burdens on many farm families and regions.

The extension of Social Security to farmers in 1955 was an important step toward separating income support payments from commodity prices. Land retirement programs, which make rental payments to farmers for removing marginal land from intensive crop production, have also served as a means of transferring income to farmers without directly affecting market prices. The Food and Agriculture Act of 1965 and the Agriculture Act of 1970 go further than earlier legislation in separating income support payment from production-incentive and price-stabilization payments.

The use of effective income protection programs that establish some socially acceptable minimum standard of living and protection against the risks imposed on individuals by the uncertainties involved in agricultural developments are particularly important in the short run. Such payments permit the program participants to achieve a level of consumption more nearly in line with American standards. But they usually do not meet the addi-

Table 29-4   MEAN INCOME OF FARM AND NONFARM FAMILIES BY REGION AND COLOR, 1970.

| | UNITED STATES | NORTHEAST | NORTH CENTRAL | SOUTH | | | WEST |
				WHITE	NONWHITE	ALL RACES	
Mean income	$11,106	$11,988	$11,358	$10,634	$6,441	$9,949	$11,538
Nonfarm	11,254	12,006	11,593	NA	NA	10,129	11,611
Farm	7,983	10,542	8,331	NA	NA	7,027	8,604
Families with incomes of $10,000 or more	49.1%	54.8%	52.3%	44.9%	17.5%	40.4%	51.8%
Nonfarm	50.0%	54.9%	53.6%	NA	NA	41.5%	52.2%
Farm	29.7%	48.9%	32.9%	NA	NA	21.6%	33.7%
Families with incomes under $5,000	19.3%	15.5%	16.9%	20.7%	47.5%	25.0%	17.6%
Nonfarm	18.3%	15.2%	15.7%	NA	NA	23.9%	17.2%
Farm	36.2%	21.2%	32.0%	NA	NA	44.0%	31.7%

Source: From Bureau of Census, *Current Population Reports,* Series P-60, No. 80, and unpublished data.

NA = Not Available.

distribution of values

| 0–14.3 | 14.4–19.3 | 19.4–25.4 | 25.5–33.3 | 33.4–71.9 |

percent

FIGURE 29-2. *The percent of all families with incomes of less than $4,000, by county, 1970.* [*Courtesy USDA.*]

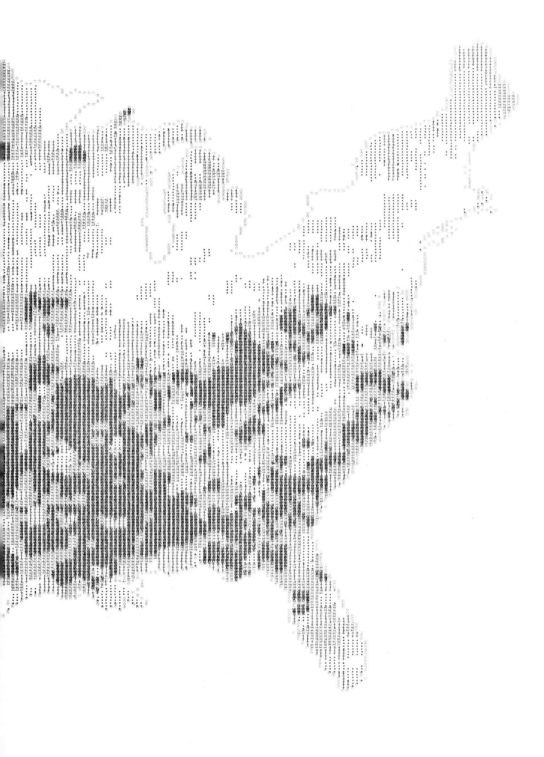

tional objective of enabling the participant to contribute effectively to the further growth of the American economy.

To meet this latter objective, a stronger emphasis should be placed on investment in the human agent of production—in man. The traditional pattern of underinvestment in rural health, rural education, and other rural social services will have to be corrected.

THE DEVELOPMENT OF A LABOR-INTENSIVE AGRICULTURAL SYSTEM: JAPAN

Japan was the first Asian country to succeed in bringing about a striking transformation in the productivity of its agriculture. In contrast to the United States, agricultural development in Japan occurred within a framework of increasing labor intensity. The average size of the Japanese farm was approximately 1.0 hectare (2.2 acres) in 1878 and 0.8 hectare (1.7 acres) in 1962.

In many respects the Japanese experience was the outcome of factors peculiar to the Japanese environment. Japan is an island economy, and social institutions had been cultivated and refined over generations to permit an efficient balance between population and a limited resource base.

FIGURE 29-3. *Current rice yields (rough rice) in selected countries related to Japan's historical trend. Yields based on historical estimates from Japanese Ministry of Agriculture.* [*After Brown,* Increasing World Food Output, *USDA,* Foreign Agricultural Economic Report No. 25, *April 1965.*]

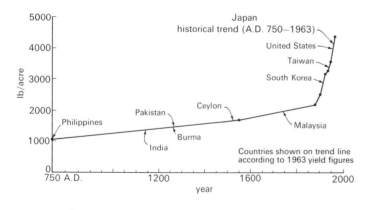

Despite other differences, at the time of the "take-off," Japanese agricultural development was similar in many respects to that of other countries in Asia—a traditional agriculture characterized by small-scale subsistence farms dominated by a hierarchial social structure. The level of productivity in Japanese agriculture prior to the Meiji Restoration (1868), as measured by yields per hectare or per man, were probably only slightly higher than the levels of productivity that persist today throughout Asia (Fig. 29-3).

The remarkable feature of the Japanese agricultural transformation, and the feature that makes it of special interest in the rest of Asia today, is that it took place within a traditional framework of small-scale agriculture. Although the average farm was declining in size, the average rice yield rose from 1.8 metric tons per hectare (1868 to 1882) to 4.0 metric tons in the late 1950's and 4.1 metric tons in the early 1970's.

During this development Japanese agriculture went through three phases (Table 29-5).

1. A period of rapid growth lasting from approximately the time of the Meiji Restoration (1868) until the end of World War I.
2. A period of slower growth lasting from the end of World War I through World War II.
3. A new period of rapid growth, starting almost immediately after the end of World War II and continuing to the present time.

Productivity growth from intensification of traditional agriculture, 1870–1920

The rapid increase in Japanese agricultural productivity between 1870 and 1920 was due to the diffusion of the superior practices already in use in Japanese traditional agriculture combined with the limited adoption of Western methods, primarily soil and fertilizer science adopted from Germany.

The basic agricultural policy of the new Meiji Government, established in 1868, developed around the ideas of Lord Iwakura, who visited North America and Europe in 1871 and 1873. Iwakura's visits were followed by tours made by other Japanese officials, and western agricultural experts were invited to Japan. At first most of these experts came from the United States and England. Later, experts from Germany were invited. The Japanese officials who returned to Japan from western countries stressed the need of raising the level of Japanese agriculture to that of the West by adopting western style extensive farming in place of small-scale intensive farming. It became apparent, however, that instead of the large-scale farm-management techniques developed in North America and England, the knowledge and techniques of Germany, particularly the new knowledge of soils and fertilizers, would become more realistic and practical in application to the small-scale Japanese agriculture. Only in the northern island of Hokkaido did extensive farming based on livestock production and western style "horse mechanization" find a permanent home.

In 1887 a new agricultural policy was adopted that shifted emphasis away from introducing western farming methods to bolstering traditional farming methods. There were three major elements in this new policy: (1) selection and diffusion of high-yielding varieties of rice; (2) the establishment of a fertilizer-consuming agriculture; (3) introduction and diffusion of new cultural practices and implements.

These three policies were closely interrelated. The new rice varieties selected were those that responded well to nitrogen fertilizer. Such cultural practices as deep plowing, double cropping, straight-row planting, revision of field layouts, irrigation and drainage, and others were also oriented toward obtaining a favorable response from increased fertilization. Table 29-6 traces the evolution of the use of fertilizer in one Japanese prefecture from 1877 to 1957.

These policies were effective for several reasons. Over the preceding three hundred years of the Tokugawa period agricultural techniques had been slowly improving, but the restraints of the feudal system had suppressed the diffusion of new techniques. Under the feudal system peasants were bound to their land and not allowed to leave their villages except for religious pilgrimages. Nor were they free to choose what crops to plant or varieties to sow. The feudal lords were anxious to raise agricultural productivity in their own territories,

Table 29-5 AVERAGE ANNUAL CHANGE IN TOTAL OUTPUTS, INPUTS, AND PRODUCTIVITY IN JAPANESE AGRICULTURE, 1880–1965.[1]

ITEM	1880–1920	1920–1935	1935–1955	1955–1965
Farm output	1.8	0.9	0.6	3.6
Total inputs	0.5	0.5	1.2	0.7
Total productivity	1.3	0.4	−0.6	2.9
Labor inputs	−0.3	−0.2	0.6	−3.0
Labor productivity	2.1	1.1	0.0	6.6
Land inputs	0.6	0.1	−0.1	0.1
Land productivity	1.2	0.8	0.7	3.5

Source: Saburo Yamada and Yujiro Hayami, "Growth Rates of Japanese Agriculture, 1880–1965," (revised). Paper presented at the Conference on *Agricultural Growth in Japan, Korea, Taiwan, and the Philippines* at the East-West Center, Honolulu, Hawaii, February 5–9, 1973.
[1] Based on five-year averages centered on the years shown.

but they frequently prohibited the transfer of techniques or varieties outside their bounds. It is even recorded that one village placed a guard at its border to prevent a variety of seed selected in the village from being taken out.

A second factor in the success of the Meiji agricultural policy was the pattern of investment in agricultural education by the government. Agricultural schools were organized as early as 1876. An agricultural experiment station and seed breeding station was established in 1877—ten years before the United States Congress passed the Hatch Act, which established experiment stations in each state. By 1900 Japan had developed a number of national research institutions as well as a network of experiment stations at the prefectural level.

Initially, the most successful "veteran farmers" were used to carry improved techniques to other farmers in their own and in other prefectures. By 1893, eleven years before the establishment of the Federal-State extension service in the United States, the prefectural experiment stations were given responsibility for formal extension activity.

Thus the prefectural experiment stations became extension centers for the dissemination of new knowledge to associations of village farmers. Failure to adopt improved technologies was frequently punished by fines or arrest. This obviously created a somewhat different level of receptivity

Table 29-6 CHANGES IN APPLICATION OF FERTILIZERS IN KANAYA-MACHI, SHIZUOKA PREFECTURE, JAPAN, (1877–1967).

YEAR	MAJOR FERTILIZER	RICE		BARLEY		TEA	
		PLAINS	MOUNTAINS	PLAINS	MOUNTAINS	PLAINS	MOUNTAINS
1877	NATURAL ORGANICS						
1887							
1897	GREEN MANURE and ORGANIC BY-PRODUCTS (FISH, SOYBEAN, RAPESEED MEAL)	Grass	Grass	Night soil	Unknown	Soybean meal, shrimp	Unknown
1907		Grass, compost, soybean and fish meal	Grass, compost, soybean and fish meal	Night soil, compost, soybean meal, superphosphate	Night soil, compost	Soybean meal, shrimp, organic mixes	
		Grass, compost, soybean and rapeseed meal, spent distillery mash, green soybean	Grass, soybean and rapeseed meal, seaweed	Night soil, compost, soybean meal, superphosphate	Night soil, compost, superphosphate	Night soil, shrimp, organic mixes, soybean meal, green soybean	
1917		Grass, compost, soybean and rapeseed meal, sodium nitrate, ammoniated superphosphate	Grass, soybean and rapeseed meal, ammonium sulfate, superphosphate	Night soil, compost, superphosphate, soybean meal, ammonium sulfate	Night soil, compost, superphosphate	Night soil, meal, organic mixes, shrimp	Soybean and fish meal
1927	CHEMICAL	Grass, compost, fish meal, ammonium sulfate, sodium nitrate, mixed	Grass, compost, night soil, organic mixed fertilizers, superphosphate, ammonium sulfate, sodium nitrate	Night soil, compost, meal, ammonium sulfate, superphosphate, mixed	Night soil, compost, meal, ammonium sulfate, superphosphate, mixed	Night soil, meal, green soybean, organic mixes	Grass organic mixes, superphosphate, ammonium sulfate
1937		Compost, fish meal, ammonium sulfate, superphosphate, potassium chloride, calcium cyanamide, ammonium phosphate, mixed	Grass, compost, wood ash, night soil, ammonium sulfate, calcium cyanamide, superphosphate, potassium chloride, mixed	Night soil, compost, ammonium sulfate, superphosphate	Night soil, compost, ammonium sulfate, superphosphate, mixed	Superphosphate, ammonium sulfate, mixed	Rapeseed meal, grass, ammonium sulfate, mixed
1947		Rapeseed meal, ammonium sulfate, calcium cyanamide, potassium chloride, fused phosphate, mixed	Night soil, compost, ammonium sulfate, potassium chloride, urea, fused phosphate, mixed	Night soil, compost, potassium chloride, ammonium sulfate, calcium cyanamide, mixed	Night soil, compost, ammonium sulfate, superphosphate, mixed	Grass, ammonium sulfate, calcium cyanamide, mixed	Fish and rapeseed meal, ammonium sulfate, calcium cyanamide, urea, mixed
1957							
1967		High synthetic, low synthetic compound, calcium cyanamide (granular, dust), calcium cyanamide compound, potassium phosphate ammonium mixed, special compound for rice seedling bed, silic calcium, ammonium chloride, magnesium lime, mixed, urea intermixed compound, ammonium potassium compound°	Low synthetic compound	None grown		Superphosphate of lime, fused phosphate, potassium chloride, potassium sulfate, low synthetic compound, high synthetic compound, organic matter intermixed compound, multiphosphate, phosphate potassium ammonium mixed, phosphate ammonium nitrate, potassium mixed, magnesium lime, fish meal, seed cake, chicken droppings, ground bonemeal, ground composts°	

Source: From Takekazu Ogura (ed.) *Agricultural Development in Japan,* Japan FAO Association, 1963, pp. 372, 373 (after Y. Ienaga).; unpublished data for 1957–1967 from Y. Ienaga.
° The distinction between plains and mountains has tended to disappear as a result of continued modernization.

than the American county agent faced when he began his work in 1914.

Stagnation of traditional agriculture, 1920–1946

Shortly after World War I, Japan appeared to have reached the limit of agricultural development that could be attained by using traditional methods. The rate of growth of agricultural output declined. Food shortages developed in the face of a growing urban population. Whereas agriculture had been a major source of support for Japan's industrial revolution during the Meiji era it now became a depressed area in the economy. The land tenure system placed increased burdens on the tenants and dampened incentives to produce.

A number of significant changes took place during this period. Agricultural land development was pushed into marginal areas. Farmers began to shift their production away from rice into the even more labor intensive livestock production, silkworm raising (sericulture), and fruit and vegetable production. The shift away from rice was partly a response to the lowering of prices that resulted from the importation of rice from Formosa (Taiwan) and Korea.

The rapid shift of farm workers from agricultural to nonagricultural employment during the period was not sufficient to reduce the size of the agricultural labor force. Farm employment did decline, however, from roughly 40 percent of the labor force at the beginning of the period to less than 20 percent by the beginning of World War II.

By the beginning of World War II Japanese agriculture was again beginning to experience rapid economic growth. During the war, however, agriculture suffered from shortages of labor and materials. Availability of commercial fertilizer dropped sharply. The yield of rice fell in 1945 to only about 70 percent of the prewar level.

The modernization of Japanese agriculture,

Following World War II Japanese agriculture experienced a new burst of productivity. Two factors appear to have been particularly important. One was the increased incentive to produce, which

resulted from the land reform of 1947–1950. A second was the backlog of modern technology resulting from the increased sophistication of experiment station research and the increased industrial capacity that emerged from World War II.

The postwar land reform represented the culmination of land improvement that began during the Meiji restoration rather than any sharp break with the past. All farmland owned by absentee landlords and all farmland leased by resident landlords in excess of 4 hectares was appropriated by the government and sold to the actual tenants. About 2 million hectares, approximately 80 percent of the tenant-cultivated land, was acquired by tenant farmers. The result was a major improvement in incentives to adopt new technology and increase production (Table 29-5).

The change in agricultural technology since World War II can be illustrated by the following. The new rice varieties being planted in Japan today are the result of experiment station breeding programs designed to produce high-yielding, disease-resistant varieties that respond well to fertilization. This is in contrast to the basis of selection that prevailed during the Meiji period when the best varieties were selected from the many already in existence. The new breeding programs, however, did not depart in their objective from that of the older selection programs. The varietal improvements and associated cultural practices continued to be directed toward the development of a "fertilizer-consuming rice culture."

The increased capacity of Japanese industry to produce fertilizers and other agricultural chemicals has complemented this traditional objective. There has, however, been one major change. As a result of Japan's continued rapid industrial development, the agricultural labor force is finally beginning to decline, and agricultural wage rates have been rising. Small-scale mechanization, including the use of power sprayers and dusters and the use of electric motors and internal combustion engines for threshing and pumping irrigation water, has expanded rapidly. Most striking of all has been the rapid small-scale mechanization of plowing and other field operations. The use of mechanized equipment for field operations was trivial before World War II, but by 1960 there were half a million small tractors or cultivators in use.

*The significance of Japanese
agricultural development*

This history of Japanese agricultural development
illustrates how the agricultural sector of the economy was able to fulfill its traditional role in the
strategy of overall development. Japanese agriculture, in the course of its transformation, was able
to earn foreign exchange, provide savings and
investment for a developing urban-industrial sector, and supply raw materials and foodstuffs for
the rest of the economy.

Most significant of all, this was achieved within
a system of small-scale labor-intensive farming
made possible by placing greater emphasis upon
the "biological revolution" than upon the "mechanical revolution." In the next several decades,
as the Japanese agricultural labor force declines,
it seems likely that Japanese agriculture will successfully complement the "biological revolution"
of the last 100 years with a "mechanical revolution," leading to a fully modern system of agriculture. The Japanese example, with its initial stress
on the "biological revolution," represents a more
valid model for many of the less developed economies than the United States model.

THE TAKE-OFF THAT FAILED:
ARGENTINA

Agricultural development in Argentina presents an
interesting contrast to both the United States and
Japan. Like the United States, Canada, and Australia, Argentina developed, during the nineteenth
century, an agricultural system based on an abundance of land and a relative scarcity of labor. By
the late 1920's Argentina ranked among the leading countries in the world with respect to per
capita income. Its prosperity was based on an
economic system that involved (1) exportation of
agricultural raw materials, particularly food and
feed grains, oil crops, and meat and wool, and (2)
importation of manufactured capital equipment
and consumer goods.

The disorganization of international markets
during the world depression of the 1930's and
during World War II impressed upon Argentina's
policy-makers the desirability of developing a more
diversified economy. Industrialization became a
major policy goal.

In an effort to increase the foreign exchange
earnings needed to finance industrialization, Argentina established a government monopoly to handle
the export of primary commodities. The monopoly
paid low prices to Argentina's agricultural producers and attempted to bargain for the highest possible prices for the export commodities on world
markets. The prices received by Argentine farmers
were kept far below world market prices. This was
equivalent to a heavy tax on agricultural production.

The effect was to dampen incentives to either
produce or export agricultural commodities. Agricultural production increased by less than 10
percent between the periods 1945–1949 and
1959–1961. Argentina's share of world exports of
a number of major agricultural commodities declined sharply from prewar levels (Table 29-7).
This weakened Argentina's capacity to import. By
the mid-1950's imports of capital goods needed for
industrialization had dropped below prewar
levels.

The lack of price incentives might have been
partially offset if Argentina had given support to
technological change in agriculture by building
agricultural research and extension programs in a
manner comparable to the United States and Japan. Even today, however, Argentina has only a
relatively weak agricultural research and extension
establishment. Corn yields remain relatively low
compared to those of the United States and a
number of other countries (Fig. 29-1).

Thus the heavy taxation of Argentina's agricultural technology killed the goose that laid the
golden eggs that were expected to pay for the
program of industrialization. Today Argentina's
per capita income ranks her either as one of the
less prosperous developed nations or one of the
more prosperous underdeveloped nations, rather
than in the position among world leaders that she
held in the 1920's.

Argentina's heavy taxation of agricultural exports, in such a way as to reduce directly the prices
received by farmers, was in sharp contrast to the
policies adopted by the Japanese during the last
third of the nineteenth century. Japan successfully

utilized the export earnings from tea and silk to finance industrialization. Distortion of production incentives was avoided by using a heavy land tax to raise revenue from agriculture instead of directly interfering with prices. The favorable prices for silk and tea were allowed to act as incentives to produce supplies for export, and the land tax, which was not affected by the level of production, was used to finance industrialization.

DEVELOPMENT IMPERATIVES

There are numerous examples of situations in which rapid progress in agricultural development did not lead to sustained economic growth because of the absence of other elements critical to the process of development. Many examples of this type can be drawn from the former colonial economies. The Dutch experience in Java is particularly striking. The Dutch were remarkably successful in their efforts to develop improved crop varieties and cultural practices for a number of tropical export crops. The Agricultural Research Institutes that they developed at Bogor were among the best in the world.

The colonial system was not, however, successful in developing the capacity of the Javanese population. The result was a dual economy that provided high incomes for the expatriate owners and employees who ran the plantation and for the firms than handled the export commodities. Production of food for local consumption continued to be produced by traditional methods by the local population. Gains in productivity per hectare were absorbed by population growth with no rise in the standard of living of the indigenous population.

The same pattern was repeated, although perhaps not as strikingly, in India, in the Philippines, and in many other former colonial economies in Asia and Africa.

Successful economic development involves a complex of technological, economic, and cultural changes. Rapid productivity changes in agriculture are becoming an increasingly essential component in this complex. In the past countries like the United Kingdom, the United States, and Australia were able to achieve rising incomes in advance

Table 29-7 SHARE OF ARGENTINE EXPORTS IN WORLD EXPORTS OF SELECTED COMMODITIES (EXPRESSED AS PERCENTAGES).

COMMODITY	1934–1938	1959–1962	1968–1970
Wheat and wheat flour	19.3	5.5	4.6
Maize	64.0	18.5	14.2
Greasy wool	} 11.7	9.9	5.8
Degreased wool		13.9	13.8
Linseed and linseed oil	67.6	49.3	29.3
Linseed, cake and meal	NA	71.6	59.5
Sunflower seed and its oil	NA	14.1	5.3
Sunflower seed, cake and meal	NA	95.2	75.0
All fresh, chilled, frozen meat	39.7	17.5	11.8

Sources: From *Trade Yearbook, 1963*, FAO, United Nations, Rome; and *Analisis y Proyecciones del Desarrollo Economico, V, El Desarrollo Economico de la Argentina*, Naciones Unidas, Mexico, 1959, Part 2, p. 48. (It has been assumed that three tons of linseed and sunflower seed are equivalent to one ton of linseed oil and sunflower seed oil.)

NA = Not available.

of growth in land productivity (Fig. 29-4). The United Kingdom was able to postpone the yield takeoff because it was more profitable to capitalize on its early technological lead by exporting manufactured goods and importing foodstuffs and

FIGURE 29-4. *Take-off dates for income per person and yield per acre in selected countries.* [*After Brown,* Increasing World Food Output, USDA, *Foreign Agricultural Economics Report No. 25, April 1965.*]

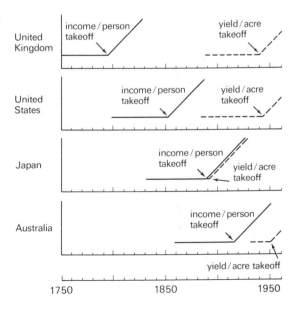

raw materials. The United States and Australia, because of a relatively favorable ratio of land resources to labor were able to concentrate on achieving rapid growth of labor productivity prior to emphasis on land productivity.

Japan's experience in generating an income takeoff and a yield takeoff at the same time is unique. This is clearly the optimum pattern for most of the presently less developed countries, particularly those that already have a high population relative to land resources. For many of the less developed countries, particularly those that are unable to limit the rate of population growth, it may be necessary for a yield takeoff to precede an income takeoff.

Important as yield increases are, they are but one of a number of closely interrelated factors involved in agricultural development. The development of the countries studied in this chapter—the United States, Japan, and Argentina—illustrates the workings of five "essentials" and five "accelerators."

The five essentials are: (1) new farm technology, (2) availability of purchasable inputs, (3) markets for products, (4) transportation, and (5) incentives for agricultural producers. The adequacy of these five determine the possibilities of agricultural development. They are like the parts of a wheel (Fig. 29-5); none is useful without the others.

The accelerators are those factors which, although not absolutely essential for agricultural growth, can contribute to speeding up the rate of growth once the essentials are met. The five accelerators are: (1) education, (2) production credit, (3) effective farm organizations or associations, (4) improving or expanding the land base, and (5) effective agricultural planning.

There can and will be some growth in agricultural productivity wherever all of the essentials are present, but without all of them there will be none. The situation is different with the accelerators. Each of them is important but none are indispensable. But in many countries it is necessary that agricultural development proceed as rapidly as possible. Recent studies have documented that investment in agricultural research often gives returns 5 to 10 times higher than conventional returns on capital investment. By placing em-

Markets for farm products provide one spoke.

New farm technology adds a second.

The local availability of farm supplies and equipment is a third.

Adequate incentives for farmers provide the fourth.

Transportation facilities complete the wheel.

FIGURE 29-5. *The five essentials for agricultural development are like the parts of a wheel.* [*After Mosher, Getting Agriculture Moving: Essentials for Development and Modernization, New York: Praeger, 1966.*]

phasis on the accelerators, as well as on the essentials, the pace of progress can be accelerated; and the return on the investment in the essentials can be raised.

DEVELOPMENT CONSTRAINTS

It is conventional to emphasize the social and economic gains brought about by agricultural development. These gains have been substantial. In short, productivity growth in agriculture has been an essential component of the total development process.

The harmful effects of agricultural development are frequently ignored. They are primarily of two types. *First,* there is the environmental stress that occurs as a by-product of productivity growth. *Second,* there is the social stress resulting from inequities in the partitioning of the costs and benefits of agricultural development.

Environmental stress

The environmental damages caused by productivity growth were discussed in Chapters 5, 24 and

28. They are largely the by-products of the increases in intensity of cropping systems associated with the use of insecticides and herbicides, and of nitrogen and phosphate fertilizers, and of the processing of agricultural and forest products. Concern has also been expressed about the loss of genetic diversity in crops that have been bred for high yields and about the effects of deforestation on climate and on oxygen production.

These social costs have not been adequately reflected in most calculations of the economic returns from investment in research (Table 27-2) or in trends in output per unit of input (Tables 29-3 and 29-5). There can be no question that the problems of environmental congestion and pollution, from both agricultural and industrial sources, have reached serious dimensions in specific localities and regions. The casual use and diffusion of certain materials, such as chlorinated hydrocarbons, pose a serious threat to environmental stability, public health, and economic activity.

Recent debate on environmental policy has tended to polarize around two alternatives. One is the antigrowth movement. It is founded on the view that the relationship between man and the environment is so delicate that the effects of economic growth on the natural world may seriously impair the capacity of the earth to support life. The capacity of the ecosystem to sustain production and to absorb the by-products of productive activity is regarded as finite; the limits to growth are being approached exponentially; and the conclusion is that, if present growth trends continue, the world will face ecological disaster within a matter of decades.

An alternative view is that scientific and technical effort can be redirected to permit the reduction of environmental stress and the continued acceleration of the performance of the ecosystem.

In the past the capacity of the environment to absorb pollutants or residuals, seemed to be infinite. Consequently the tendency was to bias the direction of technical change toward excessive production of residuals. This process has been clearly apparent in agriculture. Although one effect of the agricultural commodity programs has been to make the use of land in agricultural production "expensive" by restricting its use, at the

same time, the ability of the environment to absorb the residuals from crop and livestock production has been assumed to be limitless. As a result, in the development of agricultural innovations in the United States, too much emphasis has been placed on the development of land substitutes—plant nutrients, plant protection chemicals, crop varieties, management systems that reflected the overvaluation of land and the undervaluation of the social consequences of the absorption or disposal of the residuals from agricultural production processes. It also seems apparent that these biases in resource pricing have led to underinvestment in scientific and technological efforts directed toward those pest and soil management systems that would be more compatible with current efforts to preserve the quality of the environment for the people that must live in it.

The implication of this perspective is that the appropriate response to the environmental crises is the redirection of scientific and technical effort to reduce the environmental stress caused by agricultural commodity production. This move will also require the innovation of social institutions having the authority to establish and regulate both private and public property rights in environmental resources. One possibility might be the establishment of private firms or public authorities with appropriate incentives to manage environmental resources. Another alternative is to design market or market-like mechanisms to direct the production and use of environmental commodities and services.

The basic limitation of the first, or "crisis," approach to environmental stress is that too much preoccupation with the threat of ultimate disaster diverts attention from the efforts needed for solution of the problems that are of immediate concern. The skills attained in scientific, technical, and institutional innovation required to solve the more immediate, if less dramatic problems, will add to man's capacity to solve the more distant, though more dramatic, problems as they emerge.

Social stress

Agricultural and economic development places great stress on the social systems as well as on the ecosystem. Technological change widens the options available to a society. It makes available to society new income streams that can be used to support a wide range of individual and social objectives. Economic units such as firms, households, and public agencies are engaged in a continuous struggle to capture or internalize the new income streams resulting from economic growth and to avoid the costs associated with growth. The broader society is simultaneously engaged in a struggle to force the economic units to bear the costs of growth and to diffuse or externalize the benefits.

In agriculture the socialization of much of agricultural research, particularly the research leading to advances in biological technology, is an example of public sector institutional innovation designed to realize for society the potential gains from advances in agricultural technology. The political and legislative history of farm price programs, from the mid-1920's to the present, can be viewed as a struggle between agricultural producers and society generally regarding the partitioning of the new income streams between agricultural producers and consumers.

The gains from agricultural development are also distributed unequally among the several social classes engaged in agricultural production. For example, when the benefits of technical change are computed, usually little consideration is given to the losses undergone by displaced workers. In a study of the returns from the invention, development, and use of the tomato harvester in the U.S. it was found that the social returns, assuming that workers would be compensated for their loss of jobs, were several orders of magnitude below the level of returns when such losses are not included.° Effective social institutions to facilitate such compensation are not available. Therefore, the adoption of the tomato harvester provided a net gain to society but also imposed substantial uncompensated losses on workers. Part of this gain, then, was actually a subsidy to the producers by the workers.

In the United States the effects of such displace-

ment are often of relatively short duration because of the capacity of economic growth in other sectors of the economy to absorb the labor displaced by technological change in the agricultural sector. Where the capacity to absorb displaced workers does not exist, the intensity of the social-political struggle over the partitioning of the gains—the new income streams—resulting from agricultural development may be quite intense. Some of the conflict is between land owners, tenants, and landless laborers over the distribution of the increase in farm income. Some is between the rural and urban sectors regarding how much of the gains will be transferred to urban workers in the form of lower prices. There is a struggle between the private and public sectors concerning control of the uses of the growth "dividends." Also, within the public sector, the "military" and the "development" bureaucracies frequently struggle over the uses of new resources available to the public sector. The effects of agricultural development, and of economic development generally, may be to generate more social tension than the political systems of many countries seem able to absorb.

A viable sociopolitical environment is essential for the successful solution of the problems of environmental stress accompanying economic growth. The evolution of a viable sociopolitical system, capable of introducing institutional innovations that effectively allocate the gains of growth among classes and sectors, is fundamental to the development process.

SCIENCE, TECHNOLOGY, AND AGRICULTURAL PROGRESS

This book has given major attention to the scientific and technological foundations that underlie the creation of new agricultural practices. New technology, particularly new crop production technology resulting in higher yields, is one of the five essentials for agricultural progress.

The biological science foundation on which this new crop production technology rests is becoming increasingly sophisticated. It requires greater depth in the training of biological and agricultural scientists in their respective disciplines. If the new

° Andrew Schmitz and David Seckler, "Mechanized Agriculture and Social Welfare: The Case of the Tomato Harvester," *American Journal of Agricultural Economics*, Vol. 52, No. 4, November 1970, pp. 569–577.

agricultural technology is to exert an impact on agricultural production, increasing sophistication is also required in the agribusiness and public sectors of the economy, since these sectors organize and make available to agricultural producers the other four essentials—purchasable inputs, markets, transportation, and incentives.

It is also true that the biological and agricultural scientists who develop the new science and technology can make the greatest contribution to agricultural progress if they can achieve a broad understanding of the process of agricultural development itself. Agricultural progress depends not only on the specialized skills of a wide variety of persons engaged in many occupations and activities in each country attempting to develop its

agriculture but also on an understanding of the nature of agricultural development. Such understanding is needed not only by agricultural scientists, educators, planners, and administrators but by legislators and editors, merchants and bankers, and many others.

For this reason the scope of this book is considerably wider than the typical introduction to a field of science. It provides a broad international and interdisciplinary orientation to the problems of crop science and production. There are, however, many chapters that remain unwritten or incomplete. Their completion will depend on the potential scientists, educators, farmers, and citizens who will be organizing and carrying out the agricultural development efforts of the future.

Selected References

Brubaker, Sterling. *To Live on Earth: Man and His Environment in Perspective.* The Johns Hopkins University Press, Baltimore, 1972. (A well balanced analysis of the burdens that population, economic growth, and technological advancement impose on our water, air, and land. Environmental threats are classified according to their gravity—from aesthetic nuisance to genetic damage or impairment of the earth's life support system. The immeidate and long-range choices that must be made if man is to live on earth are outlined.)

Cochrane, Willard W. *The World Food Problem: A Guardedly Optimistic View.* Thomas Y. Crowell, New York, 1969. (An analysis of the factors—political, social, and economic—that created the world food problem. Discussion of the modern food dilemma—the failure to achieve improved levels of food consumption rather than mass starvation. Discussion of the part that America can play in helping less developed countries meet their food demands.)

Hayami, Yujiro, and Vernon W. Ruttan. *Agricultural Development: An International Perspective.* The Johns Hopkins University Press, Baltimore, 1971. (A theoretical and empirical analysis of the role of technical and institutional change in the agricultural development of the United States and Japan. Implications of the U.S. and Japanese models of agricultural development for the less developed countries and for agricultural development policy are examined.)

Mosher, Arthur T. *Getting Agriculture Moving.* Praeger, New York, 1966. (This book deals with the needs and problems at early stages of agricultural development and its principal purpose is to serve as a framework for systematic discussion of the development process in agriculture in in-service programs.)

Mosher, Arthur T. *To Create a Modern Agriculture: Organization and Planning.* Agricultural Development Council, New York, 1971. (An excellent introduction to the problems of planning for agricultural development. Emphasis is on the public policies that can enable the farmer, agribusiness, the local community to respond to modernization opportunities.)

Owens, Edgar, and Robert Shaw. *Development Reconsidered: Bridging the Gap Between Government and People.* D. C. Heath, Lexington, 1972. (This book sets forth a strategy of development in which participation by the mass of small producers—farmers, artisans, entrepreneurs—is both the means and end for development. This approach is contrasted to the capital intensive dualistic strategy based on capital intensive development in the big cities and on big farms.)

APPENDIX

CONVERSION TABLES

To convert a temperature, in either Centigrade (Celsius) or Fahrenheit, to the other scale, find that temperature in the center column, and then find the equivalent temperature in the other scale either in the Centigrade column to the left or in the Fahrenheit column to the right. For example, if the Centigrade temperature is 44°, the equivalent Fahrenheit temperature (in the righthand column) is 111.2°. If the *Fahrenheit* temperature is 44°, the equivalent Centigrade temperature (in the lefthand column) is 6.67°.

On the Centigrade scale the temperature of melting ice is 0° and that of boiling water is 100° at normal atmospheric pressure. On the Fahrenheit scale, the equivalent temperatures are 32° and 212° respectively. The formula for converting Centigrade to Fahrenheit is C = 5/9 F − 32 , and the formula for converting Fahrenheit to Centigrade is F = 9/5 C + 32.

C	C or F	F	C	C or F	F	C	C or F	F	C	C or F	F
− 73.33	− 100	− 148.0	− 6.67	20	68.0	15.6	60	140.0	43	110	230
− 70.56	− 95	− 139.0	− 6.11	21	69.8	16.1	61	141.8	49	120	248
− 67.78	− 90	− 130.0	− 5.56	22	71.6	16.7	62	143.6	54	130	266
− 65.00	− 85	− 121.0	− 5.00	23	73.4	17.2	63	145.4	60	140	284
− 62.22	− 80	− 112.0	− 4.44	24	75.2	17.8	64	147.2	66	150	302
− 59.45	− 75	− 103.0	− 3.89	25	77.0	18.3	65	149.0	71	160	320
− 56.67	− 70	− 94.0	− 3.33	26	78.8	18.9	66	150.8	77	170	338
− 53.89	− 65	− 85.0	− 2.78	27	80.6	19.4	67	152.6	82	180	356
− 51.11	− 60	− 76.0	− 2.22	28	82.4	20.0	68	154.4	88	190	374
− 48.34	− 55	− 67.0	− 1.67	29	84.2	20.6	69	156.2	93	200	392
− 45.56	− 50	− 58.0	− 1.11	30	86.0	21.1	70	158.0	99	210	410
− 42.78	− 45	− 49.0	− 0.56	31	87.8	21.7	71	159.8	100	212	414
− 40.0	− 40	− 40.0	0	32	89.6	22.2	72	161.6	104	220	428
− 37.23	− 35	− 31.0				22.8	73	163.4	110	230	446
− 34.44	− 30	− 22.0	0.56	33	91.4	23.3	74	165.2	116	240	464
− 31.67	− 25	− 13.0	1.11	34	93.2	23.9	75	167.0	121	250	482
− 28.89	− 20	− 4.0	1.67	35	95.0	24.4	76	168.8	127	260	500
− 26.12	− 15	5.0	2.22	36	96.8	25.0	77	170.6	132	270	518
− 23.33	− 10	14.0	2.78	37	98.6	25.6	78	172.4	138	280	536
− 20.56	− 5	23.0	3.33	38	100.4	26.1	79	174.2	143	290	554
− 17.8	0	32.0	3.89	39	102.2	26.7	80	176.0	149	300	572
			4.44	40	104.0	27.2	81	177.8	154	310	590
− 17.2	1	33.8	5.00	41	105.8	27.8	82	179.6	160	320	608
− 16.7	2	35.6	5.56	42	107.6	28.3	83	181.4	166	330	626
− 16.1	3	37.4	6.11	43	109.4	28.9	84	183.2	171	340	644
− 15.6	4	39.2	6.67	44	111.2	29.4	85	185.0	177	350	662
− 15.0	5	41.0	7.22	45	113.0	30.0	86	186.8	182	360	680
− 14.4	6	42.8	7.78	46	114.8	30.6	87	188.6	188	370	698
− 13.9	7	44.6	8.33	47	116.6	31.1	88	190.4	193	380	716
− 13.3	8	46.4	8.89	48	118.4	31.7	89	192.2	199	390	734
− 12.8	9	48.2	9.44	49	120.2	32.2	90	194.0	204	400	752
− 12.2	10	50.0	10.0	50	122.0	32.8	91	195.8	210	410	770
− 11.7	11	51.8	10.6	51	123.8	33.3	92	197.6	216	420	788
− 11.1	12	53.6	11.1	52	125.6	33.9	93	199.4	221	430	806
− 10.6	13	55.4	11.7	53	127.4	34.4	94	201.2	227	440	824
− 10.0	14	57.2	12.2	54	129.2	35.0	95	203.0	232	450	842
− 9.44	15	59.0	12.8	55	131.0	35.6	96	204.8	238	460	860
− 8.89	16	60.8	13.3	56	132.8	36.1	97	206.6	243	470	878
− 8.33	17	62.6	13.9	57	134.6	36.7	98	208.4	249	480	896
− 7.78	18	64.4	14.4	58	136.4	37.2	99	210.2	254	490	914
− 7.22	19	66.2	15.0	59	138.2	37.8	100	212.0	260	500	932

AREA

Metric

1 square centimeter	=	0.155 sq inch
	=	100 sq millimeters
1 square meter	=	1,550 sq inches
	=	10.764 sq feet
	=	1.196 sq yards
	=	10,000 sq centimeters
1 square kilometer	=	0.3861 sq mile
	=	1,000,000 sq meters
1 hectare	=	2.471 acres
	=	10,000 sq meters

Imperial

1 square inch	=	6.452 sq centimeters
	=	1/144 sq foot
	=	1/1296 sq yard
1 square foot	=	929.088 sq centimeters
	=	0.0929 sq meter
1 square yard	=	8,361.3 sq centimeters
	=	0.8361 sq meter
	=	1,296 sq inches
	=	9 sq feet
1 square mile	=	2.59 sq kilometers
	=	640 acres
1 acre	=	0.4047 hectare
	=	43,560 sq feet
	=	4,840 sq yards
	=	4,046.87 sq meters

LENGTH

Metric

1 millimicron	=	0.001 micron
1 micron	=	0.001 millimeter
1 millimeter	=	0.001 meter
	=	0.0394 inch
1 centimeter	=	10 millimeters
	=	0.3937 inch
	=	0.01 meter
1 meter	=	39.37 inches
	=	3.281 feet
	=	1,000 millimeters
	=	100 centimeters
	=	1.2 varas
1 kilometer	=	3,281 feet
	=	1,094 yards
	=	0.621 mile
	=	1,000 meters

Imperial

1 inch	=	25.4 millimeters
	=	2.54 centimeters
1 foot	=	30.48 centimeters
	=	0.3048 meter
	=	12 inches
1 yard	=	0.9144 meter
	=	91.44 centimeters
	=	3 feet
1 mile	=	1,609.347 meters
	=	1.609 kilometers
	=	5,280 feet
	=	1,760 yards

WEIGHT

Metric

1 milligram	=	0.001 gram
	=	0.0154 grain
1 centigram	=	0.01 gram
	=	0.1543 grain
1 gram	=	0.0353 avoirdupois ounce
	=	15.4324 grains
1 kilogram	=	1,000 grams
	=	353 avoirdupois ounces
	=	2.2046 avoirdupois pounds
1 metric ton	=	1,000 kilograms
	=	2,204.6 pounds
	=	1.102 short tons
	=	0.984 long ton

Imperial

1 grain	=	1/7000 avoirdupois pound
	=	0.064799 gram
1 ounce (avoirdupois)	=	28.3496 grams
	=	437.5 grains
	=	1/16 pound
1 pound (avoirdupois)	=	453.593 grams
	=	0.45369 kilograms
	=	16 ounces
1 short ton	=	907.184 kilograms
	=	0.9072 metric ton
	=	2,000 pounds

YIELD

Metric

1 kilogram per hectare	=	0.89 pound per acre
1 cubic meter per hectare	=	14.2916 cubic feet per acre

Imperial

1 pound per acre	=	1.121 kilograms per hectare
1 ton (2,000 lb) per acre	=	2.242 metric tons per hectare
1 cubic foot per acre	=	0.0699 cubic meter per hectare
1 bushel (60 lb) per acre	=	67.26 kilograms per hectare

VOLUME

Metric

1 liter	=	1.057 U.S. quarts liquid
	=	0.9081 quart, dry
	=	0.2642 U.S. gallon
	=	0.221 Imperial gallon
	=	1,000 milliliters or cc
	=	0.0353 cubic foot
	=	61.02 cubic inches
	=	0.001 cubic meter
1 cubic meter	=	61,023.38 cubic inches
	=	35.314 cubic feet
	=	1.308 cubic yards
	=	264.17 U.S. gallons
	=	1,000 liters
	=	28.38 U.S. bushels
	=	1,000,000 cu. centimeters
	=	1,000,000,000 cu. millimeters

Imperial

1 fluid ounce	=	1/128 gallon
	=	29.57 cubic centimeters
	=	29.562 milliliters
	=	1.805 cubic inches
	=	0.0625 U.S. pint (liquid)
1 U.S. quart liquid	=	946.3 milliliters
	=	57.75 cubic inches
	=	32 fluid ounces
	=	4 cups
	=	1/4 gallon
	=	2 U.S. pints (liquid)
	=	0.946 liter
1 quart dry	=	1.1012 liters
	=	67.20 cubic inches
	=	2 pints (dry)
	=	0.125 peck
	=	1/32 bushel
1 cubic inch	=	16.387 cubic centimeters
1 cubic foot	=	28,317 cubic centimeters
	=	0.0283 cubic meter
	=	28.316 liters
	=	7.481 U.S. gallons
	=	1,728 cubic inches
1 U.S. gallon	=	16 cups
	=	3.785 liters
	=	231 cubic inches
	=	4 U.S. quarts liquid
	=	8 U.S. pints liquid
	=	8.3453 pounds of water
	=	128 fluid ounces
	=	0.8327 British Imperial gallon
1 British Imperial gallon	=	4.546 liters
	=	1.201 U.S. gallons
	=	277.42 cubic inches
1 U.S. bushel	=	35.24 liters
	=	2,150.42 cubic inches
	=	1.2444 cubic feet
	=	0.03524 cubic meter
	=	2 pecks
	=	32 quarts (dry)
	=	64 pints (dry)

Index